HANDBOOK OF ATMOSPHERIC SCIENCE

Handbook of Atmospheric Science

Principles and Applications

EDITED BY

C.N. Hewitt
Department of Environmental Science
Lancaster University

AND

Andrea V. Jackson
The School of the Environment
University of Leeds

Blackwell
Publishing

© 2003 by Blackwell Science Ltd
a Blackwell Publishing company

350 Main Street, Malden, MA 02148-5020, USA
108 Cowley Road, Oxford OX4 1JF, UK
550 Swanston Street, Carlton, Victoria 3053, Australia

The right of C.N. Hewitt and A.V. Jackson to be identified as the Authors of the Editorial Material in this Work has
been asserted in accordance with the UK Copyright, Designs, and Patents Act 1988.

First published 2003

Library of Congress Cataloging-in-Publication Data

Handbook of atmospheric science / edited by C.N. Hewitt and A.V. Jackson.
p.   cm.
Includes index.
ISBN 0-632-05286-4 (alk. paper)
1. Atmospheric physics—Handbooks, manuals, etc.   I. Hewitt, C.N.   II. Jackson, A.V. (Andrea V.)
QC864 .H36 2003
551.5—dc21
2002028285

ISBN 0-632-05286-4 (hardback)

A catalogue record for this title is available from the British Library.

Set in 9/11$\frac{1}{2}$ pt Trump Mediaeval
by SNP Best-set Typesetter Ltd., Hong Kong
Printed and bound in the United Kingdom
by TJ International Ltd, Padstow, Cornwall

For further information on
Blackwell Publishing, visit our website:
http://www.blackwellpublishing.com

Contents

List of contributors, ix
Preface, xi

Part 1: Principles of Atmospheric Science

1   CHEMICAL EVOLUTION OF THE ATMOSPHERE, 3
Richard P. Wayne
1.1   Introduction, 3
1.2   Creation of the planets and their earliest atmospheres, 4
1.3   The Earth's atmosphere before life began, 9
1.4   Comparison of Venus, Earth, and Mars, 11
1.5   Life and the Earth's atmosphere, 14
1.6   Carbon dioxide in Earth's atmosphere, 18
1.7   The rise of oxygen concentrations, 20
1.8   Protection of life from ultraviolet radiation, 25
1.9   Conclusions, 29
1.10   Further reading, 29
References, 29

2   ATMOSPHERIC ENERGY AND THE STRUCTURE OF THE ATMOSPHERE, 35
Hugh Coe and Ann R. Webb
2.1   Introduction, 35
2.2   The structure of the Earth's atmosphere, 35
2.3   Solar and terrestrial radiation, 37

2.4   Absorption of radiation by trace gases, 39
2.5   Solar radiation, ozone, and the stratospheric temperature profile, 40
2.6   Trapping of longwave radiation, 42
2.7   A simple model of radiation transfer, 42
2.8   A brief overview of more complex radiative transfer, 45
2.9   Conduction, convection, and sensible and latent heat, 46
2.10   The energy budget for the Earth's atmosphere, 52
2.11   Energy transfer in the atmosphere and ocean, 55
2.12   Solar radiation and the biosphere, 56
References, 58

3   THE EARTH'S CLIMATES, 59
John G. Lockwood
3.1   Introduction, 59
3.2   Polar climates, 62
3.3   Temperate latitude climates, 69
3.4   Tropical climates, 75
3.5   Closing remarks, 87
References, 87

4   BIOGEOCHEMICAL CYCLES AND RESIDENCE TIMES, 90
Dudley E. Shallcross, Kuo-Ying Wang, and Claudia H. Dimmer
4.1   Introduction, 90
4.2   The global carbon cycle, 90
4.3   The global nitrogen cycle, 96

4.4   The global sulfur cycle, 99
4.5   The global halogen cycle, 105
4.6   Conclusions, 117
References, 117

5   SOURCES OF AIR POLLUTION, 124
     Andrea V. Jackson
     5.1   Introduction, 124
     5.2   Primary pollutants, 124
     5.3   Long-lived pollutants, 138
     5.4   Secondary gaseous pollutants, 140
     5.5   Other hazardous air pollutants, 143
     5.6   Particulate material, 145
     References, 149

6   TROPOSPHERIC
     PHOTOCHEMISTRY, 156
     Paul S. Monks
     6.1   Introduction, 156
     6.2   Initiation of photochemistry by light,
            158
     6.3   Tropospheric oxidation chemistry,
            159
     6.4   Nitrogen oxides and the
            photostationary state, 161
     6.5   Production and destruction of ozone,
            164
     6.6   The tropospheric ozone budget, 168
     6.7   The role of hydrocarbons, 169
     6.8   Urban chemistry, 171
     6.9   The spring ozone maximum, 173
     6.10   Nighttime oxidation chemistry, 176
     6.11   Ozone–alkene chemistry, 180
     6.12 NO_2–diene chemistry, 182
     6.13   Sulfur chemistry, 182
     6.14   Halogen chemistry, 184
     6.15   Conclusions, 184
     References, 184

7   STRATOSPHERIC CHEMISTRY AND
     TRANSPORT, 188
     A. Robert Mackenzie
     7.1   Introduction, 188
     7.2   The structure of the stratosphere, 190
     7.3   Gas-phase chemistry of the
            stratosphere, 196
     7.4   Aerosols and clouds in the
            stratosphere, 201

7.5   Heterogeneous chemistry of the
       stratosphere, 202
7.6   Future perturbations to the
       stratosphere, 204
References, 206

8   AQUEOUS PHASE CHEMISTRY OF
     THE TROPOSPHERE, 211
     Peter Brimblecombe
     8.1   The aqueous phase in the atmosphere,
            211
     8.2   Nonvolatile solutes, 214
     8.3   Reactions and photochemistry, 219
     8.4   Conclusions, 224
     References, 225

9   ATMOSPHERIC PARTICULATE
     MATTER, 228
     *Urs Baltensperger, Stefan Nyeki, and
     Markus Kalberer*
     9.1   Introduction, 228
     9.2   Size distribution, composition, and
            concentration, 230
     9.3   Aerosol sources, 230
     9.4   Heterogeneous chemistry, 238
     9.5   Climate forcing, 240
     9.6   Tropospheric and stratospheric
            aerosols: remote sensing, 248
     Appendix: nomenclature, 250
     References, 251

10   ATMOSPHERIC DISPERSION AND
     AIR POLLUTION METEOROLOGY,
     255
     David Carruthers
     10.1   Introduction, 255
     10.2   The atmospheric boundary layer, 255
     10.3   Atmospheric dispersion, 266
     10.4   Mean concentrations, 270
     10.5   Conclusions, 272
     References, 274

11   SYNOPTIC-SCALE METEOROLOGY,
     275
     Douglas J. Parker
     11.1   Introduction, 275
     11.2   Basic physical descriptions and
            models, 275

11.3 Applications to weather systems, 299
11.4 On airmasses, 307
11.5 Practicalities: how to perform a synoptic analysis, 308
11.6 Conclusions, 311
References, 312

12 ATMOSPHERIC REMOVAL PROCESSES, 314
Brad D. Hall
12.1 Introduction, 314
12.2 Dry deposition of gases, 314
12.3 Bulk resistance ("big leaf") model, 315
12.4 Dry deposition of particles, 324
12.5 Measuring dry deposition, 327
12.6 Wet deposition, 329
References, 334

Part 2: Problems, Tools, and Applications

13 GLOBAL AIR POLLUTION PROBLEMS, 339
Atul K. Jain and Katharine A.S. Hayhoe
13.1 Introduction, 339
13.2 Historical evidence of the impact of human activities on climate, 341
13.3 Future outlook of climate and ozone changes, 355
13.4 Potential impacts of stratospheric ozone and climate changes, 364
13.5 Pathways to policy considerations, 367
13.6 Conclusions, 370
References, 371

14 REGIONAL-SCALE POLLUTION PROBLEMS, 376
Crispin J. Halsall
14.1 Introduction, 376
14.2 Monitoring frameworks, 377
14.3 The regional ozone problem, 378
14.4 Deposition of nitrogen and sulfur across Europe: acidification and eutrophication, 384
14.5 Arctic haze, 388
14.6 Current trends and uncertainties in regional air pollution, 395
References, 395

15 URBAN-SCALE AIR POLLUTION, 399
Jes Fenger
15.1 Introduction, 399
15.2 Pollutants and sources, 402
15.3 From emission to pollution levels, 405
15.4 Urban-scale impacts, 411
15.5 Means of mitigation, 418
15.6 Case studies, 422
15.7 Conclusions, 432
References, 433

16 ATMOSPHERIC MONITORING TECHNIQUES, 439
Rod Robinson
16.1 Introduction, 439
16.2 Requirements, 439
16.3 Standardized methods, 440
16.4 Sampling techniques, 440
16.5 Expression of results, 441
16.6 Monitoring air quality, 441
16.7 Monitoring meteorological parameters, 461
16.8 Monitoring of the middle to upper atmosphere, 463
16.9 Conclusions, 470
References, 470

17 EMISSION INVENTORIES, 473
David Hutchinson
17.1 Introduction, 473
17.2 Emission inventory procedures, 477
17.3 Emissions from road traffic, 483
17.4 Emissions from rail transport, 489
17.5 Emissions at airports, 489
17.6 Emissions from shipping, 490
17.7 Area emission sources, 491
17.8 Point source emissions, 494
17.9 A comparison of the London and Tokyo atmospheric emissions inventories, 494
17.10 Conclusions, 499
References, 500

18 POLLUTANT DISPERSION MODELING, 503
Yasmin Vawda
18.1 Introduction, 503

Contents

18.2 Emission sources recognized by atmospheric dispersion models, 503

18.3 Assessment criteria for the results of dispersion models, 504

18.4 Meteorological data requirements of atmospheric dispersion models, 504

18.5 Types of atmospheric dispersion model, 509

18.6 Input data requirements, 510

18.7 Output data and interpretation, 516

18.8 Background air quality, 516

18.9 Choice of dispersion model, 516

18.10 Accuracy of dispersion modeling predictions, 520

Appendix: list of models, 520

References, 523

19 CLIMATE MODELING, 525
William Lahoz

19.1 Introduction, 525

19.2 The modeling tools, 529

19.3 Evaluation of the modeling tools, 547

19.4 Use of the modeling tools, 551

19.5 Latest results, 554

19.6 Future avenues, 557

References, 557

20 CRITICAL LEVELS AND CRITICAL LOADS AS A TOOL FOR AIR QUALITY MANAGEMENT, 562
Wim de Vries and Maximilian Posch

20.1 Introduction, 562

20.2 Critical levels of air pollutants, 564

20.3 Methods to derive critical loads for terrestrial ecosystems, 565

20.4 Critical loads of nitrogen, 570

20.5 Critical loads of acidity, 577

20.6 Critical loads for heavy metals, 585

20.7 The use of critical loads in policy assessments, 591

20.8 Discussion and conclusions, 597

References, 597

21 THE PRACTICE OF AIR QUALITY MANAGEMENT, 603
Bernard E.A. Fisher

21.1 Introduction to air quality management, 603

21.2 Carbon monoxide, 607

21.3 Benzene, 609

21.4 1,3-Butadiene, 611

21.5 Lead, 611

21.6 Nitrogen dioxide, 612

21.7 Sulfur dioxide, 614

21.8 Particles, 616

21.9 Ozone, 618

21.10 Polycyclic aromatic hydrocarbons, 619

21.11 Dispersion models for local air quality management, 620

21.12 Accuracy of air pollution models, 624

21.13 Example of a method of estimating uncertainty, 625

21.14 Carbon dioxide, 626

References, 627

Index, 629

Color plates facing p. 180

Contributors

URS BALTENSPERGER *Laboratory for Atmospheric Chemistry, Paul Scherrer Institute, CH-5232 Villigen-PSI, Switzerland* (urs.baltensperger@psi.ch)

PETER BRIMBLECOMBE *School of Environmental Sciences, University of East Anglia, Norwich NR4 7TJ, UK* (p.brimblecombe@uea.ac.uk)

DAVID CARRUTHERS *Cambridge Environmental Research Consultants, 3 Kings Parade, Cambridge CB2 1SJ, UK* (David.carruthers@cerc.co.uk)

HUGH COE *Department of Physics, UMIST, PO Box 88, Manchester M60 1QD, UK* (Hugh.coe@umist.ac.uk)

CLAUDIA H. DIMMER *NERC, Polaris House, North Star Avenue, Swindon SN2 1EU, UK*

JES FENGER *National Environmental Research Institute, Department of Atmospheric Environment, Frederiksborgvej 399, PO Box 358, DK-4000 Roskilde, Denmark* (JFE@DMU.dk)

BERNARD E.A. FISHER *Environment Agency, Centre for Risk and Forecasting, Kings Meadow House, Kings Meadow Road, Reading RG1 8DQ, UK* (Bernard.fisher@environment-agency.gov.uk)

BRAD D. HALL *Climate Monitoring and Diagnostics Laboratory, National Oceanic and Atmospheric Administration, 325 Broadway, Boulder, CO 80305, USA* (Bradley.Hall@noaa.gov)

CRISPIN J. HALSALL *Department of Environmental Science, Institute of Environmental and Natural Sciences, Lancaster University, Lancaster LA1 4YQ, UK* (c.halsall@Lancaster.ac.uk)

KATHARINE A.S. HAYHOE *Department of Atmospheric Sciences, University of Illinois at Urbana-Champaign, 105 S Gregory Street, Urbana, IL 61801, USA* (hayhoe@atmos.uiuc.edu)

DAVID HUTCHINSON *Greater London Authority, City Hall, The Queen's Walk, London SE1 2AA, UK*

ANDREA V. JACKSON *Institute for Atmospheric Science, School of the Environment, University of Leeds, Leeds, LS2 9JT* (andrea@env.leeds.ac.uk)

ATUL K. JAIN *Department of Atmospheric Sciences, University of Illinois at Urbana-Champaign, 105 S Gregory Street, Urbana, IL 61801, USA* (jain@atmos.uiuc.edu)

MARKUS KALBERER *Department of Chemistry, ETH Zürich, 8092 Zürich, Switzerland*

WILLIAM LAHOZ *Data Assimilation Research Centre, Department of Meteorology, University of Reading, P.O. Box 243, Earley Gate, Reading RG6 6BB, UK* (wal@met.reading.ac.uk)

JOHN G. LOCKWOOD *4 Woodthorne Croft, Leeds LS17 8XQ, UK* (jglockwood@clara.co.uk)

A. ROBERT MACKENZIE *Department of Environmental Science, Institute of Environmental and Natural Sciences, Lancaster University, Lancaster LA14YQ, UK (r.Mackenzie@lancaster.ac.uk)*

PAUL S. MONKS *Department of Chemistry, University of Leicester, Leicester LE1 7RH, UK (p.s.monks@le.ac.uk)*

STEFAN NYEKI *Laboratory for Atmospheric Chemistry, Paul Scherrer Institute, CH-5232 Villigen-PSI, Switzerland and Institute for Environmental Research, Department of Chemistry, University of Essex, Colchester, Essex CO4 3SQ, UK*

DOUGLAS J. PARKER *Institute for Atmospheric Science, School of the Environment, University of Leeds, Leeds LS2 9JT (doug@env.leeds.ac.uk)*

MAXIMILIAN POSCH *National Institute of Public Health and the Environment (RIVM), P.O. Box 1, NL-3720 BA Bilthoven, The Netherlands*

ROD ROBINSON *National Physical Laboratory, Teddington, Middlesex TW11 0LW, UK (Rod.robinson@npl.co.uk)*

DUDLEY E. SHALLCROSS *Biogeochemistry Research Centre, School of Chemistry, Cantock's Close, University of Bristol, Bristol BS8 1TS, UK (d.e.shallcross@bristol.ac.uk)*

YASMIN VAWDA *Casella Stanger, Great Guildford House, 30 Great Guildford Street, London SE1 0ES, UK (yasminvawda@casellagroup.com)*

WIM DE VRIES *ALTERRA Green World Research, P.O. Box 47, NL-6700 AA Wageningen, The Netherlands (w.devries@alterra.wag-ur.nl)*

KUO-YING WANG *Department of Atmospheric Science, National Central University, Chung-Li, Taiwan*

RICHARD P. WAYNE *Physical and Theoretical Chemistry Laboratory, University of Oxford, South Parks Road, Oxford OX1 3QZ, UK (wayne@physchem.ox.ac.uk)*

ANN R. WEBB *Department of Physics, UMIST, PO Box 88, Manchester M60 1QD, UK (ann.webb@umist.ac.uk)*

Preface

Understanding and predicting the behavior of the Earth's atmosphere, as it manifests itself in the endless variations of weather and climate, has exercised Man since the beginning of time, and indeed the systematic study of the atmosphere is one of the oldest sciences. From Man's first tentative, and maybe unintentional, sea voyages, to the present time, understanding the behavior of the atmosphere has literally meant the difference between life and death. In the past few decades, the study of the atmosphere has gained a new dimension, as it has become apparent that Man is no longer merely a passive observer of weather and climate, but that our activities actively affect the behavior of the atmosphere.

The severity of the effects of Man's activities on the atmosphere were amply demonstrated in 1987 by the Montreal Protocol on Substances that Deplete the Ozone Layer and its subsequent amendments; for the first time governments around the world accepted that the atmosphere is a delicate veil that requires legislative protection from the excesses of industrialization. The time between ozone depletion being first observed, the role of cholorofluorocarbon compounds in ozone depletion being explained, and international controls on the emissions of these polluting gases being agreed was remarkably short, and was an excellent example of the critical importance of the atmospheric sciences. Without measurement and observation techniques being developed and deployed, without an excellent underpinning knowledge of gas phase and heterogeneous chemical reactions, and without the scientific vision to link the observed ozone depletion to the emissions of the CFC gases, such rapid protection of the population through legislation would not have been possible.

In many respects, ozone depletion is a problem solved. It will take time for the ozone-depleting substances already in the atmosphere to be removed, and although there may be surprises in store, essentially ozone depletion is a twentieth-century problem. Of much more immediate concern, and in many respects of much greater concern, is global climate change. Governments around the world now accept that emissions of carbon dioxide, methane, and other radiatively active gases are changing, and will continue to change, the radiative balance of the Earth's atmosphere, and hence are changing, and will continue to change, the Earth's climate. Of particular concern is the possibility that some of the resultant effects may be large and unpredictable. In 2002, legislative controls on greenhouse gas emissions look some way away; until the single most profligate country on Earth, the United States, accepts the economic necessity of controls, meaningful reductions in emissions on a global scale look unlikely. But a better understanding of the underpinning science, and a greater awareness among the population of the realities of this science, can only enhance the possibility of controls being agreed.

In conceiving this book, and in putting it together, we draw upon our experiences as university teachers of atmospheric science. We see great enthusiasm on the part of our students to understand these major environmental issues. Ozone depletion, and air pollution in general, and climate change, excite the imagination and stir the conscience. However, teaching the underpinning, and all important, basic scientific principles is much more difficult, and hence one of our aims with this project was to provide both the underpinning science and the more appealing applications.

The first 12 chapters of the book provide an up-to-date account of the background, the principles of atmospheric science. They cover the chemical composition of the atmosphere and the chemistry that occurs within the atmosphere. They also emphasize the physical processes at work in the atmosphere, the energy balance, and the Earth's resultant climate and meteorology. The sources of air pollution and the ways in which pollutants are transformed and removed from the atmosphere are also covered.

The second half of the book tackles the major problems current in the atmosphere and describes the tools used to understand these problems and their applications. There are chapters on air pollution at the urban, regional, and global scales, measuring and monitoring techniques, the modeling of air pollution and of climate, and the use of critical loads as a tool for air quality management.

We hope this book allows the reader to obtain a clear and firm understanding of the principles of the atmospheric sciences, both physical and chemical, and of their application to understanding and predicting the Earth's atmosphere, weather, and climate and Man's pollution and modification of it.

We are grateful to all the experts who have contributed to this book, and thank them for all their efforts and work. We thank the many students who have passed through our classes over the years for their enthusiasm and comments, and who have demonstrated to us the need for this volume. We also thank our colleagues at Lancaster and Leeds for their support. One of us (NH) thanks Roy Harrison of Birmingham University for his mentoring and support over many years.

C.N. Hewitt, *Lancaster University*
Andrea V. Jackson, *University of Leeds*

Part 1

Principles of Atmospheric Science

1 Chemical Evolution of the Atmosphere

RICHARD P. WAYNE

1.1 INTRODUCTION

This chapter is concerned with how the planet Earth comes to have an atmosphere, and how that atmosphere has been modified by chemical, physical, and biological processes to move toward its present-day composition. The story begins with the "Big Bang" in which the universe was created, and we leave it some hundreds of millions of years before present (BP). Other chapters of this book discuss more recent changes in composition, especially in connection with climate. As we approach our own era, within a million years or so, the record of atmospheric composition and climate becomes richer and more detailed.

Particularly fruitful sources of information have proved to be the examination of cores of rock drilled deep into the ocean floors and cores obtained from the ice sheets covering Greenland and the Antarctic. Analysis has been conducted on icecore samples from Greenland and from some mountain glaciers, as well as from Antarctica. Virtually no melting of snow occurs even in the summer, so that each year a new layer of snow is added to the ice caps and compressed into solid ice. As the snow falls, it scavenges aerosols from the atmosphere, and these aerosols are trapped together with bubbles of air in the ice. Chemical or biological alteration to the trapped material is not expected, so that a core taken from the ice provides a stratigraphic record of the atmosphere with a high degree of resolution and stability. A limitation in the time-span of the record is ultimately imposed by the compression and resulting horizontal flow of the lowest layers of ice under the weight of newer ice deposited on top. Nevertheless, the oldest reliable samples date back to 250 kyr BP in cores from Greenland and 500 kyr BP in cores from Antarctica. The cores are up to 2500 m long. Composition and behavior of the atmosphere hundreds of thousands of years ago can thus be discussed with comparative certainty. That is a luxury that we are denied. All information about atmospheric composition in this period must perforce be obtained by inference—for example, from geological or paleontological evidence. Nevertheless, some clear principles emerge, and it is worth stating those at the outset to act as a guide to what will follow.

The first main point is that the Earth and its neighbors Venus and Mars must have early lost any primordial atmosphere with which they might have been born. Instead, a secondary atmosphere was formed from volatile materials trapped within the solid body when it was formed, or brought in later by impacting solar-system debris (comets and meteors). Life on Earth has had an enormous effect in bringing about subsequent changes to the composition of our own atmosphere, especially in terms of the relative abundances of CO_2, N_2, and O_2. Carbon dioxide, which is present at less than 0.04% in our atmosphere, makes up more than 95% of the atmospheres of Venus and Mars. Conversely, the N_2 and O_2 that make up the bulk of our atmosphere are only minor components of the other two atmospheres. Yet it is likely that all three planets acquired initially similar secondary atmospheres: biological or biologically mediated processes have modified our atmosphere. A link must therefore be sought between the evolution of life and the evolution of the Earth's atmosphere. What is more, we shall see later that the oxygen may have a critical role to play in protecting organisms on land from destructive ultraviolet radiation from the Sun. Molecular oxygen itself and its atmospheric product ozone (O_3) are the only known absorbers of such radiation in the contemporary atmosphere. There is thus a further link between life and the atmosphere.

Figure 1.1 provides a clear illustration of some of the statements just made. The pie charts show the fractional amounts of CO_2, N_2, O_2, and Ar (where they are large enough to be seen) in the atmospheres of Earth, Venus, and Mars. The left-hand column shows present-day compositions. The Earth's atmosphere immediately stands out as different from that of its inner and outer neighbors in the solar system. The right-hand column shows a composition for Earth "reconstructed" as though life were not present (Morrison & Owen 1996). Mars, being a small planet with relatively low gravitational attraction, has undergone much loss of its atmosphere as a result of erosion by impacts (Ahrens 1993; Newman et al. 1999) and by escape. The right-hand diagram for Mars gives the composition "reconstructed" as though these losses had not occurred. Venus has probably not lost substantial amounts of any of the gases under discussion; the pie chart for the present day is thus repeated on the right hand. There is now a remarkable similarity between the three charts, indicating that the assumptions about the effects of life on Earth and of loss from Mars really can explain an evolution of the atmospheres from the same starting compositions to what is found at the present day. Furthermore, although the total pressures on the three planets are now very different, as indicated on the figure, the "reconstructed" (Morrison & Owen 1996) pressures are much more closely similar. The remaining differences might well come about because of the differences in sizes of the planets and in their positions in the solar system. It is therefore time to trace the development of the atmospheres from the creation of the solar system, first of all to the compositions suggested by the right-hand column of the figure.

1.2   CREATION OF THE PLANETS AND THEIR EARLIEST ATMOSPHERES

Hydrogen and helium were present in the universe almost immediately after the "Big Bang," some 10–20 Gyr BP (current best estimates, based on Hubble space-telescope data, are 12 Gyr). Nuclear fusion transformed these elements into heavier ones such as carbon, nitrogen, oxygen, magnesium, silicon, and iron. Elements of atomic number higher than that of iron are formed in supernova explosions that scatter material through the galaxies as tiny dust grains 1–1000 nm in diameter and probably composed mainly of graphite, H_2O ice, and iron and magnesium silicates. Dust and gas in the universe (Frayer & Brown 1997) are concentrated in the arms of spiral galaxies in which new stars, such as our Sun, are formed (Brush 1990; Bachiller et al. 1997; Lunine 1997). Radiodating of meteorites and lunar samples provides ample evidence that the solar system was formed 4.6×10^9 years ago (Cameron 1988). Accretion of the primitive gases and dusts lies behind virtually all models for the origin of the solar system. In some regions, temperatures are low enough to permit crystallization of the metal silicates, but not of the ices of volatile elements such as H, C, or N. According to what has now become the standard model (Wetherill 1990), planets grow by agglomeration of rocky "planetesimals," with diameters of up to a few kilometers, that form in the solar nebula. "Planetary embryos," larger bodies of the size of the Moon or Mercury, with masses of 10^{22}–10^{23} kg, grow where the planetesimals are sufficiently closely packed to allow collisions between them. The largest of these bodies could form in ~10^5 yr. The growth of planets the size of Earth or Venus would require the merger of about 100 of these bodies. Completion of the process is estimated to occur on a time scale of about 10^7–10^8 yr, much longer than that thought

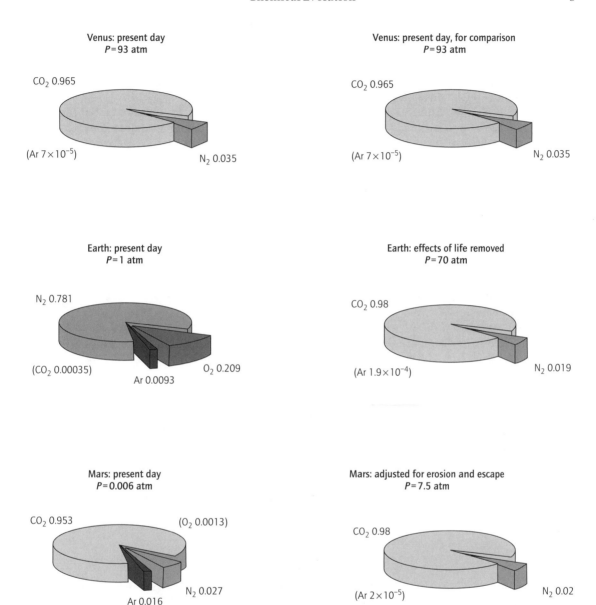

Fig. 1.1 Abundances of gases in the atmospheres of Venus, Earth, and Mars. The pie charts show the fractional abundances of the dominant gases, while those of other key components are shown in parentheses. The right-hand set of charts show what would happen on Earth if the effects of life were removed, and on Mars if the atmosphere were adjusted for loss by erosion and escape. (Data for contemporary atmospheres are taken from Wayne 2000, who cites the original references, while the reconstructions are reported by Morrison & Owen 1996, p. 347.)

likely ($\lesssim 3 \times 10^6$ yr) for the loss of the gaseous solar nebula. The most significant variant of the standard model is one in which the loss of the nebular gas occurred after the formation of the planets. Detailed accounts of the origin of the elements and of the planets and their atmospheres can be found in articles and books by Lewis and Prinn (1984), Atreya *et al.* (1989), Cox (1989, 1995), Pepin (1989), Javoy (1997, 1998), Lewis (1997), and Yung and DeMore (1999), among others.

Four possible mechanisms could account for the existence of atmospheres on Venus, Earth, and Mars (Cameron 1983). These mechanisms can be broadly classified as (i) capture of gases from the primitive solar nebula; (ii) capture of some of the solar wind; (iii) collision with volatile-rich comets and asteroids; and (iv) release ("outgassing") of volatile materials trapped together with the dust grains and planetesimals as the planets accreted (Wanke & Dreibus 1988; Wetherill 1990). One pointer to the origins of the atmospheres of the inner planets comes from the noble gases present (Pollack & Black 1982; Lupton 1983; Pepin 1989, 1991, 1992; Zhang & Zindler 1989; Atreya *et al.* 1995; Farley & Neroda 1998; Kamijo *et al.* 1998; Tolstikhin & Marty 1998), because chemical inertness prevents loss to surface rocks, and, except in the case of helium, they cannot readily escape (Hunten 1990) to space (but see the next paragraph). Figure 1.2 shows the patterns of abundances of the noble gases and carbon on the inner planets and in the Sun; to give a clearer view, the abundances are normalized to those of

(nonvolatile) silicon on the planet and in the Sun. Table 1.1 provides more detailed data for isotopes of neon and argon. Two types of noble gas can be distinguished: primordial and radiogenic. Primordial isotopes such as ^{20}Ne, ^{36}Ar, ^{38}Ar, ^{84}Kr, and ^{132}Xe were present in the solar system from the time of its creation. Radiogenic isotopes, however, have built up from the decay of radioactive nucleides: ^{40}Ar from the decay of ^{40}K, and ^{4}He from the decay of ^{232}Th, ^{235}U, and ^{238}U. It is clear from

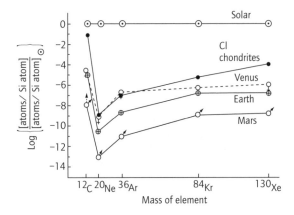

Fig. 1.2 Abundances of noble gases and carbon in the atmospheres of Venus, Earth, and Mars, and in CI chondrites. The abundances are subject to a double normalization: first, they are given as the atomic ratio to the abundance of Si on the body of interest, and then that ratio is divided by the ratio appropriate to the Sun. (Data from Hunten 1993, with permission from American Association for the Advancement of Science.)

Table 1.1 Abundance of noble gases. (Compiled from data given by Pepin 1989.) Pepin quotes errors and indicates the uncertainties in measurements and interpretations. Argon isotope ratios are taken from Pollack and Black (1982).

Object	kg per kg of object		Number ratios			
	^{20}Ne	^{36}Ar	^{36}Ar/^{12}C	^{36}Ar/^{38}Ar	^{40}Ar/^{36}Ar	^{20}Ne/^{22}Ne
Sun	2.2×10^{-3}	9.0×10^{-5}	2.3×10^{-2}	5.6	<1.0	13.7
CI*	2.9×10^{-10}	1.3×10^{-9}	3.4×10^{-8}	5.3	–	8.9
Venus	2.9×10^{-10}	2.5×10^{-9}	9.7×10^{-5}	5.6	1.0	11.8
Earth	1.0×10^{-11}	3.5×10^{-11}	2.3×10^{-6}	5.3	296	9.8
Mars	4.4×10^{-14}	2.2×10^{-13}	1.9×10^{-5}	4.1	2840	10.1

* CI, carbonaceous chondrites, a class of meteorite.

the entries in the table that the relatively sizable (about 1%) concentrations of Ar shown in Fig. 1.1 for the present-day atmospheres of Earth and Mars are composed almost entirely of the radiogenic ^{40}Ar isotope.

In comparison with solar abundances, on a mass per unit mass basis, the Earth's atmosphere is depleted of ^{36}Ar by a factor of more than two million, while for ^{20}Ne the depletion is 220 million. The depletions are even more strongly emphasized by Fig. 1.2, in which the normalization allows for the enormous solar masses of hydrogen and helium. On Earth, ^{20}Ne and ^{36}Ar are depleted by a factor of nearly 10^{11} and 10^9, respectively, on this measure, but for carbon, much of which is bound up in involatile compounds, the depletion by a factor of 10^5 is much less. Evidence of this kind is taken as clear proof that Earth has lost almost all its primordial atmosphere, if such an atmosphere existed at all, and that the present atmosphere has been acquired later. It was once supposed that the depletion of noble gases would be even more marked for the hotter planet Venus, but that Mars, having formed in a cooler part of the solar system, might have retained more of its primordial components. The Viking mission (1976) dispelled that idea for Mars, as can be seen in Table 1.1. Natural isotopes of argon and neon are even more deficient on Mars than on Earth, and the radiogenic isotopes relatively more important. Pioneer–Venus (1978) showed that the Mars results were not a freak, but that there was a real tendency for there to be greater abundances of the noble gases in the atmospheres of the planets closer to the Sun. Table 1.1 shows that the abundance of ^{20}Ne relative to ^{36}Ar is comparable on the three planets (in the range 0.1 to 0.3), but much less than the solar ratio (about 25). The planets' patterns of abundances of the noble gases seen in Fig. 1.2 are very obviously different from the Sun's; on the other hand, the planetary atmospheres show similar patterns, but again demonstrate the decreasing primordial gas residue on the planets at greater distances from the Sun. The depletions relative to solar abundances are greatest for the lightest elements. Carbon is much less depleted on the planets than are the noble gases.

The atmospheres of Venus, Earth, and Mars are thus clearly shown not to be primordial remnants. One remaining question, however, is why the patterns of rare gas abundances are so different on the planets compared with the solar abundances. There is apparently some discrimination in favor of the higher mass elements on the planets. What is more, there is also a strong hint of mass discrimination even between isotopes, as in the case of the ^{20}Ne/^{22}Ne ratios (Table 1.1), where the planetary ratios are distinctly smaller than the solar ones. Has there, after all, been escape of the noble gases from the planets that has favored the retention of the heavier elements and the heavier isotopes? Shizgal and Arkos (1996) have reviewed the ways in which gases escape from the atmospheres of Venus, Earth, and Mars. Ordinary thermal (Jeans) escape does not appear to be able to explain the observations. Planetary atmospheres can be eroded by impacts with bolides (Ahrens 1993; Newman *et al.* 1999). Another highly plausible idea (Hunten 1993; Pepin 1997) is that a rapid hydrodynamic outflow (or blowoff) of a light gas can drag with it quantities of a heavier gas. The rate of such a process depends (in a negative sense) linearly on the mass of the heavier gas, rather than exponentially as in thermal escape. Suitable driver gases might be H or H_2, but there is no certainty that the required driver flow ever existed. One possibility involves photodissociation of water vapor by short-wavelength ultraviolet light. Young stars (the "T-Tauri" stars), which resemble the Sun at the age of a few million years, emit 10^3–10^4 times as much ultraviolet radiation as the present Sun (Canuto *et al.* 1982, 1983b; Zahnle and Walker 1982), so that the Sun could have provided sufficient energy at the appropriate period. Furthermore, there may well have been much water vapor in the atmosphere at the same time, as we discuss at the end of the present section.

Hypotheses (i) and (ii) presented above argue for gravitational capture and retention by the planets after their formation either of gases of the primordial solar nebula or of the solar wind that has flowed over them during their lifetimes. The differences in ratios of noble gases between the Sun and the planets are evidence against these mechanisms. Substantial numbers of small bodies have impacted with the planets of the inner solar system over the planetary lifetimes, and the comet–asteroid hypothesis (iii) proposes that atmospheres were brought to the planets as a result of such impacts (Deming 1999). However, Venus and Earth have a roughly equal chance of encountering comets and asteroids, yet Earth has nearly two orders of magnitude less ^{36}Ar on a mass for mass basis than Venus, thus suggesting that the comet–asteroid hypothesis cannot account for a substantial proportion of the present-day atmospheres (but see later). The remaining hypothesis (iv), which is that volatile materials were incorporated into the planet as it accreted, thus seems the most probable. If the planetesimals that formed the planets contained small amounts (perhaps a fraction 10^{-4} by mass) of volatile materials, then gases could be released from within the planet as it heated up (as a result of the accretion process itself, of impacts of infalling bodies on the unprotected surface, and of decay of short-lived radioactive elements). Minerals containing bound H_2O would dissociate, and physically trapped components would become degassed (Pepin 1987; Zahnle et al. 1988; Tajika 1998). One explanation for the large excess of nonradiogenic noble gases on Venus compared to Earth could be that the planetesimals that formed Venus were exposed to an intense solar wind that was absorbed before it reached the part of the solar system where Earth (or Mars) formed. In this context, it is interesting that the pattern of noble-gas abundances for Venus (Fig. 1.2) shows a hint of a solar modification of the Earth's pattern; too much emphasis should not be given to this observation, however, since it hinges on a disputed abundance for ^{84}Kr.

An objection to the accretion hypothesis (iv) arises because of one of the most satisfactory theories (Benz et al. 1986, 1987) on the origin of the Earth–Moon system, according to which a body of nearly the size of Mars collided with the proto-Earth to melt both the impactor and most of the Earth's mantle. This very early catastrophic event would necessarily mean that almost all the water and other moderately volatile compounds that accreted with the Earth would have been lost to space. In these circumstances, a variant of the comet–asteroid hypothesis has received renewed support. The gaseous components of present-day meteorites are of interest because they may reflect the composition of the primitive materials out of which the planets accreted, as well as providing an indicator of the materials present in the solar system that are available for impact degassing (as required by the comet–asteroid hypothesis). For this reason, gas composition data for one important class of meteorite are presented in Table 1.1. The striking similarity in the $^{20}Ne/^{36}Ar$ ratios for the CI chondrites and the planetary atmospheres has prompted speculation that there may have been a single type of parent gas reservoir, with the present small spread of abundances determined by evolutionary processes. One group of models (Javoy 1998) based on this idea envisages volatile-rich planetesimals accreting relatively late to form a "veneer," resembling the composition of carbonaceous chondritic meteorites, on the planetary surface, although the details of the true origin of the veneer are not firmly established (Lewis and Prinn 1984). However, the measured isotopic ratios pose severe constraints on a single type of volatile mass distribution. The ratio of $^{20}Ne/^{22}Ne$ is significantly higher in the Venusian than the terrestrial atmosphere (and than the ratio for CI chondrites), as shown in Table 1.1, and the ratio of $^{36}Ar/^{38}Ar$ on Mars is anomalously low compared with that found on other bodies of the solar system. It seems, then, that the similar elemental but disparate isotopic compositions cannot be a result of accretion of planetesimals with constant inventories of volatile species, at least if the compositions were to resemble those of present-day meteorites. The explanation of the similarities, as well as the differences, of atmospheric composition must therefore be a coincidental result of the fractionation and mixing processes that operated, both before and after accretion. Escape of early solar-composition atmospheres from planetesimals and planets is currently thought to be the most likely mechanism that could have achieved the requisite fractionation. Whatever the detailed processes turn out to be, it is clear that the new atmospheric measurements obtained by the planetary missions have provided the basis for reasonable speculation about the origins of the planets and their atmospheres.

The way in which the inner planets heated up after they had accreted has a bearing on the composition of the earliest atmospheres if the atmospheres were primarily outgassed (Tajika 1998) from the planets themselves (hypothesis iii). If the rate of accretion was sufficient that melting occurred as the planet was forming, then, at least on Earth, iron could migrate to the core, leaving an iron-free silicate mantle in an inhomogeneous process (Walker 1976). The alternative, the homogeneous accretion model, proposes that the heating and differentiation of the Earth took place after the accretion itself. At present, the inhomogeneous model is more widely accepted than the homogeneous model. The importance for the atmosphere is that the presence or absence of iron in the mantle would determine the oxidation state and composition of the volatiles that outgassed (Levine 1985). With iron present, reduced compounds such as CH_4, NH_3, and H_2 would be expected, whereas with outgassing through a mantle that was already differentiated, or from a veneer, the probable species would be H_2O, CO_2, and N_2.

Regardless of which gases were initially released, it does not seem likely that the reduced gases could survive for long in the early atmosphere. Molecular hydrogen, like helium, is able to escape the gravitational attraction of the Earth (Hunten 1990) and this process could account for the loss of almost all H_2 as long as there were adequate heat sources in the upper atmosphere (Walker 1982). Photochemical processes are likely to transform both NH_3 and CH_4, the primary steps being

$$NH_3 + h\nu \rightarrow NH_2 + H, \quad \lambda \leq 230\,nm \quad (1.1)$$

$$CH_4 + h\nu \rightarrow CH_2 + H_2, \quad \lambda \leq 145\,nm \quad (1.2)$$

In addition, OH radicals would be formed photochemically if any H_2O were present in the atmosphere, and both NH_3 and CH_4 are attacked by OH (to yield NH_2 and CH_3). The lifetime of NH_3 against photolysis is very short (Kuhn & Atreya 1979), ranging from less than a day to a few years, depending on the assumed mixing ratio. Above an altitude of about 100 km, the lifetime of CH_4 against photolysis is also only a few days (Levine *et al.* 1982), but other gases (particularly H_2O) shield CH_4 from photodissociation at lower altitudes. Reaction with OH is then the major loss process for CH_4, and lifetimes even for trace quantities of CH_4 are estimated (Levine *et al.* 1982) to be about 50 years. Thus, in the absence of a continuous source of the gases, the abundances of these reduced species must have been low. Rocks from as long ago as 3.8×10^9 years (from Isua, in West Greenland) consist of highly metamorphosed sediments that show that abundant CH_4 was not present when they were formed. Sagan and Chyba (1997), on the other hand, argue that atmospheric methane could have been kept at relatively high concentrations, either as a result of a recycling process or, after life had been established, by the formation of the gas by methanogenic microorganisms. These workers pursue the argument further, by suggesting that CH_4 photochemistry could have led to the formation of a high-altitude absorbing layer of organic solid aerosol that in turn could have protected NH_3 from rapid photolysis. While this view is not widely accepted, it must be considered as a possibility, especially as it has implications for both greenhouse heating and the intensities of ultraviolet radiation reaching the surface of the Earth. Both these topics are touched on again later in this chapter (Sections 1.6 and 1.8).

1.3 THE EARTH'S ATMOSPHERE BEFORE LIFE BEGAN

Section 1.2 outlines the principles that are thought to have applied to the acquisition of primitive secondary atmospheres by Venus, Earth, and Mars. For Earth, the most likely scenario, according to Kasting (1993), is that it formed relatively rapidly (over a period of 10^7–10^8 yr), that its interior was hot initially as a result of a large number of impact events, and that the core was probably formed as the planet accreted. Metallic iron could have been removed from the upper mantle, and volcanic gases could have been relatively oxidized as early as 4.5 Gyr BP (Kasting *et al.* 1993). Impact probably led to the release of many of the Earth's volatile materials, and thus to a transient atmosphere very heavily laden with water vapor. Infalling iron-rich planetesimals in this phase of planetary accretion would have led to the reduction of water to form copious amounts of H_2, as required by the postulate presented earlier of hydrodynamic-outflow fractionation of the noble gases. After the end of the main accretionary phase, the water-vapor atmosphere would have condensed out to form the oceans, thus making possible the conversion of CO_2 in the atmosphere to carbonate rocks such as limestone and dolomite, to be discussed shortly. Most current models predict that the early atmosphere consisted mostly of CO_2, N_2, and H_2O, along with traces of H_2 and CO. Such models are based on the assumption that the redox state of the upper mantle has not changed, so that volcanic gas composition has remained approximately constant with time. Kasting *et al.* (1993) argue that this assumption is probably incorrect. They believe that the upper mantle was originally more reduced than today, although not as reduced as the metal-arrest level, and has become progressively more oxidized as a consequence of the release of reduced volcanic gases and the subduction of a hydrated, oxidized seafloor. Data on the redox state of sulfide and chromite inclusions in diamonds imply that the process of mantle oxidation was slow, so that reduced conditions could have prevailed for as much as half of the Earth's history. Other oxybarometers of ancient rocks give different results, so that the question of when or if the mantle redox state changed remains unresolved. Mantle redox evolution is intimately linked to the oxidation state of the primitive atmosphere: a reduced Archean atmosphere would have had a high hydrogen escape rate and should correspond to a changing mantle redox state, while an oxidized Archean atmosphere should be associated with a constant mantle redox state.

The estimated abundance of carbon that accumulated in the crust of the Earth would be sufficient to produce a partial pressure of 60–80 atm if it were all in the atmosphere in the form of CO_2 (Kasting 1993). Speculative reconstructions of the climates of the past have been used to infer what the partial pressure might really have been (Walker 1985; Durham & Chamberlain 1989; Worsley & Nance 1989; Morrison & Owen 1996). The "reconstruction" in Fig. 1.1, reported by Morrison and Owen (1996), puts the entire burden into the atmosphere, while Durham and Chamberlain (1989) suggest a partial pressure of 14 atm. Whatever the original load, there is only about 3.5×10^{-4} atm in the contemporary atmosphere. The remainder is now incorporated mainly in the carbonate rocks, with a little dissolved in the oceans: the partitioning between atmosphere, hydrosphere, and lithosphere is now roughly $1 : 50 : 10^5$. A most important aspect of atmospheric evolution on the Earth is thus how atmospheric CO_2 might be converted to solid carbonates. Inorganic weathering reactions (Wayne 2000) can convert silicate rocks such as diopside ($CaMgSi_2O_6$) to carbonate

$$CaMgSi_2O_6 + CO_2 \rightleftharpoons MgSiO_3 + CaCO_3 + SiO_2 \tag{1.3}$$

Although this aspect is getting rather ahead of the story, we ought to note here that even apparently abiological changes such as this weathering can be modulated by biological influences. Partial pressures of CO_2 in the soil where weathering occurs are 10–40 times higher than the atmospheric pressure, and these high partial pressures are maintained by soil bacteria. Thus, when appropriate organisms had developed, conversion rates could be greatly enhanced. In addition, of course, a further most important source of carbonate minerals is secretion by animals and plants, and the deposition of calcite ($CaCO_3$) shells. Life thus exerts an extremely significant influence on the removal of CO_2 from the atmosphere. Formation of carbonates, and the evolution of life itself, have in common the need for liquid water, so the nonatmospheric reservoirs for CO_2 on the Earth seem linked to the presence of the liquid phase over geological time. We return later to the question of the evolution of CO_2 concentrations.

1.4 COMPARISON OF VENUS, EARTH, AND MARS

The composition and the chemical and physical behavior of the atmospheres of Venus (Schubert & Covey 1981; Krasnopolsky 1986; Yung & DeMore 1999) and Mars (McElroy *et al.* 1977; Krasnopolsky 1986; Yung and DeMore 1999) provide further evidence about the evolution of our own atmosphere. Comparisons (Dreibus & Wanke 1987; Prinn & Fegley 1987; Kasting 1988; Durham & Chamberlain 1989; Hunten 1993) of the three atmospheres, in particular, show how the emergence of life on Earth modified our atmosphere dramatically from the atmospheres of our two planetary neighbors.

We have already seen how the compositions and surface pressures of the atmospheres of the three planets differ markedly (Fig. 1.1). The surface temperatures are also very different: 732, 288, and 223 K, for Venus, Earth, and Mars respectively.

The temperatures can be interpreted in terms of the atmospheric compositions and pressures. Carbon dioxide and water vapor are both "greenhouse" gases. That is, they trap infrared radiation that would otherwise escape to space, and raise the temperature at the surface of the planet (Goody & Yung 1989; Wayne 2000). An estimate can be made of the temperatures that would be experienced on the planets without any atmosphere at all. A simple radiative equilibrium calculation, taking into account differing reflectivities of the planets, would suggest values of 227, 256, and 217 K, some 505, 32, and 6 K less than found, listed in the order Venus, Earth, Mars.

Because of its relative closeness to the Sun, Venus may have suffered from a "runaway greenhouse effect" in which a positive feedback process ultimately led to vaporization of all surface water (Pollack 1969; Kasting 1988). Water vapor makes a sizable contribution to atmospheric heating. Since vapor pressures rapidly increase with increasing temperature, thus further increasing trapping, there exists a mechanism for positive feedback in the greenhouse effect. Evaporation from a planetary surface will proceed either until the atmosphere is saturated with water or until all the

available water has evaporated. What happens on any particular planet will depend on the starting temperature in the absence of radiation trapping, since that will decide whether the vapor ever becomes saturated at the temperatures reached. On Mars and the Earth, the additional heating due to liberation of water vapor is not sufficient to prevent the vapor reaching saturation as ice or liquid. However, on Venus there comes a critical vapor pressure (~10 mbar) when the rate of heating begins to increase dramatically: that vapor pressure is never reached at the lower temperatures on Earth or Mars. As a result, the *P–T* curve for the atmospheric water vapor increases more slowly than that for the vapor–liquid phase-equilibrium curve. Condensation never occurs on Venus, and additional burdens of H_2O serve to increase the temperature even further. Certainly this positive feedback mechanism would explain why there is no surface water on Venus at the present day. Large amounts of water vapor could have been the dominant species in the early Venusian atmosphere, but photodissociation and escape of hydrogen to space would have removed most of the H_2O to leave the rather dry atmosphere now found (Hunten 1990, 1993). Venus contains quantities of carbon and nitrogen similar to those on Earth, but hydrogen is deficient. Water abundance on Venus is about $42 \, kg \, m^{-2}$ compared with $2.7 \times 10^6 \, kg \, m^{-2}$ on Earth. There is certainly no liquid water on the surface of Venus today, and the mixing ratio for water in the atmosphere is probably not more than 2×10^{-4}.

Most mechanisms identified as potentially important for escape of hydrogen from Venus discriminate strongly against loss of deuterium, because of the large escape velocity ($10.3 \, km \, s^{-1}$) from that planet. Enrichment of deuterium might therefore be expected if Venus had originally possessed a water-rich atmosphere. Several pieces of evidence support deuterium enrichment, although they are not unequivocal. The ion mass spectrometer on Pioneer–Venus detected a signal at $m/e = 2$ from the upper atmosphere that can be attributed to D^+, and interpretation of the intensity data would require D/H in the bulk atmosphere of ~10^{-2}. Mass peaks at $m/e = 18.01$ and 19.01

obtained in the lower atmosphere (below 63 km) with the large-probe neutral mass spectrometer may be caused by H_2O and HDO (although the m/e = 19 ion could be H_3O^+). If HDO is the source of the heavier ion, then D/H on Venus is $(1.6 \pm 0.2) \times 10^{-2}$, in agreement with the upper-atmospheric ion data. On Earth, D/H $\sim 1.6 \times 10^{-4}$ overall (and perhaps twice that value in the upper atmosphere, according to Spacelab 1 observations), so that the deuterium enrichment on Venus is 50–100, implying large quantities of water in the early history of the planet. This enrichment factor is the maximum that could arise, as we shall show shortly.

Outgassing might be expected to release materials with oxidation states similar to those for terrestrial volcanic gases ($[CO]/[CO_2] \sim 10^{-2}$; $[H_2]/[H_2O] \sim 10^{-2}$). However, at high Venusian temperatures, the gas-phase equilibrium

$$CO + H_2O \rightleftharpoons CO_2 + H_2 \qquad (1.4)$$

and reactions such as

$$2FeO + H_2O \rightarrow Fe_2O_3 + H_2 \qquad (1.5)$$

at the planetary surface would have increased the H_2 content relative to H_2O. Molecular hydrogen would thus have been the dominant gas in the early upper atmosphere of Venus.

Supersonic hydrodynamic outflow, powered by solar ultraviolet heating, would have resulted in the loss of H_2 to space. Interestingly, this flow would have entrained HD, thus sweeping deuterium away, until the mixing ratio of H_2 dropped below $\sim 2 \times 10^{-2}$. Only after this limit was passed would deuterium enrichment begin, regardless of how much water was originally present. Hydrogen, in the form of water, is now present at a mixing ratio of $\sim 2 \times 10^{-4}$, according to the Venera spectrophotometer data for 54 km altitude. Deuterium enrichment is thus limited to a factor of ~ 100, in accordance with the apparent measured value. The escape rate calculated for loss of H_2 would have exhausted the equivalent of the Earth's oceans in about 280 million years.

As the Venusian atmosphere progressed toward its contemporary water vapor content, additional hydrogen loss processes probably began to operate. Translationally "hot" hydrogen atoms can escape if their velocities exceed $10.3 \, km \, s^{-1}$, and can be generated on Venus by elastic collision between "hot" O^* and ambient H

$$O^* + H \rightarrow O + H^* \qquad (1.6)$$

A source of O^* on Venus is dissociative recombination of molecular oxygen ions, which can provide $239 \, kJ \, mol^{-1}$ excess translational energy, corresponding to a velocity of $\sim 5.5 \, km \, s^{-1}$. Approximately 15% of collisions between H possessing thermal velocities (at 300 K) and O^* will produce H^* in reaction (1.6) with speeds in excess of the $10.3 \, km \, s^{-1}$ escape velocity. Mariner 5 Lyman-α (H resonance) airglow observations showed that there is an H-atom component with an effective temperature of 1000 K in addition to the atoms that are thermally equilibrated at 300 K. Escape via this collisional mechanism could have reduced the hydrogen content from 2% to the contemporary 0.02% in about 4.2 Gyr. Probably both hydrodynamic and ionic-collisional mechanisms operated at the higher hydrogen abundances, with the hydrodynamic loss becoming less important as the water vapor content approached its present level. Whatever the detailed mechanism, the deuterium enhancement suggests that Venus was once much moister than it is now. The contemporary D/H ratio does not provide evidence for the loss of several oceans' worth of water, although detailed models of escape of hydrogen suggest that large quantities of water might once have been present. Massive loss of hydrogen from water brings with it the problem of disposal of the oxygen. It may be that the oxygen escaped to space along with the hydrogen; alternatively, oxidation of surface material would provide a plausible sink if the surface were molten. Another problem concerns the present-day escape of hydrogen from the atmosphere of Venus. Calculations put the time taken to exhaust hydrogen from the atmosphere at the contemporary escape rate at between 500 and 1500 Myr.

The longer time is perhaps compatible with a gradual depletion of water over the life of the planet, but if the shorter time is correct, then the implication is that water is being replenished as fast as it escapes. Such replenishment could be provided by outgassing from the planetary interior (and possibly by cometary impacts). Mixing ratios for water vapor drop by a factor of about five between 10 km altitude and the surface, suggesting that there is a large flux of water from the atmosphere into the surface, which could nearly balance a relatively large flux of juvenile water from the interior. Substantial oxidation of the surface would be expected with large water fluxes through it, and some results (e.g. from Venera 13) indicate the presence of Fe(III) minerals that are consistent with a relatively highly oxidized surface.

Mars presents a sharp contrast. Surface channeling features suggest strongly that there was once liquid water on the surface. There seems to be no liquid water now, but there is water vapor in the atmosphere and clouds are observed. The polar caps (and much of the winter hemisphere) are covered with water ice, and the winter polar cap probably contains solid CO_2 as well. The escape velocity for relatively small Mars is less than one-half (and the energy required for escape thus less than one-quarter) of that for either of the two larger planets. A large proportion of the outgassed species can therefore have been lost from the Martian atmosphere.

Mariner airglow data showed exceptionally large scale-heights in the upper atmosphere of Mars of oxygen (at mixing ratios of 5 to 10×10^{-3}) and hydrogen (at an almost constant concentration of 3×10^4 atom cm^{-3}), indicating clearly that these atomic species are escaping the gravitational field of the planet at the present day. The exobase lies at ~230 km on Mars. Although temperatures are relatively low (~320 K), thermal escape of hydrogen is possible because of the small value of g, and hence escape velocity, on Mars (~5 km s^{-1}). Escape fluxes of about 1.2×10^8 atom cm^{-2} s^{-1} of hydrogen are predicted for the measured concentration of 3×10^4 atom cm^{-3}. That any hydrogen remains in the exosphere therefore implies the existence of an equivalent source, presumably dissociation of a hydrogen-bearing molecule such as H_2O, that is currently operating in the Martian atmosphere. The four H atoms are liberated through the intermediacy of ionic processes. Furthermore, O-atom escape also involves ions, and the rates of oxygen and hydrogen escape processes are self-regulating to be equivalent to loss of H_2O. Indeed, ionic processes are capable of forming not just H and O atoms, but C and N atoms as well, that possess more than the Martian escape velocity. These processes are all "dissociative recombinations" of the type

$$AB^+ + e \rightarrow A + B \qquad (1.7)$$

in which the ionization energy of the AB molecule is released to both break the A–B bond and drive apart the newly formed fragments.

On Mars, escape of the ^{14}N isotope is slightly faster than that of ^{15}N because of the lower mass (McElroy *et al.* 1977). With the atmosphere as the reservoir of nitrogen, the N_2 remaining will have become slowly richer in ^{15}N over the life of the planet. In comparison with nitrogen on Earth, where neither isotope escapes, the Martian ^{15}N is enriched by a factor of 1.6. It follows that Mars once had ten or more times as much nitrogen in its atmosphere as it has now. Continuous degassing of nitrogen from the planet's interior would tend to sustain the original isotopic ratio, so that the observed enrichment favors an evolutionary model in which Mars acquired its nitrogen atmosphere early in its history, with relatively little degassing in later epochs. By way of contrast to the nitrogen isotopes, the Martian $^{16}O/^{18}O$ ratio is almost exactly the same as that for Earth, and ^{18}O has been enriched by less than 5%. Yet Mars is losing O atoms at present at the rate of 6×10^7 atoms s^{-1} for every square centimeter of surface. Lack of ^{18}O enrichment implies a source of "new" oxygen, in a reservoir holding at least 4.5×10^{25} atoms cm^{-2}, presumably in the form of H_2O, since the escape of hydrogen and oxygen from Mars is constrained to have a 2:1 stoichiometry. There is, however, apparently an enrichment of D over H by a factor of about five over the terrestrial value. If D/H for juvenile water on Mars is the same as that for the Earth, the observed enhancement must be explained by a divergent history of atmospheric evolution on the two planets. Several steps in the escape of hydrogen favor loss of H over loss of D, but, even so, the observed enrichments can only be attained if some of the (D-enriched) atmospheric water can exchange back with the condensed phase. Model calculations, based on the assumption that D/H in primordial Martian H_2O is the same as the terrestrial value, imply an initial reservoir of hydrogen equivalent to a layer of water about 3.6 m thick, most of which has escaped, to leave a present-day residue that is 0.2 m thick. The calculations assume, amongst other things, that the escape rate has remained constant over geological time. Nevertheless, the required exchangeable surface layer thickness is almost two orders of magnitude less than geological inventories of subsurface water, so that the postulated loss is not unreasonable. Presumably, the much weaker fractionation between ^{18}O and ^{16}O compared with that between D and H has prevented a measurable enhancement of the heavier oxygen isotope even in the presence of exchange with a modest surface reservoir.

Water on Earth is particularly interesting. Geological evidence (Worsley & Nance 1989) suggests that the waters of the ocean have neither completely frozen nor completely evaporated for at least the past 3.5×10^9 years. Yet the infrared luminosity of the Sun has increased substantially over that period (Gilliland 1989; Sagan & Chyba 1997), leading to the so-called "faint young Sun paradox" to which we return in Section 1.6 when we examine CO_2 concentrations in more detail. Thermostatic control seems to have been exerted by a reduction of the concentration of greenhouse gases in the Earth's atmosphere, probably effected directly or indirectly by the biota. Such control is, of course, looked at closely by the proponents of the Gaia hypothesis (Section 1.5). In the absence of water on Venus, carbonate rocks have probably never been formed, and most of the carbon dioxide remains in the atmosphere. The total burden of CO_2 is not much different on the Earth from that on Venus; it is just the distribution between reservoirs that differs. The total amounts of atmospheric nitrogen in the atmospheres of Venus

and the Earth are not dissimilar, either, if the total pressures of the atmospheres are taken into account. The much greater solubility of nitrates compared with carbonates means that much less nitrogen than CO_2 has been deposited as solid minerals on the Earth, and the atmospheric amounts are closer to those released by outgassing. Some CO_2 on Mars may be stored as solid at the winter poles, but the atmosphere also seems to have undergone much more physical evolution over geological time than the atmospheres of Venus or Earth as a result of the relatively small escape velocity.

1.5   LIFE AND THE EARTH'S ATMOSPHERE

Our atmosphere appears not to obey the laws of physics and chemistry: it is a disequilibrium mixture of chemical species (Lovelock 1979; Wayne 2000). Indeed, the atmosphere is like a low-temperature combustion system, and it contains a variety of easily oxidizable substances, such as H_2, CH_4, CO, H_2S, and hundreds of other minor constituents in the presence of molecular oxygen. Even the nitrogen and oxygen, the major components of the atmosphere, are not in thermodynamic equilibrium.

The key to the apparently anomalous composition of the atmosphere is the existence of living organisms: the biota bring about the entropy reduction by utilizing solar energy. Photosynthesis, as we shall see in Section 1.7, is the only mechanism that can account for the relative abundance of oxygen in our atmosphere. Other biological processes, most often microbiological, release many fully or partially reduced compounds to the atmosphere. Within the atmosphere itself, photochemical transformations convert the substances released biologically, so that photochemistry plays a central part in maintaining the disequilibrium of the atmospheric gases. It is highly instructive to examine what the probable composition of our atmosphere would be like if life were not present. The present-day and "reconstructed" atmospheres for the Earth displayed in Fig. 1.1 provide

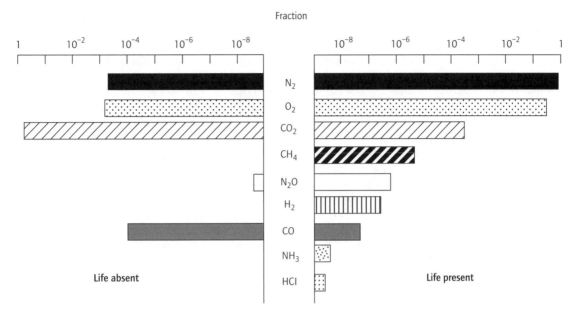

Fig. 1.3 Life's influence on Earth's atmosphere. The diagram shows, on a logarithmic scale, the mixing ratios for the major gases and some significant trace species found in our atmosphere in the presence of life and those expected in its absence. (From IGBP 1992, adapted from Margulis & Lovelock 1974.)

the comparison for CO_2, N_2, and O_2. Figure 1.3 illustrates the differences in more detail: it gives the mixing ratios for several gases actually found in our atmosphere, and those that might be expected if life were absent. Bearing in mind that the scale of this diagram is logarithmic, the enormous influence of the biota is immediately evident. Oxygen concentrations are at least one thousand times smaller than they are in the contemporary atmosphere (they are probably much lower again, as discussed in Section 1.7). The reduced compounds such as CH_4 and H_2 are virtually absent, and the only readily oxidizable species is CO, which on this view is more abundant in the absence of life (low O_2) than in its presence.

Earth's atmosphere has a composition that is influenced by the biota, as just outlined. Conversely, atmospheric composition is of evident importance for the biota. Feedbacks of this kind are frequently found in atmospheric chemistry. The composition, temperature, and pressure of the atmosphere can each be modulated in response to

biological activity, and these atmospheric parameters themselves have consequences for that activity. Temperature, for example, is determined in part by the "greenhouse" heating afforded by atmospheric CO_2; the partial pressure of CO_2 in the atmosphere is strongly dependent on photosynthetic activity (short term) and chemical weathering of silicate rocks (long term), which are themselves sensitive to temperature.

It is observations of these feedbacks that have led Lovelock and coworkers (Lovelock & Margulis 1974; Lovelock 1979, 1988, 1989; Volk 1998) to their "Gaia hypothesis." They see the interaction between life and the atmosphere as so intense that the atmosphere can be regarded as an extension of the biosphere: although the atmosphere is not living, it is a construction maintained by the biosphere. The Gaia hypothesis postulates that the climate and chemical composition of the Earth's surface and its atmosphere arc kept at an optimum by and for the biosphere. Naturally, this idea does not receive universal acceptance, and many workers have argued (Kirchner 1989) that the close links between atmosphere, oceans, and biosphere do not necessarily imply the existence of an adaptive control system. Nevertheless, the concept of Gaia forms an interesting framework for the discussion of the evolution of the atmosphere, the evolution of life, and the possible relationship between them.

Life is unlikely to have originated, or at least persisted, on Earth immediately after the main accretionary phase had ended (Kasting 1993). Significant numbers of impactors as large as 100 km in diameter continued to hit the Earth during the period of "heavy bombardment" which lasted from 4.5 to 3.8 Gyr BP, and, until 3.8 Gyr BP, the uppermost layers of the ocean would probably have been vaporized repeatedly by the impacts (Sleep *et al.* 1989). Sterilization of the planet would therefore have precluded the survival of life on Earth, even if it had appeared any earlier (Oberbeck & Fogelman 1989). Nevertheless, by 3.5 Gyr BP life was almost certainly extant (Schopf 1983, 1999; Schopf & Packer 1987; Oró *et al.* 1990; Kasting 1993). There is ample morphological evidence that documents the existence of microbiota in sediments that have not been metamorphosed (Oró *et al.* 1990). According to Schopf (1999), the oldest fossils meeting the criteria for dating authenticity are from 3.465 Gyr BP, and were found in the Apex chert of northwestern Australia. They consist of 11 types of prokaryotic threadlike organisms, and include several types of cyanobacteria that produce and consume oxygen. The extraordinary conclusion is that these most ancient of identified fossils are of organisms that were already surprisingly advanced, and they certainly suggest that life was established well before their formation. Circumstantial support for the evidence is given by several discoveries of stromatolites from the same period: these organo-sedimentary structures are similar to mats produced by present-day cyanobacteria, and microbial communities seem to have played an active role in their formation (Margulis *et al.* 1980; Schopf 1999). There is even indirect evidence (Schopf 1983) that life existed 3.8 Gyr BP, although supposed microfossils from

the Isua sedimentary deposits of this date may have nonbiological origins, and carbon-isotope indicators of biological activity in the same deposit may have been affected by metamorphic processes (Oró *et al.* 1990). Notwithstanding these reservations, Mojzsis and coworkers (Mojzsis *et al.* 1996, 1997; Eiler *et al.* 1997; Nutman *et al.* 1997) argue that their ion-microprobe measurements of carbon isotopes are indeed consistent with graphitic microdomains of bio-organic origin. The oldest samples come again from Isua in West Greenland (3.8 Gyr old) and from the neighboring island of Akilia (3.85 Gyr old: the oldest sediments yet documented). Holland (1997) has provided a useful critique of the use of geochemical data of this kind in tracing the evidence for life on Earth so far back. However, it does seem possible that the record of life on Earth may extend back more than 3.85 Gyr.

Because the origin of life must have depended on the prior existence of organic species, there has long been intense interest in the possible conversion of atmospheric gases to simple organic molecules. Among the early experiments, those of Miller (1953) aroused much interest, because it was shown that simulated lightning discharges passed through mixtures of methane and ammonia produced a wide range of organic compounds that included amino acids. The less reducing early atmosphere now thought probable (N_2, CO_2, and H_2O, together with traces of volcanic H_2 and CO: see first paragraph of Section 1.3) has more recently been shown to yield a variety of organic compounds when subject to ultraviolet irradiation, although the more reducing atmospheres generate a greater variety of compounds in larger yield (Oró *et al.* 1990). Bar-Nun and Chang (1983) found that continuous irradiation at $\lambda = 184.9$ nm of mixtures of CO and H_2O gave CO_2 and H_2 as the major products, and smaller quantities of CH_3OH, HCHO, and CH_4. Some C_2 molecules (C_2H_5OH, CH_3CHO, and C_2H_6) were also observed. Wen *et al.* (1989) have analyzed these experiments in terms of a photochemical kinetic scheme, and the simulated abundances of most of the products are in surprisingly good accord with the experimental findings. The main steps in the main pathway involve photolysis of water

$$H_2O + h\nu \rightarrow H + OH \qquad (1.8)$$

followed by termolecular addition of atomic hydrogen to CO to yield the formyl radical (HCO), which is then the precursor of the more complex organic molecules. Photolysis of carbon dioxide

$$CO_2 + h\nu \rightarrow CO + O \qquad (1.9)$$

is a continuous source of CO. Wen *et al.* (1989) have employed their scheme to predict photochemical production rates of the organic molecules in the prebiotic atmosphere to show that substantial quantities of organic material could be formed in this way.

As just noted, synthesis of organic compounds is less efficient in an environment dominated by N_2 rather than by NH_3. Brandes *et al.* (1998) have demonstrated the mineral-catalyzed reduction of N_2 to NH_3 at temperatures and pressures typical of crustal and oceanic hydrothermal conditions, and they speculate that, even though the prebiotic atmosphere was present predominantly as N_2, exchange with a mineral-catalyzed oceanic source might have provided substantial amounts of NH_3 to the atmosphere. Another, rather unusual, mechanism for the fixation of nitrogen in the early atmosphere has been proposed by Navarro-Gonzalez *et al.* (1998). Lightning discharges inside explosive volcanic clouds are estimated, on the basis of experimental laboratory simulations, to have produced as much as 10^9–10^{10} kg yr^{-1} of nitric oxide (NO) at 4 Gyr BP.

Methane seems an unavoidable requirement in any atmospheric source of prebiotic compounds. The CH_4 formed in the photolysis of the CO—CO—H$_2$O system could have supplemented that outgassed from the Earth. As explained at the end of Section 1.2, CH_4 is unstable against photolysis and, especially, attack by OH radicals (Levine *et al.* 1982; Levine 1985). However, the radical products of photolysis and attack (CH_2 and CH_3) can themselves participate in processes that yield more

complex organic species. Zahnle (1986) has pointed out an especially interesting reaction for CH_2 in the prebiotic atmosphere. Atomic nitrogen must have been abundant in the anaerobic middle atmosphere as a result of the occurrence of several processes involving photodissociation and photoionization. Nitrogen atoms react rapidly with (triplet) CH_2 to form hydrogen cyanide

$$N + {}^3CH_2 \rightarrow HCN + H \qquad (1.10)$$

thus opening up new vistas in organic chemistry. The chemistry of HCN in the contemporary atmosphere is not yet completely understood (Cicerone & Zellner 1983), but it seems that the molecule is relatively unreactive. Addition of OH and reaction of the adduct with O_2 presumably leads to complete oxidation in an oxygen-rich atmosphere. Photolysis of HCN

$$HCN + h\nu \rightarrow H + CN \qquad (1.11)$$

is followed in an oxygen atmosphere by reaction of CN with atomic or molecular oxygen and loss of the C—N bond. However, the CN radical is partially protected in a more reducing atmosphere because hydrogen abstraction regenerates HCN. Under these circumstances, reactions such as

$$CN + HCN \rightarrow C_2N_2 \qquad (1.12)$$

$$CN + C_2H_2 \rightarrow HCCCN \qquad (1.13)$$

offer routes to other interesting nitrogen-containing organic species. One of the most fascinating aspects of these speculations is the way in which they parallel our understanding of the chemistry of Titan's atmosphere (Wayne 2000), where organic photochemical aerosols and hazes are formed in an atmosphere consisting mainly of molecular nitrogen.

The possibility that atmospheric CH_4 fluxes were sufficient to permit occurrence of the chemistry described has recently received encouragement from suggestions that there was a significant extraterrestrial source of CO in the period of the Earth's history earlier than 3.8×10^9 yr BP. Kasting (1990) has estimated the effects of incoming comets and carbonaceous asteroids, whose rate of impact would have been high early on. These bolides bring in CO-ice and/or organic carbon that can be oxidized to CO in the impact plume. The elemental iron in ordinary chondritic impactors could further enhance CO by reducing CO_2. Nitric oxide (NO) is likely to be formed in a high-temperature reaction between N_2 and CO_2, and this gas also indirectly increases the [CO] to [CO_2] ratio. A photochemical model shows that, for a total atmospheric pressure of roughly 2 atm, the [CO] to [CO_2] ratio might even have exceeded unity at times more than 4.0×10^9 years BP.

Comets, invoked here as a source of CO, and other extraterrestrial sources have, of course, long been looked on as potential carriers of organic molecules to the Earth (Chyba *et al.* 1990; Shimoyama 1997; Irvine 1998), and even of life (Whittet 1997). This alternative view finds its most extreme expression in the suggestions of Hoyle and Wickramasinghe, who propose that interstellar molecules accumulated within the heads of comets as explained, for example, by Hoyle (1982) and by Wickramasinghe and Hoyle (1998). Chemical evolution occurring a few hundred meters below the cometary surface is seen as progressing as far as biopolymers and microorganisms. Hoyle and Wickramasinghe even contend that some past and present epidemics were initiated by the viruses and bacteria falling to Earth in cometary dust (Hoyle & Wickramasinghe 1977, 1983). However, there seem to be perfectly plausible ways in which organic molecules can be formed from the precursors already present in the atmosphere, and there are doubts about the survivability of amino acids and nucleobases subject to the pyrolytic conditions to which they would be exposed (Basiuk & Douda 1999). The stages between the appearance of the organic molecules on Earth and the appearance of life cannot concern us here, however fascinating those steps might be (Maurel & Decout 1999). An excellent review of the facts and speculations about the origin of life is provided by Orgel (1998). Our immediate interest here is the

evolution of life after it appeared, and the influence that it had on our atmosphere.

1.6 CARBON DIOXIDE IN EARTH'S ATMOSPHERE

Both the geochemical record and the persistence of life itself indicate that the oceans can never have either frozen or boiled in their entirety (Sagan and Mullen 1972; Owen *et al.* 1979; Kasting *et al.* 1988; Nutman *et al.* 1997). Mean surface temperatures have probably never departed from the range of 5–50°C, and may have been highest at very early periods. But this conclusion leads to a riddle! Standard stellar evolutionary models predict that at 4.5 Gyr BP the Sun's luminosity was lower than it is today by 25–30 per cent, as mentioned earlier (Gilliland 1989). We should note before proceeding further that speculative new models of solar physics have led to suggestions that the Sun may have lost mass as it evolved. A larger initial mass and luminosity would then result in larger radiative fluxes at the Earth's surface than predicted by the "standard" Sun model. The validity of such a model has not, however, been established. On the other hand, the standard model of evolution of solar intensity has created the "faint young Sun paradox," mentioned earlier, because it translates into a decrease of the Earth's effective temperature by 8%, low enough to keep sea-water frozen for ~2 Gyr, if the atmosphere possessed its present-day composition and structure. The paradox may be only illusory if atmospheric behavior was different 3–4 Gyr ago from what it is today. Explanations proposed include changes in albedo or increases in the greenhouse efficiency. Sagan and Chyba (1997), for example, argue that NH_3 concentrations might have been much higher than usually believed (see final paragraph of Section 1.2 and the discussion of mineral-catalyzed NH_3 formation in Section 1.5), and that this gas could have contributed to radiation trapping. Alterations in clouds could exert a negative-feedback, stabilizing, effect, since lower temperatures would mean decreased cloud cover and reduced reflection away of solar radiation (Rossow *et al.* 1982). Water vapor makes the largest contribution to the greenhouse effect in the contemporary atmosphere, but it is unlikely to be the agent of long-term temperature control. Its relatively high freezing and boiling points render its blanketing effect prone to unstable positive feedbacks by increasing ice and snow albedo at low temperatures (further reducing temperature), but by increasing water vapor content at high temperatures (yet further increasing the greenhouse effect). Whatever greenhouse gas or other mechanism kept the Earth warm, it must have been smoothly reduced to avoid exceeding the high-temperature limit for life. Carbon dioxide seems the most likely greenhouse gas to have exerted thermostatic control of our climate (Kasting 1993). Negative feedback mechanisms can be identified for this gas. Nonbiological control might include acceleration of the weathering of silicate minerals to carbonate deposits in response to increased temperatures. However, as discussed in Section 1.3, present-day weathering is biologically determined, and the biota both sense and amplify temperature changes. This feedback regulation of climate is seen by its proponents as evidence in support of the Gaia hypothesis (Section 1.5). The biota both increase the partial pressure of CO_2 in soils and generate humic acids. Each of these effects increases the rate of weathering, and thus of CO_2 loss.

Temperatures were likely to be higher in the very early history of the Earth than at present, perhaps by as much as 60°C as a global mean (Kasting 1993). Nevertheless, firm evidence exists for glaciations around 2.5–2.3 Gyr BP, and again at about 0.8 Gyr BP: according to some reports (Young *et al.* 1998) the earliest glaciation may even be dated back to ~2.9 Gyr BP. Figure 1.4 shows one back-projection of the history of atmospheric CO_2 concentrations using this kind of information (Kasting 1993). A radiative–convective climate model was employed in the calculations, in which radiation trapping was brought about by CO_2 and by H_2O (with a temperature-dependent vapor pressure). The constraint on the model was that it was required to maintain the Earth's mean surface temperature at 5–20°C at all times, and the entire surface kept at >0°C (ice free) at all times except

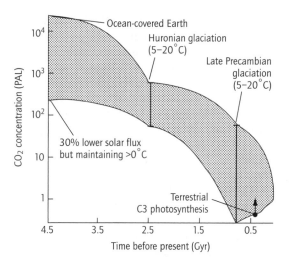

Fig. 1.4 Evolution of atmospheric carbon dioxide concentrations, given relative to the Present Atmospheric Level (PAL). The concentrations are estimated using a radiative–convective climate model to keep the Earth's temperatures within the limits marked on the diagram, and explained in the text. More [CO_2] than at present is needed early on to compensate for a Sun whose infrared intensity had not yet built up to its full intensity. The shading represents the range of concentrations permitted by the various indicators discussed. (Adapted with permission from Kasting 1993. Copyright American Association for the Advancement of Science.)

during the glaciations at 2.5 and 0.8 Gyr. Unfortunately, the probability zone (indicated by the shading) is rather wide, and encompasses concentration ranges of more than a factor of 100 at worst, and the zone narrows only as we approach, within a few hundred million years, the present day. The concentration soon after the formation of the Earth must have been >0.1 atm, or 300 PAL (present atmospheric level) to maintain a mean surface temperature in excess of freezing point, but may have been as high as the 70 atm (2×10^5 PAL) suggested by Fig. 1.1. Such concentrations seem reasonable if we accept that in the Earth's prebiological atmosphere CO_2 was a major component, as it is on Venus or Mars today. Although these calculations assume a constant cloud and surface albedo throughout, similar results are obtained using models that allow for changes in cloud albedo consequential on changes in temperature. Rye *et al.* (1995) have examined the weathering of paleosols (ancient soils: these were 2.75–2.2 Gyr old) in an attempt to assess atmospheric CO_2 levels. Iron lost from the tops of the sample profiles was precipitated lower down as silicate minerals rather than carbonate, indicating that $[CO_2]$ was roughly 100 PAL at 2.75–2.2 Gyr BP, just in the middle of the range shown in Fig. 1.4 for that period.

The model results just described assume that CO_2 and H_2O were the only important greenhouse gases present in the atmosphere. This is, indeed, likely to have been the case in the prebiotic atmosphere, as discussed earlier. After life appeared on the planet, gases like NH_3 and CH_4 could have been generated by several microbiological processes, and trapped additional radiation. If substantial quantities of these gases were present, then CO_2 concentrations might have been lower than those represented in Fig. 1.4 (Kasting 1993). However, this effect would be limited to the period before 2 Gyr BP, because after then concentrations of NH_3 and CH_4 would be expected to drop dramatically in response to the oxygen that then became abundant in the atmosphere. Such oxygen changes are the subject of the next section.

1.7   THE RISE OF OXYGEN CONCENTRATIONS

Oxygen concentrations can only have increased from their prebiotic levels to their present-day levels through photosynthetic generation of the molecule. The growth in atmospheric $[O_2]$ is thus clearly linked to the evolution of life on the planet, and in this section we attempt to summarize the evidence that can be used to infer $[O_2]$ at particular periods in the past.

In the absence of life, O_2 could have been formed only by inorganic photochemistry. Photolysis of water vapor to form OH and H (reaction 1.8) and of CO_2 to form CO and O (reaction 1.9) are the main primary steps, and secondary chemistry can lead to the formation of free O_2

$$OH + OH \rightarrow O + H_2O \qquad (1.14)$$

$$O + OH \rightarrow O_2 + H \qquad (1.15)$$

$$O + O + M \rightarrow O_2 + M \qquad (1.16)$$

In each case, the final stage is the combination of oxygen atoms in reaction 1.16. Two important limitations are placed on the amount of oxygen that can be produced. It has long been recognized that H_2O and CO_2 are photolysed by ultraviolet radiation in a spectral region that is absorbed by O_2 (say at $\lambda \leq 240$ nm for H_2O, and $\lambda \leq 230$ nm for CO_2). "Shadowing" by the O_2 thus self-regulates photolysis at some concentration. Secondly, photolyses of H_2O vapor or CO_2 do not, on their own, constitute net sources of O_2. Reactions 1.8 and 1.14–1.16 have the effect of converting two H_2O molecules to one O_2 molecule and four H atoms. Only if atomic hydrogen is lost by exospheric escape is there a gain in O_2, because otherwise H_2O is re-formed. Addition of CO_2 photolysis (reaction 1.9) to the scheme does not alter this conclusion, since the CO product interacts rapidly with OH

$$CO + OH \rightarrow CO_2 + H \qquad (1.17)$$

Water-vapor photolysis now follows a new route that involves the reaction sequence $3 \times (1.8) + 1 \times (1.9)$, $1 \times (1.17)$, $1 \times (1.14)$, and the net result is

$$\left(2H_2O + 3h\nu_{H_2O}\right) + \left(CO_2 + h\nu_{CO_2}\right) \rightarrow 2O + 4H \qquad (1.18)$$

However, the outcome is still that for every O_2 molecule formed, four H atoms must be lost. Escape is thus the crucial event, and the rate is determined by the transport of all hydrogen species through lower levels of the atmosphere to the exosphere. Loss of O_2, for example by reaction with crustal or oceanic Fe^{2+}, or with volcanic H_2, competes with production, and so further limits the amount of free O_2 that can build up without the help of photosynthesis.

Considerable difficulties arise in giving quantitative expression to the prebiological formation of oxygen because of uncertainties in the concentrations of precursor molecules (H_2O and CO_2), temperatures, and solar ultraviolet intensities. Concentrations of CO_2 might have been much greater before the gas was converted to carbonate deposits, and water-vapor levels would have been elevated had surface and atmospheric temperatures been higher than they are now. Young stars (the "T-Tauri" stars), which resemble the Sun at the age of a few million years, emit 10^3 to 10^4 times as much ultraviolet radiation as the present Sun. If enhanced solar ultraviolet intensity was available during the prebiological evolutionary period of our atmosphere, then the rates of photolysis of H_2O and CO_2 are greatly enhanced, and become a significant source of O_2, especially if $[CO_2]$ is high. Photochemical models developed for interpretation of the modern atmosphere can be adapted for the paleoatmosphere by incorporating appropriate source terms, temperature profiles, and boundary conditions. Such models suggest (Levine 1985) that prebiological O_2 at the surface would have been limited to about 2.5×10^{-14} of the present atmospheric level (PAL) had both $[CO_2]$ and solar ultraviolet intensities been at their current values. With 100 times more CO_2, and 300 times more ultraviolet radiation from the young Sun, the surface $[O_2]$ calculated is $\sim 5 \times 10^{-9}$ PAL. The geological record provides some further information about oxygen concentrations. The simultaneous existence of oxidized iron and reduced uranium deposits in early rocks (>2.2 Gyr old) requires $[O_2]$ to be more than 5×10^{-12} PAL, but less than 10^{-3} PAL; the values accommodated both by the model and by geochemistry thus seem to be roughly in the range 5×10^{-12} to 5×10^{-9} PAL.

Prebiological oxygen concentrations in the paleoatmosphere are of importance in two ways connected with the emergence of life. Organic molecules are susceptible to thermal oxidation and photooxidation, and are unlikely to have accumulated in large quantities in an oxidizing atmosphere. Living organisms can develop mechanisms that protect against oxidative

degradation, but they are still photochemically sensitive to radiation at $\lambda \leq 290$ nm. Life-forms known to us depend on an ultraviolet screen provided by atmospheric oxygen and its photochemical derivative, ozone, because DNA and nucleic acids are readily destroyed. Biological evolution therefore seems to have proceeded in parallel with the changes in our atmosphere from an oxygen-deficient to an oxygen-rich one. We return to this theme in Section 1.8.

It is worth noting that, although the tiny prebiological concentrations of oxygen preclude the existence of a useful oxygen and ozone shield against ultraviolet radiation, the low concentrations were probably essential in the early stages of the synthesis of complex organic molecules that became the basis of life. Organic molecules are susceptible to thermal oxidation and photooxidation, and could not have accumulated in large quantities in a strongly oxidizing atmosphere. Living organisms are known to develop mechanisms and structures that protect against oxidative degradation, and so are able to survive in atmospheres containing large amounts of oxygen. At about 1% of PAL, organisms can derive energy from glucose by respiration rather than by anaerobic fermentation, and they gain an energy advantage of a factor of 16. However, the fact remains that oxygen is toxic, and organisms have to trade off the energy advantage against the need to protect themselves from the oxidant.

Photosynthesis is now the dominant source of O_2 in the atmosphere. For the purposes of the present discussion, this complex and fascinating piece of chemistry can be represented by the simplified equation

$$nCO_2 + nH_2O + mh\nu \rightarrow (CH_2O)_n + nO_2$$
$$(1.19)$$

The essence of the process is the use of photochemical energy to split water and thus to reduce CO_2 to carbohydrate, shown here as $(CH_2O)_n$. Isotope experiments show that photosynthetically produced O_2 comes exclusively from the H_2O and not from the CO_2. According to one estimate (Wayne 2000), photosynthesis releases 400×10^{12} kg of oxygen annually at present. Since the atmosphere contains about 1.2×10^{18} kg of O_2, the oxygen must cycle through the biosphere in roughly 3000 years.

It is important to realize that substantial concentrations of oxygen can build up in the atmosphere only if the carbohydrate formed in the photosynthetic process is removed from contact with the atmospheric oxygen by some form of burial. Without such burial, spontaneous oxidation would rapidly reverse the changes brought about by photosynthesis. In the contemporary atmosphere, marine organic sediment deposition buries about 0.12×10^{12} kg yr^{-1} of carbon, and so releases about 0.32×10^{12} kg yr^{-1} of O_2. At that rate, atmospheric oxygen could therefore double in concentration in about 4×10^6 years. There are balancing processes, including geological weathering (e.g. of elemental carbon to CO_2, sulfide rocks to sulfate, and iron(II) rocks to iron(III)) and oxidation of reduced volcanic gases (e.g. H_2 and CO). Nevertheless, marked variations in atmospheric O_2 are likely to have occurred over geological time, and such changes are a central theme of this survey.

The rise in $[O_2]$ from its prebiotic levels has generally been linked to the geological time scale either on the basis of the stratigraphic record (see, for example, Fleet 1998; Lecuyer & Ricard 1999; Rasmussen & Buick 1999) of oxidized and reduced mineral deposits (including information about isotope ratios) or from fossil evidence combined with estimates of the oxygen requirements of ancient organisms. Evolution of the climate system (Walker 1990) is likely to have influenced the composition and mineralogy of sedimentary rocks. In Section 1.5, we saw that attempts had been made (Walker 1985; Durham & Chamberlain 1989; Worsley & Nance 1989) to use paleoclimatological evidence to suggest the history of $[CO_2]$ in the Earth's atmosphere. It has become increasingly apparent (Kasting 1987) that changes in atmospheric $[O_2]$ and $[CO_2]$ must be studied alongside each other, and that it is necessary to have a reliable estimate of past $[CO_2]$ in order to correctly infer past values of $[O_2]$.

As suggested in Section 1.5, living organisms have been present on our planet from a very early stage, and perhaps from as long ago as 3.85 Gyr BP. From that time on, there has thus been a potential photosynthetic source of O_2 which could lead to an increase of atmospheric concentrations over their very low prebiotic levels. Cloud (1972, 1983) noted that redbeds, which contain some iron in the higher oxidation state, III, are absent before roughly 2 Gyr BP, and that reduced minerals, such as uraninite, were generally formed before this date. Banded-iron formations (BIFs), which contain iron (II) rather than iron (III), were also formed only up until about 1.85 Gyr BP. An anoxic deep ocean is apparently required for the deposition of BIFs in order that iron can be transported over large distances as iron (II). Much evidence has now accumulated to confirm that there was a large change in atmospheric O_2 level just before 2 Gyr BP (Holland et al. 1989; Holland & Beukes 1990). Karhu and Holland (1996) have examined the isotopic composition of carbon in carbon sediments deposited between 2.6 and 1.6 Gyr BP. There is a large excursion in the amount of the ^{13}C isotope between 2.22 and 2.06 Gyr BP, which was probably related to an abnormally high rate of organic carbon deposition, and thus of O_2 production. There is thus clear evidence that atmospheric O_2 concentrations increased very significantly over this time period. A number of investigators have used chemical profiles of paleosols to reconstruct the evolution of atmospheric oxygen levels in the Earth's early history. Rye and Holland (1998) have provided a critical review of the data. Part of the problem lies in the identification of authentic paleosol material, but where the evidence seems firm, the profiles suggest a dramatic change in atmospheric oxygen concentrations during the period 2.2 to 2.0 Gyr BP. Every true paleosol older than 2.44 Gyr suffered significant loss of iron during weathering, indicating that the atmospheric pressure of O_2 was at $\lesssim 5 \times 10^{-4}$ atm (2.5×10^{-3} PAL) before that date. Indeed, iron loss from paleosol of age 2.245–2.203 Gyr is consistent with a partial pressure of $<4 \times 10^{-3}$ PAL. Nevertheless, the presence of redbeds, containing some Fe(III), immediately overlying these paleosols suggests that by about 2.2 Gyr BP there was a substantial (but unquantified) amount of O_2 present in the atmosphere. Iron loss is

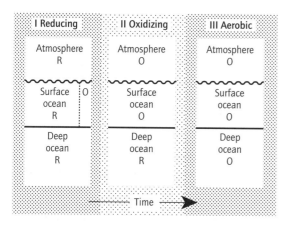

Fig. 1.5 The "three-stage–three-box" model used to represent the increase in oxygen concentrations in the Earth's atmosphere. The three boxes are for the atmosphere, the surface ocean, and the deep ocean: R shows that the conditions were reducing, and O that they were oxidizing. There may have been oxygen-rich "oases" in the surface ocean during stage I. (Adapted with permission from Kasting 1993. Copyright American Association for the Advancement of Science.)

negligible in paleosols aged 2.2–2.0 Gyr and in all younger samples.

The evidence just presented does suggest strongly that O_2 increased rather rapidly at around 2 Gyr BP. Rye and Holland (1998) even suggest that the concentration had reached as much as 0.15 PAL some time between 2.2 and 2.0 Gyr BP, although other studies would put the level nearer 0.01 PAL. Kasting (1993) explains the arguments in terms of his "three-stage–three-box" model (itself a simplification of an earlier suggestion of Walker *et al.* (1983), who proposed a four-stage model). Figure 1.5 shows the outline of the model. The three boxes are the atmosphere, the surface ocean, and the deep ocean, and the separate stages of the model correspond to one after the other of these compartments becoming oxidized or oxidizing, rather than being reducing. Stage I has all three compartments reducing (anoxic), with the possible exception of some regions of the surface ocean where oxygen production might be especially favorable. By stage II, the upper two boxes were

oxidizing, but the deep ocean remained anoxic. This arrangement accommodates a temporal overlap between BIF and redbed formation. Finally, in stage III, all parts of the oceans and atmosphere contained abundant free oxygen, as at present. Mass-balance considerations show that O_2 concentrations during stage II were ≤ 0.03 PAL, since otherwise O_2 conveyed to the deep ocean would have been lost more rapidly than it was produced in the upper levels. An oxic deep ocean in stage III implies an atmospheric O_2 concentration $\geq 2 \times 10^{-3}$ PAL in order to compensate for the influx of reductants from hydrothermal vents. Anoxic bottom waters may have persisted until well after the deposition of BIFs ceased (Canfield 1998) 1.85 Gyr BP, if sulfide rather than oxygen removed iron from deep-ocean water, and the aerobic deep-ocean may not have developed until 1.0–0.54 Gyr BP. Another line of evidence involves geologically stable biomarker molecules that can act as "molecular fossils." Biomarker lipids discovered in northwestern Australia have been shown (Brocks *et al.* 1999) to be almost certainly contemporaneous with the 2.7 Gyr old shales in which they are found, and thus provide fairly secure evidence for the presence of life. Some of the compounds are formed only by cyanobacteria ("blue-green algae"). Photosynthetic bacterial cells of this kind initially changed the atmosphere from its oxygen content of $<10^{-8}$ PAL toward much higher concentrations. The earliest cells lacked a nucleus, and are classified as prokaryotic; bacteria fall into this category, and can operate anaerobically. Larger cells that are almost certainly eukaryotic, or possessing a nucleus, are found in the fossil record. Vellai and Vida (1999) discuss the difference between prokaryotic and eukaryotic cells, and the origin of the eukaryotes. It was formerly thought that eukaryotic organisms first appeared about 1.4 Gyr BP. However, Han and Runnegar (1992) propose that the discovery of the corkscrew-shaped organism *Grypania* puts the fossil record of eukaryotic cells back to 2.1 Gyr BP. Furthermore, the Australian biomarkers include compounds most plausibly formed from eukaryotic membranes, so that these more complex cells may already have been present by 2.7 Gyr BP. The importance of this

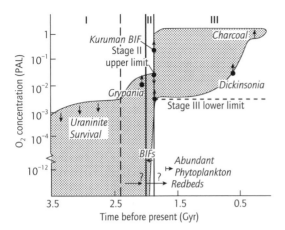

Fig. 1.6 The proposed timing for the three stages of Fig. 1.5. Various indicators of concentration are marked on the figure, and discussed in the text. The dashed vertical line indicates a possible earlier (2.4 Gyr BP) onset for stage II than is usually accepted. (Adapted with permission from Kasting 1993. Copyright American Association for the Advancement of Science.)

finding for the interpretation of atmospheric evolution is that almost all eukaryotic cells require large quantities of oxygen ($\geq$0.01 PAL) to function (Runnegar 1991). Cell division is preceded by a clustering and splitting of the chromosomes within the nucleus (mitosis), a process dependent on the protein actomyosin, which cannot form in the absence of oxygen. According to Kasting (1993), [O_2] ~ 0.01 PAL is below the upper limit for stage II of his model. It is even conceivable that such concentrations could have arisen during stage I, because the possibility exists of localized oxygen oases containing as much as 0.08 PAL. The argument is, then, that the presence of the very early eukaryotic cells does not necessarily imply that the atmosphere contained abundant (>0.01 PAL) free oxygen much before 2 Gyr BP.

The data admittedly permit alternative interpretations, but accepting the general approach adopted by Kasting (1993) leads to an evolutionary picture for O_2 such as that shown in Fig. 1.6. This illustration is in the same form as that in Fig. 1.4, and once again suffers from very wide probability ranges. On the figure are mapped out the several

mineralogical and biological markers that place constraints on the oxygen concentrations, and the three stages of Kasting's model are also indicated. One of the outstanding questions about the interpretation that we have adopted is the reason for the apparently sudden change in atmospheric O_2 concentrations at about 2 Gyr BP. Kasting (1993) reviews the evidence, and cites the original literature. Perhaps the transition from stage I to stage II in the model came about when the reduced and lower oxidation state mineral sinks on the surface had finally been almost all consumed. Or perhaps the transition in O_2 concentrations reflects a time when the net rate of photosynthetic production of O_2 became greater than the rate of oxygen consumption in reaction with volcanic H_2 and other reduced gases. A reduction in volcanic outgassing or an increased rate of carbon burial as the continental shelf area expanded are both possible explanations. These two scenarios would each be accompanied by a changing $^{13}C/^{12}C$ isotopic ratio. Until 1992, it was generally believed that the ratio remained constant over the critical period, in apparent conflict with either of the hypotheses. In that year, however, DesMarais et al. (1992) found that the isotope ratio in certain marine kerogens increased in two distinct steps, between 2.5–2.0 Gyr BP and 1.1–0.8 Gyr BP. The first of these increases might thus be linked to an increase in carbon burial rates and a concomitant increase in atmospheric oxygen levels at about 2 Gyr BP, in accordance with the evidence already presented. Of course, the data do not reveal whether the rise in atmospheric O_2 was brought about by an increased carbon burial rate, or whether it caused it (Kasting 1993). Measurements of the $^{15}N/^{14}N$ ratio yield broadly similar results (Beaumont & Robert 1999). The $^{13}C/^{12}C$ isotope ratio certainly increased more substantially in the period 600–550 Myr BP, and the effect has been ascribed (Derry et al. 1992) to a marked rise in the rate of carbon burial. At least some of the additional burial would have been accompanied by an increase in atmospheric oxygen concentrations, which is represented in the lower bound of the area of Fig. 1.6.

One important biological indicator at the "transition stage" for O_2 at 2 Gyr BP is the appearance of enlarged thick-walled cells on filaments of cyanobacteria found as fossils. The significance of this observation is that cell structures were developing in response to the need to protect the organisms against the quantities of O_2 by then appearing in the atmosphere. As the atmospheric oxygen reached about 10^{-2} PAL, a revolution occurred, because eukaryotic cells, and then animal and plant life, emerged. Respiration and large-scale photosynthesis became of importance, enough free oxygen became available for the fibrous protein collagen to be formed, and the scene was set for the appearance of metazoans, or multicelled species. About 550 million years ago, the Cambrian period opened. According to earlier ideas, this period heralded an "evolutionary explosion." In many ways, the real significance of the Cambrian period is that the first animals with clear external skeletons are preserved as fossils whose identity has been recognized for centuries, while remains of earlier life forms had not yet been discovered or understood. Metazoan fossils from the preceding 120 million years, the "Ediacarian" period, are now known. Many of these are from species resembling jellyfish. Such organisms can absorb their oxygen through the external surfaces at concentrations of about 7% of PAL. A reasonable estimate for when this level of oxygen was reached in the atmosphere can thus be set at about 670 Myr BP. The relatively impervious surface coverings of the Cambrian metazoans suggest that 120 million years later the oxygen concentration was approaching 10^{-1} PAL. Following the opening of the Cambrian, the complexity of life is known to have multiplied rapidly and the foundations for all modern phyla were laid. "Advanced" life forms (i.e. nonmicroscopic) were found ashore by the Silurian age (420 Myr BP), and by the Early Devonian, only 30 Myr later, great forests had appeared. Soon afterwards, amphibian vertebrates ventured onto dry land.

There is a further aspect of the rise of oxygen concentrations that will serve as an introduction to the next section. One view that will be presented is that ozone formed from oxygen in the atmosphere acts, together with the O_2 itself, as a screen for biologically damaging ultraviolet radiation. Until the atmospheric attenuation of ultraviolet intensity became sufficient, photosynthetic organisms such as the eukaryotic phytoplankton would not have been present in the oceans, and the rates of both carbon burial and oxygen generation would have been less than they are at present. The atmospheric shield would have become effective for oceanic organisms with $[O_2]$ in the range 0.01–0.1 PAL (Kasting 1987). These concentrations correspond with the most rapid rate of rise in the lower bound of $[O_2]$ shown in Fig. 1.6 (and, for that matter, in the upper bound). It seems, therefore, that a part of this inquiry into the rise of O_2 concentrations should include an examination of possible ways in which organisms could be protected from the adverse effects of exposure to ultraviolet radiation.

1.8 PROTECTION OF LIFE FROM ULTRAVIOLET RADIATION

The macromolecules, such as proteins and nucleic acids, that are characteristic of living cells are damaged by radiation of wavelength shorter than about 290 nm. Ozone in the atmosphere has been seen by many as the key species in reducing mid-ultraviolet intensities to a level at which life can survive on the surface of the planet. One outstanding feature of the properties of ozone is the relationship between its absorption spectrum and the protection of living systems from the full intensity of solar ultraviolet radiation. Major components of the atmosphere, especially O_2, filter out solar ultraviolet with wavelengths <230 nm; at that wavelength, only about one part in 10^{16} of the intensity of an overhead sun would be transmitted through the molecular oxygen. But at wavelengths longer than ~230 nm, the only species in the present-day atmosphere capable of attenuating the Sun's radiation is ozone. Ozone has an unusually strong absorption just at the critical wavelengths (230–290 nm), so that it is an effective filter despite its relatively small concentration. For example, at $\lambda = 250$ nm less than one part in 10^{30} of the incident (overhead) solar radiation penetrates the ozone layer.

Ozone is formed in the atmosphere from molecular oxygen, the necessary energy being supplied by the absorption of solar ultraviolet radiation. The oxygen in the Earth's contemporary atmosphere is largely biological in origin, and ozone, needed as an ultraviolet filter to protect life, is itself dependent on the atmospheric oxygen. These links further emphasize the special nature of Earth's atmosphere. Actually, the interactions are even more subtle than we have suggested. Absolute concentrations, and, indeed, the height distribution of ozone, depend on a competition between production and loss. Loss of ozone is regulated by chemistry involving some of the other trace gases of the atmosphere, such as the oxides of nitrogen, which are themselves at least partly of biological origin. Biological processes thus influence both the generation and destruction of ozone.

The starting point for many discussions of ozone concentrations in the paleoatmosphere is the work of Berkner and Marshall (1965). These workers suggested that life was able to evolve in response to increasing protection from solar ultraviolet radiation as the atmospheric ozone shield developed. In its original form, the thesis propounded that life would initially develop in stagnant pools where liquid water of 10 m depth or more would be able to filter out the damaging radiation. At this stage, life in the oceans would be unlikely since organisms would be brought too close to the surface to survive. As the atmospheric content of oxygen, and thus of protective ozone, increased, life could migrate from the safety of the pools and lakes to the oceans and, finally, to dry land. Accompanying these changes would be a greatly enhanced photosynthetic and, indeed, evolutionary activity. Marine biota certainly seem to have paved the way, in terms of modification of the atmosphere, for the evolution of the terrestrial biota (Raven 1997). At one time, the opening of the Cambrian period was thought to be characterized by an "explosion" of evolution, and it was an attractive idea that the dawn of the Cambrian was linked to the attainment in the atmosphere of an adequate ozone shield. However, as the discussion of Section 1.5 will have made clear, life was

abundant long before the Cambrian, and much of the reasoning for the evolutionary explosion was a result of earlier fossils not being recognized for what they were. All the same, complexity and diversity did increase rapidly after the opening of the Cambrian, and the possibility cannot be discounted that increased mobility following sufficient protection might have been responsible.

A major question attached to the interpretation just presented is whether the biological evolutionary events were linked causally to the atmospheric changes that undoubtedly occurred. If they were, then some kind of feedback mechanism may have been in operation, since the atmospheric evolution was certainly mediated by the biota. Resolution of this question will require further information: in the first place, it is necessary to put the time-history of growth of O_2 and O_3 in the atmosphere on a firmer footing than it seems to be at present. What we can do is to use atmospheric photochemical models to calculate the ozone concentrations that accompanied smaller O_2 levels in the early atmosphere. Figure 1.7 shows the results of one such calculation, in which the full chemistry of catalytic cycles (Levine 1982) was incorporated.

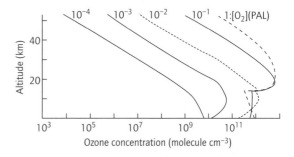

Fig. 1.7 Vertical distribution of ozone for different total atmospheric oxygen contents ranging from 10^{-4} to 1 PAL. The model used to obtain these results includes nitrogen, hydrogen, carbon, and chlorine chemistry; and allowance is made for ozone loss at the planetary surface. (Copyright Richard P. Wayne, 1985, 1991, 2000. Reprinted from *Chemistry of Atmospheres: An Introduction to the Chemistry of the Atmospheres of Earth, the Planets and Their Satellites*, by Richard P. Wayne (3rd edition, 2000) by permission of Oxford University Press.)

One interesting feature of the evolution of ozone concentrations in our atmosphere is immediately apparent. At low $[O_2]$, maximum ozone concentrations were found near the surface, but as oxygen concentrations increased, an ozone layer developed with its peak at successively higher altitudes.

Whatever shield is needed by the developing organisms, it is still difficult to specify what flux of radiation is dangerous to microorganisms. Experiments performed in space (Rettberg *et al.* 1998) provide a way of examining biological response to the full spectrum of solar ultraviolet radiation. Kasting *et al.* (1989) report a maximum allowable dose for stable heredity and clone survival that would be acquired in approximately 0.3 s on an unprotected Earth. Absorption of radiation by DNA falls off at wavelengths longer than the broad maximum at $\lambda = 240$–280 nm, but most genetic damage may be caused at the longest wavelengths absorbed (about 302 nm). Damage would occur primarily at the few hours near midday, when the Sun is most nearly overhead. For organisms whose generation span is one day or less, the maximum exposure is thus about 4 h of high-intensity light. Experiments on genetic damage to corn pollen suggest that the maximum acceptable dose over the period of 4 h would be $0.1 \, J \, m^{-2} s^{-1}$ for $\lambda \leq 302$ nm. An adequate UV screen would be provided by an ozone column density of 7×10^{18} molecule cm^{-2}. Other calculations of François and Gérard (1988) indicate that, at these densities, roughly 10% of a colony of cyanobacteria would survive. If a survival rate of 10^{-3} is acceptable, then the critical column density of ozone is 4.5×10^{18} molecule cm^{-2} for the most sensitive organisms. According to the relation between $[O_3]$ and $[O_2]$ given by François and Gérard, this limit is set at $[O_2]$ rather below 0.01 PAL. The calculations of Levine (1982), on the other hand, would require $[O_2]$ near 0.1 PAL. Whichever number is closer to the truth, it is abundantly clear that there would be insufficient ozone in the atmosphere to afford adequate protection for land-based life during the initial stages of oxygen growth. The largest prebiotic levels of ozone, based on the most extreme scenario (Canuto *et al.* 1982, 1983a, b), come nowhere near the lowest quantities needed. Enough ozone would not be present

until at least approximately 2 Gyr BP, and possibly not until the opening of the Cambrian (say 550 Myr BP). Figure 1.8 summarizes the material presented so far in this section. The growth in surface $[O_2]$ is identified on this plot by markers denoting the geochemical and fossil evidence; it corresponds roughly to the higher limits represented in Fig. 1.6, but with a less abrupt rise at the 2 Gyr transition. Column ozone abundances calculated using the model just described (Levine 1982) are also displayed in the figure. One problem is to decide what flux of ultraviolet radiation is tolerable to life, and hence what column density of ozone furnishes an adequate screen. Levine's own estimates and his O_3–O_2 relationship suggest that protection would be sufficient at $[O_2] \sim 0.1$ PAL.

It is also interesting to examine the emergence of life onto dry land in relation to the growth in protection by atmospheric ozone. For $[O_2] \leq 0.01$ PAL, a layer of water of thickness >4–5 m is needed for additional protection; by $[O_2] \sim 0.1$ PAL, water is probably no longer required. Does the transition to $[O_2] \geq 0.1$ PAL then really explain the appearance of life on dry land? Shelled organisms require dissolved oxygen that would be in equilibrium with >0.1 PAL in the atmosphere, so that the critical level of O_3 for biological protection would have already been passed when the organisms appeared abundantly in the Cambrian period (550 Myr BP). According to the arguments presented in this chapter, even in the worst case, there must have been an adequate atmospheric filter by 550 Myr BP at the latest to protect life on dry land from solar UV radiation. Yet life was apparently not firmly established on land until some 170 Myr later, toward the end of the Silurian. Thus the possibility exists that the development of the ozone shield was not directly linked to the spread of life onto land. Such a linkage would certainly be regarded favorably by the supporters of the Gaia hypothesis (see Section 1.5), and the connection between the emergence of life out of water and the filtering of ultraviolet radiation by the atmosphere remains tantalizing.

Some researchers (Rambler & Margulis 1980) have argued that an early atmospheric ultraviolet screen was not needed at all. Many present-day

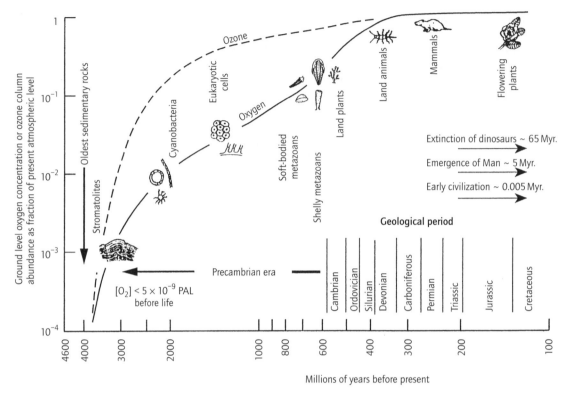

Fig. 1.8 Evolution of ozone (dashed line) on Earth in responses to the changes in oxygen levels. The oxygen concentration (continuous line) follows similar trends to those shown in Fig. 1.6, with a rather less abrupt increase at ~2 Gyr BP. Various geological and biological indicators of O_2 concentration shown on the diagram are discussed in the text. (Copyright Richard P. Wayne, 1985, 1991, 2000. Reprinted from *Chemistry of Atmospheres: An Introduction to the Chemistry of the Atmospheres of Earth, the Planets and Their Satellites*, by Richard P. Wayne (3rd edition, 2000) by permission of Oxford University Press.)

bacteria seem rather resistant to ultraviolet sterilization techniques, and it is known that other organisms can protect themselves by producing coatings that absorb in the ultraviolet. However, such sophisticated defense mechanisms were unlikely to be available to the simplest early cells. Perhaps we must look to the cyanobacteria that can protect themselves by forming covering mats (Schopf 1999), or to protection by prebiotic organic polymers and inorganic absorbers in the oceans (Cleaves & Miller 1998). Lovelock (1988) has proposed that an atmospheric screen could have been afforded by a hydrocarbon-particle smog, initiated by photolysis of CH_4, well before 2.5 Gyr BP, and

far earlier than the ozone screen is generally thought to have developed (although Towe (1990, 1996), for example, has argued for an ozone screen by this period). This idea of a hydrocarbon aerosol screen has subsequently been taken up again by Sagan and Chyba (1997), as discussed at the end of Section 1.2. Yet another suggestion originated with Kasting *et al.* (1989), who proposed that SO_2 and H_2S, released as volcanic gases, could be photolyzed to yield oligomers of sulfur (especially S_8) within the atmosphere, and that these particles could absorb and scatter solar ultraviolet radiation to an extent adequate for the survival of life on the planetary surface.

1.9   CONCLUSIONS

Despite these various imaginative and speculative ideas, it is widely accepted that ozone is at present the key atmospheric filter for ultraviolet radiation in the region $\lambda \sim 200\text{--}300\,\text{nm}$. Indeed, much of the research described in this chapter owes its existence to the perceived threat of destruction of the ozone layer as a consequence of Man's release of various species that can act as catalysts for removal of stratospheric ozone (Wayne 2000). Lovelock and Whitfield (1982) looked far into the future to raise the question of how much longer the biosphere could survive. They argue that the Sun's luminosity is likely to have increased by 25–30% over the past 4.5 Gyr, and that there is no reason to suppose that the trend has ceased. Whether by active or passive control, changes in carbon dioxide concentration appear to have compensated for the increasing solar intensity in such a way as to keep the planetary temperature very nearly constant. But the capacity for control might now nearly be exhausted, because $[CO_2]$ is approaching the lower limit tolerable for photosynthesis. If that limit is taken to be 150 p.p.m., then the CO_2 control of a climate favorable for life can continue for another 30–300 Myr. Some adaptation to lower CO_2 concentrations and to higher temperature is possible, but according to Lovelock and Whitfield it would not buy much time. Caldeira and Kasting (1992) are more encouraging! They employed a more elaborate model that treats greenhouse heating more accurately, includes biologically mediated weathering, and allows for some plant photosynthesis to persist to $[CO_2] < 10\,\text{p.p.m.}$ This treatment gives the biosphere another 0.9–1.5 Gyr of life, after which Earth might lose all its water to space within a further 1 Gyr. The outcome will then be that the Earth's atmosphere will perhaps once again come to resemble that of its sister, Venus, as it did before the evolution of life modified it.

1.10   FURTHER READING

The References provide literature citations for specific points raised in the text. More generally, there are several excellent sources of information about the origin of the atmosphere, its evolution, and the connection with the origin and evolution of living organisms. The books by Levine (1985), Lewis (1997), Wayne (2000), and Yung and DeMore (1999) are primarily about the present-day solar system and the atmospheres of the planets and satellites in it, but they do contain fairly extended treatments of the origin and evolution of the atmospheres. Lewis and Prinn (1984) and Walker (1977) are specifically concerned with the origin and evolution of atmospheres; although the books are now rather old, the principles presented still guide current thinking on the subject. Lovelock (1979, 1988) emphasizes the interaction between life and the atmosphere. Schopf and Klein (1992) edited a multi-author volume that contains contributions that remain authoritative guides to many of the topics touched upon in the present chapter. The more recent book by Schopf (1999) provides a very readable introduction to several matters of interest here, including the origin of life, the oldest fossils and their interpretation, stromatolites, and the importance of the cyanobacteria. Shorter papers of relevance to our inquiry include those by Blake and Carver (1977), Canuto *et al.* (1983a), Nunn (1998), Pepin (1991), Prinn (1982), and Walker (1980). To round off this list, we should mention two very accessible articles published in *Science* that provide an excellent overview of our subject: those by Hunten (1993), where the emphasis is on Venus and Mars; and by Kasting (1993), which provides the complement to Hunten's presentation.

REFERENCES

Ahrens, T.J. (1993) Impact erosion of terrestrial planetary atmospheres. *Annual Review of Earth and Planetary Sciences* **21**, 525–55.

Atreya, S.K., Pollack, J.B. & Matthews, M.S. (eds) (1989) *Origin and Evolution of Planetary and Satellite Atmospheres.* University of Arizona Press, Tucson.

Atreya, S.K., Edgington, S.G., Gautier, D. & Owen, T.C. (1995) Origin of the major planet atmospheres—clues from trace species. *Earth, Moon and Planets* **67**, 71–5.

Bachiller, R., Forveille, T., Huggins, P.J. & Cox, P. (1997)

The chemical evolution of planetary nebulae. *Astronomy and Astrophysics* **324**, 1123–34.

Bar-Nun, A. & Chang, S. (1983) Photochemical reactions of water and CO in Earth's primitive atmosphere. *Journal of Geophysical Research* **88**, 6662–72.

Basiuk, V.A. & Douda, J. (1999) Pyrolysis of simple amino acids and nucleobases: survivability limits and implications for extra-terrestrial survivability. *Planetary and Space Science* **47**, 577–84.

Beaumont, V. & Robert, F. (1999) Nitrogen isotope ratios of kerogens in Precambrian cherts: a record of the evolution of atmospheric chemistry? *Precambrian Research* **96**, 63–82.

Benz, W., Slattery, W.L. & Cameron, A.G.W. (1986) The origin of the moon and the single-impact hypothesis, 1. *Icarus* **66**, 515–35.

Benz, W., Slattery, W.L. & Cameron, A.G.W. (1987) The origin of the moon and the single-impact hypothesis, 2. *Icarus* **71**, 30–45.

Berkner, L.V. & Marshall, L.C. (1965) On the origin and rise of oxygen concentration in the Earth's atmosphere. *Journal of Atmospheric Science* **22**, 225–61.

Blake, A.J. & Carver, J.H. (1977) The evolutionary role of atmospheric ozone. *Journal of Atmospheric Sciences* **34**, 720–8.

Brandes, J.A., Boctor, N.Z., Cody, G.D., Cooper, B.A., Hazen, R.M. & Yoder, H.S. (1998) Abiotic nitrogen reduction on the early Earth. *Nature* **395**, 365–7.

Brocks, J.J., Logan, G.A., Buick, R. & Summons, R.E. (1999) Archean molecular fossils and the early rise of eukaryotes. *Science* **285**, 1033–6.

Brush, S.G. (1990) Theories of the origin of the solar system 1956–1985. *Reviews of Modern Physics* **62**, 43–112.

Caldeira, K. & Kasting, J.F. (1992) The life span of the biosphere revisited. *Nature* **360**, 721–3.

Cameron, A.G.W. (1983) Origin of the atmospheres of the terrestrial planets. *Icarus* **56**, 195–201.

Cameron, A.G.W. (1988) Origin of the solar system. *Annual Reviews of Astronomy and Astrophysics* **26**, 441–72.

Canfield, D.E. (1998) A new model for Proterozoic ocean chemistry. *Nature* **396**, 450–3.

Canuto, V.M., Levine, J.S., Augustsson, T.R. & Imhoff, C.L. (1982) UV radiation from the young Sun and oxygen and ozone levels in the prebiological palaeoatmosphere. *Nature* **296**, 816–20.

Canuto, V.M., Levine, J.S., Augustsson, T.R. & Imhoff, C.L. (1983a) Oxygen and ozone in the early Earth's atmosphere. *Precambrian Research* **20**, 109–20.

Canuto, V.M., Levine, J.S., Augustsson, T.R., Imhoff, C.L. & Giampapa, M.S. (1983b) The young Sun and the atmosphere and photochemistry of the early Earth. *Nature* **305**, 281–6.

Chyba, C.F., Thomas P.J., Brookshaw, L. & Sagan, C. (1990) Cometary delivery of organic molecules to the early Earth. *Science* **249**, 366–73.

Cicerone, R.J. & Zellner, R. (1983) The atmospheric chemistry of hydrogen cyanide (HCN). *Journal of Geophysical Research* 88, 689–96.

Cleaves, H.J. & Miller, S.L. (1998) Oceanic protection of prebiotic compounds from UV radiation. *Proceedings of the National Academy of Sciences of the USA* **95**, 7260–3.

Cloud, P.E. (1972) A working model of the primitive Earth. *American Journal of Science* **272**, 537–48.

Cloud, P.E. (1983) The biosphere. *Scientific American* **249**(September), 132–89.

Cox, P.A. (1989) *The Elements: Their Origin, Abundance, and Distribution.* Oxford University Press, Oxford.

Cox, P.A. (1995) *The Elements on Earth.* Oxford University Press, Oxford.

Deming, D. (1999) On the possible influence of extraterrestrial volatiles on Earth's climate and the origin of oceans. *Palaeogeography, Palaeoclimatology, and Palaeoecology* **146**, 33–51.

Derry, L.A., Kaufman, A.J. & Jacobsen, S.B. (1992) Sedimentary cycling and environmental-change in the late Proterozoic—evidence from stable and radiogenic isotopes. *Geochimica et Cosmochimica Acta* **56**, 1317–29.

DesMarais, D.J., Strauss, H., Summons, R.E. & Hayes, J.M. (1992) Carbon isotope evidence for the stepwise oxidation of the Proterozoic environment. *Nature* **359**, 605–9.

Dreibus, G. & Wanke, H. (1987) Volatiles on Earth and Mars—a comparison. *Icarus* **71**, 225–40.

Durham, R. & Chamberlain, J.W. (1989) A comparative study of the early terrestrial atmospheres. *Icarus* **77**, 59–66.

Eiler, J.M., Mojziz, S.J. & Arrhenius, G. (1997) Carbon isotope evidence for early life. *Nature* **386**, 665.

Farley, K.A. & Neroda, E. (1998) Noble gases in the Earth's mantle. *Annual Reviews of Earth and Planetary Science* **26**, 189–218.

Fleet, M.E. (1998) Detrital pyrite in Witwatersrand gold reefs: X-ray diffraction evidence and implications for atmospheric evolution. *Terra Nova* **10**, 302–6.

François, L.M. & Gérard, J.-C. (1988) Ozone, climate, and biospheric environment in the ancient oxygen-poor atmosphere. *Planetary and Space Science* **36**, 1391–414.

Frayer, D.T. & Brown, R.L. (1997) Evolution of the abundance of CO, O_2, and dust in the early universe. *Astrophysical Journal Supplement Series* **113**, 221–43.

Gilliland, R.L. (1989) Solar evolution. *Global and Planetary Change* **75**, 35–55.

Goody, R.M. & Yung, Y.L. (1989) *Atmospheric Radiation*. Oxford University Press, Oxford.

Han, T.-M. & Runnegar, B. (1992) Megascopic eukaryotic algae from the 2.1-billion-year-old Negaunee iron formation, Michigan. *Science* **257**, 232–5.

Holland, H.D. (1978) *The chemistry of the atmosphere and oceans.* John Wiley, Chichester.

Holland, H.D. (1997) Evidence for life on Earth more than 3850 million years ago. *Science* **275**, 38–9.

Holland, H.D. & Beukes, N.J. (1990) A paleoweathering profile from Griqualand West, South Africa—evidence for a dramatic rise in atmospheric oxygen between 2.2 and 1.9 by BP. *American Journal of Science* **290A**, 1–34.

Holland, H.D., Feakes, C.R. & Zbinden, E.A. (1989) The Flin Flon paleosol and the composition of the atmosphere 1.8 BY BP. *American Journal of Science* **289**, 362–89.

Hoyle, F. (1982) Comets—a matter of life and death. *Vistas in Astronomy* **24**, 123–39.

Hoyle, F. & Wickramasinghe, N.C. (1977) Does epidemic disease come from space? *New Scientist* **76**, 402–4.

Hoyle, F. & Wickramasinghe, N.C. (1983) Bacterial life in space. *Nature* **306**, 420.

Hunten, D.M. (1990) Escape of atmospheres—ancient and modern. *Icarus* **85**, 1–20.

Hunten, D.M. (1993) Atmospheric evolution of the terrestrial planets. *Science* **259**, 915–20.

IGBP (1992) *Global Change: Reducing Uncertainties.* International Geosphere–Biosphere Programme, Stockholm.

Irvine, W.M. (1998) Extraterrestrial organic matter: a review. *Origins of Life and Evolution of the Biosphere* **28**, 365–83.

Javoy, M. (1997) The major volatile elements of the Earth: their origin, behavior, and fate. *Geophysical Research Letters* **24**, 177–80.

Javoy, M. (1998) The birth of the Earth's atmosphere: the behavior and fate of its major elements. *Chemical Geology* **147**, 11–25.

Kamijo, K., Hashizume, K. & Matsuda, J. (1998) Noble gas constraints on the evolution of the atmosphere–mantle system. *Geochimica et Cosmochimica Acta* **62**, 2311–21.

Karhu, J.A. & Holland, H.D. (1996) Carbon isotopes and the rise of atmospheric oxygen. *Geology* **24**, 867–70.

Kasting, J.F. (1987) Theoretical constraints on oxygen and carbon dioxide concentrations in the Precambrian atmosphere. *Precambrian Research* **34**, 205–39.

Kasting, J.F. (1988) Runaway and moist greenhouse atmospheres and the evolution of Earth and Venus. *Icarus* **74**, 472–94.

Kasting, J.F. (1990) Bolide impacts and the oxidation state of carbon in the Earth's early atmosphere. *Origins of Life* **20**, 199–231.

Kasting, J.F. (1993) Earth's early atmosphere. *Science* **259**, 920–6.

Kasting, J.F., Toon, O.B. & Pollack, J.B. (1988) How climate evolved on the terrestrial planets. *Scientific American* **258**(February), 46–53.

Kasting, J.F., Young, A.T., Zahnle, K.J. & Pinto, J.P. (1989) Sulfur, ultraviolet radiation, and the early evolution of life. *Origins of Life* **19**, 95–108.

Kasting, J.F., Eggler, D.H. & Raeburn, S.P. (1993) Mantle redox evolution and the oxidation state of the Archean atmosphere. *Journal of Geology* **101**, 245–57.

Kirchner, J.W. (1989) The Gaia hypothesis—can it be tested? *Reviews of Geophysics* **27**, 223–35.

Krasnopolsky, V.A. (1986) *Photochemistry of the Atmospheres of Mars and Venus.* Springer-Verlag, Berlin.

Kuhn, W.R. & Atreya, S.K. (1979) Ammonia photolysis and the greenhouse effect in the primordial atmosphere of the Earth. *Icarus* **37**, 207–13.

Lecuyer, C. & Ricard, Y. (1999) Long-trem fluxes of ferric iron: implication for the redox states of the Earth's mantle and atmosphere. *Earth and Planetary Science Letters* **165**, 197–211.

Levine, J.S. (1982) The photochemistry of the paleoatmosphere. *Journal of Molecular Evolution* **18**, 161–72.

Levine, J.S. (ed.) (1985) *The Photochemistry of Atmospheres.* Academic Press, Orlando, FL.

Levine, J.S., Augustsson, T.R. & Natarajan, M. (1982) The prebiological paleoatmosphere: stability and composition. *Origins of Life* **12**, 245–59.

Lewis, J.S. (1997) *Physics and Chemistry of the Solar System*, rev. edn. Academic Press, San Diego.

Lewis, J.S. & Prinn, R.G. (1984) *Planets and Their Atmospheres: Origin and Evolution.* Academic Press, Orlando, FL.

Lovelock, J.E. (1979) *Gaia: A New Look at Life on Earth.* Oxford University Press, Oxford.

Lovelock, J.E. (1988) *The Ages of Gaia: A Biography of Our Living Earth.* Oxford University Press, Oxford.

Lovelock, J.E. (1989) Geophysiology, the science of Gaia. *Reviews of Geophysics* **27**, 215–22.

Lovelock, J.E. & Margulis, L. (1974) Atmospheric homeostasis by and for the biosphere. *Tellus* **26**, 2–9.

Lovelock, J.E. & Whitfield, M. (1982) Life-span of the biosphere. *Nature* **296**, 561–3.

Lunine, J.I. (1997) Physics and chemistry of the solar nebula. *Origins of Life* **27**, 205–24.

Lupton, J.E. (1983) Terrestrial inert gases: isotope tracer studies and clues to primordial components in the mantle. *Annual Reviews of Earth and Planetary Science* **11**, 371–414.

McElroy, M.B., Kong, T.Y. & Yung, Y.L. (1977) Photochemistry and evolution of Mars' atmosphere: a Viking perspective. *Journal of Geophysical Research* **82**, 4379–88.

Margulis, L. & Lovelock, J.E. (1974) Biological modulation of the Earth's atmosphere. *Icarus* **21**, 471–89.

Margulis, L., Barghoorn, E.S., Asherdorf, D., Banerjee, S. & Chase, D. (1980) The microbial community in the layered sediments at Laguna Figueroa, Baja California, Mexico: does it have Precambrian analogs? *Precambrian Research* **11**, 93–123.

Maurel, M.C. & Decout, J.L. (1999) Origins of life: molecular foundations and new approaches. *Tetrahedron* **55**, 3141–82.

Miller, S.L. (1953) A production of amino acids under possible primitive Earth conditions. *Science* **117**, 528–9.

Morrison, D. & Owen, T. (1996) *The Planetary System*. Addison-Wesley, Reading, MA.

Mojzsis, S.J., Arrhenius, G., McKeegan, K.D., Harrison, T.M., Nutman, A.P. & Friend, C.R.L. (1996) Evidence for life on Earth before 3800 million years ago. *Nature* **384**, 55–9.

Mojzsis, S.J, Arrhenius, G., McKeegan, K.D., Harrison, T.M., Nutman, A.P. & Friend, C.R.L. (1997) Evidence for life on Earth before 3800 million years ago. *Nature* **386**, 738.

Navarro-Gonzalez, R., Molina, M.J. & Molina, L.T. (1998) Nitrogen fixation by volcanic lightning in the early Earth. *Geophysical Research Letters* **25**, 3123–6.

Newman, W.I., Symbalisty, E.M.D., Ahrents, T.J. & Jones, E.M. (1999) Impact erosion of planetary atmospheres. *Icarus* **138**, 224–40.

Nunn, J.F. (1998) Evolution of the atmosphere. *Proceedings of the Geologists' Association* **109**, 1–13.

Nutman, A.P., Mojzsis, S.J. & Friend, C.R.L. (1997) Recognition of ≥ 3850 Ma water-lain sediments in West Greenland and their significance for the early Archean Earth. *Geochimica et Geologica Acta* **61**, 2475–84.

Oberbeck, V.R. & Fogelman, O. (1989) Estimates of the maximum time required to originate life. *Origins of Life* **19**, 549–60.

Orgel, L.E. (1998) The origin of life — a review of facts and speculations. *Trends in Biochemical Sciences* **23**, 491–5.

Oró, J., Miller, S.L. & Lazcano, A. (1990) The origin and early evolution of life on Earth. *Annual Reviews of Earth and Planetary Science* **18**, 317–56.

Owen, T., Cess, R.D. & Ramanathan, V. (1979) Enhanced greenhouse effect to compensate for reduced solar luminosity on early Earth. *Nature* **277**, 640–2.

Pepin, R.O. (1987) Volatile inventories of the terrestrial planets. *Reviews of Geophysics* **25**, 293–6.

Pepin, R.O. (1989) Atmospheric compositions: key similarities and differences. In Atreya, S.K., Pollack, J.B. & Matthews, M.S. (eds) *Origin and Evolution of Planetary and Satellite Atmospheres*. University of Arizona Press, Tucson.

Pepin, R.O. (1991) On the origin and early evolution of terrestrial planet atmospheres and meteoritic volatiles. *Icarus* **92**, 2–79.

Pepin, R.O. (1992) Origin of noble gases in the terrestrial planets. *Annual Reviews of Earth and Planetary Science* **20**, 389–430.

Pepin, R.O. (1997) Evolution of Earth's noble gases: Consequences of assuming hydrodynamic loss driven by giant impact. *Icarus* **126**, 148–56.

Pollack, J.B. (1969) A non-gray CO_2—H_2O greenhouse model of Venus. *Icarus* **10**, 314–41.

Pollack, J.B. & Black, D.C. (1982) Noble gases in planetary atmospheres: implications for the origin and evolution of atmospheres. *Icarus* **51**, 169–98.

Prinn, R.G. (1982) Origin and evolution of atmospheres: an introduction to the problem. *Planetary and Space Science* **8**, 741–53.

Prinn, R.G. & Fegley, B. (1987) The atmospheres of Venus, Earth, and Mars — a critical comparison. *Annual Review of Earth and Planetary Science* **15**, 171–212.

Rambler, M. & Margulis, L. (1980) Bacterial resistance to ultraviolet radiation under anaerobiosis: Implications for pre-Phanerozoic evolution. *Science* **210**, 638–40.

Rasmussen, B. & Buick, R. (1999) Redox state of the Archean atmosphere: Evidence from detrital minerals in ca. 3250–2750 Ma sandstones from the Pilbara Craton, Australia. *Geology* **27**, 115–18.

Raven, J.A. (1997) The role of marine biota in the evolution of terrestrial biota: Gases and genes–atmospheric composition and evolution of terrestrial biota. *Biogeochemistry* **39**, 139–64.

Rettburg, P., Horneck, G., Strauch, W., Facius, R. & Seckmeyer, G. (1998) Simulation of planetary UV radiation climate on the example of the early Earth. *Advances in Space Research* **22**, 335–9.

Rossow, W.B., Henderson-Sellers, A. & Weinreich, S.K. (1982) Cloud feedback—a stabilizing effect for the early Earth. *Science* **217**, 1245–7.

Runnegar, B. (1991) Precambrian oxygen levels estimated from the biochemistry and physiology of early eukaryotes. *Global Planetary Change* **97**, 97–111.

Rye, R. & Holland, H.D. (1998) Paleosols and the evolution of atmospheric oxygen: a critical review. *American Journal of Science* **298**, 621–72.

Rye, R., Kuo, P.H. & Holland, H.D. (1995) Atmospheric carbon dioxide concentrations before 2.2 billion years ago. *Nature* **378**, 603–5.

Sagan, C. & Chyba, C. (1997) The early faint Sun paradox: organic shielding of ultraviolet-labile greenhouse gases. *Science* **276**, 1217–21.

Sagan, C. & Mullen, G. (1972) Earth and Mars: Evolution of atmospheres and surface temperatures. *Science* **177**, 52–6.

Schopf, J.W. (ed.) (1983) *The Earth's Earliest Biosphere.* Princeton University Press, Princeton, NJ.

Schopf, J.W. (1999) *Cradle of Life: The Discovery of Earth's Earliest Fossils.* Princeton University Press, Princeton, NJ.

Schopf, J.W. & Klein, C. (eds) (1992) *The Proterozoic Biosphere: A Multidisciplinary Study.* Cambridge University Press, Cambridge.

Schopf, J.W. & Packer, B.M. (1987) Early Archean (3.3 billion to 3.5 billion year old) microfossils from Warrawoona Group, Australia. *Science* **237**, 70–3.

Schubert, G. & Covey, C. (1981) The atmosphere of Venus. *Scientific American* **245**(July), 44–52.

Shimoyama, A. (1997) Complex organics in meteorites. *Advances in Space Research* **19**, 1045–52.

Shizgal, B.D. & Arkos, G.G. (1996) Nonthermal escape of the atmospheres of Venus, Earth, and Mars. *Reviews of Geophysics* **34**, 483–505.

Sleep, N.H., Zahnle, K.J., Kasting, J.F. & Morowitz, H.J. (1989) Annihilation of ecosystems by large asteroid impacts on the early Earth. *Nature* **342**, 139–42.

Tajika, E. (1998) Mantle degassing of major and minor volatile elements during the Earth's history. *Geophysical Research Letters* **25**, 3991–4.

Tolstikhin, I.N. & Marty, B. (1998) The evolution of terrestrial volatiles: a view from helium, neon, argon and nitrogen isotope modelling. *Chemical Geology* **147**, 27–52.

Towe, K.M. (1990) Aerobic respiration in the Archean. *Nature* **348**, 54–6.

Towe, K.M. (1996) Environmental oxygen conditions during the origin and early evolution of life. *Advances in Space Research* **18**, 7–15.

Vellai, T. & Vida, G. (1999) The origin of eukaryotes: the difference between prokaryotic and eukaryotic cells. *Proceedings of the Royal Society of London* **B266**, 1571–7.

Volk, T. (1998) *Gaia's body: Toward a Physiology of Earth.* Springer-Verlag, New York.

Walker, J.C.G. (1976) Implications for atmospheric evolution of the inhomogeneous accretion model of the origin of the Earth. In Windley, B.F. (ed.), *The Earliest History of the Earth.* Wiley, New York.

Walker, J.C.G. (1977) *Evolution of the Atmosphere.* Macmillan Publishing Co., New York.

Walker, J.C.G. (1980) Atmospheric constraints on the evolution of metabolism. *Origins of Life* **10**, 93–104.

Walker, J.C.G. (1982). The earliest atmosphere of the Earth. *Precambrian Research* **17**, 147–71.

Walker, J.C.G. (1985) Carbon dioxide on the early Earth. *Origins of Life* **16**, 117–27.

Walker, J.C.G. (1990) Precambrian evolution of the climate system. *Palaeogeography, Palaeoclimatology, and Palaeoecology (Global Planetary Change)* **82**, 261–89.

Walker, J.C.G., Klein, C., Schidlowski, M., Schopf, J.W., Stevenson, D.J. & Walter, M.R. (1983) Earth's earliest biosphere: its origin and evolution. In Schopf, J.W. (ed.), *The Earth's Earliest Biosphere.* Princeton University Press, Princeton, NJ.

Wanke, H. & Dreibus, G. (1988) Chemical composition and accretion history of terrestrial planets. *Philosophical Transactions of the Royal Society of London* **A325**, 545–57.

Wayne, R.P. (2000) *Chemistry of Atmospheres*, 3rd edn. Oxford University Press, Oxford.

Wen, J.-S., Pinto, J.F. & Yung, Y.L. (1989) Photochemistry of CO and H_2O: analysis of laboratory experiments and applications to the prebiotic Earth's atmosphere. *Journal of Geophysical Research* **94**, 14957–70.

Wetherill, G. (1990) Formation of the Earth. *Annual Reviews of Earth and Planetary Science* **18**, 205–56.

Whittet, D.C.B. (1997) Is extraterrestrial organic matter relevant to the origin of life on Earth? *Origins of Life and Evolution of the Biosphere* **27**, 249–62.

Wickramasinghe, N.C. & Hoyle, F. (1998) Infrared

evidence for panspermia: an update. *Astrophysics and Space Science* **259**, 385–401.

Worsley, T.R. & Nance, R.D. (1989) Carbon redox and climate control through Earth history. *Global Planetary Change* **75**, 259–82.

Young, G.M., von Brunn, V., Gold, D.J.C. & Minter, W.E.L. (1998) Earth's oldest reported glaciation: Physical and chemical evidence from the Archean Mozaan Group (similar to 2.9 Ga) of South Africa. *Journal of Geology* **106**, 523–38.

Yung, Y.L. & DeMore, W.B. (1999) *Photochemistry of Planetary Atmospheres*. Oxford University Press, Oxford.

Zahnle, K.J. (1986) Photochemistry of methane and the formation of hydrocyanic acid (HCN) in the Earth's early atmosphere. *Journal of Geophysical Research* **91**, 2819–34.

Zahnle, K.J. & Walker, J.C.G. (1982) The evolution of solar ultraviolet intensity. *Reviews of Geophysics and Space Physics* **20**, 280–92.

Zahnle, K.J., Kasting, J.F. & Pollack, J.B. (1988) Evolution of a steam atmosphere during Earth's accretion. *Icarus* **74**, 62–97.

Zhang, Y. & Zindler, A. (1989) Noble-gas constraints on the evolution of the Earth's atmosphere. *Journal of Geophysical Research* **94**, 3719–37.

2 Atmospheric Energy and the Structure of the Atmosphere

HUGH COE AND ANN R. WEBB

2.1 INTRODUCTION

The Sun provides almost all of the energy input to the Earth, its oceans and atmosphere, a massive 5×10^{24} J yr^{-1}; in contrast the internal energy of the Earth generates only around 10^{21} J yr^{-1}. The solar input is responsible for driving the atmospheric circulation, maintaining the temperature structure of the atmosphere and evaporating water into the atmosphere to initiate the hydrological cycle. In addition, it initiates many chemical processes in the atmosphere and is central to photosynthesis, the process by which the biosphere reduces carbon dioxide (CO_2) to carbohydrates. It is clearly important to understand the radiation balance of the Earth and its atmosphere since the transfer of solar radiation underpins so many of the important processes in the atmosphere, biosphere, and oceans.

The average solar flux reaching the top of the Earth's atmosphere and the average temperature of the Earth vary by only fractions of a percent from one year to the next. This indicates that, although the Earth receives a huge amount of energy each year, it does not retain it and loses the same amount to space. The system is in balance and we must be able to understand the way this balance is maintained if we are to predict the effect future changes to parts of the system will have on the whole. The Earth's atmosphere is not totally transparent to either incoming or outgoing radiation and the interaction between the atmosphere and radiation controls the surface radiation budget, the physics of the atmosphere, and its chemical composition.

2.2 THE STRUCTURE OF THE EARTH'S ATMOSPHERE

Before we discuss atmospheric radiation in more detail we will give a brief overview of the vertical structure of the atmosphere. Figure 2.1 shows the change in temperature and density with height. There is of course considerable variability from day to day, seasonally and from one location to another. However, the main features in the profile are typical of the vertical structure of the atmosphere in general. The pressure profile of the atmosphere can be calculated from the change in pressure, dp, experienced in a small change of height, dz:

$$dp = -g\rho dz \qquad (2.1)$$

where g is the acceleration due to gravity and ρ is the density of air. The acceleration due to gravity can be treated as constant as the atmosphere is thin with respect to the radius of the Earth but the density clearly varies with altitude. However, air behaves more or less as an ideal gas, so:

$$\rho = \frac{Mp}{RT} \qquad (2.2)$$

where M is the molar mass of air, R is the gas constant (8.314 J K^{-1} mol^{-1}) and T is the temperature in Kelvin. This gives:

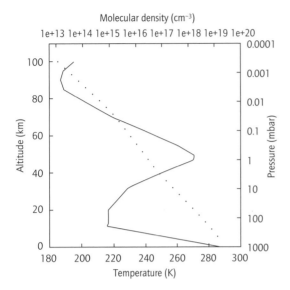

Fig. 2.1 The average vertical temperature and molecular density structure of the Earth's atmosphere. The data are taken from the US standard atmosphere and represent a time and spatial average. A local instantaneous sounding will vary considerably from this profile.

$$\frac{dp}{p} = -\frac{gM}{RT}dz \qquad (2.3)$$

which can be integrated from the surface $(z = 0, p = p_0)$:

$$p = p_0 \exp\int_0^z \frac{-dz}{(RT/gM)} \qquad (2.4)$$

This is known as the hydrostatic equation. The denominator, RT/gM, has units of length and is often referred to as the *scale height*. The scale height is the vertical distance over which the pressure falls to $1/e$ of its initial value.

As can be seen from Fig. 2.1, the temperature does not remain constant throughout the atmosphere and so the scale height also changes with height. This analysis assumes that the atmosphere is composed of a gas of single molar mass, M. In reality the atmosphere is composed of several gases and so we might expect a different scale

height for each gas at any altitude. If this was the case then the composition of the atmosphere in the absence of sources and sinks would vary with height as the heavier molecules have smaller scale heights. This is not observed below 100 km as the mean free path of molecules, or molecular mixing length, is much smaller than the macroscopic mixing length resulting from turbulence and convection. As a result macroscopic mixing processes act equally on all molecules and dominate the molecules' specific diffusion processes. Above 100 km this is not the case and molecular separation with height is observed. The mean molar mass of air in the lower part of the atmosphere is determined by the mix of molecular nitrogen and oxygen and is 28.8 g.

The atmosphere can be subdivided into layers based on its thermal structure. Closest to the surface the temperature reduces with height to a minimum at around 10 km, known as the **tropopause**. The region closest to the surface is known as the **troposphere** (*tropos* — Greek for "turning"). At the surface the temperature varies from minima of around −50°C at the wintertime poles to maxima of 40°C over the continents close to the Equator. The temperature in the troposphere decreases by, on average, 6.5 K km⁻¹ from the surface to the tropopause. The **stratosphere** (*stratus* — Greek for "layered") is a region between 10 and 50 km in which the temperature profile of the atmosphere increases to a maximum at the **stratopause**. This inversion is caused, as we shall see below, by absorption of solar ultraviolet radiation by a layer of ozone (O_3). Increasing temperature with height suppresses vertical mixing through the stratosphere and so causes its layered structure.

Above 50 km warming by ultraviolet absorption can no longer compete with the cooling processes and temperatures begin to decrease. This region is called the **mesosphere** and extends to around 90 km when the atmospheric temperature reaches a second minimum, the **mesopause**. However, unlike the troposphere where the rate of decrease of temperature with height, the **lapse rate**, is sufficient for convection to occur, the mesospheric lapse rate is only around 2.75 K km⁻¹ and the layer remains stable. Above the mesopause the temperature again increases through the so-called **thermosphere**. At these altitudes the air becomes so thin that molecular collisions become infrequent. Thus atomic species with high translational energy cannot redistribute that energy to highly excited vibrational and rotational states in molecular species. The high temperatures in the thermosphere reflect not a large energy source but the inability of the thin atmosphere at these altitudes to lose energy via radiative transfer.

2.3 SOLAR AND TERRESTRIAL RADIATION

2.3.1 *Solar radiation*

The Sun is a middle-aged, medium-sized star with a composition of approximately 75% hydrogen and 25% helium. The Sun's energy is derived from the fusion of hydrogen into helium nuclei and is then transferred to the surface of the Sun via shortwave electromagnetic radiation. Although the Sun has a radius of 7.0×10^5 km, virtually all the energy received by the Earth is emitted by the outer 500 km, known as the photosphere. The Solar photosphere emits light across the entire electromagnetic spectrum, from gamma rays to radio waves. However, most of the radiative power incident at the top of the Earth's atmosphere is due to light of wavelength between 200 nm in the ultraviolet and 4 µm in the infrared, with a peak intensity at about 490 nm in the green part of the visible region of the spectrum.

The irradiance of the Sun as a function of wavelength is shown in Fig. 2.2. The Sun's photosphere has a temperature of approximately 5800 K and can be thought of as a blackbody emitter of this temperature. A blackbody emits the maximum amount of radiation possible at each wavelength at a given temperature. Planck related the emissive power, or intensity, of a blackbody $B(\lambda, T)$ at a given wavelength, λ, to the temperature, T, of the emitter by

$$B(\lambda, T) = \frac{2hc^2}{\lambda^5 \exp(hc/k\lambda T) - 1} \qquad (2.5)$$

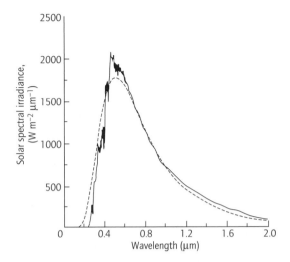

Fig. 2.2 The spectral irradiance from the Sun compared to that of a blackbody at 5777 K. (From Iqbal 1983.)

where k is the Boltzmann constant (1.381×10^{-23} J K^{-1}), c is the speed of light in vacuum (2.998×10^8 m s^{-1}), and h is Planck's constant (6.626×10^{34} J s). The blackbody curve for an emitter at a temperature of 5777 K is also shown in Fig. 2.2. The total flux emitted by a blackbody radiator, F_B, and the total emitted intensity B, can be found by integrating the Planck blackbody function (2.1) over all wavelengths

$$F_B = \pi B = \pi \int_0^\infty B(\lambda, T)\mathrm{d}\lambda = \sigma T^4 \qquad (2.6)$$

where σ is the Stefan–Boltzmann constant (5.671×10^{-8} W m^{-2} K^{-4}).

Solar radiation has an average intensity of approximately 1370 W m^{-2} at the distance of the Earth from the Sun. This value is often referred to as the solar constant, S, although it varies with time on a wide variety of scales. The Sun rotates with a period of 27 days and both active, brighter regions known as **faculae** and less active, darker regions known as **sunspots** face the Earth during each rotation. The output from these different regions of the Sun varies by between 0.1 and 0.3% of the total flux. The number of sunspots on the surface of the Sun varies with a cycle of 11 years, causing variations in radiative flux at the top of the Earth's atmosphere of the order of 1%. Lower frequency variations in the solar flux, again of the order of 1–2%, have also been inferred from isotopic abundance measurements.

2.3.2 Terrestrial radiation

The Earth also acts as a blackbody radiator, but as its global mean surface temperature, T_s, is 288 K, most of the irradiance from the Earth is in the infrared part of the spectrum and peaks at about 10 μm. Figure 2.3a shows a blackbody curve for an emitter of temperature 288 K compared to one at 5777 K, representing the solar spectrum. There is very little overlap between the incoming solar radiation at ultraviolet and visible wavelengths and the outgoing infrared radiation from the Earth's surface. Thus, incoming solar radiation and outgoing terrestrial radiation are distinct from one another, separated by a gap at around 4 μm, and are often referred to as shortwave (SW) and longwave (LW) radiation respectively.

As the mean surface temperature of the Earth changes little from year to year and has varied by less than 5°C in the past 20,000 years it is clear that the system is in equilibrium and the energy inputs must be balanced by energy losses. The effective area of the Earth receiving sunlight at any one time is given by πR^2, where R is the radius of the Earth, yet the total area of the Earth is $4\pi R^2$, so the average radiant flux over the Earth is given by $S/4$. However, not all of the incoming radiation is absorbed by the surface; some is reflected back to space by the surface, clouds, aerosol particles, or scattering from molecules in the atmosphere. The fractional reflectance is known as the global mean planetary reflectance or albedo, A. The average surface albedo is around 0.15 but the high reflectivity of clouds leads to an overall planetary albedo of 0.3. Thus the incoming irradiance absorbed by the Earth's surface, F_s, is given by

$$F_s = (1-A)\frac{S}{4} \qquad (2.7)$$

and has a value of 240 W m⁻².

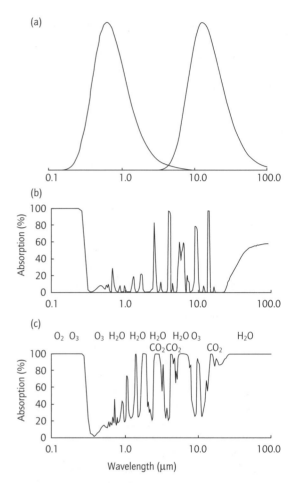

Fig. 2.3 Panel (a) shows the blackbody curves for 5777 K and 280 K, representing the solar photosphere and the Earth's surface. The Sun emits mainly in the visible, while the Earth emits predominantly in the infrared. The so-called incoming shortwave and outgoing longwave radiation is separated by a gap at around 4 μm. Panels (b) and (c) show the fractional absorption of radiation from the top of the atmosphere to 10 km and the surface of Earth respectively. The main absorbers in each wavelength region are indicated in panel (c). Most absorption of longwave outgoing radiation occurs in the troposphere and is due principally to water vapor and CO_2. However, note that the strong absorption band of O_3 that occurs at 9.6 μm in the center of the so-called atmospheric window is due mainly to stratospheric absorption.

F_s must be balanced by the outgoing blackbody radiation of the Earth given by σT_e^4, where T_e is the effective blackbody temperature of the Earth–atmosphere system. Equating incoming and outgoing fluxes gives an expression for T_e

$$T_e = \left(\frac{(1-A)S}{4\sigma} \right)^{\frac{1}{4}} \qquad (2.8)$$

that yields an equilibrium temperature of 255 K, compared to 288 K, the average surface temperature of the Earth. The fact that the Earth–atmosphere system is 33 K warmer than predicted by this simple calculation suggests that other processes act to offset the loss of heat by longwave cooling. Even if the albedo is halved to completely remove the contribution of clouds to the planetary albedo the equilibrium temperature only rises to 268 K. To understand why the Earth–atmosphere system is warmer than predicted by the simple calculation above we need to examine the interaction between trace constituents in the atmosphere and incoming and outgoing radiation.

2.4 ABSORPTION OF RADIATION BY TRACE GASES

So far we have assumed that the atmosphere acts simply to scatter and reflect incoming shortwave radiation and does not absorb light. However, this is not the case. The atmosphere interacts with both incoming solar radiation and outgoing terrestrial radiation and as we will see the strength of the interaction as a function of wavelength is responsible for the heating of the lower atmosphere.

Figure 2.3 illustrates the effect of absorption by trace gases in the atmosphere on the transmission of incoming shortwave radiation from the Sun and outgoing longwave radiation from the Earth. Figure 2.3a shows the blackbody curves for emitters at 5777 K and 280 K respectively, representing the solar and terrestrial emission spectra. Figure 2.3b and c shows the fraction of light entering the top of the Earth's atmosphere that is absorbed before reaching 10 km and sea level respectively, as a function of wavelength. At 10 km, the top of the

troposphere, virtually all radiation below 290 nm has been absorbed. All radiation below 100 nm is absorbed in the thermosphere above 100 km. Molecular oxygen absorbs strongly at wavelengths between 100 and 200 nm and also in a weaker band between 200 and 245 nm. The oxygen absorptions appreciably attenuate incoming ultraviolet radiation of wavelengths less than 200 nm above an altitude of 50 km. Radiation of wavelengths between 200 and 300 nm is strongly absorbed in the stratosphere by the oxygen trimer, ozone, and transmission of radiation of wavelengths less than 290 nm is negligible below 10 km. Between 300 and 800 nm the stratosphere is only weakly absorbing and most of the solar radiation at these wavelengths is transmitted into the troposphere. A comparison of Fig. 2.3b and c shows that there is little further absorption in the troposphere at wavelengths below 600 nm but H_2O and CO_2 absorption bands, whose abundances are dominated by their tropospheric concentrations, deplete the near infrared part of the incoming solar flux appreciably. As a result the solar irradiance at the surface is dominated by visible wavelengths.

The interaction between the outgoing longwave radiation and the atmosphere can also be seen in Fig. 2.3b and c and compared with an irradiance spectrum of a blackbody emitter of temperature 288 K, representing the radiation emitted from the surface of the Earth in Fig. 2.3a. Several different molecules are efficient absorbers of infrared radiation and many of these are most abundant in the troposphere. Consequently much of the outgoing radiation is absorbed in the lowest 10 km of the atmosphere. Much of the outgoing radiation of wavelengths less than 7 μm is absorbed by water vapor, with some contribution from methane and nitrous oxide, N_2O. Radiation of wavelengths longer than 13 μm is efficiently absorbed by CO_2, whose absorption band is centered at 15 μm. This band is particularly important as it lies close to the maximum of the longwave irradiance spectrum. At longer wavelengths water vapor is excited into many rotational states that effectively form an absorption continuum beyond 25 μm. Minor absorbers between the CO_2 and water bands are mainly N_2O and CH_4. The only fraction of the outgoing radiation that is transmitted through the troposphere without undergoing appreciable absorption lies in the so-called atmospheric window between 7 and 13 μm.

The only significant absorptions of infrared radiation in the stratosphere are due to ozone. The 9.6 μm band of ozone happens to lie in the middle of the atmospheric window and as a result means that stratospheric ozone plays a significant role in the outgoing longwave radiation budget of the Earth.

The absorption of radiation by gases in the atmosphere is complex and mainly involves several trace gas species rather than major constituents. These interactions are key to the chemical composition, thermal structure, and radiative balance of our atmosphere.

2.5 SOLAR RADIATION, OZONE, AND THE STRATOSPHERIC TEMPERATURE PROFILE

We have already seen that ozone is a very efficient absorber of solar radiation between 200 and 300 nm. Its detailed chemistry is discussed in Chapter 7. However, it is worth briefly mentioning here as its formation and presence controls the stratospheric temperature profile. Figure 2.4 shows an average vertical profile of ozone through the atmosphere compared to an average temperature profile. The main layer of ozone in the atmosphere is situated between 15 and 30 km and reaches a maximum concentration of around 5 × 10^{12} molecules cm^{-3} at 22 km. The maximum temperature at the top of the stratosphere occurs at around 50 km, well above the main ozone layer. To understand the effect of stratospheric ozone on the temperature profile we need to understand the way ozone is created and destroyed in the mesosphere and stratosphere.

Ozone is formed from the photo-dissociation of molecular oxygen but is itself removed by photo-dissociation. Photo-dissociation is the fragmentation of a molecule as a result of its absorption of a photon that is energetic enough to break its molec-

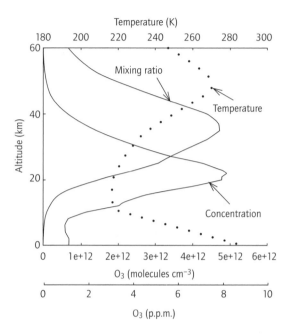

Fig. 2.4 The average vertical profile of ozone and temperature through the atmosphere. The ozone profile is represented as both a mixing ratio and a molecular concentration. Data from the AFGL standard ozone and temperature profiles.

ular bonds. Both O_2 and O_3 absorb ultraviolet light very strongly and prevent highly energetic radiation penetrating to lower altitudes. Figure 2.5 shows the extent to which ultraviolet light penetrates the Earth's atmosphere as a function of wavelength and indicates the gas species responsible for its absorption. Figure 2.6 shows the absorption cross-section of molecular oxygen. The peak in the absorption cross-section of O_2 occurs in the Schumann–Runge continuum at around 145 nm. For wavelengths less than 175 nm, O_2 is dissociated into two oxygen atoms, one of which is electronically excited. This strong absorption prevents sunlight of wavelengths below 175 nm from penetrating below around 70 km (Fig. 2.5). The oxygen atoms formed as a result of O_2 photolysis react with other molecules of O_2 to form ozone. Figure 2.7 shows the strong absorption cross-sections of ozone occurring between 240 and 300 nm with a

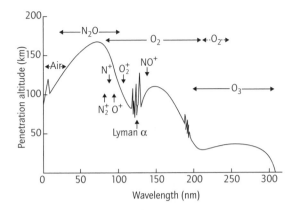

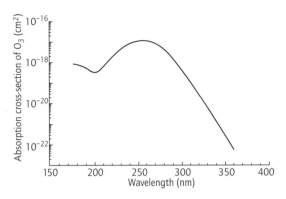

Fig. 2.5 The extent to which ultraviolet solar radiation penetrates through the atmosphere as a function of wavelength. The penetration altitude is the altitude at which the initial intensity at any wavelength is attenuated to e^{-1} of its original intensity. (From Salby 1996.)

Fig. 2.7 The absorption cross-section of ozone at 273 K as a function of wavelength between 170 and 360 nm. (From Seinfeld & Pandis 1998.)

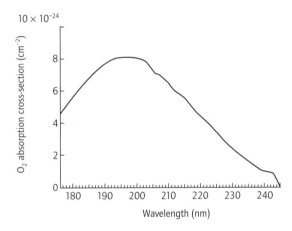

Fig. 2.6 The absorption cross-section of molecular oxygen as a function of wavelength. (Data from DeMore *et al.* 1997.)

maximum value of 1.1×10^{-17} cm^2 at 255 nm. As a result, above 60 km ozone is photolysed very efficiently back to O_2 and atomic oxygen, reducing its concentration and favoring the existence of atomic oxygen.

The Herzberg continuum between 200 and 240 nm is responsible for photolysis of O_2 below

60 km because the shorter wavelengths have already been removed (Fig. 2.5), while radiation of these wavelengths penetrates down to around 20 km.

High up in the atmosphere little O_3 is produced as the air density is low and there is little O_2 to be photolysed or to subsequently react with the atomic oxygen formed by its photolysis. As we descend through the atmosphere the density increases, favoring ozone formation via the combination of atomic and molecular oxygen, and the concentration of ozone increases. A maximum concentration in O_3 is observed at around 20–25 km. Lower in the stratosphere the overhead ozone column is now significant and absorbs much of the radiation between 200 and 290 nm, thus limiting photolysis of oxygen and slowing the rate of ozone formation. The concentration of ozone reduces and reaches a minimum by the tropopause where radiation of less than 290 nm is almost completely removed.

The absorption of ultraviolet radiation by both oxygen and ozone leads to their photolysis, and the energy involved in these sunlight-induced reactions produces local warming. The temperature at a particular altitude will then be a combination of the rates of photolysis of the two oxygen species, in particular ozone, and the air density. The rates of photolysis will depend on the local incidence of radiation and thus on the optical density of the atmosphere in the column above at a given wavelength. This in turn will be dependent on the overhead concentration profile of O_2 and O_3 themselves. As the air density increases any products of photochemical processes that remain energetically excited are deactivated more rapidly via an increased chance of collisions, leading to an increase in temperature. Although the temperature profile is strongly linked to that of ozone, its maximum occurs not at the maximum ozone concentration, but above it and close to the region where the photolytic formation and loss processes of ozone are most rapid.

2.6 TRAPPING OF LONGWAVE RADIATION

Incoming visible and ultraviolet radiation from the Sun is energetic enough to excite electrons within certain optically active molecules. We have seen that in the cases of ozone and molecular oxygen the photon energy is sufficient to fragment the molecule and cause its photolysis. Less energetic outgoing photons of infrared wavelengths induce vibrational and rotational excitations of molecules. These excitations do not cause chemical changes in the absorbing molecule; instead the excited molecule, below 100 km at least, is rapidly deactivated by collisions and the energy absorbed from the original photon is distributed thermally.

We can imagine the effect on a layer of atmosphere as a result of these interactions. Some fraction of the outgoing longwave radiation entering the base of the layer is absorbed by molecules such as CO_2, H_2O, and CH_4 in the layer (see Fig. 2.3c). The absorbed energy is transferred to kinetic energy by collisions between the absorbing molecules and others in the layer. The layer will itself act as a blackbody and re-radiate infrared radiation; however, the layer will radiate uniformly in all directions and so act to increase the longwave flux through the lower layers of the atmosphere. This process raises the local temperature in the lower layers of the atmosphere above that predicted from

a straightforward surface budget calculation of the kind described in Section 2.3.

2.7 A SIMPLE MODEL OF RADIATION TRANSFER

Several gases in the atmosphere absorb strongly in the infrared. As we shall see, each gas has a complex absorption pattern made up of many different individual vibrational and rotational transitions. The way these different absorptions interact is not straightforward and should be accounted for in a detailed description of radiative transfer through the atmosphere. We discuss some of these effects and considerations in the next section. However, first we provide a general picture of the processes taking place in the atmosphere by deriving a simple model of radiative transfer based on an atmosphere that is transparent to incoming shortwave radiation and includes only one trace gas that absorbs uniformly at all infrared wavelengths. This model atmosphere is known as a **gray atmosphere**. Our model is further simplified by the removal of scattering and by assuming that radiation is either emitted or absorbed only in the vertical direction. Lastly, we also assume that each level of the atmosphere is in local thermodynamic equilibrium.

First, we need to describe the absorption of light by an absorbing species in the atmosphere. The intensity of light of wavelength λ, $I(\lambda)$, which passes through a depth dz of an absorber with number concentration n, is reduced by an amount $dI(\lambda)$ given by:

$$dI(\lambda) = -I(\lambda)n\sigma(\lambda)dz = I(\lambda)d\chi \qquad (2.9)$$

where $\sigma(\lambda)$ is the absorption cross-section at wavelength λ and is constant for any given species, and χ is the optical depth. We can obtain the intensity of light transmitted a distance z through the absorber, $I_z(\lambda)$, by integrating eqn 2.9:

$$I_z(\lambda) = I_0(\lambda)\exp\left\{-\int_0^z n\sigma(\lambda)dz\right\} \qquad (2.10)$$

where $I_0(\lambda)$ is the initial intensity of light of wavelength λ. In cases where the concentration of the absorber is independent of the depth of the absorbing slab the above relation becomes the **Beer–Lambert Law**:

$$I_z = I_0(\lambda)\exp\{-n\sigma(\lambda)z\} \qquad (2.11)$$

This is not the case for a vertical slice through the atmosphere.

In our simplified model the single species absorbs uniformly over all wavelengths so we can simplify eqn 2.10 to give:

$$I_z = I_0\exp\left\{-\int_0^z n\sigma\mathrm{d}z\right\} = I_0\exp\left\{-\int_{\chi_0}^0 \mathrm{d}\chi\right\}, \qquad (2.12)$$

where χ_0 is by convention the optical depth at the base of the atmosphere.

So far we have only considered the absorption of light. However, we know that the layer will reemit radiation as a blackbody in a similar way so we must also include the intensity of emitted radiation, B, and assuming Kirchoff's Law:

$$\mathrm{d}I = -In\sigma\mathrm{d}z + Bn\sigma\mathrm{d}z = (I - B)\mathrm{d}\chi \qquad (2.13)$$

Furthermore, in any slice of the atmosphere there may be some downwelling longwave radiation arising from blackbody emission of the layers above, so we should treat both the upwelling and downwelling radiative fluxes, $F^\uparrow$ and $F^\downarrow$, separately:

$$\frac{\mathrm{d}F^\uparrow}{\mathrm{d}\chi} = F^\uparrow - \pi B \quad \text{and} \quad -\frac{\mathrm{d}F^\downarrow}{\mathrm{d}\chi} = F^\downarrow - \pi B \qquad (2.14)$$

The net flux through a layer is given by $F = F^\uparrow - F^\downarrow$, the difference between the upwelling and downwelling radiation. As we have assumed that the atmosphere is in local thermodynamic equilibrium the flux must not change with height and is therefore constant throughout the depth of the atmosphere. By summation and subtraction of the upward and downward fluxes we obtain:

$$\frac{\mathrm{d}F}{\mathrm{d}\chi} = \bar{F} - 2\pi B \quad \text{and} \quad \frac{\mathrm{d}\bar{F}}{\mathrm{d}\chi} = F, \qquad (2.15)$$

where $\bar{F} = F^\uparrow + F^\downarrow$ is the total flux leaving one layer.

As F is constant these expressions are easily integrated to give:

$$\bar{F} = 2\pi B \quad \text{and} \quad \bar{F} = F\chi + c \qquad (2.16)$$

The blackbody emission flux of the outermost layer of the Earth's atmosphere is given by πB_0, and this must be equal to half of the total flux from this layer, $\bar{F}$. As there are no overlying layers to supply a contribution to $F^\downarrow$, $\bar{F} = F$ and so:

$$B = \frac{F}{2\pi}\chi + B_0 = \frac{F}{2\pi}(\chi + 1) \qquad (2.17)$$

In this model the blackbody emission decreases linearly with height from the surface to the top of the atmosphere and there is a constant difference between the up and downwelling fluxes, i.e. $F = F^\uparrow - F^\downarrow$ is constant. Futhermore, as there is no heat gained or lost by the atmosphere, the upwelling longwave radiation leaving the top of the atmosphere must be equal to the solar radiation absorbed at the Earth's surface, F_s (eqn 2.7), so at the top of the atmosphere

$$F^\uparrow = F = F_s \qquad (2.18)$$

We now consider the boundary conditions at the surface. We must balance the upward flux of radiation emitted by the Earth at a temperature T_s, $\pi B(T_s)$, with the downwelling short and longwave radiation.

$$\pi B(T_s) = F_s + F^\downarrow(\chi_s) \qquad (2.19)$$

where χ_s is the optical depth of the lowest layer of the atmosphere. However,

$$\bar{F} - F = 2F^\downarrow = 2\pi B - F_s \qquad (2.20)$$

which when evaluated at the surface gives

$$\pi B(T_s) = \pi B(\chi_s) + F_s/2 \qquad (2.21)$$

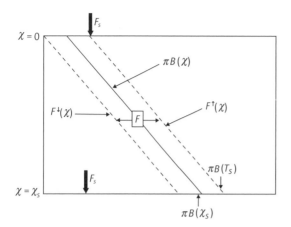

Fig. 2.8 Schematic of the variation of upwelling and downwelling radiation and blackbody emission fluxes with optical depth through the atmosphere. The difference between the upwelling and downwelling radiation fluxes is constant with height and the model predicts a discontinuity in the blackbody emission flux at the surface.

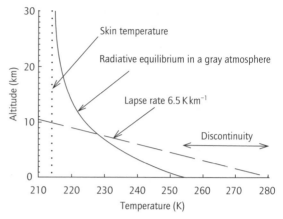

Fig. 2.9 The temperature structure of the gray atmosphere. The temperature predicted by a model of radiative equilibrium in a gray atmosphere tends to the skin temperature at high altitudes. At the surface there is a marked discontinuity. In reality the warm surface of the Earth heats the air immediately above and initiates convection, modeled by the average tropospheric lapse rate of $6.5\,\mathrm{K\,km^{-1}}$.

This expression implies that there is a temperature discontinuity between the surface and the cooler lowest layer of the atmosphere. Figure 2.8 shows schematically how the radiation fluxes vary with optical depth through the atmosphere and emphasizes the discontinuous blackbody emission flux at the surface.

If we assume that our absorber varies in concentration solely as a function of pressure then we can express its optical density in the atmosphere as

$$\chi = \chi_s \exp\{-z/H_s\} \qquad (2.22)$$

where z is the height above the surface and H_s is the scale height in the surface layers, approximately 7 km. When the atmosphere is optically thin, $\chi < 1$, then radiation traverses the level with little interaction. When $\chi > 1$ radiation is absorbed efficiently within the layer and successive layers do not strongly interact. We will therefore choose $\chi_s = 1$ to crudely illustrate how our model works. We can evaluate χ and so using eqn 2.17 we can derive B as a function height. However, from eqn 2.6:

$$F_B = \pi B = \pi \int_0^\infty B(\lambda, T)\mathrm{d}\lambda = \sigma T^4$$

and so we can calculate the temperature at any level in the atmosphere. The results of such a calculation are shown in Fig. 2.9. The discontinuity in the temperature at the surface is again obvious, reducing from 282 K to 255 K above the surface. The temperature falls rapidly with height and tends to some limit, the so-called **skin temperature**, the temperature as z tends to infinity, $(F_s/2\sigma)^{1/4}$. If the atmosphere did not interact with outgoing longwave radiation ($\chi = 0$) the temperature of the atmosphere would be constant with altitude and be equal to the skin temperature.

Clearly the temperature in the troposphere does not vary in this way. However, the temperature profile in the lower stratosphere (Fig. 2.1) is close to that described above. In Section 2.9 we discuss convection and show that the rate of decrease of temperature with height, or **lapse rate**, through most of the troposphere is determined by heat transport by convection rather than radiative transfer. The average lapse rate in the tropopause is around $6.5\,\mathrm{K\,km^{-1}}$, somewhere between the lapse rate of dry air and that of cloudy air. This lapse rate is also shown in Fig. 2.9.

In the model, the air in contact with the ground is heated by the surface, becomes buoyant and initiates convection. In this way convective processes dominate the heat transfer of the troposphere and the temperature of the overlying layers decreases with height at the average tropospheric lapse rate. The temperature is greater than that predicted by the radiative scheme to a height of 8 km, above which the radiative scheme predicts warmer temperatures than the lapse rate. So the model predicts a convective lower atmosphere that is turbulent and well mixed and a transition at around 8 km to a stable atmosphere whose temperature structure is controlled by radiative processes. The transition level in this model is a little lower than the observed tropopause but nevertheless this simple scheme predicts the broad temperature characteristics of the atmosphere below 20 km.

2.8 A BRIEF OVERVIEW OF MORE COMPLEX RADIATIVE TRANSFER

In Section 2.7 we made very simple assumptions about the nature of the absorber in our model atmosphere. In reality of course the situation is considerably more complex. The absorption spectrum of each radiatively active gas is composed of many individual lines. Figure 2.10 shows part of the $14\,\mu\mathrm{m}$ band of CO_2, and the complex structure in the band is immediately evident. At some wavelengths much of the light is transmitted, while at other wavelengths the absorption by the atmosphere is total.

Although the individual absorption lines arise from discrete transitions between particular vibrational and rotational energy levels within an absorbing molecule the individual lines are not infinitely narrow. The lifetime of the state to which the molecule has been excited can never be predicted exactly as there is always some small but finite uncertainty in the energy of the excited state. This uncertainty in the decay time leads to a broadening of the transition line over a range of frequencies. However for vibrational and rotational transitions in the infrared part of the spectrum other line broadening mechanisms are more important.

The absorbing molecules can collide with other molecules while in the excited state and this affects the re-radiation of light from the molecule. This broadening effect is known as pressure broadening. Lastly, Doppler broadening occurs as a result of the absorbing or emitting molecule moving in either the same direction as or the opposite direction to the photon of light it emits. This leads to a small frequency shift, observed as line broadening. In the troposphere the frequency of collisions is high and pressure broadening is the dominant mechanism.

When the atmospheric column of a gas absorbs only a small fraction of the incident radiation the absorption is said to be weak at that wavelength. In this case the thicker the layer of absorber the light traverses the greater the attenuation. This reduction in intensity is described by the Beer–Lambert Law (eqn 2.11). However, as can be seen in Fig. 2.10, several gases in the atmosphere absorb very

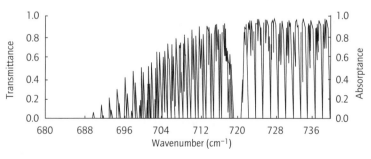

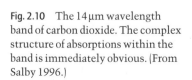

Fig. 2.10 The $14\,\mu\mathrm{m}$ wavelength band of carbon dioxide. The complex structure of absorptions within the band is immediately obvious. (From Salby 1996.)

strongly and certain individual transitions effectively absorb all light of that wavelength through the atmospheric column, and the absorptance approaches unity. This is known as saturation: radiation at these wavelengths has effectively been removed and the only further change in the absorption spectrum of the atmosphere after further passage through the absorbing atmosphere is at the edges of the line where the gas is more weakly absorbing. Saturation will therefore change the shape of the absorbing band through the depth of the absorbing column. In the case where the spectrum of an absorbing gas contains many strong absorption lines close together, saturation may lead to the merging of these lines and the absorption spectrum may become continuous.

Furthermore, unlike the simple model we discussed above, there are many different absorbing species in the real atmosphere, each with its own spectral characteristics. This can lead to some of the absorbing features of different gases overlapping and saturation at some wavelengths may occur even though the contribution from each of the individual absorbing molecules may be weak.

In theory all of these effects can be treated explicitly. However, in practice this is difficult and also computationally very expensive. Radiation schemes usually use some parameterizations of the absorption spectra, or band models that represent the general features of the absorption spectra over a range of wavelengths. The absorption spectrum of CO_2 around $14\,\mu m$ (Fig. 2.10) shows a regular pattern of absorption lines and can be modeled using an evenly spaced set of lines separated by mean line spacing with a line strength derived from the mean strength of lines in the band. Absorption lines of other absorbers may be randomly spaced and this needs to be accounted for in the band model.

2.9   CONDUCTION, CONVECTION, AND SENSIBLE AND LATENT HEAT

2.9.1   Introduction

So far we have only considered transfer of heat

through the atmosphere by radiative processes. Certainly radiative transfer of energy is very important in the atmosphere. However, energy may also be transferred through **conduction** and **convection**. The process of conduction occurs by the transfer of kinetic energy from one molecule to an adjacent one. The process will be most efficient when the molecules are tightly constrained in solids and especially when there is a defined structure to the material, such as in a metal. Gases, including air, have low thermal conductivities and so the atmosphere is a poor conductor of heat. Although conduction can be neglected in the atmosphere it is the main mechanism by which heat is transferred away from the warm surface through the underlying layers of soil or rock.

Convection occurs much more efficiently than conduction in fluids as warmer parts of the mass can mix much more rapidly with cooler parts and transfer heat. This transfer of heat on the macroscale is far faster than transfer on the molecular scale and makes this process extremely important when considering heat transfer in the atmosphere. Heat is exchanged between the Earth's surface, which is radiatively heated, and the lowest layer of the atmosphere by conduction at the molecular level. The heating of the air causes density changes in the fluid and locally the air expands. This makes the warmed parcel more buoyant and may in itself cause the parcel to mix through the bulk of the air above, a process known as **free convection**.

However, the atmosphere is continually stirred by large-scale winds generated by pressure gradients and motion around and over mountain ranges. This process forces the heated air close to the ground to mix through the air above and warm the whole air mass. Hence, this process is known as **forced convection**.

Convection then mixes parcels of warm and cold air together and so changes the temperature of the two parcels. The warm parcel loses heat as it cools and the colder parcel gains heat as it warms. Enthalpy, or specific heat, is transferred along this temperature gradient. The specific heat content of a parcel of air of unit mass is defined as $c_p T$, where

c_p is the specific heat at constant pressure and T is the temperature of the parcel.

Energy may be transferred indirectly, without changing the temperature of the air parcel, through a change in phase of water in the atmosphere, otherwise known as **latent heat**. A large amount of heat is required to change liquid water to water vapor and the same amount of energy is released when water vapor condenses and a cloud forms. Already we can see that cloud formation releases energy and so will have an effect on the temperature profile compared to the dry atmosphere. The **latent heat of vaporization**, L, is the energy required to convert 1 kg of liquid water to water vapor at the same temperature: at 0°C $L = 2.5 \times 10^6 \, \mathrm{J \, kg^{-1}}$. The latent heat of melting is the energy required to melt 1 kg of ice to form liquid water. At 0°C this is around $3.3 \times 10^5 \, \mathrm{J \, kg^{-1}}$. We will now look in more detail at the effects sensible and latent heats have on the temperature structure of the troposphere.

2.9.2 *Sensible heat and the temperature structure of the dry atmosphere*

We will now consider the processes acting on a rising parcel of air in the atmosphere from a thermodynamic perspective to obtain a vertical temperature profile assuming that the atmosphere is well mixed. This is shown schematically in Fig. 2.11. A rising parcel of air will expand because the pressure exerted on the parcel by the surrounding air reduces with height. The work done by the air parcel in expanding is at the expense of its own internal energy and so the temperature of the parcel falls. We can consider the effect of a small change in altitude on a dry air parcel of unit mass by considering small perturbations to the pressure,

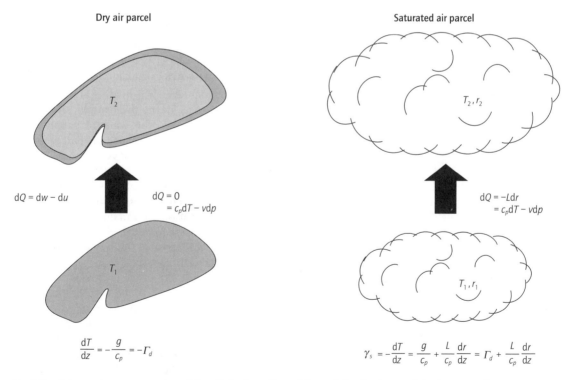

Dry air parcel

T_2

$dQ = dw - du$ $dQ = 0$
  $= c_p dT - v dp$

T_1

$$\frac{dT}{dz} = -\frac{g}{c_p} = -\Gamma_d$$

Saturated air parcel

T_2, r_2

$dQ = -L dr$
  $= c_p dT - v dp$

T_1, r_1

$$\gamma_s = -\frac{dT}{dz} = \frac{g}{c_p} + \frac{L}{c_p}\frac{dr}{dz} = \Gamma_d + \frac{L}{c_p}\frac{dr}{dz}$$

Fig. 2.11 Schematic representation of the adiabatic cooling of dry and cloudy air parcels in the atmosphere.

dp, volume, dv, and temperature, dT, of that parcel using the ideal gas law ($pv = RT$).

$$(p + dp)(v + dv) = R(T + dT) \qquad (2.23)$$

where R is the molar gas constant $(8.314\,\mathrm{J\,mol^{-1}}$ $\mathrm{K^{-1}})$. If we expand and assume that the products of the increments are negligible:

$$pdv + vdp = RdT \qquad (2.24)$$

The work done by the parcel of gas at pressure p in expanding by a volume increment dv is

$$dw = pdv \qquad (2.25)$$

From the first law of thermodynamics we know that the quantity of heat, dQ, supplied to a unit mass of a gas is balanced by an increase in the internal energy of the gas, du, and the external work done by the gas, dw:

$$dQ = du + dw = du + pdv \qquad (2.26)$$

We should now introduce the specific heats at constant pressure and volume, c_p and c_v, which are defined in the following way:

$$c_p = \left(\frac{dQ}{dT}\right)_p \quad \text{and} \quad c_v = \left(\frac{dQ}{dT}\right)_v \qquad (2.27)$$

Let us first consider the case when there is no volume change, dv = 0. In this case

$$dQ = du \qquad (2.28a)$$

and so from eqn 2.27:

$$du = c_v dT \qquad (2.28b)$$

In general then:

$$dQ = c_v dT + pdv = c_v dT + RdT - vdp \qquad (2.29)$$

We may now consider the particular case when there is no pressure change, dp = 0.

$$dQ = (c_v + R)dT \qquad (2.30a)$$

and so from eqn 2.27

$$c_p = \left(\frac{dQ}{dT}\right)_p \qquad (2.30b)$$

we have

$$dQ = c_p dT - vdp \qquad (2.31)$$

In most situations in the troposphere vertical motion of air is rapid enough to far outweigh any heat transfer by conduction or radiation. Under these circumstances there is no net exchange of heat and dQ = 0, such processes are known as **adiabatic** changes and

$$c_p dT = vdp = RT\left(\frac{dp}{p}\right) \qquad (2.32a)$$

or

$$\frac{dT}{T} = \frac{Rdp}{c_p p} \qquad (2.32b)$$

On integrating, we can see that

$$\frac{T}{p^K} = \text{constant}, \quad \kappa = \frac{R}{c_p} = \frac{c_p - c_v}{c_p} = 0.288 \qquad (2.33)$$

This sole dependence of the temperature of an air parcel on pressure during an adiabatic process means that we can find the temperature an air parcel would have if it was moved from some arbitrary pressure level to 1000 mbar adiabatically. This temperature is known as the **potential temperature**, θ, where

$$\theta = T\left(\frac{1000}{p}\right)^\kappa \qquad (2.34)$$

If θ is constant with height then the atmosphere is said to be in convective equilibrium and the fall of temperature with height, or **lapse rate**, is found by substituting the hydrostatic equation into eqn 2.32. Remembering that we have considered a parcel of unit mass, so $v\rho = 1$:

$$c_p dT + g dz = 0 \qquad (2.35)$$

or

$$\frac{dT}{dz} = -\frac{g}{c_p} = -\Gamma_d = -9.8\,\text{K km}^{-1} \qquad (2.36)$$

Γ_d is known as the **dry adiabatic lapse rate** and is the rate of reduction of temperature with height through the atmosphere assuming that the air is unsaturated and in convective equilibrium. That is to say that if a parcel of air is displaced vertically its temperature at the new pressure will be the same as the surrounding air at that level. Though not true locally, on average the atmosphere would show a lapse rate close to this value as long as the air is dry and there is no contribution from phase transitions of water to the energy budget.

2.9.3 Stability of dry air

The measured lapse rate in the atmosphere, γ, is the observed rate of change of temperature with height. We can compare this value with the dry adiabatic lapse rate to investigate the likely extent of vertical motion of an air parcel in that layer. The two environment curves in Fig. 2.12, marked A and B, have lapse rates that are respectively greater and less than Γ_d. An environment curve shows the vertical temperature structure of the real environment. First consider environment curve A. If a parcel of air at point O is displaced upwards its temperature reduces along the dry adiabatic lapse rate and will therefore be higher than that of its surroundings. Since the pressure is the same, the density of the parcel must be less than that of its surroundings. The air parcel will then have positive buoyancy and will be accelerated upwards. Similarly, if the parcel is displaced downwards then it becomes cooler and denser than its surroundings and sinks further. An atmosphere under these conditions is said to be **unstable**.

Conversely, when we consider a parcel of air at point O on environment curve B, an upward dis-

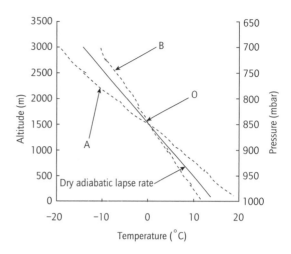

Fig. 2.12 The lapse rates of the two environment curves A and B are respectively greater and less than the dry adiabatic lapse rate of 9.8 K km^{-1}. If an air parcel at O ascends adiabatically its temperature changes at a rate Γ_d. If the surrounding environment has a lapse rate greater than this, A, then the ascending parcel is warmer than the air around it, and is buoyant and unstable. If the lapse rate of the surrounding air is less than that of the ascending parcel, B, then the displaced parcel is cooler than its surroundings and the air is stable.

placement of the air parcel along the dry adiabatic lapse rate curve will result in the temperature of the parcel displacement being cooler than that of its surroundings and more dense. Likewise, a downward displacement will result in the parcel being warmer and more buoyant than its surroundings. In both cases the parcel is subjected to a restoring force that tends to return it back to point O. Vertical columns of air with temperature profiles similar to the environment curve B are said to be **stable**.

2.9.4 Latent heat and the effect of clouds on the vertical temperature structure

We have so far assumed that the atmosphere is dry: that is to say, no water exists in the atmosphere in the liquid or solid phase. However, water can change phase readily under atmospheric conditions and such changes of state produce large

changes in the energy budget of the system. It is its
ability to act as an energy store that makes water so
important in the troposphere.

The amount of water vapor in a parcel of air may
be expressed as the mass mixing ratio, r, where
$r = \rho_v/\rho_a$ and ρ_v and ρ_a are the densities of water
vapor and dry air in the parcel. The mixing ratio is
usually expressed in units of grams of water per
kilogram of air. The maximum amount of water
vapor an air parcel can hold at any given temperature is given by the saturation mixing ratio, r_w. The
saturation mixing ratio can be thought of as the
amount of water in the vapor phase above a
thermally isolated body of pure liquid water at
equilibrium. Any further increase in water into
the already saturated air parcel will lead to condensation. The saturation mixing ratio is solely a
function of temperature: as the parcel warms its
capacity to hold water vapor increases and conversely at colder temperatures air can hold less
moisture before condensation occurs. The Antarctic continent may therefore be thought of as a
desert because there is little moisture available in
the air at such low temperatures, reducing the
possibility of significant precipitation.

The adiabatic changes experienced by a cloudy
air parcel are shown schematically in Fig. 2.11.
Consider an unsaturated air parcel close to the
Earth's surface. As the air parcel rises it will cool
and its saturation mixing ratio will decrease. If
there is no exchange between the air parcel and its
environment the water vapor mixing ratio of the
parcel remains constant. If the parcel continues to
rise it will eventually reach a level at which the
ambient mixing ratio of the parcel is equal to the
saturation mixing ratio. This point is known as the
condensation level and is often observed by a clearly defined base to a layer of clouds. As the air parcel
continues to rise the water vapor mixing ratio is at,
or slightly above, r_w and so condensation occurs.
During condensation, heat is released into the air
parcel. Some of the energy required to expand the
parcel as it rises into a lower pressure environment
is met by this energy release and so the parcel no
longer needs to meet all of the work of expansion
from its own internal energy. The effect of condensation within a cloud is therefore to offset some of

the temperature reduction with height of a rising
parcel under dry adiabatic conditions by releasing
energy through the phase transition of water.

We can modify our mathematical description of
the dry atmosphere to account for these changes in
the following way. Unlike the dry atmosphere
where there is no change in heat of the parcel with
height, heat is provided to the parcel by the mass of
water condensed. If we are considering a parcel of
unit mass the latent heat released will supply a
change in heat $dQ = -Ldr$, where dr is the change in
mixing ratio of water vapor in the parcel. We can
therefore modify eqn 2.31 to give:

$$dQ = -Ldr = c_p dT - v dp \qquad (2.37)$$

and by substituting the hydrostatic equation:

$$-Ldr = c_p dT + g dz \qquad (2.38)$$

As a result we can see that the reduction in temperature with height of a cloudy air parcel under adiabatic conditions, the **saturated adiabatic lapse rate**,
γ_s, is given by:

$$\gamma_s = -\frac{dT}{dz} = \frac{g}{c_p} + \frac{L}{c_p}\frac{dr}{dz} = \Gamma_d + \frac{L}{c_p}\frac{dr}{dz} \qquad (2.39)$$

As the cloudy air parcel ascends the water vapor
mixing ratio decreases with height as water is
condensed onto cloud droplets so dr/dz is negative.
This reduces γ_s below the dry adiabatic lapse rate,
Γ_d, by an amount that is directly proportional to
the mass of water vapor condensing in the rising air
over a fixed height interval.

It is clear that as the amount of water vapor a
parcel can hold at saturation is a strong function of
temperature, the saturated adiabatic lapse rate will
vary markedly depending on the temperature of
the air parcel. At high latitudes, or altitudes with
very cold temperatures, an air parcel can hold very
little water vapor at saturation. As a result most of
the energy for expansion of a rising cloudy air parcel must still come from the internal energy of the
parcel and the lapse rate is little different from the
parcel in dry conditions. However, at low latitudes
and altitudes air temperatures are higher and there

is a considerable amount of water vapor held in the parcel at saturation. Under these circumstances, condensation may contribute significantly to the energy budget of the parcel. At high ambient temperatures it may be that γ_s may be as low as $0.35\Gamma_d$.

It should also be pointed out that the extremely low temperatures at the tropopause act as a cold trap. At these very low temperatures air can sustain very little water in the vapor phase before reaching saturation. As water vapor is condensed and precipitates out of the parcel the remaining parcel is very dry and little water vapor is transferred into the stratosphere.

Clearly the average lapse rate of the troposphere will be less than the dry adiabatic lapse rate but will be considerably more than the $3.5\,\mathrm{K\,km^{-1}}$ experienced in the lower levels of equatorial cumulonimbus clouds. In fact the average lapse rate in the troposphere is around $6.5\,\mathrm{K\,km^{-1}}$. This lapse rate was used in the consideration of the temperature structure of the lower atmosphere in Section 2.7.

2.9.5 Stability in cloudy air

It follows from the above that the rates of change of temperature of an ascending air parcel will be different in dry and cloudy air. We can imagine then that an air parcel will respond differently to small vertical perturbations in position depending on whether cloud is present in the air parcel or not. What is certainly true is that if an air parcel is unstable in dry air (see discussion in Section 2.9.3) then it must be unstable in cloudy conditions. Put another way, if the ambient temperature reduces with height more steeply than the dry adiabatic lapse rate it must fall faster than the saturated lapse rate.

However, the converse clearly does not hold. Consider an air parcel at point O in Fig. 2.13. The ambient fall in temperature with height is less than the dry adiabatic lapse rate and in dry air this will lead to any vertical motions of a parcel being suppressed as the air parcel is stable. However, if the temperature structure of the atmosphere is the same but the air parcel at O has a water vapor mixing ratio of $6\,\mathrm{g\,kg^{-1}}$ then the parcel at point O will be

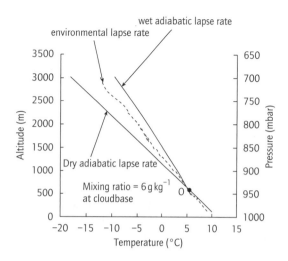

Fig. 2.13 The effect of moisture on the atmospheric lapse rate. A parcel of air at O has a mixing ratio of 6 g kg^{-1} of water vapor and so is just at saturation. As the air parcel rises adiabatically its temperature will decrease at a rate γ_s as some of the energy required to expand the parcel is supplied by the condensation of water vapor. If the environment curve is greater than γ_s and less than Γ_d the parcel is stable in dry air but becomes unstable when cloud is present; in other words, conditionally unstable.

just saturated and so will be at cloudbase. The reduction in the adiabatic lapse rate caused by heat released during condensation is sufficient to reduce the saturated adiabatic lapse rate to less than the ambient temperature profile in the cloudy column above point O. Any vertical displacement of an air parcel subjected to these conditions will lead to it becoming more buoyant than the air around it and hence the parcel will be unstable. This is known as **conditional instability** and a stable column of air that is forced to rise over orography or a frontal zone may cool until condensation occurs. This may release enough latent heat to make the air column positively buoyant and hence unstable.

Table 2.1 shows the five different stability criteria possible from absolutely unstable to absolutely stable. Neutral stability occurs in dry air when any vertical movement of an air parcel neither increases nor decreases its buoyancy relative to the surrounding air. The lapse rate under these conditions

Table 2.1   Stability criteria for a moist air parcel.

1	$\gamma < \gamma_s$	Absolutely stable
2	$\gamma = \gamma_s$	Saturated neutral
3	$\gamma_s > \gamma > \Gamma_d$	Conditionally unstable
4	$\gamma = \Gamma_d$	Dry neutral
5	$\gamma > \Gamma_d$	Absolutely unstable

The ambient reduction in temperature with height, γ, may be greater or less than the change in temperature with height induced by either a dry or a saturated adiabatic process, Γ_d or γ_s.

will be equal to Γ_d. Similarly neutral stability can also occur in cloudy air, though of course the lapse rate under cloudy conditions will then be equal to γ_s.

2.10   THE ENERGY BUDGET FOR THE EARTH'S ATMOSPHERE

We have seen that the input of energy to the Earth and its atmosphere comes from the Sun in the form of shortwave visible and ultraviolet radiation. However, the temperature of the Earth and its atmosphere is not changing considerably with time so the system appears to be close to steady state overall and energy inputs balance energy losses. We have discussed the re-radiation of longer-wavelength infrared radiation from the Earth and its atmosphere and also the transfer of energy through the atmosphere through convection and latent heat release. We should now consider the contributions these processes make to the overall energy budget of the Earth atmosphere system. First, we will consider the energy budget averaged over the whole globe and simply consider the relative contributions of the various processes as global means. Although this gives a good indication of which of the processes are most important it does not give us any idea of the behavior of the atmosphere at a particular location. We will therefore consider how the energy budget varies from one part of the globe to another and discuss the effect of these inhomogeneities on the atmosphere of the Earth.

2.10.1   The average energy budget

Over one year around 5.5×10^{24} J of solar energy is received at the top of the Earth's atmosphere in the form of visible and ultraviolet radiation. If, as in Fig. 2.14, we assume this input to be 100 units we can compare the other parts of the energy budget of the Earth–atmosphere system with respect to this total. The planetary albedo is approximately 0.3 and so 30% of the incoming shortwave radiation is reflected back to space with no interaction. Two-thirds of this reflection is from cloud tops, one-fifth is from molecular scattering in the atmosphere, and the remainder is from reflection at the Earth's surface. Of the remaining 70 units of energy, 19 are absorbed by the atmosphere and 51 are absorbed by the surface.

We have seen that the Earth itself acts as a blackbody radiator and emits radiation from its surface, mainly at infrared wavelengths. Overall, the number of units of energy lost from the surface as longwave radiation is 117, greater than the energy input from the Sun. However, of this large loss from the surface only 6 units escape to space directly, while 111 units are absorbed by the atmosphere, mainly by water vapor, CO_2, and clouds. In addition to the absorption of shortwave and longwave radiation, the atmosphere gains a further 23 units from evaporation of surface water into the atmosphere in the hydrological cycle and also 7 units from sensible heat transfer by convective processes. In total, then, the atmosphere gains 160 units of energy per year.

Clearly the atmosphere is not undergoing a very large warming and this energy gain is balanced overall by loss of the same amount of energy in the form of longwave radiation. Of the 160 units lost by the atmosphere 96 units are re-radiated back to the surface and 64 units are radiated out to space. This longwave re-radiation by the atmosphere also balances the energy budget at the Earth's surface, where in total 147 units of energy are received in the form of shortwave and longwave radiation and lost by longwave radiation, evaporation, and convective transfer. Likewise, the incoming 100 units of solar energy are balanced by a loss of 100 units

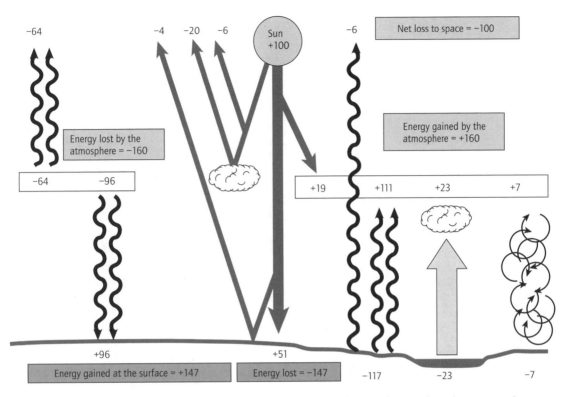

Fig. 2.14 The average energy budget of the Earth–atmosphere system. The contributions from shortwave and longwave radiation to the energy budget are indicated by straight and wavy arrows respectively. One-fifth of the loss of energy from the surface is in the form of sensible and latent heat, the remainder is due to longwave radiation.

from reflection and longwave radiative emission from both the Earth and the atmosphere.

2.10.2 Variations in the heat budget across the globe

The above energy budget is averaged over all latitudes and seasons. Clearly, the shortwave radiation input is not uniform over the entire globe. Low latitudes receive considerably more energy from solar radiation than higher latitudes. Furthermore, the albedo of the surface determines the fraction of sunlight reflected away from the surface and so affects the amount of shortwave radiation absorbed. For example, the albedo of fresh snow is very high (0.8–0.9), while that of the ocean is as low as 0.08. The heating rates therefore vary greatly as a function of latitude, but this does not lead to a concomitant increase in temperature with time. One explanation of this is that the outgoing longwave radiation flux also varies in a similar way. However, satellite data show that this is not the case. The average albedo of the Earth's land surfaces is higher than that of the oceans and as most of the land surface is in the northern hemisphere and the southern hemisphere is mostly ocean there are differences in the heating rates of the two hemispheres and hence in their dynamical circulation.

Figure 2.15a shows the calculated shortwave

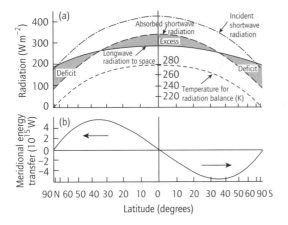

Fig. 2.15 (a) The calculated shortwave and longwave energy budgets of the Earth–atmosphere system averaged over time as a function of latitude. There is an excess of shortwave solar radiation at low latitudes relative to the longwave losses in the same region of the planet. This increase in heat flux is balanced by a net loss of radiation at latitudes above 40°. As the temperatures of the lower atmosphere in both the tropical and polar regions are approximately constant with time there must be a poleward transfer of energy, shown in (b). (From Wells 1997, © John Wiley & Sons Limited. Reproduced with permission.)

and longwave energy budgets of the Earth–atmosphere system averaged over time as a function of latitude. There is an excess of incoming shortwave radiation between 35° S and 40° N and a deficit at higher latitudes compared with the outgoing longwave radiation budget. If equilibrium were to be maintained at every latitude the shortwave and longwave radiation should balance locally and the two curves in Fig. 2.15a would be identical. Because they are not, and local mean temperatures close to the Equator are not increasing with time and those close to the poles are not decreasing, heat energy must be transported from low latitudes poleward. This is achieved by circulation within both the ocean and the atmosphere, transporting heat away from the Equator toward the pole and maintaining a higher temperature at latitudes greater than 50° than would be possible

from a system in radiative equilibrium, illustrated by the thin broken curve in Fig. 2.15a.

The difference between the incoming short-wave and outgoing terrestrial longwave radiation is known as the **net radiation** and is shown as the shaded area in Fig. 2.15a. Net radiation is positive close to the Equator indicating energy gain and negative above 40°. The energy transfer from Equator to pole can be calculated for each hemisphere by integrating the net radiation from the Equator toward the pole. The result of this calculation is shown in Fig. 2.15b. The energy transfer reaches a maximum around 40°, the latitude below which the Earth–atmosphere system on average gains energy. The energy is transported to higher latitudes where the system on averages loses more energy than it receives. This heat pump supplies the energy required to drive the ocean circulation and global wind patterns in the atmosphere.

Figure 2.16 shows the spatial variability of net radiation over the globe for February 2002 that has been calculated from the combined information from several different satellites during the Earth Radiation Budget Experiment (ERBE). The striking feature of the global variability of net radiation is the zonal homogeneity. The shortwave radiative flux is largely dependent on the albedo and this varies markedly from one location to another at similar latitudes. The atmosphere above the oceans at low latitudes is characterized by clear skies and so has a low albedo of around 0.1, whereas over the equatorial land masses large-scale organized convection leads to large cumulonimbus clouds and hence increased albedos greater than 0.3. However, the cloud tops over land are at around 10–12 km and consequently have low temperatures. There is therefore significant longwave trapping over low-latitude land masses and this approximately offsets the change in albedo and leads to the lack of zonal variability in the net radiation over the globe.

Detailed satellite measurements of net radiation can be used to deduce the size of the energy transfer from the Equator to the poles. However, this gives us no indication of the relative size of the contributions made to this transfer by the circulation of the ocean and atmosphere.

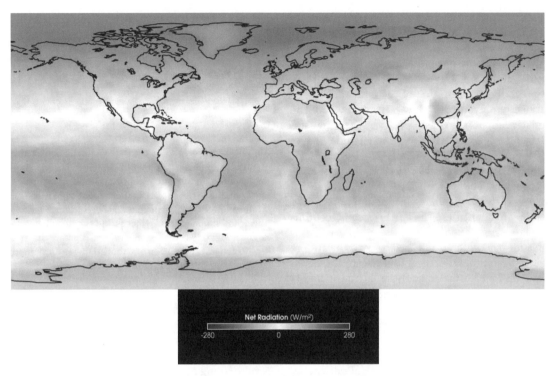

Fig. 2.16 A false color map of the spatial variability of net radiation over the globe for February 2002. The data were obtained from several different satellites and combined during the Earth Radiation Budget Experiment (ERBE). (Data courtesy of the ERBE and CERES Projects, NASA LaRC: http://www.earthobservatory.nasa.gov)

2.11 ENERGY TRANSFER IN THE ATMOSPHERE AND OCEAN

As we have seen, for the Earth's ocean–atmosphere system to be in equilibrium large quantities of heat must be transferred from the Equator toward the poles. Circulation systems in both the atmosphere and the ocean are significant in achieving this redistribution of heat. In the atmosphere, the high levels of net radiation at the Equator warm the surface and drive large-scale convection. The rising air cools and condensation enhances instability and fuels further convection by very large releases of latent heat. Penetration of the rising air into the stratosphere is greatly suppressed by the large temperature inversion at the tropopause and the air flows poleward. As the air moves away from

the Equator the air cools as it transports heat poleward. The air is no longer buoyant and sinks, providing a return flow at low level towards the Equator. This circulation is known as the Hadley circulation and extends to around 30° latitude, north and south of the Equator. A small fraction of the descending air flows poleward and rises in the mid-latitudes before a weak upper-level mean flow returns the air toward the Equator. This second cell is known as a Ferrel cell.

The Hadley cell has a strong meridional component as it is driven by convection. In contrast the mean circulation in the mid-latitudes is driven by large latitudinal pressure differences and the flow is largely westerly with only small mean meridional flow. However, this situation is unstable and synoptic-scale transient perturbations to the

westerly flow, in the form of cyclonic and anticyclonic weather systems, very efficiently transport
heat poleward.

Radiation transfers more heat to the ocean than
to the land because the albedo of the sea surface
is generally less than that of the land and the sea
surface temperatures at low latitudes, where the
radiative input is highest, are lower than the land
surface temperatures at the same latitude, reducing longwave cooling. The largest net loss of heat
from the ocean is by evaporation rather than sensible heat loss. The heat capacity of water is large and
so, unlike the atmosphere, the ocean can store very
significant quantities of heat. The oceans accumulate large amounts of heat in the mid-latitudes in
the summer, which is released back to the atmosphere during the winter, giving rise to increased
cloud and precipitation at these latitudes at this
time of year. However, the wintertime heat loss
from the mid-latitude oceans is larger than their
heat gain in summer and a poleward flux of heat is
required to maintain the equilibrium.

Figure 2.17 shows the contributions made by
the atmosphere and ocean to the total net energy
transport in the northern hemisphere as a function
of latitude. To more clearly see the contributions
of the processes described above the energy transport of the atmosphere is split into its mean and
transient components. At low latitudes heat is
transported approximately equally by the oceans
and the atmosphere, whose contribution is almost
entirely due to the Hadley circulation. The ocean
heat flux dominates over the atmosphere between
15 and 30°N. This is the region south of the mid-
latitude circulation where the Hadley circulation
is descending. The maximum heat flux occurs at
around 30°N and at higher latitudes the atmospheric transport is once more larger than that in
the ocean. However, unlike the equatorial regions
the main transport mechanism is not the mean circulation but the transient synoptic-scale perturbations to the mean zonal flow.

2.12 SOLAR RADIATION AND
THE BIOSPHERE

In addition to providing the energy to our Earth-

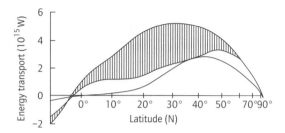

Fig. 2.17 The total poleward longitudinally averaged
net transport of heat as a function of latitude. The total
has been subdivided into the atmospheric component
and that due to the ocean circulation (shaded area). The
atmospheric contribution has been divided into its
mean and transient components (lower and upper parts
of the non shaded area respectively) to separate the flux
due to general circulation and that arising from synoptic
scale mixing processes. (From Gill 1982.)

atmosphere system and providing a climatology in
which life can flourish, solar radiation has many
direct influences on the biosphere (the narrow
band of the atmosphere–Earth–ocean system
where living organisms are found). There are also
indirect effects and complex feedback systems
between the atmosphere, biosphere, and radiative
transfer, especially when man's activity is considered as part of the biosphere: for instance, anthropogenic emissions of carbon dioxide and methane
and ozone depletion. Here we will concentrate on
the more direct effects of solar radiation on life
forms.

In many respects the behavior of living organisms in sunlight is not an effect, in the sense of an
existing organism responding to an externally imposed stimulus, but an example of the evolutionary adaptation of the organism to utilize available
resources. Many of the so-called effects of radiation (usually detrimental) come from an upsetting
of the balance between the radiation climate in
which the organism evolved and that to which it is
currently exposed; this change might be considered as an external stimulus.

The radiation balance of the Earth–atmosphere
system discussed in earlier sections is only concerned with the total energy in the complete solar
waveband (0.3–4 μm). However, many of the pho-
toreactions initiated by sunlight are wavelength

dependent and different parts of the solar spectrum have to be considered for different reactions. In addition, the simple physical energy contained in a waveband is not always a good indicator of its potential to induce the desired reaction; the action spectrum is also required. The action spectrum, or response spectrum, for a given reaction describes the wavelength-dependent sensitivity of the reaction or target body to the incident radiation, often normalized to unity at the wavelength of maximum response. Thus if the incident solar spectral intensity is $I(\lambda)$ and the action spectrum of interest is denoted by $R(\lambda)$ then the intensity of biologically effective radiation (i.e. the physical energy weighted with its effectiveness in producing the specified reaction) will be:

$$\int I(\lambda)R(\lambda)\mathrm{d}\lambda \qquad (2.40)$$

Many of the important biological action spectra are in the ultraviolet and visible portions of the solar spectrum, where the individual photons have most energy. Approximately half of the total solar energy is in the visible part of the spectrum, with a small amount in the ultraviolet and the rest at infrared wavelengths (see Fig. 2.3).

Two fundamental uses of solar radiation by inhabitants of the biosphere are vision and photosynthesis. Respectively they allow the majority of mobile creatures to see, and plants to convert solar energy into a usable form (sugar), initially for themselves and then, as plant matter, for other levels of the food chain. Our optical system (the eye) responds to visible radiation (wavelengths between 400 and 700 nm), and the photopic response peaks in the middle of this range (green light). It is no coincidence that the solar spectrum, both extraterrestrially and at the surface, is a maximum in the same wavelength region. Understanding the illuminance (visually effective radiation with $R(\lambda)$ equal to the photopic response in eqn 2.40) is important in, for example, building design and is measured with a luxmeter that has a response spectrum very similar to the photopic response of the eye.

Systems that photosynthesize also make good use of the same waveband. Photosynthesis by chlorophyll-containing plants is the process by

which water and atmospheric CO_2 are converted into simple sugars (and thence more complex compounds), oxygen, and water. The absorptance of typical green leaves exceeds 90% at blue and red wavelengths, but decreases to less than 80% in the green waveband, where reflectance and transmittance increase (hence the observed green color). Absorptance drops precipitously at the longwave end of the visible spectrum and is less than 5% between 0.7 and 1 µm, thereafter increasing again. The reflection and transmission of infrared radiation helps to prevent the plant from overheating. Given the comparatively constant spectral composition to photosynthetically active radiation (PAR), the photosynthesis rate will depend on the intensity of the radiation, plus water availability, carbon dioxide concentration, and the presence or lack of other environmental stresses (e.g. temperature). In the absence of other limiting factors the photosynthesis rate increases almost linearly with increasing incident radiation up to a limiting value. At this point the plant is light saturated and further increases in radiation are not beneficial. The light saturation point occurs at different irradiance and photosynthesis rates depending on the type of plant: both are low for shade-loving plants and increase until, for some plants, it is difficult to reach light saturation in sunlight.

At ultraviolet wavelengths photon energies become sufficient to cause damage and it is the detrimental effects of ultraviolet that are most often cited, although there are beneficial effects as well. Prominence has been given to the UVB waveband as it is radiation in this waveband that is most affected by changes in stratospheric ozone. Ozone depletion leads to increased UVB at the surface and a shift of the short-wavelength limit of the solar spectrum to shorter (more damaging) wavelengths. In humans and animals UVB radiation is necessary for skeletal health as it initiates the cutaneous synthesis of vitamin D, but it also damages DNA, produces sunburn/tanning, affects skin-mediated immunosuppression, and causes damage to the eye. Sunburn is probably the best known detrimental effect, with an action spectrum that includes both UVB and UVA radiation. It is also associated with increased risk of skin cancer, particularly the most fatal variety, malignant melanoma. Skin cancers, like chronic eye damage resulting in cataracts, are diseases whose risks increase with accumulated lifetime exposure to ultraviolet, moderated for skin cancers by the skin's natural sensitivity to ultraviolet radiation. Fair skinned people are at greater risk than those with naturally high levels of pigmentation. The relation between skin color and ultraviolet effects is a good example of an evolutionary balancing act. People originating from high latitudes (low levels of sunlight and ultraviolet) have fair, sun-sensitive skins with little melanin (a competing absorber for ultraviolet photons and responsible for color in tanned, brown, or black skin), enabling them to take advantage of available ultraviolet for vitamin D synthesis. Movement to higher radiation environments (equatorwards) increases the risks of sunburn and skin cancer. Conversely, highly pigmented peoples from low latitudes have natural protection against high levels of ultraviolet there, but if they move polewards where ultraviolet is reduced they can become susceptible to vitamin D deficiency.

REFERENCES

De More, W.B., Sander, S.P., Golden, D.M. *et al.* (1997) *Chemical Kinetics and Photochemical Data for Use in Stratospheric Modeling. Evaluation No. 12.* Jet Propulsion Laboratory, Pasadena, CA.

Gill, A.E. (1982) *Atmosphere–Ocean Dynamics.* Academic, London.

Iqbal, M. (1983) *An Introduction to Solar Radiation.* Academic Press, Toronto.

Salby, M.L. (1996) *Fundamentals of Atmospheric Physics.* Academic Press, San Diego.

Seinfeld, J.H. & Pandis, S.N. (1998) *Atmospheric Chemistry and Physics: From Air Pollution to Climate Change.* Wiley, New York.

Wells, N. (1997) *The Atmosphere and Ocean: A Physical Introduction,* 2nd edn. Wiley, New York.

3 The Earth's Climates

JOHN G. LOCKWOOD

3.1 INTRODUCTION

3.1.1 Basics

Climate is the average state of the atmosphere observed as the weather over a finite period (e.g. a season) for a number of different years. The distribution of climates across the Earth's surface is determined by its spherical shape, its rotation, the tilt of the Earth's axis of rotation at 23.5° in relation to a perpendicular line through the plane of the Earth's orbit round the Sun and the nature of the underlying surface. The spherical shape creates sharp north–south temperature differences, while the tilt is responsible for month-by-month changes in the amount of solar radiation reaching each part of the planet, and hence the variations in the length of daylight throughout the year at different latitudes and the resulting seasonal weather cycle. The intensity of solar radiation in space at a given distance from the Sun is, of course, constant throughout the year. The 23.5° tilt also accounts for the position of the Tropics: the Tropic of Cancer at 23° 27′ N and the Tropic of Capricorn at 23° 27′ S. Here the Sun is overhead at midday on the solstices, June 21 or 22 and December 22 or 23 respectively. Indeed, the length of daylight does not vary significantly from 12 hours throughout the year over the whole of the area between the two Tropics. At the spring (March 21 or 22) and autumnal (September 22 or 23) equinoxes, when the noon Sun is vertically overhead at the Equator, day and night are of equal length everywhere. This is not so at the winter and summer solstices, since each year the areas lying poleward of the Arctic (at 66° 33′ N) and Antarctic (at 66° 33′ S) circles have at least one complete 24-hour period of darkness at the winter solstice and one complete 24-hour period of daylight at the summer solstice. At the poles themselves there is almost six months of darkness during winter followed by six months of daylight in summer.

Over time scales of around a year or longer, the complete Earth–atmosphere system is almost in thermal equilibrium. That is, the total global absorption of solar radiation by the Earth's surface and atmosphere is balanced by the total global emission to space of infrared radiation. While most of the solar radiation is absorbed during daylight in the tropical regions of the world, infrared radiation is lost continually to space from the whole of the Earth's surface and atmosphere, including both the night side and the middle and polar latitudes. Heat energy is therefore transferred from the Tropics where it is in surplus, to middle and high latitudes where it is in deficit, by the north–south circulations in the atmosphere and oceans. In very general terms, these circulation patterns take the form of a number of meridional overturning cells in the atmosphere, with separate zones of rising air motion at low and middle latitudes, and corresponding sinking motions in subtropical and polar latitudes. This very simple meridional circulation both implies a rotating Earth and is also further modified by the Earth's rotation.

If the Earth did not rotate relative to the Sun — that is, it always kept the same side toward the

Sun—the most likely atmospheric circulation would consist of rising air over an extremely hot, daylight face and sinking air over an extremely cold, night face. The diurnal cycle of heating and cooling obviously would not exist since it depends on the Earth's rotation. Surface winds everywhere would blow from the cold night face toward the hot daylight face, while upper flow patterns would be the reverse of those at lower levels. Whatever the exact nature of the atmospheric flow patterns, the climatic zones on a nonrotating Earth would be totally different from anything observed today. Theoretical studies suggest that if this stationary Earth started to rotate, then as the rate of rotation increased, the atmospheric circulation patterns would be progressively modified until they resembled those observed today.

The most obvious influence of the surface on climate is seen in the distribution of continents and oceans. This significantly modifies a simple north–south distribution of climates, causing marked differences in characteristics between the eastern and western sides of the continents because there are marked differences between oceanic and continental climates. The oceans are a very large reservoir of heat energy and this damps

the rate of change of oceanic air temperatures. They are also a major source of atmospheric water vapor. In contrast, the heat capacity of land surfaces is low, so they warm rapidly during the day, helping to induce convection that both vertically mixes the atmosphere and is also a cause of precipitation. Vegetation growth reduces the albedo of bare land surfaces, alters the moisture balance of both the lower atmosphere and the soil, and often lowers the maximum temperature. In contrast, polar snow and ice covers reflect much of the incident solar radiation, causing temperatures to be much lower than they would be in the absence of ice.

3.1.2 *General atmospheric circulation*

A schematic representation of the mean meridional circulation between Equator and pole is shown in Fig. 3.1. A simple direct circulation cell, known as the Hadley cell, is clearly seen equatorward of about 30° latitude. Eastward angular momentum is transported from the equatorial latitudes to the middle latitudes by nearly horizontal eddies, 1000 km or more across, moving in the upper troposphere and lower stratosphere. This

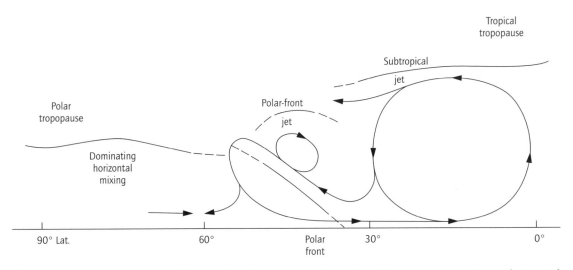

Fig. 3.1   Schematic representation of the meridional circulation and associated jet stream cores in winter. The tropical Hadley and middle latitude Ferrel Cells are clearly visible. (From Palmen 1951.)

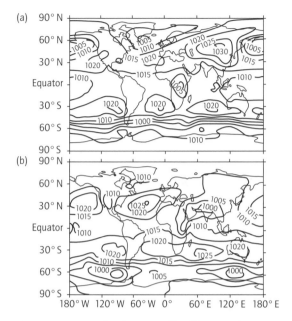

(a) 90° N
60° N
30° N
Equator
30° S
60° S
90° S

(b) 90° N
60° N
30° N
Equator
30° S
60° S
90° S
180° W 120° W 60° W 0° 60° E 120° E 180° E

Fig. 3.2 Mean sea-level pressure (hPa) (1963–73) averaged for: (a) December, January, February and (b) June, July, August. (From Henderson-Sellers & Robinson 1986; Oort 1983.)

transport, together with the dynamics of the middle latitude atmosphere, leads to an accumulation of eastward momentum between 30 and 40° latitude, where a strong meandering current of air, generally known as the subtropical westerly jet stream, develops. The cores of the subtropical westerly jet streams in both hemispheres and throughout the year occur at an altitude of about 12 km. The air subsiding from the jet streams forms the belts of subtropical anticyclones at about 30 to 40° N and S (Fig. 3.2). More momentum than is necessary to maintain the subtropical jet streams against dissipation through internal friction is transported to these zones of upper strong winds. The excess is transported downwards and polewards to maintain the eastward-flowing surface winds (temperate latitude westerlies) against ground friction. The middle latitude westerly winds are part of an indirect circulation cell known as the Ferrel cell. The supply of eastward momentum to the Earth's surface in middle latitudes tends to speed up the Earth's rotation. Coun-

teracting such potential speeding up of the Earth's rotation, air flows from the subtropical anticyclones toward the equatorial regions, forming the so-called trade winds. The trade winds, with a strong flow component directed toward the west (easterly winds), tend to retard the Earth's rotation, and in turn gain eastward momentum.

The greatest atmospheric variability occurs in middle latitudes, from approximately 40 to 70° N and S, where large areas of the Earth's surface are affected by a succession of eastward moving cyclones (frontal depressions) and anticyclones or ridges. This is a region of strong north–south thermal gradients with vigorous westerlies in the upper air at about 10 km, culminating in the polar-front jet streams along the polar edges of the Ferrel cells. The zone of westerlies is permanently unstable and gives rise to a continuous stream of large-scale eddies near the surface, the cyclonic eddies moving eastward and poleward and the anticyclonic ones moving eastward and equatorward. In contrast, at about 10 km, in the upper westerlies, smooth wave-shaped patterns are the general rule. Normally there are four or five major waves around the hemisphere, and superimposed on these are smaller waves that travel through the slowly moving train of larger waves. The major waves are often called Rossby waves, after Rossby who first investigated their principal properties. Compared to the Hadley cells the middle latitude atmosphere is highly disturbed and the suggested meridional circulation shown in Fig. 3.1 is largely schematic.

Both polar regions are located in areas of general atmospheric subsidence, though the climate is not particularly anticyclonic nor are the winds necessarily easterly. The moisture content of the air is low because of the intense cold, and horizontal thermal gradients are normally weak, with the result that energy sources for major atmospheric disturbances do not exist and are rarely observed. Vowinckel and Orvig (1970) suggest that the Arctic atmosphere can be defined as the hemispheric cap of fairly low kinetic energy circulation lying north of the main courses of the planetary westerlies, which places it roughly north of 70° N. The situation over the south polar regions is more complex because of the presence of the Antarctic ice sheet, and the boundary of the Antarctic atmosphere is less clear.

3.1.3 Climatic classifications

To bring some order and understanding to the great variety of climates observed across the Earth's surface it is useful to use a systematic classification system. Since natural vegetation tends to reflect the local climate, a number of schemes have been proposed that reflect vegetation zones. One of the most widely used systems of climatic classification is that of Wladimir Köppen (1846–1940), a German biologist who devoted most of his life to climatic problems. His first classification of 1900 was based largely on vegetation zone, but without the use of formulae involving climatological data. In 1918 he revised his first classification, paying greater attention to temperature and rainfall, together with their seasonal characteristics. Later, he used simple formulae to compare temperature and precipitation (Köppen 1923). He used temperature as a substitute for evaporation data, which were too scanty and unreliable in his time to be used on a world basis. Köppen's system is strictly empirical, in that each climatic zone is defined according to temperature and precipitation; no consideration is given to climatic causes or to pressure systems, wind belts, etc. Each climatic type was designated by a shorthand notation which replaced imprecise and rather lengthy general descriptions. Revised versions of this shorthand notation are widely used today, particularly in European countries; therefore they are summarized in Table 3.1 and Fig. 3.3.

In 1948 the American climatologist Charles Warren Thornthwaite introduced a climatic classification based on a water balance approach. He used two climatic factors, a moisture index and a thermal efficiency index. The former is obtained by determining water surpluses and deficits for each year (the difference between precipitation and potential evapotranspiration), while the latter is equivalent to annual potential evapotranspiration. Because evaporation data were not widely available when Thornthwaite (1948) produced his classification, he used mean monthly temperature as a substitute. As in Köppen's classification, each climatic type is designated by a shorthand notation. While Thornthwaite's classification has some use today, particularly in North America, Köppen's classification is very widely used and is followed in this chapter.

3.2   POLAR CLIMATES

3.2.1 Introduction

The most distinctive climatic features in both north and south polar regions are the presence of ice, snow, deeply frozen ground and a long winter period of continuous darkness. The radiation budget of the polar surfaces is nearly always negative. This is because of the absence in winter of any solar radiation, the low angle of solar incidence in the summer, and the high albedo of the ice fields. Indeed, the polar regions serve as global sinks for the energy that is transported poleward from the Tropics by warm ocean currents and by atmospheric circulation systems, particularly travelling cyclones and blocking anticyclones. However, there are major differences between the Arctic and Antarctic regions in terms of the distribution of land, sea, and ice. The Arctic is a largely ice-covered, landlocked ocean, whereas Antarctica consists of a continental ice sheet over 2000 m thick surrounded by ocean. Surrounding the Antarctic continent is a zone of floating sea ice, about 1.5 m thick, which undergoes a marked annual cycle from a minimum in February/March to a maximum in September/October. In the summer the Antarctic sea ice melts almost back to the coastline of the continent, so only slightly more than 10% survives the summer. The winter advance of the sea ice is not restricted by land at its equatorward boundary and it crosses the Antarctic circle. In contrast, in winter the southern limits of the Arctic sea ice are constrained by the northern coastlines of Asia and North America, but pack ice extends into middle latitudes off eastern Canada and eastern Asia where northerly winds and cold currents transport ice far southward. West of Norway there is open water to about 78° N because

Table 3.1 Köppen climate classification system. Köppen's climate classification is strictly empirical in that each climate is defined according to mean annual temperature and precipitation, no consideration is given to climate causes. The system allows for five major climate categories that are further subdivided at the secondary and tertiary levels.

Primary level

The five major categories are designated by capital letters

A	Tropical rainy climates. Average temperature of every month above 18°C. No winter season. Annual rainfall is large and exceeds annual evapotranspiration
B	Dry climates. No temperature limitation *per se*, but on average evapotranspiration exceeds precipitation throughout the year
C	Humid mesothermal climates. Coldest month has average temperature under 18°C but above −3°C; at least one month has an average temperature above 10°C
D	Humid microthermal climates. Warmest month is above 10°C. Coldest month below −3°C
E	Ice climates. Mean temperature of warmest month is less than 10°C

Secondary level

Af	Tropical rainforest climate. All months >61 mm of rain per month
Am	Monsoon climate with excessively heavy precipitation in some months in order to compensate for one or more months <61 mm
Aw	Savanna climate with winter dry period
BS	Steppe
BW	Desert climates. Drier than BS
Cf and Df	Moist with precipitation well distributed throughout the year
Cs and Ds	Summer dry period. Driest month receives <30.5 mm of rain
Cw and Dw	Winter dry period
ET	Tundra: warmest month <10°C but >0°C
EF	Ice cap: warmest month <0°C; perpetual frost

Tertiary level

Each tertiary letter deals with some further aspect of temperature

For B climates	
h	Hot: mean annual temperature >18°C
k	Cold: mean annual temperature <18°C
For C and D climates	
a	Hot summer: mean temperature of warmest month >22°C
b	Warm summer: mean temperature of the warmest month <22°C but at least four warmest months each with a mean temperature >10°C
c	Cool summer: mean temperature of warmest month <22°C; from 1 to 3 months with a mean temperature >10°C
For D climates alone	
d	Severe winter: mean temperature of the coldest months <−38°C

of the warm waters of the North Atlantic Drift and thermohaline circulation. The North Atlantic thermohaline circulation is particularly vigorous and its climatic effect is often illustrated by comparing the surface temperature of the northern Atlantic with comparable latitudes of the Pacific, since the former is 4–5°C warmer. During summer the sea ice melts back toward the pole, the boundary being around 80°N in the European sector and 70°N in the North American sector. Thus much of the Arctic ice has survived at least one summer melt season, and as a consequence its mean thickness is 3–4 m. The thickness is highly variable locally, with narrow openings (leads) throughout

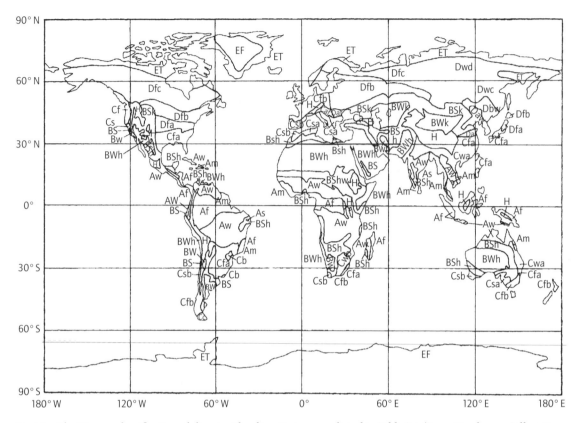

Fig. 3.3 The Köppen classification of climate. The climatic types are listed in Table 3.1. (From Henderson-Sellers & Robinson 1986.)

the pack ice even in winter, and larger openings (polynyas) adjacent to coastal areas where winds blow offshore. More than 80% of Greenland is covered by an extensive ice sheet that rises above 3000 m in elevation.

Cooling and sinking of the air raises pressure at the surface, but there is no permanent dome of high pressure to be observed over the poles, neither synoptically nor on the scale of means. This is because cold polar air is regularly but discontinuously transported away. Over Greenland, and most notably over Antarctica, the outflow of cold air is accelerated by the slope of the ice domes. Antarctica experiences katabatic winds as a constant feature, often of extreme violence, as air flows off the ice dome. Leroux (1993, 1998) has introduced the concept of Mobile Polar Highs to describe

these outbreaks of cold air, which can have an important influence on middle latitude circulation patterns.

3.2.2 Arctic climates

In a climatological sense, the Arctic corresponds approximately to the land areas beyond the northern limits of forests and ocean areas with at least a seasonal sea-ice cover. The boundary between Boreal forest and tundra vegetation corresponds closely to the 10°C mean isotherm of the warmest month and to the average July location of the Arctic Front separating Arctic and temperate air masses. Indeed, in the Köppen climatic classification system tundra climates (ET) are those with the warmest month below 10°C but above 0°C, while in ice cap climates (EF) the warmest monthly average is less than 0°C (i.e. perpetual frost).

At high latitudes, the total daily solar radiation depends largely on day length, which in turn varies widely with the season of the year. The radiation climate produced with continuous darkness in winter and continuous daylight in summer is distinctly different from that found at lower latitudes. It is also observed that the distribution of global radiation is controlled not only by astronomical considerations, but also by the distribution of cloud, which is closely related to the atmospheric circulation and weather systems. Thus the most cloudy regions of the Arctic are the North Atlantic and European sectors, where, because of the intense cyclonic activity and the warm North Atlantic Drift, the frequencies of overcast skies are very high. The least cloudy regions in the Arctic are the northeastern parts of the Canadian archipelago and North Greenland, which are subject to the prolonged influence of the polar and Greenland anticyclones. A second minimum is observed in eastern Siberia, where cloudiness is high along the coast and decreases inland.

There are two main areas of semipermanent low-level inversions (atmospheric temperature increases with height) in the world: the subtropical belt and the polar regions. The inversions found in the subtropical belt are largely dynamic in origin and are separated from the surface by an unstable layer dominated by cumulus convection. The main cause of the polar inversions is the negative energy balance at the surface, but the inversion is also maintained by warm air advection aloft with associated subsistence. The strongest vertical temperature gradients are therefore often found near the surface, and the gradients are very much steeper than in the subtropical inversions. Over the Central Polar Ocean the temperature inversion persists for at least 60% of the days in all months, reaching a maximum in late winter of 100%. Because of this the surface air temperature is primarily dependent on the nature of the surface conditions. The prevailing summer melting of snow and ice holds the surface air temperature close to 0°C, but slightly positive temperatures are usually

observed near the pole in the second half of June (Fig. 3.4). Relatively warm water below the sea ice keeps winter temperatures around –30 to –35°C, since sensible heat is conducted upwards through the sea ice to warm the lowest air layers. Indeed, the lowest winter surface temperatures are not found over the Central Polar Ocean, but over the land areas of Northeast Siberia and the Yukon (Fig. 3.5). Here, under the influence of the continental anticyclones, calm, clear, subsiding air becomes trapped in hollows and the temperature falls below –50°C on occasion and winter monthly means below –40°C are common. Rapid loss of heat during the long winter nights in the form of longwave radiation cools the ground surface but, in contrast to the sea ice, this heat cannot be replaced. Similar low temperatures are found on the high plateau of the Greenland ice sheet. Solar forcing of the diurnal temperature cycle is at a minimum in polar regions, making the polar diurnal temperature cycle extremely weak and difficult to compare directly with cycles at locations nearer to the Equator with a stronger diurnal solar radiation signal. During the winter months of December to February there is almost continuous darkness in the Arctic; therefore the conventional concept of solar radiative forced maximum and minimum temperatures does not apply and diurnal temperature range is governed by the atmospheric circulation.

Precipitation over the Central Polar Ocean (Fig. 3.6) is mainly associated with the frontal zones of depressions, with annual amounts decreasing northward from over 200 mm around the margins to close to 100 mm at the North Pole. Mean wind speeds are not high (4–5 m s⁻¹), but blowing snow occurs on over 100 days per year.

From December to April the Arctic air mass is polluted (Arctic haze) by man-made mid-latitude emissions from fossil fuel combustion, smelting, and industrial processes. For the rest of the year pollution levels are much lower. This is due in winter to less efficient processes of pollutant removal and better south-to-north transports (Barrie 1986). The winter Arctic air mass consists of a dome 7–8 km deep over the pole with shallow tongues of air 0–5 km deep spilling southward over the land masses. In summer it is confined to the

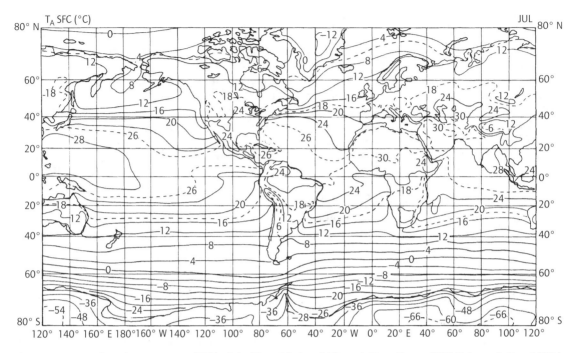

Fig. 3.4 Mean surface air temperature (°C) for July. (From National Climatic Data Center 1987; Peixoto & Oort 1992.)

northern polar regions. Low cloud varies seasonally from about 75% between May and September to about 35% between December and March. Clear skies are very rare in summer and more frequent in winter. Summer low stratus is often formed by warm air advection over a cold ice pack, and it is frequently associated with drizzle, which is important in the removal of air pollution. Main pollution source regions for Arctic air pollution are eastern North America, Southeast Asia and Europe. Much of the pollution from North America and Southeast Asia is washed out by precipitation over the Atlantic and Pacific Oceans. In contrast, European pollution in winter travels northeast over snow-covered surfaces and is strongly transported into the Arctic by the persistent Siberian high-pressure region.

3.2.3   Antarctic climates

The Antarctic continent consists of two domed ice sheets covering about 97.6% of its area at an average elevation of about 2200 m, compared to Asia's, the second highest continent, at 800 m. The high plateau of East Antarctica has extensive areas above 3000 m. This desolate world of ice is the coldest and windiest of the world's great land masses (Allison *et al.* 1993). On August 24, 1960, Vostock (78.5° S, 106.9° E; 3488 m) recorded a temperature of −88.3°C, which is probably the lowest temperature recorded on the Earth's surface. Antarctica and its environs play an important role in the workings of climate on a global scale, since the region acts as a large sink in the global energy cycle (King & Turner 1997). It is clear that the region is intimately tied to global climate and its variations, and that it is involved in teleconnections with remote regions (Simmonds 1998). For example, Savage *et al.* (1988), Carleton (1989), Karoly (1989), Smith and Stearns (1993), and Simmonds and Jacka (1995) have found evidence of connections between the Southern Oscillation Index and aspects of the Antarctic environment, at a variety of temporal leads and lags.

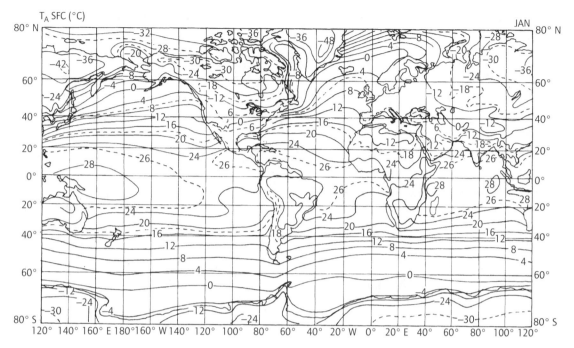

Fig. 3.5 Mean surface air temperature (°C) for January. (From National Climatic Data Center 1987; Peixoto & Oort 1992.)

Mean surface temperatures at coastal stations typically range, depending on location and other factors, between 0 and –5°C in January and –15 and –25°C in July. The temperatures decrease with latitude south and elevation. At the South Pole (2800 m) they are –28.9°C in January and –59.2°C in July, while at Vostock (3488 m) they are –33.4 and –66.7°C respectively. (Schwerdtfeger 1970, 1984; Simmonds 1998). Radiation balance is the principal climate-causing factor in the Antarctic, where there are strong seasonal variations in incoming solar radiation as well as variations in elevation and surface characteristics (Schwerdtfeger 1970, 1984). In summer very large amounts of solar energy reach the plateau surface, where the high surface albedo and thin clouds enhance the importance of diffuse sky radiation, so that global radiation on cloudy days differs little from that on clear days. The annual accumulation of snow over the continent appears to vary from about 600 mm yr⁻¹ (water equivalent) along the coast to about 50 mm yr⁻¹ in the center of the plateau. Since evaporation rates are normally very low, even these meager precipitation rates maintain the ice sheet. Mean annual snow accumulation for the continent is around 150–170 mm yr⁻¹ water equivalent (Turner *et al.* 1999).

The seasonal cycle of surface air temperature at South Pole Station (Table 3.2) shows a number of interesting features that are peculiar to Antarctica. The temperature rises rapidly in the southern spring (October and November), then averages –28°C for the summer months of December and January. The extreme shortness of the polar summer, not more than about 30 days between mid-December and mid-January, is a characteristic of a wide area of the high plateau. In February and March the temperature falls rapidly by nearly half a degree Celsius per day, reaching winter levels by the March equinox. The normal temperature peak comes only eight days after the summer solstice, a very short lag in comparison with

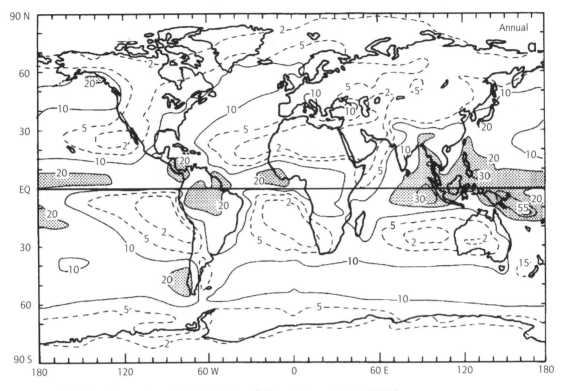

Fig. 3.6 Global distribution of precipitation in dm yr^{-1}. (From Peixoto & Oort 1992.)

Table 3.2 Mean temperature data for South Pole, 1957–66, latitude 90° S, elevation 2800 m (data from Orvig 1970).

Month	Daily mean (°C)
January	−28.8
February	−40.1
March	−54.5
April	−58.5
May	−57.4
June	−56.5
July	−59.2
August	−58.9
September	−59.0
October	−51.3
November	−38.9
December	−28.1

other continents, for which the lag is closer to 30 days.

The shortness of the Antarctic summer is caused by the variations of global radiation and the surface albedo. The maximum global radiation is on average around the time of the summer solstice. In the weeks proceeding the solstice the surface albedo decreases, due to the increasing solar elevation above the horizon and a slight metamorphosis of the snow cover. It is normal for the albedo of many surfaces to decrease with increasing solar elevation; thus more solar radiation can be absorbed at the snow surface. After the solstice, the global radiation decreases, and the first light snowfall or influx of drifting snow restores surface conditions to those favoring higher albedo values. Hence there are good reasons to expect absorbed solar radiation to decrease rapidly in January and the surface temperature to fall quickly. During the Antarctic winter the intense radiational cooling and the highly transmissive atmosphere generates a persistent low-level temperature inversion that reaches its greatest depth over the higher elevations of the ice sheet, where it is present almost the entire year (Phillpot & Zillman 1970). The inversion is no stronger at the end than at the beginning of the polar night. This is because an equilibrium is reached between the surface longwave radiation loss, which decreases as temperature falls, and the downward radiation from the inversion layer, which changes relatively little with time (Schwerdtfeger 1970, 1984). The absence of a clearly defined time for the occurrence of the surface minimum temperature is a feature of the Antarctic, comprising the so-called "coreless winter" (van Loon 1967).

Katabatic winds occur when cooling causes a shallow blanket of air adjacent to the surface to become colder and therefore heavier than the atmosphere above; the air then drains downslope under the influence of gravity. These winds are most persistent where the ground is covered by ice and snow and they dominate the surface wind regime of Antarctica. Nearly everywhere on the ice sheet the surface winds are directed downslope from the interior toward the surrounding ocean. The surface katabatic winds are related to the energy balance of the low-level inversion over Antarctica. Radiosonde measurements have been made of the net radiation and cooling rate in the lower troposphere at the South Pole (Schwerdtfeger 1970, 1984). There are difficulties with the radiosonde measurements, but an average radiative cooling rate of 4°C per day at the inversion level is acceptable. This is the cooling rate that appears to be occurring from radiative balance considerations alone; the energy is lost by longwave radiation to space. The real cooling rate in the atmosphere obtained from temperature observations is 1°C or less per month. Thus other atmospheric processes must be compensating for the longwave radiative loss to space. These additional atmospheric processes are identified as horizontal advection and adiabatic sinking (Simmonds & Law 1995). Air converges over Antarctica at middle levels, and

sinks and warms due to adiabatic compression, causing the low-level inversion over the ice surface. The sinking air cools rapidly by infrared emission in the inversion layer, as explained above. The radiatively cooled air sinks through the inversion layer, is replaced by sinking air from above, and flows away as the surface katabatic wind.

3.3 TEMPERATE LATITUDE CLIMATES

Daily weather charts for any extensive region in middle latitudes reveal well defined synoptic systems that normally move from west to east with a speed that is considerably smaller than their constituent air currents. At the surface the predominant features are closed cyclonic and anticyclonic systems of irregular shape, while at higher altitudes in the atmosphere smooth wave-shaped patterns are the general rule. These upper Rossby waves are important because surface synoptic systems tend to move in the direction of the broad upper flow with a velocity that is proportional to the intensity of the upper flow. So an intense upper flow pattern is associated with the rapid passage eastward of frequent surface frontal depressions and anticyclones. Surface frontal depressions also tend to form slightly downwind of upper troughs and similarly surface anticyclones tend to form slightly downwind of upper ridges. Major temperate climatological cyclone development regions are therefore located just downwind of upper climatological troughs.

In a simple ideal atmosphere, Rossby waves could arise anywhere in the middle latitude atmosphere, as is observed in the predominantly ocean-covered southern temperate latitudes. In contrast the northern temperate latitude Rossby waves tend to be locked in preferred locations which are shown in Fig. 3.7. These preferred locations may arise because the atmospheric circulation is influenced not only by the thermal properties of land and sea, but also by high mountain ranges and highlands in general. When a westerly air stream crosses a north–south aligned mountain range such as the Rocky Mountains, anticyclonic defor-

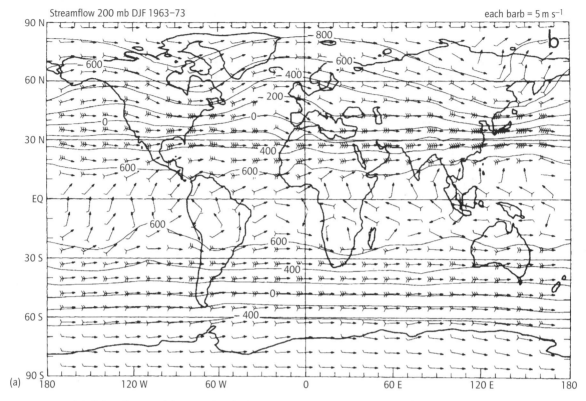

Streamflow 200 mb DJF 1963–73   each barb = 5 m s⁻¹

Fig. 3.7 Global distribution of the mean height (1963–73) of the 200 hPa pressure field represented as mean height minus 11,784 gpm for: (a) December, January, February; (b) June, July, August. Wind speed and direction shown by arrows. Each barb on the tail of an arrow represents a wind speed of 5 m s⁻¹. (From Peixoto & Oort 1992.) (*Continued*)

mation of flow is found over the mountain range and cyclonic deformation to leeward. These orographically generated ridges and troughs propagate downstream and can reinforce similar features generated by the direct thermal effects of land and sea. In January the northern hemisphere shows two dominant upper troughs near the eastern extremities of the two continental land masses, while ridges lie over the eastern parts of the oceans. A third weak trough extends from north Siberia to the eastern Mediterranean. Climatologically, the positions of the main upper troughs may be associated with cold air over the winter land masses, and the ridges with relatively warm sea surfaces. This so-called continentality is most marked in the middle-latitude continents, where

the low heat capacity of the land allows temperatures to fall to very low values in winter but reach equally high values in summer. The large heat capacity of the oceans generates a more maritime climate with relatively small annual temperature ranges. Thus, in July, the mean upper ridge found in January over the Pacific has moved about 25° west, and now lies over the warm North American continent, while there is a definite trough over the east Pacific. Patterns elsewhere are less marked, but a weak trough does appear over Europe, and may perhaps be connected with the coolness of the North Sea, the Baltic, the Mediterranean, and the Black Sea. The winter troughs found over the northeastern seaboards of the two major continental land masses are associated with vigorous

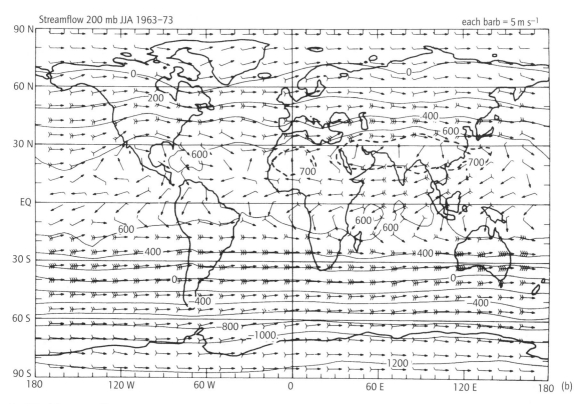

Fig. 3.7 *(Continued)*

frontal depression formation over the neighboring oceans. The depressions newly formed just downwind of the upper trough are steered to the northeast and then east by the upper flow, until they decay near the climatological upper troughs over the northeastern parts of the oceans, forming the so-called Icelandic and Aleutian climatological low-pressure areas.

The upper troughs may also be related to the mean trajectories of Mobile Polar Highs (MPH) moving out of the polar regions (Leroux 1998). These are masses of cold air of polar origin that tend to subside as they move equatorward because of the conservation of potential vorticity. They are very cold and dry, unless they are warmed by convection over a relatively warm ocean. Greenland, together with the high ground of Ellesmere Island, funnels Arctic MPHs on average every two days preferentially towards North America, and feeds

them into the Atlantic only south of latitude 60° N. In the European/western Asiatic sector, MPHs emerging from the Arctic follow a "Scandinavian" trajectory, moving along either the western or eastern slopes of the Scandinavian mountains. MPHs channeled to the east of Greenland and along the western slopes of the Scandinavian mountains move directly across western Europe. MPHs moving along the eastern flank of the Scandinavian mountains are directed either toward the Black Sea (channeled by the Urals) and the eastern Mediterranean basin or toward the Turin basin occupied by the Caspian and Aral seas, from which it is difficult for them to escape since this low-lying area is surrounded by high relief on three sides. In winter, continental cold air, relief, and MPHs maintain a powerful and highly stable atmosphere over the interior of Asia.

Leroux (1998) comments that in the southern

hemisphere, rapid and violent moment of cold air is facilitated by the slope of the Antarctic ice sheet, and dispersion occurs in all directions. The orography of the southern continents has a marked effect on the development of MPH trajectories. The Andes become impassable to MPHs north of latitude 40°S, and the southern African plateau divides MPHs approaching the Cape into two, one section flowing northwards along the foot of the Namibian escarpment, and the other following the coastline of Natal toward the Mozambique Channel.

3.3.1 Europe

Most of Europe is within the Cfb zone of the Köppen climate classification system, but this changes gradually to the east and northeast into the Dfb and Dfc climates that cover much of northern Asia (Fig. 3.3). Northwest Europe has a humid temperate climate, but the southern rim of the continent is characterized by mild winters and hot dry summers, a regime often referred as a Mediterranean type climate (Csa, Csb).

Important climate forming factors in Europe include the north–south gradient of solar radiation; the east–west alignment of mountain ranges along the southern rim; the large expanse of ocean to the west, with its associated winter warmth; the extensive continental land mass to the east, with its associated winter cold; and the indented shoreline. The open North European Plain allows the mid-latitude westerly winds, with their associated eastward-moving frontal depressions to penetrate far inland, while the exchange of air masses between the Mediterranean Sea and the North European Plain is restricted by the Alps and their associated mountain ranges. This is in marked contrast to North America, where north–south mountain ranges along the Pacific coast restrict the passage of mild air masses inland, but the interior of the continent is open to the unrestricted exchange of air masses in a north–south direction. Europe is mostly open to the north, but is shielded from very cold Arctic air by the relatively warm North Atlantic Ocean. Indeed, the mildness of the winter climate of western Europe is one of the

more spectacular latitudinal anomalies in the world climatic pattern; in January on the Norwegian coast the temperature anomaly may be as high as 22–26°C, gradually decreasing to 1°C southward across the continent. This mildness is related to the abnormally vigorous thermohaline circulation in the North Atlantic.

Of the elements governing the distribution of climate, the most basic is radiation from the Sun and its balance with outgoing longwave radiation from the Earth. The southernmost rim of Europe lies at about 35°N, almost within the subtropics, and its northernmost island extensions reach 78°N, almost in the Arctic. One important consequence of this large latitudinal spread is that the length of daylight at the solstices varies significantly over the region. Except along the southern rim, the net radiation balance of the continental surface in winter is negative, but the energy lost by the continual cooling of the land is replaced by vigorous warm air flow from the Atlantic Ocean to the west. Variations in the vigor of this westerly flow because of the North Atlantic Oscillation are of great significance in relation to the level of European winter temperatures and are discussed in detail below. Because the warm air masses from the Atlantic Ocean cool as they pass eastward over the winter continental surface, the mean isotherms in January run north–south, with temperatures falling both eastward away from the Atlantic and northward toward the Arctic (Fig. 3.5). In contrast, in summer the net radiation balance over the continental surface is positive: values fall toward the north, so isotherms are aligned east–west (Fig. 3.4).

The atmospheric circulation over Europe is controlled by an almost permanent cyclonic center over the North Atlantic, the Icelandic low, which is deepest in winter, and by the Azores high, which is most intense and extensive in summer. Because Europe is open to the resulting invasions of the maritime westerlies, the change eastward from maritime to continental climates is very gradual, and maritime effects are still apparent in eastern Europe. Indeed, the climate is dominated by continental air masses only east of a line running from around St Petersburg to Vienna. Along the Atlantic coasts of Europe, precipitation (Fig. 3.6) is evenly distributed through the year, with a tendency to a maximum in the winter half, when frontal depressions are most frequent and intense. Frontal depressions lose their intensity as they move inland, so a higher proportion of the precipitation in the interior comes from convective showers, which are most frequent in summer, producing a summer rainfall maximum. During the summer months, the prevailing westerlies retreat northward from southern Europe, leaving it under the influence of the subsiding air of the subtropical anticyclones and creating an intense dry season with only occasional rainfall from convective showers. In winter the westerlies return to southern Europe, bringing frequent depressions that cause a precipitation maximum in the winter half of the year.

A major source of interannual variability in the atmospheric circulation over the North Atlantic and western Europe is the so-called "North Atlantic Oscillation" (NOA), which is associated with changes in the strength of the oceanic surface westerlies. Its influence extends across much of the North Atlantic and well into Europe and it is usually defined through the regional sea-level pressure field, although it is readily apparent in mid-troposphere height fields. The NAO's amplitude and phase vary over a range of time scales from intraseasonal to interdecadal; the NAO is present throughout the year but the largest amplitudes typically occur in winter (Commission on Geosciences, Environment, and Resources 1998). The NAO is often indexed by the standardized difference of December to February sea-level atmospheric pressure between Ponta Delgado, Azores (37.8° N, 25.5° W) or Lisbon, Portugal (38.8° N, 9.1° W) and Stykkisholmur, Iceland (65.18° N, 22.7° W). Statistical analysis reveals that the NAO is the dominant mode of variability of the surface atmospheric circulation in the Atlantic and accounts for more than 36% of the variance of the mean December to March sea-level pressure field over the region from 20 to 80° N and 90° W to 40° E, during 1899 through to 1994. Marked differences are observed between winters with high and low values of the NAO. Typically, when the index is high the

Icelandic low is strong, which increases the influence of cold Arctic air masses on the northeastern seaboard of North America and enhances westerlies carrying warmer, moister air masses into western Europe (Hurrell 1995). During high NAO winters, the westerlies directed onto northern Britain and southern Scandinavia are over $8 \, \text{m s}^{-1}$ stronger than during low NAO winters, with higher than normal pressures south of 55° N and a broad region of anomalously low pressure across the Arctic. In winter, western Europe has a negative radiation balance and mild temperature levels are maintained by the advection of warm air from the Atlantic. Thus strong westerlies are associated with anomalously warm winters, weak westerlies with anomalously cold winters, and NAO anomalies are related to downstream wintertime temperature and precipitation anomalies across Europe, Russia, and Siberia (Hurrell 1995; Hurrell & van Loon 1996). They have been linked to changes in the thermohaline circulation in the North Atlantic (Lazier 1988; Dickson et al. 1997), mass balance of European Glaciers (Pohjola & Rogers 1997), the Indian monsoon (Dugam et al. 1997), and the atmospheric export of North African dust (Moulin et al. 1997).

Sea surface temperature (SST) anomalies in the North Atlantic subpolar basin varied concurrently over the twentieth century with the atmospheric surface wind anomalies (Deser & Blackmon 1993). Warm periods are characterized by positive SST anomalies around southern Greenland and negative anomalies along the northeastern United States seaboard. Deser and Blackmon also noted that the atmosphere and ocean anomalies display a roughly 10-year period and are consistent with what could be expected theoretically if the phenomenon were inherently due to coupling between the atmosphere and oceans. They also comment that there is a high negative correlation between the SST anomalies and anomalies in sea-ice extent in the Baffin Bay/Labrador Sea region. These sea-ice anomalies lead the SST and wind anomalies by a few years. A significant surface freshwater anomaly appeared in about 1969 in the Labrador Sea. Now known as the "Great Salinity Anomaly" (GSA), this feature can be traced moving eastward across the subpolar gyre, into the Norwegian Sea and ending up near Fram Strait more than 10 years later (Commission on Geosciences, Environment, and Resources 1998). The GSA was probably caused by an increase in sea-ice export from the Arctic and there is evidence that its stabilization of the upper water column interrupted the thermohaline circulation and deepwater production in the North Atlantic. Variations in the strength of the North Atlantic thermohaline circulation probably influence the NAO and can have a profound influence on the climate of western Europe.

Overlying the interannual variability there have been four main phases of the NAO index during the historical record: prior to the 1900s the index was close to zero; between 1900 and 1930, strong positive anomalies were evident; between the 1930s and 1960s, the index was low; and since the 1980s, the index has been strongly positive (Wilby *et al.* 1997). The recent persistent high positive phase of the NAO index, extending from about 1973 to 1995, is the most persistent and high of the historical record. Preliminary reconstructions from tree ring data of a 1000-year long record suggest that the recent extended period of high values in the NAO index series may not be unique (Stockton & Glueck 1999). During each positive phase, higher than normal winter temperatures prevail over much of Europe, culminating in the unprecedented strongly positive NAO index values and mild winters of 1989 and 1990. During high NAO index winters, drier conditions also prevail over much of central and southern Europe and the Mediterranean, while enhanced rainfall occurs over the northwestern European seaboard (Hurrell 1995). The recent high positive phase of the NAO index has resulted in an extended dry period in Morocco, with as much as 35% reduction in runoff from major river catchments and serious reduction in cereal crop production.

3.3.2   Interior North America and Asia

The interiors of both continents are far removed from oceanic influences, and since they are located in middle latitudes experience extreme continen-

tality of climate (Köppen's Dfc, Dwd, Bsk, and Bwk climates). The main effect of extreme continentality is to produce large seasonal temperature variations (Borisov 1965; Lockwood 1974). Thus Bergen (60°24′N, 5°19′E, 44m) on the Atlantic coast of Norway has a mean annual temperature of 7.8°C and an annual range of 13.7°C, while at approximately the same latitude the central Asian city of Omsk (54°56′N, 73°24′E, 105m) has a mean annual temperature of 0.4°C and a range of 38.4°C. Since the oceans are the main atmospheric moisture source, the continental interiors also tend to be dry, allowing the subtropical deserts to extend into temperate latitudes (Fig. 3.6). In the United States the region between the Rockies and 100°W has annual precipitation means between 300 and 500mm, while vast areas of central Asia receive less than 500mm, but in both cases amounts increase toward the east coast. In North America, the mountain ranges along the west coast curtail the penetration of moist Pacific air masses inland and have the same drying effect as the whole of lowland western Europe has on air masses entering Asia. The interiors of both continents are open to invasions by air currents of Arctic origin, which can travel far without much modification. North America is also open to invasion by air currents of subtropical origin that contrast strongly with the cold northern air masses, generating at times extremely active cold fronts, convective activity and tornadoes. The Midwestern and western plains states of the USA are visited by more severe tornadoes than any other area in the world; they are most frequent during spring and summer months. Moist tropical air is blocked from penetrating into central Asia by the Tibetan Plateau and its associated mountain ranges.

While the upper winter ridge over western North America directly influences the weather over the interior of that continent, a similar relationship does not exist between the ridge over western Europe and central Asia (Fig. 3.7). This is because a further winter trough–ridge pair of small amplitude are located between the main ridge over western Europe and the east Asian trough, with the secondary ridge located about 85°E. In January, almost the whole of Canada, the north interior, and northeast have snow covers; similarly all Asia north of about 40°N is snow-covered. In winter, the interiors of both continents are dominated by large anticyclones (Fig. 3.2), the result of cold air ponding over the cold continental surfaces. The most intense is over Asia and is located on average to the east of the 85°E meridian, which is just downwind of the upper climatological ridge. The anticyclone over North America is a weak and unstable feature that can be interpreted as a statistical average rather than as a semipermanent and quasi-stationary system. Both systems act as the sources of continental polar air, which is intensely cold and dry. In summer, the continental anticyclones vanish and are replaced by shallow heat lows.

Winter surface radiation balances are largely negative over the interior and northern parts of both continents, so unless temperatures are to fall to very low values the energy loss to space must be replaced by atmospheric heat advection. In January, almost the whole of Canada and the northern interior and northeastern United States have mean temperatures below 0°C (Fig. 3.5), the 10°C isotherm not being reached until just short of the Gulf of Mexico. Similarly, all Asia north of about 40°N has mean January temperature below 0°C, and in China the 0°C isotherm reaches 33°N, nearer to the Equator than anywhere else in the world. Indeed, it has already been noted that the lowest winter temperatures are not found over the Arctic but over northeastern Siberia and the Yukon. The intense winter cold has a marked influence upon soil temperatures in the northern parts of the continents. This can result in the soil becoming cemented by frozen water, and this is known as permafrost if the soil layers remain in this state throughout the year. Permafrost occurs in Alaska and over much of Canada, but its greatest extent is found in Siberia.

July is when the surface radiation balance is positive over the whole of the northern hemisphere and this is reflected by a nearly zonal trend in the isotherms over central Asia and North America, where the mean temperature is above 10°C everywhere except in the Arctic and mountainous regions (Fig. 3.4). The positive surface radi-ation balance leads to relatively uniform temperatures over the interiors of the two continents. Cool water along the coasts of Labrador and Newfoundland cause the July mean isotherms to dip southward, and a similar trend is also observed along the eastern coast of Asia. Extreme summer temperatures of 30°C are recorded even in the north of central Asia and similar temperatures are observed in Canada, where the extreme maximum temperature at Winnipeg has exceeded 40°C.

The Pacific Ocean can influence the climate of North America. The Pacific-North American (PNA) pattern represents a large-scale atmospheric teleconnection between the north Pacific Ocean and North America. It appears as four distinct cells in the 500 hPa geopotential height field near Hawaii, over the north Pacific, over Alberta in Canada, and over the Gulf Coast of the United States. Wallace and Gutzler (1981) define an index for the phase of this teleconnection pattern through a weighted average of 500 hPa normalized-height anomaly difference between the centers of the four cells. The PNA is also reflected in the sea-level pressure (Rogers 1990) and can be depicted by the North Pacific Index (NPI) (Trenberth 1990). The NPI is defined as the average pressure over a large area of the North Pacific Ocean near the center of the Aleutian low.

3.4 TROPICAL CLIMATES

3.4.1 *Introduction*

The tropical world may be considered as bounded at the surface by the two belts of subtropical high pressure, at about 30°N and S (Fig. 3.2), and in the upper air by the corresponding subtropical westerly jet streams. Under this definition tropical climates are found in those areas dominated by the tropical Hadley cell circulations or tropical monsoon circulations.

The tropical atmosphere differs meteorologically from the middle latitude atmosphere in a number of important aspects (Lockwood 1974; McGregor & Nieuwolt 1998). In the Tropics temperatures tend to be uniform in the horizontal over vast areas, so contrasting air currents are rare. In middle latitudes there are marked north–south temperature gradients, and air currents with differing origins therefore have contrasting temperatures and humidities. That is, the middle-latitude atmosphere is strongly baroclinic with large horizontal temperature gradients and therefore strong thermal winds (measure of wind shear with height). The strongly baroclinic nature of the middle latitude atmosphere explains the formation of frontal depressions and jet streams, with the latent heat released by the condensation of water vapor only exerting a modifying influence. In the Tropics, where atmospheric moisture values are often considerable, rainfall and therefore condensation rates can be high, resulting in the release of large amounts of latent heat, which can be of importance in the development of tropical circulation systems. Because the Coriolis parameter is small in the Tropics, small pressure gradients can generate air currents as intense as those found in middle latitudes.

In the tropical world the Sun is nearly always almost overhead at midday, with little seasonal variation in day length from about 12 hours. Seasonal temperature variations in the humid Tropics are therefore generally small, particularly near the Equator (Figs 3.4 and 3.5). The annual temperature range (mean January minus mean July) is very small at 3°C or less over the oceans in the equatorial zone, and is only 2 or 3°C greater at 30°N and S. While values of annual temperature range are equally small over the equatorial continents, values increase rapidly toward the subtropics to reach, for example, 20°C in the Sahara Desert and 15°C in central Australia. Since horizontal temperature gradients in the tropical atmosphere are small, heating of the lower atmosphere often causes convective overturning because it cannot be compensated by horizontal cold air advection from elsewhere. Therefore, tropical circulation patterns are particularly influenced by heat inputs from such sources as warm ocean surfaces acting through latent heat released in deep cumulus convection. Other heat sources, such as high tropical plateaus and equatorial rainforest, are also important. These heat sources show marked latitudinal and longitudinal variations in their distributions,

and also a marked tendency to vary on both annual and, in the case of the tropical oceans, greater than annual scales. One consequence of this is that tropical rainfall patterns show both annual and greater than annual variations and also marked teleconnections with distant locations.

Mean winds in the tropical world reflect the mean pressure patterns (Fig. 3.2). In the centers of the subtropical anticyclones and in the equatorial trough, winds are normally light and variable. Between the equatorial trough and the subtropical anticyclones there is a region of easterly winds, with a small deflection toward the Equator, usually known as the "trade winds." A seasonal reversal of wind direction takes place over southern Asia and the northern Indian Ocean; these are the Asian summer monsoon circulations.

3.4.2 Subtropical anticyclones

The subtropical anticyclones are normally attributed to the permanent subsidence associated with the upper-level westerly subtropical jet streams and the descending limbs of the Hadley cells. The most notable high-pressure cells (Fig. 3.2) are found over the eastern parts of the oceans; the Atlantic and Pacific oceans in both the northern and southern hemispheres and the Indian Ocean in the southern hemisphere. They are often absent at the surface though present in the upper air over the tropical land masses because of strong solar heating, and in particular over southern Asia in summer where they are replaced by deep monsoon low-pressure systems. Subtropical anticyclones show great permanence, for they tend to be located at fixed positions on the globe and undergo only slight seasonal variations, amounting on average to about 5° of latitude. The high-pressure belts are nearest to the Equator in winter, but there is a slight asymmetry with regard to the geographic Equator, with the southern ridge situated about 5° of latitude closer to it in the mean than the northern one. At the subtropical ridge lines, pressure is practically equal in both hemispheres, varying on average from 1015 hPa in summer to 1020 in winter. The continual subsidence in the subtropical anticyclones generates a low-level temperature in-

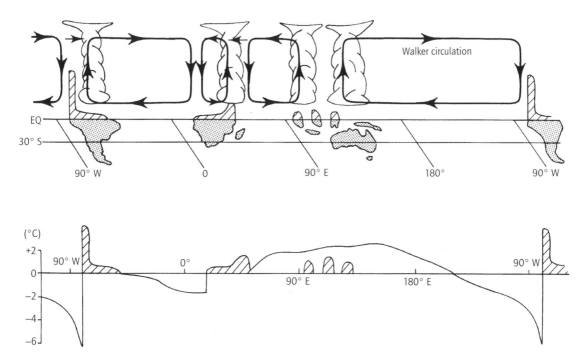

Fig. 3.8 Schematic representation of the Walker circulation along the Equator during non-ENSO conditions. The sea surface temperature departures from the zonal-mean along the Equator are shown in the lower part of the figure. (From Peixoto & Oort 1992; Wyrtki 1982.)

version and makes it almost impossible for extensive clouds to form and for precipitation to fall; therefore, they are almost rainless. The cool air below the inversion is continually renewed over the oceans by mobile polar highs, which manage to penetrate into the Tropics. Most of the extensive hot deserts (e.g. Sahara) of the world are situated in the latitudes of the subtropical anticyclones (Fig. 3.6).

3.4.3 The trade winds

The trade winds occur on the equatorial sides of the subtropical anticyclones (Fig. 3.2), and are found over the greater part of the tropical oceans. In general the trade winds blow from ENE in the northern hemisphere and from ESE in the southern, and are noted for extreme constancy in both speed and direction. They are characterized by generally fine sunny weather with only a small annual temperature range. Many locations in this climatic region show a summer rainfall maximum, because this is when sea surface temperatures are highest and tropical weather systems most frequent. While pressure is high along the subtropical high-pressure ridge lines, a broad zone of low pressure is found in the region of the Equator. In no other climatic regime do winds blow so steadily, for this steadiness reflects the permanence of the subtropical anticyclones; normally interruptions in the flow occur only with the formation of a major atmospheric disturbance such as a tropical cyclone. As winter is the season when the subtropical highs tend to be most intense, the trade winds are strongest in the winter hemisphere and weakest in the summer.

A distinctive feature of the trade winds is the presence of a dynamic temperature inversion at about 1000–2000 m. This forms in the subtropical anticyclones, probably from the remains of mobile polar highs that have penetrated into the Tropics, and is carried toward the Equator by the trade winds. Air streams flowing toward the Equator tend to subside and diverge because of dynamic considerations such as the conservation of potential vorticity. From these considerations it might be expected that the trade-wind inversion should decrease in height downwind toward the Equator. This is not observed, since over vast ocean areas trade-wind cumulus convection "diffuses" energy and water vapor gained from the oceans to higher levels, and causes the trade-wind inversion to rise, even though the mean vertical motion along trajectories is downward. The trade-wind inversion is not therefore a material surface, and subsiding air above the inversion is slowly incorporated through the inversion into the moist, relatively cool, layer below. The trade-wind inversion can have unfortunate effects on pollution levels in cities in the trade-wind zone, because it very effectively stops any vertical exchange of air from taking place, trapping pollutants below the inversion layer.

3.4.4   *The equatorial trough*

The northern and southern trade winds meet in the equatorial trough, which is a zone of relatively low pressure situated near the Equator. Here the winds normally become weak and can be replaced by calms or by equatorial westerlies. The equatorial trough is not a region of continuous cloud and rain, since very large areas are dominated by the trade-wind inversion and experience fine, dry weather and even a near desert climate. This is because, while the ascending limbs of the Hadley cells are located in the equatorial trough zone, the actual ascent is localized in a limited number of extremely active cumulonimbus convective clouds, which may be organized into synoptic systems. Ascent is absent away from the active cumulonimbus clouds, in contrast to the descending subtropical limbs of the Hadley cells where subsidence occurs over wide areas.

3.4.5   *Tropical storms*

Atmospheric disturbances are frequent over the

Table 3.3   Tropical storm development regions (data from Lockwood 1974).

Area location	Average number of tropical storms per year
NE Pacific	10
NW Pacific	22
S Pacific	7
NW Atlantic (including W Caribbean and Gulf of Mexico)	7
Bay of Bengal	6
S Indian Ocean	6
Arabian Sea	2
Off NW Australian coast	2

western parts of the subtropical oceans and since these disturbances drift westward, they bring ample rainfall to the eastern coasts of the subtropical land masses. The most frequent disturbances are the easterly waves and tropical storms (cyclones, hurricanes, and typhoons) (Pielke & Pielke 1997). The major tropical storm development regions are listed in Table 3.3. Tropical storms are driven largely by latent heat released by condensing water vapor associated with intense rainfall. As rainfall rates near the center of a tropical cyclone can exceed $500\,mm\,day^{-1}$, a continuous inflow of water vapor into the storm is necessary. This is supplied by warm ocean surfaces with temperatures above about 26.5°C, which are normally found in the tropical oceans in the late summer and autumn. Tropical storms are therefore most frequent over warm tropical oceans in the late summer and autumn. They decay rapidly when the supply of water vapor is cut off by the passage across cooler water surfaces or land areas. In summer and autumn they frequently bring high winds, storm surges, and intense rainfall to the eastern coasts of the United States and Central America, subtropical southeast Asia, the Indian subcontinent and northern Australia.

3.4.6   *Walker circulations, the Southern Oscillation, and El Niño*

Satellite imagery and rainfall data clearly show

(Fig. 3.6) three equatorial regions of maximum cloudiness and rainfall: the so-called "Maritime Continent" of the Indonesian archipelago, the Amazon river basin in South America, and the Zaire river basin in Africa. The rest of the equatorial region is comparatively dry, and some, like the coasts of Peru, even desert. These longitudinal variations in rainfall are associated with east–west regional circulations along the Equator, the most important being the Walker circulation, which involves rising air motion over the Indonesian archipelago and sinking over the eastern Pacific (Fig. 3.8). The rising air motion takes place mostly in deep convective clouds and is associated with intense convective rainfall and therefore the wet humid climates of the Indonesian archipelago. The subsiding air suppresses cloud formation and rainfall, and is therefore associated with the coastal deserts of Peru. This circulation is named after Sir Gilbert Walker, who in 1910 first started to document the global-scale correlations between variations in tropical weather features. The Hadley cell circulation refers to the north–south component of these circulations: equatorward motion at low levels, rising in the convective regions near the Equator, and poleward flow aloft. The Walker circulation refers to the east–west component, which is particularly prominent in the equatorial plane. Both circulations are driven by the release of latent heat in deep convective shower clouds.

The Walker circulation is closely coupled with the sea surface temperature distribution over the Pacific, with relatively cool water in the east and warm in the west. When the Pacific Ocean off the coast of South America is particularly cold, the air above is too stable to take part in the ascending motion of the Hadley cell circulation. Instead, the equatorial air flows westward between the Hadley cell circulations of the two hemispheres to the warm west Pacific where, having been heated and supplied with moisture from the warmer waters, the equatorial air can take part in large-scale ascent (Fig. 3.8). The easterly winds that blow along the Equator and the northeasterly winds that blow along the Peru and Ecuador coasts both tend to drag the surface water along with them. The Earth's rotation then deflects the resulting surface currents

toward the right (northward) in the northern hemisphere and to the left (southward) in the southern hemisphere. The surface waters are therefore deflected away from the Equator in both hemispheres and also away from the coastline. Where the surface water moves away under the influence of the trade winds, colder, nutrient-rich water upwells from below to replace it. Since the newly upwelled water is colder than its surroundings, its signature in infrared satellite images takes the form of a distinctive "cold tongue" extending westward along the Equator from the South American coast. The winds that blow along the Equator also affect the ocean thermocline, i.e. the boundary between the warm surface water and the deep cold water. In the absence of the wind the thermocline would be nearly horizontal; but the trade winds drag the surface water westward, raising the thermocline nearly to the ocean surface in the east and depressing it in the west. The situation in the equatorial Atlantic is analogous to the equatorial Pacific in that the warmest part is in the west, at the coast of Brazil, but west–east contrasts of water temperature are much smaller than in the Pacific. However, in January a thermally driven Walker circulation may operate from the Gulf of Guinea to the Andes, with the axis of the circulation near the mouth of the Amazon.

Associated with the Walker circulation is the El Niño phenomena, when every few years the tropical Pacific Ocean off the coasts of Peru and Ecuador occasionally becomes much warmer than average for periods of several months. Under El Niño conditions the Walker circulations become reversed, resulting in heavy rain in the normally arid areas of Peru, and drought in the western Pacific. Sir Gilbert Walker described in his papers in the 1920s and 1930s what he named the Southern Oscillation (SO). The Southern Oscillation is dominated by an exchange of air between the southeast Pacific subtropical high and the Indonesian equatorial low, with a period that varies between roughly one and five years. During one phase of this oscillation, the trade winds are intense and converge into the warm western tropical Pacific, where rainfall is plentiful and sea-level pressures are low. At such times the atmosphere over the eastern tropical Pacific is cold and dry. During the complementary phase, the trade winds relax, the zone of warm surface waters and heavy precipitation shifts eastwards, and sea-level pressure rises in the west while it falls in the east. The latter phase is the more unusual and in the eastern Pacific has become known as El Niño, while vigorous episodes of the former are often termed La Niña. The combined atmospheric/oceanic conditions that give rise to these changes in rainfall across the Pacific and neighboring areas are referred to as El Niño Southern Oscillation (ENSO) events. ENSO is important climatologically (Glantz 1996) for two main reasons. First, it is one of the most striking examples of interannual climatic variability on an almost global scale. Second, in the Pacific it is associated with considerable fluctuations in rainfall and sea surface temperature, and also with extreme weather events around the world.

The Southern Oscillation may be defined in terms of the difference in sea level pressure between Darwin in Australia and Tahiti. Records are available, with the exception of a few years and occasional months, from the late nineteenth century. The El Niño phenomena are associated with extreme negative Southern Oscillation values, but for much of the time the series exhibits continuous transitions from high to low values, with most values being positive. Of great importance to the development of the Southern Oscillation is the difference in the ways the ocean and atmosphere respond to changes in the winds and sea surface temperature patterns, respectively. During La Niña, intense trade winds drive warm surface waters of the equatorial Pacific westward while exposing cold water at the surface in the east. When the winds relax during El Niño, the warm waters move back eastward, overflowing the cold water. The response of the ocean to changing winds involves not only changes in currents but also the excitation of waves that travel back and forth across the Pacific. These waves have a large signature, not at the surface, but at the thermocline, which separates the warm waters of the upper ocean from the colder water at depth. These waves eventually bring the ocean to a new equilibrium after a change in the winds, and the time it takes them to propagate across the Pacific is important in determining the time scale of the Southern Oscillation.

3.4.7 The monsoon circulations of Southern Asia and Eastern Africa

The characteristics of the monsoon climate are to be found mainly in the Indian subcontinent (Pant & Rupa Kumar 1997), where over much of the region the annual changes may conveniently be divided into the northeast (dry) and southwest (wet) seasons. Over many tropical oceans, the atmospheric circulation undergoes very little seasonal variation. In contrast, over tropical continents the atmospheric circulation displays distinct seasonal rhythms and variations. Over the oceans, evaporation consumes a high proportion of the incident radiation, which is also rapidly absorbed by the water surface and spread over great depths by waves and turbulence, and dissipated to other latitudes by ocean currents. Over the tropical continents, particularly if they are dry because of low rainfall, solar radiation is used mostly to warm the Earth's surface, so that surface temperatures in these regions reach very high values owing to the much lower heat capacity of the soil compared to water. This warming over the tropical continents in early summer produces thermal lows which gradually take on some functions of the equatorial trough, thus forming a new region of convergence. The trade winds from the winter hemisphere cross the Equator, slow down and create a minor area of convergence owing to their change in direction, caused by the reversal of the Coriolis force. In winter, the tropical continents experience relatively low temperatures and high surface pressure; winds are reestablished flowing toward the Equator that are similar to the trade winds. Thus the tropical continents and adjacent oceans can experience a semiannual reversal in wind direction characteristic of monsoons.

When upper winds are taken into account, it is found that the Asian monsoon is a fairly complex system. During the northern winter season the subtropical westerly jet stream crosses southern Asia, with its core located at about 12,000 m altitude. It divides in the region of the Tibetan Plateau, with one branch flowing to the north of the plateau and the other to the south. The two branches merge to the east of the plateau and form an immense upper convergence zone over China. In May and June the subtropical jet stream over northern India slowly weakens and disintegrates, causing the main westerly flow to move north into central Asia. While this is occurring, an easterly jet stream, mainly at about 14,000 m, builds up over the equatorial Indian Ocean and expands westward into Africa. The formation of the equatorial easterly jet stream is connected with the formation of an upper-level high-pressure system over Tibet. In October the reverse process occurs; the equatorial easterly jet stream and the Tibetan high disintegrate, while the subtropical westerly jet stream reforms over northern India. The Himalayan–Tibetan plateau is of importance because it appears to accelerate the onset of the Asian monsoon and to increase its ultimate intensity (Hahn & Manabe 1975). Central and southeastern parts of Tibet remain free of snow throughout most of the year; hence, the plateau must heat rapidly during the northern spring. This direct warming of the middle troposphere creates an upper-level anticyclone, which is readily observed on synoptic charts, with upper-level divergence and low-level convergence. Thus, suitable conditions are produced for the Asian monsoon in the northern spring. Latent heat released in intense tropical storms over India keeps the system functioning during the northern summer. Since a complex feedback system produces the southwest monsoon, failures in the system are common and produce extensive breaks in the monsoon rains when the whole system shows signs of collapse. Variations in winter snow cover over Tibet will influence the start and intensity of the southwest monsoon. General cooling over southern Asia at the end of the northern summer causes its collapse.

The southwest monsoon current in the lower 5000 m near India consists of two main branches: the Bay of Bengal branch, influencing the weather over the northeast part of India and Burma; and the Arabian Sea branch, dominating the weather over the west, central, and northwest parts of India. The low-level flow across the Equator during the southwest monsoon is not evenly distributed between the latitudes 40 and 80° E, but has been found by Findlater (1971) to take the form of low-level high-speed southerly currents, which are concentrated between about 39 and 55° E. A particularly important feature of this flow is the strong southerly current with a mean wind speed of about 14 m s^{-1} observed at the Equator over eastern Africa from April to October. The strongest flow occurs near the 1000–5000 m level, but it often increases to more than 25 m s^{-1} and occasionally to more than 45 m s^{-1} at heights between 1000–2000 and 2000–5000 m. This high-speed current flows intermittently during the southwest monsoon from the vicinity of Mauritius through Madagascar, Kenya, eastern Ethiopia, and Somalia, and then across the Indian Ocean toward India.

Nearly half of the Indian subcontinent is arid or semiarid (Fig. 3.3), and by far the larger part of this dry area is located in northwest India and southwestern Pakistan. The Arabian Sea summer monsoon circulation is linked with a heat low over Arabia, Pakistan, and northwest India. This heat low, which develops during May, establishes the low-level westerly monsoon wind regime a full month before heavy monsoon rains start over western India. In mid-July the heat low is deepest, the southwest monsoon is strongest, and the west Indian rains are heaviest. The question then arises as to why the heat low remains cloud-free while rain falls to the south. According to Flohn (1964), the unique summer aridity of the desert belt from the western Sahara to Pakistan is strongly correlated with the forced descending motion on the northern side of the exit region of the equatorial easterly jet stream. During the northern summer, the equatorial easterly jet stream extends at about 14,000 m in the latitude belt 5–20° N from the Philippines across southern Asia and northern Africa to almost the western Atlantic. Over the whole exit region from India westward the very gradual deceleration of the jet stream core results in widespread sinking motions on the northern side and rising motions on the southern. Since the Hadley cell circulations already generate desert conditions over north Africa and the Middle East, the deceleration of the equatorial easterly jet stream intensifies the aridity of these deserts. Similarly, numerical simulations by Hahn and Manabe (1975) suggest that large areas of northeast Africa would have considerably more rainfall if the Tibetan Plateau and the equatorial easterly jet stream were absent. The reverse flow is observed in the entrance region over southeast Asia, where air sinks to the south and rises to the north, with a rainy area over southern Asia. Radiosonde ascents at Karachi and Jodhpur indicate the frequent presence of a low-level inversion, with moist air originating over the Arabian Sea below the inversion. Subsidence limits the height to which the surface air from the Arabian Sea can ascend, restricting cloud development, and thus favoring strong solar heating.

About 80% of the annual rainfall over a large part of India occurs during the summer monsoon period (June to September). The variation in the all-India average seasonal rainfall may be considered as a measure of the intensity of the planetary-scale monsoon over the India region. A time series of the all-India summer monsoon rainfall (AISMR), has been devised by Parthasarathy et al. (1994) using the area-weighed rainfall at 306 well distributed rain gauges across the country. The mean AISMR is 852 mm and the standard deviation from year to year is 84 mm. The analysis of the short-term fluctuations in the AISMR time series shows that there are epochs of above- and below-normal rainfall. There appears to be an inherent internal epochal variability in the rainfall series. The periods 1880–95 and 1930–63 were characterized by above-normal rainfall with very few droughts. In contrast, the periods 1895–1930 and 1963–90 were characterized by below-normal rainfall with very frequent droughts. The turning points are noted around 1880, 1895, 1930, 1963, and 1990. The fall from an extreme state of above-normal to an extreme state of below-normal rainfall occurs in a short span of about a decade (1890–1900, 1955–65). However, the rise above normal state is gradual and may take about five decades. The epochal behavior in the rainfall record can be broken by strong external forcing such as El Niño. The major extreme events of rainfall (severe floods/droughts) are due to the phase-locking between the epochal variability and the external forcing; thus the impact of El Niño on the AISMR is more severe during the below-normal epochs, while La Niña has a more severe impact during the above-normal ones. It should be noted that in some years a monsoon drought occurs without the occurrence of a El Niño event, one of the worst being 1979. Other factors are also believed to be important for the interannual behavior of the AISMR. For example, a recent study by Kripalani et al. (1996) suggests that the Eurasian snow mass is related to the Indian monsoon.

3.4.8 Australia

The chief determinant of the climate (Fig. 3.2) and wind field over Australia is the subtropical anticyclonic belt. In general the continent is affected by mid-latitude westerlies on the southern fringe, tropical convergence on the northern fringe in summer, and stable subsiding air under the subtropical anticyclones over the interior (Sturman & Tapper 1996). The result is that most of the continent is covered by arid or semiarid climates. The average altitude of Australia is only 300 m, with 87% of the continent less than 500 m and 99.5% less than 1000 m. In general the low relief of Australia does not significantly obstruct the movement of the atmospheric systems that control the climate.

Most of the interior of Australia lies within the BWh and BSh divisions of Köppen's climatic classification (Fig. 3.3); that is, much of the interior is hot with either a desert or a steppe climate. Tropical cyclones develop over the oceans around northern Australia between November and April. Much of northern Australia therefore lies in Köppen's Aw or Am divisions; that is, tropical summer wet and winter dry climate. Parts of the coastal fringes of southern Australia enjoy a Mediterranean type climate with hot, dry summers and therefore fall in the Cs climatic divisions. Many eastern coastal fringes have moist climates with mild winters and therefore fall within the Ca and Cb climatic divisions.

Most of Australia is warm to hot (Figs 3.4 and 3.5), with the exception of the alpine area in the southeast where there is seasonal snow. The month with the highest temperature varies from November in the far north to February in the south. In the north the buildup of monsoon clouds cools the latter part of the summer, while in the south the time taken to warm the ocean delays the peak temperature until late summer. July is the coldest month throughout the country. Australia is the driest continent, excluding Antarctica, and no continent has less runoff from its rivers. More than a third of the country receives on average less than 250 mm of rainfall annually, and only 9% receives more than 1000 mm. Most of the area south of 35° S receives mainly winter rains, while north of 25° S, most rain falls in summer, associated with the summer monsoon and tropical cyclones. Australian rainfall averages disguise an extremely variable rainfall, with droughts and flooding being very common.

Australian rainfall is more variable than could be expected from similar climates elsewhere in the world, mainly due to the impact of ENSO. Conrad (1941) examined the relationship between interannual rainfall variability and long-term mean annual rainfall, using data from across the globe. He defined the relative variability of annual rainfall as the mean of the absolute deviations of annual rainfalls from the long-term mean, expressed as a percentage of the long-term mean. Conrad found that the relative variability decreased, in general, as the mean precipitation increased. Nicholls (1988) compared the relationship between relative variability and mean rainfall in areas affected by ENSO with the relationship elsewhere. The relative variability was typically one-third to one-half higher for these stations, compared with stations with the same mean rainfall in areas not affected by ENSO.

Australian rainfall fluctuations, as well as being more severe because of ENSO's influence, also operate on very large spatial scales (Nicholls 1991). High rainfall totals in Australia occur when the Southern Oscillation Index (SOI) is large and positive (La Niña events). In contrast, when the SOI is strongly negative (El Niño years) drought occurs over much of the continent. Thus the continental scale of the 1982/3 drought is typical of many years, although it was more severe than most. Among others, Ropelewski and Halpert (1987, 1989) have demonstrated that Australian rainfall fluctuations tend to coincide over large areas, during both ENSO and anti-ENSO events. There have been long-term variations in this relationship; since the early 1970s rainfall appears to have been greater, relative to the SOI, than was the case in earlier years (Opoku-Ankomah & Cordery 1993; Nicholls *et al.* 1996).

Extended periods of drought or extensive rains in Australia do not occur randomly in time, in relation to the annual cycle (e.g. Ropelewski & Halpert 1987). The ENSO phenomenon, and Australian rainfall fluctuations associated with it, are phase-locked with the annual cycle. Thus the heavy rainfall of an anti-ENSO event tends to start early in the calendar year and finish early in the following year. The dry periods associated with ENSO events tend to occupy a similar time period. For example, the 1982/3 drought started about April 1982 and broke over much of the country in March and April 1983. Nicholls (1991) comments that rainfall is not the only aspect of Australian climate affected by ENSO. Frosts tend to be more common in inland Australia during ENSO events, because low rainfall is associated with decreased cloud cover, allowing increased radiative cooling at night. The decreased cloud cover also causes higher maximum temperatures during ENSO events. Drosdowsky (1996) and Nicholls (1991) also report that north of about 25° S low-level winds in winter during ENSO events tend to be about three to four times stronger than during anti-ENSO events. In the southeast the winds in summer tend to be two to three times stronger in ENSO events. The higher maximum temperatures and stronger winds associated with ENSO-related droughts increase the likelihood of wildfires. The latitudinal variations in the season in which the ENSO-amplified winds occur (winter in the north, summer in the south) means that the stronger winds occur at the time of year when fires are most likely.

Westoby (1980) noted that "climates with the same general level of aridity can offer very different mixtures of growth opportunities, because of the patterning of rainfall in time; accordingly different mixtures of growth forms are found." Thus Australian vegetation should be suited to an environment of highly variable rainfall with frequent severe droughts or pluvials extending, typically, for about 12 months. It should also be able to survive frequent fires, since the coincidence of strong winds with dry periods was noted above. Nicholls (1991) lists among others the following characteristics of Australian vegetation that may be, at least in part, attributable to ENSO's influence on climate.

1 Absence of succulents. Succulents are almost totally absent from Australian arid and semiarid regions, because although adapted to arid climates and requiring little moisture, they need regular rainfall. Such plants are therefore unsuited to the high rainfall variability ENSO produces over much of Australia.

2 Vegetation height. Australia has more trees at a given level of aridity than elsewhere. This is because intense rainfall events accompanying anti-ENSO periods produce deep water penetration into the soil. Larger trees, with large root structures able to remove the stored rainwater from far below the surface, are favored by the intense rainfall, relative to an area where rainfall was lighter and more frequent.

3 Fire resistance/dependence. Much of the Australian flora is fire-resistant or even dependent on fire for successful reproduction.

3.4.9 *South America*

Of particular climatological significance in South America are the Andes Mountains along the west coast and the tapered shape of the continent, with much of its area lying in tropical latitudes. The latter causes the property of continentality, so significant in other continents that extend into middle latitudes, to be totally absent from South America. The oceans are therefore of major and immediate importance in the climates over large areas of the continent (Fig. 3.3).

Cerveny (1998) notes that three circulation regimes dominate the climate of the continent:

1 The prevailing southern westerly winds of the extreme southern latitudes of the continent.
2 The semipermanent subtropical high-pressure cells positioned over the South Atlantic and South Pacific oceans.
3 The location of the ITCZ (Intertropical Convergence Zone), a migrating band of maximum convergence, convection cloudiness, and rainfall.

There are several regional climates of South America that are of particular interest. The movements of the ITCZ strongly influence the climates of tropical South America. The ITCZ reaches its northernmost location in June and September/October, causing the season of greatest rainfall for northern South America and the Caribbean. At the height of the northern winter, the ITCZ extends southward into the central Amazon Basin. Consequently, the months of January and February mark the dry season for much of northern South America and the tropical Atlantic, while the Amazon basin receives much of its annual rainfall in this season. In general, the onset of the rainy season occurs first in southeast Amazonia, with onset dates occurring progressively later toward the northwest. The demise of the rainy season occurs first in the southeast and progressively later to the northwest. Associations with the extremes of the SO are most marked in central Amazonia and near the mouth of the Amazon. Years with the low/high SO phase are consistent with anomalously dry/wet rainy seasons in these two regions, due mainly to a late/early onset of the rainy season (Marengo *et al.* 1999b). During the El Niño, the near surface Atlantic trade winds are weaker, consistent with an anomalously northward displaced ITCZ; thus the moisture input from the Atlantic into the Amazon basin is weak, the moisture convergence and convection are also weak over Amazonia, and lower rainfall is observed specially in the central and mouth of Amazon River regions. The sea surface temperature dipole in the tropical Atlantic indicates anomalously warmer surface water to the north of the Equator. During La Niña, these patterns are reversed, with a southward displacement of the ITCZ, strong Atlantic trades winds and moisture transport into Amazonia, and an inverted sea surface temperature dipole (anomalously warmer surface water south of the Equator). Over the western part of the basin there is a descending branch of the Walker cell during El Niño, with subsidence affecting a zonal band from the Andes to the Atlantic. There is also strong upward flow, convection, and rainfall over northern Peru to the west of the Andes, associated with anomalously warm Pacific sea surface temperatures, implying compensatory subsidence and reduced rainfall over western Amazonia. This flow pattern is inverted during La Niña. The upper level Bolivian tropospheric high is weaker during El Niño, with the subtropical westerly jet stronger and located anomalously northward of its average position, while during La Niña the jet is weaker and anomalously southward of its average position (Marengo *et al.* 1999a).

During the winter months of April to October cold fronts invade Matto Grosso and the southern Amazon basin one to three times a year. Such invasions cause a drastic fall in temperature called **friagem** and give rise to widespread showers, often accompanied by hail. They can sometimes result in damaging frost affecting the agricultural areas of southern and southeastern Brazil. Strong cold air outbreaks are generally associated with a strongly amplified wave pattern in southern middle latitudes, with an amplified upper tropospheric ridge situated near or just to the west of the Andes mountains and a trough near the east coast of South America. The enhanced northward flow between the ridge and the trough facilitates the rapid northward advance of cold air from Argentina to southern Brazil that can eventually result in the occurrence of frost at very low latitudes (Cavalcanti *et al.* 1999).

Northeast Brazil is an exceptional area in which rainfall diminishes from in excess of 1000 mm along the coast to less than 400 mm inland (Fig. 3.3). During the southern winter almost all of Brazil south of the Equator is under the influence of a greatly enlarged subtropical anticyclone, and rainfall is at a minimum over wide areas. In the summer, the anticyclone weakens and equatorial air invades much of southern Brazil, with the major exception of the northeastern portion, which remains under the influence of the high

pressure. The relative dryness of northeast Brazil is caused primarily by the flow patterns of the general atmospheric circulation, particularly by thermally driven circulations of the Hadley–Walker types. To the immediate west of northeast Brazil is the Amazon basin, where the high rainfall is associated with vigorous upward convection in cumulonimbus clouds. The water vapor condensing in these clouds releases latent heat, which warms the air and thereby maintains the ascending motion. The air rising over Amazonia descends in a Walker-type circulation over the eastern subtropical Atlantic Ocean, including northeast Brazil and possibly the western coast of Africa. The circulation patterns are reflected in the rainfall and cloud distributions. Similarly, the north–south circulation patterns of the Hadley cells, with ascending motion in the convective clouds of the ITCZ and descending motion over the subtropical Atlantic of both hemispheres, also contributes to the aridity of northeast Brazil. The interannual variability in both strength and geographic position of the Hadley–Walker circulations results in the high interannual variability of precipitation in northeast Brazil.

Observational studies indicate the existence of a link between rainfall anomalies in northeast Brazil and anomalies in the atmospheric circulation on a semiglobal scale. These studies show that in years with large rainfall fluctuations about the mean the atmospheric circulation over the entire Atlantic Ocean was altered (Namias 1972; Hastenrath & Heller 1977; Hastenrath *et al.* 1984; Nobre *et al.* 1985). In wet years, the Azores anticyclone over the subtropical North Atlantic is stronger than normal and displaced southward of its usual position. The North Atlantic trade winds are stronger, and the ITCZ is displaced further south. A mirror image of this atmospheric circulation about the Equator is found in dry years, with the South Atlantic anticyclone now more intense, stronger South Atlantic trade winds, and the ITCZ north of its normal position.

In oceanic regions the ITCZ generally lies over or near the highest sea surface temperatures. Therefore, a relationship should be expected to exist between the general sea surface temperature distribution in the tropical Atlantic and the rain-fall over northeast Brazil. Warmer (colder) sea surface temperatures in the tropical South Atlantic and colder (warmer) ones in the tropical North Atlantic are associated with wet (dry) years in northeast Brazil (Hastenrath & Heller 1977; Markham & Mclain 1977; Moura & Shukla 1981; Hastenrath & Greischar 1993; Hastenrath 1995). Thus, in wet years (e.g. 1964, 1967, 1984, and 1985) the sea surface temperature anomalies were positive in the tropical South Atlantic and negative or near zero in the tropical North Atlantic. In contrast, in dry years (e.g. 1951, 1953, 1958, 1970, and 1979) the sea surface anomalies were negative in the tropical South Atlantic and positive in the tropical North Atlantic.

Another unusual arid region occurs along the northern coasts of Venezuela and Colombia, a location that could be expected to receive copious rainfall. The explanation of this lack of rainfall requires an analysis of local patterns of divergence and subsidence. In spring, the driest season, a deep easterly flow prevails, with divergence and local subsidence up to at least 2000 m. In summer, the ITCZ is absent and subsidence continues, although it is not as severe as in spring and some rainfall does occur. Winter, like autumn, is relatively rainy but analysis still indicates divergence at the surface. Part of the explanation of this anomalously dry coastal strip is found in the east–west orientation of the coast. Where easterly flow predominates along an east–west coast with water (low-friction surface) to the north and land (high-friction surface) to the south, divergent flow and subsidence result over the coastal region. Two other factors may enhance the subsidence. The first is the cold water upwelling found offshore of the dry zone; and the second is the modification of the normal land–sea breeze by inclination of the land surface.

3.4.10 Africa

Of all the continents, Africa is the most symmetrically located with regard to the Equator, and this is reflected in the climatic zonation (Fig. 3.3). The continent may be regarded as a giant plateau, for there is a relative absence of very pronounced topography, although some high mountains exist, especially in the East African region. The latitudinal position of Africa means that the continent is influenced by tropical, subtropical and mid-latitude pressure and wind systems. As the near-equatorial trough migrates with the seasons, the ITCZ lies south of the Equator in the northern winter and north of the Equator in the summer. Connections between the equatorial easterly jet stream and north African rainfall were discussed in the section on monsoons. The ITCZ in southern Africa is neither spatially continuous nor aligned east–west, particularly in summer when it is displaced southward as far as 20° S.

Of particular interest in Africa is the sub-Saharan region (Sahel), since it shows a dramatic decrease in rainfall since the late 1960s and continued severe drought conditions from the early 1970s up to at least the late 1980s. As in the case of northeast Brazil, this is probably connected with regional circulation changes and North Atlantic sea surface temperatures. Indeed, colder than average tropical North Atlantic sea surface temperatures appear to be strongly associated with drought in the African Sahel (Folland *et al.* 1986, 1991; Hastenrath 1990; Rowell *et al.* 1992; Semazzi *et al.* 1993). Marine temperatures over the period 1856–1981 show a worldwide fluctuation of about 0.6°C, with the coldest period being centered on 1905–10 and the warmest occurring in the 1940s (Folland *et al.* 1984) The major climatic fluctuations in the first half of the twentieth century occurred simultaneously in both hemispheres, with the greater amplitude in the northern hemisphere. The mid-latitudes of the two hemispheres now appear to be fluctuating out of phase, as they were to some extent before 1900.

The tropical Atlantic Ocean shows a coherent structure in SST variability. The dominant pattern of SST often shows a warm pool in the tropical North Atlantic and a complementary cool pool in the tropical South Atlantic, or vice versa. These centers of action seem to vary coherently over decadal time scales but independently on shorter time scales (Hastenrath 1990; Chang *et al.* 1997). As already shown, this low-frequency SST phenomenon shows concurrently anomalies in the rainfall over Brazil and northern Africa.

3.5 CLOSING REMARKS

Within one short chapter it is not possible to describe in detail every aspect of every climate found on the Earth's surface. For example, no mention has been made of the role of vegetation changes in helping to prolong droughts in tropical areas or of boreal forest in warming the polar climates. Similarly, it is not possible to describe the substantial climate changes that have occurred over the past 20,000 years or are likely to occur in the future as a result of global warming due to the anthropogenic emissions of radiatively active trace gases.

REFERENCES

Allison, I., Wendler, G. & Radock, U. (1993) Climatology of the East Antarctic ice sheet (100° E to 140° E) derived from automatic weather stations. *Journal of Geophysical Research* **98**, 8815–23.

Barrie, L.A. (1986) Arctic air chemistry: an overview. In Stonehouse, B. (ed.), *Arctic Air Pollution*. Cambridge University Press, Cambridge.

Borisov, A.A. (1965) *Climates of the USSR*, 2nd edn. Oliver and Boyd, London.

Carleton, A.M. (1989) Antarctic sea-ice relationships with indices of the atmospheric circulation of the Southern Hemisphere. *Climate Dynamics* **3**, 207–20.

Cavalcanti, I.F.A., Paulista, C. & Kousky, V. (1999) Interannual variability of cold air outbreaks over southern and southeastern Brazil from 1979 to 1997 and sensitivity of the CPTEC/COLA GCM in predicting extreme cases. *Abstract, 10th Symposium on Global Change Studies*. American Meteorological Society, Boston.

Cerveny, R.S. (1998) Present climates of South America. In Hobbs, J.E., Lindesay, J.A. & Bridgman, H.A. (eds), *Climates of the Southern Continents*. Wiley, Chichester.

Chang, P., Ji, L. & Li, H. (1997) A decadal climate variation in the tropical Atlantic Ocean from thermodynamic air-sea interactions. *Nature* **385**, 516–18.

Commission on Geosciences, Environment, and Resources, National Research Council. (1998) *Decade-to-century-scale Climate Variability and Change: A Science Strategy*. National Academy Press, Washington, DC.

Conrad, V. (1941) The variability of precipitation. *Monthly Weather Review* **69**, 5–11.

Deser, C. & Blackmon, M.L. (1993) Surface climate variations over the North Atlantic Ocean during winter 1900–1989. *Journal of Climate* **6**, 1743–53.

Dickson, R., Lazier, J.R.N., Meincke, J., Rhines, P. & Swift, J. (1997) Longterm coordinated changes in the convective activity of the North Atlantic. *Progress in Oceanography* **38**, 241–95.

Drosdowsky, W. (1996) Variability of the Australian summer monsoon at Darwin: 1957–1992. *Journal of Climate* **9**, 85–96.

Dugam, S.S., Kakade, S.B. & Verma, R.K. (1997) Interannual and long-term variability in the North Atlantic Oscillation and Indian summer monsoon rainfall. *Theoretical and Applied Climatology* **58**, 21–9.

Findlater, J. (1971) Mean monthly air-flow at low levels over the western Indian Ocean. *Geophysical Memoirs* **115**. HMSO, London.

Flohn, H. (1964) Investigations on the tropical easterly jet. *Bonner Meteorologische Abhandlugen* **4**. Bonn.

Folland, C.K., Parker, D.E. & Kates, F.E. (1984) Worldwide marine temperature fluctuations 1856–1981. *Nature* **310**, 670–3.

Folland, C.K., Palmer, T.N. & Parker, D.E. (1986) Sahel rainfall and worldwide sea temperatures 1901–85. *Nature* **320**, 602–7.

Folland, C.K., Owen, J., Ward, M.N. & Colman, A. (1991) Prediction of seasonal rainfall in the Sahel region using empirical and dynamical methods. *Journal of Forecasting* **10**, 21–56.

Glantz, M.H. (1996) *Currents of Change: El Niño's Impact on Climate and Society*. Cambridge University Press, Cambridge.

Hahn, D.G. & Manabe, S. (1975) The role of mountains in the south Asian monsoon circulation. *Journal of Atmospheric Science* **32**, 1515–41.

Hastenrath, S. (1990) Decadal scale changes of the circulation in the tropical Atlantic sector associated with Sahel drought. *International Journal of Climatology* **10**, 459–72.

Hastenrath, S. (1995) Recent advances in tropical climate prediction. *Journal of Climate* **8**, 1519–32.

Hastenrath, S. & Greischar, L. (1993) Circulation mechanisms related to northeast Brazil rainfall anomalies. *Journal of Geophysical Research* **98**, 14,917–23.

Hastenrath, S. & Heller, L. (1977) Dynamics of climate

hazards in Northeast Brazil. *Quarterly Journal Royal Meteorological Society* **103**, 77–92.

Hastenrath, S., Wu, M.C. & Chu, P.S. (1984) Towards the monitoring and prediction of Northeast Brazil droughts. *Quarterly Journal Royal Meteorological Society* **110**, 411–25.

Henderson-Sellers, A. & Robinson, P.J. (1986) *Contemporary Climatology*. Longman, Harlow.

Hurrell, J.W. (1995) Decadal trends in the North Atlantic Oscillation: regional temperature and precipitation. *Science*, **269**, 676–9.

Hurrell, J.W. & van Loon, H. (1996) Decadal variations in climate associated with the North Atlantic oscillation. *Climatic Change* **36**, 301–26.

Karoly, D.J. (1989) Hemisphere circulation features associated with El Niño-Southern Oscillation events. *Journal of Climate* **2**, 1239–52.

King, J.C. & Turner, J. (1997) *Antarctic Meteorology and Climatology*. Cambridge University Press, Cambridge.

Köppen, W. (1923) *Die klimate der Erde: Grundkriss der Klimakunde*. Degruyter, Berlin.

Kripalani, R.H., Singh, S.V., Vernekar, A.D. & Thapliyal, V. (1996) Empirical study on Nimbus-7 snow mass and Indian summer monsoon rainfall. *International Journal of Climatology* **16**, 23–34.

Lazier, J.R. (1988) Temperature and salinity changes in the deep Labrador Sea, 1962–1986. *Deep Sea Research* **18**, 1247–53.

Leroux, M. (1993) The Mobile Polar High: a new concept explaining present mechanisms of meridional airmass and energy exchanges. *Global and Planetary Change* **7**, 69–73.

Leroux, M. (1998) *Dynamic Analysis of Weather and Climate: Atmospheric Circulation, Perturbations, Climate Evolution*. Wiley, Chichester.

Lockwood, J.G. (1974) *World Climatology: An Environmental Approach*. Edward Arnold, London.

McGregor, G.R. & Nieuwolt, S. (1998) *Tropical Climatology: An Introduction to the Climates of the Low Latitudes*. Wiley, Chichester.

Marengo, J.A., Grimm, A.M. & Zaratini, P. (1999a) Impacts of the extremes of the Southern Oscillation in Amazonia. Part 2. Circulation, convection and SSTS. *Abstract, 10th Symposium on Global Change Studies*. American Meteorological Society, Boston.

Marengo, J.A., Liebmann, I., Wainer, I. & Kousky, V. (1999b) On the characteristics of the onset and demise of the rainy season in Amazonia. *Abstract, 10th Symposium on Global Change Studies*. American Meteorological Society, Boston.

Markham, C.G. & McLain, D.R. (1977) Sea surface temperatures related to rain in Ceara, Northeastern Brazil. *Nature* **265**, 320–3.

Moulin, C., Lambert, C.E., Dulac, F. & Dayan, U. (1997) Control of atmosphere export of dust from North Africa by the North Atlantic Oscillation. *Nature* **387**, 691–4.

Moura, A.D. & Shukla, J. (1981) On the dynamics of droughts in Northeast Brazil: observations, theory and numerical experiment with a general circulation model. *Journal of Atmospheric Science* **38**, 2653–75.

National Climatic Data Center (1987) *Monthly Climatic Data for the World, Volume 40, Nos 1 and 7*. National Oceanic and Atmospheric Administration, Asheville, NC.

Namias, J. (1972) Influence of northern hemisphere general circulation on drought in Northeast Brazil. *Tellus* **24**, 336–43.

Nicholls, N. (1988) El Niño–Southern Oscillation and rainfall variability. *Journal of Climate* **1**, 418–21.

Nicholls, N. (1991) The El Niño–Southern Oscillation and Australian vegetation. *Vegetatio* **91**, 23–36.

Nicholls, N., Lavery, B., Frederiksen, C., Drosdowsky, W. & Torok, S. (1996) Recent apparent changes in relationships between the El Niño–Southern Oscillation and Australian rainfall and temperature. *Geophysical Research Letters* **23**, 3357–60.

Nobre, P., Nobre, C.A. & Moura, A.D. (1985) *Large-scale Circulation Anomalies and Prediction of Northeast Brazil Drought*. Proceedings of the Sixteenth Conference on Hurricanes and Tropical Meteorology. American Meteorological Society, Boston.

Oort, A.H. (1983) *Global Atmospheric Circulation Statistics, 1958–1973*. National Oceanic and Atmospheric Administration, Washington, DC.

Opoku-Ankomah, Y. & Cordery, I. (1993) Temporal variation of relations between New South Wales rainfall and the Southern Oscillation. *International Journal Climatology* **13**, 51–64.

Orvig, S. (ed.) (1970) *Climate of the Polar Regions. Volume 14, World Survey of Climatology*. Elsevier, Amsterdam.

Pant, G.B. & Rupa Kumar, K. (1997) *Climates of South Asia*. Wiley, Chichester.

Palmen, E. (1951) The role of atmospheric disturbances in the general circulation. *Quarterly Journal Royal Meteorological Society* **77**, 337–54.

Parthasarathy, B., Munot, A.A. & Kothawale, D.R. (1994) All-India monthly and seasonal rainfall series 1871–1993. *Theoretical and Applied Climatology* **49**, 217–24.

Peixoto, J.P. & Oort, A.H. (1992) *Physics of Climate.* American Institute of Physics, New York.

Phillpot, H.R. & Zillman, J.W. (1970) The surface temperature inversion over the Antarctic continent. *Journal of Geophysical Research* **75**, 4161–9.

Pielke, R.A. Jr & Pielke, R.A. Sr (1997) *Hurricanes: Their Nature and Impacts on Society.* Wiley, Chichester.

Pohjola, V.A. & Rogers, J.C. (1997) Atmospheric circulation and variations in the Scandinavian glacier mass balance. *Quaternary Research* **47**, 29–36.

Ropelewski, C.F. & Halpert, M.S. (1987) Global and regional scale precipitation patterns associated with the El Niño/Southern Oscillation. *Monthly Weather Review* **115**, 1606–26.

Ropelewski, C.F. & Halpert, M.S. (1989) Precipitation patterns associated with the high index phase of the Southern Oscillation. *Journal of Climate* **2**, 268–84.

Rogers, J.C. (1990) Patterns of low-frequency monthly sea level pressure variability (1899–1986) and associated wave cyclone frequencies. *Journal of Climate* **3**, 1364–79.

Rowell, D.P., Folland, C.K., Maskell, K., Owen, J.A. & Ward, M.N. (1992) Modelling the influence of global sea surface temperatures on the variability and predictability of seasonal Sahel rainfall. *Geophysical Research Letters* **19**, 905–8.

Savage, M.L., Stearns, C.R. & Weidner, G.A. (1988) The Southern Oscillation signal in Antarctica. *Proceedings of the Second Conference on Polar Meteorology and Oceanography.* American Meteorological Society, Boston.

Schwerdtfeger, W. (1970) The climate of the Antarctic. In Orvig, S. (ed.), *World Survey of Climatology. Volume 14, Climate of the Polar Regions.* Elsevier, Amsterdam.

Schwerdtfeger, W. (1984) *Weather and Climate of the Antarctic.* Elsevier, Amsterdam.

Semazzi, F.H.M., Neng, L.-H., Neng, L.-L. & Giorgi, F. (1993) A nested model study of the Sahelian climate response to sea-surface temperature anomalies. *Geophysical Research Letters* **20**, 2897–900.

Simmonds, I. (1998) The climate of the Antarctic Region. In Hobbs, J.E., Lindesay, J.A. & Bridgman, H.A. (eds), *Climates of the Southern Continents: Present, Past and Future.* Wiley, Chichester.

Simmonds, I. & Jacka, T.H. (1995) Relationships between the interannual variability of Antarctic sea ice and the Southern Oscillation. *Journal of Climate* **8**, 637–47.

Simmonds, I. & Law, R. (1995) Associations between Antarctic katabatic flow and the upper level winter vortex. *International Journal of Climatology* **15**, 403–21.

Smith, S.R. & Stearns, C.R. (1993) Antarctic pressure and temperature anomalies surrounding the minimum in the Southern Oscillation Index. *Journal of Geophysical Research* **98**, 13,071–83.

Stockton, C.W. & Glueck, M.F. (1999) Long-term variability of the North Atlantic Oscillation (NAO). *Abstracts, 10th Symposium on Global Change Studies.* American Meteorological Society, Boston.

Sturman, A.P. & Tapper, N.J. (1996) *The Weather and Climate of Australia and New Zealand.* Oxford University Press, Melbourne.

Thornthwaite, C.W. (1948) An approach to a rational classification of climate. *Geographical Review* **38**, 55–94.

Trenberth, K.E. (1990) Recent observed interdecadal climate changes in the Northern Hemisphere. *Bulletin American Meteorological Society* **71**, 988–93.

Turner, J., Coonlley, W.M., Leonard, S., Marshall, G.J. & Vaughan, D.G. (1999) Spatial and temporal variability of net snow accumulation over the Antarctic from ECMWF reanalysis project data. *International Journal of Climatology* **19**, 697–724.

Van Loon, H. (1967) The half-yearly oscillations in middle and high southern latitudes and the coreless winter. *Journal of Atmospheric Science* **24**, 472–86.

Vowinckel, E. & Orvig, S. (1970) The climate of the North Polar Basin. In Orvig, S. (ed.), *Climates of the Polar Regions.* Elsevier, Amsterdam.

Wallace, J.M. & Gutzler, D.S. (1981) Teleconnections in the geopotential height field during the Northern Hemisphere winter. *Monthly Weather Review* **109**, 784–812.

Westoby, M. (1980) Elements of a theory of vegetation dynamics in arid rangelands. *Israel Journal of Botany* **28**, 485–97.

Wilby, R.L., O'Hare, G. & Barnsley, W. (1997) The North Atlantic Oscillation and British Isles climate variability, 1865–1996. *Weather* **52**, 266–76.

Wyrtki, K. (1982) The Southern Oscillation, ocean–atmosphere interaction, and El Niño. *Marine Technology Society Journal* **16**, 3–10.

4 Biogeochemical Cycles and Residence Times

DUDLEY E. SHALLCROSS, KUO-YING WANG,
AND CLAUDIA H. DIMMER

4.1 INTRODUCTION

The cycling of the elements carbon, nitrogen, sulfur, and the halogens, chlorine, bromine, and iodine between the atmosphere, the lithosphere, and the hydrosphere is of extreme importance to the life forms that inhabit the Earth. Human activities are increasing the loading of so-called greenhouse gases (most of which contain one or more of these elements) in the atmosphere (see Chapter 1 of this handbook), altering the state of the atmosphere through modification of the Earth's radiative budget. It is of prime importance to understand how biogeochemical cycling of these elements may respond to a changing climate, and much effort is now being focused on this very area. It is the purpose of this chapter to briefly describe the main atmospheric reservoirs for each of these elements, their sources, be they natural (biogenic) or via human activities (anthropogenic), and their sinks, and hence to provide an estimate of their atmospheric residence times where possible. Although it is beyond the scope of this chapter to discuss in detail the potential effects of climate change on individual biogeochemical cycles, it is instructive to assess the perceived most likely response from terrestrial and oceanic flora and fauna. It should be noted, however, that our knowledge is far from complete and that systems may respond in a very different way to that currently expected. Since CO_2 plays such a dominant role in understanding the Earth's climate it is sensible to begin this chapter by considering the atmospheric carbon cycle.

4.2 THE GLOBAL CARBON CYCLE

All living matter is carbon based and the element is an essential part of life on Earth. The major reservoirs of carbon in the atmosphere are CO_2 (360 p.p.m.), CH_4 (1.8 p.p.m.) and CO (~0.1 p.p.m.), however myriad other volatile organic compounds are emitted into the Earth's atmosphere from a variety of sources and constitute a large pool of atmospheric carbon.

4.2.1 Carbon dioxide

Measurements of levels of carbon dioxide (CO_2) over the past 50 years or so (Fig. 4.1) show a dramatic rise from around 315 p.p.m. in the late 1950s to the present-day value of 360 p.p.m. Analysis of ice cores has facilitated an assessment of the changes to atmospheric levels of CO_2 over the past 1000 years (Barnola *et al.* 1987). The analysis shows that until 1800 levels of CO_2 were constant at 280 p.p.m., but have risen dramatically since the onset of the Industrial Revolution. The cause of this rapid change is a combination of the burning of fossil fuel, deforestation, and changes in land use. It is well known that CO_2 is a potent greenhouse gas, trapping outgoing terrestrial radiation in the infrared region of the spectrum, and that an increase in its atmospheric mixing ratio is expected to lead to a warming of the Earth's surface. The extent of the warming predicted is currently of considerable debate. However, estimates based on future global population and land use, suggest that

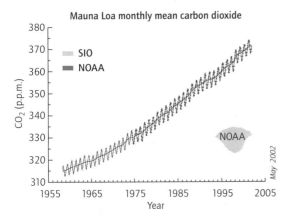

Mauna Loa monthly mean carbon dioxide

Fig. 4.1 Measurements of the change in atmospheric CO_2 within the past 50 years. The National Oceanic and Atmospheric Administration (NOAA), Climate Monitoring and Diagnostics Laboratory (CMDL), Carbon Cycle Greenhouse Gases Group is gratefully acknowledged for these data.

Table 4.1 Global budget for an anthropogenic perturbation of CO_2 (based on IPCC 1996).

	CO_2 $(GtC\,yr^{-1})$
Source	
Fossil fuel combustion and related processes	5.5 ± 0.5
Deforestation (Tropics) and land-use change	1.6 ± 1.0
Total	7.1 ± 1.1
Sinks	
Retained in the atmosphere	3.3 ± 0.2
Oceanic uptake	2.0 ± 0.8
Uptake by forest regrowth (northern hemisphere)	0.5 ± 0.5
Total	5.8 ± 1.0
Imbalance	1.3 ± 1.5

levels of CO_2 may double from preindustrial levels to around 700 p.p.m., and that global surface temperatures may increase by around 6°C over the next century (IPCC 1996). Current estimates for the "budget" of CO_2 (see Table 4.1) based on meas-urements over the period 1980–9 show that for a perturbation of $7.1\,GtC\,yr^{-1}$, nearly half is retained in the atmosphere, with major known sinks being the ocean and uptake by regrowth of trees in the northern hemisphere. However, a substantial missing sink ($1.3\,GtC\,yr^{-1}$) is unaccounted for at present.

If CO_2 levels rise as expected and the associated warming of the surface is realized, this will have potential consequences for the CO_2 sink terms. CO_2 is less soluble in warmer water and global warming will lead to a gradual rise in sea surface temperatures, hence reducing the effectiveness of the ocean as a sink for CO_2. In addition, as the amount of dissolved inorganic carbon in the ocean increases, a decrease in ocean buffering is expected, again reducing the effectiveness of the ocean in taking up CO_2. Nevertheless, biological processes in the ocean may themselves counteract these physical changes. For example, favorable changes in the external supply of biologically limiting nutrients such as Fe, Si, P, and N could increase the strength of biological production within the ocean surface and enable more CO_2 to be dissolved in the ocean.

It is readily apparent that the climate system is extremely complex. The exchange of CO_2 with terrestrial plants is another important sink. Plants assimilate CO_2 in the process of photosynthesis, releasing O_2, and the gain in carbon is known as the gross primary production (GPP). However, plants also release CO_2 through respiration processes (taking up O_2) and the difference between assimilation and release of CO_2 gives the net primary production (NPP), i.e. the amount of new stored carbon. The largest NPP is found in savannas (32%) and tropical rainforests (18%), with cultivated land (11%), temperate forests (8%), wetlands (7%), grassland (7%), boreal forests (6%), and evergreen tropical forests (6%) all making significant contributions to CO_2 uptake (Ajtay *et al.* 1979). Figure 4.2 shows the distribution of NPP derived from a land surface model (Wang & Shallcross 2000).

Assessing the response of plants to increasing CO_2 levels is far from simple. McKee and Woodward (1994), for example, have shown that

Mean net primary production (Gt C per month)

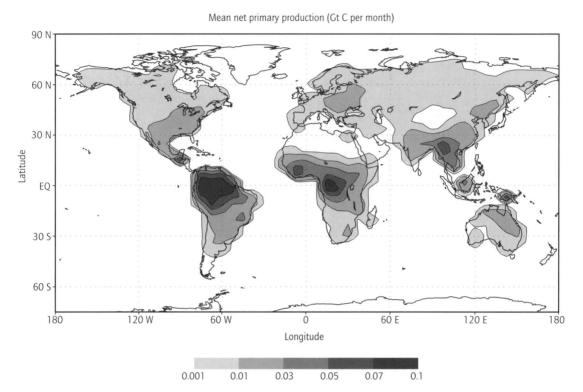

Fig. 4.2 Distribution of NPP derived from a land-surface vegetation model (see Wang & Shallcross 2000).

growth promotion occurs under elevated CO_2 levels, and in general studies suggest that an increase in growth of more than 30% is typical for a doubling of CO_2 concentration. However, growth temperature is very important, and Moya *et al.* (1998) found that simultaneous increases in air temperature and CO_2 concentration offset the stimulation of biomass and grain yield in rice compared with increases in CO_2 alone. Elevated CO_2 does seem to raise the maximum temperature at which plants can survive. However, for any plant, exceeding their optimal growing temperature will have serious consequences on their growth. Therefore, the response of a particular species to changes in CO_2 levels will depend on how far away from its optimal growing temperature it is. Again in studies on rice, Baker *et al.* (1996) found that season-long enrichments of CO_2 from 300 to 700 p.p.m. resulted in a 21–27% increase in net canopy photo-synthesis and a 10% reduction in total evapotranspiration. Higher levels of CO_2 lead to partial closure of stomata, reducing the stomatal conductance and leading to increasing water use efficiency. Therefore, increasing CO_2 levels may have an overall positive effect on terrestrial plant growth, but there are caveats, which are discussed in a later section. The overall lifetime of CO_2 is around 120 years, and it is therefore well mixed in the atmosphere.

4.2.2 Methane

Methane (CH_4) is the most abundant organic species in the atmosphere. In a similar way to CO_2, levels of methane have risen dramatically over the past 200 years. Preindustrial levels were around 750 p.p.b., but now stand at 1760 p.p.b. in the northern hemisphere and 1630 p.p.b. in the southern hemisphere. Methane was increasing at around 1.3% per year for most of the twentieth century until the early 1990s (Blake & Rowland 1988), although since this time the rate of growth of methane has slowed to 0.6% per year (Steele *et al.* 1992). It is not known with certainty why there has been a slowdown in the rate of increase: reduction in sources or an increase in the concentration of the hydroxyl radical, the major sink for methane, are possibilities. On a molecule per molecule basis, methane is a far more effective greenhouse gas than CO_2 (factor of 20 based on a 100-year time horizon) and therefore understanding its budget is of great importance. The current estimated budget for methane is shown in Table 4.2, where it is apparent that anthropogenic sources dominate, being double natural sources. The anthropogenic sources are spread between fossil fuel related release (27% of the total anthropogenic source), waste management (24%), enteric fermentation of cattle (23%), biomass burning (11%), and rice paddies (15%). Natural sources of methane are dominated by wetland emissions, particularly in the Tropics. The decomposition of organic matter under oxygen-deficient conditions leads to the production of methane and it is no surprise that where temperature is highest and microbial activity most intense, the largest natural source of methane is found. Termites also produce a significant quantity of methane, and other insects may well do the same. The dominant loss process for methane in the atmosphere is via reaction with the hydroxyl radical, OH ($k_{298} = 6 \times 10^{-15}\,\mathrm{cm^3\,molecule^{-1}\,s^{-1}}$: DeMore *et al.* 1997), with minor contributions from stratospheric removal and consumption of methane in soils by methanotrophic bacteria. The global residence time of methane in the atmosphere is approximately 10 years, leading to a slight inter-hemispheric gradient.

4.2.3 Carbon monoxide

The budget of carbon monoxide (CO) is dominated by the atmospheric oxidation of methane and other VOCs (see next section), initiated by the OH radical (see Chapter 6) and from incomplete com-

Table 4.2 Sources and sinks for methane in the atmosphere (data taken from IPCC 1996).

	Likely (Tg yr^{-1})	Range (Tg yr^{-1})
Natural sources		
Wetlands		
Tropics	65	30–80
Northern latitude	40	20–60
Others	10	5–15
Termites	20	10–50
Ocean	10	5–50
Freshwater	5	1–25
Geological	10	5–15
Total	160	
Anthropogenic sources		
Fossil fuel related		
Coal mines	30	15–45
Natural gas	40	25–50
Petroleum industry	15	5–30
Coal combustion	15	5–30
Waste management system		
Landfills	40	20–70
Animal waste	25	20–30
Domestic waste treatment	25	15–80
Enteric fermentation	85	65–100
Biomass burning	40	20–80
Rice paddies	60	20–100
Total	375	
Total sources	535	
Sinks		
Reaction with OH	490	405–575
Removal in the stratosphere	40	32–48
Removal by soils	30	15–45
Total sinks	560	
Atmospheric increase	37	35–40

bustion associated with biomass burning and fossil fuels. Vegetation and oceans are also a nonnegligible source of CO. Hence, anthropogenic sources dominate, although it should be noted that the oxidation of natural VOCs released from vegetation provides over half the source from VOC oxidation. Despite having no significant direct impact on global warming, CO is an extremely

Table 4.3 Global budget for carbon monoxide in the atmosphere (data taken from Graedel & Crutzen 1993).

	CO (Tg yr^{-1})
Sources	
Technological sources	440 ± 150
Biomass burning	700 ± 200
Vegetation	75 ± 25
Ocean	50 ± 40
CH$_4$ oxidation	600 ± 200
NMHC oxidation	800 ± 400
Total sources	2700 ± 1000
Sinks	
Reaction with OH	2000 ± 600
Uptake by soils	250 ± 100
Flux into stratosphere	110 ± 30
Total sinks	2400 ± 750

important atmospheric species, whose major loss process is reaction with OH ($k_{298} = 2.4 \times 10^{-13}$ cm^3 molecule^{-1} s^{-1}: DeMore *et al.* 1997):

$$CO + OH \rightarrow CO_2 + H \qquad (4.1)$$

In fact, reaction with CO constitutes ~70% of the total sink for OH; hence the importance of CO in the Earth's atmosphere. Current levels of CO are 60–70 p.p.b. in the southern hemisphere and 120–180 p.p.b. in the northern hemisphere, with extremely high levels experienced in urban and industrialized areas (p.p.m. levels). The global atmospheric residence time of CO is approximately two months, giving rise to the hemispheric asymmetry. The trend in CO is difficult to estimate, but analysis by Novelli *et al.* (1994) suggests that CO has been decreasing recently. It is possible that either the level of OH has increased or emissions from combustion processes, for example, have decreased. A summary of the global atmospheric budget is given in Table 4.3.

4.2.4 *Volatile organic compounds*

Myriad volatile organic compounds (VOCs) are emitted into the atmosphere by both biogenic and anthropogenic sources. These VOCs have varying residence times depending on their structure, with the vast majority undergoing photooxidation initiated by the OH radical (Chapter 6), ultimately leading to the production of CO$_2$ and H$_2$O. Photooxidation may lead to the production of intermediate compounds such as carboxylic acids, alcohols and nitrates which can then be removed by wet deposition (Chapter 12). Globally, natural sources of VOCs are believed to dominate (Singh & Zimmerman 1992), constituting around 90% of the carbon flux. Apart from methane and dimethyl sulfide (see Section 4.4), the major source of biogenic VOCs is terrestrial plants (Fall 1999). The reasons why plants manufacture certain VOCs are well known. For example, ethene (C$_2$H$_4$) plays a key role in growth and development and its production is greatly enhanced following plant wounding, exposure to chemicals, and infection (Fall 1999). However, for other VOCs, such as isoprene (2 methyl 1,3 butadiene, C$_5$H$_8$), the reasons for their production are less clear. Emission of isoprene is in fact the single largest source of organic carbon to the atmosphere after methane. In general, deciduous trees and most woody plants emit isoprene, although ferns, vines, and some herbaceous plants are also emitters. Emission of isoprene is light-dependent, dropping to zero flux in the dark, and is also temperature-dependent, with the flux increasing rapidly with temperature (Guenther *et al.* 1995). Figure 4.3 shows the distribution of isoprene emissions derived from a land-surface vegetation model (Wang & Shallcross 2000) using emission algorithms from Guenther *et al.* (1995). Singsaas *et al.* (1997) have suggested that isoprene acts as a thermal protectant, which is consistent with the observation that emissions are elevated as temperature increases. However, there are problems with this explanation, in that many species in hot desert environments do not emit isoprene at all, whereas plants of the same genus in more temperate climates do (Fall 1999). Another possible explanation includes neutralization of free radicals within the plant.

Isoprene is extremely short-lived: its rate coefficient for reaction with OH is near the gas collision limit ($k_{298} = 1.01 \times 10^{-10}$ cm^3 molecule^{-1} s^{-1}: Atkinson *et al.* 1992), giving rise to a residence

Annual mean C_5H_8 emissions (mg m^{-2} month^{-1})

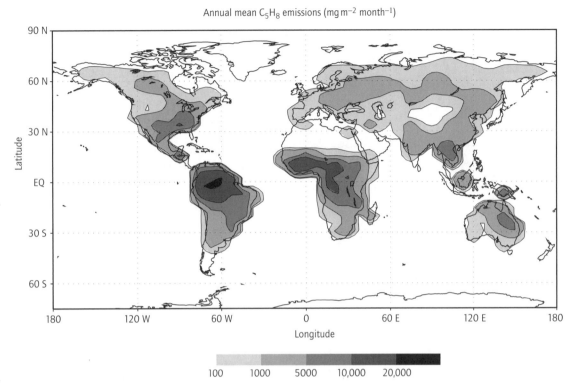

Fig. 4.3 Distribution of isoprene emissions derived from a land-surface model (see Wang & Shallcross 2000) using emission algorithms from Guenther *et al.* (1995).

time of a few hours. Consequently, levels of isoprene in the atmosphere are very variable and display a strong diurnal cycle, peaking in mid-afternoon and rapidly decaying at night, as well as a strong seasonal cycle, peaking in late spring. In forests, levels of isoprene can reach 15 p.p.b., although p.p.t. levels are more typical elsewhere.

In Chapter 6 a detailed discussion is presented of the role played by VOCs in the presence of NO_x (NO and NO_2) in the production of tropospheric ozone. At the Earth's surface ozone is deleterious to plants, entering the stomata and destroying cell lining, while being itself a greenhouse gas. As levels of CO_2 increase it is anticipated that global emissions of isoprene will also increase, driven mainly by increasing surface temperature (Constable *et al.* 1999). Horowitz *et al.* (1998) and Wang and Shallcross (2000), among others, have

noted that isoprene may be responsible for a significant proportion of ozone production in the troposphere, and speculate that increasing emissions of isoprene in the future may exacerbate the rise in tropospheric ozone. However, very high surface ozone in the future will lead to a reduction in plant NPP, due to the retardation of growth, and it is not clear at present at what level of ozone such retardation will occur. Once again, natural biogeochemical cycles are intricately coupled with the climate system.

Monoterpenes ($C_{10}H_{16}$) are another major class of VOC manufactured by plants. A selection of the most common monoterpenes is shown in Fig. 4.4. Their ecological role includes herbivore defense, attraction of pollinators, and allelopathic effects on competing plants. Monoterpenes rapidly react with the hydroxyl radical, and also with the nitrate

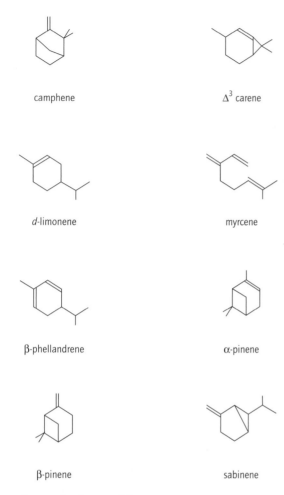

camphene

Δ^3 carene

d-limonene

myrcene

β-phellandrene

α-pinene

β-pinene

sabinene

Fig. 4.4 A selection of the most common monoterpenes found in the atmosphere.

radical (NO_3), and like isoprene have very short residence times, of the order of a few hours, and also have a strong diurnal and seasonal cycle. Plants also emit C_6 aldehydes and alcohols, known as the hexenal family, which have antibiotic properties (Hatanaka 1993).

Anthropogenic emissions of VOCs are dominated by combustion of fossil fuel, with alkanes, aromatics, and alkenes being the major VOC classes. In urban areas, emissions of VOCs follow a typical diurnal cycle, tracking rush hour traffic flow. Photochemical smog formation in the summer months is driven by the emission of these VOCs, in the presence of NO_x and sunlight (Chapter 6). In urban areas, individual VOC levels can be of the order of some p.p.b., while in background air, levels of the more reactive species can be in the p.p.t. range or less.

Table 4.4 lists some of the major classes of VOC, their sources, estimated emission strength, and approximate residence time. Table 4.4 illustrates the point that unsaturated species such as alkenes (e.g. ethene) and dialkenes (e.g. isoprene) react more rapidly with the OH radical than do alkanes (e.g. ethane) and aromatics (e.g. benzene), yielding a shorter residence time.

4.3 THE GLOBAL NITROGEN CYCLE

N_2 represents nearly 80% of the Earth's atmosphere, and is chemically inert throughout the lower atmosphere. Only in the upper atmosphere, where significant fluxes of very short wavelength radiation can be found, can N_2 be broken down by photolysis. Hence 99.9999% of the nitrogen budget is well known, and given that nitrous oxide (N_2O) represents around 99% of the remainder of the budget, one would imagine that the global nitrogen cycle is well understood. However, despite their small and highly variable sources, the other nitrogen-containing species present in the atmosphere at significant levels—ammonia (NH_3), nitric oxide (NO), and nitrogen dioxide (NO_2)—play a disproportionately large role in determining atmospheric behavior. Ammonia, for example, is the only basic gas in the atmosphere and is important in the neutralization of acid aerosols. Nitrogen oxides—NO and NO_2—play a vital role in the production and destruction of ozone in both the troposphere (Chapter 6) and the stratosphere (Chapter 7) and therefore affect the radiative budget of the Earth and the oxidizing capacity of the atmosphere.

4.3.1 Nitrous oxide

Nitrous oxide (N_2O) is an important greenhouse gas, whose current atmospheric mixing ratio is

Table 4.4 Estimates of the global source strengths for some VOCs and their residence times in the atmosphere (emission data and source assignment are derived from Singh & Zimmerman 1982; lifetimes are from estimates by Seinfeld 1999).

VOC	Emissions (Tg C year^{-1})	Sources	Residence time (days)
Ethane	10–15	Natural gas, biomass burning, oceans, vegetation	78
Ethene	20–45	Combustion of fuel, biomass burning, plants	2
Ethyne	3–6	Combustion of fuel, biomass burning	23
Propene	7–12	Combustion of fuel, biomass burning, oceans	0.8
Benzene	4–5	Combustion of fuel, biomass burning	17
Toluene	4–5	Combustion of fuel, biomass burning, solvents	3
Isoprene	175–500	Plants and trees	0.2
Monoterpenes	125–350	Plants and trees	0.2

about 310 p.p.b., compared with an estimated preindustrial level of around 285 p.p.b. N_2O has an extremely long residence time (~150 years) and is virtually inert in the troposphere, being destroyed in the stratosphere by direct photolysis and reaction with $O(^1D)$ atoms:

$$N_2O + hv \rightarrow N_2 + O(^1D) \qquad (4.2)$$

$$N_2O + O(^1D) \rightarrow N_2 + O_2 (\sim 40\%) \qquad (4.3a)$$

$$N_2O + O(^1D) \rightarrow NO + NO (\sim 60\%) \qquad (4.3b)$$

Nitrous oxide is released to the atmosphere from both soils and aquatic systems, with undisturbed soils (40%), cultivated soils (24%), and oceans (20%) making up the bulk of the known sources. Denitrifying bacteria transform nitrate to N_2 and some N_2O under anaerobic conditions, which can then evade to the surface and enter the atmosphere. A budget for N_2O is presented in Table 4.5 based on the IPCC (1996) assessment. There are many uncertainties, including the estimation of the ocean source, since N_2O is both lost to and emitted from the oceans. Since it delivers NO_x to the stratosphere, N_2O plays an important role in controlling the abundance of stratospheric ozone.

4.3.2 *Ammonia*

Ammonia (NH_3) has a short residence time in the

Table 4.5 Estimates of the global sources and sinks for N_2O (data taken from IPCC 1996).

	Likely (Tg N yr^{-1})	Range (Tg N yr^{-1})
Natural sources		
Oceans	3.0	1–5
Tropical soils		
Wet forests	3.0	2.2–3.7
Dry savannas	1.0	0.5–2.0
Temperate soils		
Forests	1.0	0.5–2.3
Grasslands	1.0	
Total	9.0	
Anthropogenic sources		
Cultivated soils	3.5	1.8–5.3
Biomass burning	0.5	0.2–1.0
Industrial sources	1.3	0.7–1.8
Cattle and feed lots	0.4	0.2–0.5
Total	5.7	
Total sources	14.7	10–17
Sinks		
Stratospheric *hv*	12.3	9–16
Total sinks	12.3	9–16
Imbalance (increase)	3.9	3.1–4.7

atmosphere of approximately 10 days and is removed mainly by both wet and dry deposition processes, with some additional loss via reaction with OH radicals. The main sources arise from

Table 4.6 Global atmospheric budget for ammonia (data from the compilation of Brasseur *et al.* 1999).

	Ammonia (Tg N yr^{-1})	Range
Sources		
Domestic animals	21.3	20–40
Human excrement	2.6	2.6–4
Soil emissions	6	6–45
Biomass burning	5.7	1–9
Wild animals	0.1	0.1–6
Industry	0.2	
Fertilizer use	9	5–10
Fossil fuel	0.1	0.1–2.2
Ocean	8.2	5–15
Sinks		
Wet deposition, land	11	11–80
Wet deposition, ocean	10	6–26
Dry deposition, land	11	10–150
Dry deposition, ocean	5	
Reaction with OH	3	1–9

biological activity, such as the decomposition of urea in animal urine by enzymes, the decomposition of excrement, and the release from soils and the ocean following mineralization of organic material. Anthropogenic sources center on its use in fertilizers and as a byproduct of waste production. Since deposition processes dominate its loss and sources are diverse, levels of ammonia are highly variable, ranging from 0.1 to 10 p.p.b. over continental regions. It has already been noted that ammonia is the only basic gas in the atmosphere, and it forms ammonium sulfate aerosols. Deposition of ammonium sulfate to the soil decreases its pH, leading to a decline in growth. A global budget for ammonia is summarized in Table 4.6.

4.3.3   *Oxides of nitrogen*

In the atmosphere, NO and NO_2 (NO_x) are extremely tightly coupled during sunlit hours and rapidly interconvert with one another in the presence of ozone:

$$NO + O_3 \rightarrow NO_2 + O_2 \qquad (4.4)$$

$$NO_2 + hv \rightarrow NO + O \qquad (4.5)$$

$$O + O_2 + M \rightarrow O_3 + M \qquad (4.6)$$

Individually, NO and NO_2 have extremely short residence times, of the order of seconds. However, if we consider the two compounds together as NO_x, the residence time is lengthened to many hours. Hence, NO and NO_2 display strong diurnal cycles and their concentrations display a seasonal cycle. In urban areas, NO_x can reach hundreds of p.p.b., and in particularly polluted environments p.p.m. levels, whereas clean maritime levels are only 5–10 p.p.t. The major loss processes for NO_x are conversion to HNO_3 via reaction with OH

$$NO_2 + OH + M \rightarrow HNO_3 + M \qquad (4.7)$$

and dry deposition of NO_x. HNO_3 can be removed by wet and dry deposition, constituting a loss of NO_x from the atmosphere. In the troposphere, in the presence of VOCs, NO_x can promote the formation of ozone (Chapter 6) and can also be sequestered to form temporary nitrate reservoirs such as peroxyacetylnitrate (PAN: $CH_3C(O)O_2NO_2$). PAN can allow NO_x to be transported away from source regions and influence chemistry on regional and global scales. NO_x is also the major natural catalytic cycle operating in the stratosphere. NO_x is mostly emitted in the form of NO. Natural sources of NO are from soil processes and lightning discharge, while an ever increasing source is the combustion of fossil fuels. At the high temperatures inside an internal combustion engine the Zeldovitch mechanism

$$O + N_2 \rightarrow NO + N \qquad (4.8)$$

$$N + O_2 \rightarrow NO + O \qquad (4.9)$$

leads to the formation of NO. There are considerable uncertainties within the budget for NO_x, such as the soil and lightning source strength. However, the burden from fossil fuels is reasonably well defined and set to increase globally, despite the growing use of control technologies.

4.4 THE GLOBAL SULFUR CYCLE

Sulfur is an important secondary constituent of amino acids and proteins, and is therefore an essential element for living organisms on Earth. Cross-linking in proteins via sulfur–sulfur bonds is of great importance: intermolecular "disulfide" linkages can lead to large-scale structures such as nails, while intramolecular linkages allow the protein to adopt specific configurations, as in the case of an enzyme. On decomposition by bacteria, organic sulfur compounds usually release hydrogen sulfide (H_2S). However, H_2S is but one of many sulfur compounds released into the atmosphere, which include sulfur dioxide (SO_2), carbonyl sulfide (OCS), carbon disulfide (CS_2), and dimethyl sulfide (CH_3SCH_3, also known simply as DMS). Since the onset of the Industrial Revolution the sulfur burden in the atmosphere has increased dramatically, due to the burning of fossil fuels that inevitably contain some sulfur. Hence, the anthropogenic contribution to the total sulfur emission budget, mainly in the form of SO_2, approaches 75%, with the bulk of these emissions (around 90%) emanating from the northern hemisphere. Natural emission sources, which make up the remaining 25%, are distributed over the two hemispheres, with a slight bias towards the northern hemisphere (Brasseur *et al*. 1999).

It should be noted that a very large amount of sulfate is released into the atmosphere from the oceans in the form of sea salt. However, these very coarse particles are rapidly deposited back to the ocean and play no further role in the global sulfur cycle. In addition, sulfur-containing minerals are transported around the globe following wind-driven erosion of soils (Aneja 1990).

A summary of the most recent estimates of the global budgets for each of the five main sulfur species is presented in Table 4.8, and these sources and the residence time of each sulfur species will be discussed. It is apparent that each of these budgets has a large error associated with it, despite intensive research over the past 25 years. The diverse range of sources and their frequent inaccessibility to study means that the sulfur budget is still only known approximately, and new sources and sinks

Table 4.7 Global atmospheric budget for NO_x (data from the compilation of Brasseur *et al*. 1999 and IPCC 1996).

	NO_x (Tg N year^{-1})	Range (Tg N year^{-1})
Sources		
Fossil fuel combustion	20	14–28
Biomass burning	12	4–24
Soil emissions	20	4–40
Lightning	8	2–20
NH_3 oxidation	3	0–10
Aircraft	0.5	
Stratosphere	0.6	5–10
Total	64	25–122
Sinks		
Wet deposition, land	19	8–30
Wet deposition, ocean	8	4–12
Dry deposition, land	16	12–22
Total	43	24–64

for each compound may well become apparent in the future.

4.4.1 Sulfur dioxide

The burning of fossil fuels is the major source of sulfur dioxide (SO_2) in the atmosphere, with volcanoes being another significant source. The contribution to the SO_2 budget from the oxidation of natural reduced sulfur compounds (see below) is quite minor by comparison. SO_2 emissions are strongly linked with acid rain formation and are also known to lead to respiratory problems in humans and animals. The presence of aerosols in the atmosphere has an influence on the Earth's climate, resulting from both direct and indirect effects on the radiation budget (see Chapter 9). The direct effect arises because aerosols scatter and absorb incoming solar radiation, thereby reducing the energy reaching ground level. The indirect effect results from the role of aerosols in cloud formation, since clouds reflect incoming radiation. Both effects are influenced by the number, size distribution, and chemical composition of the aerosol, and are currently believed to lead to

Table 4.8 Global annual sources and sinks of OCS, CS_2, H_2S, DMS, and SO_2 in $Tg\,yr^{-1}$.

	OCS	CS_2	H_2S	DMS	SO_2
Sources					
Open ocean	0.10 ± 0.15	0.11 ± 0.04	1.50 ± 0.60	20.70 ± 5.20	
Coastal ocean	0.10 ± 0.05	0.04 ± 0.02	0.30 ± 0.10		
Salt marshes	0.10 ± 0.05	0.03 ± 0.02	0.50 ± 0.35	0.07 ± 0.06	
Anoxic soils	0.02 ± 0.01	0.07 ± 0.06			
Vegetation		0.37 ± 0.07	1.58 ± 0.86		
Tropical forests		0.42 ± 0.12	1.60 ± 0.50		
Soils				0.29 ± 0.17	
Wetlands	0.03 ± 0.03	0.02 ± 0.02	0.20 ± 0.21	0.12 ± 0.07	
Volcanism	0.05 ± 0.04	0.05 ± 0.04	1.05 ± 0.94		20
Precipitation	0.13 ± 0.08				
OCS + OH			0.08 ± 0.07		
DMS oxidation	0.17 ± 0.04				
CS_2 oxidation	0.42 ± 0.12				
Biomass burning	0.07 ± 0.05				4
Anthropogenic	0.12 ± 0.06	0.34 ± 0.17	3.30 ± 0.33	0.13 ± 0.04	176
Total sources	1.31 ± 0.68	0.66 ± 0.37	7.72 ± 2.79	24.49 ± 5.30	200
Sinks					
Oxic soil	0.92 ± 0.78	0.44 ± 0.38			
Vegetation	0.56 ± 0.10				
Reaction with OH	0.13 ± 0.10	0.57 ± 0.25	8.50 ± 2.80	25.00 ± 1.30	20
Reaction with O	0.02 ± 0.01				
Photolysis	0.03 ± 0.01				
Cloud scavenging					180
Total sinks	1.66 ± 1.00	1.01 ± 0.63	8.50 ± 2.80	25.00 ± 1.30	200
Total imbalance	0.35 ± 1.68	0.35 ± 1.00	0.78 ± 5.59	0.51 ± 6.60	–

Sources: Möller (1984), Andreae (1990), Fried *et al.* (1992), Chin and Davis (1993), Barnes *et al.* (1994), Nguyen *et al.* (1995), Xie *et al.* (1997), Watts and Roberts (1999), Watts (2000).

atmospheric cooling, which offsets the warming influence of radiatively active trace gases such as CO_2. The majority of aerosols in the atmosphere are generated as a result of gas-to-particle conversion processes, although there are substantial contributions from other sources, such as resuspended mineral dust and sea-salt aerosols. An essential prerequisite for gas-to-particle conversions to occur is the presence of a species in the gas phase at a partial pressure in excess of its saturation vapor pressure with respect to the condensed phase (i.e. condensable material).

It is generally accepted that the most significant condensable molecule formed in the troposphere is sulfuric acid (H_2SO_4), which has also been long recognized as the most important from the point of view of the nucleation of new particles. The major source of H_2SO_4 results from the oxidation of anthropogenically derived SO_2, for which the predominant gas-phase oxidation pathway is initiated by reaction with the hydroxyl radical (Stockwell & Calvert 1983):

$$OH + SO_2 + M \rightarrow HOSO_2 + M \qquad (4.10)$$

$$HOSO_2 + O_2 \rightarrow SO_3 + HO_2 \qquad (4.11)$$

$$SO_3 + H_2O + M \rightarrow H_2SO_4 + M \qquad (4.12)$$

Once formed, sulfuric acid will either be taken up by existing aerosol or create new ones. SO_2 is sufficiently soluble for aqueous-phase oxidation to be another important route for its conversion to sulfuric acid ($K_H \sim 21.5 \times 10^{-3}$ mol dm^{-3} atm^{-1} at 25°C, Brasseur *et al.* 1999). Once taken up into the aqueous phase, SO_2 establishes an equilibrium with the bisulfite ion and the sulfite ion

$$SO_2(aq) \leftrightarrow H^+ + HSO_3^- \qquad (4.13)$$

$$HSO_3^- \leftrightarrow H^+ + SO_3^{2-} \qquad (4.14)$$

where K_1 is the equilibrium constant for reaction 4.13 and K_2 is the equilibrium constant for reaction 4.14. The Henry's Law coefficient for SO_2, $K_H(SO_2)$, can be expressed as:

$$K_H(SO_2) = [SO_2(aq)]/P(SO_2) \qquad (4.15)$$

where [SO_2(aq)] is the aqueous concentration of SO_2 and $P(SO_2)$ is the partial pressure of SO_2 in the gas phase. However, because of the formation of bisulfite (the dominant species in HSO_3^- in aqueous media) and sulfite ions, it is more common to use an effective Henry's Law coefficient, termed $k_{Heff}(SO_2)$ and defined as

$$
\begin{aligned}
&k_{Heff}(SO_2) \\
&= \left([SO_2(aq)] + [HSO_3^-] + [SO_3^{2-}]\right)/P(SO_2)
\end{aligned}
\qquad (4.16)
$$

Equation 4.16 can then be rearranged in the form:

$$
\begin{aligned}
&K_{Heff}(SO_2) \\
&= k_H(SO_2)\left(1 + K_1/[H^+] + K_1 K_2/[H^+]^2\right)
\end{aligned}
\qquad (4.17)
$$

Inspection of eqn 4.17 shows that solubility of SO_2 is dependent on pH, decreasing at low solution pH, where the $k_{Heff}(SO_2)$ then approaches the value of $k_H(SO_2)$. The aqueous-phase oxidation of the bisulfite ion to the sulfate ion is dominated by H_2O_2 for

pH less than 5, with oxidation by dissolved ozone becoming the main contributor for pH greater than 4.5. Other mechanisms for aqueous-phase oxidation exist (see Hoffmann & Jacob 1984) but they are generally less important. Depending on the amount of moisture in the atmosphere, 20–80% of the SO_2 emitted is oxidized to sulfate, with the remainder being removed by dry deposition. The mixture of SO_2 and sulfate has a lifetime of between two and six days before being lost via wet or dry deposition. Highest SO_2 mixing ratios are found over the major industrial regions of the world, the eastern United States of America, Europe, and the Far East, where levels are as high as a few p.p.b. Reliable measurements of SO_2 in remote regions are problematic due to the problem of loss of SO_2 on moist surfaces, such as instrument inlets; however, mixing ratios drop in such areas to below 100 p.p.t.

4.4.2   Carbonyl sulfide

The lifetime of carbonyl sulfide (OCS) in the atmosphere is approximately 18 months, making it the most long-lived of the sulfur species considered. OCS is relatively insoluble ($K_H \sim 21.5 \times 10^{-3}$ mol dm^{-3} atm^{-1} at 25°C: Watts 2000) where the major loss processes are uptake by oxic soils (such as aridsols) and vegetation. Recent studies have shown that the plant enzyme higher plant carbonic anhydrase (HPCA) processes not only CO_2 but also OCS (Protoschill-Krebs *et al.* 1996) and that HPCA seems to preferentially take up OCS over CO_2 (Kesselmeier & Merk 1993). The reaction of OCS with the OH radical is slow, where $k_{298} = 1.96 \times 10^{-15}$ cm^3 molecule^{-1} s^{-1} (De More *et al.* 1997), and therefore gas-phase removal in the troposphere is a minor loss process. Since OCS is so long lived it can be transported up to the stratosphere, where it is photolyzed by ultraviolet radiation, or reacts with O atoms to subsequently act as a source of SO_2 and ultimately sulfate particles (known as the Junge layer). The oxidation of OCS to SO_2 in the stratosphere is summarized by reactions 4.18 to 4.22.

$$OCS + h\nu \rightarrow CO + S \qquad (4.18)$$

$$O + OCS \rightarrow CO + SO \qquad (4.19)$$

$$S + O_2 \rightarrow SO + O \qquad (4.20)$$

$$SO + O_2 \rightarrow SO_2 + O \qquad (4.21)$$

$$SO + NO_2 \rightarrow SO_2 + NO \qquad (4.22)$$

The conversion of SO_2 to sulfate is discussed above. One of the major estimated sources of OCS is the atmospheric oxidation of CS_2 (Barnes *et al.* 1994), which is initiated by the OH radical:

$$CS_2 + OH \leftrightarrow CS_2OH \qquad (4.23)$$

$$CS_2OH + 2O_2 \rightarrow HO_2 + SO_2 + OCS \sim 85\%$$
$$(4.24a)$$

$$CS_2OH + 2O_2 \rightarrow HCO + 2SO_2 \sim 15\% \quad (4.24b)$$

and also leads to SO_2 production. The exact mechanism is unknown, but product yields suggest that OCS is a major product of the decomposition of the CS_2OH adduct in the presence of oxygen.

Other important sources of OCS include oceanic photochemical production from dissolved organosulfur species (Zepp & Andreae 1994), biomass burning, aluminum production, coal combustion, and car emissions (the last three clearly being of anthropogenic origin). Another source of OCS that has been speculated is precipitation, which is a somewhat counterintuitive source, since one would not expect OCS to be scavenged from the atmosphere given its Henry's Law coefficient. However, there are reports of precipitation being supersaturated with OCS (Belvisio *et al.* 1987), where organic matter, including amino acids containing sulfur, are believed to be responsible for the production (e.g. Mopper & Zika 1987). Although vegetation is believed to be a net sink for OCS (Watts 2000) there is evidence that spruce trees emit OCS (e.g. Berresheim & Vulcan 1992) and it may well be that trees are both a source and sink. Typical mixing ratios of OCS in the troposphere are around 500 p.p.t. (e.g. Fried *et al.* 1991), which does not vary dramatically with altitude. The combination of low solubility, low reactivity

with OH, and somewhat diffuse sources would be concomitant with these observations. Latitudinal measurements differ somewhat, but more recent studies suggest that a gradient does exist across the hemispheres that changes with season (higher concentrations in the summer hemisphere), perhaps reflecting the dependence of oceanic production on the availability of sunlight (Weiss *et al.* 1995).

4.4.3 Carbonyl disulfide

The lifetime of carbonyl disulfide (CS_2) is of the order of a week, where the major loss process is reaction with the OH radical ($k_{298} = 1.16 \times 10^{-12}$ cm^3 molecule^{-1}s^{-1}: De More *et al.* 1997), which was shown above to yield SO_2 and OCS as major products. The reaction of OCS with the OH radical is slow, with CS_2 another relatively insoluble species, where $K_H = 54.95 \times 10^{-3}$ mol dm^{-3} atm^{-1} at 25°C (DeBruyn *et al.* 1995), and therefore wet deposition is unlikely to be important. Watts (2000) has suggested that oxic soils may also be an important sink for CS_2. However, these estimates are based on a relatively limited amount of data and it is possible that this figure is overestimated. Anthropogenic sources such as chemical processing dominate for CS_2, being over 50% of the estimated source, most notably in the production of cellulose. Rotting organic matter in oceans, soils, and marshes is thought to be the main natural source, with possible contributions from wetlands and anoxic soils. The significantly shorter lifetime for CS_2, compared with OCS, results in a highly nonuniform distribution in the troposphere, with very little penetration into the stratosphere expected. Observed levels range from 2 p.p.t. under clean marine conditions up to approximately 300 p.p.t. in areas heavily influenced by anthropogenic emissions (Bandy *et al.* 1993).

4.4.4 Hydrogen sulfide

Hydrogen sulfide (H_2S) reacts rapidly with the OH radical to form the HS radical ($k_{298} = 4.7 \times 10^{-12}$ cm^3 molecule^{-1}s^{-1}: De More *et al.* 1997), giving rise to a lifetime of just two to three days.

The HS radical is in turn rapidly oxidized in the atmosphere to SO_2. Although the precise details of the mechanism are uncertain, the suggested route is summarized by reactions 4.25 to 4.27.

$$H_2S + OH \rightarrow HS + H_2O \qquad (4.25)$$

$$HS + O_3 \rightarrow HSO + O_2 \qquad (4.26)$$

$$HSO + O_3 \rightarrow SO + OH + O_2 \qquad (4.27)$$

$$SO + O_2 \rightarrow SO_2 + O \qquad (4.21)$$

$$SO + NO_2 \rightarrow SO_2 + NO \qquad (4.22)$$

The budget of H_2S is the least well characterized of all the sulfur compounds considered here. However, it is widely agreed that H_2S is the major reduced sulfur compound released from soils and vegetation and is also a significant component of the marine budget (Möller 1984). Like the other sulfur compounds considered thus far it is quite insoluble ($K_H \sim 95 \times 10^{-3}\,mol\,dm^{-3}\,atm^{-1}$ at 25°C: DeBruyn *et al.* 1995), and in conjunction with available seawater measurements (Andreae 1990) it would appear that H_2S is supersaturated in seawater. Hydrolysis of OCS appears to be the main source of H_2S in seawater (Elliot *et al.* 1989), although there is evidence suggesting that H_2S formation is related to primary production in the oceans (Radford-Knoery & Cutter 1994). Earlier in this chapter it was noted that decomposition of organic matter by bacteria would lead to H_2S production, and estuaries, mudflats, salt marshes, and swamps are all active H_2S emission regions for this reason (e.g. Bates *et al.* 1992). Higher plants are known to release H_2S when the enzyme HPCA acts on OCS instead of CO_2 (Protoschill-Krebs *et al.* 1996) and may be the cause of H_2S emissions from lichens, for example (Gries *et al.* 1994). A clear link has been established between OCS assimilation in plants and H_2S emissions (Bartell *et al.* 1993), but Materna (1966) and Rennenberg (1991) have noted that H_2S can also be emitted by plants as a detoxification mechanism, i.e. to remove excess sulfur. Hence H_2S can also be emitted independent of ambient OCS levels. Another major natural source of H_2S is of course volcanoes: some OCS and CS_2 is also released but H_2S and SO_2 are the major sulfur species emitted from this source. Like CS_2, anthropogenic sources of H_2S, such as the combustion of fossil fuel, dominate, contributing nearly half of the total burden. Although the removal of OCS by OH is a small loss process for OCS, it is believed to lead to the formation of H_2S and is a minor contributor to the H_2S budget, but provides an *in situ* atmospheric source. The short lifetime of H_2S inevitably means that its distribution in the troposphere is highly variable, where marine levels can vary between 5 and 100 p.p.t., rising to as high as many hundreds of p.p.t. in wetland regions (Brasseur *et al.* 1999).

4.4.5 Dimethyl sulfide

Dimethyl sulfide (CH_3SCH_3), commonly known as DMS, is the most abundant natural sulfur compound emitted into the atmosphere. Haas first discovered the release of DMS from phytoplankton in 1935 and in 1948 Challenger and Simpson showed that DMS was generated from the Zwitter ion dimethyl-sulfonio-propionate (DMSP), also known as dimethyl-β-propiothetin. It is believed that DMSP is produced by phytoplankton as an osmoregulating substance, which is released by grazing from zooplankton, cell leakage, sensescence, or viral infection (Watts 2000). Once in the water, DMSP can be metabolized by the enzyme DMSPase, found intracellularly and in DMS-producing bacteria (Ansede *et al.* 2001), leading to DMS production. DMS is insoluble ($K_H \sim 4.74 \times 10^{-1}\,mol\,dm^{-3}\,atm^{-1}$ at 25°C and 32.5 salinity units: DeBruyn *et al.* 1995), and therefore degasses from the water column. Although other sources of DMS have been identified, such as from vegetation, production from the oceans is believed to dominate. Estimates of the oceanic source have been wide ranging, from around 10 Tg per year up to 110 Tg per year (Watts 2000). Two approaches have been used to estimate the oceanic flux, one using seawater measurements and a mass transfer coefficient from sea to air (Liss & Slater 1974), the other using air concentrations of DMS and its residence time.

Both methods have their flaws. For the former the actual mass transfer coefficient used will depend on which tracer, such as CO_2 or Rn, the measurement is based on (e.g. Liss & Merlivat 1986) and will vary with wind speed, for which several parameterizations exist (e.g. Wanninkhof 1992). Air–sea transfer is also affected by the composition of the microlayer at the interface, which is poorly understood. For the latter approach a sure knowledge of the residence time for DMS is required, while both methods require a compilation of many seawater and atmospheric observations over the whole globe throughout the year. The residence time of DMS is assumed to be very short— of the order of a day—as it is rapidly removed by reaction with both OH ($k_{298} = 6.6 \times 10^{-12}\,cm^3$ molecule^{-1} s^{-1}: Atkinson *et al.* 1992) and the NO_3 radical ($k_{298} = 1.1 \times 10^{-12}\,cm^3$ molecule^{-1} s^{-1}: Atkinson *et al.* 1992). However, removal of DMS via reaction with halogen radicals such as Cl or BrO has been speculated and may well be a significant additional loss processes in the marine environment (e.g. James *et al.* 2000). The estimate for DMS production adopted here (Watts 2000) is based on the latter method of atmospheric measurements and an assessment of residence times. If additional loss processes do exist then this estimate will of course be a lower limit. The oxidation mechanism for DMS is extremely complex and has been the focus of numerous studies.

Reaction of DMS with OH is thought to proceed via two channels, abstraction 4.28a and addition 4.28b:

$$CH_3SCH_3 + OH \rightarrow CH_3SCH_2 + H_2O \quad (4.28a)$$

$$CH_3SCH_3 + OH + M \rightarrow CH_3S(OH)CH_3 + M \quad (4.28b)$$

The subsequent fate of the CH_3SCH_2 radical and $CH_3S(OH)CH_3$ adduct in the atmosphere is an area of considerable debate, with new laboratory studies constantly refining the assumed mechanism. Here, the mechanism adopted is a simple one based on the work of Jenkin (1996), and assumes that peroxy radicals react with NO only (see Chapter 6 for a fuller discussion on the reactions of

peroxy radicals). CH_3SCH_2 is therefore oxidized in the following manner:

$$CH_3SCH_2 + O_2 + M \rightarrow CH_3SCH_2O_2 + M \quad (4.29)$$

$$CH_3SCH_2O_2 + NO \rightarrow CH_3SCH_2O + NO_2 \quad (4.30)$$

$$CH_3SCH_2O \rightarrow CH_3S + HCHO \quad (4.31)$$

$$CH_3S + NO_2 \rightarrow CH_3SO + NO \quad (4.32)$$

$$CH_3SO + NO_2 \rightarrow CH_3SO(O) + NO \quad (4.33)$$

The $CH_3SO(O)$ radical, now referred to as CH_3SO_2, is thought to be the common intermediate for both the abstraction and addition mechanisms leading to the degradation of DMS. The oxidation of the $CH_3S(OH)CH_3$ adduct may proceed as follows:

$$CH_3S(OH)CH_3 + O_2 \rightarrow CH_3S(O)CH_3 + HO_2 \quad (4.34)$$

where $CH_3S(O)CH_3$ is a stable compound dimethyl sulfoxide (DMSO). DMSO can itself react with OH:

$$CH_3S(O)CH_3 + OH + M$$
$$\rightarrow CH_3SO(OH)CH_3 + M \quad (4.35)$$

$$CH_3SO(OH)CH_3 + O_2 \rightarrow CH_3SO(O)CH_3 + HO_2 \quad (4.36)$$

where $CH_3SO(O)CH_3$ is the stable compound dimethyl sulfone (DMSO$_2$). DMSO$_2$ can itself react with OH:

$$CH_3SO(O)CH_3 + OH \rightarrow CH_3SO(O)CH_2 + H_2O \quad (4.37)$$

$$CH_3SO(O)CH_2 + O_2 + M$$
$$\rightarrow CH_3SO(O)CH_2O_2 + M \quad (4.38)$$

$$CH_3SO(O)CH_2O_2 + NO$$
$$\rightarrow CH_3SO(O) + HCHO + NO_2 \quad (4.39)$$

Hence $CH_3SO(O)$ or CH_3SO_2 is formed once more. The fate of CH_3SO_2 in the atmosphere determines whether SO_2 or $CH_3SO(O)OH$, known as methane sulfonic acid (MSA), is formed. CH_3SO_2 can thermally decompose

$$CH_3SO(O) + M \rightarrow CH_3 + SO_2 \qquad (4.40)$$

to produce SO_2 directly. However, it can also react with either NO_2 or O_3 to form $CH_3SO(O)O$ or CH_3SO_3, which is the precursor to MSA formation

$$CH_3SO(O) + NO_2 \text{ (or } O_3)$$
$$\rightarrow CH_3SO(O)O + NO \text{ (or } O_2) \qquad (4.41)$$

$$CH_3SO(O) + O_3 \rightarrow CH_3SO(O)O + O_2 \qquad (4.42)$$

$$CH_3SO(O)O + HCHO$$
$$\rightarrow CH_3SO(O)OH + HCO \qquad (4.43)$$

The importance of the oxidation of SO_2 to H_2SO_4 has been discussed in detail. H_2SO_4 can act as a condensation nuclei, whereas, MSA tends to stick to existing aerosol surfaces and does not lead to the formation of new particles (Brasseur *et al.* 1999). A summary of the DMS oxidation scheme is depicted in Fig. 4.5. The distribution of DMS reflects the fact that it is so short lived and has a strong biological source from the oceans. Therefore DMS concentrations are highest around high-productivity coastal and open ocean regions, and display a strong seasonal cycle, reflecting the seasonal pattern of phytoplankton growth and decay. Since DMS is also so short lived, it also displays a very strong diurnal cycle. Some measured values for boundary layer DMS concentrations are given in Table 4.9.

4.4.6 *The CLAW hypothesis*

In 1987 a link between phytoplankton and cloud albedo was proposed. In their seminal paper, Charlson *et al.* suggested that phytoplankton are able to modify the climate through the generation of clouds via CCN (cloud condensation nuclei) derived from the oxidation of DMS emissions from

Table 4.9 Measured concentrations of DMS at various locations within the marine boundary layer.

Location	Minimum concentration $(10^{-9}\,g\,m^{-3})$	Reference
Antarctic	806	Shooter *et al.* (1992)
Arctic	830	Ferek *et al.* (1995)
Atlantic (North)	152	Andreae *et al.* (1985)
Atlantic (Tropical)	177	Andreae *et al.* (1985)
Pacific (North)	47	Andreae *et al.* (1985)
Pacific (South)	335	Andreae *et al.* (1985)
Pacific (Tropical)	301	Andreae *et al.* (1985)

the ocean. Over the oceans the numbers of sea-salt particles produced are insufficient to account for the abundance of CCN at cloud height in the marine boundary layer (MBL). Numerous studies have suggested that the main source of CCN in the MBL must be from non-sea-salt sulfate (NSSS). Size distribution measurements of NSSS (Andreae 1982) show that they are appropriate for supersaturation between 0.1 and 1%, which is correct for marine clouds (relative humidities of between 100.3 and 101%). Other studies, such as correlation between mass concentration of NSSS and the CCN population, and correlation between CCN and NSSS lifetimes and light scattering data, further support the notion that NSSS is the major CCN in the MBL (see Charlson *et al.* 1987). Since the dominant NSSS source in the MBL is from DMS it seems likely that DMS is responsible for the generation of CCN. This is supported by the fact that there is a seasonal relationship in CCN number and density, and cloudiness, which in turn is related to the times when phytoplankton are producing DMS. The CLAW hypothesis could provide a route by which the biosphere may offset human activities and provide a negative climate feedback.

4.5 THE GLOBAL HALOGEN CYCLE

The halogens chlorine, bromine, and iodine are present in the Earth's atmosphere in both

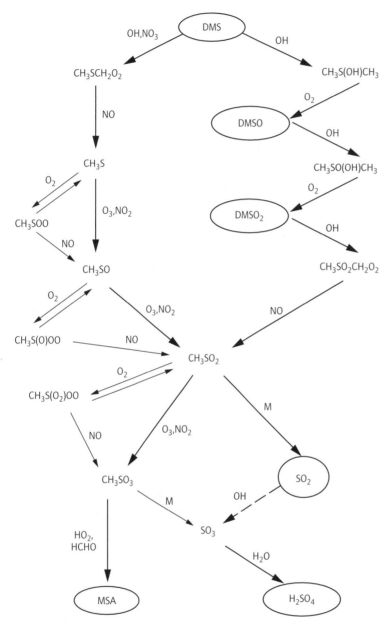

Fig. 4.5 A simplified DMS oxidation scheme.

inorganic and organic forms. Natural ecosystems use halogen-containing compounds for myriad purposes (Gribble 1994), while the use of halogen-containing species, such as chlorofluorocarbons (CFCs), hydrochlorofluorocarbons (HCFCs), hy-

droflurocarbons (HFCs), and halons (containing bromine as well as fluorine and possibly chlorine) in a variety of industrial applications has led to a dramatic increase in the halogen burden in the atmosphere. In Chapter 7, the impact of the CFCs and their replacement compounds on stratospheric ozone is discussed in detail, and hence it is only briefly mentioned here. The total atmospheric budget for the halogens is far from complete. However, in this section the sources, sinks, and lifetimes of the major known species are presented.

4.5.1 Organohalogens

Gschwend and MacFarlane (1986) have found that two distinct pools of natural halocarbons exist, with two different production mechanisms, namely monohalomethanes and polyhalomethanes. The synthesis of monohalomethanes in nature occurs via the enzyme-catalyzed methylation of halide ions, where the enzyme is methyl transferase. The enzymatic synthesis of CH_3Cl and CH_3Br through an S-adenosyl methionine (SAM) transfer mechanism has been observed in marine species such as red macroalgae (Harper 1995) and a number of terrestrial organisms. White rot fungi (Harper 1995), brassica leaves (Attieh *et al.* 1995), and higher plants (Saini *et al.* 1995) have all been shown to be capable of biosynthesizing monohalomethanes in this way. Polyhalomethanes are believed to be produced by the successive enzymatic halogenation of ketone metabolites (Beissner *et al.* 1981), catalyzed by haloperoxidase mechanisms. In seawater, $CHBr_3$ is the dominant product of the haloperoxidase reaction, although other products (e.g. $CHBr_2Cl$, $CHBrCl_2$, and $CHCl_3$) are formed due to the replacement of Br by Cl.

A variety of studies have begun to elucidate the potential roles of organohalogens in natural ecosystems. Nightingale *et al.* (1995) showed that levels of CH_3I, CH_2Br_2, $CHBr_3$, $CHBr_2Cl$, $CHBrCl_2$, and $CHCl_3$ were elevated in beds of the marine macroalgae *Laminaria digitata*. Release rates were influenced by partial desiccation, light availability, tissue age, tissue wounding, and grazing. These compounds may therefore act as antimicrobial agents or grazing deterrents, since the herbivory of marine macroalgae is intense and is often the primary factor affecting their distribution and abundance. The simpler organohalogens may act as intermediates in the synthesis of more complex halogenated antigrazing compounds, or could result from the breakdown of such compounds following plant death. In soils (Laturnus 1995) very high haloperoxidase activity has been reported in the surface organic layer, where high microbial activity (e.g. fungi and bacteria) is known to be involved in the degradation of organic matter. The function of haloperoxidase activity in soil may be intra- or extracellular formation of organohalogens for the chemical defense of the producing microorganisms. Alternatively, haloperoxidases could be released by microorganisms in order to produce hypohalites. Formation of hypohalites *in situ* in the soil will cause oxidation of the refractory organic matter in the organic layer to produce more soluble and accessible organic compounds, but also organohalogens such as chloroform ($CHCl_3$). Harper (1985) showed that methyl halide formation occurs in fungi and in further studies Harper and coworkers (1990) showed that CH_3Cl was an effective precursor to the biosynthesis of veratryl alcohol (3,4-dimethoxybenzyl alcohol). Veratryl alcohol is a secondary metabolite which is biosynthesized by many white rot fungi and plays a central role in lignin degradation. A further possibility for the role of CH_3Cl is associated with the high methoxyl content of lignin, which can be as much as 20% (Chen & Chang 1995). The diversion of excess one-carbon fragments from this source into volatile CH_3Cl that will rapidly diffuse from the cell may represent a means of overcoming the biochemical handicap likely to be imposed by a very large one-carbon pool in the cell. The sources, sinks, and residence times of a range of organohalogens will now be discussed.

4.5.2 Chlorofluorocarbons

Table 4.10 shows the uses of the major chlorofluorocarbons (CFCs) and other chlorinated halocarbons prior to controls on their use introduced by the Montreal Protocol. Originally thought to be inert in the atmosphere, CFCs are now known to lead to the destruction of stratospheric ozone. Provided that the implementation of the Montreal Protocol and subsequent legislation is adhered to, the stratospheric abundance of halogenated ozone-

Table 4.10 Historical applications of CFCs (prior to controls over use).

Sector	Compound	Application
Aerosols	CFC-11 ($CFCl_3$) CFC-12 (CF_2Cl_2) CFC-114 ($C_2F_4Cl_2$) HCFC-22 (CHF_2Cl)	Used as pure fluids or mixtures for aerosol propellants
Foam-blowing	CFC-11	Blowing polyurethane and phenolic foams
	CFC-12	Blowing polystyrene and polyethylene foams
Solvents	CFC-113 ($C_2F_3Cl_3$) CH_3CCl_3 CFC-11 CCl_4	Solvents used for electronics, precision cleaning, and dry cleaning
Refrigeration and air-conditioning	CFC-11 CFC-12 HCFC-22	Used for wide range of refrigeration and air conditioning applications

depleting substances is expected to return to its pre-1980 level of 2 p.p.b. chlorine equivalent by about 2050. Total atmospheric chlorine grew at 7 p.p.t. Cl per year between mid-1992 and mid-1994. While CFC-12 (CF_2Cl_2) was still increasing at 9 p.p.t. Cl per year, CFC-11 ($CFCl_3$) showed a decline of 2 p.p.t. Cl per year, with CFC-113 ($C_2F_3Cl_3$) approximately constant over the time period (WMO 1998). Prior to the introduction of the Montreal Protocol, CFC-11 was increasing at 9.3–10.1 p.p.t. per year, CFC-12 at 16.9–18.2 p.p.t. per year and CFC-113 at 5.4–6.2 p.p.t. per year. The CFCs are destroyed by photolysis in the stratosphere or via reaction with $O(^1D)$ atoms, e.g. for CFC-11:

$$CFCl_3 + hv \rightarrow CFCl_2 + Cl \qquad (4.44)$$

$$CFCl_3 + O(^1D) \rightarrow CFCl_2 + ClO \qquad (4.45)$$

Reaction 4.45 may well produce other products but the ClO radical is the dominant product. The effectiveness of such species in destroying ozone, i.e. their ozone depletion potential, is discussed in Chapter 7. Residence times for these compounds

are long: 50 years for CFC-11, 102 years for CFC-12, and 110 years for CFC-113 (WMO 1998).

4.5.3 *"Replacement" compounds: HCFCs and HFCs*

The phase-out of CFC production generated an immediate need to find replacements to meet the continued demand. However, scope for direct substitution is limited, and some historical CFC requirements are being superseded by alternative technology, such as the use of pump-action aerosols or the use of hydrocarbons as refrigeration (McCulloch 1994). Alternative replacement compounds must not only match the properties needed to provide the specific effect desired but must also fulfill other criteria, such as energy efficiency, safety (nontoxic and nonflammable), and environmental acceptability.

HCFCs were introduced as an interim CFC replacement (Oram *et al.* 1996), since the presence of a labile C–H bond allows the compound to be oxidized via reaction with the hydroxyl radical in the troposphere. Tropospheric removal via reaction with OH considerably shortens the atmospheric residence time of the HCFC compared with the CFC it replaces. HCFC-22 (CHF_2Cl) is currently the most widely used HCFC and has a history of use in the refrigeration sector. Production of HCFC-22 increased steadily from $56 \times 10^6 \, kg \, yr^{-1}$ in 1970 to $245 \times 10^6 \, kg \, yr^{-1}$ in 1992 (WMO 1998). Analyses of archived air samples collected in Tasmania between 1978 and 1996 (Miller 1998) indicated a mixing ratio of 116.7 p.p.t. in mid-1996, increasing at 6 p.p.t. per year. The abundance at this site in 1978 was suggested to be approximately 35 p.p.t. HCFC-22 has a residence time of 11.5 years, which is still long and permits some penetration of this compound into the stratosphere.

HCFC-141b (CH_3CFCl_2) and HCFC-142b (CH_3CF_2Cl) are used predominantly as replacement compounds in foam-blowing and solvent cleaning. Industrial production of HCFC-141b and HCFC-142b increased markedly in the late 1980s and early 1990s. By 1995 nearly three times as much HCFC-141b was sold into dispersive uses as HCFC-142b. Sales figures were $113 \times 10^6 \, kg \, yr^{-1}$

for HCFC-141b and $39 \times 10^6 \, \text{kg yr}^{-1}$ for HCFC-142b. Data from a global ground-based flask sampling network indicates that the mean global tropospheric mixing ratio of HCFC-141b increased from 0.7 p.p.t. in mid-1993 to 3.5 p.p.t. in mid-1995, with the 1995 rate of increase estimated at 1.9 p.p.t. per year (Montzka *et al.* 1996). The residence times are 9.4 years for HCFC-141b and 19.5 years for HCFC-142b.

4.5.4 Methyl chloroform

Methyl chloroform (CH_3CCl_3) is used primarily as a cleaning solvent. Industrial sales have declined rapidly in recent years. In 1995 global sales of $220 \times 10^6 \, \text{kg}$ were less than one-third of the amount reported in 1990, the year of peak production (McCulloch & Midgley 1996). Total consumption of CH_3CCl_3 in developing countries was approximately $30 \times 10^6 \, \text{kg}$ in 1994. Phase-out in the developed world occurred in 1996, with developing world phase-out set for 2015 (WMO 1998). Data from two ground-based global sampling networks (AGAGE and NOAA/CMDL) have been updated, indicating dramatic declines in the global tropospheric abundance of CH_3CCl_3 (Prinn *et al.* 1995; Montzka *et al.* 1996). The rate of decline of methyl chloroform since 1991 was between 13 and 14 p.p.t. per year by mid-1996. Interestingly, measurements from the two networks disagreed by about 10%, with estimates ranging from 89 (AGAGE) to 97 p.p.t. (NOAA/CMDL). As industrial sources of CH_3CCl_3 become less significant, sources such as biomass burning will become significant contributors to the global budget and may well give rise to differences in measurements. The residence time of CH_3CCl_3 is 5.4 years, where reaction with OH is the dominant loss process, but uptake by the oceans may also be a nonnegligible loss. Prinn *et al.* (1995) have used CH_3CCl_3 measurements to infer global OH concentrations.

4.5.5 Carbon tetrachloride

In 1931, carbon tetrachloride (CCl_4) began to be used as a chemical intermediate for the production of CFCs, with usage increasing rapidly from the 1950s to the 1980s. Since the introduction of the Montreal Protocol, which listed CCl_4 as a controlled substance, large-scale production has declined rapidly, and been accelerated by the global phase-out of CFC production, which consumed 80–90% of the total CCl_4 production. Atmospheric concentrations of CCl_4 measured at the five remote AGAGE surface stations (Barbados, Tasmania, Oregon, American Samoa, and Ireland) reached a peak in 1989–90 of 104.4 p.p.t. and have since been decreasing by 0.7 p.p.t. a year (Simmonds *et al.* 1998). CCl_4 is removed by photodissociation in the stratosphere, where it eventually yields phosgene ($COCl_2$). The ocean is also a sizable sink for CCl_4, with approximately 20% of atmospheric CCl_4 being consumed by ocean mixing and hydrolysis (Wallace *et al.* 1994). The combined loss processes lead to a residence time for CCl_4 of 42 years.

4.5.6 Halons (1211 and 1301)

Halon-1211 (CF_2BrCl) and Halon-1301 (CF_2Br_2) are very effective fire extinguishers. Halon-1211 is used mainly in mobile hand-held equipment, whereas Halon-1301 tends to be employed in fixed installation: for example, to protect computers or flammable storage areas from fire, or in aircraft applications. Halons act by destroying hydrogen atoms which sustain combustion

$$CF_2BrCl + H \rightarrow CF_2Cl + HBr \qquad (4.46)$$

$$HBr + H \rightarrow H_2 + Br \qquad (4.47)$$

$$Br + H \rightarrow HBr \qquad (4.48)$$

Halon production in the developed world was banned in 1994, under the Montreal Protocol, two years earlier than the CFCs, due to the exceptionally high ODP (ozone depletion potential) of these brominated species. Hence the rate of increase in atmospheric abundance of halons has decreased, and because no suitable replacement has been approved there is now an emphasis on conservation of existing stocks and recycling. Halons can enter the atmosphere either through leakage from installations or through direct emission in firefighting, and although production is banned, the use of existing equipment is not, so emissions are expected to continue for the next 20 years. China now produces around 90% of the world's halon-1211, and global emissions of halon-1211 have been found to be 50% greater than previously thought. Halon-1211 is now believed to be responsible for up to 20% of global ozone destruction. Photolysis is the major loss process for both halons, giving rise to residence times of 20 years (Halon-1211) and 65 years (Halon-1301).

4.5.7 Dichloromethane

Dichloromethane (CH_2Cl_2) is a man-made solvent, and is used in a wide range of applications, such as the decaffeination of coffee and paint stripping. McCulloch and Midgely (1996) estimated that the annual global emission was 592×10^6 kg for 1988 and 527×10^6 kg for 1996, with over 90% of this emission occurring in the northern hemisphere. Northern hemisphere and southern hemisphere values of 40–50 p.p.t. and 15–20 p.p.t. respectively were obtained during cruises in the Atlantic and Pacific (Atlas *et al.* 1993; Koppmann *et al.* 1993). There is no direct evidence for substantial natural sources, although atmospheric measurements of background concentrations in the southern hemisphere are double those that have been calculated using known emission fields. Gribble (1994) notes that CH_2Cl_2 has been observed in the oceans from marine algae and may well have a seasonal maritime source. The residence time for CH_2Cl_2 is around 0.4 years, with reaction with OH being the dominant loss process, where $k_{298} = 1.15 \times 10^{-13}$ cm^3 molecule^{-1} s^{-1} (DeMore *et al.* 1997).

4.5.8 Chloroform

Chloroform ($CHCl_3$) is known to have both anthropogenic and natural atmospheric sources, which are approximately of equal magnitude. Anthropogenic emissions arise from the use of $CHCl_3$ as a chemical intermediate in industrial processes, as a solvent, as a secondary product during chlorination processes, and from coal combustion and

waste incineration. Chlorination processes that lead to the production of chloroform include sewage treatment, paper pulp bleaching, and the chlorination of drinking water. Aucott *et al.* (1999) estimated that industrial emissions amounted to 60 Gg per year, of which paper manufacture contributed about 30 Gg per year, and waste water treatment a further 20 Gg per year. Natural production of chloroform appears to be very widespread among different biota, and known sources include rice fields (Khalil *et al.* 1990), soil ecosystems (Hoekstra & de Leer 1995), termites (Khalil *et al.* 1990), fungi (Hoekstra *et al.* 1998), and moss (Gribble 1994), although marine algae are believed to represent the dominant natural source. The average fresh-weight $CHCl_3$ production rates from macroalgae were 12 ng g^{-1} day^{-1} for kelps and 8 ng g^{-1} day^{-1} for nonkelp species (Nightingale *et al.* 1995). The yearly global production of $CHCl_3$ is therefore estimated to be 0.17 Gg for kelps and 0.06 Gg for nonkelps. The total estimated yearly release by macroalgae is just 0.23 Gg. However, release is species-dependent. If release rates by the common kelp *L. saccharina* were typical, the global flux to the atmosphere from macroalgae would be the same order of magnitude as the total estimated emissions from global oceans. Biomass burning is also another significant source of chloroform (Graedel & Keene 1995). Elkins *et al.* (1996) reported a seasonal cycle for $CHCl_3$ in phase with CH_2Cl_2, PCE (C_2Cl_4), and CH_3Cl, where northern hemisphere concentrations were 10–15 p.p.t. and southern hemisphere concentrations were 5–7 p.p.t. The residence time for chloroform is about six months, where reaction with the OH radical is the dominant loss process ($k_{298} = 1.0 \times 10^{-13}$ cm^3 molecule^{-1} s^{-1}: DeMore *et al.* 1997).

4.5.9 Perchloroethene and trichloroethene

Perchloroethene (PCE, C_2Cl_4) and trichloroethene (TCE, C_2HCl_3) are used as industrial solvents and degreasers. Industrial emission estimates (McCulloch & Midgely 1996) suggest that source regions are broadly similar to CH_2Cl_2. There has been a suggestion that certain types of macroalgae in the subtropical oceans may be capable of synthesizing chlorinated alkenes such as TCE and PCE (Abrahamsson *et al.* 1995a, b). However, Scarratt and Moore (1999) detected no production of TCE or PCE from any cultures of the red microalgae *Porphyridium purpureum*, used in the work of Abrahamsson *et al.* Tropospheric measurements made by Quack and Seuss (1999) on a western Pacific cruise in September 1994 suggested a natural source of TCE in the area of the Indonesian archipelago, as did TCE and PCE ocean water supersaturations measured on a cruise in 1981 between 40°N and 30°S (Khalil *et al.* 1998). However, measurements made on the Polarstern cruise across the Atlantic Ocean between 45°N and 30°S in August and September 1989 found that the PCE and TCE concentrations were highest between 40 and 45°N (Koppmann *et al.* 1993). Furthermore there was a distinct interhemispheric gradient. Such observations indicate that TCE and PCE are not well mixed and that the dominant source regions are located within the northern hemisphere.

Consumption of PCE has been declining steadily since the 1970s with the introduction of more efficient cleaning equipment, which has enabled a greater proportion of solvent to be recovered and recycled rather than released into the environment. The primary loss process for both species in the atmosphere is reaction with the OH radical, where residence times are 96 days for PCE ($k_{298} = 1.72 \times 10^{-13}\,\text{cm}^3\,\text{molecule}^{-1}\,\text{s}^{-1}$: DeMore *et al.* 1997) and 5 days for TCE ($k_{298} = 2.2 \times 10^{-12}\,\text{cm}^3\,\text{molecule}^{-1}\,\text{s}^{-1}$: DeMore *et al.* 1997). Ocean supersaturation of PCE and TCE have led to the suggestion that the ocean is a source of these compounds. However, the TCE and PCE profiles with depth were unlike those for CH_3I or isoprene, showing increases down to 500 m. Moore (1999) postulated that seasonal variations in PCE and TCE could drive a wintertime flux from the atmosphere to the oceans and a reverse flux in the summer.

4.5.10 Chloromethane or methyl chloride

Chloromethane or methyl chloride (CH_3Cl) is by far the most abundant organohalogen in the atmosphere, with man-made emissions being negli-gible in comparison with natural ones (Harper 1994). As CFC levels have risen, so the relative contribution of CH_3Cl to the total stratospheric loading has decreased from greater than 80% to less than 25%. A short residence time (1–1.5 years) following reaction with OH ($k_{298} = 3.76 \times 10^{-14}\,\text{cm}^3\,\text{molecule}^{-1}\,\text{s}^{-1}$: DeMore *et al.* 1997), and the absence of an interhemispheric gradient is consistent with widespread natural sources, not concentrated within the industrialized northern hemisphere (Rasmussen 1977). Khalil and Rasmussen (1999) have used a model to estimate that a global source of about 3.7 Tg per year is needed to explain observed atmospheric levels. Including model uncertainty widens the range from 2.8 to 4.6 Tg per year. Anthropogenic sources identified include biomass burning and other combustion processes. CH_3Cl is formed from Cl^- during combustion of vegetation, such as in forest fires and slash-and-burn agriculture. Mixing ratios as high as 2 p.p.b. have been observed in forest fire smoke. Andreae *et al.* (1996) estimated that biomass burning produced 1000 Gg per year. Current evidence would tend to indicate that the overwhelming proportion of the atmospheric CH_3Cl burden is of direct biological origin, and biosynthesis has been demonstrated in a wide variety of organisms, such as polypore fungi (Watling & Harper 1998), macroalgae (Laturnus *et al.* 1998), and higher plants (Saini *et al.* 1995).

Studies by Singh *et al.* (1983) and Rasmussen *et al.* (1980) concluded that elevated mixing ratios of CH_3Cl in the boundary layer typified air masses of marine origin, consistent with the idea of a widespread natural flux from the oceans. Photochemical production via the reaction of Cl^- with CH_3I in seawater (Zafirou 1974) has been suggested, but a failure to find a strong relationship between these two compounds in seawater has tended to negate this idea.

Manley and Dastoor (1987) estimated that the giant kelp *Macrocystos*, a dominant primary producer of Californian coastal waters, released CH_3Cl at a rate of 160 ng g^{-1} per day. However, assuming that this release rate observed is typical of all marine macroalgae, the global standing crop of 58×10^6 tonnes can at most be responsible for the production of only 3000 tonnes of CH_3Cl per year, an insignificant contribution to the total global flux.

The first direct evidence of CH_3Cl release by phytoplankton cultures was obtained by Moore and coworkers (Tait *et al.* 1994; Moore *et al.* 1995), who demonstrated emission by nonaxenic cultures of both warm-water and cold-water diatoms. Overall rates of biosynthesis were very low and when scaled to phytoplankton abundance in the oceans they cannot account for more than 20,000 tonnes per year: 0.5% of the global flux. Extrapolation from measurements in the northwest Atlantic and Pacific oceans suggests emission of CH_3Cl at lower latitudes from warmer waters (400–600 $Gg yr^{-1}$), but uptake at higher latitudes by colder waters (100–300 $Gg yr^{-1}$), yielding a global net oceanic source strength of 200–400 $Gg yr^{-1}$. Based on these estimates, the net oceanic emission would only account for some 7–13% of the sources needed to balance the calculated sink.

Large quantities of gaseous CH_3Cl can be released during the growth of certain wood-rotting fungi and such emissions may therefore make an important contribution to the atmospheric burden. CH_3Cl acts as a methyl donor in the biosynthesis of certain methyl esters and O-methyl phenols by lignin-degrading fungi. Hutchinson (1971) and Cowan *et al.* (1973) initially discovered that CH_3Cl was released from several species of *Phellinus*, a widespread group of bracket fungi. *Phellinus* species are widely distributed in both temperate and tropical regions of the world, showing vegetative growth not only on wood, but on plant litter and even on soil organic matter. Even in situations where environmental concentrations of chloride are relatively low, significant amounts of CH_3Cl can be generated because of the high affinity of the methylating system for halide ion. If a significant proportion of atmospheric CH_3Cl arises from fungi, the extensive global deforestation that has occurred over the past 200 years may have significantly altered the flux of CH_3Cl into the environment, and thus altered the prevailing concentration of CH_3Cl in the atmosphere.

Higher plants represent a possible terrestrial source of CH_3Cl. A survey of 60 cultivars of potato has indicated that CH_3Cl emissions by tubers are as high as 590 $ng g^{-1}$ day^{-1} within 48 hours of harvest (Varns 1982). Saini *et al.* (1995) presented evidence that plants can produce CH_3Cl, CH_3Br, CH_3I, and CH_3SH through the methyltransferase reaction discussed above, which may provide a way to eliminate halide ions, which are known to be phytotoxic. The highest activities were found in brassicas (cabbage family). Leaf disks were also observed to emit approximately tenfold more CH_3I than roots. High concentrations of organohalogens in rain falling through spruce trees (Asplund & Grimvall 1991) have led to the assertion that biohalogenation in the needles themselves may be taking place. Although reaction with OH is the major loss process for CH_3Cl in the troposphere, photolysis becomes dominant in the stratosphere. CH_3Cl is the main natural chlorine carrier to the stratosphere.

4.5.11 *Methyl bromide*

Atmospheric methyl bromide (CH_3Br) is derived from both anthropogenic and biogenic sources. The budget of CH_3Br has been the subject of intense debate recently and it is still not fully resolved. CH_3Br is believed to be the main carrier of bromine to the stratosphere, where bromine is extremely efficient at promoting ozone destruction (see Chapter 7). Current mean global tropospheric background concentrations are around 10 p.p.t. (Shauffler *et al.* 1998), with a north–south interhemispheric gradient of 1.3 p.p.t. (Grosko & Moore 1998). The suggestion is therefore that there are either greater sources in the northern hemisphere or greater sinks in the southern hemisphere. The oceans are now believed to be a net sink for tropospheric CH_3Br, although parts are a source. Early studies suggested the oceans were a net source (Singh *et al.* 1983). However, Lobert *et al.* (1997) derived a net flux of −21 Gg per year from data collected on two extensive cruises and one regional cruise in the Southern Ocean. Grosko and Moore (1998) derived a net flux of −10 Gg per year with data from cruises of the Pacific and northwest Atlantic oceans. They observed undersaturations in cold, high-latitude waters and some supersaturated waters in the Atlantic Gulf stream and around 35° S in the South Pacific. It has been noted that if the atmospheric concentration of CH_3Br dropped to much lower than 7 p.p.t. (Elkins *et al.* 1996) then the oceans would become a net source. It is extremely important to determine the magnitude of the oceanic flux in order to regulate industrial use of CH_3Br effectively; loss to the oceans is a significant sink and reduces the residence time of CH_3Br and hence affects the calculation of its ozone depleting potential (ODP).

CH_3Br is used as a pesticide in the cultivation of high-value agricultural produce, although its use is being phased out as part of the Montreal Protocol. CH_3Br is decomposed in soils at a rate of 5–10% a day via processes of hydrolysis and demethylation and it has been calculated that around 42% of the CH_3Br applied may be ultimately released to the atmosphere. Soils themselves are a sink for CH_3Br, and Shorter *et al.* (1995) have estimated a total sink of around $42 \, Gg \, yr^{-1}$. Equatorial regions are the strongest sinks for CH_3Br. Vegetation may be the important missing source suggested by Lee-Taylor *et al.* (1998). Saini *et al.* (1995) and Gan *et al.* (1998) have shown that CH_3Br is produced by vegetables, oil crops, pastures, and ornamental crops. Rapeseed alone could be responsible for up to $7 \, Gg \, yr^{-1}$ of CH_3Br production. Rice plants may also be an important source of CH_3Br (Muramatsu & Yoshida 1995). Chen *et al.* (1999) have estimated that around $5 \, Gg$ of CH_3Br is generated from vehicle emissions, and there is the suggestion that vehicles running on leaded gasoline produce more than unleaded. Mano and Andreae (1994) have shown that biomass burning is another potentially large source of CH_3Br, although the current uncertainties in the emission rates are large. The combined residence time of CH_3Br, allowing for loss to the oceans and soils and by reaction with the OH radical ($k_{298} = 2.98 \times 10^{-14} \, cm^3 \, molecule^{-1} \, s^{-1}$: DeMore *et al.* 1997), is currently estimated to be 0.7 years.

4.5.12 *Bromoform and other brominated species*

The anthropogenic source of bromoform ($CHBr_3$)

and bromochloromethanes is primarily related to water treatment and disinfection. Gschwend *et al.* (1985) estimated a production rate of organic bromine compounds from water chlorination of $0.6 \, Gg \, Br \, yr^{-1}$ from seawater and $4 \, Gg \, yr^{-1}$ from fresh water. Emission of ethylene dibromide from gasoline additives and fumigation has been significant in past decades, but there are less emissions today due to regulation. Bromoform is normally the major compound produced by macroalgae and microalgae (Sturges *et al.* 1992) and has been observed in polar ice microalgae in both the Arctic and the Antarctic. These microalgae colonize the underside of sea ice and are highly photoadapted to low light levels, and undergo an intense bloom during the spring. Unlike the macrophytes, they are not restricted to shallow water and narrow coastal margins but are believed to inhabit the underside of all annual and pack ice in the polar seas — an area of $7 \times 10^6 \, km^2$ in the Arctic and $16 \times 10^6 \, km^2$ in the Antarctic (6% of the world ocean). A bromoform rich layer at the top of the sea ice, under the snow pack, has been observed and is believed to slowly leak into the atmosphere. Satellite observations of tropospheric air masses enriched in BrO are always situated close to sea ice (Wagner & Platt 1998) and typically extend over areas of about $300–2000 \, km^2$.

The loss of bromoform from seawater can follow different mechanisms, such as hydrolysis, bacterial degradation, bioaccumulation, adsorption onto particles followed by sedimentation, and evaporation to the atmosphere. Chemical degradation by hydrolysis is a slow process in cold seawater but increases dramatically in warmer water. Loss of $CHBr_3$ is dominated by photolysis in the tropopause region, with OH radical initiated degradation playing a minor role ($k_{298} = 1.5 \times 10^{-13} \, cm^3 \, molecule^{-1} \, s^{-1}$: DeMore *et al.* 1997). The overall residence time for $CHBr_3$ is around 36 days. Interestingly, Dvortsov *et al.* (1999) have shown that $CHBr_3$ is likely to contribute more organic bromine than all the conventional long-lived sources (halons and CH_3Br) to the lower most stratosphere, where the observed mid-latitude ozone trend maximizes.

4.5.13   Methyl iodide

Although present at low levels (0.2–5 p.p.t.), methyl iodide (CH_3I) is believed to be the main gaseous iodine species in the troposphere and as such plays an important role in the natural iodine cycle. The C–I bond is easily cleaved and CH_3I has the longest residence time of any organoiodine compound. CH_3I is photolyzed, giving rise to a residence time of between 2 and 10 days depending on location and season. CH_3I is reported to be almost exclusively of marine origin (Chameides & Davis 1980), with some minor contribution from biomass burning (Andreae *et al.* 1996), rice paddies (Redeker *et al.* 2000), peatland ecosystems, and wetlands (Dimmer *et al.* 2000). There have been no enhanced levels of CH_3I observed in urban areas (e.g. Singh *et al.* 1983), but it has been suggested as a replacement fumigant for CH_3Br. Rasmussen *et al.* (1982) observed that CH_3I levels exhibited a seasonal cycle, with highest levels in summer and early autumn, and was found to be more abundant in the Tropics and over oceanic regions of high biomass productivity. Happell and Wallace (1996) suggested that in the warm tropical surface waters, subject to high solar irradiance, photochemical production of CH_3I would be significant. Such an assertion is backed up by positive saturation anomalies (1.5–7.7 pmol kg^{-1}), indicating a sea–air flux. This contrasted with negative saturation anomalies (–0.65 pmol kg^{-1}) measured in cold surface waters of the open ocean around Greenland, subject to low light levels. Therefore, high-latitude oceans may be a significant sink during autumn and winter. Photochemical production of CH_3I would go some way to explaining the lack of correlation between CH_3I and CH_3Cl, CH_3Br, CH_2Br_2, $CHBr_3$, $CHBr_2Cl$, and chlorophyll. Kelp, bacteria (Manley & Dastoor 1988), phytoplankton (Manley & de la Cuesta 1997), and macroalgae (Giese *et al.* 1999) are all believed to be small but significant sources of oceanic CH_3I.

4.5.14   Other organoiodine compounds

Atmospheric measurements of other organoiodine compounds are sparse. A number of short-lived organoiodine species have been observed, including C_2H_5I, CH_2ICl, CH_2IBr, and CH_2I_2, whose residence times are approximately four days, a day, an hour, and a few minutes respectively. Carpenter *et al.* (1999) measured an extensive suite of alkyl iodides and bromides at Mace Head in Ireland. Positive correlations were between $CH_2IBr/CHBr_3$ and CH_3I/C_2H_5I, which was interpreted as signifying common or linked marine sources. The rapid photolysis of these compounds can lead to IO formation (Alicke *et al.* 1999) via the reaction of iodine radicals with ozone. Measurements of these organoiodine compounds (Klick & Abrahamsson 1992; Moore & Tokarczyk 1993) in coastal and open oceans suggest that the organoiodine flux to the atmosphere may be considerably underestimated if CH_3I alone is considered. However, due to the short residence times, it is believed that very small amounts of iodine precursors enter the stratosphere (Wennberg *et al.* 1997). Table 4.11 summarizes the current best estimates for the budgets of CH_3Cl, CH_3Br and CH_3I.

It is worth noting that several thousand naturally occurring organohalogen compounds have been identified in aquatic and terrestrial systems. The edible seaweed *Asparagopsus taxiformis*, for example, contains over 100 such organohalogen compounds (Moore 1977), including mixed chlorinated, brominated, and iodinated ketones and alkenes. They are used as feeding deterrents, as irritants, as a pesticide, in food gathering, and possibly as antifreeze in types of thermophiles. Whether there is a significant flux of these compounds to the atmosphere is still unknown.

4.5.15   Inorganic halogens

The single most important direct emission of inorganic halogens arises from the volatilization of sea-salt aerosol. In the case of chlorine, HCl is liberated (Graedel & Keene 1995):

$$NaCl(s) + HNO_3(g) \rightarrow HCl(g) + NaNO_3(s) \tag{4.49}$$

$$2NaCl(s) + H_2SO4(g) \rightarrow 2HCl(g) + Na_2SO_4(s) \tag{4.50}$$

Table 4.11 Estimated budgets for CH_3Cl, CH_3Br and CH_3I.

	CH_3Cl ($Tg\,yr^{-1}$)	CH_3Br ($Gg\,yr^{-1}$)	CH_3I ($Gg\,yr^{-1}$)
Sources			
Oceans	3 (±1)	56 (5–130)	
Kelp			0.2–0.6
Bacteria			300
Phytoplankton			1.2
Photochemical			200–500
Terrestrial			
Fumigation, soils		26.5 (16–48)	
Fumigation, durables		6.6 (4.8–8.4)	
Fumigation, perishables		5.7 (5.4–6.0)	
Fumigation, structures		2	
Gasoline		5 (0–10)	
Industrial	0.3 (±0.1)		
Biomass burning	0.5 (±0.2)	20 (10–40)	3.4
Wetlands		4.6	7.3
Salt marshes		14 (7–29)	
Plants, rapeseed		6.6 (4.8–8.4)	
Rice fields		1.5 (0.5–2.5)	20–71
Fungus		1.7 (0.5–5.2)	
Total	3.7 (±1.3)	151 (56–290)	~700
Sinks			
Oceans	0.3 (±0.6)	77 (37–133)	
Reaction with Cl^-			~300
OH and *hv*	3.3 (±1)	86 (65–107)	~400
Soils		46.8 (32–154)	
Total	3.6 (±1.6)	210 (134–394)	~700
Typical mixing ratio (p.p.t.)	650	10	2

Source: Singh *et al.* (1983), Manley and Dastoor (1987, 1988), Graedel and Keene (1995), Muramatsu and Yoshida (1995), Shorter *et al.* (1995), Blake *et al.* (1996), Manley and de la Cuesta (1997), Gan *et al.* (1998), WMO (1998), Moore and Grosko (1999), Varner *et al.* (1999), Dimmer *et al.* (2000), Keppler *et al.* (2000), Redeker *et al.* (2000), Rhew *et al.* (2000), N. Bell, Harvard (personal communication).

Direct volcanic emission can also be an important source for HCl. For bromine, it is believed that N_2O_5 can initiate the release of bromine nitrite, which is itself rapidly photolyzed, yielding Br atoms (Finlayson-Pitts *et al.* 1990)

$$NaBr(s) + N_2O_5(g)$$
$$\rightarrow BrNO_2(g) + NaNO_3(s) \quad (4.51)$$

$$BrNO_2 + hv \rightarrow Br + NO_2 \quad (4.52)$$

Further sources of inorganic bromine involve hydrolysis of bromine nitrate on sulfate aerosol (Fan & Jacob 1992). Mechanisms for the generation of X_2 or XY, where X and Y are Cl, Br, or I, has been speculated by a number of workers. Oum *et al.* (1998) has shown that on photolysis of ozone in the presence of sea-salt particles, Cl_2 can be released. It is argued that photolysis of ozone in the gas phase gives rise to the formation of H_2O_2, which can be taken up in the aqueous phase, where its photolysis leads to the formation of OH in the aqueous phase. OH radicals can then oxidize Cl^-:

$$OH(aq) + Cl^-(aq) \rightarrow HOCl^-(aq) \qquad (4.53)$$

$$HOCl^- + H^+ \rightarrow Cl + H_2O \qquad (4.54)$$

$$Cl + Cl^- \rightarrow Cl_2^- \qquad (4.55)$$

$$Cl_2^- + Cl_2^- \rightarrow Cl_2 + 2Cl^- \qquad (4.56)$$

Very high levels of Cl_2 (300 p.p.t.) have been reported by Spicer *et al.* (1998), and may well arise via ozone mediated release. Sander and Crutzen (1996) and Vogt *et al.* (1996) developed a mechanism for the release of BrCl in the aqueous phase:

$$HOBr(aq) + Cl^- + H^+ \rightarrow BrCl(aq) + H_2O \qquad (4.57)$$

Since BrCl is only sparingly soluble it can transfer into the gas phase, where it is rapidly photolyzed to yield Br and Cl atoms. Vogt *et al.* (1999) have also modeled the analogous reaction for aqueous HOI, yielding aqueous ICl, which will also preferentially partition into the gas phase:

$$HOI(aq) + Cl^- + H^+ \rightarrow ICl(aq) + H_2O \qquad (4.58)$$

Vogt *et al.* (1999) also note that since the analogous reaction involving Br^- with HOI is calculated to be two orders of magnitude faster than reaction 4.58, the formation of IBr may also be important, i.e.

$$HOI(aq) + Br^- + H^+ \rightarrow IBr(aq) + H_2O \qquad (4.59)$$

Hence, if HOI levels are elevated, they may induce the release of Br and Cl from sea salt. Organoiodine species such as CH_2I_2, CH_2ICl, and CH_2IBr are known to be photolyzed rapidly (Rattigan *et al.* 1997; Mössinger *et al.* 1998), releasing iodine atoms, which can rapidly form HOI in the gas-phase by the following reaction sequence:

$$I + O_3 \rightarrow IO + O_2 \qquad (4.60)$$

$$IO + HO_2 \rightarrow HOI + O_2 \qquad (4.61)$$

Hence it is speculated that the emission of short-lived organoiodine precurors can lead to an increase in inorganic halogen, particularly within the marine boundary layer (McFiggans *et al.* 2000) via aerosol processing of HOI.

In general, once released, halogen atoms will react with O_3 (X is Cl, Br, and I)

$$X + O_3 \rightarrow XO + O_2 \qquad (4.62)$$

forming XO. XO can react with itself, where multiple product channels are possible:

$$XO + XO \rightarrow X + X + O_2 \qquad (4.63a)$$

$$XO + XO + M \rightarrow XOOX + M \qquad (4.63b)$$

$$XO + XO \rightarrow OXO + X \qquad (4.63c)$$

$$XO + XO \rightarrow X_2 + O_2 \qquad (4.63d)$$

However, at 298 K and at surface pressure (1 bar) channel 4.63b dominates for Cl, channel 4.63a dominates for Br, and channel 4.63a represents around 60% of the total reaction for I, with channel 4.63c appearing to represent the remainder, forming OIO (DeMore *et al.* 1997). The molecule OIO can also be formed by cross reaction between IO and BrO (DeMore *et al.* 1997). Considerable interest surrounds OIO, which has been observed in the marine boundary layer (Stutz *et al.* 1999; Allan *et al.* 2001), although its chemistry is uncertain at this time. XO can also form temporary reservoir species, HOX and $XONO_2$, via reaction with HO_2 and NO_2:

$$XO + HO_2 \rightarrow HOX + O_2 \qquad (4.64)$$

$$XO + NO_2 + M \rightarrow XONO_2 + M \qquad (4.65)$$

For Br and particularly I, these reservoirs are quite short-lived in the troposphere, being either taken up into aerosol or destroyed by photolysis in the case of $IONO_2$. The role of halogen reservoirs in the stratosphere is discussed in Chapter 7. XO can also react with NO to regenerate X:

$$XO + NO \rightarrow X + NO_2 \qquad (4.66)$$

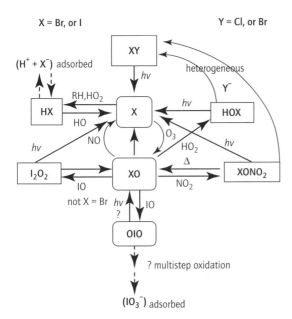

X = Br, or I   Y = Cl, or Br

Fig. 4.6 Cycling of the halogens in the troposphere.

summary of inorganic halogen chemistry is shown in Fig. 4.6.

4.6 CONCLUSIONS

The biogeochemical cycling of the elements carbon, nitrogen, sulfur, chlorine, bromine, and iodine has been investigated, with the main sources and their residence times discussed. For all elements the natural flux is substantial, although additional sources derived from human activities dominate for certain reservoir species. The complex interactions between climate and the cycling of these elements has been noted, and further research is required before we will be able to predict with any confidence how these cycles may be perturbed by a changing climate.

Chlorine can react with hydrocarbons RH to form HCl, which limits its effectiveness in destroying ozone in the troposphere, but Br can only abstract hydrogens from labile species such as aldehydes and HO_2, although addition to unsaturated species is also possible. Iodine atoms are unable to react with any hydrocarbons and reaction with ozone is by far the dominant channel in the troposphere. Iodine chemistry is therefore the most likely to lead to significant ozone destruction in the troposphere, although, under perturbed conditions, such as in the boundary layer in the Arctic spring, bromine chemistry can initiate rapid surface ozone destruction (e.g. Finlayson-Pitts *et al.* 1990). It is worth noting that, once released, fluorine can react with O_3, but on reaction with RH forms the very stable HF. Since release of fluorine from its major reservoir compounds, such as the CFCs, HCFCs, HFCs, and PFCs (perfluorocarbons), occurs in the stratosphere, HF is transported back down to the troposphere, where it is removed by wet deposition. The lifetime of inorganic halogens and their reservoirs is highly variable and depends strongly on the aerosol loading present. A simple

ACKNOWLEDGMENTS

The authors would like to thank the following people for helpful discussions that aided the preparation of this manuscript: Drs Carlos Canosa-Mas, Glenn Carver, Gordon McFiggans, Graham Nickless, and Carl Percival, and Professor John Pyle.

REFERENCES

Abrahamsson, K., Ekdahl, A., Collen, J., Falstrom, E. & Pedersen, M. (1995a) The natural formation of trichloroethylene and perchloroethylene in sea water. In Grimvall, A. & de Leer, E.W.B. (eds), *Naturally Produced Organohalogens*.

Abrahamsson, K., Ekdahl, A., Collen, J. & Pedersen, M. (1995b) Marine-algae: a source of trichloroethylene and perchloroethylene. *Limnology and Oceanography* **40**, 1321–6.

Ajtay, G.L., Ketner, P. & Duvigneaud, P. (1979) Terrestrial primary production and phytomass. In Bolin, B., Degens, E.T., Kempe, S. & Ketner, P. (eds), *The Global Carbon Cycle*. Wiley, Chichester.

Alicke, B., Hebestreit, K., Stutz, J. & Platt, U. (1999) Iodine monoxide in the marine boundary layer. *Nature* **397**, 572–3.

Allan B.J., McFiggans, G., Plane, J.M.C. & Coe, H. (2001) Observations of OIO in the remote marine boundary layer. *Geophysical Research Letters* **28**, 1945–8.

Andreae, M.O. (1982) Marine aerosol chemistry at Cape Grim, Tasmania and Townsville, Queensland. *Journal of Geophysical Research* **87**, 8875–85.

Andreae, M.O. (1990) Ocean-atmosphere interactions in the global biogeochemical sulphur cycle. *Marine Chemistry* **30**, 1–29.

Andreae, M.O., Ferek, R.J., Bermond, F. *et al.* (1985) DMS in the marine atmosphere. *Journal of Geophysical Research* **90**, 12,891–900.

Andreae, M.O., Atlas, E., Harris, G.W. *et al.* (1996) Methyl halide emissions from savanna fires in southern Africa. *Journal of Geophysical Research* **101**, 23,603–13.

Aneja, V.P. (1990) Natural sulfur emissions into the atmosphere. *Journal of the Air and Waste Management Association* **40**, 469–76.

Ansede, J.H., Friedman, R. & Yoch, D.C. (2001) Phylogenetic analysis of culturable dimethyl sulfide-producing bacteria from a Spartina-dominated salt marsh and estuarine water. *Applied Environmental Microbiology* **67**, 1210–17.

Asplund, G. & Grimvall, A. (1991) Organohalogens in nature. *Environmental Science and Technology* **25**, 1346–50.

Atkinson, R., Baulch, D.L., Cox, R.A., Hampson, R.F., Kerr, J.A. & Troe, J. (1992) Evaluated kinetic and photochemical data for atmospheric chemistry, Supplement IV: IUPAC subcommittee on gas kinetic data evaluation for atmospheric chemistry. *Journal of Physical Chemistry Reference Data* **21**, 1125–568.

Atlas, E., Pollock, W., Greenberg, J., Heidt, L. & Thompson, A.M. (1993) Alkyl nitrates, nonmethane hydrocarbons, and halocarbon gases over the equatorial Pacific-Ocean during SAGA-3. *Journal of Geophysical Research* **98**, 16,933–47.

Attieh, J.M., Hanson, A.D. & Saini, H.S. (1995) Purification and characterisation of a novel methyl transferase responsible for biosynthesis of halomethanes and methanethiol in Brassica oleracea. *Journal of Biological Chemistry* **270**, 9250–7.

Aucott, M.L., McCulloch, A., Graedel, T.E., Kleiman, G., Midgely, P.M. & Li, Y-F. (1999) Anthropogenic emissions of trichloromethane and chlorodifluoromethane: Reactive Chlorine Emissions Inventory. *Journal of Geophysical Research* **104**, 8405–15.

Baker, J.T., Allen, L.H., Boote, K.J. & Pickering, N.B. (1996) Rice responses to drought under carbon dioxide enrichment. *Global Change Biology* **2**, 101–10.

Bandy, A.R., Thornton, A.C. & Jonson, J.E. (1993) CS_2 measurements in the atmosphere of the western North Atlantic and Northwestern South Atlantic Oceans. *Journal of Geophysical Research* **98**, 23,449–57.

Barnes, I., Becker, K.H. & Patroescu, I. (1994) The tropospheric oxidation of DMS: a new source of OCS. *Geophysical Research Letters* **21**, 2389–92.

Barnola, J.M., Raynaud, D., Korotkevich, Y.S. & Lorius, C. (1987) Vostok ice core provides 160,000 year record of atmospheric CO_2. *Nature* **329**, 408–14.

Bartell, U., Hoffman, U., Hoffman, R., Kreuzburg, B. & Andreae, M.O. (1993) OCS and H_2S fluxes over a wet meadow in relation to photosynthetic activity: an anlysis of measurements made on 6th September 1990. *Atmospheric Environment* **27A**, 1851–64.

Bates, T.S., Lamb, B., Guenther, A., Dignon, A. & Stoiber, R.E. (1992) Sulfur emissions to the atmosphere from natural sources. *Journal of Atmospheric Chemistry* **14**, 315–17.

Beissner, R.S., Guildford, W.J., Coates, R.M. & Hager, L.P. (1981) Synthesis of brominated heptanones and bromoform by a bromoperoxidase of marine origin. *Biochemistry* **20**, 3724–31.

Berresheim, H. & Vulcan, V.D. (1992) Vertical distributions of COS, CS_2, DMS, and other sulfur compounds in a loblolly pine forest. *Atmospheric Environment* **26**, 2031–6.

Belvisio, S., Mihalopoulos, N. & Nguyen, B. (1987) The supersaturation of OCS in rainwaters. *Atmospheric Environment* **21**, 1361–7.

Blake, D.R. & Rowland, F.S. (1988) Continuing worldwide increase in tropospheric methane, 1978–1987. *Science* **239**, 1129–31.

Blake, N.J., Blake, D.R., Sive, B.C. *et al.* (1996) Biomass burning emissions and vertical distributions of atmospheric methyl halides and other reduced carbon gases in the South Atlantic region. *Journal of Geophysical Research* **101**, 24,151–64.

Brasseur, G.P., Orlando, J.J. & Tyndall, G.S. (1999) *Atmospheric Chemistry and Global Change*. Oxford Univeristy Press, New York.

Carpenter, L.J., Sturges, W.T., Penkett, S.A. *et al.* (1999) Short-lived alkyl iodides and bromides at Mace Head, Ireland: links to biogenic sources and halogen oxide production. *Journal of Geophysical Research* **104**, 1679–89.

Challenger, F. & Simpson, M.I. (1948) Studies on biological methylation. Part XII. A precursor of dimethyl sulphide evolved by Polysiphonia fastigiata. Dimethyl-2-carboxyethylsulphonium hydroxide and its salts. *Journal of the Chemical Society* 2, 1591–7.

Chameides, W.L. & Davis, D.D. (1980) Iodine: its possible role in tropospheric photochemistry. *Journal of Geophysical Research* 85, 7383–98.

Charlson, R.J., Lovelock, J.E., Andreae, M.O. & Warren, S.G. (1987) Oceanic phytoplankton, atmospheric sulfur, cloud albedo and climate. *Nature*, 326, 655–61.

Chen, C.L. & Chang, H.M. (1995) Chemistry of lignin degradation. In Higuchi, T. (ed.), *Biosynthesis and Biodegradation of Wood Components*. Academic Press, Orlando, FL.

Chen, T.-Y., Blake, D.R., Lopez, J.P. & Rowland, F.S. (1999) Estimation of global vehicular methyl bromide emissions: Extrapolation from a case study in Santiago, Chile. *Geophysical Research Letters* 26, 283–6.

Chin, M. & Davis, D.D. (1993) Global sources and sinks of OCS and CS_2, and their distribution. *Global Biogeochemical Cycles* 7, 321–37.

Constable, J.V.H., Guenther, A.B., Schimel, D.S. & Monson, R.K. (1999) Modelling changes in VOC emission in response to climate change in the continental United States. *Global Change Biology* 5, 791–806.

Cowan, M.I., Glen, A.T., Hutchinson, S.A., MacCartney, M.E., MacKintosh, J.M. & Moss, A.M. (1973) Production of volatile metabolites by species of fomes. *Transactions of the British Mycological Society* 60, 347–360.

DeBruyn, W.J., Swartz, E., Hu, J.H., Shorter, J.A. & Davidovits, P. (1995) Henry's law solubilities and Setchenow coefficients for biogenic reduced sulfur species obtained from gas–liquid uptake measurements. *Journal of Geophysical Research* 100, 7245–51.

DeMore, W.B., Sander, S.P., Golden, D.M. *et al.* (1997) Chemical kinetics and photochemical data for use in stratospheric modeling. Evaluation 12. *JPL Publications* 97–4.

Dimmer, C.H., Simmonds, P.G., Nickless, G. & Bassford, M.R. (2000) Biogenic fluxes of halomethanes from Irish peatland ecosystems. *Atmospheric Environment* 35, 321–30.

Dvortsov, V.L., Geller, M.A., Solomon, S., Schauffler, S.M., Atlas, E.L. & Blake, D.R. (1999) Rethinking reactive halogen budgets in the midlatitude lower stratosphere. *Geophysical Research Letters* 26, 1699–702.

Elkins, J.W., Butler, T.M., Thompson, A.M. *et al.* (1996) Nitrous oxide and halocompounds. In Hofmass, D.J., Peterson, J.T. & Rosson, R.M. (eds), *Climate Monitoring and Diagnostics Laboratory No. 23 Summary Report 1994–1995*. US Department of Commerce, Boulder, CO.

Elliot, S., Lu, E. & Rowland, S. (1989) Rates and mechanisms for the hydrolysis of OCS in natural waters. *Environmental Science and Technology* 23, 458–61.

Fall, R. (1999) Biogenic emissions of volatile organic compounds from higher plants. In Hewitt, C.N. (ed.), *Reactive Hydrocarbons in the Atmosphere*. Academic Press, San Diego.

Fan, S.M. & Jacob, D.J. (1992) Surface ozone depletion in Arctic spring sustained by bromine reactions on aerosols. *Nature* 359, 522–4.

Ferek, R.J., Hobbs, P.V., Radke, L.F., Herring, J.A., Sturges, W.T. & Cota, G.F. (1995) Dimethyl sulfide in the arctic atmosphere. *Journal of Geophysical Research* 100, 26,093–104.

Finlayson-Pitts, B.J., Livingston, B.J. & Berko, H.N. (1990) Ozone destruction and bromine photochemistry at ground level in the Arctic spring. *Nature* 343, 622–5.

Fried, A.J., Drummond, R., Henry, B. & Fox, J. (1991) Versatile integrated tunable diode laser system for ambient measurements of OCS. *Applied Optics* 30, 1916–32.

Fried, A., Henry, B., Ragazzi, R.A. *et al.* (1992) Measurements of OCS in automotive exhausts and an assessment of its importance to the global sulfur cycle. *Journal of Geophysical Research* 97, 14,621–34.

Gan, J., Yates, S.R., Ohr, H.D. & Sims, J.J. (1998) Production of methyl bromide by terrestrial higher plants. *Geophysical Research Letters* 25, 3595–8.

Giese, B., Laturnus, F., Adama, F.C. & Wiencke, C. (1999) Release of volatile iodinated C_1–C_4 hydrocarbons by marine macroalgae from various climate zones. *Environmental Science and Technology* 33, 2432–9.

Graedel, T.E. & Crutzen, P.J. (1993) *Atmospheric Change: An Earth System Perspective*. W.H. Freeman, New York.

Graedel, T. & Keene, W. (1995) Tropospheric budget of reactive chlorine. *Global Biogeochemical Cycles* 9, 47–77.

Gribble, G.W. (1994) Natural organohalogens. *Chemistry in Britain* 71, 907–11.

Gries, C., Nash, T.H. III & Kesselmeier, J. (1994) Exchange of reduced sulfur gases between lichens and the atmosphere. *Biogeochemistry* 26, 25–39.

Grosko, W. & Moore, R.M. (1998) Ocean–atmosphere exchange of methyl bromide: NW Atlantic and Pacific Ocean studies. *Journal of Geophysical Research* 103, 16,737–41.

Gschwend, P.M., MacFarlane, J.K. & Newman, K.A. (1985) Volatile halogenated organic compounds released to seawater from temperate marine macroalgae. *Science* **227**, 1033–5.

Gschwend, P.M. & MacFarlane, J.K. (1986) Polybromomethanes—a year round study of their release to seawater from ascophyllum nodosum and focus vesiculosis. *Organic Marine Chemistry*, ACS series, 314–22.

Guenther, A., Hewitt, C.N., Erickson, D. *et al.* (1995) A global-model of natural volatile organic-compound emissions. *Journal of Geophysical Research* **100**, 8873–92.

Haas, P. (1935) The liberation of methyl sulphide by seaweed. *Biochemistry Journal* **29**, 1297–9.

Happell, J.D. & Wallace, D.W.R. (1996) Methyl iodide in the Greenland/Norwegian Seas and the tropical Atlantic Ocean: evidence for photochemical production. *Geophysical Research Letters* **103**, 2105–8.

Harper, D.B. (1985) Halomethane from halide ion—a highly efficient fungal conversion of environmental significance. *Nature* **315**, 55–7.

Harper, D.B. (1994) Biosynthesis of halogenated methanes. *Biochemical Society Transactions* **2**, 1007–11.

Harper, D.B. (1995) *Naturally-produced Organohalogens*. Kluwer Academic Publishers, Dordrecht.

Harper, D.B., Buswell, J.A., Kennedy, J.T. & Hamilton, J.T.G. (1990) Chloromethane, methyl donor in veratryl alcohol biosynthesis in Phanerochaete chrysosporium and other lignin degrading fungi. *Applied Environmental Microbiology* **56**, 3450–7.

Hatanaka, A. (1993) The biogeneration of odour by green leaves. *Phytochemistry* **34**, 1201–18.

Hoekstra, E.J. & de Leer, E.W.B. (1995) Organohalogens—the natural alternatives. *Chemistry in Britain* **31**, 127–31.

Hoekstra, E.J., de Leer, E.W.B. & Brinkman, U.A.T. (1998) Natural formation of chloroform and brominated trihalomethanes in soil. *Environmental Science and Technology* **32**, 3724–9.

Hoffmann, M.R. & Jacob, D.J. (1984) Kinetics and mechanisms of the catalytic oxidation of dissolved sulfur dioxide in aqueous solution: an application to nighttime fog water chemistry. In Calvert, J.G. (ed.), *SO₂, NO and NO₂ Oxidation Mechanisms*. Butterworth, London.

Horowitz, L.W., Liang, J.Y., Gardner, G.M. & Jacob, D.J. (1998) Export of reactive nitrogen from North America during summertime: sensitivity to hydrocarbon chemistry. *Journal of Geophysical Research* **103**, 13,451–76.

Hutchinson, S.A. (1971) Biological activity of volatile fungal metabolites. *Transactions of the British Mycological Society* **57**, 185–200.

IPCC (1996) *Climate Change 1995: The Science of Climate Change*. Cambridge University Press, Cambridge.

James, J.D., Harrison, R.M., Savage, N.H. *et al.* (2000) Quasi-Lagrangian investigation into dimethyl sulfide oxidation in martime air using a combination of measurements and model. *Journal of Geophysical Research* **105**, 26,379–92.

Jenkin, M.E. (1996) Chemical mechanism forming condensable material. AEA Technology report, AEA/RAMP/20010010/002.

Keppler, F., Elden, R., Niedan, V., Pracht, J. & Schöler, H.F. (2000) Halocarbons produced by natural oxidation processes during degradation of organic matter. *Nature* **403**, 298–301.

Kesselmeier, J. & Merk, L. (1993) Exchange of OCS between agricultural plants and the atmosphere: studies on the deposition of OCS to peas, corn and rapeseed. *Biogeochemistry* **23**, 47–59.

Khalil, M.A.K. & Rasmussen, R.A. (1999) Atmospheric methyl chloride. *Atmospheric Environment* **33**, 1305–21.

Khalil, M.A.K., Rasmussen, R.A., Wang, M.X. & Ren, L. (1990) Emissions of trace gases from Chinese rice fields and biogas generators—CH_4, N_2O, CO, CO_2, chlorocarbons, and hydrocarbons. *Chemosphere* **20**, 207–26.

Khalil, M.A.K., Rasmussen, R.A., Shearer, M.J., Chen, Z.-L., Yao, H. & Yang, J. (1998) Emissions of methane, nitrous oxide and other trace gases from rice fields in China. *Journal of Geophysical Research* **103**, 25,241–50.

Klick, S. & Abrahamsson, K. (1992) Biogenic volatile iodated hydrocarbons in the ocean. *Journal of Geophysical Research* **97**, 12,683–7.

Koppmann, R., Johnen, F.J., Plass-Dulmer, C. & Rudolph, J. (1993) Distribution of methyl chloride, dichloromethane, trichloroethene and tetrachloroethene over the North and South Atlantic. *Journal of Geophysical Research* **98**, 20,517–26.

Laturnus, F. (1995) Release of volatile halogenated organic compounds by unialgal cultures of polar macroalgae. *Chemosphere* **31**, 3387–95.

Laturnus, F., Adams, F.C. & Wiencke, C. (1998) Methyl halides from Antarctic macroalgae. *Geophysical Research Letters* **25**, 773–6.

Lee-Taylor, J.M., Doney, S.C., Brasseur, G.P. & Muller,

J.-F. (1998) A global three-dimensional atmosphere–ocean model of methyl bromide distributions. *Journal of Geophysical Research* **103**, 16,039–57.

Liss, P.S. & Merlivat, L. (1986) Air–sea gas exchange rates: introduction and synthesis. In Menard, P.B. (ed.), *The Role of Air–Sea Exchange in Geochemical Cycling*. D. Riedall. Norwell.

Liss, P.S. & Slater, P.G. (1974) Flux of gases across the air–sea interface. *Nature* **247**, 181–4.

Lobert, J.M., Yvon-Lewis, S.A., Butler, J.H., Montzka, S.A. & Myers, R.C. (1997) Undersaturation of CH_3Br in the Southern Ocean. *Geophysical Research Letters* **24**, 171–2.

McCulloch, A. (1994) Sources of hydrochlorofluorocarbons, hydrofluorocarbons and fluorocarbons and their potential emissions during the next 25 years. *Environmental Monitoring Assessement* **31**, 167–74.

McCulloch, A. & Midgely, P.M. (1996) The production and global distribution of emissions of trichloroethene, tetrachlororethene and dichloromethane over the period 1988–1992. *Atmospheric Environment* **30**, 167–74.

McFiggans, G., Plane, J.M.C., Allan, B.J., Carpenter, L.J., Coe, H. & O'Dowd, C. (2000) A modelling study of iodine chemistry in the marine boundary layer. *Journal of Geophysical Research* **105**, 14,371–85.

McKee, I.F. & Woodward, F.I. (1994) CO_2 enrichment responses of wheat: interactions with temperature, nitrate and phosphate. *New Phytologist* **127**, 447–53.

Manley, S.L. & Dastoor, M.N. (1987) Methyl halide production from the giant kelp, Macrocystis and estimates of global CH_3X production by kelp. *Limnology and Oceanography* **32**, 709–15.

Manley, S.L. & Dastoor, M.N. (1988) Methyl iodide production by kelp and associated microbes. *Marine Biology* **98**, 477–82.

Manley, S.L. & de la Cuesta, J.L. (1997) Methyl iodide production from marine phytoplankton cultures. *Limnology and Oceanography* **42**, 142–7.

Mano, S. & Andreae, M.O. (1994) Emission of methyl bromide from biomass burning. *Science* **263**, 1255–8.

Materna, M. (1966) Die ausschiedung des durch die fichtennadeln absorbierten schwefeldioxids. *Archiv für das Forstwesen* **15**, 691–2.

Miller, B.R. (1998) Abundances and trends of atmospheric chlorodifluoromethane and bromomethane. PhD thesis, Scripps Institute of Oceanography, University of California.

Möller, D. (1984) On the global natural sulfur emission. *Atmospheric Environment* **18**, 29–39.

Montzka, S.A., Butler, J.H., Myers, J.C. *et al.* (1996) Decline in the tropospheric abundance of halogen from halocarbons: implications for stratospheric ozone depletion. *Science* **272**, 1318–22.

Moore, R.E. (1977) Volatile compounds from marine algae. *Accounts of Chemical Research* **10**, 40–7.

Moore, R.M. (1999) On the fluxes of short-lived organochlorine compounds between the ocean and atmosphere. *IUGG XXII General Assembly Abstracts*, A110.

Moore, R.M. & Groszko, W. (1999) Methyl iodide distribution in the ocean and fluxes to the atmosphere. *Journal of Geophysical Research – OCEANS* **104**(C5), 11,163–71.

Moore, R.M. & Tokarczyk, R. (1993) Volatile biogenic halocarbons in the northwest Atlantic. *Global Biogeochemical Cycles* **7**, 195–210.

Moore, R.M., Tokarczyk, R., Tait, V.K., Poulin, M. & Green, C. (1995) Marine phytoplankton as a natural source of volatile organohalogens. In Grimvall, A. & de Leer, E.W.B. (eds), *Naturally Produced Organohalogens*.

Mopper, K. & Zika, R.G. (1987) Free amino acids in marine rains: evidence for oxidation and potential role in nitrogen cycling. *Nature* **325**, 246.

Mössinger, J.C., Shallcross, D.E. & Cox, R.A. (1998) UV-visible absorption cross-sections and atmospheric lifetimes of CH_2Br_2, CH_2I_2 and CH_2BrI. *Journal of the Chemical Society Faraday Transactions* **94**, 1391–6.

Moya, T.B., Ziska, L.H., Namuco, O.S. & Olszyk, D. (1998) growth dynamics and genotypic variations in tropical, field-grown paddy rice (*Oryza sativa* L.) in response to increasing carbon dioxide and temperature. *Global Change Biology* **4**, 657–66.

Muramatsu, Y. & Yoshida, S. (1995) Volatilization of methyl iodide from the soil plant system. *Atmospheric Environment* **29**, 21–5.

Nguyen, B.C., Mihalopoulos, N., Putard, J.P. & Bonsang, B. (1995) OCS emissions from biomass burning in the Tropics. *Journal of Atmospheric Chemistry* **22**, 55–65.

Nightingale, P.D., Malin, G. & Liss, P.S. (1995) Production of chloroform and other low-molecular weight halocarbons by some species of macroalgae. *Limnology and Oceanography* **40**, 680–9.

Novelli, P.C., Masarie, K.A., Tans, P. & Lang, P.M. (1994) Recent changes in atmospheric carbon monoxide. *Science* **263**, 1587–90.

Oram, D.E., Reeves, C.E., Sturges, W.T., Penkett, S.A., Fraser, P.J. & Langenfelds, R.L. (1996) Recent tropospheric gowth rate and distribution of HFC-134a (CF_3CH_2F). *Geophysical Research Letters* **23**, 1949–52.

Oum, K.W., Lakin, M.J., DeHaan, O., Brauers, T. & Finlayson-Pitts, B.J. (1998) Formation of molecular chlorine from the photolysis of ozone and aqueous sea-salt particles. *Science* **279**, 74–6.

Prinn, R.G., Weiss, R.F., Miller, B.R. *et al.* (1995) Atmospheric trends and lifetime of CH_3CCl_3 and global OH concentrations. *Science* **269**, 187–92.

Protoschill-Krebs, G., Wilhelm, C. & Kesselmeier, J. (1996) Consumption of OCS by higher plant carbonic anhydrase (CA). *Atmospheric Environment* **30**, 3151–6.

Quack, B. & Seuss, E. (1999) Volatile halogenated hydrocarbons over the Western pacific between 43° and 4° N. *Journal of Geophysical Research* **104**, 1663–78.

Radford-Knoery, J. & Cutter, G.A. (1994) Biogeochemistry of dissolved H_2S species and OCS in the western North Atlantic Ocean. *Geochimica et Cosmichimica* **58**, 5421–31.

Rasmussen, R.A. (1977) *Methyl Chloride in the Air of Kenya.* Final report, Manufacturing Chemical Association.

Rasmussen, R.A., Rasmussen, L.E., Khalil, M.A.K. & Dalluge, R.W. (1980) Concentration distribution of methyl chloride in the atmosphere. *Journal of Geophysical Research* **85**, 7350–6.

Rasmussen, R.A., Khalil, M.A.K., Gunawardena, R. & Hoyt, S.D. (1982) Atmospheric methyl iodide. *Journal of Geophysical Research* **87**, 3086–90.

Rattigan, O.V., Shallcross, D.E. & Cox, R.A. (1997) UV absorption cross-sections and atmospheric photolysis rates of CF_3I, CH_3I, C_2H_5I and CH_2ICl. *Journal of the Chemical Society Faraday Transactions* **93**, 2839–46.

Redeker, K.R., Wang, N-Y., Low, J., Macmillan, A., Tyler, S.C. & Cicerone, R. (2000) Emissions of methyl halides from a CA rice field. *Science* **290**, 966–8.

Rennenberg, H. (1991) The significance of higher plants in the emission of sulfur compounds from terrestrial ecosystems. In Sharkey, Th.D. (ed.), *Trace Gas Emission by Plants.* Academic Press, New York.

Rhew, R.C., Miller, B.R. & Weiss, R.F. (2000) Natural methyl bromide and methyl chloride emissions from coastal salt marshes. *Nature* **403**, 292–5.

Saini, H.S., Attieh, J.M. & Hanson, A.D. (1995) Biosynthesis of halomethanes and methanethiol by higher plants via a novel methyl transferase reaction. *Plant, Cell and Environment* **18**, 1027–33.

Sander, R. & Crutzen, P.J. (1996) Model study indicating halogen activation and ozone destruction in polluted air masses transported to the sea. *Journal of Geophysical Research* **101**, 9129–38.

Scarratt, M.G. & Moore, R.M. (1999) Production of chlorinated hydrocarbons and methyl iodide by the red microalga *Porphyridium purpureum. Limnology and Oceanography* **44**, 703–7.

Schauffler, S.M., Atlas, E.L., Klocke, F., Lueb, R.A., Stroud, V. & Travnicek, W. (1998) Measurements of bromine-containing organic compounds at the tropical tropopause. *Geophysical Research Letters* **25**, 317–20.

Seinfeld, J.H. (1999) Global atmospheric chemistry of reactive hydrocarbons. In Hewitt, C.N. (ed.), *Reactive Hydrocarbons in the Atmosphere.* Academic Press, San Diego.

Shooter, D., Demora, S.J., Grout, A., Wylie, D.J. & He, Z.Y. (1992) The chromatographic analysis of reduced sulfur gases in Antarctic waters following preconcentration onto Tenax. *International Journal of Environmental Analytical Chemistry* **47**, 239–49.

Shorter, J.H., Kolb, C.E., Crill, P.M. *et al.* (1995) Rapid degradation of atmospheric methyl bromide in soils. *Nature* **377**, 717–19.

Simmonds, P.G., Cunnold, D.M., Weiss, R.F. *et al.* (1998) Global trends and emissions of CCl_4 from in-situ background observations from July 1978 to June 1996. *Journal of Geophysical Research* **103**, 16,017–27.

Singh, H.B. & Zimmerman, P.R. (1992) Atmospheric distribution and sources of nonmethane hydrocarbons. In Nriagu, J.O. (ed.), *Gaseous Pollutants: Characterization and Cycling.* Wiley, New York.

Singh, H.B., Salas, L.J. & Stiles, R.E. (1983) Methyl halides in and over the eastern Pacific (40° N–32° S). *Journal of Geophysical Research* **88**, 3684–90.

Singsaas, E.L., Lerdau, M., Winter, K. & Sharkey, T.D. (1997) Isoprene increases thermotolerance of isoprene-emitting species. *Plant Physiology* **115**, 1413–20.

Spicer, C.W., Chapman, E.G., Finlayson-Pitts, B.J. *et al.* (1998) Observations of molecular chlorine in coastal air. *Nature* **394**, 353–6.

Steele, L.P., Dlugokencky, E.L., Lang, P.M., Tans, P.P., Martin, R.C. & Masarie, K.A. (1992) Slowing down of the global accumulation of atmospheric methane during the 1980s. *Nature* **358**, 313–16.

Stockwell, W.R. & Calvert, J.G. (1983) The mechanism of the HO–SO_2 reaction. *Atmospheric Environment* **17**, 2231–5.

Sturges, W.T., Cota, G.F. & Buckley, P.T. (1992) Bromoform emission from Arctic ice algae. *Nature* **358**, 660–2.

Stutz, J., Hebestreit, K., Alicke, B. & Platt, U. (1999) Chemistry of halogen oxides in the troposphere: comparison of model calculations with recent field data. *Journal of Atmospheric Chemistry* **34**, 65–85.

Tait, V.K., Moore, R.M. & Tokarczyk, R. (1994) Measurements of methyl chloride in the NW Atlantic. *Journal of Geophysical Research* **99**, 7821–33.

Varner, R.K. Crill, P.M. & Talbot, R.W. (1999) Wetlands: a potentially significant source of atmospheric methyl bromide and methyl chloride. *Geophysical Research Letters* **26**, 2433–6.

Varns, J.L. (1982) The release of methyl chloride from potato tubers. *American Potato Journal* **59**, 593–604.

Vogt, R., Crutzen, P.J. & Sander, R. (1996) A mechanism for halogen release from seasalt aerosol in the remote marine boundary layer. *Nature* **383**, 327–30.

Vogt, R., Sander, R., von Glasow, R. & Crutzen, P.J. (1999) Iodine chemistry and its role in halogen activation and ozone loss in the marine boundary layer: a model study. *Journal of Atmospheric Chemistry* **32**, 375–95.

Wagner, T. & Platt, U. (1998) Satellite mapping of enhanced BrO concentrations in the troposphere. *Nature* **395**, 486–9.

Wallace, D.W.R., Beining, P. & Putzka, A. (1994) Carbon tetrachloride and chlorofluorocarbons in the South Atlantic Ocean, 19°S. *Journal of Geophysical Research* **99**, 7803–19.

Wang, K.-Y. & Shallcross, D.E. (2000) Modelling terrestrial biogenic isoprene fluxes and their potential impact on global chemical species using a coupled LSM-CTM model. *Atmospheric Environment* **34**, 2909–25.

Wanninkhof, R.J. (1992) Relationship between wind-speed and gas-exchange over the ocean. *Journal of Geophysical Research* **97**, 7373–82.

Watling, R. & Harper, D.B. (1998) Chloromethane production by wood-rotting fungi and an estimate of the global flux to the atmosphere. *Mycology Research* **102**, 769–87.

Watts, S.F. (2000) The mass budgets of carbonyl sulfide, dimethyl sulfide, carbon disulfide and hydrogen sulfide. *Atmospheric Environment* **34**, 761–79.

Watts, S.F. & Roberts, C.N. (1999) H_2S from car catalytic converters. *Atmospheric Environment* **33**, 169–70.

Weiss, P.S., Jonson, J.E., Gammon, R.H. & Bates, T.S. (1995) Reevaluation of the open ocean source of carbonyl sulfide to the atmosphere. *Journal of Geophysical Research* **100**, 23,083–92.

Wennberg, P.O., Brault, J.W., Hanisco, T.F., Salawitch, R.J. & Mount G.H. (1997) The atmospheric column abundance of IO: implications for stratospheric ozone. *Journal of Geophysical Research* **102**, 8887–98.

WMO (1998) *Global Ozone Research and Monitoring Project.* Report No. 44, Scientific Assessment of Ozone Depletion. World Meteorological Organization.

Xie, H., Moore, R.M., Miller, W.L. & Scarratt, M.G. (1997) A study of the ocean source of CS_2. *Abstracts of the Papers of the American Geophysical Union*, **214**(1), 86-GEOC.

Zafirou, O.C. (1974) Reaction of methyl halides with seawater and marine aerosols. *Journal of Marine Research* **33**, 75–81.

Zepp, R.G. & Andreae, M.O. (1994) Factors affecting the production of OCS in seawater. *Geophysical Research Letters* **21**, 2813–16.

5 Sources of Air Pollution

ANDREA V. JACKSON

5.1 INTRODUCTION

The atmosphere can be divided into several regions. The region closest to the Earth's surface, typically around 12 to 15 km depth, is the troposphere. Although it represents only a small fraction of the whole atmospheric volume, this is the region into which most of the anthropogenic and all of the biogenic emissions are released. It can essentially be regarded as a continually mixing chemical vessel where its chemical composition is dependent on factors such as temperature, solar aspect, weather conditions, and location. It was originally thought that the composition of the atmosphere was on the whole inactive (Junge 1963) and derived solely from volcanic emissions, outgassing from the oceans, and transport from the stratosphere. However, it has become increasingly apparent in recent years that human activities have a significant effect upon both the troposphere and the stratosphere.

Air pollution has been defined (Weber 1982) as "the presence of substances in the ambient atmosphere, resulting from the activity of man or from natural processes, causing adverse effects to man and the environment." If we ignore the highly variable amounts of water vapor, we find that more than 99.9% of the troposphere is composed of nitrogen, oxygen, and trace levels of the chemically inert noble gases (largely argon) and carbon dioxide. The combined concentration of all the remaining "reactive" trace gases in the lower atmosphere rarely rises above 0.001% (10 p.p.m.), even in the most polluted areas. Despite their small contribution, the trace species in the atmosphere have disproportionately large effects as air pollutants.

This chapter reviews the natural and anthropogenic sources of air pollutants including carbon-, sulfur-, nitrogen-, and halogen-containing primary pollutants; secondary pollutants such as ozone and peroxyacetyl nitrate (PAN); hazardous air pollutants such as benzene and 1,3-butadiene; persistent organic pollutants such as polycyclic aromatic hydrocarbons, polycyclic biphenyls (PCBs), and dioxins; and particulate material. Long-term trends as well as the spatial and temporal distribution of their emissions are also included. Table 5.1 summarizes the sources and sinks of the majority of pollutants discussed in this chapter. The chemistry associated with the majority of these pollutants is discussed in other chapters of the handbook and when used in conjunction with this one will provide a comprehensive insight into air pollution chemistry.

5.2 PRIMARY POLLUTANTS

5.2.1 *Carbon dioxide*

Carbon dioxide as a greenhouse gas

Carbon dioxide (CO_2) is the most important of all radiately active or "greenhouse" (apart from water vapor) gases, and is considered to contribute 70% of the enhanced greenhouse effect that has been experienced to date (Houghton 1997). This contrasts

Table 5.1 Natural and anthropogenic sources of a selection of trace gases (from Cox & Derwent 1981).

Compound	Natural sources	Anthropogenic sources
Carbon-containing compounds		
Carbon dioxide (CO_2)	Respiration; oxidation of natural CO; destruction of forests	Combustion of oil, gas, coal and wood; limestone burning
Methane (CH_4)	Enteric fermentation in wild animals; emissions from swamps, bogs etc., natural wet land areas; oceans	Enteric fermentation in domesticated ruminants; emissions from paddy fields; natural gas leakage; sewerage gas; colliery gas; combustion sources
Carbon monoxide (CO)	Forest fires; atmospheric oxidation of natural hydrocarbons and methane	Incomplete combustion of fossil fuels and wood, in particular motor vehicles, oxidation of hydrocarbons; industrial processes; blast furnaces
Light paraffins, C_2–C_6	Aerobic biological source	Natural gas leakage; motor vehicle evaporative emissions; refinery emissions
Olefins, C_2–C_6	Photochemical degradation of dissolved oceanic organic material	Motor vehicle exhaust; diesel engine exhaust
Aromatic hydrocarbons	Insignificant	Motor vehicle exhaust; evaporative emissions; paints, gasoline, solvents
Terpenes ($C_{10}H_{16}$)	Trees (broadleaf and coniferous); plants	
CFCs & HFCs	None	Refrigerants; blowing agents; propellants
Nitrogen-containing trace gases		
Nitric oxide (NO)	Forest fires; anaerobic processes in soil; electric storms	Combustion of oil, gas, and coal
Nitrogen dioxide (NO_2)	Forest fires; electric storms	Combustion of oil, gas, and coal; atmospheric transformation of NO
Nitrous oxide (N_2O)	Emissions from denitrifying bacteria in soil; oceans	Combustion of oil and coal
Ammonia (NH_3)	Aerobic biological source in soil Breakdown of amino acids in organic waste material	Coal and fuel oil combustion; waste treatment
Sulfur-containing trace gases		
Dimethyl sulfide (DMS)	Phytoplankton	Landfill gas
Sulfur dioxide (SO_2)	Oxidation of H_2S; volcanic activity	Combustion of oil and coal; roasting sulfide ores
Other minor trace gases		
Hydrogen	Oceans, soils; methane oxidation, isoprene and terpenes via HCHO	Motor vehicle exhaust; oxidation of methane via formaldehyde (HCHO)
Ozone	In the stratosphere; natural NO–NO_2 conversion	Man-made NO–NO_2 conversion; supersonic aircraft
Water (H_2O)	Evaporation from oceans	Insignificant

with much smaller enhancements from compounds such as methane (24%) and nitrous oxide (6%). The turnover of carbon is vast in scale, with an estimated 60 Gt yr^{-1} processed via photosynthesis and respiratory processes and around 90 Gt yr^{-1} via ocean–atmosphere exchange. This cycle results in an atmospheric loading of around 750 Gt of carbon, comprised almost entirely of CO_2.

Anthropogenic emissions to the atmosphere are only small in comparison. However, the fine balance of the carbon cycles means that at current rates of emission, there is an estimated increase in the atmospheric carbon load of 3.5 Gt yr^{-1} (Houghton 1997).

Historical records of the Earth's atmospheric CO_2 levels constructed from studies of gases within ice cores indicate that a very stable concentration was maintained at approximately 280 p.p.m.v. before significant human activity occurred (Raynaud et al. 1993). Since the Industrial Revolution, however, concentrations have been observed to steadily increase, and currently stand approximately 30% higher than preindustrial values at around 360 p.p.m.v.

Sources of CO_2

There are many sources of direct anthropogenic CO_2 emissions to the atmosphere, but most originate from the combustion of fossil fuels as biomass, coal, oil, or gas (see Table 5.2). Fossil fuel burning (plus a small contribution from cement production) released on average 5.4 ± 0.3 Pg C yr^{-1} during 1980–9, and 6.3 ± 0.4 Pg C yr^{-1} during 1990–9 (Prentice et al. 2001). Land use change is

Table 5.2 Global CO_2 budgets (in Pg C yr^{-1}) based on intradecadel trends in CO_2 and O_2. Positive values are fluxes to the atmosphere; negative values represent uptake from the atmosphere. Error bars denote uncertainty ($\pm 1\sigma$), not interannual variability. (From Prentice et al. 2001.)

	1980s	1990s
Atmospheric increase	3.3 ± 0.1	3.2 ± 0.1
Emissions (fossil fuel, cement)	5.4 ± 0.3	6.3 ± 0.4
Ocean–atmosphere flux	-1.9 ± 0.6	-1.7 ± 0.5
Land–atmosphere flux (partitioned as follows)	-0.2 ± 0.7	-1.4 ± 0.7
Land-use changes	1.7 (0.6 to 2.5)	n/a
Residual terrestrial sink	-1.9 (-3.8 to 0.3)	n/a

n/a, not available.

responsible for the rest of the emissions. Almost all this combustion is related to the production of energy for human needs: for example, in Europe 32% of CO_2 is emitted from the energy sector (mainly power and heat generation), followed by transport (22%) and industry (21%) (EEA 2000).

Reducing CO_2

At the Third Conference of the Parties of United Nations Framework Convention on Climate Change (UNFCCC), held in Kyoto in 1997, developed countries agreed to reduce their emissions of carbon dioxide, methane (CH_4), nitrous oxide (N_2O), hydrofluorocarbons (HFCs), perfluorocarbons (PFCs), and sulfur hexafluoride (SF_6) by an overall 5% from 1990 levels by 2008–12, expressed as carbon dioxide equivalents. The EU and its member states are committed to a reduction of 8% below the 1990 level in the period 2008–12, while Central and Eastern European (CEE) countries are committed to reduction of 5–8%. The UNFCCC's aim is to reach atmospheric concentrations that would prevent dangerous anthropogenic interference with the climate system but that would allow sustainable economic development.

5.2.2 Volatile organic compounds

Volatile organic compounds (VOC) are ubiquitous in the atmosphere and are central to atmospheric chemistry from the urban to the global scales. The term "VOCs" denotes vapor-phase atmospheric organics excluding CO and CO_2 (Lee & Nicholson 1994). They include pure hydrocarbons containing C and H only (e.g. alkanes, alkenes, alkynes, and aromatics) as well as species that contain other elements, such as oxygen and chlorine (e.g. aldehydes, ethers, alcohols, ketones, esters, chlorinated alkanes and alkenes, chloroflurocarbons (CFCs), and hydrofluorocarbons). Hence, VOC species display many different physical and chemical behaviors.

The most abundant of the organic compounds present in ambient air are the alkanes and aromatics, with mean concentrations ranging from <0.11 to 10 p.p.b.v. (Derwent 1993).

Table 5.3 Estimated sources and sinks of methane (Tg yr^{-1}). (From Houghton *et al.* 1995, Prather *et al.* 1995, Schimel *et al.* 1995, Seinfeld & Pandis 1998.)

Identified sources	Individual estimate	Total
Natural		
Wetlands	115 (55–150)	
Termites	20 (10–50)	
Oceans	10 (5–50)	
Other	15 (10–40)	
Total identified natural sources		160 (110–210)
Anthropogenic		
Fossil fuel related sources		
Natural gas	40 (25–50)	
Coal mines	30 (15–45)	
Petroleum industry	15 (5–30)	
Coal combustion	? (1–30)	
Total fossil fuel related		100 (70–120)
Biospheric carbon		
Enteric fermentation	85 (65–100)	
Rice paddies	60 (20–100)	
Biomass burning	40 (20–80)	
Landfill	40 (20–70)	
Animal waste	25 (20–30)	
Domestic sewage	25 (15–80)	
Total biospheric		275 (200–350)
Total anthropogenic sources		375 (300–450)
Total identified sources		535 (410–660)
Sinks		
Tropospheric OH	445 (360–530)	
Stratosphere	40 (32–48)	
Soils	30 (15–45)	
Total sinks		515 (430–600)
Total global burden 4850 Tg CH_4		

5.2.3 Methane

Methane (CH_4) is the most simple VOC in terms of structure. It is also the most abundant VOC in the atmosphere (~1800 p.p.b.v.) and, as such, has a significant effect on atmospheric chemistry. Table 5.3 shows the global sources of CH_4 to the atmosphere.

Sources of CH_4

Wetlands, mainly tropical and subtropical, are the largest natural source of CH_4 to the atmosphere (Khalil & Rasmussen 1983; Cicerone & Oremland 1988; Crutzen 1991; Fung *et al.* 1991; Houghton 1997) producing it by bacterial degradation (by methanogens) of organic matter in these anoxic environments. The majority of CH_4 emissions from rice cultivation occur in the 15–35°N latitude region. The role of rice paddies in the increase of atmospheric CH_4 is thought to have been important from the middle of the tentieth century to present but the rate of increase has slowed in the past decade (Neue & Roger 2000). However, global population levels indicate that the demand for rice will still increase over the next 20 years (IRRI 1989) and it is predicted that global CH_4 emissions from wetland rice agriculture are likely to increase. Domestic ruminants account for approximately 95% of animal CH_4 emissions (mainly from cattle and buffalo), produced from both microbial digestion of feeds in the animals' digestive tract and from microbial degradation of excreted residues in manure (Johnson *et al.* 2000).

There are many more minor sources, including biomass burning, coal mines, landfills, urban areas, sewage disposal, natural gas leakage, lakes, oceans, termites, and tundra, and collectively these contribute significantly to the global budget. Biomass material (living and dead material in forests, savannas, and agricultural wastes, fuel wood) contains about 40% carbon by weight and its incomplete combustion simultaneously produces carbon species such as CH_4, CO, NMHCs, and particulate carbon (Lobert *et al.* 1991). Methane is formed in coal during the process of coalification. Generally, more CH_4 is formed during coalification than can be stored within the coalbed itself, so excess CH_4 migrates into, and can be stored in, the surrounding strata, where it is retained until mining releases this pressure, allowing the CH_4 to be released.

In Europe the majority of CH_4 is emitted from ruminant animals (45%) and from processes such as waste treatment and disposal, mainly landfill (36%), coal mining, and leakage from natural gas distribution networks (EEA 2000). Emissions fell by 12% between 1990 and 1996 in the EU, mainly as a result of economic restructuring, the switch from coal to natural gas, the decline of deep mining in the UK, and the replacement of old gas-distribution pipework. Agricultural emissions also fell, mainly due to a reduction in the number of dairy cows (AEA 1998).

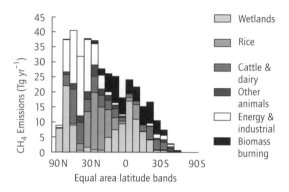

Fig. 5.1 The spatial CH_4 emission rates as a function of latitude. (From Khalil & Shearer 2000.)

Spatial and temporal distribution of CH_4 emissions

Figure 5.1 shows, as expected, that the majority of CH_4 emissions are confined to biologically productive land, although the oceans are a relatively small source (Lambert & Schmidt 1993; Bates *et al.* 1996).

Seasonal cycles of CH_4 concentration, particularly at middle and higher northern latitudes, demonstrate substantial variations of CH_4 emissions to the atmosphere (Khalil & Rasmussen 1983), which is substantiated by measurements of CH_4 emissions from rice fields, wetlands, or swamps. These show that emissions are confined to specific periods of the year (for example, Harriss *et al.* 1982; Seiler *et al.* 1984; Walen & Reeburgh 1992).

The imbalance between the current sources and sinks of CH_4 means there is an accumulation in the atmosphere, confirmed by data from ice cores. These show that for at least two thousand years before 1800 the concentration of CH_4 in the atmosphere was about 0.7 p.p.m.v. (Khalil & Rasmussen 1987). Since then CH_4 concentration has more than doubled and is increasing on average at about 0.6% per year (Watson *et al.* 1990). Its concentration now stands at 1.7 p.p.m.v. and is continuing to increase, although the rate of acceleration has slowed over the past decade (Khalil & Rasmussen 1990; Steele *et al.* 1992). Studies of geographic and agricultural data suggest that the increases of CH_4 over the past 100–300 years have been caused by increasing emissions as opposed to being caused by decreasing OH (see Khalil & Rasmussen 1985; Levine *et al.* 1985; Thompson & Cicerone 1986; Lu & Khalil 1991; Pinto & Khalil 1991; Krol *et al.* 1998), predominantly as a result of the increase in cattle population and increased area of rice harvested (Chappellaz *et al.* 1993; Kammen & Marino 1993; Khalil *et al.* 1996).

This increase is of concern for several reasons. First, CH_4 chemistry affects the production of tropospheric ozone (O_3), and computer models have predicted that CH_4 may account for approximately 20% of the O_3 formed photochemically over northwest Europe (Hough & Derwent 1987). Second, 80–90% of the CH_4 destruction that occurs in the troposphere is through reaction with the hydroxyl radical (Cicerone & Oremland 1988). A rise in the background level of CH_4 would result in the reduction of the hydroxyl radical, which in turn would result in a further increase in CH_4 concentration. This positive feedback would lead to an overall decrease in the oxidizing capacity of the troposphere. Finally, CH_4 is a greenhouse gas whose warming efficiency is up to 30 times that of CO_2 (Dickinson & Cicerone 1986) and it contributes about 24% to the enhanced greenhouse effect to date (Houghton 1997). Through chemical interactions CH_4 can also indirectly influence climate. For example, oxidation of CH_4 eventually produces CO_2 and stratospheric water vapor, and leads to changes in tropospheric and stratospheric O_3.

5.2.4 *Nonmethane volatile organic compounds (NMVOC)*

There are two major reasons for the interest in the NMVOCs in the atmosphere. The first is the direct toxicity of some compounds and the second is their role as precursors of photochemical ozone. Vast quantities of NMHC are emitted into the atmosphere from many different sources (Lamb *et al.* 1986) and the global emission rate is estimated at $\sim 1 \times 10^{14}\,g\,C\,yr^{-1}$ for anthropogenically derived compounds (stationary and mobile sources)

(Piccot *et al.* 1992; WMO 1992) and $\sim 8 \times 10^{14}\,g\,C\,yr^{-1}$ for biogenic compounds (Zimmerman *et al.* 1978; Guenther *et al.* 1995). The development of accurate emission inventories for NMVOCs is difficult due to the diversity of processes from which they are emitted. In addition to this, there are a great many species of NMVOCs in the atmosphere. Almost all VOCs are precursors of ozone through photochemical oxidation, but they differ greatly in their potential to promote the production of ozone which has led to a system of classifying hydrocarbons according to their photochemical ozone creation potential.

Anthropogenic NMVOCs

As shown in Table 5.4, motor vehicles are the dominant source of anthropogenic NMVOCs on a global scale, predominantly in the form of alkanes and aromatics, emitted mainly as a result of the incomplete combustion of fuel or from its vaporization. Solvent use is the second largest source (Middleton

Table 5.4 Estimated global anthropogenic emissions of NMVOCs (1990) (from Middleton 1995).

Activity	Emission ($Tg\,yr^{-1}$)
Fuel production/distribution	
Petroleum	8
Natural gas	2
Oil refining	5
Gasoline distribution	2.5
Fuel consumption	
Coal	3.5
Wood	25
Crop residues (including waste)	14.5
Charcoal	2.5
Dung cakes	3
Road transport	36
Chemical industry	2
Solvent use	20
Uncontrolled waste burning	8
Other	10
Total	142

Table 5.5 UK Emissions of NMVOC for 1999 by UNECE category (from DEFRA, *Digest of Environmental Statistics* 2001).

Source	kt	% of total
Power stations	8	0
Refineries	1	0
Combustion in fuel extraction and transformation	1	0
Domestic	42	2
Commercial, public, and agricultural combustion	4	0
Iron and steel	1	0
Other industrial combustion	5	0
Production processes	212	12
Extraction and distribution of fossil fuels	259	15
Solvent use	472	27
Road transport	473	27
Off-road sources	56	3
Military	1	0
Railways	2	0
Shipping	2	0
Civil aircraft	3	0
Waste treatment and disposal	24	1
Agriculture and managed forestry	0	0
Forests	178	10
Total	1744	100

1995). The global distribution of NMVOCs is relatively well described for a few species, such as ethane, propane, and acetylene. A strong inter-hemispheric gradient is seen for these compounds with highest concentrations observed in the northern hemispheric regions at around 50° N decreasing southwards (Ehhalt *et al.* 1985; Rudolph 1988; Boissard *et al.* 1996). Higher mixing ratios are observed over America than over Europe.

Table 5.5 shows that the main sources of CH_4 to the UK atmosphere are processes and solvents, the next largest source being road transport. This is a similar situation to that in the EU overall, where the main emitters are transport, industrial and household solvent use, and the storage and distribution of fossil fuels. Maps of the UK and the rest of Europe show high NMVOC emission from areas with high traffic density and where industrial processes and refineries are located.

Emissions from transport have declined as a result of the increased use of catalytic converters and the switch from gasoline to diesel cars. It is expected that emissions from this source will continue to reduce in the future as more new cars with increasingly stringent emission controls replace old polluting vehicles. Emissions from solvent use and industrial processes have also decreased as water-based alternatives and pollutant abatement technology are introduced. Despite these emissions reductions, the EU has so far not seen fewer exceedances of critical O_3 levels or O_3 concentration thresholds, and substantial reductions are necessary from most EU countries to meet targets for the proposed CLRTAP NMVOC protocol targets for 2010.

Higher VOC concentrations are observed in the wintertime in urban areas due to the much shallower boundary layer and reduced solar flux, reducing the production of hydroxyl radicals (responsible for removing VOCs). Diurnal variations are also observed. VOCs such as benzene and 1,3-butadiene show very similar diurnal variation to carbon monoxide and NO_x, confirming that motor vehicular emissions are the major source of these species. However, diurnal variation of the lower alkanes, such as ethane and propane, in urban areas are characteristically different from those for the motor vehicle derived pollutants, which is consistent with a pollutant that has a surface source and a constant emission rate. Concentrations exhibit a tendency to be elevated at night, when vertical dispersion is least effective, decreasing to a minimum mid-afternoon as the planetary boundary layer rises and wind speeds increase, and then rising again to the nighttime maximum.

Biogenic NMVOCs

Organic compounds are released naturally to the atmosphere from many types of vegetation and it was Went (1960) who first put forward the suggestion that natural foliar emissions of VOCs from trees and other vegetation could have a significant effect on the chemistry of the Earth's atmosphere. There have since been many investigations of the speciation of natural VOCs, as well as the atmospheric distribution of these compounds and their oxidation products. Over 1000 natural VOCs are known to enter the atmosphere and there are potentially thousands of others known to exist in plants that have not yet been reported in atmospheric samples. The main biogenic compounds released to the atmosphere (excluding CH_4) are: isoprene and the monoterpene family emitted from plants; and dimethyl sulfide emitted from marine phytoplankton. It has been found that natural emissions from terrestrial biogenic sources can sometimes be so large that they can contribute to ozone episodes (Chameides *et al.* 1988).

Not all plants emit the same set of VOCs. Deciduous trees typically emit isoprene (C_5H_8), while conifers typically emit terpenes (e.g. alpha- and beta-pinene). Isoprene exhibits a strong increase in emission as temperature increases and is only emitted in the presence of light. Clear diurnal cycles can be therefore observed for isoprene, with maximum emissions during the daytime and minimum ambient levels during the night. In contrast, monoterpenes appear to be the result of actual biophysical processes, with emissions independent of the amount of light, but dependent upon temperature. Their diurnal cycle can therefore display maximum ambient concentrations during the nighttime. Natural VOC emissions are therefore generally highest on hot summer days.

Hydrocarbon emission inventories show that biogenic hydrocarbon emissions dominate the global flux of reactive VOCs to the atmosphere (Müller 1992; Guenther *et al.* 1995). Their sources are widely spread over the Earth's surface and there is much variability in their ambient concentrations, but the largest emissions (predominantly in the form of isoprene) occur in the Tropics as a result of the high temperatures and large biomass densities. Strong local sources such as biomass burning also significantly contribute to the concentration of NMVOCs in this region.

Significant concentrations of isoprene have recently been measured in urban areas and correlation with exhaust gases such as 1,3-butadiene or carbon monoxide suggests a traffic-related source (McLaren *et al.* 1996; Christensen *et al.* 1999). Direct vehicle emission analysis showed clearly that isoprene is a component of exhaust gases. This is further supported by the lack of seasonal variation in isoprene concentrations in urban atmospheres, compared to strong seasonal variation observed in rural and nonurban atmospheres as a result of variation in biogenic activity (Jobson *et al.* 1994; Hagerman *et al.* 1997; Young *et al.* 1997). The amount of isoprene that can be attributed to traffic in urban areas compared to its biogenic emission has been the focus of recent studies. A study by Borbon *et al.* (2001) found that in urban areas isoprene of biogenic origin was most dominant in summer (traffic-related isoprene varied from 10 to 50%), while vehicle exhausts were mainly responsible for its levels from late September to April. It has been suggested that due to their high reactivity this source of VOCs could be a significant contributor to O_3 formation (Chameides *et al.* 1988).

5.2.5 Carbon monoxide

Sources and sinks of CO

Carbon monoxide (CO) is released as a by-product of the incomplete combustion of carbon-containing materials and from the photochemical conversion of atmospheric CH_4 and other hydrocarbons. Table 5.6 shows the IPCC range of estimates for sources and sinks of CO, as well as those calculated by the IMAGES three-dimensional global chemistry-transport model (Granier *et al.* 2000). Chemical production from the oxidation of CH_4 and other VOCs represents a significant contribution on the global scale. This production mechanism was first discussed by Zimmermann *et al.* (1978), and many subsequent studies have attempted to quantify the yield of CO from hydrocarbon oxidation (Brewer *et al.* 1984; Paulson & Seinfeld 1992; Miyoshi *et al.* 1994). Results from the IMAGES model (Müller & Brasseur 1995; Pham *et al.* 1995; Granier *et al.* 1996, 1999, 2000) show the significance of this source, with CH_4 oxidation representing 53% and the oxidation of

Table 5.6 Global CO budget estimates and uncertainties (from Houghton *et al.* 1995 and Granier 2000).

	Range from IPCC (1995) compilation (Tg CO yr^{-1})	Estimates from IMAGES model (Tg CO yr^{-1})
Surface emissions		
Anthropogenic	300–550	476
Biomass burning	300–700	675
Oceans and soils	80–360	186
Chemical production		
CH$_4$ oxidation	400–1000	721
Isoprene oxidation		273
Terpenes oxidation		60
Other NMHCs oxidation	200–600	
Natural NMHCs		208
Anthropogenic NMHCs		106
Destruction		
Chemical destruction	1400–2600	2380
Surface deposition	200–640	290

isoprene, terpenes, and other hydrocarbons representing 20, 4, and 23% respectively of this chemical production.

The IMAGES model also shows the contribution of human activities to the CO surface concentration, and unexpectedly this was seen to be highest over the polluted areas of the northern hemisphere. However, it also showed high CO concentrations over the Northern Atlantic. These have been explained by the long global lifetime of hydrocarbons such as ethane, and their oxidation products, so that they can impact far away from the location of their emission or chemical production.

The influence of anthropogenic sources is reflected by the contribution of road transport to total national emissions for the UK and the USA. Road transport, in particular gasoline vehicles, accounted for almost 75% of the UK total in 1997 (DETR 2000) and for 57% of the US total in 1998 (US EPA 2000). Diesel vehicle emissions of CO are

relatively small, and in 1997 contributed only 3% of the UK national total. Maximum ambient concentrations tend to be recorded at kerbside and roadside sites, reflecting the dominant influence of motor vehicle exhaust emissions on ambient CO concentrations. Other sources of CO in the UK and USA include industrial processes (such as metals processing and chemical manufacturing), solvent use, storage and transport, public power, waste disposal, and recycling. The highest levels of CO in the outside air typically occur during the colder months of the year when inversion conditions are more frequent: air pollutants become trapped near the ground beneath a layer of warm air.

Reaction with the hydroxyl radical (OH) accounts for 80–90% of the CO sink (Prather *et al.* 1994). It is an important reaction as an increase in CO levels may suppress OH, decreasing the oxidizing capacity of the troposphere and thus decreasing the oxidation rates of other reduced trace gases, such as CH$_4$ (Sze 1970; Thompson & Cicerone 1986). Remaining CO is lost to the stratosphere and to the Earth's surface by deposition, where it is oxidized via biological consumption in soils and via the ocean (Warneck 1988).

Trends in CO emissions

Studies during the 1980s showed that atmospheric CO concentrations were increasing at ~1.2 ± 0.6% per year (Khalil & Rasmussen 1987). This increase continued to around 1987, after which concentrations started to fall, with a particularly rapid decline in the southern hemisphere (Khalil & Rasmussen 1994; Novelli *et al.* 1994). This trend is thought by many to be the result of decreasing anthropogenic emissions. For example, annual CO emissions in the UK have been declining rapidly since 1990, falling by 33% between 1990 and 1997, and projections show that annual urban road transport emissions of CO are expected to decline by 59% between 1995 and 2005. In the USA, emissions peaked in 1970 and decreased rather steadily thereafter. Although there are no internationally agreed emission reduction targets or emission ceilings for CO, it is subject to control under O$_3$ strategies. As such, CO is classed as a "priority" pollutant in many countries, meaning that it is subject to stringent controls. The observed decreases are therefore largely a result of these controls, through the introduction of catalysts on gasoline-engined vehicles, new vehicle technologies, and clean fuels programs.

Alternative explanations, such as an increase in OH resulting from stratospheric ozone decreases, have been put forward to explain the global decrease in CO concentration (Madronich & Granier 1992; Prinn *et al.* 1992) but the decreases of CO are too rapid (1.5% per year in the northern hemisphere and >5% per year in the southern hemisphere) compared to the estimated increase of OH from stratospheric ozone depletion.

The rapid decline observed in the middle and higher latitudes of the southern hemisphere is thought to be mostly due to the decrease of tropical biomass burning (Khalil & Rasmussen 1994), which is the dominant source of CO in the Tropics (Logan *et al.* 1981). However, this is as yet unconfirmed.

Spatial and temporal variation in CO

Carbon monoxide has a short lifetime, of around two months. This, together with its varied sources, means that there is much regional variation of CO concentration. For example, biomass burning can increase the concentration of CO above background levels over large areas (Fishman *et al.* 1991) and oceanic emissions are also regionally important (Erickson & Taylor 1992). The total source strength in the northern hemisphere is estimated to be about two times more than that in the southern hemisphere due to the large influence of anthropogenic sources (Logan *et al.* 1981; Dianov-Klokov *et al.* 1989; Khalil & Rasmussen 1994), which leads to an annually averaged northern hemisphere abundance of around 100–125 p.p.b.v. compared to the southern hemisphere with an abundance of 50–65 p.p.b.v. (Novelli *et al.* 1992, 1994).

Mixing ratios of CO are highest during winter and lowest in the summer in both hemispheres, but the amplitude of this cycle is greatest in the high northern latitudes, decreasing toward the Equator (Novelli *et al.* 1992; Derwent *et al.* 1998). In contrast, mixing ratios are relatively constant with increasing latitude south of 10–20° S (Seiler *et al.* 1984). The difference in seasonal cycles between the hemispheres reflects the seasonal differences in the imbalance of sources and sinks and the effects of CO transport (Novelli *et al.* 1992). For example, in the middle and high latitudes OH levels are typically highest in summer, leading to increased CO oxidation, and lowest in the winter, resulting in decreased CO loss (Logan *et al.* 1981; Spivakovsky *et al.* 1990). In the middle and high latitudes of the northern hemisphere, and to a lesser extent in the southern hemisphere, anthropogenic CO emissions reach their maximum during winter, coincident with the minimum in loss via OH oxidation. In contrast, in the Tropics, irradiation and OH levels vary much less than at higher latitudes, and consequently CO oxidation also varies less on a seasonal basis (Novelli *et al.* 1998).

Effects of CO on human health

Carbon monoxide is a toxic gas detrimental to human health if exposure is sufficient. It enters the bloodstream through the lungs and combines preferentially with hemoglobin to produce carboxyhemoglobin (COHb), which displaces oxygen and reduces oxygen delivery to the body's organs and tissues (Varon & Marik 1997). Carbon monoxide binds reversibly to hemoglobin with an affinity 200–230 times that of oxygen (Rodkey *et al.* 1974). Consequently, relatively minute concentrations of the gas in the environment can result in toxic concentrations in human blood. In the UK, a National Air Quality Standard of 11.6 mg m^{-3} (10 p.p.m.) as an eight-hour running mean, to be achieved by the end of 2005, has been assigned, based on scientific and medical evidence regarding the environmental and health effects of CO.

However, the greatest concerns regarding health effects of CO are those associated with exposure to CO in indoor air from sources such as heaters, appliances that use carbon-based fuels, household fires, tobacco smoke, or wood stoves.

5.2.6 Oxides of nitrogen

Role of NO_x in the atmosphere

The most abundant oxide of nitrogen in the atmosphere is nitrous oxide (N_2O), produced by natural microbiological processes in the soil. Although it is relatively unreactive in the troposphere, it does have an effect on stratospheric O_3 concentration. It is also a radiately active greenhouse gas. Oxidized nitrogen species of tropospheric significance (those excluding N_2O) are collectively referred to as NO_y and are taken to consist of NO, NO_2, higher oxides (NO_3 and N_2O_5), oxyacids (HNO_3, HO_2NO_2, and HONO), organic peroxy nitrates (RO_2NO_2), organic nitrates ($RONO_2$), and aerosol nitrate.

The pollutant nitrogen oxides of particular concern are nitric oxide, NO, and nitrogen dioxide, NO_2, the sum of the two compounds being referred to as NO_x. They play an important role in the atmospheric chemical processes that govern the "quality" of the air surrounding us on local, regional, and global scales. First, reactions involving NO_x generate secondary pollutants, the most important being ozone (O_3). Second, although the direct greenhouse effect of NO_x is negligible, the global warming potential (GWP) of NO_x via the formation of O_3 is comparable to that of CH_4. Third, NO_x is oxidized to produce acids, which are removed from the atmosphere by both wet and dry deposition, the overall process being termed acid deposition. This is known to have detrimental effects on watercourses, vegetation, materials, and buildings. Finally, NO_x has several adverse effects on human health and is known to cause diseases such as pulmonary edema and to damage central nervous system tissue. NO_2 on its own in the 1 p.p.m. range can cause a whole host of reversible and irreversible effects on humans arising from structural changes in the cells of the respiratory system (Brimblecombe 1996). With these and other health effects in mind, the EPAQS has set health objectives for the UK for NO_2 as $200\,\mu g\,m^{-3}$ (105 p.p.b.) one-hour mean, not to be exceeded more than 18 times a year, and $40\,\mu g\,m^{-3}$ (21 p.p.b.) annual mean, both to be achieved by the end of 2005. For these reasons, there is considerable interest in estimating sources of NO_x emissions and their fate in the atmosphere.

Global sources of NO_x

Nitrogen oxides are released into the troposphere from a variety of biogenic and anthropogenic sources (Logan 1983; Lee *et al.* 1997), as shown in Table 5.7. The main sources are production by industry and traffic, biomass burning, microbiological emission by soil, exchange with the stratosphere, lightning, and air traffic. Although current production estimates for some sources are very uncertain, it is estimated that the majority of the total NO_x emissions are anthropogenic (Lee *et al.* 1997).

NO_x from fossil fuel combustion. The NO_x produced through combustion, whether from stationary or mobile sources, can be formed by the oxidation of nitrogen-containing compounds in the fuel (fuel NO_x), by reaction of air-derived nitrogen and oxygen (thermal NO_x), and by reaction between radicals in the combustion flame (prompt NO_x) (Bachmaier *et al.* 1973; Barnard & Bradley

Table 5.7 Global NO_x budget (from Logan 1983).

	Tg N yr^{-1}
Sources	
Fossil fuel combustion	21 (14–28)
Biomass burning	12 (4–24)
Lightning	8 (2–20)
Microbial activity in soil's	8 (4–16)
Oxidation of ammonia	0–10
Photolytic/biological processes in ocean	<1
Input from the stratosphere	≈0.5
Total	25–99
Sinks	
Precipitation	12–42
Dry deposition	11–22
Total	23–64

1985; Glassman 1987; Baulch *et al.* 1992). Under high-temperature combustion conditions, thermodynamics favor NO formation and this results in less than 10% of the NO_x in typical exhausts being in the form of NO_2. NO, however, converts to NO_2 in the atmosphere. The amount of NO_x produced depends on the type and composition of the fuel and on combustion conditions; a good review can be found in Colbeck and Mackenzie (1994).

Lightning production. There is a large uncertainty surrounding the role of lightning in the NO_x global budget but best estimates converge in the 2–$6\,TgN\,yr^{-1}$ range (Lawrence *et al.* 1995; Levy *et al.* 1996; Lee *et al.* 1997; Price *et al.* 1997; Bradshaw *et al.* 2000). Data gathered from NASA's DC-8 aircraft during the Subsonics Assessment Ozone and Nitrogen Dioxides Experiment (SONEX) showed evidence of a substantial contribution from lightning to the NO_x budget over the North Atlantic (Singh *et al.* 1999), and other studies have shown that there are strong variations in lightning emissions in the atmosphere (Crawford *et al.* 1999; Thompson *et al.* 1999; Allen *et al.* 2000).

Production of NO_x in soils. Microorganisms make a significant contribution to the global NO_x budget through the processes of nitrification and denitrification. In developed countries, well aerated N fertilized arable soils can be the largest sources of soil NO emission. In some cases soil emissions can be comparable to those from combustion and other nonbiological processes (Stohl *et al.* 1996). Estimates of the soil contribution to the European total NO_x budget range from 2 to 23% (Simpson *et al.* 1999), with increased contributions on hot summer days (Stohl *et al.* 1996).

Briefly, nitrification is the production of nitrate ions through the oxidation of ammonium ions in the soil, mediated by chemoautotrophic bacteria such as *Nitrosomonas* spp. and *Nitrobacter* spp. (Haynes 1986). The following net chemical transformation occurs under aerobic conditions, with the rate of nitrification dependent upon the availability of NH_4^+ and O_2 and other factors (Haynes 1986; Firestone & Davidson 1989).

$$NH_4^+ \rightarrow NH_2OH \rightarrow [HNO]$$
$$\rightarrow NO_2^- \rightarrow NO \text{ or } N_2O \qquad (5.1)$$

Denitrification reactions occur under anaerobic conditions in the soil, mediated by heterotrophic bacteria such as *Pseudomonas* and *Achromobacter*, in the presence of a plentiful supply of easily decomposable organic matter. One denitrification reaction is the reduction of nitrate to form nitrogen gas, a process involving several steps and the production of nitrite ion and nitric oxide intermediates. Denitrification can act as a source or sink of these intermediates depending on the relative rates of reduction and intermediate release (Firestone & Davidson 1989).

$$4NO_3^-(aq) + 5\{CH_2O\} + 4H_3O^+(aq)$$
$$\rightarrow 2N_2 + 5CO_2 + 11H_2O \qquad (5.2)$$

NO_x can also be produced in soils by purely chemical means. Sources include the photolysis of HONO and aqueous nitrite and the disproportionation of nitrous acid (Haynes 1986; Galbally 1989).

Biomass burning, although an important direct source of NO_x emissions to the troposphere, can also significantly enhance soil emissions of NO from nitrification (Anderson *et al.* 1988).

Regional NO_x emissions. In the UK almost half of all annual NO_x emissions are from road transport, as shown in Table 5.8. The electricity supply industry and the industrial and commercial sectors also make significant contributions. Almost similar contributions are seen on a European scale and in the USA (EEA 2000; US EPA 2000). The dominance of road transport in NO_x emissions is reflected in the annual concentrations of NO_2 in urban areas, which are generally highest at roadside and kerbside sites. In fact, major routeways (e.g. motorways and primary routes) are clearly defined on maps of UK NO_x emission (DETR 2000). Conurbations and city centers also show high emissions resulting from large volumes of road transport, and residential and commercial combustion.

Table 5.8 Estimated emissions of NO$_x$ in the UK by UNECE source category (from DEFRA, *Digest of Environmental Statistics* 2001).

Source	kt	% of total
Power stations	338	21
Refineries	29	2
Combustion in fuel extraction and transformation	55	3
Domestic	71	4
Commercial, public, and agricultural combustion	32	2
Iron and steel	24	2
Other industrial combustion	145	9
Production processes	6	0
Extraction and distribution of fossil fuels	1	0
Road transport	714	44
Off-road sources	82	5
Military	23	1
Railways	12	1
Shipping	56	3
Civil aircraft	12	1
Waste treatment and disposal	3	0
Agriculture & managed forestry	0	0
Total	1605	100

Trends in NO$_x$ emissions

In the UK total NO$_x$ emissions increased from 1970 to 1989 as a result of increased road transport. They subsequently declined by 30% between 1990 and 1998 as a result of the introduction of catalytic converters on gasoline vehicles. The UK has agreed to cut the emission of NO$_x$ to below 1181×10^3 t by 2010 under the UNECE Gothenburg Protocol to the Convention on Long Range Transboundary Air Pollution.

Total EU annual emissions of NO$_x$ have declined rapidly in recent years, largely due to the introduction of three-way catalysts to gasoline-engined cars, but also due to improved abatement in the energy and industry sectors (EEA 2000). However, reducing emissions of NO$_x$ is proving more difficult than it is for SO$_2$. Switching to natural gas from oil only provides small NO$_x$ emission reductions and the low-NO$_x$ burners fitted to many large power plants are not as effective in reducing emissions as flue gas desulfurization is for SO$_2$. Furthermore, increasing road travel, whose emissions are more difficult to control, has partly offset reductions achieved by emission abatement. It is forecast that NO$_x$ emissions in the UK from road transport in urban areas will decrease by 59% between 1995 and 2005 and they are projected to continue to decrease until they bottom out between 2015 and 2020, before increasing again after 2020 (DETR 2000). Although the total number of vehicles in cities of developed countries is much higher than that of developing countries, vehicular density is much lower in developed countries due to better road network and infrastructure. In developing countries the growing number of running vehicles, poor vehicle maintenance, and poor infrastructure of roads are resulting in an increase in pollution from these areas (Shishar & Patil 2001).

5.2.7 Sulfur dioxide

Role of sulfur dioxide in the atmosphere

Sulfur dioxide (SO$_2$) is formed when fuel containing sulfur is burned. Interest in this pollutant gas lies in its detrimental effects on the environment and on human health. High exposures to SO$_2$ in the ambient air result in breathing problems, respiratory illness, alterations in the lung's defenses, and aggravation of existing respiratory and cardiovascular disease. People most sensitive to SO$_2$ include asthmatics and individuals with chronic lung disease (such as bronchitis and emphysema) or cardiovascular disease, as well as children and the elderly (Lippman 1992). The EPAQS has set the following SO$_2$ air quality objectives for the protection of human health in the UK: $350 \mu g\,m^{-3}$ (132 p.p.b.) one-hour mean, not to be exceeded more than 24 times a year, to be achieved by the end of 2004; and $266 \mu g\,m^{-3}$ (100 p.p.b.) 15-minute mean, not to be exceeded more than 35 times a year, to be achieved by the end of 2005.

Sulfur dioxide and NO$_x$ are the major precursors of acid rain. Briefly, once SO$_2$ is emitted into the atmosphere it is oxidized to sulfuric acid (H_2SO_4) and sulfate, which can be transported large distances before it is deposited to the surface. The effects of this acid deposition on the environment include acidification of soils, lakes, and streams, and the accelerated corrosion of buildings and monuments. Sulfur dioxide is also a precursor to $PM_{2.5}$, which is a significant health concern as well as a main pollutant that impairs visibility.

Sources of SO_2

The major source of SO_2 is the combustion of fossil fuels, predominantly coal and fuel oil. Natural gas and gasoline have a relatively low sulfur content. Until recently diesel engine emissions led to small elevations in SO_2 alongside busy roads, but regulations have significantly reduced the sulfur content of this fuel. The main sources of SO_2 in Europe are energy (60%), industry (25%), and transport (6%). Combustion of coal in power stations is by far the major source of SO_2 emissions in the UK (as shown in Table 5.9), as is the case in other countries.

Trends in SO_2 levels

In the first half of the century, SO_2 emissions in the UK were predominantly a result of the combustion of coal in domestic, commercial, and industrial premises, and in power stations (at this time these were predominantly within towns and cities). Following the smogs of the 1950s and the Clean Air Act of 1956, efforts were made to reduce levels of smoke and SO_2 in air, by means of controlling chimney heights to ensure adequate dispersion of SO_2, by control of domestic smoke emissions by local authorities through the promotion of the use of alternative fuels and by the introduction of "smoke control zones." Cleaner fuels replaced coal in the domestic sector and power generation was concentrated in much larger, more efficient stations in rural areas. Today, although there are significant emissions from the industrial sector, including refineries and the iron and steel industry, emissions are now dominated by a relatively small number of large emitters, such as fossil-fuelled power stations (which in 1997 accounted

Table 5.9 Estimated UK emissions of SO_2 for 1999 by UNECE source category (from DEFRA, *Digest of Environmental Statistics* 2001).

Source	kt	% of total
Power stations	776	65
Refineries	93	8
Combustion in fuel extraction and transformation	15	1
Domestic	53	4
Commercial, public, and agricultural combustion	22	2
Iron and steel	42	4
Other industrial combustion	114	10
Production processes	17	1
Extraction and distribution of fossil fuels	1	0
Solvent use	4	0
Road transport	12	1
Off-road sources	5	0
Military	6	1
Railways	1	0
Shipping	22	2
Civil aircraft	1	0
Waste treatment and disposal	4	0
Total	1187	100

for 62% of the national total). However, even power generating companies began to switch to gas for new power stations during the 1990s. All of these measures have resulted in a steady decline in national emissions of SO_2 since the peak in concentration in 1970, almost to the point where urban and rural concentrations are virtually indistinguishable.

Under the United Nations Economic Commission for Europe's (UNECE) Second Sulphur Protocol, the UK must achieve a 70% reduction in its total SO_2 emission by 2005 and 80% by 2010 from a 1980 baseline. By the end of 1998 the UK had achieved a 67% reduction from 1980 baseline levels. The UK has also agreed under the UNECE Gothenburg Protocol to the Convention on Long-range Transboundary Air Pollution that its emissions will be below 625,000 t by 2010.

Emissions decreased in the USA by 28% from 1980 to 1999 and 21% from 1990 to 1999. Reductions in SO_2 concentrations and emissions since 1994 are due, in large part, to controls implemented under the Environmental Protection Agency's Acid Rain Program, which began in 1995.

5.3 LONG-LIVED POLLUTANTS

5.3.1 Halocarbons

Sources of halocarbons to the atmosphere

Atmospheric halocarbons (halogen-containing organic compounds) are produced naturally by biological processes in the oceans, from sea salt and from biomass burning. However, for many of these species, industrial synthesis for use as refrigerant gases, propellants, and blowing agents, i.e. human activities, is their sole source. It is these man-made halocarbons that are of concern with regard to atmospheric chemistry, in particular those that contain chlorine (chlorofluorocarbons, CFCs, and carbon tetrachloride) and bromine (halons). Table 5.10 shows the UK supply and use of some CFCs and substitute HCFCs.

Environmental effects of halocarbons

One of the unique properties of these compounds is their inertness in the troposphere, meaning they have very long lifetimes (see Table 5.11). As a result of the increased use of CFCs through the 1980s this has led to an accumulation of these compounds in the atmosphere, with several environmental consequences.

Halocarbons as greenhouse gases. Halocarbons are greenhouse gases. They possess absorption bands in the region known as the longwave atmospheric window, where few other gases absorb. A CFC molecule has a greenhouse effect 5000 to 10,000 thousand times greater than an added molecule of carbon dioxide. Thus, despite their very small concentration, compared, for instance, with carbon dioxide, they have a significant greenhouse effect.

Table 5.10 The supply and use of CFCs and substitute HCFCs in the UK (tonnes) (from DEFRA, *Digest of Environmental Statistics* 2001).

	1986	1990	1994	1995
CFC-11	26,910	7,221	2,057	293
CFC-12	18,769	8,263	3,068	1,112
CFC-113	7,389	7,212	2,752	30
CFC-114	1,232	295	427	70
CFC-502	2,054	2,376	1,131	172
Other CFCs	3	43	42	25
Refrigeration	5,296	5,416	2,504	597
Solvents	7,139	6,793	2,689	30
Aerosols	36,771	7,395	2,653	995
Foams	8,651	5,456	1,531	0
Other uses	500	350	100	50
Total CFCs	58,357	25,410	9,477	1,702
HCFC-22	4,372	6,031	6,636	6,817
HCFC-141b	0	7	1,855	3,761
HCFC-142b	1	1,294	1,474	1,316
Other HCFCs	0	87	388	559
Refrigeration	4,142	5,121	5,600	6,176
Solvents	0	0	458	740
Aerosols	150	377	395	412
Foams	81	1,921	3,900	5,112
Other uses	0	0	10	13
Total HCFCs	4,373	7,419	10,353	12,453

Role of halocarbons in depletion of the ozone layer. Halocarbons play an important role in the depletion of the stratospheric O_3 layer. Briefly, several years ago scientists in Antarctica observed significant reductions in stratospheric ozone; such reductions are now routinely observed during early spring (Farman *et al.* 1985). These observations were explained by the catalytic depletion of O_3 by chlorine, released from CFCs which had entered the stratosphere from the troposphere as a result of their long lifetimes (Molina & Rowland 1974). During the Antarctic winter a strong circumpolar wind develops in the middle to lower stratosphere, producing what is known as the Polar Vortex, isolating air above the polar region. Cold air descends from the upper stratosphere/

Table 5.11 Lifetimes for radiatively active gases and halocarbons (from Schimel *et al.* 1995).

Species	Formula	Lifetime	Conc. (p.p.b.v.) 1992	Conc. (p.p.b.v.) preindustrial	Current growth (p.p.b.v. yr^{-1})
Gases phased out before 2000 under Montreal Protocol and its amendments					
CFC-11	CCl_3F	50	0.268	0	0.000
CFC-12	CCl_2F_2	102	0.503	0	0.007
CFC-113	CCl_2FCClF_2	85	0.082	0	0.000
Carbon tetrachloride	CCl_4	42	0.132	0	0.0005
Chlorinated hydrocarbons controlled by the Montreal Protocol and its amendments					
HCFC-22	$CHClF_2$	12.1	0.100	0	0.005
HCFC-141b	CH_3CFCl_2	9.4	0.002	0	0.001
HCFC-142b	CH_3CF_2Cl	18.4	0.006	0	0.001
Perfluorinated compounds					
Sulfur hexafluoride	SF_6	3,200	0.032	0	0.0002
Perfluoromethane	CF_4	50,000	0.070	0	0.0012

lower mesosphere and frozen clouds known as polar stratospheric clouds (PSC) are formed in the very low temperatures (−80°C). These provide active surfaces for the conversion of the reservoir species into more active forms of chlorine. In more recent times reductions in stratospheric ozone concentrations during the Arctic spring have also been detected (Müller *et al.* 1997).

Reduction of ozone-depleting substances. International agreement to limit the production and consumption of O_3-depleting substances was reached in 1987 through the Montreal Protocol on Substances that Deplete the Ozone Layer and its Adjustments and Amendments (UNEP 1987), and this has led to substantial reductions. In Europe, EC Regulation 3093/94 on Substances that Deplete the Ozone Layer resulted in the phase-out of CFCs and carbon tetrachloride (CCl_4) a year ahead of the Montreal Protocol deadline, and introduced tighter controls on HCFCs. Data on the consumption of CFCs globally and within the EU over this period are given in Table 5.12.

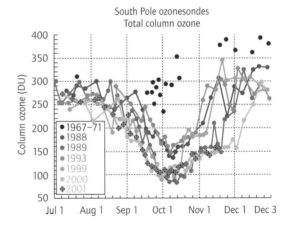

Fig. 5.2 Total ozone column as measured by balloon flights over the South Pole. (From http://www. cmdl.noaa.gov/ozwv/ozsondes/spo/spototal.html, courtesy of Climate Monitoring and Diagnostics Laboratory, National Oceanic and Atmospheric Administration, Boulder, CO.)

Table 5.12   CFC production and consumption in the EU and the world 1986–98, in thousand (ODP) tonnes. Production and consumption figures have been multiplied by the ozone depletion factor (ODP) to reflect the potential damage to the ozone layer (from DEFRA, *Digest of Environmental Statistics* 2001).

	1986	1987	1988	1989	1990	1991	1992	1993	1994	1995	1996	1997	1998
EU													
Production	429	435	413	355	269	246	219	190	85	0	0	0	0
Consumption	304	317	299	223	175	154	129	114	47	6	6	9	4
World													
Production/ consumption	932	1012	1020	907	618	570	500	411	225	139	78	56	51

CFC replacements. CFCs have been replaced by hydrochlorofluorocarbons (HCFCs) and hydrofluorocarbons (HFCs). These compounds contain C–H bonds that allow them to be attacked by OH radicals in the troposphere, and the products are removed via wet deposition processes (precipitation) before they reach the stratosphere. HFCs have the added advantage that they do not contain the chlorine or bromine atoms primarily responsible for the catalytic destruction of ozone. While being less destructive to O_3 than the CFCs, HCFCs, and HFCs are strong absorbers of terrestrial infrared radiation and are therefore greenhouse gases, and it was decided in Copenhagen in 1992 that HCFCs would also be phased out by the year 2030. However, their shorter lifetime does mean that their concentration in the atmosphere, and therefore their con-tribution to global warming for a given rate of emission, will be less than for the CFCs.

Other replacements, such as perfluorocarbons (PFCs, e.g. CF_4, C_2F_6) and sulfur hexafluoride (SF_6), are also greenhouse gases. PFCs are produced as a by-product of aluminum smelting, while SF_6 is produced during magnesium manufacture, from electrical equipment, and from atmospheric tracers for scientific studies. These compounds possess very long atmospheric lifetimes, some more than 1000 years, meaning that all emissions are accumulating in the atmosphere and will continue to influence climate for a long period. If emissions of these gases increase substantially from their present low levels, their effect on climate could become significant.

5.4   SECONDARY GASEOUS POLLUTANTS

5.4.1   Ozone

Photochemical smog

The phenomenon of photochemical smog was first recognized in Los Angeles in the mid-1940s; it was noted for the plant damage, eye irritation, and decrease in visibility it caused. Because of the region where it was initially observed, this smog quickly became known as "Los Angeles smog" but was soon found to occur in cities worldwide and was renamed "photochemical smog." The historical word "smog" implies that this pollution is formed from the condensation of smoke and fog. However, it is in fact formed as a result of high temperatures and bright sunlight promoting the rapid conversion of primary compounds such as carbon monoxide, nitric oxide, aromatics, other unburnt hydrocarbons (e.g. alkanes and alkenes), and partially oxidized fuel (e.g. aldehydes), emitted from motor vehicles, into secondary pollutants, the most well known of which is ozone (O_3). Other secondary pollutants include nitrogen dioxide, ozone, peroxyacyl nitrates, aldehydes, and alkyl nitrates. It also consists of organic nitrates, oxidized hydrocarbon, and secondary aerosol.

Ozone formation

Ozone is not directly emitted from human sources. Although stratospheric intrusions of ozone to the troposphere occur at certain times of the year, tropospheric O_3 is predominantly formed by photochemical reactions, primarily involving NO_x and VOCs. These chemical reactions take place over periods of several hours or even days depending on the VOCs; therefore, once O_3 has been produced it may persist for several days. As a result, O_3 measured at a particular location may have arisen from VOC and NO_x emitted large distances away. This means that maximum O_3 concentrations are generally found downwind of the source areas of precursor pollutant emissions. The high NO_x concentrations found in urban areas tend to reduce O_3 concentrations, but as the air masses move away from these areas, more O_3 is generated and may lead to very high O_3 concentrations downwind.

In severe photochemical smogs, such as in Los Angeles where the basin-like topography and high temperatures and high solar flux in this region are ideal for promoting the photochemical processes, O_3 concentrations may exceed 400 p.p.b. Although the same chemical processes result in elevated concentrations of ground-level ozone on a regional scale in Europe (extending over hundreds of miles simultaneously) (PORG 1997) the problem is not as severe as that experienced in Los Angeles. However, similarly, occurrence of these episodes is dependent on the meteorology. Anticyclonic conditions produce high pressures and stagnant air masses, which effectively act as a huge chemical reactor for the trapped pollutants, producing elevated concentrations of O_3. Concentrations commonly exceed 100 p.p.b. during these episodes, which are superimposed on a baseline that averages between 20 and 50 p.p.b. in the UK. There is evidence that this baseline has roughly doubled since 1900, largely due to the increase in man-

made NO_x emissions in the whole of the northern hemisphere. Taking into account the factors producing elevated O_3 levels, they are observed to occur more frequently in the UK: in summer; in the south versus the north; and in rural and suburban areas versus city centers (PORG 1997).

The time it takes for O_3 to form and then be destroyed in the atmosphere, and hence the distance it can travel, makes it a transboundary pollutant (i.e. O_3 precursors emitted in one country can influence O_3 levels in another). For example, measures taken by the UK to reduce NO_x and VOC emissions will have a significant impact on domestically generated O_3, but some of the ozone problem within the UK stems from pollutants generated outside the UK. Action is therefore required on a European scale in order to reduce ambient levels. Action by one country alone would be of limited effectiveness in the overall reduction of O_3 levels on the regional scale.

Effects of O_3

Health effects of tropospheric O_3. The detrimental effects of increased tropospheric O_3 concentration on lung function and respiratory symptoms have long been known, although until recently no clear association had been proved. Studies now suggest that there is a direct link between summertime hospital visits for respiratory problems, linked with O_3 pollution (e.g. Walters & Ayres 1996; Burnett *et al.* 1997; Sartor *et al.* 1997). Exposure to high concentrations of ozone may cause slight irritation to the eyes and nose. However, if very high levels of exposure (500–1000 p.p.b.) are experienced over several hours, damage to the airway lining followed by inflammatory reactions may occur, although such concentrations do not occur in ambient air, except in the stratosphere. There is also evidence that minor changes in the airways may occur at lower concentrations, down to about 80 p.p.b. The UK air quality objective for the protection of human health, as recommended by the UK EPAQS for O_3, is 50 p.p.b., not to be exceeded more than 10 times per year and measured as the daily maximum of running eight-hour mean, to be achieved by the end of 2005.

Effects of O_3 on materials and vegetation. The occurrence of material damage associated with exposure to ozone was originally reported in the 1960s (Jaffe 1967). Scaling from US estimates made in the 1960s, the estimated cost of damage to materials such as rubber, metal, and stone in the U.K. is £170–345 million per year (Lee *et al.* 1996). Costs in prevention of such damage are estimated at £25–63 million per year to consumers and £25–189 million per year to manufacturing industry. Damage is mainly caused by the attack of unsaturated molecules in polymeric substances such as natural rubber (Norton 1940; Newton 1945; Lewis 1986). It has also been shown that the amount of SO_2-induced corrosion of metals is enhanced in the presence of O_3 (Svensson & Johansson 1993).

Ozone is known to be detrimental to many plant species (Taylor *et al.* 1958; Oshima *et al.* 1975); in particular, the root/shoot ratio is significantly reduced under high ambient ozone concentrations (Hofstra *et al.* 1981).

5.4.2 *Peroxyacetylnitrate (PAN)*

Formation of PAN

Peroxyacetyl nitrate is the first compound in the peroxyacyl nitrate series, whose general formula is $RC(O)OONO_2$. It was first discovered in the 1950s as a component of photochemical smog (Stephens 1969, 1987) but it is now also known to be present in rural regions (Nielson *et al.* 1981; Brice *et al.* 1984). There is interest in this compound through its role as a reservoir species of NO_x, its extreme phytotoxicity (Taylor 1969), and its detrimental effects on human health (it is a powerful lachrymator and is also thought to play a possible role in the growing incidences of skin cancer) (Lovelock 1977).

PAN is formed in the atmosphere via acetyl radicals (CH_3CO) which are formed predominantly from the oxidation of acetaldehyde:

$$CH_3CHO + OH \rightarrow CH_3CO + H_2O \quad (5.3)$$

$$CH_3CO + O_2 \rightarrow CH_3C(O)O_2 \quad (5.4)$$

$$CH_3C(O)O_2 + NO_2 + M$$
$$\rightarrow CH_3C(O)O_2NO_2 + M \quad (5.5)$$

Reaction 5.5 must compete with the following reaction converting NO to NO_2:

$$CH_3C(O)O_2 + NO \rightarrow NO_2 + CH_3C(O)O \quad (5.6)$$

PAN is not readily lost in the troposphere through aqueous-phase scavenging or photodissociation. It exists in thermal equilibrium and the principal loss mechanism for PAN is thermal decomposition back to the peroxy radical and NO_2. Its lifetime is strongly controlled by tropospheric temperatures (~1 day lifetime at 4°C, ~20 days at −12°C, and 100 days at −30°C: Moxim *et al.* 1996), meaning that it is unstable close to the surface but highly stable in the colder regions of the middle and upper troposphere. It was this property that originally led to the speculation that PAN could serve as a stable form of NO_x, transporting it long distances from source regions (Crutzen 1979), and this has been the subject of many investigations (e.g. Heikes *et al.* 1996; Jacob *et al.* 1996; Schultz *et al.* 1999). For example, when studying the impact of fossil fuel combustion emissions on the distribution of reactive nitrogen compounds, Kasibhatla *et al.* (1993) noted that if PAN chemistry was neglected in their studies, the amount of NO_x in remote regions was underestimated. Measurements at the Jungfraujoch show PAN to be the most abundant reactive nitrogen compound present (average of 36%) (Zellweger *et al.* 2000).

Spatial and temporal distribution of PAN

A modeling study by Moxim *et al.* (1996) found that PAN is mostly found in the northern hemisphere, concentrated over continental sites of NO_x emissions in the boundary layer, and measurements confirm that concentrations are significantly higher in urban areas. Concentrations in excess of 50 p.p.b.v. have been recorded in southern California during severe smog episodes (Grosjean

1993). These studies also show that, similarly to O_3, PAN is formed photochemically during transport. Concentrations recorded during European photochemical episodes are significantly lower than those reached in southern California at around 1–4 p.p.b.v., and they show a background concentration of 0.2–0.5 p.p.b.v. (Tsani-Bazaca *et al.* 1988; Tsalkani *et al.* 1987; Wunderli & Gehrig 1991).

PAN accumulates in the coldest regions of the northern hemisphere, generally the highest latitudes and altitudes, but in the southern free troposphere, the maximum PAN levels are found in an Equator to 30° S belt stretching from South America to Australia. PAN has been found to dominate the NO_y budget in the Arctic (Beine & Krognes 2000) and concentrations in the Antarctic (Jacobi *et al.* 2000) are an order of magnitude lower than those measured in high northern latitudes (Jacobi *et al.* 1999). This would be expected, as the Arctic is surrounded by continents that are strong sources of PAN precursors.

5.5 OTHER HAZARDOUS AIR POLLUTANTS

Hazardous air pollutants are emitted from thousands of sources and are known to cause, or are suspected of causing, cancer or other serious human health effects, such as neurological, cardiovascular, and respiratory effects, effects on the liver, kidney, immune, and reproductive system, and effects on child development. They are also known to cause damage to ecosystems. As well as those dealt with above, there are many of these pollutants, but examples include heavy metals such as lead, mercury, and chromium and organic chemicals such as benzene and 1,3-butadiene, polycyclic aromatic hydrocarbons (PAHs), dioxins, and polycyclic biphenyls (PCBs). Many of these substances are capable of accumulating in the environment and can be transported over long distances, and humans become exposed to them not only by breathing polluted air but by ingesting foodstuffs and from polluted waters. Ambient data for individual hazardous air pollutants are presently limited (both spatially and temporally) in comparison to data for other air pollutants.

5.5.1 Benzene and 1,3-butadiene

Benzene and 1,3-butadiene are among the priority pollutants routinely monitored in the UK, as they are recognized genotoxic human carcinogens. Studies of industrial workers exposed in the past to high levels of benzene have demonstrated an excess risk of leukemia, which increased in relation to their working lifetime exposure, and exposure to 1,3-butadiene is thought to induce cancers of the lymphoid system and blood-forming tissues, lymphomas, and leukemia. The UK air quality objectives, based on health advice from the EPAQS for benzene and 1,3-butadiene, are $16.25\,\mu g\,m^{-3}$ (5 p.p.b.) and $2.25\,\mu g\,m^{-3}$ (1 p.p.b.) respectively, both measured as running annual means to be achieved by the end of 2003.

The combustion and distribution of gasoline, of which it is a minor constituent, is the most important source of benzene to the UK atmosphere, as shown in Tables 5.13 and 5.14. It is also formed during the combustion of aromatics in the gasoline. Exhaust gases and evaporation from gasoline vehicles accounted for 56 and 4% of total benzene emissions respectively in 1997. Similarly, the

Table 5.13 Estimated emissions of 1,3-butadiene for 1999 by UNECE source category (from DEFRA, *Digest of Environmental Statistics* 2001).

Source	t	% of total
Combustion in fuel extraction and transformation	3	0
Commercial, public, and agricultural combustion	0	0
Other industrial combustion	0	0
Production processes	372	6
Extraction and distribution of fossil fuels	28	0
Road transport	5260	85
Other transport	492	8
Waste treatment and disposal	13	0
Total	6167	100

Table 5.14 Estimated emissions of benzene for 1999 by UNECE source category (from DEFRA, *Digest of Environmental Statistics* 2001).

Source	kt	% of total
Refineries	7	0
Combustion in fuel extraction and transformation	148	0
Domestic	2,971	10
Commercial, public, and agricultural combustion	146	0
Iron and steel	251	1
Other industrial combustion	589	2
Production processes	1,792	6
Extraction and distribution of fossil fuels	930	3
Road transport	20,963	71
Other transport	1,822	6
Waste treatment and disposal	71	0
Total	29,690	100

major source of 1,3-butadiene in the UK is from road transport, accounting for some 68% of the total UK emissions. The majority of these emissions are derived from gasoline combustion, with the only other major sources arising from the chemical industry through the manufacture and use of 1,3-butadiene and in the production of synthetic rubber for tyres.

Benzene emissions fell by 39% between 1990 and 1998 and 1,3-butadiene emissions fell by 46% over the same period. Projections show that annual urban road emissions of both compounds will continue to decrease as new emission limits for cars are agreed and new fuel quality standards are introduced (DETR 2000).

5.5.2 *Polycyclic aromatic hydrocarbons*

PAHs are hydrocarbons that are composed of two or more fused benzene rings. More than 100 species have been identified in the atmosphere, ranging from naphthalene (two benzene rings; RMM 128) to coronene (seven rings; RMM 300). The larger PAHs, containing five or more aromatic rings, are found in the atmosphere predominantly as aerosols, while naphthalene exists exclusively in the gas phase (Wayne 2000). PAHs are very persistent organic pollutants in the environment, the importance of which resides in their impact upon human health. The toxicity of PAHs varies widely but some of them are carcinogenic, the most potent including benzo(a)pyrene, benzofluoranthenes, benz(a)anthracene, dibenzo(ah)anthracene, and indeo (1,2,3-cd) pyrene. Such PAHs are recognized by the EU and the United States Environmental Protection Agency (USEPA) as priority pollutants.

The major source of PAHs is the incomplete combustion of organic materials. A small contribution may come from natural sources such as forest fires or volcanoes, but the predominant sources are anthropogenic and include motor vehicles (both diesel and gasoline), stationary power plants (coal- and oil-fired), and domestic (coal and wood burning, tobacco smoke), as well as deliberate biomass burning.

Road transport is the predominant source of PAHs in the UK, accounting for 28% of total emissions (Salway *et al.* 1997). Other major sources include combustion of domestic coal and wood, accounting for 18 and 23% of the total emission respectively in 1994. The relatively inefficient combustion conditions of domestic fires and boilers compared to power stations and industrial combustion accounts for the higher emissions from domestic sources. Aluminum and iron and steel production also contribute. Emissions have decreased substantially over time, as shown in Table 5.15 for selected persistent organic pollutants.

5.5.3 *Dioxins*

Dioxin refers to polychlorinated dibenzo-p-dioxins (PCDD) and polychlorinated dibenzofurans (PCDF), and emissions monitored in the UK cover 17 PCDD and PCDF congeners, including 2,3,7,8-TCDD, the most toxic. PCDD/Fs possess a number of toxicological properties, but the main concern is regarding their possible role in immunologic and reproductive effects. Their main sources are thermal processes where chlorine is

Table 5.15 Estimated total emissions of persistent organic pollutants in the UK, 1990–9 (from DEFRA, *Digest of Environmental Statistics* 2001).

	Units	1990	1991	1992	1993	1994	1995	1996	1997	1998	1999
Selected PAHs											
Benz[a]anthroacene	t	83	81	77	69	65	53	30	24	22	21
Naphthalene	t	2074	2003	1827	1310	1162	949	745	677	616	645
Phenanthrene	t	1368	1316	1258	1178	1123	894	350	242	225	185
Pyrene	t	387	374	358	335	318	253	103	72	66	56
PCBs	kg	6976	6397	5901	5407	4846	4292	3750	3248	2747	2071
Dioxins	g	1142	1123	1098	1049	953	819	589	384	361	346

present. They can also be emitted from the chemical production and use of polychlorinated aromatic pesticides and herbicides, many of which are now controlled.

The largest source of dioxins in the UK is incineration of municipal solid waste and clinical waste, which accounted for around 73% of the total in 1993 (Salway *et al.* 1997). Combustion of coal and other solid fuels under certain conditions can result in dioxin formation, but power stations contribute only a small percentage of dioxins to the atmosphere (2% to the UK in 1993), as a large proportion of any dioxins formed are destroyed by high temperatures in the boilers. Emissions from domestic coal combustion are significantly higher (4% in 1993) despite lower coal consumption due to inefficient combustion conditions. Table 5.15 shows the decrease in UK dioxin emissions since 1990.

5.5.4 *Polychlorinated biphenyls (PCBs)*

PCBs are classified as probably carcinogenic to humans and have been linked with subtle chronic effects such as reduced male fertility and long-term behavioral and learning impairment. They are extremely persistent in the environment and possess the ability to accumulate in the food chain—their lipophilic properties mean that they become concentrated in the fatty tissues of animals.

PCBs are synthetic organic compounds that were originally manufactured for use in electrical transformers and capacitors, and as plasticizers.

They have not been manufactured in the UK since the mid-1970s and emissions to the atmosphere are now associated with leaks from poor maintenance of in-service appliances. Leakage from large capacitors and transformers comprised around 92% of the total PCB emissions to the UK atmosphere in 1993 (Salway *et al.* 1997). There is also a small emission from the fragmentation of small capacitors used in cars and household appliances, contributing around 7% to the UK total. Large quantities of PCBs in old appliances have been disposed of to landfill, but now such equipment is disposed of by chemical incineration, where high temperatures and adequate oxygen ensure the destruction of PCBs. There still remain large uncertainties on all emission estimates for PCBs.

5.6 PARTICULATE MATERIAL

Particulate matter is the general term used for a mixture of solid particles and liquid droplets found in the air. Particles can vary widely in size and composition. Some particles are large or dark enough to be seen as soot or smoke, others are much smaller and can only be detected with an electron microscope.

5.6.1 *Sources of global particulate material*

On a global scale particulate material plays a major role in influencing climate: directly through scattering and absorbing radiation and indirectly

Table 5.16 Source strength of various types of aerosol particles. Range reflects estimates reported in the literature. The actual range of uncertainty may encompass values larger and smaller than those reported here (from Penner *et al.* 2001).

	Global flux (Tg yr⁻¹)	Low	High
Primary source			
Carbonaceous aerosols			
Organic matter (0–2 μm)			
Biomass burning	54	45	80
Fossil fuel	28	10	30
Biogenic (>1 μm)	56	0	90
Black carbon (0–2 μm)			
Biomass burning	5.7	5	9
Fossil fuel	6.6	6	8
Aircraft	0.006		
Industrial dust	100	40	130
Soil dust (mineral aerosol)	2150	1000	3000
Sea salt	3340	1000	6000
Secondary source			
Sulfate (as NH₄HSO₄)			
Anthropogenic	122	69	214
Biogenic	57	28	118
Volcanic	21	9	48
Nitrate (as NO₃⁻)			
Anthropogenic	14.2	9.6	19.2
Natural	3.9	1.9	7.6
Organic compounds			
Anthropogenic	0.6	0.3	1.8
VOC			
Biogenic VOC	16	8	40

Soil dust is a major contributor to aerosol loading, especially in subtropical and tropical regions. Estimates of its global source strength range from 1000 to 5000 Mt yr⁻¹ (Duce 1995), with high spatial and temporal variability. Major dust sources are found in the desert regions of the northern hemisphere, while dust emissions in the southern hemisphere are relatively small. Sea-salt aerosols are generated by various physical processes, especially the bursting of entrained air bubbles during whitecap formation (Blanchard 1983; Monahan *et al.* 1986), and the total sea-salt flux from the ocean to the atmosphere is estimated to be 3300 Tg yr⁻¹ (Penner *et al.* 2001 and references therein). Anthropogenic primary particles originate from a wide variety of sources, including transportation, coal combustion, cement manufacturing, metallurgy, and waste incineration. It is estimated that 100 (Andraea 1995) to 200 Tg yr⁻¹ (Wolf & Hidy 1997) are emitted through anthropogenic processes.

Carbonaceous aerosols (consisting of predominantly organic substances and various forms of black carbon) make up a large, but highly variable, proportion of global particulate matter. The main sources of carbonaceous particles are biomass and fossil fuel burning, and the oxidation of biogenic and anthropogenic volatile organic compounds (see Table 5.16). Primary biogenic aerosols consist of plant debris, humic matter, and microbial particles, but there is presently insufficient information available to make a reliable estimate of the contribution of primary biogenic particles to the atmospheric aerosol.

Secondary particulates include sulfate, nitrate, and organic compounds. Sulfate aerosols are produced by chemical reactions in the atmosphere from gaseous precursors such as sulfur dioxide (from fossil fuel burning and volcanoes) and dimethyl sulfide (from marine plankton). Aerosol nitrate is closely tied to the relative abundance of ammonia and sulfate. For example, if ammonia is available in excess of the amount required to neutralize sulfuric acid, nitrate can form small aerosols. In the presence of accumulation mode sulfuric acid containing aerosols, however, nitric acid deposits onto larger, alkaline mineral or salt particles.

through modifying the optical properties and lifetime of clouds (Penner *et al.* 2001). The sources of primary and secondary particulate material are shown in Table 5.16. Although natural sources represent around 90% of the total mass emission to the atmosphere of aerosols and their precursors, it is anthropogenic emissions that are estimated to result in almost half the direct aerosol effect, the major contribution arising from SO₂. On a regional scale, natural sources are often dwarfed by particles arising from human activity.

Volcanic emissions are also important contributors to global particulate matter through direct emission of primary volcanic dust, or through emission of gaseous sulfur species capable of reacting to form secondary sulfate particles. Estimates from this source are highly uncertain, as only a few potential sources have been measured. There is also much variability between sources and between different stages of volcanic activity (Penner *et al.* 2001).

5.6.2 Black smoke

On a regional scale, particulate material can have detrimental effects on the environment and on human health. The largest sets of particulate measurements in the UK are those for black smoke, a measurement related to the blackness or soiling capacity of particles. These show that road transport, particularly the diesel fuel component, and coal burning in domestic premises are now the main sources of black smoke emissions. Burning bituminous coal is prohibited in domestic premises within urban areas, and therefore road transport is by far the major source of black smoke emissions in these areas.

During the London smog of December 1952, airborne concentrations of black smoke exceeded $1500 \mu g\,m^{-3}$, accompanied by massive concentrations of SO_2. The daily death rate increased greatly during this time and it is thought that there were around 4000 premature deaths. This led to legislation to control urban air pollution by smoke and SO_2. The smogs were caused primarily by low-level emissions from coal combustion during periods of meteorology unsuitable for effective pollutant dispersal (low wind speeds and shallow mixing depth). The combination with fog led to dramatic losses in visibility.

5.6.3 *PM₁₀ and PM₂.₅*

Particles are now frequently measured by a method that determines the mass of that fraction which is considered most likely to be deposited in the lung, the most common of which relies on the use of a size-selective sampler that preferentially collects particles smaller than $10 \mu m$. The resultant mass of material is known as PM_{10}. Black smoke and PM_{10} are different and no simple relationship exists between them either for emissions or for ambient concentrations. PM_{10} includes both primary and secondary particulate material, which can be further divided into "fine" (aerodynamic diameters less than or equal to $2.5 \mu m$, often referred to as $PM_{2.5}$) and coarse (aerodynamic diameter greater than $2.5 \mu m$, but less than or equal to $10 \mu m$ material), and it can generally be considered as being composed of three main categories of source:

1 Primary particles emitted directly by combustion processes. These are generally less than $2.5 \mu m$ and often less than $1 \mu m$.
2 Secondary particles formed in the atmosphere from gas-to-particle conversion processes in the atmosphere, consequently composed substantially of sulfates and nitrates formed from the reactions of emissions of SO_2 and NO_x, together with organic and elemental carbon and a range of trace metals (QUARG 1993). These secondary particles are also generally less than $2.5 \mu m$, but the size could vary depending on humidity.
3 Primary particles formed from a variety of primarily noncombustion sources. These include natural events such as wind-blown dusts and soils, forest and other natural fires, and human-influenced sources such as resuspended road dust and tire debris, and construction and mining/quarrying activity. The particles generated by these sources mostly arise from mechanical attrition and are thus relatively large, generally greater than $2.5 \mu m$.

Health effects of PM_{10}

Epidemiological studies have indicated that exposure to airborne particulate matter is connected with an increased incidence in both respiratory and cardiac diseases (Walters & Ayres 1996), and current ambient concentrations are even thought to be sufficient to lead to increased mortality and morbidity. Sensitive groups that appear to be at greatest risk to such effects include the elderly, individuals with cardiopulmonary disease such as asthma, and children. The similarity of exposure–response coefficients from cities in different parts of the world indicates that the chemical composition of the particles is unlikely to be a major driver of these effects on health (Harrison *et al.* 2000). Present-day UK urban concentrations of PM_{10} are orders of magnitude below those found in the major smog episodes of the 1950s and 1960s. For many years after these smogs, UK pollution control policy focused on smoke and SO_2, and it is only in recent years that the focus has transferred to motor traffic as the major source of urban air pollution. In view of the health risks posed by PM_{10} there are two air quality objectives for the UK to be achieved by the end of 2004: $50\,\mu g\,m^{-3}$ not to be exceeded more than 35 times a year, measured as a 24-hour mean, and $40\,\mu g\,m^{-3}$ measured as an annual mean.

Sources of primary PM_{10}

The estimated emissions of anthropogenic primary PM_{10} to the UK atmosphere by UNECE source category are given in Table 5.17.

The main source of primary PM_{10} in the UK is

Table 5.17 UK emissions of PM_{10} for 1999 by source category (from DEFRA, *Digest of Environmental Statistics* 2001).

Source	kt	% of total
Power stations	19	10
Refineries	3	2
Combustion in fuel extraction and transformation	4	2
Domestic	38	20
Commercial, public, and agricultural combustion	5	3
Iron and steel	1	0
Other industrial combustion	18	10
Production processes	11	6
Construction, mining, and quarrying	25	13
Road transport	36	20
Other transport	11	6
Waste treatment and disposal	1	1
Nonlivestock agriculture	14	8
Total	186	100

road transport, predominantly from diesel vehicles. Emissions also arise from brake and tire wear and from the reentrainment of dust on the road surface (QUARG 1996; NETCEN 2001).

Power stations are the largest users of coal in the UK, but all coal-fired stations are equipped with electrostatic precipitators (ESP), which remove on average 99.5% of the particulate matter. The main stationary combustion source of PM_{10} is from domestic premises and this has traditionally been from the use of coal burnt in open fires. However, a combination of the restrictions on coal burning since the smogs of the 1950s and the substantial switch to electricity and gas for domestic heating over much of the UK has meant a decrease from this source. Domestic coal burning can still be a significant source in some smaller towns and villages: for example, in Northern Ireland and in areas associated with the coal industry.

There are a large number of noncombustion industrial sources, including iron and steel, pesticide production, construction, and quarrying, but these tend to dominate locally and so do not contribute a large fraction of the emissions nationally. Agricultural sources such as land preparation, fertilizer application, and harvesting also contribute; these tend to be very seasonal.

Trends in primary PM_{10} emissions

National PM_{10} emissions have declined since 1970 in the UK, mainly due to the reduction in coal use (Salway *et al.* 1997). Emissions from electricity generation have declined dramatically, due to the move away from coal to natural gas for electricity generation, the move toward more efficient power stations, improvements in the performance of electrostatic precipitators at coal-fired power stations, and the increased use of nuclear generation capacity. National emissions from road transport grew steadily, in line with increasing traffic. However, emissions decreased by 30% between 1991 and 1997 due to the introduction into the fleet of new diesel vehicles meeting tighter particle emission regulations. While road transport accounts for 26% of national emissions, it accounts for over 80% of primary emissions inside cities. Among the noncombustion and nontransport sources, the major emissions are from a range of industrial processes and quarrying, and these have remained fairly constant over the period.

Secondary PM$_{10}$

Sulfate produced by the photochemical oxidation of SO$_2$ emitted from combustion sources is perhaps the most well known secondary particulate, in the form of particulate sulfuric acid and ammonium sulfate. This is due to its role in health impacts, reduction in visibility, acid rain, and environmental acidification. Trends of particulate sulfate concentrations in the UK show a steady rise through the 1950s and 1960s to a maximum in the 1970s, after which concentrations started to fall through to the present day. Measurements across Europe show that there is a clear gradient in particulate sulfate, with concentrations decreasing from east to west. Across the UK there is a gradient in levels from about 1.5 μg S m^{-3} in the south and east to about 0.5 μg S m^{-3} in the north and west. In the USA "secondary" fine particles are thought to comprise as much as half of the PM$_{2.5}$ measured (USEPA 2000), although there is much regional variation in concentrations, with higher annual averages in the east compared to the west. This has been mainly attributed to the high amounts of sulfate at eastern sites.

The contribution to urban PM$_{10}$ from particulate nitrate is somewhat smaller than that of particulate sulfate, although it is likely that it is also formed photochemically. Long-term time series (Lee & Atkins 1994) show a steady rise throughout the period from 1954 onwards, without the recent decrease observed for particulate sulfate. Some organic compounds emitted from high-temperature processes are capable of rapidly cooling and condensing onto preexisting particles in the atmosphere, contributing to urban PM$_{10}$. As molecular weight increases and volatility decreases, the organic compounds become increasingly attached to particles rather than remaining in the gas phase.

When one considers the sources of particulate material to the UK atmosphere, while primary particle emissions generally make their largest proportional contribution to local air quality, secondary particles are formed in the atmosphere via processes that are relatively slow compared with airborne travel times. This means that they must be considered a transboundary pollutant and European sources as well as UK precursor emissions must be considered when developing policy surrounding the control of UK particle concentrations.

REFERENCES

AEA (1998) *Options to Reduce Methane Emissions.* Report prepared for the Commission (DGXI). AEA, London.

Allen, D.J., Pickering, K.E., Stenchikov, G., Thompson, A.M. & Kondo, Y. (2000) A three-dimensional total odd nitrogen (NO$_x$) simulation during SONEX using a stretched-grid chemical transport model. *Journal of Geophysical Research* **105**, 3851–76.

Anderson, I.C., Levine, J.S., Poth, M.A. & Riggan, P.J. (1988) Enhanced biogenic emissions of nitric oxide and nitrous oxide following biomass burning. *Journal of Geophysical Research* **93**, 3893–8.

Andraea, M.O. (1995) Climatic effects of changing atmospheric aerosol levels. In Henderson-Sellers, A. (ed.), *World Survey of Climatology. Volume 16: Future Climates of the World.* Elsevier, Amsterdam.

Bachmaier F., Eberius, K.H. & Just, T. (1973) Formation of nitric oxide and the detection of hydrogen cyanide in premixed hydrocarbon–air flames at 1 atmosphere. *Combustion Science and Technology* **7**, 77–84.

Barnard, J.A. & Bradley, J.N. (1985) *Flame and Combustion*, 2nd edn. Chapman and Hall, New York.

Bates, T.S., Kelly, K.C., Johnson, J.E. & Gammon, R.H. (1996) A reevaluation of the open ocean source of methane to the atmosphere. *Journal of Geophysical Research* **101**, 6953–61.

Baulch, D.L., Cobos, C.J., Cox, R.A. *et al.* (1992) Evaluated kinetic data for combustion modelling. *Journal of Physical and Chemical Reference Data* **21**, 411–734.

Beine, H.J. & Krognes, T. (2000) The seasonal cycle of peroxyacetyl nitrate (pan) in the European Arctic. *Atmospheric Environment* **34**, 933–40.

Blanchard, D.C. (1983) The production, distribution and bacterial enrichment of the sea-salt aerosol. In Liss, P.S. & Slinn, W.G.N. (eds), *Air–Sea Exchange of Gases and Particles.* Reidel, Boston.

Boissard, C., Bonsang, B., Kanakidou, M. & Lambert, G. (1996) TROPOZ II: global distributions and budgets of methane and light hydrocarbons. *Journal of Atmospheric Chemistry* **25**, 115–48.

Borbon, A., Fontaine, H., Veillerot, M., Locoge, N., Galloo, J.C. & Guillermo, R. (2001) An investigation into the traffic-related fraction of isoprene at an urban location. *Atmospheric Environment* **35**, 3749–60.

Bradshaw, J.D., Davis, D., Grodzinsky, G. *et al.* (2000) Observed distributions of nitrogen oxides in the remote free troposphere from the NASA Global Tropospheric Experiment Programs. *Review Geophysical* **38**, 61–116.

Brewer, D.A., Ogliaruse, M.A., Augustsson, T.R. & Levine, J.S. (1984) The oxidation of isoprene in the troposphere: mechanism and model calculations. *Atmospheric Environment* **18**, 2723–44.

Brice, K.A., Penkett, S.A., Atkins, D.H.F. *et al.* (1984) Atmospheric measurements of peroxyacetylnitrate (pan) in rural, south-east England: seasonal variations, winter photochemistry and long-range transport. *Atmospheric Environment* **18**, 2691–702.

Brimblecombe, P. (1996) *Atmospheric Composition and Chemistry*, 2nd edn. Cambridge University Press, Cambridge.

Burnett, R.T., Brook, J.R., Yung, W.T., Dales, R.E. & Krewski, D. (1997) Association between ozone and hospitalisation for respiratory diseases in 16 Canadian cities. *Environmental Research* **72**, 24–31.

Calpaldo, K., Corbett, J.J., Kasibhatla, P., Fischbeck, P. & Pandis, S.N. (1999) Effects of ships' emissions on sulphur cycling and radiative climate forcing over the ocean. *Nature* **400**, 743–6.

Chameides, W.L., Lindsay, R.W., Richardson, J. & Kiang, C.S. (1988) The role of biogenic hydrocarbons in urban photochemical smog; Atlanta as a case study. *Science* **241**, 1473–5.

Chappellaz, J., Fung, I.Y. & Thompson, A.M. (1993) The atmospheric CH_4 increase since the last glacial maximum. *Tellus* **45**, 228–41.

Christensen, C.S., Skov, H. & Palmgren, F. (1999) C_5–C_8 non-methane hydrocarbon measurements in Copenhagen: concentrations, sources and emission estimates. *The Science of the Total Environment*, **236**, 163–71.

Cicerone, R.J. & Oremland, R.S. (1988) Biogeochemical aspects of atmospheric methane. *Global Bioceochemical Cycles* **2**, 299–327.

Colbeck, I. & Mackenzie, A.R. (1994) *Air Pollution by Photochemical Oxidants*. Air Quality Monographs, Volume 1. Elsevier Science, Amsterdam.

Corbett, J.J. & Fischbeck, P.S. (1997) Emissions from ships. *Science* **278**, 823–4.

Corbett, J.J., Fischbeck, P.S. & Pandis, S.N. (1999) Global nitrogen and sulfur emissions inventories for ocean-going ships. *Journal of Geophysical Research* **104**, 3457–70.

Cox, R.A. & Derwent, R.G. (1981) Gas kinetics and energy transfer. Specialist *Periodical Reports Chemistry Society* **4**, 189.

Crawford, J., Davis, D., Olson, J. *et al.* (1999) Assessment of upper tropospheric HO_x sources over the tropical Pacific based on NASA/GTE PEM data: net effect on HO_x and other photochemical parameters. *Journal of Geophysical Research* **104**, 16,255–73.

Crutzen, P.J. (1979) The role of NO and NO_2 in the chemistry of the troposphere and stratosphere. *Annual Review of Earth Planet Science* **7**, 443–72.

Crutzen, P.J. (1991) Methane sources and sinks. *Nature* **350**, 380–1.

Department for Environment, Food and Rural Affairs (2001) *Digest of Environmental Statistics, 2001.* Internet version (www.defra.gov.uk/environment/statistics/des/index.htm).

Department of the Environment, Transport and the Regions (2000) *The Air Quality Strategy for England, Scotland, Wales and Northern Ireland. Working Together for Clean Air.* HMSO, London.

Derwent, R.G. (1993) Hydrocarbons in the atmosphere: their sources, distributions and fates. Proceedings of the International Conference on Volatile Organic Compounds in the Environment, London, October 27–28.

Derwent, R.G., Simmonds, P.G., Seuring, S. & Dimmer, C. (1998) Observations and interpretation of the seasonal cycles in the surface concentrations of ozone and carbon monoxide at Mace Head, Ireland from 1990 to 1994. *Atmospheric Environment* **32**, 145–57.

Dianov-Klokov, V.I.L., Yurganov, N., Grechko, E.I. & Dzola, A.Z. (1989) Spectroscopic measurements of atmospheric carbon monoxide and methane 1: latitudinal distribution. *Journal of Atmospheric Chemistry* **8**, 139–51.

Dickinson, R.E. & Cicerone, R.J. (1986) Future global warming from atmospheric trace gases. *Nature* **319**, 109–15.

Duce, R. (1995) Distributions and fluxes of mineral aerosol. In Charlson, R.J. & Heintzenberg, J. (eds), *Aerosol Forcing of Climate*. Wiley, Chichester.

Ehhalt, D.H., Rudolph, J., Meixner, F. & Schmidt, U. (1985) Measurements of selected C_2–O_5 hydrocarbons in the background troposphere: vertical and latitudinal variations. *Journal of Atmospheric Chemistry* **3**, 29–52.

Erickson, D.J. III & Taylor, J.A. (1992) 3-D tropospheric co modelling: the possible influence of the ocean. *Geophysical Research Letters* **19**, 1955–8.

EEA (2000) *Emissions of Atmospheric Pollutants in Europe, 1980–1996.* Topic report No. 9/2000. European Environment Agency, Copenhagen.

Farman, J.C., Gariner, B.G. & Shanklin, J.D. (1985) Large losses of total ozone in Antarctica reveal seasonal ClO_x/NO_x interaction. *Nature* **315**, 207–10.

Firestone, M.K. & Davidson, E.A. (1989) Microbiological basis of NO and N_2O production and consumption in soil. In Andreae, M.O. & Schimel, D.S. (eds), *Exchange of Trace Gases between Terrestrial Ecosystems and the Atmosphere.* Wiley, Chichester.

Fishman, J., Fakhruzzaman, K., Cros, B. & Nganga, D. (1991) Identification of widespread pollution in the southern hemisphere deduced from satellite analysis. *Science* **252**, 1693–6.

Fung, I., John, J., Lerner, J., Matthews, E., Prather, M., Steele, L.P. & Fraser, P.J. (1991) Three-dimensional model synthesis of the global methane cycle. *Journal of Geophysical Research* **96**, 13,033–65.

Galbally, I.E. (1989) Factors controlling NO_x emissions from soils. In Andreae, M.O. & Schimel, D.S. (eds), *Exchange of Trace Gases between Terrestrial Ecosystems and the Atmosphere.* Wiley, Chichester.

Glassman, I. (1987) *Combustion*, 2nd edn. Academic Press, Orlando, FL.

Granier, C., Müller, J.F., Madronich, S. & Brasseur, G.P. (1996) Possible causes for the 1990–1993 decrease in the global tropospheric CO abundance: a three-dimensional study. *Atmospheric Environment* **30**, 1673–82.

Granier, C., Müller, J.F., Petron, G. & Brasseur, G. (1999) A three-dimensional study of the global CO budget. *Chemosphere: Global Change Science* **1**, 255–61.

Granier, C., Peutron, G., Müller, J.F. & Brasseur, G. (2000) The impact of natural and anthropogenic hydrocarbons on the tropospheric budget of carbon monoxide. *Atmospheric Environment* **34**, 5255–70.

Grosjean, D. (1993) Distribution of atmospheric nitrogenous pollutants at Los Angeles area smog receptor sites. *Environmental Science and Technology* **17**, 13–19.

Guenther, A., Hewitt, C.N., Erickson, D. *et al.* (1995) A global model of natural volatile organic compound emissions. *Journal of Geophysical Research* **100**, 8873–92.

Hagerman, L.M., Aneja, V.P. & Lonneman, W.A. (1997) Characterisation of non-methane hydrocarbons in the rural southeast United States. *Atmospheric Environment* **31**, 4017–38.

Harrison, R.M., Yin, J., Mark, D. *et al.* (2000) Studies of the coarse particle (2.5–10 µm) component in the UK urban atmospheres. *Atmospheric Environment* **35**, 3667–79.

Harriss, R.C., Sebacher, D.I. & Day, F.P. Jr (1982) Methane flux in the Great Dismal Swamp. *Nature* **297**, 673–4.

Haynes, R.J. (1986) *Mineral Nitrogen in the Plant–Soil System.* Academic Press, London.

Heikes, B., Lee, M., Jacob, D. *et al.* (1996) Ozone, hydroperoxides, oxides of nitrogen and hydrocarbon budgets in the marine boundary layer over the South Atlantic. *Journal of Geophysical Research* **101**, 24,221–34.

Hofstra, G., Ali, A., Wukasch, R.T. & Fletcher, R.A. (1981) The rapid inhibition of root respiration after exposure of bean (*Phaseolus vulgaris* L.) plants to ozone. *Atmospheric Environment* **15**, 483–7.

Hough, A.M. & Derwent, R.G. (1987) Computer modelling studies of the distribution of non-methane hydrocarbons from the biosphere to the atmosphere in the UK: present knowledge and uncertainties. *Atmospheric Environment* **26**, 3069–77.

Houghton, J. (1997) *Global Warming. The Complete Briefing*, 2nd edn Cambridge University Press, Cambridge.

Houghton, J.T., Meira, L.G., Callander, B.A., Harris, N., Kattenberg, A. & Maskell, K. (eds) (1995) *The Science of Climate Change.* Contribution of Working Group I to the Second Assessment Report of the Intergovernmental Panel on Climate Change. Cambridge University Press, Cambridge.

IRRI (1989) *IRRI toward 2000 and Beyond.* Manila, International Rice Research Institute.

Isodorov, V.A., Zenkevich, I.G. & Ioffe, B.V. (1985) Volatile atmospheric compounds in the atmosphere of forests. *Atmospheric Environment* **19**, 1–8.

Jacob, D.J., Heikes, B.G., Fan, S.-M. *et al.* (1996) Origin of ozone and NO_x in the tropical troposphere: a photochemical analysis of aircraft observations over the South Atlantic Basin. *Journal of Geophysical Research* **101**, 24,235–50.

Jacobi, H.-W., Weller, R., Bluszez, T. & Schrems, O. (1999) Latitudinal distribution of peroxyacetyl nitrate (PAN) over the Atlantic Ocean. *Journal of Geophysical Research* **104**, 26,901–12.

Jacobi, H.-W., Weller, R., Jones, A.E., Anderson, P.S. & Schrems, O. (2000) Peroxyacetyl nitrate (PAN) concentrations in the Antarctic troposphere measured during the Photochemical Experiment at Neumayer (PEAN'99), *Atmospheric Environment* **34**, 5235–47.

Jaffe, L.S. (1967) The effects of photochemical oxidants on material. Air Pollution Control Association **17**, 375–8.

Jobson, B.T., Wu, Z. & Niki, H. (1994) Seasonal trends of isoprene, C_2–C_5 alkanes, and acetylene at a remote boreal site in Canada. *Journal of Geophysical Research* **99**, 1589–699.

Johnson, D.E., Johnson, K.A., Ward, G.M. & Branine, M.E. (2000) Ruminants and other animals. In Khalil, M.A.K. (ed.), *Atmospheric Methane. Its Role in the Global Environment*. Springer-Verlag, Berlin.

Junge, C.E. (1963) *Air Chemistry and Radioactivity*. Academic Press, New York.

Kammen, D.M. & Marino, B.D. (1993) On the origin and magnitude of pre-industrial anthropogenic CO_2 and CH_4 emissions. *Chemosphere* **26**, 69–86.

Kasibhatla, P.S., Levy, H. II & Moxim, W.J. (1993) Global NO_x, HNO_3, PAN and NO_y distributions from fossil-fuel combustion emissions: a model study. *Journal of Geophysical Research* **98**, 7165–80.

Khalil, M.A.K. & Rasmussen, R.A. (1983) Sources, sinks and seasonal cycles of atmospheric methane. *Journal of Geophysical Research* **88**, 5131–44.

Khalil, M.A.K. & Rasmussen, R.A. (1985) Causes of increasing atmospheric methane: depletion of hydroxyl radicals and the rise of emissions. *Atmospheric Environment* **19**, 397–407.

Khalil, M.A.K. & Rasmussen, R.A. (1987) Atmospheric methane: trends over the last 10,000 years. *Atmospheric Environment* **21**, 2445–52.

Khalil, M.A.K. & Rasmussen, R.A. (1990) Atmospheric methane: recent global trends. Environmental Science and Technology **24**, 549–53.

Khalil, M.A.K. & Rasmussen, R.A. (1994) Global decrease in atmospheric carbon monoxide concentration. *Nature* **370**, 639–41.

Khalil, M.A.K. & Shearer, R.A. (2000) Sources of methane: an overview. In Khalil M.A.K. (ed.), *Atmospheric Methane. Its Role in the Global Environment*. Springer-Verlag, Berlin.

Khalil, M.A.K., Shearer, M.J. & Rasmussen, R.A. (1996) Atmospheric methane over the last century. *World Resource Review* **8**, 481–92.

Krol, M., van Leeuwen, P.J. & Lelieveld, J. (1998) Global OH trend inferred from methylchloroform measurements. *Journal of Geophysical Research* **103**, 10,697–711

Lamb, B., Westberg, H. & Allwine, G. (1986) Isoprene emission fluxes determined by an atmospheric tracer technique. *Atmospheric Environment* **20**, 1–8.

Lambert, G. & Schmidt, S. (1993) Reeavluation of the oceanic flux of methane: uncertainties and long term variations. *Chemosphere* **26**, 95–109.

Lawrence, M.G., Chameides, W.L., Kasibhatla, P.S., Levy, H. & Moxim, W. (1995) Lightning and atmospheric chemistry: the rate of atmospheric NO production. In *Handbook of Electrodynamics, Volume 1*. CRC Press, Boca Raton, FL.

Lee, D.S. & Atkins, D.H.F. (1994) Atmospheric ammonia emissions from agricultural waste combustion. *Geophysical Research Letters* **21**, 281–4.

Lee, D.S. & Nicholson, K.W. (1994) The measurement of atmospheric concentrations and deposition of semi-volatile organic compounds. *Environment Mon. Assessment* **32**, 59–91.

Lee, D.S., Holland, M.R. & Falla, N. (1996) The potential impact of ozone on materials in the UK. *Atmospheric Environment* **30**, 1053–65.

Lee, D.S., Köhler, I., Grobler, E. *et al.* (1997) Estimations of global NO_x emissions and their uncertainties. *Atmospheric Environment* **31**, 1735–49.

Levy, H., Moxim, W.J. & Kasibhatla, P.S. (1996) A global three-dimensional time-dependent lightning source of tropospheric NO_x. *Journal of Geophysical Research* **101**, 22,911–22.

Levine, J.S., Rinsland, C.P. & Tennille, G.M. (1985) The photochemistry of methane and carbon monoxide in the troposphere in 1950 and 1985. *Nature* **318**, 254–7.

Lewis, P.M. (1986) Effect of ozone on rubbers: countermeasures and unsolved problems. *Polymer Degradation and Stability* **15**, 33–66.

Lippman, M. (1992) Health effects of atmospheric acidity. In Radojevic, M. & Harrison, R.M. (eds), *Atmospheric Acidity: Sources, Consequences and Abatement*. Elsevier Applied Science, London.

Lobert, J.M., Scharffe, W.-M., Kuhlsbusch, R., Seuwen, P. & Warneck, P.J. (1991) Experimental evaluation of biomass burning emissions: nitrogen and carbon containing compounds. In Levine, J.S. (ed.), *Global Biomass Burning: Atmospheric, Climatic, and Biospheric Implications*. MIT Press, Cambridge, MA.

Logan, J.A. (1983) Nitrogen oxides in the troposphere: global and regional budgets. *Journal of Geophysical Research* **88**, 10,785–807.

Logan, J.A., Prather, M., Wofsy, S.C. & McElroy, C. (1981) Tropospheric chemistry: a global perspective. *Journal of Geophysical Research* **86**, 7210–54.

Lovelock, J.E. (1977) PAN in the natural environment; its possible significance in the epidemiology of skin cancer, *Ambio* **6**, 131–3.

Lu, Y. & Khalil, M.A.K. (1991) Tropospheric OH: model calculations of spatial, temporal and secular variations. *Chemosphere* **23**, 397–444.

McLaren, R., Singleton, D.L., Lai, J.Y.K. *et al.* (1996) Analysis of motor vehicle sources and their contribution to ambient hydrocarbons distributions at urban sites in Toronto during the Southern Ontario Oxidants Study. *Atmospheric Environment* **30**, 2219–32.

Madronich, S. & Granier, C. (1992) Impact of recent total ozone changes on tropospheric ozone photodissociation, hydroxyl radicals, and methane trends. *Geophysical Research Letters* **19**, 465–7.

Middleton, P. (1995) Sources of air pollutants. In Singh, H.B. (ed.), *Composition, Chemistry, and Climate of the Atmosphere*. Van Nostrand Reinhold, New York.

Miyoshi, A.S., Hatakeyama, N. & Washida, N. (1994) OH radical initiated photooxidation of isoprene. *Journal of Geophysical Research* **97**, 20,703–15.

Molina, M.J. & Rowland, F.S. (1974) Stratospheric sink for chlorofluoromethanes – chlorine catalyzed destruction of ozone. *Nature* **249**, 810–12.

Monahan, E.C., Spiel, D.E. & Davidson, K.L. (1986) A model of marine aerosol generation via whitecaps and wave disruption in oceanic whitecaps. In Monahan, E.C. & Niocaill, G.M. (eds), *Oceanic Whitecaps and Their Role in Air–Sea Exchange Processes*. D. Reidel, Dordrecht.

Moxim, W.J., Levy, H. II & Kasibhatla, P.S. (1996) Simulated global tropospheric PAN: its transport and impact on NO_x. *Journal of Geophysical Research* **101**, 12,621–38.

Müller, J.-F. (1992) Geographical distribution and seasonal variation of surface emissions and deposition velocities of atmospheric trace gases. *Journal of Geophysical Research* **97**, 3787–804.

Müller, J.-F. & Brasseur, G. (1995) IMAGES: a three-dimensional chemical-transport model of the global troposphere. *Journal of Geophysical Research* **100**, 19,015–33.

Müller R., Crutzen, P.J., Gross, J.U. *et al.* (1997) Severe chemical ozone loss in the Arctic during the winter of 1995–96. *Nature* **389**, 709–12.

NETCEN (2001) *Digest of Environmental Statistics.* Internet version (www.defra.gov.uk/environment/statistics/des/index.htm).

Neue, H.-U. & Roger, P.A. (2000) Rice agriculture: factors controlling emissions. In Khalil, M.A.K. (ed.), *Atmospheric Methane. Its Role in the Global Environment*. Springer-Verlag, Berlin.

Newton, R.G. (1945) Mechanism of exposure-cracking of rubbers (with a review of the influence of ozone). *Journal of Rubber Research* **14**, 87–90.

Nielson, T., Sammnelsson, U., Grennfelt, P. & Thompsen, E.L. (1981) Peroxyacetylnitrate in long-range transported polluted air. *Nature* **293**, 553.

Norton, F.J. (1940) Action of ozone on rubber. *Rubber Age* **47**, 27–39.

Novelli, P.C., Steele, P. & Tans, P.P. (1992) Mixing ratios of carbon monoxide in the troposphere. *Journal of Geophysical Research* **97**, 20,731–50.

Novelli, P.C., Masarie, K.A., Tans, P.P. & Lang, P.M. (1994) Recent changes in atmospheric carbon monoxide. *Science* **263**, 1587–90.

Novelli, P.C., Masarie, K. & Lang, P.M. (1998) Distributions and recent changes of CO in the lower troposphere. *Journal of Geophysical Research* **103**, 19,015–33.

Oshima, R.J., Taylor, O.C., Braegalmann, P.K. & Balwin, D.W. (1975) The effect of ozone on the yield and plant biomass of a commercial variety of tomato. *Journal of Environmental Quality* **4**, 463–4.

Paulson, S.E. & Seinfeld, J.H. (1992) Development and evaluation of a photooxidation mechanism for isoprene. *Journal of Geophysical Research* **97**, 20,703–15.

Penner, J.E., Andreae, M., Annegarn, H. *et al.* (2001) Aerosols, their direct and indirect effects. In *Climate Change 2001: The Scientific Basis*. Contribution of Working Group I to the Third Assessment Report of the Intergovernmental Panel on Climate Change. Cambridge University Press, Cambridge.

Pham, M., Müller, J.F., Brasseur, G., Granier, C. & Mégie, G. (1995) A three dimensional study of the tropospheric sulfur cycle. *Journal of Geophysical Research* **100**, 20,061–92.

Piccot, S.D., Watson, J.J. & Jones, J.W. (1992) A global inventory of volatile organic compound emissions from anthropogenic sources. *Journal of Geophysical Research* **97**, 9897–912.

Pinto, J. & Khalil, M.A.K. (1991) The Stability of tropospheric OH during ice ages, interglacial epochs and modern times. *Tellus* **43B**, 347–52.

PORG (1997) *Ozone in the United Kingdom*. Photochemical Oxidants Review Group, Department of the Environment, London.

Prather, M., Derwent, R., Ehhalt, D., Fraser, P., Sanhueza,

E. & Zhou, X. (1995) Other trace gases and atmospheric chemistry. In Houghton, J.T. , Meira, L.G., Callander, B.A., Harris, N., Kattenberg, A. & Maskell, K. (eds) *Radiative Forcing of Climate 1994*. Intergovernmental Panel on Climate Change, Report to the IPCC from the Scientific Assessment Working Group. Cambridge University Press, Cambridge.

Prentice, M., Ehhalt, D., Dentener, F. *et al.* (2001) Atmospheric chemistry and greenhouse gases. In *Climate Change 2001: The Scientific Basis*. Contribution of Working Group I to the Third Assessment Report of the Intergovernmental Panel on Climate Change. Cambridge University Press, Cambridge.

Price, C., Penner, J. & Prather, M. (1997) NO_x from lightning 1. Global distribution based on lightning physics. *Journal of Geophysical Research* **102**, 5929–41.

Prinn, R.G., Cunnold, D. & Simmonds, P. (1992) Global average concentration and trend for hydroxyl radicals deduced from ale gauge trichloroethane (methyl chloroform) data for 1978–1990. *Journal of Geophysical Research* **97**, 2445–61.

QUARG (1993) *Urban Air Quality in the United Kingdom*. Quality of Urban Air Review Group, London.

QUARG (1996) *Airborne Particulate Matter in the United Kingdom*. Quality of Urban Air Review Group, London.

Raynaud, D., Jouzel, J., Barnola, J.M., Chappellaz, J., Delmas, R.J. & Lorius, C. (1993) The ice core record of greenhouse gases. *Science* **259**, 926–34.

Rodkey, F., O'Neal, J. & Collinson, H. (1974) Relative affinity of hemaglobin S and hemaglobin A for carbon monoxide and oxygen. *Clinical Chemistry* **20**, 83–4.

Rudolph, J. (1988) Two-dimensional distribution of light hydrocarbons: results from the STRATOZ III experiment. *Journal of Geophysical Research* **93**, 8367–77.

Salway, A.G., Eggleston, H.S., Goodwin, J.W.L. & Murrells, T.P. (1997) *UK Emissions of Air Pollutants, 1970–1995*. A Report of the National Atmospheric Emissions Inventory, Department of the Environment, Transport and the Regions, London.

Sartor, F., Demuth, C., Snacken, R. & Walckiers, D. (1997) Mortality in the elderly and ambient ozone concentration during the hot summer, 1994, in Belgium. *Environmental Research* **72**, 109–17.

Schimel, D., Alves, D., Enting, I. *et al.* (1995) Radiative forcing of climate change. In *Climate Change 1995: The Scientific Basis*. Contribution of Working Group I to the Second Assessment Report of the Intergovernmental Panel on Climate Change. Cambridge University Press, Cambridge.

Schultz, M.G., Jacob, D.J., Want, Y. *et al.* (1999) On the origin of tropospheric ozone and NO_x over the tropical South Pacific. *Journal of Geophysical Research* **104**, 5829–43.

Seiler, W.A., Holzapfel-Pschorn, C.R. & Scharffe, D. (1984) Methane emission from rice paddies. *Journal of Atmospheric Chemistry* **1**, 241–68.

Seinfeld, J.H. & Pandis, S.N. (1998) *Atmospheric Chemistry. From Air Pollution to Climate Change*. Wiley, New York.

Shishar, L. & Patil, R.S. (2001) Monitoring of atmospheric behavior of NO_x from vehicular traffic. *Environment Mon. Assessment* **68**, 37–50.

Simpson, D., Winiwarter, W., Borjesson, G. *et al.* (1999) Inventorying emissions from nature in Europe. *Journal of Geophysical Research* **104**, 8113–52.

Singh, H.B., Thompson, A. & Schlager, H. (1999) SONEX Airborne mission and co-ordinated POLINAT-2 activity: overview and accomplishments *Geophysical Research Letters* **26**, 3053–6.

Spivakovsky, C.M., Yevich, R., Logan, J.A., Wofsy, S.C. & McElroy, M.B. (1990) Tropospheric OH in a three-dimensional chemical tracer model: an assessment based on observations of CH_3CCl_3. *Journal of Geophysical Research* **95**, 18,441–71.

Steele, L.P., Dlugokencky, E.J., Lang, P.M., Tans, P.P., Martin, R.C. & Masarie, K.A. (1992) Slowing down of the global accumulation of atmospheric methane during the 1980s. *Nature* **358**, 313–16.

Stephens, E.R. (1969) The formation, reactions and properties of peroxyacetyl nitrates (PANs) in photochemical air pollution. *Advances in Environmental Technology* **1**, 119–46.

Stephens, E.R. (1987) Smog studies of the 1950s. EOS **61**, 89–93.

Stohl, A., Williams, E., Wotawa, G. & Kromp-Kolb, H. (1996) A European inventory of soil nitric oxide emissions and the effect of these emissions on the photochemical formation of ozone. *Atmospheric Environment* **30**, 3741–55.

Svensson, J.E. & Johansson, L.G. (1993) A laboratory study of the effect of ozone, nitrogen dioxide and sulfur dioxide on the atmospheric corrosion of zinc. *Journal of the Electrochemical Society* **140**, 2210–16.

Sze, N.D. (1970) Anthropogenic CO emissions: implications for $CO-OH-CH_4$ cycle. *Science* **167**, 984–6.

Taylor, O.C. (1969) Importance of peroxyacetyl nitrate (PAN) as a phytotoxic air pollutant. *Journal of the Air Pollution Control Association* **19**, 347–51.

Thompson, A.M. & Cicerone, R.J. (1986) Possible perturbations to atmospheric CO, CH_4, and OH. *Journal of Geophysical Research* **91**, 10,853–64.

Thompson, A.M., Sparling, L.C., Kondo, Y., Anderson, B.E., Greg, G.L. & Sachse, G.W. (1999) Perspectives on NO, NO$_y$ and fine aerosol sources and variability during SONEX. *Geophysical Research Letters* **26**, 3073–6.

Tsalkani, N., Perros, P., Dutot, A.L. & Toupance, G. (1987) One year measurements of PAN in the Paris Basin: effect of meteorological parameters. *Atmospheric Environment* **25**, 1941–9.

Tsani-Bazaca, E., Glavas, S. & Güsten, H. (1988) Peroxyacetyl nitrate (PAN) concentrations in Athens, Greece. *Atmospheric Environment* **22**, 2283–6.

UNEP (1987) *The Montreal Protocol on Substances that Deplete the Ozone Layer.* United Nations Environment Programme, Nairobi.

USEPA (2000) *National Air Pollutant Emission Trends, 1900–1998. USEPA and the States – Working Together for Cleaner Air!* EPA-454/R-00–002, United States Environmental Protection Agency, Washington, DC.

Varon, J. & Marik, P.E. (1997) Carbon monoxide poisoning. *Internet Journal of Emergency and Intensive Care Medicine* **1**(2) (www.ispub.com/journals/IJEICM/Vol1N2/CO.htm).

Walen, S.C. & Reeburgh, W.S. (1992) Interannual variations in tundra methane emission: a 4-year time series at fixed sites. *Global Biogeochemical Cycles* **6**, 139–59.

Walters, S. & Ayres, J. (1996) The health effects of air pollution. In Harrison, R.M. (ed.), *Pollution. Causes, Effects and Control,* 3rd edn. Royal Society of Chemistry, Cambridge.

Warneck, P. (1988) *Chemistry of the Natural Atmosphere.* Academic Press, San Diego.

Watson, R.T. *et al.* (1990) In Houghton, J.T., Jenkins, G.J. & Ephraums, J.J. (eds), *Climate Change: The IPCC Scientific Assessment.* Cambridge University Press.

Wayne, R.P. (2000) *Chemistry of Atmospheres,* 3rd edn. Oxford University Press, Oxford.

Weber, E. (1982) *Air Pollution: Assessment Methodology and Modelling, Volume 2.* Plenum Press, New York.

Went, F.W. (1960) Organic matter in the atmosphere and its possible relation to petroleum formation. *Proceedings of the National Academy of Science* **46**, 212–21.

WMO (1992) *Scientific Assessment of Ozone Depletion: 1991.* WMO Global Ozone Research and Monitoring Project, Report No. 25. World Meteorological Organization, Geneva.

Wolf, M.E. & Hidy, G.M. (1997) Aerosols and climate: anthropogenic emissions and trends for 50 years. *Journal of Geophysical Research Atmos.* **102**, 11,113–21.

Wunderli, S. & Gehrig, R. (1991) Influence of temperature on formation and stability of surface PAN and ozone: a two year field study in Switzerland. *Atmospheric Environment* **25**, 1599–608.

Young, V.L., Kieser, B.N., Chen, S.P. & Niki, H. (1997) Seasonal trends and local influences on nonmethane hydrocarbon concentrations in the Canadian boreal forest. *Journal of Geophysical Research* **102**, 5913–18.

Zellweger, C., Ammann, M., Buchmann, B. *et al.* (2000) Summertime NO$_y$ speciation at the Jungfraujoch, 3580 m above sea level, Switzerland. *Journal of Geophysical Research* **105**, 6655–67.

Zimmerman, P.R., Chatfield, R.B., Fishman, J., Crutzen, P.J. & Hanst, P.L. (1978) Estimates of the production of CO and H$_2$ from the oxidation of hydrocarbon emissions from vegetation. *Geophysical Research Letters* **5**, 679–82.

6 Tropospheric Photochemistry

PAUL S. MONKS

6.1 INTRODUCTION

The troposphere is the lowest region of the atmosphere, extending from the Earth's surface to the tropopause at 10–18 km. About 90% of the total atmospheric mass resides in the troposphere, and the greater part of the trace gas burden is found there. The troposphere is well mixed and its bulk composition is 78% N_2, 21% O_2, 1% Ar, and 0.036% CO_2 with varying amounts of water vapor depending on temperature and altitude. The majority of the trace species found in the atmosphere are emitted into the troposphere from the surface and are subject to a complex series of chemical and physical transformations. It is becoming apparent that human activities are beginning to change the composition of the troposphere over a range of scales, leading to increased acid deposition, local and regional ozone episodes, and potentially climate change. In order to understand both the "natural" troposphere and the more perturbed state, there is a requirement to understand the fundamentals of tropospheric chemistry and its interactions.

In general, tropospheric chemistry is analogous to a low-temperature combustion system, i.e.

$$CH_4 + O_2 \rightarrow CO_2 + 2H_2 \qquad (6.1)$$

Unlike combustion, this is not a thermal process but a radical mediated process initiated and propagated by photochemistry. The chemistry that takes place in the troposphere, and in particular the photochemistry, is intrinsically linked to the chemistry of ozone. Tropospheric ozone acts as initiator, reactant, and product in much of the oxidation chemistry that takes place in the troposphere, and stratospheric ozone determines the amount of short-wavelength radiation available to initiate photochemistry. Figure 6.1 shows a typical ozone profile through the atmosphere illustrating a number of interesting points. First, 90% of atmospheric ozone can be found in the stratosphere; on average about 10% can be found in the troposphere. Second, the troposphere, in the simplest sense, consists of two regions. The lowest kilometer or so contains the planetary boundary layer (Stull 1988) and inversion layers, which can act as pre-concentrators for atmospheric emissions from the surface and hinder exchange to the so-called **free** troposphere, the larger part by volume, which sits above the boundary layer.

For a long time, transport from the stratosphere to the troposphere was thought to be the dominant source of ozone in the troposphere (Junge 1962; Fabian & Pruchniewicz 1977). Early in the 1970s Crutzen (1973) and Chameides and Walker (1973) first suggested that tropospheric ozone originated mainly from production within the troposphere by photochemical oxidation of CO and hydrocarbons catalyzed by HO_x and NO_x. These sources are balanced by *in situ* photochemical destruction of ozone and by dry deposition at the Earth's surface. Figure 6.2 gives a schematic representation of the tropospheric ozone budget. Many studies, both experimental- and model-based, have set about determining the contribution of both chemistry and

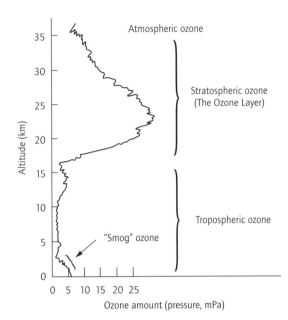

Fig. 6.1 A typical atmospheric ozone profile.

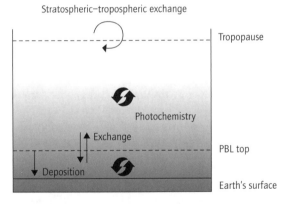

Fig. 6.2 Schematic representation of the physical and chemical factors that effect the tropospheric ozone budget.

transport to the tropospheric ozone budget on many different spatial and temporal scales.

There is growing evidence that the composition of the troposphere is changing. For example, analysis of historical ozone records has indicated that tropospheric ozone levels in both hemispheres has increased by a factor of three to four over the past century (Volz & Kley 1988; Anfossi *et al.* 1991; Sandroni *et al.* 1992) (see Fig. 6.3). Methane concentrations have effectively doubled over the past 150 years (Etheridge *et al.* 1992; Lelieveld *et al.* 1998). N_2O levels have risen by 15% since preindustrial times (Machida *et al.* 1995). Measurements of halocarbons have shown this group of chemically and radiatively important gases to be increasing in concentration (e.g. Prinn *et al.* 2000) until relatively recently.

One of the difficulties about discussing tropospheric chemistry in general terms is that by the very nature of the troposphere being the lowest layer of the atmosphere it has complex multiphase interactions with the Earth's surface, which can vary considerably between expanses of ocean to deserts. The fate of any chemical species (C_i) in the atmosphere can be represented as a continuity or mass balance equation such as

$$\frac{dC_i}{dt} = \frac{duC_i}{dx} - \frac{dvC_i}{dy} - \frac{dwC_i}{dz} + K_z\frac{dC_i}{dz}$$
$$+ P_i - L_i + S_i + \left(\frac{dC_i}{dt}\right)_{clouds} \quad (6.2)$$

where t is the time, u, v, and w are the components of the wind vector in x, y, and z accounting for the horizontal and vertical large-scale transport. Small-scale turbulence can be accounted for using K_z, the turbulent diffusion coefficient, P_i and L_i are the chemical production and loss terms, and S_i are the sources owing to emissions. Cloud processes (vertical transport, washout, and aqueous phase chemistry) are represented in a cloud processing term. The application of this type of equation is the basis of chemical modeling. In this chapter we concentrate mainly on the chemical terms in this equation and the processes that control them, but inherently, as the study of tropospheric photochemistry is driven by observations, these must be placed within the framework of eqn 6.2.

The following chapter describes the role of light in the initiation of photochemistry, as it is this process that is mainly responsible for driving the

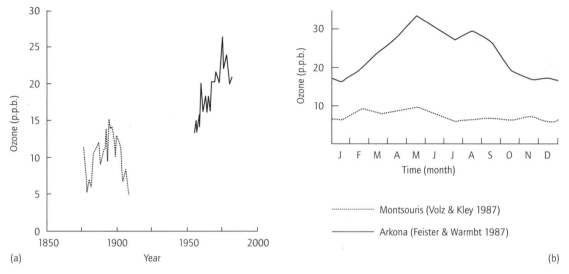

Fig. 6.3 (a) A comparison of the annual mean ozone concentrations at Montsouris (1876–86) and at Arkona (1956–83). (b) Average seasonal variation of ozone at Montsouris (1876–86, SW sector only) and at Arkona. (Reprinted with permission from *Nature* (Volz & Kley 1988). Copyright 1988 Macmillan Magazines Limited.)

chemistry. The following sections detail the basics of the oxidation chemistry, with more detailed discussion of production and destruction of ozone, the photostationary state, the tropospheric ozone budget, the role of hydrocarbons, the formation of smog, the spring ozone maximum, and nighttime oxidation chemistry, in particular the chemistry of the NO_3 radical. Further discussions cover the role of ozone–alkene and NO_2–diene chemistry in the oxidation of trace species. The chapter concludes with a brief overview of sulfur and halogen chemistry.

6.2   INITIATION OF PHOTOCHEMISTRY BY LIGHT

Photodissociation of atmospheric molecules by solar radiation plays a fundamental role in atmospheric chemistry. The photodissociation of trace species such as ozone and formaldehyde contributes to their removal from the atmosphere, but probably the most important role played by these photoprocesses is the generation of highly reactive atoms and radicals. Photodissociation of trace species and the subsequent reaction of the photoproducts with other molecules is the prime initiator and driver for the bulk of tropospheric chemistry.

The light source for photochemistry in the atmosphere is the Sun. At the top of the atmosphere there is *c*.1370 W m^{-2} of energy over a wide spectral range, from X-rays through the visible to longer wavelength. By the time the incident light reaches the troposphere much of the more energetic, shorter-wavelength light has been absorbed, by molecules such as oxygen, ozone, and water vapor, or scattered higher in the atmosphere. Typically, in the surface layers, light of wavelengths longer than 290 nm is available (see Fig. 6.4). In the troposphere, the wavelength at which the intensity of light drops to zero is termed the **atmospheric cut-off**. For the troposphere, this wavelength is determined by the overhead stratospheric ozone column (absorbs *c*. $\lambda = 310$ nm) and the aerosol loading.

The light capable of causing photochemical reactions is termed the actinic flux, $F_\lambda(\lambda)$ (cm^{-2} s^{-1} nm^{-1}), which is also known as the scalar intensity or spherical radiant flux:

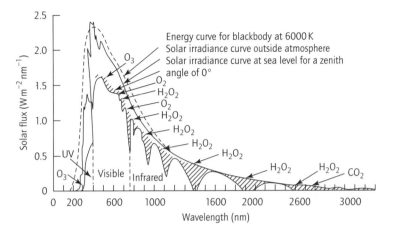

Fig. 6.4 Solar flux outside the atmosphere and at sea level, respectively. The emission of a blackbody at 6000 K is included for comparison. The species responsible for light absorption in the various regions (O_2, H_2O, etc.) are also shown. (From Howard *et al.* 1960.)

$$F_\lambda(\lambda) = \int L_\lambda(\lambda, \vartheta, \varphi) d\omega \qquad (6.3)$$

where $L_\lambda(\lambda)$ $(cm^{-2} s^{-1} sr^{-1} nm^{-1})$ denotes the spectral photon radiance, ω is the solid angle and (ϑ, φ) are the polar and azimuthal angles of incidence of the radiation interacting with the molecule of interest. The actinic flux is related but not equal to irradiance (see Madronich 1987). In essence, all angles of incident light must be considered when measuring or calculating the actinic flux $(\vartheta = 0{-}180°$, $\omega = 0{-}360°)$. Photolysis rates are often expressed as a first-order loss process, e.g. in the photolysis of NO_2

$$NO_2 + h\nu \rightarrow NO + O(^3P) \qquad (6.4)$$

i.e.

$$-\frac{d[NO_2]}{dt} = j_4[NO_2] \qquad (6.5)$$

where the photolysis frequency, j, can be expressed as

$$j = \int_{\lambda_{min}}^{\lambda_{max}} \sigma(\lambda, T)\phi(\lambda, T)F_\lambda(\lambda)d\lambda \qquad (6.6)$$

where σ is the absorption cross-section (cm^2), ϕ is the quantum yield of the photoproducts and T is the temperature.

There are two main experimental methods used for the determination of photolysis rates: chemical actinometry, which is a direct measure of a photolysis rate (e.g. Shetter *et al.* 1996); and radiometers, which measure the incident radiation, from which the photolysis rate can be calculated (e.g. Hofzumahaus *et al.* 1999). Radiative transfer through the atmosphere can be modeled using radiative transfer models (Goody & Yung 1989) that normally divide the atmosphere into layers and the radiative spectrum in wavelength intervals. The vertical distribution of trace gases and particles and the surface albedo serve as model inputs.

6.3 TROPOSPHERIC OXIDATION CHEMISTRY

Though atmospheric composition is dominated by both oxygen and nitrogen, it is not the amount of oxygen that defines the capacity of the troposphere to oxidize a trace gas. The **oxidizing capacity** of the troposphere is a somewhat nebulous term probably best described by Thompson (1992):

The total atmospheric burden of O_3, OH and H_2O_2 determines the "*oxidizing capacity*" of the atmosphere. As a result of the multiple interactions among the three oxidants and the multiphase activity of H_2O_2, there is no single expression that defines the earth's oxidizing capacity. Some researchers take the term to mean the total global OH, although even this parameter is not defined unambiguously.

Atmospheric photochemistry produces a variety of radicals that exert a substantial influence on the ultimate composition of the atmosphere. Probably the most important of these in terms of its reactivity is the hydroxyl radical, OH. The formation of OH is the initiator of radical-chain oxidation. Photolysis of ozone by ultraviolet light in the presence of water vapor is the main source of hydroxyl radicals in the troposphere:

$$O_3 + h\nu(\lambda < 340\,nm) \rightarrow O(^1D) + O_2(^1\Delta_g) \quad (6.7)$$

$$O(^1D) + H_2O \rightarrow OH + OH \quad (6.8)$$

The fate of the bulk of the $O(^1D)$ atoms produced via reaction 6.7 is collisional quenching back to ground state oxygen atoms:

$$O(^1D) + N_2 \rightarrow O(^3P) + N_2 \quad (6.9)$$

$$O(^1D) + O_2 \rightarrow O(^3P) + O_2 \quad (6.10)$$

The fraction of $O(^1D)$ atoms that form OH is dependent on pressure and the concentration of H_2O; typically, in the marine boundary layer, about 10%

of the $O(^1D)$ generates OH. Reactions 6.7 and 6.8 are the primary source of OH in the troposphere, but there are a number of other reactions and photolysis routes capable of forming OH directly or indirectly. As these compounds are often products of OH radical initiated oxidation they are often termed secondary sources of OH, and include the photolysis of HONO, HCHO, H_2O_2, and acetone and the reaction of $O(^1D)$ with methane (see Fig. 6.5). Table 6.1 illustrates the average contribution of various formation routes with altitude in a standard atmosphere (Cantrell 1998).

Two important features of OH chemistry make it critical to the chemistry of the troposphere. The first is its inherent reactivity; the second is its relatively high concentration given its high reactivity. The hydroxyl radical is ubiquitous throughout the troposphere owing to the widespread nature of ozone and water. In relatively unpolluted regimes (low NO_x) the main fate for the hydroxyl radical is reaction with either carbon monoxide or methane to produce peroxy radicals such as HO_2 and CH_3O_2:

$$OH + CO \rightarrow H + CO_2 \quad (6.11)$$

$$H + O_2 + M \rightarrow HO_2 + M \quad (6.12)$$

and

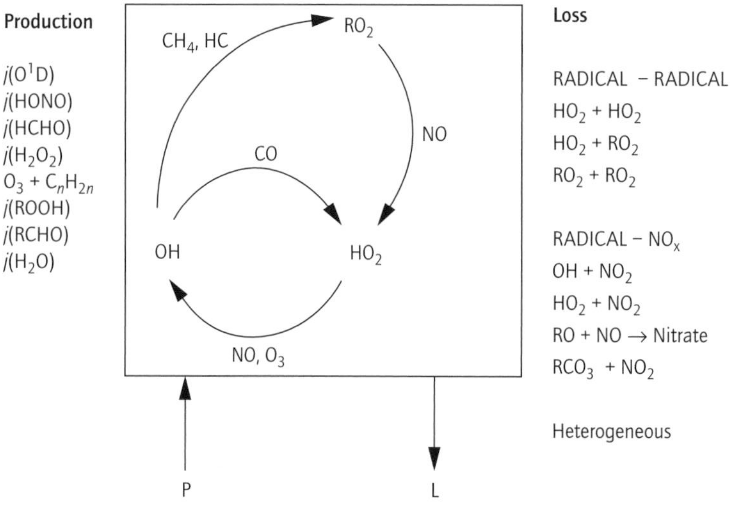

Production

$j(O^1D)$
$j(HONO)$
$j(HCHO)$
$j(H_2O_2)$
$O_3 + C_nH_{2n}$
$j(ROOH)$
$j(RCHO)$
$j(H_2O)$

Loss

RADICAL − RADICAL
$HO_2 + HO_2$
$HO_2 + RO_2$
$RO_2 + RO_2$

RADICAL − NO_x
$OH + NO_2$
$HO_2 + NO_2$
$RO + NO \rightarrow$ Nitrate
$RCO_3 + NO_2$

Heterogeneous

Fig. 6.5 The sources, interconversions, and sinks for HO_x (and RO_x) in the troposphere.

Table 6.1 Calculated fractional contribution of various photolysis rates to radical production with altitude (from Cantrell 1998).

Altitude	$j(O(^1D)) + H_2O$	$j(O(^1D)) + CH_4$	j(Acetone)	$j(H_2O_2)$	j(HCHO)
Ground	0.68	0.0	Neg.	0.15	0.17
Mid-troposphere	0.52	Neg.	0.03	0.20	0.25
Upper-troposphere	0.35	0.02	0.1	0.25	0.28
Lower stratosphere	0.40	0.1	0.25	0.1	0.15

Neg., negligible.

$$OH + CH_4 \rightarrow CH_3 + H_2O \qquad (6.13)$$

$$CH_3 + O_2 + M \rightarrow CH_3O_2 + M \qquad (6.14)$$

In low-NO_x conditions, HO_2 can react with ozone, leading to further destruction of ozone in a chain sequence involving production of hydroxyl radicals:

$$HO_2 + O_3 \rightarrow OH + 2O_2 \qquad (6.15)$$

$$OH + O_3 \rightarrow HO_2 + O_2 \qquad (6.16)$$

Alternatively, it can recombine to form hydrogen peroxide (H_2O_2):

$$HO_2 + HO_2 \rightarrow H_2O_2 + O_2 \qquad (6.17)$$

or react with organic peroxy radicals such as CH_3O_2 to form organic hydroperoxides:

$$CH_3O_2 + HO_2 \rightarrow CH_3O_2H + O_2 \qquad (6.18)$$

The formation of peroxides is effectively a chain termination reaction, as under most conditions these peroxides can act as effective sinks for HO_x. In more polluted conditions (high-NO_x), peroxy radicals catalyze the oxidation of NO to NO_2:

$$HO_2 + NO \rightarrow OH + NO_2 \qquad (6.19)$$

leading to the production of ozone from the subsequent photolysis of nitrogen dioxide and reaction of the photoproducts, i.e.

$$NO_2 + h\nu(\lambda < 420\,nm) \rightarrow NO + O(^3P) \qquad (6.4)$$

$$O + O_2 + M \rightarrow O_3 + M \qquad (6.20)$$

Hydroxyl radicals produced in reaction 6.19 can go on to form more peroxy radicals (e.g. via reactions 6.11 and 6.13). Similarly to HO_2, CH_3O_2 can also oxidize NO to NO_2:

$$CH_3O_2 + NO \rightarrow CH_3O + NO_2 \qquad (6.21)$$

The resulting methoxy radical reacts rapidly with O_2 to form formaldehyde and HO_2:

$$CH_3O + O_2 \rightarrow HCHO + HO_2 \qquad (6.22)$$

The oxidation of methane is summarized schematically in Fig. 6.6. The OH radical may have another fate. Depending on the concentration of NO_2, it can react with NO_2 to form nitric acid:

$$OH + NO_2 + M \rightarrow HNO_3 + M \qquad (6.23)$$

The formation of HNO_3 represents an effective loss mechanism for both HO_x and NO_x.

6.4 NITROGEN OXIDES AND THE PHOTOSTATIONARY STATE

From the preceding discussion it can be seen that the chemistry of nitrogen oxides is an integral part of tropospheric oxidation and photochemical processes. Nitrogen oxides are released into the troposphere from a variety of biogenic and anthropogenic sources, including fossil fuel combustion, biomass burning, microbial activity in soils, and lightning discharges (see Logan 1983). There is still

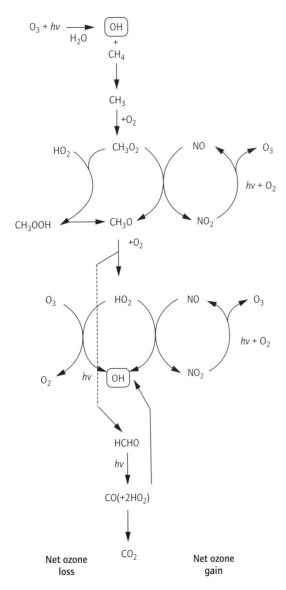

Fig. 6.6 Simplified mechanism for the photochemical oxidation of CH_4 in the troposphere. (Reprinted from *Atmospheric Environment* (Lightfoot *et al.* 1992). Copyright 1992, with permission from Elsevier Science.)

some debate about the magnitude of the various sources and sinks for NO_x (e.g. Lerdau *et al.* 2000). About 30% of the global budget of NO_x comes from fossil fuel combustion, with almost 86% of the NO_x emitted in one form or the other into the planetary boundary layer from surface processes. Typical NO/NO_2 ratios in surface air are 0.2–0.5 during the day, tending to zero at night. Over the time scales of hours to days NO_x is converted to nitric acid (reaction 6.23) and nitrates, which are subsequently removed by rain and dry deposition.

The photolysis of NO_2 to NO and the subsequent regeneration of NO_2 via reaction of NO with ozone is sufficiently fast, in the moderately polluted environment, for these species to be in dynamic equilibrium:

$$NO_2 + h\nu \rightarrow NO + O(^3P) \quad (6.4)$$

$$O(^3P) + O_2 + M \rightarrow O_3 + M \quad (6.20)$$

$$O_3 + NO \rightarrow NO_2 + O_2 \quad (6.24)$$

Therefore, at suitable concentrations, ambient NO, NO_2, and O_3 can be said to be in a photochemical steady state or photostationary state (PSS) (Leighton 1961), provided that they are isolated from local sources of NO_x and that sunlight intensity is relatively constant. Therefore,

$$[O_3] = \frac{j_4[NO_2]}{k_{24}[NO]} \quad (6.25)$$

Reactions 6.4, 6.20, and 6.24 constitute a cycle with no net chemistry. The PSS expression is sometimes expressed as a ratio:

$$\phi = \frac{j_4[NO_2]}{k_{24}[NO][O_3]} \quad (6.26)$$

If ozone is the sole oxidant for NO to NO_2 then $\phi = 1$. This situation often pertains in urban areas where NO_x levels are high and other potential oxidants of NO to NO_2, such as peroxy radicals, are suppressed. In the presence of peroxy radicals, eqn

6.25 has to be modified, as the NO/NO_2 partitioning is shifted to favor NO_2:

$$\frac{[NO_2]}{[NO]} = (k_{24}[O_3] + k_{19}[HO_2] + k_{21}[RO_2])/j_4$$

$$(6.27)$$

Though the radical concentrations are typically about 1000 times smaller than the $[O_3]$, the rate of the radical oxidation of NO to NO_2 is about 500 times larger than the corresponding oxidation by reaction with O_3. We return to the significance of the peroxy radical catalyzed oxidation of NO to NO_2 when considering photochemical ozone production and destruction. From the preceding discussion it can be seen that the behaviors of NO and NO_2 are strongly coupled through both photolytic and chemical equilibria. Because of their rapid interconversion they are often referred to as NO_x. NO_x, i.e. $(NO + NO_2)$, is also sometimes referred to as "active nitrogen."

Another major role for NO_x in the atmosphere is in the determination of the odd-hydrogen partitioning between OH and HO_2. The $[NO_x]$ controls the partitioning of HO_x between OH and HO_2 and is integrally linked to the production of ozone in the unpolluted atmosphere (see Section 6.5). This role in HO_x partitioning can be illustrated by consideration of the simple interconversion ratio (ICR) between OH and HO_2 (see also Fig. 6.5)

$$\frac{[HO_2]}{[OH]} = \frac{k_{11}[CO] + k_{13}[CH_4] + k_{16}[O_3]}{k_{19}[NO] + k_{15}[O_3]} \quad (6.28)$$

where the reaction of OH with CO, CH_4, or O_3 converts OH into HO_2 and the reaction of HO_2 with NO or O_3 converts HO_2 to OH. Figure 6.7 shows the ICR plotted against NO_x for free tropospheric conditions (Zanis *et al.* 2000). As $[NO_x]$ increases the HO_2/OH ratio decreases; [OH] increases with increasing NO_x, as the reaction between HO_2 and NO (reaction 6.19) becomes an important secondary source for OH, as much as the primary production via reaction of $O(^1D)$ with water (reactions 6.7 and 6.8). For HO_2, as $[NO_x]$ increases loss via reaction with NO (reaction 6.19) becomes dominant

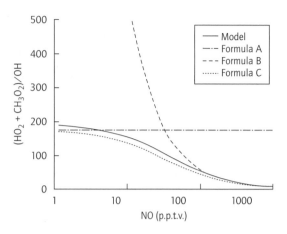

Fig. 6.7 Modeled interconversion ratio (i.e. $(HO_2 + CH_3O_2)/OH$; see text) for varying concentrations of NO. Formula A takes into account only O_3 photolysis, HO_2, and OH recycling through reaction with O_3. Formula B takes into account recycling via NO transformations. Formula C takes into account all the relevant interconversion processes and radical cross- and self-reaction termination at low NO. (From Zanis *et al.* 2000.)

over that from the self- and cross-reactions (reactions 6.17 and 6.18). This is somewhat of a simplification owing to the role of RO_2 as a source of HO_2 (for full details see Zanis *et al.* 2000). However, it is clear that $[NO_x]$ plays a critical role in the partitioning of OH and HO_2.

The extent of the influence of NO_x in any given atmospheric situation depends on its sources, reservoir species and sinks. Therefore, an important atmospheric quantity is the lifetime of NO_x. If nitric acid formation is considered to be the main loss process for NO_x (i.e. NO_2), then the lifetime of NO_x (τ_{NO_x}) can be expressed as the time constant for reaction 6.23, the NO_2 to HNO_3 conversion

$$\tau_{NO_x} = \frac{1}{k_{23}[OH]}\left(1 + \frac{[NO]}{[NO_2]}\right) \quad (6.29)$$

Therefore, using this simplification, the lifetime of NO_x is dependent on the [OH] and $[NO]/[NO_2]$ ratio. Calculating τ_{NO_x} under typical upper tropospheric conditions gives lifetimes in the order of 4–7 days, with lifetimes in the order of days in the lower free troposphere. In the boundary layer, the situation is more complex as there are NO_x loss and transformation processes other than those considered in eqn 6.29, which can make τ_{NO_x} as short as 1 hour. Integrally linked to the lifetime of NO_x and therefore the role of nitrogen oxides in the troposphere is its relation to odd nitrogen reservoir species, i.e. NO_y. The sum of total reactive nitrogen or total odd nitrogen is often referred to as NO_y and can be defined as $NO_y = NO_x + NO_3 + 2N_2O_5 + HNO_3 + HNO_4 + HONO + PAN + MPAN +$ nitrate + alkyl nitrates. NO_y can also be thought of as NO_x and all the compounds that are products of the atmospheric oxidation of NO_x. NO_y is not a conserved quantity in the atmosphere owing to the potential for some of its constituents (e.g. HNO_3) to be efficiently removed by deposition processes. Mixing of air masses may also lead to dilution of NO_y. The concept of NO_y is useful for considering the budget of odd nitrogen and evaluating the partitioning of NO_x and its reservoirs in the troposphere (Roberts 1995).

In summary, the concentration of NO_x in the troposphere:

1 Determines the catalytic efficiency of ozone production.
2 Determines the partitioning of OH and HO_2.
3 Determines the amount of HNO_3 and nitrates produced.
4 Determines the magnitude and sign of net photochemical production or destruction of ozone (see Section 6.5).

6.5   PRODUCTION AND DESTRUCTION OF OZONE

From the preceding discussion of atmospheric photochemistry and NO_x chemistry it can be seen that the fate of the peroxy radicals can have a marked effect on the ability of the atmosphere either to produce or to destroy ozone. Photolysis of NO_2 and the subsequent reaction of the photoproducts with O_2 (reactions 6.4 and 6.20) are the only known way of producing ozone in the tropo-

sphere. In the presence of NO_x the following cycle for the production of ozone can take place:

$$NO_2 + h\nu \rightarrow O(^3P) + NO \qquad (6.4)$$

$$O(^3P) + O_2 + M \rightarrow O_3 + M \qquad (6.20)$$

$$OH + CO \rightarrow H + CO_2 \qquad (6.11)$$

$$H + O_2 + M \rightarrow HO_2 + M \qquad (6.12)$$

$$HO_2 + NO \rightarrow OH + NO_2 \qquad (6.19)$$

$$\textbf{Net: } CO + 2O_2 + h\nu \rightarrow CO_2 + O_3 \qquad (6.30)$$

Similar chain reactions can be written for reactions involving RO_2. In contrast, when relatively little NO_x is present, as in the remote atmosphere, the following cycle can dominate over ozone production, leading to the catalytic destruction of ozone:

$$HO_2 + O_3 \rightarrow OH + 2O_2 \qquad (6.15)$$

$$OH + CO \rightarrow H + CO_2 \qquad (6.11)$$

$$H + O_2 + M \rightarrow HO_2 + M \qquad (6.12)$$

$$\textbf{Net: } CO + O_3 \rightarrow CO_2 + O_2 \qquad (6.31)$$

Clearly, there is a balance between photochemical ozone production and ozone loss dependent on the concentrations of HO_x and NO_x. Figure 6.8 shows the dependence of the production of ozone on NO_x taken from a numerical model. There are distinct regions in terms of $N(O_3)$ versus $[NO_x]$ in Fig. 6.8. For example, in region A the loss of ozone $(L(O_3))$ is greater than the production of ozone $(P(O_3))$; hence the net product of this process, i.e.

$$N(O_3) = P(O_3) - L(O_3), \qquad (6.32)$$

leads to a net ozone loss. The photochemical loss of ozone can be represented as

$$L(O_3) = (fj_7(O^1D)[O_3]) + k_{15}[HO_2] + k_{16}[OH] \qquad (6.33)$$

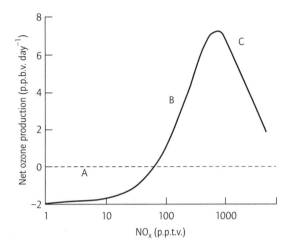

Fig. 6.8 Schematic representation of the dependence of the net ozone $(N(O_3))$ production or destruction on the concentration of NO_x.

where f is the fraction of $O(^1D)$ atoms that react with water vapor (reaction 6.8), rather than being deactivated to $O(^3P)$ (reactions 6.9 and 6.10). Evaluation of eqn 6.33 is effectively a lower limit for the ozone loss rate, as it neglects any other potential chemical loss processes for ozone, such as cloud chemistry (Lelieveld & Crutzen 1990), NO_3 chemistry (e.g. Volz *et al.* 1989), or halogen chemistry (see Section 6.15). The balance point, i.e. where $N(O_3) = 0$, is often referred to, somewhat misleadingly, as the compensation point, and occurs at a critical concentration of NO_x. Above the compensation point $P(O_3) > L(O_3)$ and therefore $N(O_3)$ is positive and the system is forming ozone. The *in situ* formation rate for ozone is approximately given by the rate at which the peroxy radicals (HO_2 and RO_2) oxidize NO to NO_2. This is followed by the rapid photolysis of NO_2 (reaction 6.4) to yield the oxygen atom required to produce O_3:

$$P(O_3) = [NO] \cdot \left(k_{19}[HO_2] + \sum k_i[RO_2]_i\right) \quad (6.34)$$

It is also worth noting that $P(O_3)$ can be expressed in terms of the concentrations of NO_x, $j_4(NO_2)$, and O_3 and temperature by substitution of eqn 6.27 into eqn 6.34 to give

$$P(O_3) = j_4[NO_2] - k_{24}[NO][O_3] \quad (6.35)$$

At some concentration of NO_x the system reaches a maximum production rate for ozone at $dP(O_3)/d(NO_x) = 0$, and even though $P(O_3)$ is still significantly larger than $L(O_3)$ the net production rate begins to fall off with increasing NO_x. Until this maximum is reached the system is said to be NO_x-limited with respect to the production of ozone. The turnover, i.e. $dP(O_3)/d(NO_x) = 0$, is caused by the increased competition for NO_x by the reaction

$$OH + NO_2 + M \rightarrow HNO_3 + M \quad (6.22)$$

In reality, the situation is somewhat more complicated owing to the presence at high concentrations of NO_x of increased levels of nonmethane hydrocarbons (NMHCs), especially in places such as the urban atmosphere. The oxidation of NMHCs, in common with much of tropospheric oxidation chemistry, is initiated by reaction with OH, leading to the rapid sequence of chain reactions:

$$OH + RH \rightarrow R + H_2O \quad (6.36)$$

$$R + O_2 + M \rightarrow RO_2 + M \quad (6.37)$$

$$RO_2 + NO \rightarrow RO + NO_2 \quad (6.38)$$

$$RO + O_2 \rightarrow \text{carbonyl products} + HO_2 \quad (6.39)$$

$$HO_2 + NO \rightarrow OH + NO_2 \quad (6.19)$$

This cycle is similar to the preceding one for the oxidation of CO, in that it is catalytic with respect to OH, R, RO, and RO_2, with HO_2 acting as the chain propagating radical. The mechanism of reaction 6.39 is strongly dependent on the structure of RO.

With the involvement of volatile organic compounds (VOCs) in the oxidation chemistry, Fig. 6.8 represents a slice through an n-dimensional surface where there should be a third axis to represent the concentration of VOCs. The peak initial concentrations of ozone generated from various initial concentrations of NO_x and VOCs are usually rep-

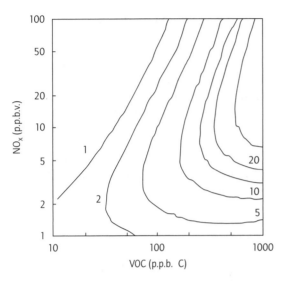

Fig. 6.9 Isopleths giving net rate of ozone production (p.p.b. h^{-1}) as a function of VOC (p.p.b. C) and NO$_x$ (p.p.b. v) for mean summer daytime meteorology and clear skies. (Reprinted from *Atmospheric Environment* (Sillman 1999). Copyright 1999, with permission from Elsevier Science.)

resented as an "O$_3$ isopleth diagram," an example of which is shown in Fig. 6.9 (Sillman 1999). In an isopleth diagram, initial mixture compositions giving rise to the same peak O$_3$ concentration are connected by the appropriate isopleth. An isopleth plot shows that ozone production is a highly nonlinear process in relation to NO$_x$ and VOC, but picks out many of the features already highlighted in Fig. 6.8, i.e. when NO$_x$ is "low" the rate of ozone formation increases with increasing NO$_x$ in a near-linear fashion. On the isopleth, the local maximum in the ozone formation rate with respect to NO$_x$ is the same feature as the turn over in $N(O_3)$ in Fig. 6.8. The ridgeline along the local maximum separates two different regimes, the so-called NO$_x$-*sensitive* regime, i.e. $N(O_3) \propto (NO_x)$, and the VOC-*sensitive* (or NO$_x$-*saturated* regime), i.e. $N(O_3) \propto$ (VOC), and decreases with increasing NO$_x$. The relationship between NO$_x$, VOCs, and ozone embodied in the isopleth diagram indicates one of the problems in the development of air quality policy with respect to ozone. Reductions in VOC are only

effective in reducing ozone under a VOC-sensitive chemistry, and reductions in NO$_x$ will only be effective if NO$_x$-sensitive chemistry predominates and may actually increase ozone in VOC-sensitive regions. In general, as an air mass moves away from emission sources, e.g. in an urban region, the chemistry tends to move from VOC-sensitive to NO$_x$-sensitive chemistry.

Another way of thinking of the ability of the atmosphere to make ozone is that the propensity for ozone formation is essentially proportional to the chain-length of the cycle, i.e. the number of free radical propagated cycles before termination (see Fig. 6.5). Therefore, the ozone production efficiency becomes a balance between propagation of the free radical interconversion cycle and the rate of termination of the cycle. Under VOC-limited conditions, i.e. when [NO$_x$] versus [VOC] is high, the competing chain propagation and termination reactions are going to be

$$OH + RH \rightarrow R + H_2O \qquad (6.36)$$

and

$$OH + NO_2 + M \rightarrow HNO_3 + M \qquad (6.23)$$

where reaction 6.36 will lead to production of ozone and reaction 6.23 will lead to termination of the HO$_x$ (and NO$_x$) chemistry. In the balance between these two reactions lies the problem of VOC versus NO$_x$ control, in that reduction of VOC may move the system into an ozone production regime. At high [VOC]/[NO$_x$] ratios or NO$_x$-limited regimes, the dominant chain terminating reactions are the peroxy radical self- and cross-reactions to form peroxides (reactions 6.17 and 6.18). Therefore, the ozone formation chain length is determined by the competition between the reactions of peroxy radicals and NO and the peroxy radical termination reactions. As shown in Fig. 6.8, this illustrates the direct dependence of $N(O_3)$ on [NO$_x$].

An example of the experimental determination of these relationships is shown in Fig. 6.10: a comparison of observed ozone production rates ($P(O_3)$) and concentrations of HO$_2$ and OH from the NASA

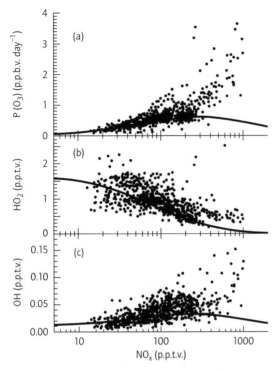

Fig. 6.10 Observed ozone production rates (P(O₃)) and concentrations of HO₂ and OH during SONEX (8–12 km altitude, 40–60° N latitude) plotted a function of the NO$_x$ concentration (NO$_x$ = observed NO + modeled NO₂). The observed rates and concentrations are averaged over 24 hours, using diel factors obtained from a locally constrained box model. The lines on the three panels correspond to model-calculated values for median upper tropospheric conditions as encountered during SONEX. (From Jaeglé *et al.* 1999.)

SONEX mission (Jaeglé *et al.* 1999) plotted as a function of NO$_x$. The data were taken from a suite of aircraft measurements between 8 and 12 km altitude at latitudes between 40 and 60° N. The model data suggest that P(O₃) becomes independent of NO$_x$ above 70 p.p.t.v. and the turnover point into a NO$_x$-saturated regime occurs at about 300 p.p.t.v. The bulk of the experimental observations below [NO$_x$] < 300 p.p.t.v. show the P(O₃) dependency predicted by the model, but above [NO$_x$] ≈ 300 p.p.t.v. P(O₃), computed from the measured

HO₂ and NO, continues to increase with NO$_x$, suggesting a NO$_x$-limited regime.

An elegant piece of experimental evidence for the photochemical destruction of ozone comes from studies in the remote marine boundary layer over the Southern Ocean at Cape Grim, Tasmania (41°S) (Ayers *et al.* 1992). In the marine boundary layer (MBL), the photochemical processes are coupled to physical processes that affect the observed ozone concentrations, namely deposition to the available surfaces and entrainment from the free troposphere. The sum of these processes can be represented in the form of an ozone continuity equation (a simplified version of eqn 6.2):

$$\frac{d[O_3]}{dt} = C + \frac{E_v([O_3]_{ft} - [O_3])}{H} + \frac{v_d[O_3]}{H} \quad (6.40)$$

where C is a term representative of the photochemistry (the net result of production, P(O₃), minus destruction, L(O₃); eqn 6.32), E_v is the entrainment velocity, $[O_3]_{ft}$ is the concentration of free-tropospheric ozone, v_d is the dry deposition velocity, and H is the height of the boundary layer. In general, the marine boundary layer is particularly suitable for making photochemical measurements owing to its stable and chemically simple nature. Figure 6.11 shows the average diurnal cycle of ozone and total peroxide (mainly H₂O₂) in clean oceanic air as measured at Cape Grim during January 1992. During the sunlit hours an ozone loss of about 1.6 p.p.b.v. occurs between mid-morning and late afternoon. This loss of ozone is followed by an overnight replenishment to a similar starting point. In contrast, the peroxide concentration increases from 600 to 900 p.p.t.v. between mid-morning and late afternoon. It is worth noting that the magnitude of this anti-correlation of ozone and peroxide is dependent on season. The daytime anti-correlation between O₃ and peroxide can be interpreted as experimental evidence for the photochemical destruction of ozone, as the ozone is destroyed via reactions 6.7, 6.15, and 6.16, while simultaneously peroxide is formed from chemistry involving the odd-hydrogen radicals OH and HO₂ (reactions 6.11, 6.12, and 6.17). The nighttime re-

Table 6.2 Calculated average ozone removal and addition rates according to pathway p.p.b.v. day^{-1} (upper part) and fractional contributions to overall production or destruction pathways (lower part) on a seasonal basis from measurements made in the marine boundary layer at Cape Grim, Tasmania (from Monks *et al.* 2000).

Pathway	O$_3$ removal		O$_3$ addition	
	Photochemistry	Deposition	Photochemistry	Entrainment
Jan.–Feb.	1.19	0.18	0.56	2.1
Aug.–Sep.	0.61	0.35	0.29	0.1
Jan.–Feb. (%)	87	13	21	79
Aug.–Sep. (%)	64	36	74	26

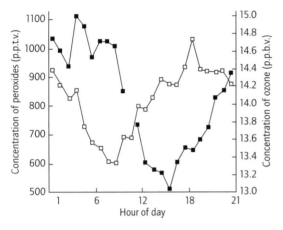

Fig. 6.11 Average diurnal cycles for peroxide (open squares) and ozone (filled) squares) in baseline air at Cape Grim, Tasmania (41° S) for January 1992. (Reprinted with permission from *Nature* (Ayers *et al.* 1992). Copyright 1992 Macmillan Magazines Limited.)

plenishment of ozone is caused by entrainment of ozone from the *free* troposphere into the boundary layer. The overnight loss of peroxide is due to deposition to the sea surface, as peroxide has a significant physical loss rate, in contrast to ozone which does not, over the sea surface. Therefore, the daytime anti-correlation of ozone and peroxide is indicative of the net photochemical destruction of ozone.

If the theoretical framework of eqn 6.40 is coupled to a suite of measured data (Monks *et al.* 2000), seasonal estimates of the main contributing processes for the control of ozone in the remote marine boundary layer can be made (see Table 6.2). There is clear evidence that in the remote MBL the degree of photochemistry determines the lower limit of the ozone concentration, while entrainment across the concentration gradient between the free troposphere and the MBL controls the upper bound. That is, in summer, when photochemistry is at a maximum, the ozone concentration is at its lowest value. By contrast, during winter, when photochemistry is at a minimum, boundary layer ozone values approach lower free tropospheric values. As can be seen from the data in Table 6.2, photochemistry is the most important ozone-destruction process in the clean MBL on a seasonal basis. Although entrainment is the dominant ozone-addition process during summer, once the differential between boundary layer and lower free troposphere ozone is reduced, the small photochemical production term becomes the dominant ozone-addition mechanism during winter.

6.6   THE TROPOSPHERIC OZONE BUDGET

The overall O$_3$ budget of the troposphere is, in the simplest sense, a combination of photochemical processes and physical processes, i.e. photochemical production or destruction of ozone, stratospheric–tropospheric exchange (STE), and destruction of ozone at the Earth's surface (see Fig. 6.2). The latest generation of three-

Table 6.3 Modeled global tropospheric ozone budget.

	Amount (Tg yr^{-1})
Input	
Photochemical production	~3500–4000
Stratospheric input	~400–850
Loss	
Photochemical loss	~3000–4000
Surface removal	~500–1200

Table 6.4 Global turnover of tropospheric gases and fraction removed by reaction with OH (from Ehhalt 1999).

Trace gas	Global emission rate (Tg yr^{-1})	Removal by OH* (%)
CO	2800	85
CH_4	530	90
C_2H_6	20	90
Isoprene	570	90
Terpenes	140	50
NO_2	150	50
SO_2	300	30
$(CH_3)_2S$	30	90
$CFCl_3$	0.3	0

* Assuming mean global [OH] = 1×10^6 molecule cm^{-3}.

dimensional global models suggest an ozone budget as shown in Table 6.3.

What is clear from Table 6.3 is that by current estimates photochemical production of ozone contributes between 82 and 90% of the total ozone inputted into the troposphere. Conversely, photochemical loss destroys between 76 and 85% of the ozone in the troposphere. Furthermore, the models seem to suggest that the tropospheric ozone budget is dominated by photochemistry. In a broader perspective, these budgets, though useful, illustrate the difficulty in estimating whether ozone is decreasing or increasing in the troposphere owing to the fine balance between net increase or decrease. Furthermore, they give a poor picture of the relative contribution of each process in the different regions of the atmosphere boundary layer versus free troposphere, northern hemisphere versus southern hemisphere, and Tropics versus extratropics (e.g. the remote marine boundary layer is a net sink for ozone, whereas the polluted continental boundary layer may be a net source of ozone). These hemispheric/regional contributions must be superimposed on the seasonal variability in a number of the sources and sinks, adding to the complexity of the problem.

6.7 THE ROLE OF HYDROCARBONS

The discussion up to this point has focused in the main on the role of CO and CH_4 as the fuels for atmospheric oxidation. It is clear that there are many more carbon compounds in the atmosphere than just these two (e.g. Guenther *et al.* 1995). One of the roles of atmospheric photochemistry is to "cleanse" the troposphere of a wide range of these compounds. Table 6.4 illustrates the global turnover of a range of trace gases, including hydrocarbons, and illustrates, for a number of trace gases, the primary role played by OH in their removal. The sources of nonmethane hydrocarbons (NMHC) are dealt with in Chapter 5. Carbon monoxide chemistry is not independent from NMHC chemistry, as 40–60% of surface CO levels over the continents, slightly less over the oceans, and 30–60% of CO levels in the free troposphere are estimated to come from NMHC oxidation (Poisson *et al.* 2000).

A major component of the reactive hydrocarbon loading is the biogenic hydrocarbons, the chemistry of which is dealt with in detail in Chapter 4. As previously indicated, hydrocarbon oxidation chemistry is integral to the production of ozone. Globally, the contribution of NMHC to net photochemical production of ozone is estimated to be about 40% (Houweling *et al.* 1998).

There are a number of inorganic molecules, such as NO_2 and SO_2 (see Table 6.4), which are also lost via reaction with OH. A number of halocarbons also exist that possess insubstantial tropospheric sinks and have importance in the chemistry of the stratosphere (see Chapter 7).

The ultimate products of the oxidation of any hydrocarbon are carbon dioxide and water vapor, but there are many relatively stable partially oxidized organic species, such as aldehydes, ketones, and carbon monoxide, that are produced as intermediate products during this process, with ozone produced as a by-product of the oxidation process. Figure 6.5 shows a schematic representation of the free radical catalyzed oxidation of methane, which is analogous to that of a hydrocarbon. As previously discussed, the oxidation is initiated by reaction of the hydrocarbon with OH and follows a mechanism in which the alkoxy and peroxy radicals are chain propagators and OH is effectively catalytic:

$$OH + RH \rightarrow R + H_2O \qquad (6.36)$$

$$R + O_2 + M \rightarrow RO_2 + M \qquad (6.37)$$

$$RO_2 + NO \rightarrow RO + NO_2 \qquad (6.38)$$

$$RO + O_2 \rightarrow carbonyl\ products + HO_2 \quad (6.39)$$

$$HO_2 + NO \rightarrow OH + NO_2 \qquad (6.19)$$

As both reactions 6.38 and 6.19 lead to the oxidation of NO to NO_2, the subsequent photolysis

leads to the formation of ozone (see reactions 6.4 and 6.20). The individual reaction mechanism depends on the identity of the organic compounds and the level of complexity of the mechanism. Though OH is the main tropospheric oxidation initiator, reaction with NO_3, O_3, or $O(^3P)$, or photolysis, may be an important loss route for some NMHCs or the partially oxygenated products produced as intermediates in the oxidation (see reaction 6.39).

Table 6.5 illustrates the impact and feedbacks inherent in NMHC oxidation chemistry on a global scale from a three-dimensional global tropospheric photochemical model (Poisson *et al.* 2000). Increasing NMHC concentrations lead to a decrease in global [OH] of about 20% (diurnally and seasonally averaged), as evidenced by the longer CH_4 lifetimes. Also, increasing NMHC oxidation increases the atmospheric burden of CO by about 40% both by increased photochemical production and, in a feedback loop, by less effective NMHC loss owing to the reduced concentration of OH. In fact, the annual secondary production of CO from CH_4 and NMHC oxidation processes is estimated to be larger than that from direct emissions. The NMHC oxidation chemistry adds about 30% to

Table 6.5 Annual global photochemical budgets in the troposphere as calculated using a three-dimensional global tropospheric model when neglecting or considering NMHC oxidation (from Poisson *et al.* 2000).

	Without NMHC	With NMHC	Impact of NMHC oxidation (%)
CH_4 lifetime owing to reaction with OH (yr)	6.5	7.4	+15
CH_4 burden (Tg)	3570	3700	+4
CH_4 oxidation by OH (Tg yr^{-1})	551	502	−9
Global O_3 photochemical budget terms (Tg yr^{-1})			
Production	3673	4834	+31
Destruction	−3260	−3816	−17
Net global photochemical O_3 production	413	668	+62
Deposition	775	1019	+31
Mean tropospheric content of O_3 (Tg)*	345	401	+16
CO global photochemical production (Tg yr^{-1})	920	1297	+41
Deposition	36	48	+33
CO burden (Tg)	224	310	+38

All averages are volume mean weighted.
* Up to 100 hPa.

the tropospheric production of ozone, but as $[O_3]$ increases the photochemical destruction of ozone increases by 17%. When these numbers are factored with the STE and increased deposition terms (see Section 6.6) for larger concentrations of O_3, they suggest an increase in tropospheric O_3 burden owing to NMHC oxidation chemistry of about 16–17% (Houweling *et al.* 1998; Poisson *et al.* 2000).

It is clear from hydrocarbon oxidation that increasing NMHC emissions and oxidation chemistry since preindustrial times have the potential to increase tropospheric O_3 concentrations (see Fig. 6.3).

In summary, the rate of oxidation of VOCs and therefore by inference the production of ozone is governed by the concentration of the catalytic HO_x radicals. There are a large variety of VOCs with a range of reactivities, and therefore this remains a complex area (Lewis *et al.* 2000).

6.8 URBAN CHEMISTRY

In some respects, the story of atmospheric chemistry, and particularly ozone photochemistry, begins with urban chemistry and photochemical smog. The term "smog" arises from a combination of the words smoke and fog. In the 1940s it became apparent that cities like Los Angeles (LA) were severely afflicted with a noxious haze (Haagen-Smit 1952). Though at the time it was thought to be a relatively local phenomenon, with the understanding of its chemistry came the development of a photochemical theory for the whole of the troposphere. The LA smog is often termed photochemical smog and is quite different in origin to the London smogs of the turn of the twentieth century, which have their origins in abnormally high concentrations of smoke particles and sulfur dioxide. The London smogs have been alleviated with the effective application of legislation that has reduced the burning of coal in the London area. The major features of photochemical smog are high levels of oxidant concentration in particular ozone and peroxidic compounds, produced by photochemical reactions. The principal effects of smog

are eye and bronchial irritation, as well as plant and material damage. The basic reaction scheme for the formation of photochemical smog is

$$VOC + h\nu \rightarrow VOC + R \qquad (6.41)$$

$$R + NO \rightarrow NO_2 \qquad (6.42)$$

$$NO_2 + h\nu(\lambda < 420\,nm) \rightarrow NO + O_3 \qquad (6.4)$$

$$NO + O_3 \rightarrow NO_2 + O_2 \qquad (6.24)$$

$$R + R' \rightarrow R'' \qquad (6.43)$$

$$R + NO_2 \rightarrow NO_y \qquad (6.44)$$

There is a large range of available VOCs in the urban atmosphere, driven by the range of anthropogenic and biogenic sources (e.g. Calvert *et al.* 2000). Table 6.6 illustrates the different loadings of the major classes of NMHC in urban and rural locations. These NMHC loadings must be coupled to measures of reactivity and the degradation mechanisms of the NMHC to give a representative picture of urban photochemistry. The oxidation of the VOCs drives, via the formation of peroxy radicals, the oxidation of NO to NO_2, where under sunlit conditions the NO_2 can be dissociated to form ozone (reaction 6.4). Preexisting ozone can also drive the NO to NO_2 conversion (reaction 6.24).

The basic chemistry responsible for urban photochemistry is essentially the same as that

Table 6.6 Percentage of NMHC classes measured in the morning at various locations (from Goldan *et al.* 1995; Fujita *et al.* 1997).

	Urban Los Angeles	Urban Boston	Rural Alabama
Alkanes	42	36	9
Alkenes	7	10	43*
Aromatics	19	30	2
Other	33	24	46†

* Large contribution from biogenic alkenes.
† Mainly oxygen containing.

which takes place in the unpolluted atmosphere (see Section 6.3). It is the range and concentrations of NMHC fuels and the concentrations of NO_x coupled to the addition of some photochemical accelerants that can lead to the excesses of urban chemistry. For example, in the Los Angeles basin it is estimated that 3333 t day^{-1} of organic compounds are emitted, as well as 890 t day^{-1} of NO_x. In addition, to reactions forming OH in the background troposphere via the reaction of $O(^1D)$ with H_2O,

$$O_3 + h\nu(\lambda < 340\,nm) \rightarrow O_2 + O(^1D) \quad (6.7)$$

$$O(^1D) + H_2O \rightarrow 2OH \quad (6.8)$$

under urban conditions OH may be formed from secondary sources such as

$$HONO + h\nu(\lambda < 400\,nm) \rightarrow OH + NO \quad (6.45)$$

where HONO can be emitted in small quantities from automobiles or formed from a number of heterogeneous pathways (Lammel & Cape 1996), as well as gas-phase routes. OH produced from HONO has been shown to be the dominant OH source in the morning under some urban conditions, where the HONO has built up to significant levels overnight (e.g. Winer & Bierman 1994). Another key urban source of OH, can be the photolysis of the aldehydes and ketones produced in the NMHC oxidation chemistry, in particular formaldehyde:

$$HCHO + h\nu(\lambda < 334\,nm) \rightarrow H + HCO \quad (6.46)$$

$$H + O_2 + M \rightarrow HO_2 + M \quad (6.12)$$

$$HCO + O_2 \rightarrow HO_2 + CO \quad (6.47)$$

Net: $HCHO + 2O_2 + h\nu \rightarrow 2HO_2 + CO \quad (6.48)$

Smog chamber experiments have shown that the addition of aldehyde can significantly increase the formation rates of ozone and the conversion rates of NO and NO_2 under simulated urban conditions (e.g. Wayne 2000).

A marked by-product of oxidation in the urban atmosphere, often associated with, but not exclusive to, urban air pollution, is peroxyacetylnitrate (PAN). PAN is formed by

$$OH + CH_3CHO \rightarrow CH_3CO + H_2O \quad (6.49)$$

$$CH_3CO + O_2 \rightarrow CH_3CO.O_2 \quad (6.50)$$

addition of NO to the peroxyacetyl radical ($RCO.O_2$) leading to the formation of peroxyacetylnitrate

$$CH_3CO.O_2 + NO_2 + M \leftrightarrow CH_3CO.O_2.NO_2 + M \quad (6.51)$$

PAN is often used as an unambiguous marker for tropospheric chemistry. The lifetime of PAN in the troposphere is very much dependent on the temperature dependence of the equilibrium in reaction 6.51, the lifetime varying from 30 minutes at $T = 298\,K$ to 8 hours at $T = 273\,K$. At mid-troposphere temperature and pressures PAN has thermal decomposition lifetimes in the order of 46 days. It is also worth noting that peroxyacyl radical like peroxy radicals can oxidize NO to NO_2:

$$CH_3CO.O_2 + NO \rightarrow CH_3CO.O + NO_2 \quad (6.52)$$

PAN can be an important component of NO_y in the troposphere (see Section 6.4) and has the potential to act as a temporary reservoir for NO_x, and in particular the potential to transport NO_x from polluted regions into the background/remote atmosphere (Singh et al. 1992).

There are other side-products of urban air pollution that are giving increasing cause for concern. These include the production of both aerosols (organic and inorganic) and particles.

The effects of photochemical smog/urban air pollution remain on the political agenda owing to their potential impact on human health and the economy. In summary, urban photochemistry is not substantially different from clean tropospheric photochemistry. It is the range and concentrations of the VOCs involved in oxidation coupled to the concentration of NO_x and other oxidants that lead to a large photochemical turnover.

6.9 THE SPRING OZONE MAXIMUM

As an example of an atmospheric conundrum that brings together many of the elements of tropospheric photochemistry discussed so far, it is worth looking at the so-called spring ozone maximum. Measurements of ozone throughout the troposphere clearly show an annual cycle. Over the past couple of decades it has become apparent that the measured annual cycle of ozone in certain locations shows a distinct maximum during spring and the magnitude of this maximum seems to have increased over time. There has been much debate as to the origins of this phenomenon (e.g. Monks 2000). There is broad agreement that much of the ozone found in the troposphere is of photochemical origin. In contrast, there is still no overarching consensus as to the mechanisms that lead to the formation of the so-called spring ozone maximum. Part of the problem would seem to lie in the interpretation of measurements and the interactions of processes occurring on differing scales, from the local to the global. The spring ozone maximum remains an interesting problem, as it may well be a proxy for the continuing change of tropospheric composition, and, at a fundamental level, it tests the basic understanding of both tropospheric chemistry and dynamics and their interrelations.

6.9.1 Spring maximum in ozone

Throughout the troposphere ozone measurements show annual cycles (e.g. Logan 1999). The shape and form of these cycles depends on location and altitude, some showing a spring maximum/summer minimum, some a broad summer maximum and some even a winter maximum/summer minimum in ozone. It is clear that there are a number of processes acting on a range of scales that contribute to the overall appearance of any individual ozone seasonal cycle. It is the identification, deconvolution, and quantification of these chemical and physical processes that lies at the heart of an understanding of the control of ozone in the troposphere and therefore the spring ozone maximum. As previously described, in the southern hemisphere in relatively remote regions, the seasonal cycle of ozone in the marine boundary layer is marked by a summer minimum and winter maximum and can be characterized by competition between photochemistry destroying ozone and physical processes entraining ozone into the boundary layer. In the northern hemisphere, it is clear that a different situation pertains, as many measurements show a spring maximum or broad spring/summer maximum. Satellite measurements of climatological tropospheric ozone (Fishman & Brackett 1997) show a widespread increase in ozone during the northern hemisphere spring as compared to the corresponding season in the southern hemisphere. Another aspect of the spring ozone maximum is the distribution of ozone with respect to altitude and season. Figure 6.12 shows a composite average of the ozone concentration as measured above Uccle in Belgium. A number of interesting features are evident in the data: the stratospheric spring ozone maximum is apparent at altitudes above 8–9 km, a lower tropospheric spring maximum is visible at lower altitudes (2–3 km) with a different timing to the stratosphere, and each of these signals seems to fade in the upper troposphere.

The appearance of the spring ozone maximum has often been rationalized in terms of increased input of stratospheric ozone into the troposphere, but it has also been suggested that tropospheric photochemical processes may make a contribution.

6.9.2 Stratospheric–tropospheric exchange

Stratospheric–tropospheric exchange (STE) of ozone has been suggested to be a major contributing factor to the observed spring ozone maximum. As exemplified in Fig. 6.12, ozone levels in the extratropical lower stratosphere, as driven by the Brewer–Dobson circulation, show a maximum in late winter/early spring. There are a number of processes occurring on a range of scales that can

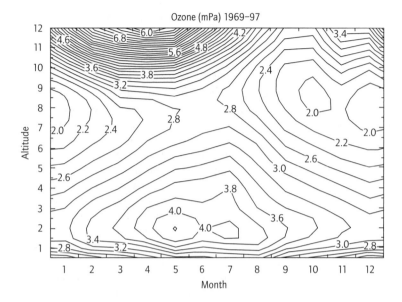

Fig. 6.12 Vertical distribution of ozone over Uccle, Belgium (1969–97). (Courtesy of Dirk DeMeur, RMI, Belgium.)

contribute to transfer of ozone from the stratosphere to the troposphere, including the global aspects of exchange coupled to synoptic and small-scale mechanisms. On the global scale, the rise and fall of the tropopause with season coupled to smaller-scale tropopause folding, convection, and streamer/col decay can lead to STE. It has been argued that there would be a spring maximum in STE even if the tropopause folding and small-scale processes remained constant all year round, owing to the seasonal build-up of tracers in the lower stratosphere over winter. A problem that remains is the quantification of these mechanisms for STE, in particular the direct versus indirect transfer of ozone, the frequency and magnitude of the events, and the large-scale transfer.

One further clue as to the contribution of STE to the spring ozone maximum may lie in the analysis of historical ozone records. It is clear from these records that the level of tropospheric ozone has approximately doubled over the past 100 years (see Fig. 6.3) and the magnitude of the spring ozone maximum has increased. It is not clear whether STE frequencies/rates have increased over the same period. Similarly, a comparison of northern and southern hemisphere STE rates might provide a clue about the hemispheric asymmetry with re-

spect to observed ozone concentrations, the lack of a spring maximum in the southern hemisphere and the contribution of STE.

6.9.3 *Photochemistry*

The role of photochemistry in the production and destruction of ozone in the troposphere is now widely accepted. The role of tropospheric photochemistry in the formation of the spring ozone maximum is less clear. Seasonal measurements of a number of tropospheric chemical tracers show a spring maximum, indicative of a significant photochemical contribution to the formation of ozone in spring (Penkett & Brice 1986). Figure 6.13 shows measurements of a number of trace gases, many of which have a spring maximum independent of whether or not they have stratospheric sources (Derwent *et al.* 1998). In order to rationalize the role of photochemistry in the spring maximum there have been various postulates with respect to reservoir mechanisms, such as accumulation of precursors to photochemical ozone production behind the polar front, and the role of convective venting and frontal uplift as mechanisms for the transport of the precursors for ozone production.

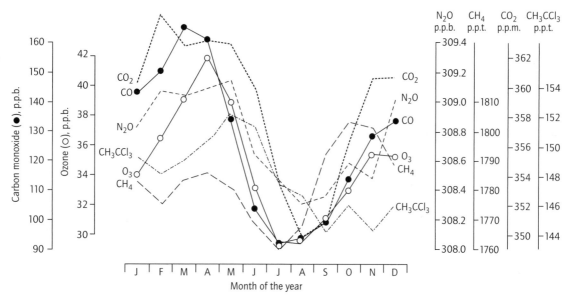

Fig. 6.13 Seasonal cycles of background ozone and carbon monoxide at Mace Head in Ireland over the period 1990–4 and the monthly mean concentrations of a wide range of trace gases in polluted air. (From Derwent *et al.* 1998.)

In order to produce ozone by photochemistry in spring there is a requirement for chemical "fuel" to drive the system. There is growing evidence that the required nitrogen oxides (NO_x) have a widespread distribution. Strong seasonal cycles in nonmethane hydrocarbons, the other precursors for photochemical O_3 production, are observed. There still remains uncertainty as to the source and sink mechanisms in relation to the spring ozone maximum for these compounds, integral in the photochemical formation of ozone. One interesting distinction in terms of the photochemical production of ozone and the contribution to the spring ozone maximum is between *in situ* photochemical production and that produced by the same mechanisms but transported long distances when the photochemical lifetimes for ozone are long. This coupling of tropospheric dynamics and chemistry is an integral part of the appearance of the spring ozone maximum.

Figure 6.14 shows the simulated and observed seasonal ozone cycles at two remote sites (Goose Bay and Mauna Loa) (Wang *et al.* 1998). An analysis of the modeling results suggests that tropospheric

photochemistry is the dominant factor affecting the appearance of the spring ozone maximum, over the stratospheric contribution, but the role of long-range transport of tropospheric ozone was highlighted as an important factor (Wang *et al.* 1998). The effect the photochemical processes that lead to the spring ozone maximum have in the overall tropospheric ozone budget remains an open question.

The appearance of a spring maximum in ozone remains an intriguing problem. To date there is still no overarching consensus on the mechanisms that lead to its formation. The spring ozone phenomenon may well be a useful proxy for the continuing changes to atmospheric composition wrought by human activities. Understanding the appearance of the spring ozone maximum and the mechanisms that lead to its formation therefore remains a fundamental issue that has potential impact on diverse areas of both policy (e.g. abatement strategies) and atmospheric science. It is an issue that lies at the heart of how we understand tropospheric chemistry and dynamics and the links between them.

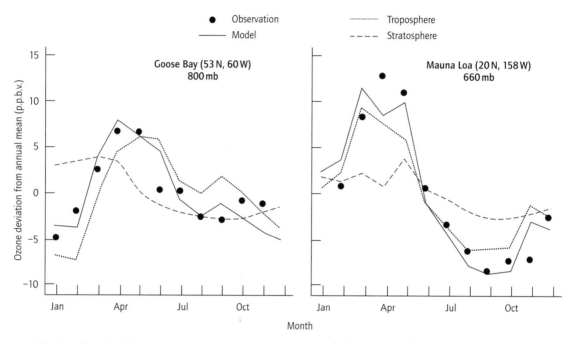

Fig. 6.14 Simulated and observed seasonal ozone at Goose Bay, Canada (800 mbar) and at the Mauna Loa Observatory, Hawaii (600 mbar). The values are mean deviations from the annual mean. Observations are from an ozonesonde climatology (Logan 1999). The seasonal variations in the model (continuous lines) are broken down into the contributions of ozone transported from the stratosphere (dashed lines) and ozone produced in the troposphere (dotted lines). (From Wang *et al.* 1998.)

6.10   NIGHTTIME OXIDATION CHEMISTRY

Although photochemistry does not take place at night, it is important to note, within the context of tropospheric oxidation chemistry, the potential for oxidation chemistry to continue at night. This chemistry does not lead to the production of ozone, in fact the opposite, but has importance owing to the potential for the production of secondary pollutants. In the troposphere, the main nighttime oxidant is thought to be the nitrate radical formed by the relatively slow oxidation of NO_2 by O_3:

$$NO_2 + O_3 \rightarrow NO_3 + O_2 \qquad (6.53)$$

The time constant for reaction 6.53 is of the order of 15 hours at an ozone concentration of 30 p.p.b.v. and $T = 290$ K. Other sources include

$$N_2O_5 + M \rightarrow NO_3 + NO_2 + M \qquad (6.54)$$

but as N_2O_5 is formed from

$$NO_3 + NO_2 + M \rightarrow N_2O_5 + M \qquad (6.55)$$

the two species act in a coupled manner. Dinitrogen pentoxide, N_2O_5, is potentially an important product as it can react heterogeneously with water to yield HNO_3. During the daytime the NO_3 radical is rapidly photolyzed as it strongly absorbs in the visible

$$NO_3 + h\nu \rightarrow NO + O_2 \qquad (6.56)$$

$$NO_3 + h\nu \rightarrow NO_2 + O(^3P) \qquad (6.57)$$

with a lifetime in the region of 5 s for overhead Sun and clear sky conditions. Further, NO_3 will react rapidly with NO

$$NO_3 + NO \rightarrow NO_2 + NO_2 \qquad (6.58)$$

which can have significant daytime concentrations in contrast to the nighttime, where away from strong source regions NO concentrations should be near zero.

The nitrate radical has a range of reactivity toward volatile organic compounds. The nitrate radical is highly reactive toward certain unsaturated hydrocarbons such as isoprene, a variety of butenes and monoterpenes, and reduced sulfur compounds such as dimethylsulfide (DMS). In the case of DMS, if the NO_2 concentration is 60% that of DMS then NO_3 is a more important oxidant than OH for DMS in the marine boundary layer (Allan *et al.* 1999). The role of NO_3 chemistry in DMS oxidation is treated in more detail in Chapter 4. In general, NO_3 abstraction reactions of the type

$$NO_3 + RH \rightarrow HNO_3 + R \qquad (6.59)$$

are relatively slow, with the alkyl radical reacting with oxygen under atmospheric conditions to form a peroxy radical. In the case of RH being an aldehyde, acyl products will form acylperoxy radicals ($R.CO.O_2$), potential sources of peroxyacylnitrates. In contrast, the reaction of NO_3 with alkenes occurs by an addition mechanism, initiating a complex chemistry involving nitro-oxy substituted organic radicals, which can either regenerate NO_2 or produce comparatively stable bifunctional organic nitrate products (Wayne *et al.* 1991). For example, the products derived from the reaction of NO_3 with propene in the presence of O_2 and NO_x include CH_3CHO, HCHO, 1,2-propanedioldinitrate (PDDN), nitoxyperoxypropylnitrate (NPPN) and α-(nitrooxy)acetone (Calvert *et al.* 2000; Wayne 2000) (see Fig. 6.15). The reaction channel that produces the nitrated acetones also yields peroxy radicals, leading to the potential for a night time source of OH, either by reaction ($HO_2 + O_3$) or by the direct reaction of the peroxy radical with NO_3. For the reaction of NO_3 with propene the initial addition can take place at either end of the double bond:

$$NO_3 + CH_2{=}CHCH_2 + M$$
$$\rightarrow CH_2CHCH_2(ONO_2) + M \qquad (6.60)$$

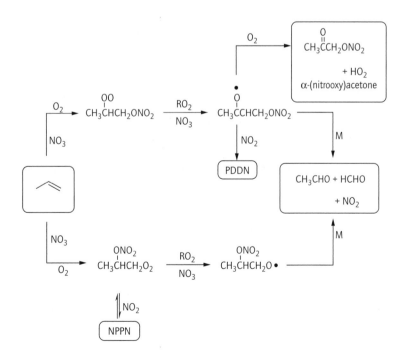

Fig. 6.15 A schematic representation of the chain propagation reactions in the NO_3 radical initiated oxidation of propene. (For details see Wayne *et al.* 1991.)

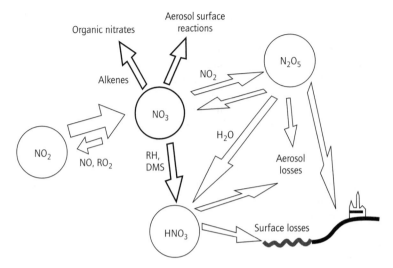

Fig. 6.16 A simplified reaction scheme for nighttime chemistry involving the nitrate radical. (From Allan *et al.* 2000.)

$$NO_3 + CH_2\!\!=\!\!CHCH_2 + M$$
$$\rightarrow CH_2CH(ONO_2)CH_2 + M \qquad (6.61)$$

The reaction can then proceed by the mechanism shown schematically in Fig. 6.15. The ratio of final products is dependent on the structure of the individual alkenes. In general, for branched alkenes, there is significant regeneration of NO_x and production of unsubstituted carbonyl products, while comparatively they are a minor source of HO_x (Wayne *et al.* 1991). For the less alkyl substituted alkenes, there is a greater yield of HO_x and bifunctional organic nitrate products but a lesser regeneration of NO_x. Therefore, depending on the mix of hydrocarbons, NO_3 chemistry can act to recycle NO_x, therefore inhibiting the formation of nitrate aerosol or HNO_3 at night and potentially leading to the generation of a nighttime source of HO_x. Figure 6.16 provides a simplified summary of the relevant nighttime chemistry involving the nitrate radical.

One important difference between NO_3 chemistry and daytime OH chemistry is that NO_3 can initiate, but not catalyze, the removal of organic compounds. Therefore, its concentration can be suppressed by the presence of fast-reacting, with respect to NO_3, organic compounds.

Evidence for the role of NO_3 in nighttime oxida-

tion chemistry has come from a number of experimentally based studies (e.g. Wayne *et al.* 1991; Allan *et al.* 1999, and references therein). Significant NO_3 concentrations have been detected over a wide range of atmospheric conditions, indicating a potential role for NO_3 over large regions of the atmosphere (Wayne *et al.* 1991). The atmospheric lifetime of NO_3 can be estimated using the steady-state approximation (compare to the photostationary state – see Section 6.4) to be

$$\tau(NO_3) = \frac{[NO_3]}{k_{53}[NO_2][O_3]} \qquad (6.62)$$

A useful quantity with which to compare this is the reciprocal of the lifetime calculated from the sum of the first-order loss processes involving NO_3 and N_2O_5:

$$\tau(NO_3)^{-1} \geq \sum_i k_{(NO_3+HC_i)}[HC_i]$$
$$+ k_{(NO_3+DMS)}[DMS] + k_{het}(NO_3)$$
$$+ \left(k^I[H_2O] + k^{II}[H_2O]^2\right.$$
$$+ k_{het}(N_2O_5)).K_{54}[NO_2] \qquad (6.63)$$

where the pseudo first-order loss rates over the *i* reactive hydrocarbons are summed and k^I and k^{II} are

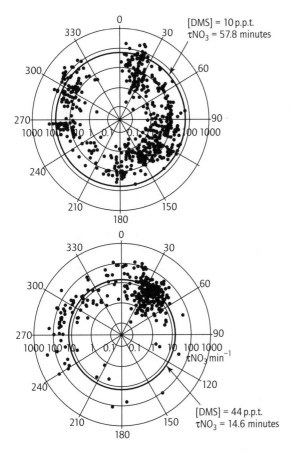

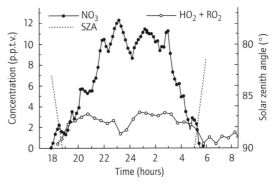

Fig. 6.18 Overnight concentrations of NO_3 and HO_2 + RO_2 as measured at the Weybourne Atmospheric Observatory on the North Norfolk Coast, UK on April 15–16, 1994. The variation of the solar zenith angles indicates the times of sunset and sunrise. (From Carslaw *et al.* 1997.)

Fig. 6.17 Polar plots showing the observed lifetime of NO_3 versus arrival wind direction for Mace Head, Ireland (top) and Tenerife (bottom). (From Allan *et al.* 2000.) The black circle represents the calculated lifetime of NO_3 from reaction with the campaign average [DMS] only.

the first- and second-order components with respect to H_2O in the reaction

$$N_2O_5 + H_2O \rightarrow 2HNO_3 \qquad (6.64)$$

$k_{het}(NO_3)$ and $k_{het}(N_2O_5)$ are the heterogeneous loss rates for these species and K_{54} is the equilibrium constant for reactions 6.54 and 6.55. Figure 6.17 shows the observed lifetime of NO_3 in the marine boundary layer at Mace Head, Ireland (Allan *et al.* 2000), segregated by arrival wind sec-

tor. Mace Head experiences a range of air masses from clean marine air to European continental outflow. The measured NO_3 lifetime varies from 2 minutes to 4 hours (see Fig. 6.17). An assessment of the parameters controlling NO_3 atmospheric lifetime (see eqn 6.63) highlights that, under the conditions encountered, the lifetime of NO_3 chemistry is very sensitive to DMS and NMHC chemistry in clean marine air. However, in more polluted air the terms involving the indirect loss of N_2O_5 either in the gas phase with H_2O or through uptake on aerosol tend to dominate.

There is a growing body of observational evidence for the generation of peroxy radicals during the night under polluted and semipolluted conditions. For example, Fig. 6.18 shows NO_3/RO_2 concentrations measured overnight at Weybourne on the north Norfolk coast of the UK (Carslaw *et al.* 1997). NO_3 and RO_2 rise simultaneously at dusk and similarly decrease at sunrise. Under the conditions encountered at Weybourne, NO_3 and RO_2 concentrations are correlated, indicative of NO_3 chemistry leading to nighttime HO_x chemistry from reactions with VOCs or through the involvement of NO_3 in radical propagation. A simplified nighttime VOC reaction scheme can be written, where NO_3 initiates the oxidation via the production of a precursor to the peroxy radical (6.65) and is also involved as a chain propagator via reaction 6.67:

$$NO_3 + \text{organic compound} \rightarrow R + \text{products} \tag{6.65}$$

$$R + O_2 + M \rightarrow RO_2 + M \tag{6.66}$$

$$RO_2 + NO_3 \rightarrow RO + NO_2 + O_2 \tag{6.67}$$

$$RO + O_2 \rightarrow R'R''CO + HO_2 \tag{6.68}$$

$$HO_2 + O_3 \rightarrow OH + 2O_2 \tag{6.15}$$

$$HO_2 + NO_3 \rightarrow OH + NO + O_2 \tag{6.69}$$

The generation of nighttime OH can take place by the reaction of HO_2 with ozone or NO_3 (reactions 6.15 or 6.69). The eventual fate of the nighttime OH is going to depend on a number of factors, but in the simplest case it can react with CO and CH_4 (reactions 6.11–6.14) to form peroxy radicals or react with NO_2 to form HNO_3 (reaction 6.23). It is worth noting that under different conditions NO_3 and RO_2 are anti-correlated, indicative of the influence of fast-reacting organics in the nighttime chemistry, or in some cases NO_3 and RO_2 are not correlated at all (e.g. Salisbury et al. 2001).

From the preceding discussion it can be seen that the involvement of NO_3 chemistry in gas-phase tropospheric chemistry has potentially six significant consequences (Wayne et al. 1991):

1 The radical can control NO_y speciation in the atmosphere at night (via reaction 6.54).

2 Nitric acid can be formed, by hydrolysis of N_2O_5, as a product of a hydrogen abstraction process or indirectly via the NO_3 mediated production of OH, which can react with NO_2 (reaction 6.23) to produce nitric acid.

3 Primary organic pollutants can be oxidized and removed at night.

4 Radicals (HO_x and RO_2) produced by NO_3 chemistry can act as initiators for chain oxidation chemistry.

5 Toxic or otherwise noxious compounds such as peroxyacylnitrates, other nitrates, and partially oxidized compounds may be formed.

6 Nitrate products or NO_3 itself may act as temporary reservoirs in the presence of NO_x.

6.11 OZONE–ALKENE CHEMISTRY

Another potential source of HO_x in the atmosphere, more particularly in the boundary layer, is the reactions between ozone and alkenes. The ozonolysis of alkenes can lead to the direct production of the OH radical at varying yields (between 7 and 100%) depending on the structure of the alkene, normally accompanied by the co-production of a (organic) peroxy radical. As compared to the reactions of both OH and NO_3 with alkenes the initial rate of the reaction of ozone with an alkene is relatively slow; this can be offset under regimes where there are high concentrations of alkenes and/or ozone. For example, under typical rural conditions the atmospheric lifetimes for the reaction of ethene with OH, O_3, and NO_3 are 20 hours, 9.7 days, and 5.2 months respectively; in contrast for the same reactants with 2-methyl-2-butene the atmospheric lifetimes are 2.0 hours, 0.9 hours, and 0.09 hours.

The mechanism for the reaction of ozone with alkenes was first suggested by Criegee in the late 1940s (for details see Criegee 1975) and involves the concerted addition of ozone to form a primary ozonide, which rapidly decomposes to form a vibrationally excited carbonyl oxide (**Criegee intermediate**) and carbonyl products. The Criegee intermediate can then either be collisionally stabilized by a third body (M) or undergo unimolecular decomposition to products (see also Section 6.13). It is now widely believed that alkyl-substituted Criegee intermediates can decompose via a vibrationally hot hydroperoxide intermediate to yield an OH radical, along with another radical species of the general form $R_1R_2CC(O)R_3$, which is expected to react rapidly with O_2 to form a peroxy radical (RO_2) in the atmosphere (e.g. Calvert et al. 2000). Figure 6.19 shows a schematic representation of the ozone–alkene reaction mechanism. The OH and peroxy radical yield is dependent on the structure and mechanism of the individual alkene–ozone reaction (see Paulson et al. 1999). Table 6.7

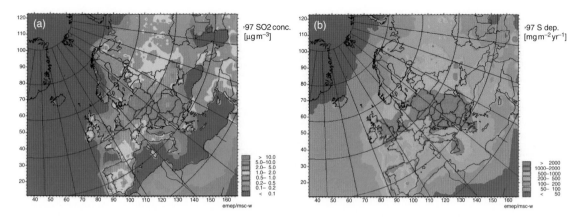

Plate 6.1 (a) Map of computed annual mean atmospheric SO_2 concentrations from the EMEP Eulerian Model for 1997 in $\mu g\, S\, m^{-3}$. (From Olendryzynski 1999.) (b) Map of computed annual mean total of dry and wet deposited oxidized sulfur from the EMEP Eulerian Model for 1997 in $\mu g\, S\, m^{-2}\, yr^{-1}$. (From Olendryzynski 1999.)

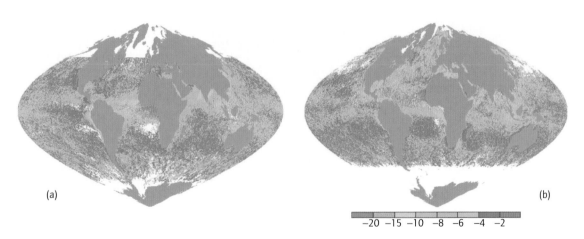

Plate 9.1 Clear-sky shortwave aerosol radiative forcing $(W\, m^{-2})$ retrieved from POLDER satellite measurements for (a) December 1996 and (b) June 1997. (From Boucher & Tanré 2000.)

(facing p. 180)

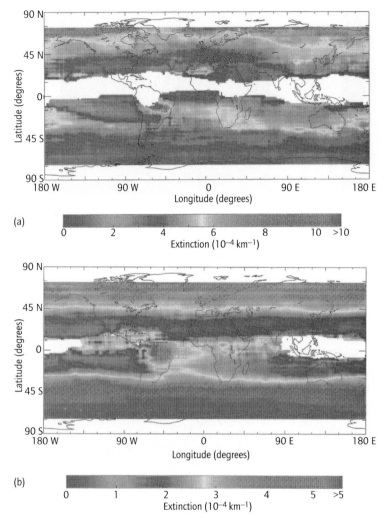

(a)

Extinction (10^{-4} km^{-1})

(b)

Extinction (10^{-4} km^{-1})

Plate 9.2 SAGE II maps of the aerosol extinction $\lambda = 1.02\,\mu$m in the troposphere averaged over the period September–November during the years 1985–90 and 1993–6. High stratospheric aerosol concentrations during and immediately after the Mount Pinatubo eruption in 1991–2 obscured the troposphere. (a) Extinction at 6.5 km. (b) Extinction at 12.5 km. (From Kent *et al.* 1998a.)

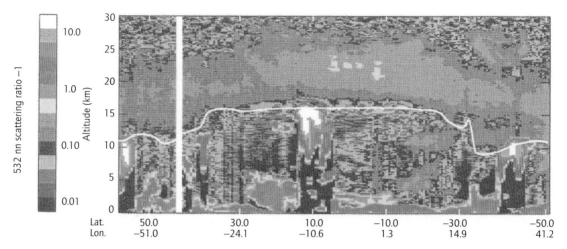

Plate 9.3 LITE profile (Orbit 115; September 17, 1994) of the lidar scattering ratio at $\lambda = 0.532\,\mu m$ from latitude 50° N to 50° S. The tropopause is indicated with a white line, above which a fairly homogeneous stratospheric aerosol layer is clearly visible. (From Osborn *et al.* 1998.)

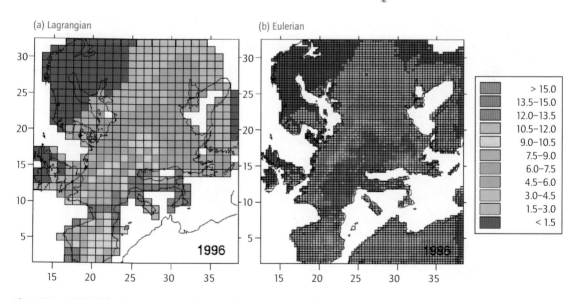

Plate 14.1 AOT 40 for the period April–September 1986 using the EMEP: (a) Lagrangian and (b) Eulerian photochemical models. (Reprinted from *Environmental Pollution* (Fowler *et al.* 1999). Copyright 1999, with permission from Elsevier Science.)

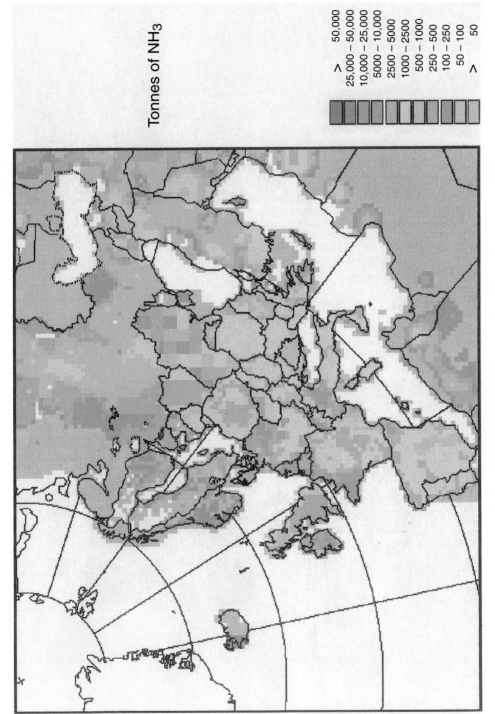

Tonnes of NH$_3$

> 50,000
25,000 – 50,000
10,000 – 25,000
5000 – 10,000
2500 – 5000
1000 – 2500
500 – 1000
250 – 500
100 – 250
50 – 100
> 50

Plate 14.2 Modeled NH$_3$ emissions from the European region for 1999 (Grid scale 50 × 50 km). (From Vestreng 2001.)

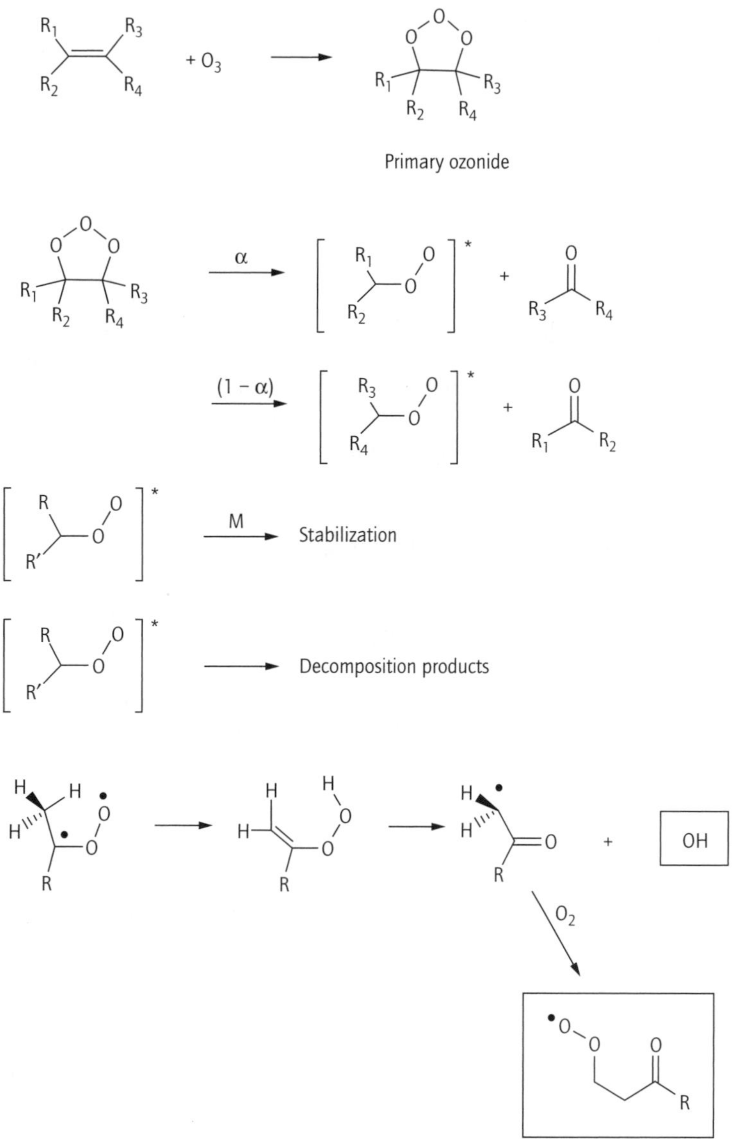

Fig. 6.19 A schematic representation of the oxidation of an alkene initiated by reaction with ozone. (From Salisbury *et al.* 2001.)

shows typical OH yields for the reaction of a range of anthropogenic and biogenic alkenes with ozone.

There has been some suggestion (e.g. Paulson & Orlando 1996; Ariya *et al.* 2000) that this chemistry could be important in the atmosphere under certain boundary layer conditions with suitable alkene/ozone loadings, e.g. the urban and rural boundary layer. To date, experimentally, the im-

portance of this chemistry in the atmosphere is still unproven, though there are indications from radical measurements (e.g. George *et al.* 1999; Salisbury *et al.* 2001) that these reactions could be important in understanding urban oxidation chemistry, marine boundary layer oxidation of alkenes, and wintertime chemistry (e.g. Ariya *et al.* 2000).

Table 6.7 Range of OH yields from the reaction of ozone with alkenes (all data taken from Paulson *et al.* 1999).

Alkene	OH yield
Ethene	0.18 ± 0.06
Propene	0.35 ± 0.07
Methylpropene	0.72 ± 0.12
Δ^3-carene	1.00

6.12 NO$_2$–DIENE CHEMISTRY

A potential source of HO$_x$ in the troposphere under heavily polluted conditions comes from the reactions between NO$_2$ and conjugated dienes (Harrison *et al.* 1998). The dienes are found in significant concentrations in vehicle exhaust emissions. Though NO$_2$–diene chemistry can play a role throughout the diurnal cycle it has been postulated to be of importance at night when concentrations of NO$_3$ and O$_3$ are suppressed by high levels of NO, such as in the polluted winter urban boundary layer. In general, the reactions of NO$_2$ with nonconjugated dienes are relatively slow ($<10^{-20}\,cm^3$ molecule^{-1} s^{-1}) (Atkinson 1997), whereas the reactions with conjugated dienes are typically $>10^{-20}\,cm^3$ molecule^{-1} s^{-1}. The reaction sequence for NO$_2$ with a conjugated diene (e.g. 1,3-butadiene; see Atkinson 1997 for details) is suggested to proceed as

$$NO_2 + RCH{=}CHCH{=}CHR'$$
$$\rightarrow RCH{=}CHCH{-}CH(NO_2){-}R' \quad (6.70)$$

$$RCH{=}CHCH{-}CH(NO_2){-}R' + O_2$$
$$\rightarrow RCH{=}CHCH(O_2){-}CH(NO_2){-}R' \quad (6.71)$$

$$RCH{=}CHCH(O_2){-}CH(NO_2){-}R' + NO$$
$$\rightarrow RCH{=}CHCH(O){-}CH(NO_2){-}R' + NO_2$$
$$(6.72)$$

The alkoxy radical formed in reaction 6.72 can react with O$_2$ (reaction 6.73) or isomerize, both of which lead ultimately to the formation of HO$_2$:

$$RCH{=}CHCH(O){-}CH(NO_2){-}R' + O_2$$
$$\rightarrow RCH{=}CHC(O){-}CH(NO_2){-}R' + HO_2$$
$$(6.73)$$

Subsequent reaction of HO$_2$ with NO (reaction 6.19) can generate OH, which can go on to react with a range of NMHC species. It is worth noting that the mechanism is catalytic with respect to NO$_2$ and can lead, via the oxidation of NO to NO$_2$ with the peroxy radicals, to an increase in the NO oxidation chain length, a key feature of the enhanced conversion of NO to NO$_2$ under heavily polluted conditions (e.g. Harrison *et al.* 1998).

6.13 SULFUR CHEMISTRY

Sulfur chemistry is an integral part of life, owing to its role in plant and human metabolism. Sulfur compounds have both natural and anthropogenic sources. In modern times, the atmospheric sulfur budget has become dominated by anthropogenic emissions, particularly from fossil fuel burning. It is estimated that 75% of the total sulfur emission budget is dominated by anthropogenic sources, with 90% of it occurring in the northern hemisphere. The natural sources include volcanoes, plants, soil, and biogenic activity in the oceans (see Chapter 4). In terms of photochemistry the major sulfur oxide, sulfur dioxide (SO$_2$) does not photodissociate in the troposphere (compare NO$_2$) i.e.

$$SO_2(X^1A_1) + h\nu(240 < \lambda < 330\,nm)$$
$$\rightarrow SO_2(^1A_2, {}^1B_1) \quad (6.74)$$

$$SO_2(X^1A_1) + h\nu(340 < \lambda < 400\,nm)$$
$$\rightarrow SO_2(^3B_1) \quad (6.75)$$

The oxidation of sulfur compounds in the atmosphere has implications in a number of different atmospheric problems, such as acidification, climate balance, and the formation of a sulfate layer in the stratosphere, the so-called Junge layer. By far the largest sulfur component emitted into the atmosphere is SO_2. Plate 6.1a (facing p. 180) shows the 1997 computed annual mean atmospheric SO_2 concentrations from EMEP (Olendryzynski 1999). In Europe, the source regions for SO_2 are quite apparent: the so-called black triangle region (southern Poland, eastern Germany, and the northern part of the Czech Republic) is the largest source of anthropogenic sulfur pollution ($>10 \mu g \, S \, m^{-3}$). There are a number of other large emission sources, including the central UK and the Kola Peninsula, also apparent in Plate 6.1a. The absolute maximum in emissions is in southern Italy around Sicily, where the largest single source of both natural (the volcano Mt Etna) and anthropogenic SO_2 is found. SO_2 can be detected from space-borne sensors (e.g. Eisinger & Burrows 1998) as a product of volcanic activity and fossil fuel burning. Plate 6.1b shows the total dry and wet deposition of oxidized sulfur in 1997 for Europe (Olendryzynski 1999). It is apparent that there are large areas of central Europe with deposition above $2000 \, mg \, S \, m^{-2} \, yr^{-1}$. It is noticeable that many of the areas with the highest deposition of oxidized sulfur are concomitant with the maxima in emission sources (compare Plate 6.1a).

The atmospheric oxidation of SO_2 can take place by a number of different mechanisms, both homogeneously and heterogeneously in the liquid and gas phases (see Fig. 6.20). The gas-phase oxidation of SO_2

$$SO_2 + OH + M \rightarrow HSO_3 + M \qquad (6.76)$$

$$HSO_3 + O_2 \rightarrow HO_2 + SO_3 \qquad (6.77)$$

$$SO_3 + H_2O + M \rightarrow H_2SO_4 \qquad (6.78)$$

can lead to the formation of sulfuric acid, which owing to its relatively low vapor pressure can rapidly attach to the condensed phase, such as aerosol particles. The bulk of the H_2SO_4 is lost via wet deposition mechanisms in cloud droplets and precipitation. There is another potential gas-phase loss route for SO_2 that can lead to the formation of sulfuric acid in the presence of H_2O: the reaction of SO_2 with Criegee intermediates (Cox & Penkett 1971). The aqueous-phase oxidation of SO_2 is more complex, depending on a number of factors such as the nature of the aqueous phase (e.g. clouds and fogs), the availability of oxidants (e.g. O_3 and H_2O_2), and the availability of light. An overview of the mechanism is given in Fig. 6.20. The key steps include the transport of the gas to the surface of a droplet, transfer across the gas–liquid interface, the formation of aqueous-phase equilibria, transport from the surface into the bulk aqueous phase, and subsequent reaction. In brief, the SO_2 gas is dissolved in the liquid phase, establishing a set of equilibria for a series of S (IV) species, i.e. $SO_2.H_2O$, HSO_3^- and SO_3^{2-}:

$$SO_2(g) + H_2O \leftrightarrow SO_2.H_2O(aq) \qquad (6.79)$$

$$SO_2.H_2O(aq) \leftrightarrow HSO_3^- + H^+ \qquad (6.80)$$

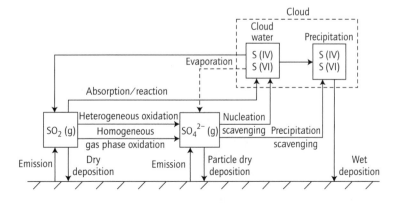

Fig. 6.20 Summary of emission, oxidation, and deposition of S(IV) and S(VI). (Reprinted from *Atmospheric Environment* (Lamb *et al.* 1987). Copyright 1987, with permission from Elsevier Science.)

$$HSO_3^- \leftrightarrow SO_3^{2-} + H^+ \qquad (6.81)$$

The solubility of SO_2 is related to the pH of the aqueous phase, decreasing at lower values of pH. The oxidation of sulfur (IV) to sulfur (VI) is a complex process dependent on many physical and chemical factors. The main oxidants seem to be O_2 (catalyzed/uncatalyzed), O_3, H_2O_2, the oxides of nitrogen, and free radical reactions in clouds and fogs (e.g. Sander *et al.* 1995). For example, H_2O_2 is highly soluble in solution, so even at relatively low gas-phase concentrations (typically about 1 p.p.b.v.) there is a significant concentration of H_2O_2 present in solution. The oxidation proceeds as

$$HSO_3^- + H_2O_2 \leftrightarrow \overset{O^-}{\underset{O}{\rangle}}S{-}OOH + H_2O \qquad (6.82)$$

$$\overset{O^-}{\underset{O}{\rangle}}S{-}OOH + HA \leftrightarrow H_2SO_4 + A^- \qquad (6.83)$$

where HA is an acid. The ubiquitous occurrence of H_2O_2, its solubility, its high reactivity, and the pH independence (under atmospheric conditions) of the rate constant for the reaction with SO_2 makes H_2O_2 one of the most important oxidants for SO_2 in the troposphere. A more detailed description of aqueous-phase oxidation of SO_2 is given in Finlayson-Pitts and Pitts (2000). The role of biogenic sulfur compounds, in particular dimethylsulfide and carbonyl sulfide, is dealt with in detail in Chapter 4.

6.14   HALOGEN CHEMISTRY

In comparison to the atmospheric chemistry taking place in the stratosphere, where halogen chemistry is well known and characterized (e.g. Wayne 2000), there has been much debate as to the role of halogen species in the oxidative chemistry of the troposphere. The halogen compounds found in the troposphere arise from chemical degradation of both natural and anthropogenic trace species. For example, many of the methyl halides (CH_3I, CH_3Br, and CH_3Cl) and higher halogenated species naturally occur in the oceans, and are also produced as a product of biomass burning. The degradation of man-made CFC replacements is a significant anthropogenic source of these compounds. Sea-salt aerosol has been implicated as a potential source of halogen species as well as a player in the recycling of halogen species in the marine atmosphere. There is growing experimental evidence as to the potential for halogen photochemistry as part of tropospheric photochemistry (Platt 2000). As many of the halogen species of interest have biogenic sources, their chemistry is dealt with in detail in Chapter 4.

6.15   CONCLUSION

The troposphere contains a diverse chemistry, driven in the main by the interaction of light with a few molecules, which drives a relatively complex array of chemistry. The main features of tropospheric chemistry have been discussed in terms of tropospheric oxidation chemistry, highlighting the key role that ozone plays in the oxidation chemistry of the troposphere. The troposphere is a region of the atmosphere rich in diversity in both chemistry and composition, and therefore in multiphase interactions. There is little doubt that, to date, the troposphere has not yielded all of its secrets.

REFERENCES

Allan, B.J., Carslaw, N., Coe, H., Burgess, R.A. & Plane, J.M.C. (1999) Observations of the nitrate radical in the marine boundary layer. *Journal of Atmospheric Chemistry* **33**, 129–54.

Allan, B.J., McFiggans, G., Plane, J.M.C., Coe, H. & McFadyen, G.G. (2000) The nitrate radical in the remote marine boundary layer. *Journal of Geophysical Research* **105**, 24,191–204.

Anfossi, D., Sandroni, S. & Viarengo, S. (1991) Tropospheric ozone in the nineteenth century: the Moncalieri series. *Journal of Geophysical Research* **96**, 17,349–52.

Ariya, P.A., Sander, R. & Cutzen, P.J. (2000) Significance

of HO_x and peroxides production due to alkene ozonolysis during fall and winter: a modelling study. *Journal of Geophysical Research* **105**, 17,721–38.

Atkinson, R. (1997) Gas-phase tropospheric chemistry of volatile organic compounds: 1. Alkanes and alkenes. *Journal of Physical and Chemical Reference Data* **26**, 217–90.

Ayers, G.P., Penkett, S.A., Gillett, R.W. *et al.* (1992) Evidence for photochemical control of ozone concentrations in unpolluted air. *Nature* **360**, 446–8.

Calvert, J.G., Atkinson, R., Kerr, J.A. *et al.* (2000) *The Mechanisms of Atmospheric Oxidation of the Alkenes.* Oxford University Press, Oxford.

Cantrell, C.A. (1998) Challenges in Photochemistry Workshop, GAAC, NCAR, Boulder, June 8–11.

Carslaw, N., Carpenter, L.J., Plane, J.M.C. *et al.* (1997) Simultaneous observations of nitrate and peroxy radicals in the marine boundary layer. *Journal of Geophysical Research (Atmosphere)* **102**, 18,917–33.

Chameides, W.L. & Walker, J.C.G. (1973) A photochemical theory for tropospheric ozone. *Journal of Geophysical Research* **78**, 8751–60.

Cox, R.A. & Penkett, S.A. (1971) Oxidation of SO_2 by oxidants formed in the ozone–olefin reaction. *Nature* **230**, 321–2.

Criegee, R. (1975) Mechanism of ozonolysis. *Agnew Chemistry International Edition* **14**, 745–52.

Crutzen, P.J. (1973) A discussion of the chemistry of some minor constituents in the stratosphere and troposphere. *Pure and Applied Geophysics* **106–8**, 1385–99.

Derwent, R.G., Simmonds, P.G., Seuring, S. & Dimmer, C. (1998) Observation and interpretation of the seasonal cycles in the surface concentrations of ozone and carbon monoxide at Mace Head, Ireland from 1990 to 1994. *Atmospheric Environment* **32**, 145–57.

Ehhalt, D. (1999) Photooxidation of trace gases in the troposphere. *Physical Chemistry Chemical Physics* **24**, 5401–8.

Eisinger, M. & Burrows, J.P. (1998) Tropospheric sulfur dioxide observed by the ERS-2 GOME instrument. *Geophysical Research Letters* **22**, 4177–80.

Etheridge, D.M., Pearman, G.I. & Fraser, P.J. (1992) Changes of tropospheric methane between 1841 and 1978 from a high accumulation-rate Antarctic ice core. *Tellus B* **44**, 282–94.

Fabian, P. & Pruchniewicz, P.G. (1977) Meridional distribution of ozone in the troposphere and its seasonal variations. *Journal of Geophysical Research* **82**, 2063–73.

Feister, U. & Warmbt, W. (1987) Long term measurements of surface ozone in the German Democratic Republic. *Journal of Atmospheric Chemistry* **5**, 1–21.

Finlayson-Pitts, B.J. & Pitts, J.N. Jr (2000) *Chemistry of the Upper and Lower Atmosphere.* Academic Press, San Diego.

Fishman, J. & Brackett, V.G. (1997) The climatological distribution of tropospheric ozone derived from satellite measurements using version 7 Total Ozone Mapping Spectrometer and Stratospheric Aerosol and Gas Experiment data sets. *Journal of Geophysical Research* **102**, 19,725–78.

Fujita, E.M., Lu, Z., Sheetz, L., Harshfeld, G. & Zielinska, B. (1997) *Determination of Mobile Source Emission Fraction Using Ambient Field Measurements.* Coordinating Research Council, Atlanta.

George, L.A., Hard, T.M. & O'Brien, R.J. (1999) Measurement of free radicals OH and HO_2 in Los Angeles smog. *Journal of Geophysical Research (Atmosphere)* **104**, 11,643–55.

Goldan, P.D., Kuster, W.C., Fehsenfeld, F.C. & Montska, S.A. (1995) Hydrocarbon measurement in the southeast United States: the Rural Oxidants in the Southern Environment (ROSE) program. *Journal of Geophysical Research* **100**, 25,945–63.

Goody, R.M. & Yung, Y.L. (1989) *Atmospheric Radiation*, 2nd edn. Oxford University Press, Oxford.

Guenther, A., Hewitt, C.N., Erickson, D. *et al.* (1995) A global model of natural volatile organic compound emissions. *Journal of Geophysical Research (Atmosphere)* **100**, 8873–92.

Haagen-Smit, A.J. (1952) Chemistry and physiology of Los Angeles smog. *Industrial Engineering Chemistry* **44**, 1342–6.

Harrison, R.M., Shi, J.P. & Grenfell, J.L. (1998) Novel nighttime free radical chemistry in severe nitrogen dioxide pollution episodes. *Atmospheric Environment* **32**, 2769–74.

Hofzumahaus, A., Kraus, A. & Müller, M. (1999) Solar actinic flux spectroradiometry: a technique for measuring photolysis frequencies in the atmosphere. *Applied Optics* **38**, 4443–60.

Houweling, S., Dentener, F. & Lelieveld, J. (1998) The impact of nonmethane hydrocarbon compounds on tropospheric photochemistry. *Journal of Geophysical Research (Atmosphere)* **103**, 10,673–96.

Howard, J.N., King, J.I.F. & Gast, P.R. (1960) Thermal radiation. In *Handbook of Geophysics.* Macmillan, New York.

Jaeglé, L., Jacob, D.J., Brune, W.H. *et al.* (1999) Ozone production in the upper troposphere and the influence of aircraft during SONEX: approach of NO_x saturated conditions, *Geophysical Research Letters* **26**, 3081–4.

Junge, C.E. (1962) Global ozone budget and exchange between and stratosphere and troposphere. *Tellus* **14**, 363–77.

Lamb, D., Miller, D.F., Robinson, N.F. & Gertler, A.W. (1987) The importance of liquid water concentration in the atmospheric oxidation of SO_2. *Atmospheric Environment* **21**, 2333–44.

Lammel, G. & Cape, J.N. (1996) Nitrous acid and nitrite in the atmosphere. *Chemical Society Review* **25**, 361–9.

Leighton, P.A. (1961) *The Photochemistry of Air Pollution*. Academic Press, New York.

Lelieveld, J. & Crutzen, P.J. (1990) Influences of cloud photochemical processes on tropospheric ozone. *Nature*, **343**, 227–233.

Lelieveld, J., Crutzen, P.J. and Dentener, F.J. (1998) Changing concentration, lifetime and climate forcing of atmospheric methane. *Tellus B* **50**, 128–150.

Lerdau, M.T., Munger, J.W. & Jacob, D.J. (2000) The NO_2 flux conundrum. *Science* **289**, 2291–3.

Lewis, A.C., Carslaw, N., Marriott, P.J., Kinghorn, R.M., Morrison, P., Lee, A.L., Bartle, K.D. & Pilling, M.J. (2000) A larger pool of ozone-forming carbon compounds in urban atmospheres. *Nature* **405**, 778–81.

Lightfoot, P.D., Cox, R.A., Crowley, J.N. *et al.* (1992) Organic peroxy radicals: kinetics, spectroscopy and tropospheric chemistry. *Atmospheric Environment* **10**, 1805–961.

Logan, J.A. (1983) Nitrogen oxides in the troposphere: Global and regional budgets. *Journal of Geophysical Research* **88**, 785–807.

Logan, J.A. (1999) An analysis of ozonesonde data for the troposphere: recommendations for testing 3-D models and development of a gridded climatology for tropospheric ozone. *Journal of Geophysical Research* **104**, 16,115–49.

Machida, T., Nakazawa, T., Fujii, Y., Aoki, S. & Watanabe, O. (1995) Increase in the atmospheric nitrous oxide during the last 250 years. *Geophysical Research Letters* **22**, 2921–4.

Madronich, S. (1987) Photodissociation in the atmosphere 1. Actinic flux and the effect of ground reflections and clouds. *Journal of Geophysical Research* **92**, 9740–52.

Monks, P.S. (2000) A review of the observations and origins of the spring ozone maximum. *Atmospheric Environment* **34**, 3545–61.

Monks, P.S., Salisbury, G., Holland, G., Penkett, S.A. & Ayers, G.P. (2000) A seasonal comparison of ozone hotochemistry in the remote marine boundary layer. *Atmospheric Environment* **34**, 2547–61.

Olendryzynski, K. (1999) *Operational EMEP Eulerian Acid Deposition Model*. EMEP/MSC-W Note 4/99. The Norwegian Meteorological Institute, Oslo.

Paulson, S.E. & Orlando, J.J. (1997) The reactions of ozone with alkenes: an important source of HO_x in the boundary layer. *Geophysical Research Letters* **23**, 3727–30.

Paulson, S.E., Chung, M.Y. & Hasson, A.S. (1999) OH radical formation from the gas-phase reaction of ozone with terminal alkenes and the relationship between structure and mechanism. *Journal of Physical Chemistry A* **103**, 8125–38.

Penkett, S.A. & Brice, K.A. (1986) The spring maximum in photo-oxidants in the northern hemisphere troposphere. *Nature* **319**, 655–7.

Platt, U. (2000) Reactive halogen species in the mid-latitude troposphere–recent discoveries. *Water, Air and Soil Pollution* **123**, 229–44.

Poisson, N., Kanakidou, M. & Crutzen, P.J. (2000) Impact of non-methane hydrocarbons on tropospheric chemistry and the oxidizing power of the global troposphere: 3-dimensional modelling results. *Journal of Atmospheric Chemistry* **36**, 157–230.

Prinn, R.G., Weiss, R.F., Fraser, P.J. *et al.* (2000) A history of chemically and radiatively important gases deduced from ALE/GAGE/AGAGE. *Journal of Geophysical Research (Atmosphere)* **105**, 17,751–92.

Roberts, J.M. (1995) Reactive odd-nitrogen (NO_y) in the atmosphere. In Singh, H.B. (ed.), *Composition and Climate of the Atmosphere*. Van Nostrand Reinhold, New York.

Salisbury, G., Rickard, A.R., Monks, P.S. *et al.* (2001) The production of peroxy radicals at night via reactions of ozone and the nitrate radical in the marine boundary layer. *Journal of Geophysical Research* **106**, 12,669–88.

Sander, R., Leleiveld, J. & Crutzen, P.J. (1995) Modelling of the nighttime nitrogen and sulfur chemistry in size resolved droplets of orographic cloud. *Journal of Atmospheric Chemistry* **20**, 89–116.

Sandroni, S., Anfossi, D. & Viarengo, S. (1992) Surface ozone levels at the end of the nineteenth century in South America. *Journal of Geophysical Research* **97**, 2535–40.

Shetter, R.E., Cantrell, C.A., Lantz, K.O. *et al.* (1996) Actinometric and radiometric measurement and modelling of the photolysis rate coefficient of ozone to $O(^1D)$ during the Mauna Loa Observatory Photochemistry Experiment 2. *Journal of Geophysical Research* **101**, 14,631–41.

Sillman, S. (1999) The relation between ozone, NO$_x$ and hydrocarbons in urban and polluted rural environments. *Atmospheric Environment* **33**, 1821–45.

Singh, H.B., O'Hara, D., Herlth, D. *et al.* (1992) Atmospheric measurements of peroxy acetyl nitrate and other organic nitrates at high latitudes possible sources and sinks. *Journal of Geophysical Research* **97**, 16,511–22.

Stull, R.R. (1988) *An Introduction to Boundary Layer Meteorology*. Kluwer Academic Publishers, London.

Thompson, A.M. (1992) The oxidizing capacity of the earths atmosphere—probable past and future changes. *Science* **256**, 1157–65.

Volz, A. & Kley, D. (1988) Evaluation of the Montsouris series of ozone measurements made in the nineteenth century. *Nature* **332**, 240–2.

Volz, A., Geiss, H., McKeen, S. & Kley, D. (1989) Correlation of ozone and solar radiation at Montsouris and Hohenpeissenberg: indications for photochemical in-

fluence. In *Tropospheric Ozone*. Proceedings of the Fourth Ozone Symposium, Göttingen 1987.

Wang, Y., Jacob, D.J. & Logan, J.A. (1998) Global simulation of tropospheric O$_3$–NO$_x$–hydrocarbon chemistry 3. Origin of tropospheric ozone and effects of nonmethane hydrocarbons. *Journal of Geophysical Research* **103**, 10,757–67.

Wayne, R.P. (2000) *Chemistry of Atmospheres*, 3rd edn. Oxford University Press, Oxford.

Wayne, R.P., Barnes, I., Biggs, P. *et al.* (1991) The nitrate radical: physics, chemistry and the atmosphere. *Atmospheric Environment* **25**, 1–203.

Winer, A.M. & Bierman, H.W. (1994) Long path differential optical absorption spectroscopy (DOAS) measurements of HONO, NO$_2$ and HCHO in the Californian south coast basin. *Res. Chem. Intermed.* **20**, 423–45.

Zanis, P., Monks, P.S., Schuepbach, E. & Penkett, S.A. (2000) The role of *in-situ* photochemistry in the control of ozone during spring at the Jungfraujoch (3580 m asl)—comparison of model results with measurements. *Journal of Atmospheric Chemistry* **37**, 1–27.

7 Stratospheric Chemistry and Transport

A. ROBERT MACKENZIE

7.1 INTRODUCTION

Ozone is formed by the photolysis of oxygen by ultraviolet radiation in the solar spectrum. Since the source of sunlight is at the top of the atmosphere, but atmospheric density increases exponentially from the top of the atmosphere to the Earth's surface, the majority of ozone is found in a layer in the middle atmosphere. Ozone itself absorbs ultraviolet radiation from the Sun. The energy trapped by this absorption is converted into heat. Hence, the almost inexorable decrease of temperature with height that characterizes the lower atmosphere, the troposphere, is halted, and is replaced by the increasing temperatures and high static stability that characterize the lower stratosphere (see Chapter 2). In absorbing solar ultraviolet, ozone not only gives the stratosphere its principal structure, it also protects living organisms from the chemical-bond-breaking danger of ultraviolet light. These roles of ozone — as controlling influence in stratospheric statics and as shade for the biosphere — explain why changes in its abundance are monitored closely.

Although chemical measurements of ozone had been available since the middle of the nineteenth century, it was the discovery of the spectroscopy of ozone that allowed it to be detected remotely. By the end of the nineteenth century, spectroscopic measurements had determined that most of the atmospheric column of ozone was in the upper atmosphere. Measurements made through the first three decades of the twentieth century identified the exact altitude of the ozone maximum, showed that a great deal of the variability in ozone column from day to day can be attributed to changes in tropopause height, and demonstrated the existence of a slow overturning circulation in the stratosphere and mesosphere.

Mean column ozone abundances as a function of latitude and season are shown in Fig. 7.1. There is a broad minimum at equatorial latitudes at all seasons, with maxima at middle-to-high latitudes in the late winter and spring of each hemisphere. At high southern latitudes, embedded in the southern hemisphere maximum, is a severe local minimum centered on September. This is the Antarctic "ozone hole," described below.

The Tropics comprise a region where ozone production is efficient. Conversely, the high latitudes are regions where ozone production is inefficient. The distribution shown in Fig. 7.1 cannot be explained, therefore, by photochemistry alone, but must also be the result of atmospheric transport. The "Brewer–Dobson," or residual, circulation deduced from consideration of the distribution of trace gases like ozone and water vapor is shown in Fig. 7.2. The circulation is not thermally direct — i.e. it is not due to hot air rising — but due to the transport of momentum by waves internal to the atmosphere (e.g. McIntyre 1992; Shepherd 2000). The details of the sources and sinks of these waves are still to be resolved, but the general sense of the residual circulation is clear. One important consequence of this circulation is the distribution of

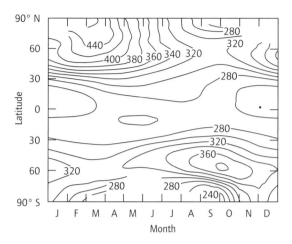

Fig. 7.1 The mean column ozone as a function of latitude and season. Note the relatively constant values in the Tropics, and the much greater variation nearer the poles. Arctic ozone columns are maximum in late winter and early spring. Antarctic ozone columns are highest in mid-winter, and show a decrease in the spring as a result of "ozone hole" chemistry. There is a marked "collar region" of higher ozone columns in the southern middle latitudes in winter and spring.

trace gases in the stratosphere. Long-lived and tropospherically inert trace gases, such as CFCs, N_2O, and water vapor, enter the stratosphere in the Tropics and are transported upwards. Exposure to shortwave, high-energy, solar radiation can convert these compounds into ozone-depleting radicals, and the reservoir compounds for these radicals (see below), as the air moves slowly poleward. In the downward leg of the circulation at high latitudes, then, the downward transport of the radical-reservoir compounds means that the potential for severe ozone depletion exists.

Early concerns about ozone depletion focused on low latitudes, and on the middle and upper stratosphere. It is here that CFCs and N_2O are broken down into ozone-depleting radicals. In the 1980s, however, attention switched abruptly to the polar lower stratosphere (Farman *et al.* 1984). Ground-based measurements showed a marked decline in the October monthly-mean ozone column over the Faraday research station, Antarctica (Fig. 7.3). Satellite measurements later confirmed that this was a continent-wide phenomenon.

Fig. 7.2 The residual circulation in the stratosphere and associated transport processes. The tropopause is shown by the thick line. Thin lines are isentropes, labeled in Kelvin. Wiggly double-headed arrows denote meridional transport by eddy motions. This eddy transport is not necessarily symmetric in and out of the stratosphere. The broad arrows show transport by the global-scale circulation. (From Holton *et al.* 1995.)

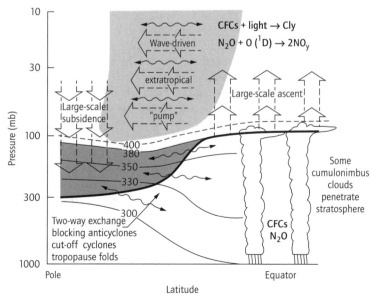

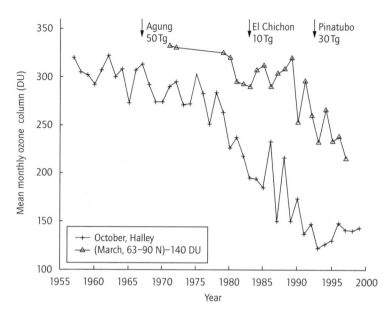

Fig. 7.3 October mean column ozone over Halley Bay (76° S) and March mean ozone columns over the Arctic (63–90 °N). Note that 140 DU have been subtracted from the Arctic means to aid graph plotting. The dates of major volcanic eruptions are also marked. (Ozone data courtesy of J. Shanklin, British Antarctic Survey, Cambridge, UK.)

Chemical ozone depletions, similar in scale to those seen in Antarctica in the early 1980s, have occurred in the Arctic, in 1994–5, 1995–6, 1996–7, and 1999–2000 particularly (e.g. Müller *et al.* 1997; Newman *et al.* 1997). Figure 7.3 also shows satellite observations of March-mean Arctic total ozone columns. The greater variability compared to the September-mean Antarctic ozone columns is clear: this is due to the greater variability in atmospheric transport in the northern hemisphere. Some recent northern hemisphere winters have shown only moderate ozone depletion. As outlined in Section 7.6, this should not be taken to mean that Arctic ozone depletion is necessarily a thing of the past.

7.2 THE STRUCTURE OF THE STRATOSPHERE

The stratosphere, as its name suggests, is a stably stratified layer of the Earth's atmosphere, extending very roughly from 10 to 50 km altitude. Temperature generally increases throughout the stratosphere, due to the absorption of solar radiation by ozone. Potential temperature, a measure of entropy, therefore increases rapidly and monotonically in the stratosphere. Because air must gain or lose energy to cross surfaces of constant potential temperature—isentropes—and because this is a process that occurs slowly relative to other transport processes occurring in the stratosphere, it is often convenient to use potential temperature as a vertical coordinate, or to follow air motion along isentropes—i.e. adiabatic air motion.

7.2.1 The residual circulation

Figure 7.2 bears closer examination. The thick line curving upwards from pole to Equator represents the tropopause, the boundary between troposphere and stratosphere. At low latitudes this boundary is best represented in the long-time mean by the mean position of the 380 K potential temperature surface. Once in the stratosphere, the slow upward transport of air carries pollutants, such as CFCs, and natural radical sources, such as N_2O and H_2O, into a regime of intense ultraviolet irradiation from the Sun. These compounds, which are chemically stable in the lower atmosphere, then begin to break down, e.g.:

$$N_2O + O(^1D) \rightarrow NO + NO \qquad (7.1)$$

$$H_2O + O(^1D) \rightarrow OH + OH \qquad (7.2)$$

$$CF_2Cl_2 + h\nu(\lambda < 227\,nm) \rightarrow CF_2Cl + Cl$$

$$(7.3)$$

where λ is the wavelength of the photon responsible for the photolysis and $O(^1D)$ is an electronically excited oxygen atom, produced from the photolysis of ozone and, to a lesser extent, molecular oxygen:

$$O_3 + h\nu(\lambda < 310\,nm) \rightarrow O_2 + O(^1D) \qquad (7.4)$$

$$O_2 + h\nu(\lambda < 175\,nm) \rightarrow O(^3P) + O(^1D) \qquad (7.5)$$

both of which processes themselves require ultraviolet sunlight. Note that the wavelength limits given here are approximate (see Wayne 2000 for details). In particular, small but significant production of $O(^1D)$ from ozone photolysis can occur at wavelengths longer than 310 nm (Ball *et al.* 1995).

The release of reactive compounds from their sources can cause ozone depletion, through the cycles outlined below, but in the upper atmosphere, these cycles are soon terminated by production of radical-reservoir compounds. A lingering doubt in predictions of upper stratospheric responses to these source-gas inputs stemmed from a consistent deficit in modeled ozone concentrations at around 40 km altitude when compared to measurements. Chemical production of ozone via vibrationally excited oxygen has been suggested as a means of correcting this ozone deficit (Miller *et al.* 1994), but recent calculations that were constrained by multiple measurements on board the UARS satellite seem to be able to account for the observed 40-km ozone concentrations without recourse to this extra production mechanism (Dessler *et al.* 1998, and references therein).

The mean upward motion in the Tropics is balanced by mean downward motion in the extratropics. Descent in the polar regions in winter is especially dramatic (Fig. 7.4), bringing upper

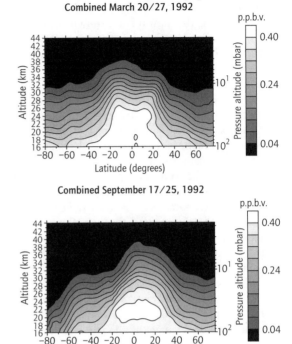

Fig. 7.4 A latitudinal cross-section of CF_2Cl_2 from the satellite instrument CLAES (data from Dessler *et al.* 1998). Note the steeply sloping isopleths in the subtropics and the wintertime polar circle. The relatively shallow slopes of the isopleths in the middle latitudes show the effects of stirring and mixing in the Rossby-wave surf zone.

stratospheric and mesospheric air down to lower stratospheric altitudes (roughly 25 km and below). The air that descends in the wintertime stratosphere is depleted in CFCs, N_2O, CH_4, and other long-lived gases, and is correspondingly enriched in chlorine, nitrogen, and hydrogen radical-reservoir compounds. With little or no warming available from the Sun, the wintertime polar regions cool, to temperatures at which aerosols take up condensable vapors (see Section 7.5). The combination of high concentrations of radical-reservoir compounds, high surface areas of aerosol, and low meridional mixing (see below) provides the ideal chemical and physical conditions for rapid ozone depletion (e.g. Schoeberl & Hartmann 1991).

Connecting the upward and downward portions of the mean circulation is a mean-meridional transport, toward the pole. The mean circulation as a whole is superimposed on much more rapid zonal transport (e.g. Andrews et al. 1987; Lindzen 1990, Chapter 5): eastward in the extratropics in winter; westward in the extratropics in summer; and oscillating between eastward and westward with a quasi-biennial period in the deep Tropics (Holton & Lindzen 1972).

Alongside the transport processes, diffusive processes mix air parcels, blurring the chemical and dynamical characteristics imparted to the air parcels by their transport through the stratosphere. Molecular diffusion is, of course, ubiquitous, and increases with altitude. Magnitudes of molecular diffusion vary by a factor of two to three for different chemical constituents, but indicative values are those for water vapor: $0.2\,cm^2\,s^{-1}$ at $0\,km$; $1.2\,cm^2\,s^{-1}$ at $16\,km$; and $75\,cm^2\,s^{-1}$ at $42\,km$ (Ghosh 1993). Molecular diffusion rates are important in the calculation of the flux of trace gases to aerosol surfaces (see below, and Turco et al. 1989; Hanson et al. 1994), but other mixing processes are much more important in the bulk mixing of stratospheric air parcels. Turbulent mixing—i.e. mixing due to unresolved eddies, hence often called "eddy diffusivity"—although much reduced compared to the troposphere, is still important in determining the fine structure of tracer distributions.

Because the winds in the stratosphere produce horizontal strain fields and strong vertical shear, mixing processes are most likely to act on vertical, cross-isentrope, contrasts in trace gas concentrations rather than on horizontal, along-isentrope, contrasts (Haynes & Anglade 1997). The exact value of vertical diffusivity, as a function of space and time, therefore becomes important. Vertical diffusivity has been estimated to reach $2000\,cm^2\,s^{-1}$ in patches of three-dimensional turbulence (Woodman & Rastogi 1984), although other analyses suggest an upper limit of $200\,cm^2\,s^{-1}$ (M. Balluch and P. H. Haynes, private communication). In any event, turbulent mixing may always be expected to be much larger than molecular diffusion.

7.2.2 Barriers to mixing: the polar vortices and the tropical "pipe"

Even with an appreciation of the basic Brewer–Dobson circulation, early models of stratospheric transport failed to reproduce the rather sharp gradients in trace gas and aerosol concentrations seen in satellite observations such as Fig. 7.4 (Nightingale et al. 1996). CF_2Cl_2 concentrations decrease with height due to photolysis (see above), as expected, but there are also striking horizontal gradients in concentration, particularly in the springtime hemisphere. At any given altitude, the sharpest changes in horizontal concentration gradient occur in the subtropics and near the polar circle. These sharp changes are indicative of mixing barriers, which inhibit the transport of air along isentropic surfaces.

High-resolution models of the stratosphere show these transport barriers particularly clearly (Juckes & McIntyre 1987; Norton 1994; Sutton et al. 1994; Waugh and Plumb 1994; Orsolini et al. 1995; and many subsequent papers). The model results have since been strongly supported by observational evidence (e.g. Plumb et al. 1994; Fairlie et al. 1997). Figure 7.5 shows one high-resolution, single-layer, model result. The polar region is bounded by a region of steep horizontal gradients. This is the wintertime polar vortex, within which severe ozone depletion can occur. Another region of steep gradients occurs in the subtropics throughout the year. This is the subtropical mixing barrier. The region bounded by the subtropical mixing barrier is sometimes known as the "tropical pipe" (Mote et al. 1996; Plumb 1996). Between the steep gradients at the polar and subtropical regions is a region composed of many interleaved filaments, known, for reasons outlined below, as the "Rossby Wave Surf Zone" or simply the "Stratospheric Surf Zone" (Juckes & McIntyre 1987).

Observations such as those shown in Fig. 7.4, and the fluid dynamical fundamentals implicit in simulations such as that which generated the model results in Fig. 7.5, present a severe difficulty

DFT 435K , 19 Sep 1999 11:59, ECMWF [lvl 2: 100.000 hPa]

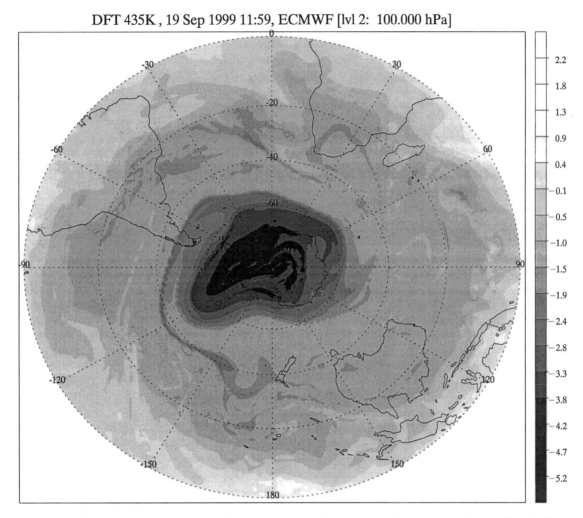

Fig. 7.5 A high-resolution simulation of the lower stratosphere showing mixing barriers. (Calculation and graphic by Gianluca Redaelli. See, for further work, Dragani *et al.* 2002, and references therein.) A regular grid of several thousand points is constructed for the domain shown. A backward air mass trajectory from each point is then calculated. The potential vorticity, for the position of the trajectory's origin, is then interpolated onto each trajectory from a meteorological analysis (from the European Centre for Medium-Range Forecasting, in this case). Finally, the potential vorticity for each trajectory is mapped onto the original regular grid, i.e. transported forward in time under the assumption that the potential vorticity is conserved.

to theories and models of stratospheric transport that assume an all-pervasive eddy diffusivity operating throughout the stratosphere (e.g. Tuck 1979). Such models are still used for long-term forecasting of stratospheric ozone (WMO 1999, Section 7), but now include analytic approximations to the large-scale behavior of the polar vortex that account to some extent for the obvious deficiencies of two-dimensionality (Garcia & Boville 1994). The realization that the eddy-diffusivity treatment is not adequate has led to the acceptance and use of models and analyses based on the interaction of internal atmospheric waves with the mean flow.

There are many, fluid-dynamically fundamental, implications arising from a view of the atmosphere based on wave–mean interaction theory. McIntyre (1992) provides an exemplary description of these implications; other highly informative reviews are provided by Hamilton (1996) and Shepherd (2000). Only a few of the practical results are dealt with here. Foremost of the practical results is that a single set of isentropic fields, that of potential vorticity, contains almost all the information on the dynamics of the stratospheric state. Vortices, mixing barriers, and the stirring of the surf zone can then all be understood as stemming from the behavior of potential vorticity waves on isentropic surfaces.

Potential vorticity—sometimes known as Rossby–Ertel potential vorticity—is defined as:

$$Q = \frac{1}{\rho}(\Omega + \zeta)\frac{\partial \theta}{\partial z}$$

where ρ is the atmospheric density, Ω is the planetary angular velocity, ζ is the relative vorticity, and the three-dimensional gradient of potential temperature has been assumed to be directed exclusively in the vertical. Note that Q has units $K\,m^2\,kg^{-1}\,s^{-1}$ (and not s^{-1}, which is the unit of vorticity, e.g. for ζ). These units are so clumsy that the "PV-unit" ($1\,PVU = 10^{-6}\,K\,m^2\,kg^{-1}\,s^{-1}$) has been coined. Thorpe and Volkert (1997) give a short history of the derivation of potential vorticity that sheds light on the quantity and its unusual units.

Q is a measure of the absolute spin of an air parcel about the vertical, including the vertical component of the Earth's rotation, and can be thought of as the "cyclonicity" of an air parcel. In the absence of frictional and other diabatic effects, Q is conserved. In the lower stratosphere, adiabatic motion is a reasonable approximation for 5–10 days, and so the conservation of Q is a reasonable assumption over similar time scales. Following the distribution of Q on isentropic surfaces gives a much clearer picture of the stirring and mixing in

the stratosphere than does following the distribution of, say, winds or geopotential heights, and this has become an important tool for visualizing stratospheric flow (Norton 1994; Waugh & Plumb 1994). The distribution of Q can cascade to very small scales in the middle latitudes in winter, when the horizontal gradients on Q are concentrated at the subtropical mixing barrier and the edge of the polar vortex (Polvani et al. 1995) (Fig. 7.5). Undulations in these mixing barriers, known as Rossby waves, can become sufficiently large that air is transported outwards in filaments across the mixing barriers into the middle latitudes. Figure 7.5 shows the superposition of wave-three (three cycles around a latitude circle) and wave-two Rossby waves, leading to a pronounced equatorward distortion of the vortex around 120° W. The formation of filaments of vortex air, or subtropical air, in the middle latitudes is called Rossby wave breaking, by analogy to the breaking of (gravity) water-waves on a beach. Transport inwards—i.e. toward Equator or pole—is much less common (e.g. Plumb et al. 1994), and this lack of symmetry explains the inability of eddy-diffusivity models to capture the mean circulation. PV visualizations can often successfully predict filamentation from Rossby wave breaking (e.g. Lee et al. 1997) but also tend to predict filaments where none are observed (Dragani et al. 2001). A parameterization of the diffusional mix-down of the filaments is required, but this is still some way off, it seems, since the sporadic occurrence of three-dimensional turbulence in the stratosphere is not well understood (see above).

7.2.3 *The tropopause*

Exchange of mass between the troposphere and the stratosphere (stratosphere–troposphere exchange, or STE) is of central importance to several problems in atmospheric pollution. The rate of CFC transport into the stratosphere affects the likely duration and extent of ozone depletion into the future. The rate of downward transport of ozone to the troposphere is also important because of the contribution of stratospheric ozone to the oxidizing capacity of the troposphere (e.g. Lelieveld et al.

1999, and references therein). As would be expected from the mean circulation shown in Fig. 7.2, downward transport occurs mainly in the middle latitudes, while upward transport occurs mostly in the Tropics.

Sudden bursts of ozone-rich, very dry, air have been measured at ground air pollution stations, particularly in spring in middle latitudes. It can be shown easily that these "ozone episodes" are not photochemical in origin, but are instead due to rapid downward transport from the stratosphere.

These rapid events are called tropopause folds, and are an extreme example of tropopause undulation. The pressure-altitude of the tropopause moves according to the weather systems in the troposphere. In particular, low-pressure systems lower the tropopause. Jet streaks are often associated with steep N–S changes in tropopause height. These streaks of high horizontal wind speeds initiate vertical circulation patterns at their beginning and end. Under particular conditions, closely related to the formation of a surface low-pressure system with its attendant fronts, the vertical circulations at the entrance to a jet bring a thin layer of stratospheric air to very low altitudes, as sketched in Fig. 7.6 (Danielsen 1968). The stratospheric air is then mixed into the lower troposphere by turbulence. Between about 1950 and 1970, rapid STE in tropopause folds constituted a potential health risk, due to the downward transport of radioactivity that had been released into the atmosphere in above-ground nuclear weapons tests. Recent advances in lidar and radar technology have provided an opportunity to study these features in much more detail (e.g. Holton *et al.* 1995; Vaughan *et al.* 2001). While tropopause folds have been recognized features of atmospheric dynamics for about half a century, it has not proved straightforward to develop a quantification of STE from a climatology of folds. New trajectory-based method—using space-filling trajectory ensembles and counting the number that cross the tropopause— offer more hope of improving estimates of hemispheric-scale STE.

The stratosphere is extremely dry. Water vapor mixing ratios in the stratosphere generally range from 1 to 6 p.p.m.v.; in the troposphere mixing ra-

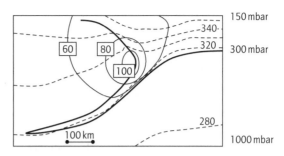

Fig. 7.6 A schematic horizontal-distance/pressure cross-section of stratosphere–troposphere exchange in a tropopause fold (adapted from Danielsen 1968). The thick line shows the tropopause. Gray continuous lines are isotachs, showing a jet that is causing the deformation of the tropopause. Broken lines show isentropes: note the rapid change in the vertical gradient of potential temperature (i.e. isentrope spacing) across the tropopause. Although the tropopause fold is reversible to some extent, turbulence in the middle and lower troposphere will cause irreversible mixing of stratospheric air into the troposphere.

tios of p.p.th.v are normal. Furthermore, most stratospheric water vapor comes from the oxidation of methane, a fact that is easily demonstrated by the near-constancy of "total hydrogen," ΣH_2, in the stratosphere:

$$\sum H_2 = 2CH_4 + H_2O + H_2$$

(H_2 itself is always a minor component of ΣH_2 in the troposphere and stratosphere). For $\Sigma H_2 \approx 7.5$ p.p.m.v., $CH_4 \approx 1.7$ p.p.m.v., and $H_2 \ll 1$ p.p.m.v., $H_2O \approx 3$ p.p.m.v. Because there are almost no chemical sinks for water capable of removing hundreds of p.p.m.v., there must be an intense drying of air as it passes upward across the tropopause. The only conceivable method for this dehydration to take place is by cloud processing.

When a cloud forms, the total mass of water substance in the air parcel is partitioned between condensed and vapor phases, according to the vapor pressure equation for water. If the ice particles formed in the cloud are large enough to precipitate out of the air mass then, on evaporation of the cloud, the resulting water vapor mixing ratio will be much less than the original total water. This process is also known as freeze-drying. Only the tropical tropopause is cold enough to dry air, in the manner discussed above, to stratospheric mixing ratios. The details of the cloud formations that are responsible for the drying are not well known, however (Holton *et al.* 1995).

7.2.4   The meso-scale and the stratosphere

In addition to the planetary-scale residual circulation, and the synoptic-scale influence of tropospheric weather systems, it has recently become apparent that smaller-scale processes—with typical horizontal scales of tens of kilometers—are also important to the distribution of trace gases in the stratosphere. The most important of these small-scale processes are gravity waves. These waves are internal to the atmosphere, generated by forced upward movement of air parcels, and have gravity as a restoring mechanism. The forced upward movement can be due to a solid obstruction to the flow, such as a mountain, or can be due to upwelling of buoyant air in convection. It is now believed that the torque applied to the mean flow in the upper stratosphere and mesosphere by dissipating gravity waves is significant, and that much of the gravity wave spectrum responsible for this torque is too small-scale to be resolved by current global climate models (Hamilton 1996).

Lower in the stratosphere, gravity waves have an impact on chemistry by subjecting air parcels to rapid, adiabatic, temperature oscillations (Murphy & Gary 1995). When the mean temperature is already low (i.e. below 200 K), the induced adiabatic oscillation can lead to uptake of water on the stratospheric aerosol, and even to cloud formation (see below). It has been hypothesized that ozone depletion by gravity-wave-induced stratospheric clouds is the largest contribution to ozone depletion in the Arctic (Carslaw *et al.* 1999). The gravity waves in this case are forced by the underlying topography: the Scandinavian Alps, the Greenland ridge, and the Urals. Studies of the distribution of gravity waves in the lower stratosphere range from relatively simple analytic approaches combined with "ridge-finding" algorithms (Bacmeister *et al.*

1994), to high-resolution modeling of mountainous regions (Dörnbrack *et al.* 1998).

Gravity waves have also been put forward as a mechanism for cloud formation, and thence dehydration, at the tropical tropopause (Potter & Holton 1995). The observational evidence for this is currently scant, but growing (Pfister *et al.* 1999; Santacesaria *et al.* 2001). The quantitative importance of small-scale gravity waves for atmospheric transport and physico-chemical transformation of trace gases remains uncertain, and so is likely to be an important area of research well into the first decade of the twenty-first century.

7.3   GAS-PHASE CHEMISTRY OF THE STRATOSPHERE

The chemical conversion of trace gases along the course of the Brewer–Dobson circulation has been mentioned in passing above. Below, a somewhat more detailed account of the major processes is given. A complete review of chemical reactions, their rate coefficients, and their mechanisms is beyond the scope of this chapter. The reader is directed to the continuing series of critical reviews under the auspices of JPL, the NASA Jet Propulsion Laboratory (DeMore *et al.* 1997) and IUPAC (Atkinson *et al.* 1999; see also the website http://www.iupac-kinetic.ch.cam.ac.uk/).

7.3.1   Oxygen-only chemistry

Given that ozone is an allotrope of oxygen, that oxygen makes up one fifth of the atmosphere, and that the upper atmosphere is bathed in photolytic radiation from the Sun, it is unsurprising that ozone is produced by photolysis of oxygen:

$$O_2 + h\nu(\lambda < 243\,nm) \rightarrow O + O \qquad (7.5a)$$

followed by

$$O + O_2 + M \rightarrow O_3 + M \qquad (7.6)$$

where M is any atmospheric molecule, acting as a heat sink for the exothermic combination of O and

O_2. Clearly, unless ozone is to accumulate in the atmosphere, there must be loss processes too. Chapman (1930) first closed the loop of the ozone life cycle by suggesting

$$O_3 + h\nu(\lambda < 1180\,nm) \rightarrow O + O_2 \qquad (7.4a)$$

$$O + O_3 \rightarrow 2O_2 \qquad (7.7)$$

(Note that a third possible oxygen-only loss, from the recombination of O, atoms is too slow to be significant in the stratosphere. Note also that reactions 7.4a and 7.5a differ from reactions 7.4 and 7.5 only in the wavelength threshold: light at the longer wavelengths can produce ground-state oxygen atoms, $O(^3P)$, from O_3 and O_2.) Reactions 7.4a and 7.6 swap oxygen rapidly between its "odd oxygen" (one-atom and three-atom) allotropes, so that it is convenient to discuss the "odd oxygen family" and to treat O and O_3 as in chemical equilibrium with each other. This grouping of reactive intermediates into families is an important concept in reducing the complexity of stratospheric chemistry, as exemplified by the use of other families below.

The oxygen-only "Chapman mechanism" for formation and destruction of ozone has the correct properties to generate an ozone layer in the stratosphere, as is observed. That is, the source of the light to photolyze oxygen originates above the atmosphere but the density of oxygen molecules per unit volume increases with decreasing altitude (e.g. Wayne 2000, pp. 160–2). However, careful measurements reveal that the rate of reaction 7.7 is too slow to account for the ozone concentrations observed, i.e. the Chapman mechanism predicts higher peak ozone concentrations than are observed.

7.3.2 *The classical radical cycles: ozone-depleting and null cycles, and termination reactions*

In the decades following the publication of the Chapman scheme, several additional loss processes for ozone have been revealed (Bates & Nicolet 1950; Crutzen 1970; Molina & Rowland 1974). In the upper stratosphere, homogeneous catalysis of reaction 7.7 is important:

$$X + O_3 \rightarrow XO + O_2 \qquad (7.8)$$

$$XO + O \rightarrow X + O_2 \qquad (7.9)$$

$$\textbf{Net}: O + O_3 \rightarrow 2O_2$$

where X is H, OH, NO, Cl, and, to a minor extent, Br. The source of these catalysts is discussed in Section 7.2.1 above. The rapid interconversion of radical intermediates—for example, the X/XO radicals above—make it convenient to consider families of radical intermediates. Hence, the HO_x family represents H, OH, and HO_2; the NO_x family represents NO, NO_2, and NO_3; and so on. The concept of chemical families makes conceptual modeling (e.g. Johnston & Podolske 1978) and numerical modeling (see Jacobson 1999 for an introduction) of the chemistry considerably simpler.

For all the radicals listed above, oxidation by ozone, and reduction of the oxidized radical by O, are extremely rapid reactions, so that, if the catalysts are present in sufficient concentration, the rate of the net reaction is increased. A second channel is available to OH radicals, since they can be reduced by O atoms, as well as oxidized by O_3:

$$OH + O \rightarrow H + O_2 \qquad (7.10)$$

$$H + O_2 + M \rightarrow HO_2 + M \qquad (7.11)$$

$$HO_2 + O \rightarrow OH + O_2 \qquad (7.12)$$

$$\textbf{Net}: O + O \rightarrow O_2$$

The catalytic reaction cycles above are in competition with null cycles, which simply tie up a proportion of the catalyst in reactions that do not destroy ozone, or with holding cycles, which remove the catalysts temporarily by forming short-lived reservoir compounds. An important null cycle is that involving NO_2:

$$NO + O_3 \rightarrow NO_2 + O_2 \qquad (7.13)$$

$$NO_2 + h\nu(\lambda < 420\,nm) \rightarrow NO + O \quad (7.14)$$

$$O + O_2 + M \rightarrow O_3 + M \quad (7.6)$$

Net: null.

Reaction 7.14 can close the null cycle channels of other radicals, since NO can be oxidized by many of the other XO species, e.g.

$$OH + O_3 \rightarrow HO_2 + O_2 \quad (7.15)$$

$$HO_2 + NO \rightarrow NO_2 + OH \quad (7.16)$$

$$NO_2 + h\nu(\lambda < 420\,nm) \rightarrow NO + O \quad (7.14)$$

$$O + O_2 + M \rightarrow O_3 + M \quad (7.6)$$

Net: null.

An important holding cycle involves the production of N_2O_5, which is stable during the night, so locking up two NO_x radicals for the hours of darkness:

$$NO_2 + O_3 \rightarrow NO_3 + O_2 \quad (7.17)$$

$$NO_2 + NO_3 + M \rightarrow N_2O_5 + M \quad (7.18)$$

$$N_2O_5 + h\nu \rightarrow NO_2 + NO_3 \quad (7.19)$$

$$NO_3 + h\nu(\lambda < 640\,nm) \rightarrow NO_2 + O \quad (7.20a)$$

$$O + O_2 + M \rightarrow O_3 + M \quad (7.6)$$

Net: null, but accumulation of N_2O_5 during darkness.

When reactions remove the radicals to form products with reactivation time scales that are comparable with transport time scales, then the radical reaction chain can be considered to be terminated. Examples are:

$$OH + NO_2 + M \rightarrow HNO_3 + M \quad (7.21)$$

$$Cl + CH_4 \rightarrow HCl + CH_3 \quad (7.22)$$

$$ClO + NO_2 + M \rightarrow ClONO_2 + M \quad (7.23)$$

$$HO_2 + NO_2 + M \rightarrow HO_2NO_2 + M \quad (7.24)$$

The efficiency of a catalytic cycle is measured by the chain length, i.e. the average number of times a particular cycle is executed before termination. However, the impact of a catalytic cycle on ozone concentrations depends not only on the efficiency of the cycle, but also on the concentration of the radicals. For example, the efficiency of the Br/BrO cycle is much larger than that of the Cl/ClO cycle, and it is only the much smaller abundance of bromine than chlorine in the atmosphere that justifies our focus on chlorine as the major threat to ozone. Were the abundance of bromine to increase in the stratosphere, the consequences for ozone would be very serious indeed.

Reactivation of the radical chemistry is by photolysis, e.g.

$$HNO_3 + h\nu \rightarrow OH + NO_2 \quad (7.25)$$

by reaction with OH, e.g.

$$HNO_3 + OH \rightarrow H_2O + NO_3 \quad (7.26)$$

or by heterogeneous processes (see Section 7.5).

The existence of important cross-family null cycles, holding cycles, and termination reactions is indicative of the nonlinear nature of the chemistry of the stratosphere with respect to changes in the concentrations of radical families. Predicting the response of ozone to changes in any one family is virtually impossible, therefore, without recourse to modeling of the full system.

7.3.3 *Ozone depletion cycles without O atoms*

The classical ozone-destroying cycles, described in the previous section, require O atoms for the reduction of the XO species. However, other catalytic cycles exist that do not depend on the presence of the O atom. Such cycles are particularly important in the lower stratosphere, where the ratio of O atoms to ozone molecules is much less than in the upper stratosphere. There are three general types of ozone-specific catalytic cycles. The first involves formation of an XO species that itself reacts with ozone, e.g.

$$OH + O_3 \rightarrow HO_2 + O_2 \qquad (7.15)$$

$$HO_2 + O_3 \rightarrow OH + 2O_2 \qquad (7.27)$$

Net: $2O_3 \rightarrow 3O_2$

or

$$NO + O_3 \rightarrow NO_2 + O_2 \qquad (7.13)$$

$$NO_2 + O_3 \rightarrow NO_3 + O_2 \qquad (7.17)$$

$$NO_3 + h\nu(584 < \lambda < 640\,nm) \rightarrow NO + O_2 \quad (7.20b)$$

Net: $2O_3 + h\nu \rightarrow 3O_2$

Note the second channel for photolysis of NO_3 (reaction 7.20b, cf. reaction 7.20a). A detailed discussion of the photolysis pathways for NO_3 is given in Wayne *et al.* (1990).

The second type of ozone-specific cycle involves formation of a compound from two XO species, with elimination of O_2, e.g.

$$Br + O_3 \rightarrow BrO + O_2 \qquad (7.28)$$

$$OH + O_3 \rightarrow HO_2 + O_2 \qquad (7.15)$$

$$HO_2 + BrO \rightarrow HOBr + O_2 \qquad (7.29)$$

$$HOBr + h\nu \rightarrow OH + Br \qquad (7.30)$$

Net: $2O_3 + h\nu \rightarrow 3O_2$

The third type of ozone-specific cycle involves formation of compounds from two XO species, and subsequent elimination of O_2 by photolysis of a different bond to the one initially formed, e.g.

$$Cl + O_3 \rightarrow ClO + O_2 \qquad (7.31)$$

$$NO + O_3 \rightarrow NO_2 + O_2 \qquad (7.13)$$

$$ClO + NO_2 + M \rightarrow ClONO_2 + M \qquad (7.32)$$

$$ClONO_2 + h\nu \rightarrow Cl + NO_3 \qquad (7.33)$$

$$NO_3 + h\nu(584 < \lambda < 640\,nm) \rightarrow NO + O_2 \quad (7.20b)$$

Net: $2O_3 + 2h\nu \rightarrow 3O_2$

An important example of this third type of ozone-specific cycle is the dimer cycle:

$$2(Cl + O_3 \rightarrow ClO + O_2) \qquad (7.31)$$

$$2ClO + M \rightarrow Cl_2O_2 + M \qquad (7.34)$$

$$Cl_2O_2 + h\nu \rightarrow Cl + ClOO \qquad (7.35)$$

$$ClOO + M \rightarrow Cl + O_2 + M \qquad (7.36)$$

Net: $2O_3 + h\nu \rightarrow 3O_2$

which is responsible for most of the ozone destruction in the springtime polar vortices, leading to the formation of ozone "holes." Note that the second and third type of ozone-specific cycles require sunlight. These cycles do not take place during polar night, therefore, even when the polar vortices are "primed" for ozone destruction by heterogeneous chemistry (see below). In the northern hemisphere, the vortex is sufficiently perturbed by planetary waves that vortex air is exposed to sunlight throughout the winter. In the southern hemisphere, the vortex is less perturbed and ozone loss can take place only at the edge of the polar vortex (Roscoe *et al.* 1997; Santacesaria *et al.* 1999) until sunlight returns to the polar regions in spring.

Figure 7.7 gives an example of a model calculation for the Arctic lower stratosphere, showing chemical conversion of chlorine from reservoirs to active radicals and concomitant ozone destruction (Becker *et al.* 2000). For this idealized, but still reasonably realistic, trajectory calculation, temperature and radiative fluxes vary as the air parcel oscillates around 70° N. This simulates the Rossby-wave dynamics of the Arctic vortex (see above). The trajectory slowly descends in altitude (and potential temperature), simulating the dia-

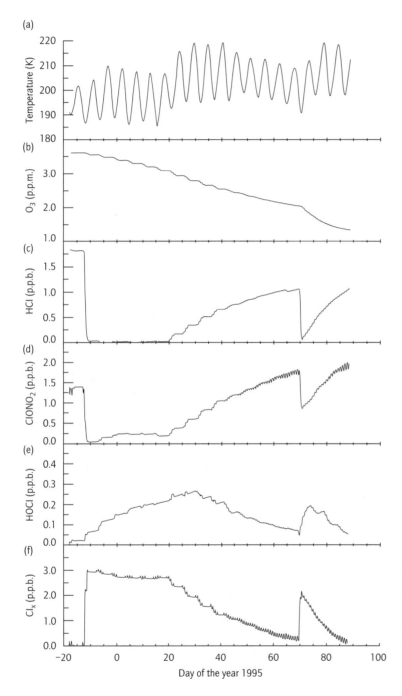

Fig. 7.7 A model of the chemical evolution in the wintertime polar lower stratosphere, following an idealized vortex air trajectory (adapted from Becker *et al.* 2000, redrawn by the authors of that work). Plots against day of the year: (a) temperature, (b) ozone, (c) HCl, (d) $ClONO_2$, (e) HOCl, (f) Cl_x.

batic cooling that occurs in the vortex. Over the three-month model run, chlorine is released rapidly from the reservoir compounds, leading to high mixing ratios of chlorine radicals (Cl_x: this process is usually called "chlorine activation"). Ozone loss occurs steadily until increased solar irradiation begins to deactivate the chlorine, converting it back into $ClONO_2$ and, more slowly, HCl. HOCl mixing ratios increase in the middle of the model run, i.e. HOCl acts like a classical reactive intermediate. By the end of the model run, ozone is reduced to about 30% of its initial value. Ozone losses of this magnitude have been deduced from observations (Müller *et al.* 1997; Rex *et al.* 1997).

7.4 AEROSOLS AND CLOUDS IN THE STRATOSPHERE

The thermal stability and dryness of the stratosphere (Section 7.2) make cloud formation unlikely, but it does occur. Nacreous clouds have been observed over the Arctic for more than a century and a quarter (Stanford & Davis 1974). Satellite observations in the past 20 years have greatly improved our understanding of the frequency and distribution of these clouds, now known as polar stratospheric clouds (PSCs) (Poole & Pitts 1994). Detailed *in situ* measurements have shown a continuum of particle sizes, from the sub-micrometer background aerosol (Junge *et al.* 1961) to ice particles with radii of many micrometers (Dye *et al.* 1992).

7.4.1 The volcanic aerosol and the Junge layer

A layer of aerosol, the Junge layer, blankets the globe. The center of mass of the Junge layer is about 5 km above the tropopause, although the aerosol is present in a wide altitude band throughout the lower stratosphere (McCormick *et al.* 1993). The stratospheric aerosol is composed primarily of concentrated sulfuric acid droplets, but can also contain mineral elements derived from meteorites (Toon & Farlow 1981). Recent *in situ* mass spectra near the tropopause detected at least 45 elements

in aerosol particles, ranging from hydrogen to lead (Murphy *et al.* 1998).

The primary source of the stratospheric aerosol is oxidation of the SO_2 injected by volcanic eruptions. An additional, very small but constant, source may be the percolation to the stratosphere of carbonyl sulfide (OCS), emitted by microbes, with subsequent oxidation to sulfuric acid. The most effective injection of volcanic aerosol in recent years came in June 1991 with the eruption of Mt Pinatubo in the Philippines (McCormick *et al.* 1993). The global enhancement of the stratospheric aerosol loading due to the eruption was estimated at 30 Tg (i.e. 30 megatonnes). Atmospheric opacity, or optical depth, peaked at values greater than 0.2 in the Tropics shortly after the eruption. This corresponds to surface area densities of up to $40 \mu m^2 cm^{-3}$ and above (Thomason *et al.* 1997). The largest injection of volcanic gas and ash into the stratosphere for which we have evidence is the eruption of Toba, in Sumatra, 73,500 years before present. It is estimated that the total stratospheric aerosol loading following Toba reached 1000 Tg (Rampino & Self 1992).

7.4.2 Polar stratospheric clouds

Polar stratospheric clouds (PSCs) catalyze the conversion of relatively stable chlorine reservoir compounds to reactive compounds that can take part in rapid ozone depletion. PSCs can also remove nitric acid from the stratosphere, as the PSC particles sediment, and so act as an important sink for nitrogen oxides in the stratosphere. Because PSCs contain nitric acid, and because their formation is a strong function of temperature, it has been suggested that PSC frequency can be affected by supersonic aircraft emissions, and by climate change. In order to quantify these sensitivities, and to quantify current ozone loss, accurate models of PSC formation and evolution are required. In the past decade, many advances have been made in the quantification of individual microphysical processes affecting PSCs (Fiocco *et al.* 1998). This includes improvements to the treatment of growth kinetics (MacKenzie & Haynes 1992), solution thermodynamics (Carslaw *et al.* 1994; Tabazadeh *et al.* 1994), and freezing kinetics (MacKenzie *et al.* 1998; Koop *et al.* 2000).

The composition of the stratospheric aerosol changes dramatically as an air mass is cooled to temperatures in the range of 180–200 K. Initially, the sulfuric acid aerosol takes up water vapor, to become more dilute. On further cooling, below about 192 K, the sulfuric acid aerosol takes up nitric acid and more water vapor, to form a PSC. For rapid temperature changes—in mountain-wave-induced clouds, for example—the combination of the Kelvin effect and the size dependence of vapor deposition rates mean that this change in composition is size-dependent. Numerical schemes have been developed to study size-dependent composition changes (e.g. Meilinger *et al.* 1995; Lowe & MacKenzie 2002). Further cooling below the frost point results in freezing of the particles.

Despite this improvement in theory and laboratory simulations, the formation and evolution of PSCs in the real atmosphere remains highly uncertain (Fiocco *et al.* 1998; Tolbert & Toon 2001, and references therein). In particular, while there is ample evidence for solid PSC particles at temperatures above the frost point, the mechanism by which they form is not known. Liquid–solid phase change is not a well quantified process in any system (Oxtoby 1992), including the concentrated aqueous solutions that are present in the lower stratosphere. Laboratory studies suggest that the trace amounts of sulfuric acid in the liquid PSC droplets substantially reduce the freezing temperature for a given cooling rate (or, equivalently, freezing time for a given temperature) compared to binary nitric acid–water droplets (e.g. Meilinger *et al.* 1995).

7.4.3 *Cirrus at the extratropical tropopause*

Important uncertainties remain about the reactivity of aerosol and cloud particles in the upper troposphere and mid-latitude lower stratosphere (e.g. Lary *et al.* 1997; Solomon *et al.* 1997). The conditions for heterogeneous chemistry to be important are that reactant compounds are present in appreciable amounts and that sufficient aerosol volume is available. It has been argued that these condi-

tions are met not only in the middle of the Junge layer, and in polar stratospheric clouds, but also at the mid-latitude tropopause (Borrmann *et al.* 1996, 1997). The hypothesis is that chlorine radicals are produced by heterogeneous chemistry at the tropopause, thus modifying the ozone budget there. The tropopause is a minimum in the temperature profile, and is often cold enough for aerosol particles present to take up water vapor and swell, and for cirrus clouds to form. For the release of chlorine radicals, the reactant compounds required are HCl, $ClONO_2$, and $HOCl$, all of which have sources predominately in the stratosphere. It is to be expected that the mixing ratios of these reactant compounds will change rapidly near the tropopause, reflecting the change in the source of air on either side of the tropopause. So, it is not clear whether the clouds and swollen aerosol particles, which require tropospheric water vapor to form after all, come into contact with the reactant compounds. Any more definitive statement must await further *in situ* observations of aerosol and chlorine radicals.

7.5   HETEROGENEOUS CHEMISTRY*
OF THE STRATOSPHERE

After the discovery of the Antarctic ozone hole (Farman *et al.* 1984), several possible mechanisms for the ozone depletion were put forward. The correct hypothesis came from Solomon *et al.* (1986): that reactions occurring on PSC particles produced chlorine radicals, and that the chlorine radicals went on to destroy ozone in a series of catalytic cycles. Crutzen and Arnold (1986) and Toon *et al.* (1986) then independently identified that nitric acid trihydrate (Taesler *et al.* 1975) was the thermodynamically stable phase for the PSCs observed

---

* In the atmospheric science community, the term "heterogeneous chemistry" is generally taken to include any reactions occurring on or in particles in the atmosphere. Reactions occurring in liquid particles are, of course, homogeneous liquid-phase reactions, *sensu stricto*, but atmospheric chemistry models usually fold the gas–particle partitioning into the rate expression (see below), so there is some justification for the terminology.

above the ice frost-point. In the decades following the discovery of the ozone hole, a substantial amount of work on particle microphysics and heterogeneous chemistry has been carried out, leading to much improved characterization of their kinetics and thermodynamics. Work on liquid particles appears to be reaching maturity—in that most of the laboratory techniques and analytic tools required have now been developed (e.g. Ravishankara & Shepherd 1999)—but there is still much to be learnt about reactions on solid particles in the atmosphere.

Table 7.1 lists some of the most important reactions occurring on sulfuric acid aerosol and PSC particles. The overall effect of heterogeneous reactions is to convert the chlorine and hydrogen in reservoir compounds into more active forms, while converting the NO_x in temporary reservoir compounds into nitric acid. Many of the heterogeneous reactions in the sulfuric acid aerosol are strongly temperature dependent, due to the variation of aerosol composition, and thus Henry's Law coefficients, with temperature. For aerosol loadings typical of a volcanically quiescent period, heterogeneous reactions begin to activate chlorine at temperatures below about 200 K. This activation

Table 7.1 Some heterogeneous reactions of importance in the lower stratosphere.

Reaction	Effect
$N_2O_5 + H_2O \rightarrow 2HNO_3$	Decreases NO_x concentration throughout lower stratosphere
$ClONO_2 + H_2O \rightarrow HOCl + HNO_3$	Chlorine activation at low temperatures, but HOCl less photolabile than Cl_2; denoxification
$ClONO_2 + HCl \rightarrow Cl_2 + HNO_3$	Chlorine activation at low temperatures; denoxification
$HOCl + HCl \rightarrow Cl_2 + H_2O$	Chlorine activation at low temperatures
$BrONO_2 + H_2O \rightarrow HOBr + HNO_3$	Indirectly affects ClO_x and HO_x concentrations
$HOBr + HCl \rightarrow BrCl + H_2O$	Indirectly affects ClO_x and HO_x concentrations

is relatively slow, however. When aerosol loadings are much increased following a volcanic eruption, the temperature at which chlorine activation begins is increased by up to 5 K, and aerosol reactions can effectively compete with reactions on type 1 PSCs. One heterogeneous reaction that is not temperature-dependent is the hydrolysis of N_2O_5. Consideration of this reaction has led to a downward revision of the importance of NO_x-related catalytic cycles in the global lower stratosphere at all temperatures (e.g. Garcia & Solomon 1994), but also led to models significantly underpredicting NO_x in the lower stratosphere. The agreement between models and measurements has now been improved once more, due to more accurate measurements of the rates of the reactions of HNO_3 and NO_2 with OH (Brown *et al.* 1999a, b), but it is safe to say that nitrogen oxide chemistry in the stratosphere is still somewhat uncertain.

7.5.1 Rates of heterogeneous reactions in liquids

Diffusion is important for reactions involving particles, and reactants partition between the gas and condensed phases according to their solubilities or adsorptivities. In liquid particles, if reaction is slow, the reactants will have time to continuously adjust to their equilibrium partitioning, and the rate of reaction has the form (e.g. Cox *et al.* 1994)

$$-d[R_1]/dt = k[R_1][R_2]$$

where k is the rate of reaction in solution ($M^{-1}s^{-1}$) for reactants R_1 and R_2. By Henry's Law and the Ideal Gas Law,

$$[R_n] = 10H^* R k_B T n_R$$

where R_n is either R_1 or R_2, H^* is the effective Henry's Law coefficient ($M\,atm^{-1}$) defining the reactant's solubility, k_B is Boltzmann's constant ($J K^{-1}$ molecule^{-1}), T is the temperature (K), and n is the gas-phase number density of the reactant (molecule cm^{-3}). So the reaction rate can be given in terms of gas-phase concentrations, which is important for coupling the heterogeneous chemistry to the gas-phase chemistry that is occurring simultaneously. The reaction of HCl and HOCl in solution can be treated this way. If the rate coefficient for reaction is faster than about $10^5 \, M^{-1} \, s^{-1}$, then the rate of reaction becomes limited by transport of reactants into the particle and the effective volume available for reaction is reduced (Hanson *et al.* 1994). The reactive uptake coefficient of a species X reacting with Y (in excess), γ_X, is then given by

$$\frac{1}{\gamma_X} = \frac{1}{\alpha_X} + \frac{\bar{\nu}}{4H_X^* RT \sqrt{k_r^v D_X c_Y}}$$

where α_X is the mass accommodation coefficient for species X, $\bar{\nu}$ is the mean molecular speed, H_X^* is the effective Henry's Law constant, R is the ideal gas constant, T is the temperature, k_r^v, is the second-order volume rate coefficient for reaction in the liquid (cm^3 s^{-1}), D_X is the liquid-phase diffusion coefficient, and c_Y is the molecular number density of compound Y in the liquid. In the limit of instantaneous reaction, the rate is dependent on the aerosol surface area rather than the volume. The reaction of $ClONO_2$ with HCl is an example of a reaction requiring this more complete treatment of multiphase transport and chemistry.

7.5.2 Rates of heterogeneous reactions on solids

For solid particles, reaction rates depend linearly on the particle surface area (but then, of course, surface area is not as easily measured or calculated as for spherical liquid drops). Assuming that one can meaningfully define the surface of solid PSCs by parameters such as σ, the surface area per adsorption site, and θ_Y, the fraction of the surface covered by adsorbate Y, and assuming that the adsorption conforms to the simple Langmuir isotherm, the reactive uptake coefficient of a species X reacting with Y (in excess), γ_X, is given by

$$\frac{1}{\gamma_X} = \frac{1}{\alpha_X} + \frac{\bar{\nu}\sigma^2}{4K_X^* k_B T k_r^s \theta_Y}$$

(Carslaw & Peter 1997), where K_X^* is the effective Langmuir constant (equal to $K_X/(1 + K_X)$), k_B is the Boltzmann constant, k_r^s is the second-order surface reaction rate for X with Y (cm^2 s^{-1}), and the other parameters are as given in Section 7.5.1. The rate of reaction is then

$$-\frac{d[X]}{dt} = \gamma \bar{\nu} \frac{A}{4}[X]$$

where A is the surface area density of the PSC cloud (cm^2 cm^{-3}).

It has yet to be established whether the surface of solid PSC particles is indeed solid, or if it consists of a "quasi-liquid layer" a few molecular radii deep (*e.g.* Hobbs 1974; MacKenzie & Haynes 1992; Pruppacher & Klett 1997). The dissolution of HCl to form hydrated Cl$^-$ ions, as seen in laboratory studies of PSC surrogates (Horn *et al.* 1992), supports the hypothesis of a quasi-liquid layer on PSC particles, but the formation of thin solid overlayers of HNO_3 coating ice (Abbatt 1997) does not support the hypothesis. Without a robust theoretical model of solid PSC particle surfaces, calculated chlorine activation rates, and hence ozone depletion rates, remain very uncertain when the chlorine activation is occurring on solid PSCs. Recent observations of a very small number of densities of large solid PSC particles in the Arctic lower stratosphere (Fahey *et al.* 2001) have refocused attention on the chemistry and microphysics of solid PSC particles.

7.6 FUTURE PERTURBATIONS TO THE STRATOSPHERE

7.6.1 Current inputs, future scenarios

The emission of ozone-depleting gases has been reduced significantly since the implementation of the Montreal Protocol (which came into force on January 1, 1989), and the implementation of amendments to the protocol in London (1990) and Copenhagen (1992). The total inorganic chlorine in the troposphere peaked at about 3.7 p.p.t. in the early 1990s. In the stratosphere the time at which total chlorine peaks is a function of height, since the transport of air to the uppermost stratosphere, via the Brewer–Dobson circulation, is slow. The peak in chlorine therefore occurred later in the stratosphere than in the troposphere. At 22 km, for example, the peak was measured in 1998.

Bromine is estimated to be about 50 times more efficient than chlorine in destroying stratospheric ozone on an atom-for-atom basis (see Section 7.3, above). Bromine is carried into the stratosphere in various forms, such as halons and substituted hydrocarbons, of which methyl bromide is the predominant form. Three major anthropogenic sources of methyl bromide have been identified: soil fumigation; biomass burning; and the exhaust of automobiles using leaded gasoline. Recent measurements have shown that there is three times as much methyl bromide in the northern hemisphere as in the southern hemisphere. Halon-1211 ($CBrClF_2$) and halon-1301 ($CBrF_3$) have been widely used in fire protection systems. Their production has now ceased but emissions from existing fire protection systems are expected to continue for decades. Global background levels are about 3.5×10^{-3} p.p.b.v. (H-1211) and 2.0×10^{-3} p.p.b.v. (H-1301). H-1211 concentrations continue to grow, but H-1301 concentrations have now begun to stabilize (WMO 1999).

In the 1970s, Johnston (1971) postulated that NO_x, emitted in the exhausts of supersonic aircraft (SST), would result in large ozone depletions. There is once more increasing concern over the possible impact of aircraft flights, both supersonic and subsonic, on ozone levels in the lower stratosphere (IPCC 1999). Production of a new generation of SSTs has not been ruled out by the aircraft industry. Total ozone changes, for cruise altitudes of 16 and 20 km, are calculated by models to be a few percent for reasonable estimates of the size of the supersonic fleet. Emissions higher in the stratosphere lead to larger local ozone losses since the NO_x emitted remains in the stratosphere for longer. Although changes in NO_x have the largest impact on ozone, the effects of H_2O emissions contribute about 20% to the calculated ozone change.

Many subsonic flights pass through the lowermost parts of the stratosphere, especially at high latitudes (recall Fig. 7.2). Subsonic aircraft flying in the North Atlantic flight corridor emit 44% of their exhaust emissions into the stratosphere. Models predict an ozone decrease in the lower stratosphere of less than 1%, but modeling the lower stratosphere is particularly difficult since a number of chemical processes are of comparable importance to each other, and to the transport processes. Industry projections are for a 5% per year increase in air traffic from 1995 and 2015, so the impact of aircraft on the atmosphere is likely to grow significantly.

7.6.2 Chemistry–climate feedbacks

The balance of expert opinion is now that human perturbation of the climate is taking place, and will increase (Houghton *et al.* 1996, 2001). Green-

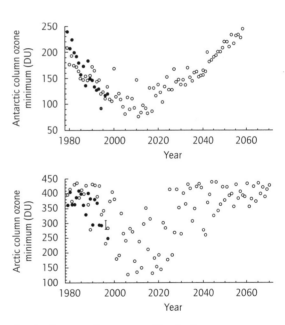

Fig. 7.8 Future column ozone. Total column ozone minima in the Arctic and Antarctic averaged over the last three days of March (Arctic) or September (Antarctic), from a GCM with simplified chemistry and microphysics. (Reprinted with permission from *Nature* (Shindell *et al.* 1998). Copyright 1998 Macmillan Magazines Limited.) Also shown are TOMS version 7 data (filled circles). Note the much greater variability in the measured and modeled Arctic data.

house gases cause a warming of the troposphere, but a cooling of the stratosphere (Fels *et al.* 1980). Changes in stratospheric temperature are likely to lead to changes in stratospheric ozone columns, and so lead to changes in ultraviolet radiation penetration of the stratosphere, to changes in the thermal structure and dynamics of the stratosphere, and, ultimately, to further changes in climate. Quantifying the effects of this climate–ozone feedback requires integration and interpretation of coupled chemistry–climate global models, which is at the limit of current resources. Nevertheless, some preliminary results have been presented (Austin *et al.* 1992, 2000; Shindell *et al.* 1998). Figure 7.8 shows Antarctic and Arctic ozone columns from such a chemistry–climate simulation, the early years of which are compared to satellite measurements. For both northern and southern hemispheres, minimum ozone columns are predicted to occur in the first two decades of the twenty-first century. This is substantially later than minimum ozone columns were predicted to occur in the absence of greenhouse cooling of the stratosphere, and about 10–15 years after chlorine loadings have peaked in the troposphere. The interplay between CFC-induced ozone depletion and greenhouse-gas-induced climate change is of central concern to policy-makers, since it demonstrates the connectedness of the Montreal and Kyoto protocols. Recent assessments aimed at policy-makers have given considerable space to this interplay, therefore (WMO 1999; Austin & Langematz 2001).

REFERENCES

Abbatt, J.P.D. (1997) Interactions of HNO_3 with water-ice surfaces at temperatures of free troposphere. *Geophysical Research Letters* **24**, 1479–82.

Andrews, D.G., Holton, J.R. & Leovy, C.B. (1987) *Middle Atmosphere Dynamics*. Academic Press, New York.

Atkinson, R., Baulch, D.L., Cox, R.A. *et al.* (1999) Evaluated kinetic and photochemical data for atmospheric chemistry. Supplement VII, organic species—IUPAC subcommittee on gas kinetic data evaluation for atmospheric chemistry. *Journal of Physical Chemistry Reference Data* 28, 191–393.

Austin, J. & Langematz, U. (2001) Stratospheric ozone and the link to climate change. In *Second European Assessment of Ozone Research*. Office for Official Publications of the European Communities, Luxembourg.

Austin, J., Butchart, N. & Shine, K.P. (1992) Possibility of an Arctic ozone hole in a doubled-CO_2 climate. *Nature* **360**, 221–5.

Austin, J., Knight, J. & Butchart, N. (2000) Three-dimensional chemical model simulations of the ozone layer: 1979–2015. *Quarterly Journal of the Royal Meteorological Society* **126**, 1533–56.

Bacmeister, J.T., Newman, P.A., Gary, B.L. & Chan, K.R. (1994) An algorithm for forecasting mountain wave-related turbulence in the stratosphere. *Weather and Forecasting* **9**, 241–53.

Ball, S.M., Hancock, G. & Winterbottom, F. (1995) Product channels in the near-uv photodissociation of ozone. *Faraday Discussions* **100**, 215–27.

Bates, D.R. & Nicolet, M. (1950) The photochemistry of atmospheric water vapor. *Journal of Geophysical Research* **55**, 301–27.

Becker, G., Müller, R., McKenna, D.S. *et al.* (2000) Ozone loss rates in the Arctic stratosphere in the winter 1994/1995: model simulations underestimate results of MATCH analysis. *Journal of Geophysical Research* **105**, 15,175–84.

Borrmann, S., Solomon, S., Dye, J.E. & Luo, B.-P. (1996) The potential of cirrus clouds for heterogeneous chlorine activation. *Geophysical Research Letters* **23**, 2133–6.

Borrmann, S., Solomon, S., Avallone, L., Toohey, D. & Baumgardner, D. (1997) On the occurrence of ClO in cirrus clouds and volcanic aerosol in the tropopause region. *Geophysical Research Letters* **24**, 2011–14.

Brown, S.S., Talukdar, R.K. & Ravishankara, A.R. (1999a) Rate constants for the reaction $OH + NO_2 + M + HNO_3 + M$ under atmospheric conditions. *Chemical Physics Letters* **299**, 277.

Brown, S.S., Talukdar, R.K. & Ravishankara, A.R. (1999b) Reconsideration of the rate constant for the reaction of hydroxyl radicals with nitric acid. *Journal of Physical Chemistry* **103**, 3031–7.

Carslaw, K.S. & Peter, Th. (1997) Uncertainties in reactive uptake coefficients for solid stratospheric particles—I, surface chemistry. *Geophysical Research Letters* **24**, 1743–6.

Carslaw, K.S., Luo, B.-P., Clegg, S.L., Peter, Th. & Brimblecombe, P. (1994) Stratospheric aerosol growth and HNO_3 gas phase depletion from coupled HNO_3 and water uptake by liquid particles. *Geophysical Research Letters* **21**, 871–4.

Carslaw, K.S., Peter, Th., Bacmeister, J.T. & Eckermann, S. (1999) Widespread solid particle formation by mountain waves in the Arctic stratosphere. *Journal of Geophysical Research* **104**, 1827–36.

Chapman, S.C. (1930) A theory of upper atmospheric ozone. *Memoirs of the Royal Meteorological Society* **3**, 103–25.

Cox, R.A., MacKenzie, A.R., Müller, R.H., Peter, Th. & Crutzen, P.J. (1994) Activation of stratospheric chlorine by reactions in liquid sulphuric acid. *Geophysical Research Letters* **21**, 1439–42.

Crutzen, P.J. (1970) The influence of nitrogen oxides on the atmospheric ozone content. *Quarterly Journal of the Royal Meteorological Society* **96**, 320–5.

Crutzen, P.J. & Arnold, F. (1986) Nitric acid cloud formation in the cold Antarctic stratosphere: a major cause for the springtime "ozone hole." *Nature* **324**, 651–5.

Danielsen, E.F. (1968) Stratospheric–tropospheric exchange based on radioactivity, ozone and potential vorticity. *Journal of the Atmospheric Sciences* **25**, 502–18.

De More, W.B., Sander, S.P., Golden, D.M. *et al.* (1997) *Chemical Kinetics and Photochemical Data for Use in Stratospheric Modeling. Evaluation No. 12.* Jet Propulsion Laboratory, Pasadena, CA.

Dessler, A.E., Burrage, M.D., Grooß, J.-U. *et al.* (1998) Selected highlights from the first five years of the upper atmosphere research satellite (UARS) program. *Reviews of Geophysics* **36**, 183–210.

Dörnbrack, A., Leutbecher, M., Volkert, H. & Wirth, M. (1998) Mesoscale forecasts of stratospheric mountain waves. *Meteorological Applications* **5**, 117–26.

Dragani, R., Redaelli, G., Visconti, G. *et al.* (2002) High resolution stratospheric tracer fields reconstructed with Lagrangian techniques: a comparative analysis of predictive skill. *Journal of Atmospheric Science* **59**, 1943–58.

Dye, J.E., Baumgardner, D., Gandrud, B.W. *et al.* (1992) Particle size distributions in Arctic polar stratospheric clouds, growth and freezing of sulfuric acid droplets, and implications for cloud formation. *Journal of Geophysical Research* **97**, 8015–34.

Fahey, D.W. *et al.* (2001) The detection of large HNO_3-containing particles in the winter Arctic stratosphere. *Science* **291**, 1026–31.

Fairlie, T.D., Pierce, R.B., Grose, W.L., Lingenfelser, G., Loewenstein, M. & Podoslke, J.R. (1997) Lagrangian forecasting during ASHOE-MAESA: analysis of predictive skill for analyzed and reverse domain filled potential vorticity. *Journal of Geophysical Research* **102**, 13,169–82.

Farman, J.C., Gardiner, B.G. & Shanklin, J.D. (1984) Large losses of total ozone reveal seasonal ClO_x/NO_x interactions. *Nature* **315**, 207–10.

Fels, S.B., Mahlman, J.D., Schwarzkopf, M.D. & Sinclair, R.W. (1980) Stratospheric sensitivity to perturbations in ozone and carbon dioxide: radiative and dynamical response. *Journal of the Atmospheric Sciences* **37**, 2265–97.

Fiocco, G., Larsen, N., Bekki, S. *et al.* (1998) Particles in the stratosphere. In *European Research on the Stratosphere: The Contribution of EASOE and SESAME to Our Current Understanding of the Ozone Layer.* Office for Official Publications of the European Communities, Luxembourg.

Garcia, R.R. & Boville, B.A. (1994) "Downward control" of the mean meridional circulation and temperature distribution of the polar winter stratosphere. *Journal of the Atmospheric Sciences* **51**, 2238–45.

Garcia, R.R. & Solomon, S. (1994) A new numerical model of the middle atmosphere, 2. Ozone and related species. *Journal of Geophysical Research* **99**, 12,937–52.

Ghosh, S. (1993) On the diffusivity of trace gases under stratospheric conditions. *Journal of Atmospheric Chemistry* **17**, 391–7.

Hamilton, K. (1996) Comprehensive meteorological modeling of the middle atmosphere: a tutorial review. *Journal of Atmospheric and Solar–Terrestrial Physics* **58**, 1591–627.

Hanson, D.R., Ravishankara, R. & Solomon, S. (1994) Heterogeneous reactions in sulphuric-acid aerosols— a framework for model-calculations. *Journal of Geophysical Research* **99**, 3615–29.

Haynes, P.H. & Anglade, J. (1997) The vertical-scale cascade of atmospheric tracers due to large-scale differential advection. *Journal of the Atmospheric Sciences* **54**, 1121–36.

Hobbs, P.V. (1974) *Ice Physics.* Oxford University Press, New York.

Holton, J.E. & Lindzen, R. (1972) An updated theory for the quasi-biennial cycle of the tropical stratosphere. *Journal of the Atmospheric Sciences* **29**, 1076–80.

Holton, J.E., Haynes, P.H., McIntyre, M.E., Douglass, A.R., Rood, R.B. & Pfister, L. (1995) Stratosphere–troposphere exchange. *Reviews of Geophysics* **33**, 403–39.

Horn, A.B., Chesters, M.A., McCoustra, M.R.S. & Sodeau, J.R. (1992) Adsorption of stratospherically important molecules on thin D_2O ice films using reflection adsorption infrared spectroscopy. *Journal of the Chemical Society, Faraday Transactions* **88**, 1077–8.

Houghton, J.T., Meiro Filho, L.G., Callander, B.A., Harris, N.R.P., Kattenberg, A. & Maskell, K. (eds) (1996) *Climate Change 1995: The Science of Climate Change.* Cambridge University Press, Cambridge.

Houghton, J.T., Ding, Y., Griggs, D.J., Nogeur, M., Van Der Linden, P.J. & Xiaosu, D. (eds) (2001) *Climate Change 2001: The Scientific Basis.* Cambridge University Press, Cambridge.

Intergovernmental Panel on Climate Change (1999) *Aviation and the Global Atmosphere.* Cambridge University Press, Cambridge.

Jacobson, M.Z. (1999) *Fundamentals of Atmospheric Modelling.* Cambridge University Press, Cambridge.

Johnston, H.S. (1971) Reduction of stratospheric ozone by nitrogen oxide catalysts from supersonic transport exhaust. *Science* **173**, 517–762.

Johnston, H.S. & Podolski, J. (1978) Interpretations of stratospheric photochemistry. *Review of Geophysics and Space Physics* **16**, 491ff.

Juckes, M.N. & McIntyre, M.E. (1987) A high resolution, one-layer model of breaking planetary waves in the stratosphere. *Nature* **328**, 590–6.

Junge, C.E., Chagnon, C.W. & Manson, J.E. (1961) Stratospheric aerosols. *Journal of Meteorology* **18**, 8.

Koop, T., Luo, B.-P., Tsias, A. & Peter, Th. (2000) Water activity as the determinant for homogeneous ice nucleation in aqueous solutions. *Nature* **406**, 611–14.

Lary, D.J., Lee, A.M., Toumi, R., Newchurch, M.J., Pirre, M. & Renard, J.B. (1997) Carbon aerosols and atmospheric photochemistry. *Journal of Geophysical Research* **102**, 3671–82.

Lee, A.M., Carver, G.D., Chipperfield, M.P. & Pyle, J.A. (1997) Three-dimensional chemical forecasting: a methodology. *Journal of Geophysical Research* **102**, 3905–19.

Lelieveld, J. *et al.* (1999) Tropospheric ozone and related processes. In *Scientific Assessment of Ozone Depletion: 1998.* World Meteorological Organisation, Geneva.

Lindzen, R.S. (1990) *Dynamics in Atmospheric Physics.* Cambridge University Press, Cambridge.

Lowe, D. & MacKenzie, A.R. (2002) A model of liquid polar stratospheric clouds with size-dependent composition and fixed size bins. *Atmospheric Chemistry and Physics.*

McCormick, M.P., Wang, P.-H. & Pitts, M.C. (1993) Background stratospheric aerosol and polar stratospheric cloud reference models. *Advances in Space Research* **13**, 7–29.

McIntyre, M.E. (1992) Atmospheric dynamics: some fundamentals, with observational implications. In Gille, J.C. & Visconti, G. (eds), *The Use of EOS for Studies of Atmospheric Physics.* North-Holland, Amsterdam.

MacKenzie, A.R. & Haynes, P.H. (1992) The influence of surface kinetics on the growth of stratospheric ice crystals. *Journal of Geophysical Research* **97**, 8057–64.

MacKenzie, A.R., Laaksonen, A., Batris, E. & Kulmala, M. (1998) The Turnbull Correlation and the freezing of stratospheric aerosol droplets. *Journal of Geophysical Research* **103**, 10,875–84.

Meilinger, S.K., Koop, T., Luo, B.-P. *et al.* (1995) Size-dependent stratospheric droplet composition in lee wave temperature fluctuations and their potential role in PSC freezing. *Geophysical Research Letters* **22**, 3031–4.

Molina, M.J. & Rowland, F.S. (1974) stratospheric sink for chlorofluoro-methanes: chlorine atom catalyzed destruction of ozone. *Nature* **249**, 810–12.

Mote, P.W., Rosenlof, K.H., McIntyre, M.E. *et al.* (1996) An atmospheric tape recorder: The imprint of tropical tropopause temperatures on stratospheric water vapor. *Journal of Geophysical Research* **101**, 3989–4006.

Miller, R.L., Suits, A.G., Houston, P.L., Toumi, R., Mack, J.A. & Wodtke, A.M. (1994) The "ozone deficit" problem. *Science* **265**, 1831–8.

Müller, R. *et al.* (1997) Severe chemical ozone loss in the Arctic during the winter of 1995–96, *Nature* **389**, 709–12.

Murphy, D.M. & Gary, B.L. (1995) Mesoscale temperature fluctuations and polar stratospheric clouds. *Journal of the Atmospheric Sciences* **52**, 1753–60.

Murphy, D.M., Thomson, D.S. & Mahoney, M.J. (1998) In situ measurements of organics, meteoritic material, mercury, and other elements in aerosols at 5 to 19 kilometres. *Science* **282**, 1664–9.

Newman, P.A., Gleason, J.F., McPeters, R.D. & Stolarski, R.S. (1997) Anomalously low ozone over the Arctic. *Geophysical Research Letters* **24**, 2689–92.

Nightingale, R.W. *et al.* (1996) Global CF_2Cl_2 measurements by UARS Cryogenic Limb Array Etalon Spectrometer: validation by correlative data and a model. *Journal of Geophysical Research* **101**, 9711–36.

Norton, W.A. (1994) Breaking Rossby waves in a model stratosphere diagnose by a vortex following coordinate system and a technique for advecting material contours. *Journal of the Atmospheric Sciences* **51**, 654–73.

Orsolini, Y., Simon, P. & Cariolle, D. (1995) Filamentation and layering of an idealized tracer by observed winds in the lower stratosphere. *Geophysical Research Letters* **22**, 839–42.

Oxtoby, D.W. (1992) Nucleation of crystals from the melt. In Henderson D. (ed.), *Fundamentals of Inhomogeneous Fluids*. Dekker, New York.

Pfister, L., Jensen, E., Selkirk, H., Browell, E.V. & Gary, B. (1999) Observations of subvisible cirrus clouds and gravity waves at the tropical tropopause and their potential significance for the stratospheric water budget. In Carslaw, K.S. & Amanatidis, G. (eds), *Proceedings of the MEPS Workshop, Bad Tölz*. Office for Official Publications of the European Union, Luxembourg.

Plumb, R.A. (1996) A "tropical pipe" model of stratospheric transport. *Journal of Geophysical Research* **101**, 3957–72.

Plumb, R.A., Waugh, D.W., Atkinson, R.J. *et al.* (1994) Intrusions into the lower stratospheric Arctic vortex during the winter of 1991–1992. *Journal of Geophysical Research* **99**, 1089–105.

Polvani, L.M., Waugh, D.W. & Plumb, R.A. (1995) On the subtropical edge of the stratospheric surf zone. *Journal of the Atmospheric Sciences* **52**, 1288–309.

Poole, L. & Pitts, M.C. (1994) Polar stratospheric cloud climatology based on Stratospheric Aerosol Measurement II observations from 1978 to 1989. *Journal of Geophysical Research* **99**, 13,083–9.

Potter, B.E. & Holton, J.R. (1995) The role of monsoon convection in the dehydration of the lower tropical stratosphere. *Journal of Atmospheric Science* **52**, 1034–50.

Pruppacher, H.R. & Klett, J.D. (1997) *Microphysics of Clouds and Precipitation*, 2nd edn. Kluwer, Dordrecht.

Rampino, M.R. & Self, S. (1992) Volcanic winter and accelerated glaciation following the Toba super eruption. *Nature* **359**, 50–52.

Ravishankara, A.R. & Shepherd, T.G. (1999) Lower stratospheric processes, in World Meteorological Organisation (WMO)/United Nations Environment Programme (UNEP). In *Scientific Assessment of Ozone Depletion: 1998*. WMO, Geneva.

Rex, M. *et al.* (1997) Prolonged stratospheric ozone loss in the 1995/96 Arctic winter. *Nature* **389**, 835–8.

Roscoe, H.K., Jones, A.E. & Lee, A.M. (1997) Midwinter start to Antarctic ozone depletion, *Science* **278**, 93–6.

Santacesaria, V., Stefanutti, L., Morandi, M., Guzzi, D. & MacKenzie, A.R. (1999) Two-year (1996/1997) ozone DIAL measurement over Dumont d'Urville (Antarctica), *Geophysical Research Letters* **26**, 463–6.

Santacesaria, V., Carla, R., MacKenzie, A.R. *et al.* (2001) Clouds at the tropical tropopause: a case study during the APE-THESEO campaign over the western Indian Ocean. *Journal of Geophysical Research*.

Schoeberl, M.R. & Hartmann, D.L. (1991) The dynamics of the stratospheric polar vortex and its relation to springtime ozone depletions. *Science* **251**, 46–52.

Shepherd, T.G. (2000) The middle atmosphere. *Journal of Atmospheric and Solar–Terrestrial Physics* **62**, 1587–601.

Shindell, D.T., Rind, D. & Lonergan, P. (1998) Increased polar stratospheric ozone losses and delayed eventual recovery owing to increasing greenhouse-gas concentrations. *Nature* **392**, 589–92.

Solomon, S., Garcia, R.R., Rowland, F.S. & Wuebbles, D.J. (1986) On the depletion of Antarctic ozone. *Nature* **321**, 755–8.

Solomon, S., Borrmann, S., Garcia, R.R. *et al.* (1997) Heterogeneous chemistry in the tropopause region. *Journal of Geophysical Research* **102**, 21,411–29.

Stanford, J.L. & Davis, J.S. (1974) A century of stratospheric cloud reports: 1870–1972. *Bulletin of the American Meteorological Society* **55**, 213–19.

Sutton, R.T., MacLean, H., Swinbank, R., O'Neill, A. & Taylor, F.W. (1994) High resolution stratospheric tracer fields estimated from satellite observations using Lagrangian trajectory calculations. *Journal of the Atmospheric Sciences* **51**, 2995–3005.

Tabazadeh, A., Turco, R.P., Drdla, K., Jacobson, M.Z. & Toon, O.B. (1994) A study of type 1 polar stratospheric cloud formation. *Geophysical Research Letters* **21**, 1619–22.

Taesler, I., Delaplane, R.G. & Olovsson, I. (1975) Hydrogen bond studies XCIV, diaquooxonium ion in nitric acid trihydrate. *Acta Crystallographica* Section B31, Structural Science, 1489–92.

Tan, D.G.H., Haynes, P.H., MacKenzie, A.R. & Pyle, J.A. (1998) Effects of fluid-dynamical stirring and mixing on the deactivation of stratospheric chlorine. *Journal of Geophysical Research* **103**, 1585–605.

Thomason, L.W., Poole, L.R. & Deshler, T. (1997) A global climatology of stratospheric aerosol surface area density produced from Stratospheric Aerosol and Gas Experiment II measurements: 1984–1994. *Journal of Geophysical Research* **102**, 8967–76.

Thorpe, A.J. & Volkert, H. (1997) Potential vorticity: a short history of its definition and uses. *Meteorologische Zeitschrift* **6**, 275–80.

Tolbert, M.A. & Toon, O.B. (2001) Solving the PSC mystery. *Science* **292**, 61–3.

Toon, O.B. & Farlow, N.H. (1981) Particles above the tropopause. *Annual Review of Earth and Planetary Sciences* **9**, 19–58.

Toon, O.B., Hamill, P., Turco, R.P. & Pinto, J. (1986) Condensation of HNO_3 and HCl in the winter polar stratospheres. *Geophysical Research Letters* **13**, 1284–7.

Tuck, A.F. (1979) A comparison of one-, two-, and three-dimensional model representations of stratospheric gases. *Philosophical Transactions of the Royal Society of London*, Ser. A. **290**, 477.

Turco, R.P., Toon, O.B. & Hamill, P. (1989) Heterogeneous physicochemistry of the polar ozone hole. *Journal of Geophysical Research* **94**, 16,493–510.

Vaughan, G.X., Gouget, H., O'Connor, F.M. & Weir, D. (2001) Fine-scale layering on the edge of a stratospheric intrusion. *Atmospheric Environment* **35**, 2215–21.

Waugh, D.W. & Plumb, R.A. (1994) Contour advection with surgery: a technique for investigating finescale structure in tracer transport. *Journal of the Atmospheric Sciences* **51**, 654–73.

Wayne, R.P. (2000) *Chemistry of Atmospheres*, 3rd edn. Oxford University Press, Oxford.

Wayne, R.P. *et al.* (1990) *The Nitrate Radical.* Air Pollution Report 31. Commission of the European Communities, Directorate-General for Science, Research and Development, Brussels, Belgium.

WMO/UNEP (1999) *Scientific Assessment of Ozone Depletion: 1998.* WMO, Geneva.

Woodman, R.F. & Rastogi, P.K. (1984) Evaluation of effective eddy diffusion coefficients using radar observations of turbulence in the stratosphere. *Geophysical Research Letters* **11**, 243–6.

At the time of going to press, the following web sites contained useful background information on stratospheric ozone:

www.atm.ch.cam.ac.uk/tour/index.html—an informal introduction to stratospheric ozone, aimed at school children, run by the Atmospheric Chemistry Modelling Unit of the University of Cambridge, UK.

www.es.lancs.ac.uk/casestud/case13.htm—an informal introduction to dehydration at the tropical tropopause, aimed at school teachers, written by myself.

www.aero.jussieu.fr/~sparc/—the home page of the *Stratospheric Processes and Climate* (SPARC) initiative of the World Climate Research Programme.

The following site contains information on reaction rate coefficients for reactions of importance in the stratosphere and troposphere:

www.iupac-kinetic.ch.cam.ac.uk

8 Aqueous Phase Chemistry of the Troposphere

PETER BRIMBLECOMBE

Liquid water is a minor component of the troposphere, yet it plays important roles in atmospheric chemistry. It is responsible for removing gases and solid trace components from air through dissolution. This process is usually followed by sedimentation or more characteristically wet removal in rainwater. In a more subtle way partition into the droplet phase can alter the rate of reactions in the gas phase via a process that might be thought as sequestration. Here species can be separated, with soluble ones going into solution and less soluble ones remaining in the gas phase. The aqueous aerosol can also allow species to reach higher concentrations or achieve faster reaction rates than are typical in the gas phase, thus allowing additional reaction pathways to become significant. More recently it has become clear that there is an active liquid-phase photochemistry in the atmosphere. This chapter reviews the chemistry of inorganic species in the atmosphere (much studied during the acid rain debate) and concludes with our developing understanding of the radical and organic chemistry of droplet phases.

8.1 THE AQUEOUS PHASE IN THE ATMOSPHERE

Water in the atmosphere can be present in all three phases, as ice, liquid water, and water vapor. Most water is present in the vapor phase, with cloud water on average being only some 4% of total available water in the atmosphere. The amount of water vapor in the atmosphere is a typically referred to by meteorologists as absolute humidity ($g\,m^{-3}$) or perhaps more conveniently as specific humidity ($g\,kg^{-1}$ (air)). More commonly we find the water vapor content expressed as relative humidity, which is the ratio of the amount of water relative to its saturation value, at the temperature under consideration. It is usually expressed as a percentage.

The solid phase, ice, is present as in the atmosphere as snow, hail, graupel, and rime (on surfaces). The chemistry of ice is more difficult to describe, because the composition may be heterogeneous and the extent to which it is in equilibrium with the atmosphere is often uncertain.

Liquid water can be present as rain, cloud, mist, fog, sleet, and dew. The amount of liquid water and other characteristics of these forms of precipitation are listed in Table 8.1. This quantity is important because it controls the amount available as a solvent. The absolute amounts are rather small, and if we consider that even in large cumulus clouds phase volume ratios of anything greater than one in a million are difficult to achieve (i.e. $1\,g\,water\,m^{-3}$).

Fog is typically defined in terms of visibility, often where visibility is less than a kilometer. Such systems are sometimes called mists and hazes, where the visibility is greater. The term "mists" tends to be applied to liquid particles, while "haze" is often used for solid aerosols. Sleet is melting snow. In the case of rainfall we should note that the number of drops in rainfall does not change markedly with intensity, but droplet size

Table 8.1 Some typical size and water contents of liquid water in the atmosphere (data for precipitation from Barry & Chorley 1998; for clouds from Mason 1975; Heitnzenberg 1998; for dew from Brimblecombe & Todd 1977; Hughes & Brimblecombe 1994).

Water	Particle radius	Liquid water content (g m^{-3})	Ionic strength molal
Precipitation			10^{-4}
Rain (0.1 cm hour)	0.1 cm		
Rain (1.3 cm hour)	0.2 cm		
Rain (10 cm hour)	0.3 cm		
Cloud, mist, and aerosol droplets			
Cumulonimbus	20 μm {2–4000}	2	
Cumulus congestus	10 μm	1	10^{-3}
Continental cumulus	6 μm {2–30}	0.45	
Stratus (Hawaii)	–{0.5–30}	0.35	
Mist, fog	–{0.5–30}	0.05–0.5	5×10^{-3}
Hygroscopic aerosols	<1 μm	10^{-5}–10^{-4}	>1
Leaf wetness			
Dew (over grass)	0.2 mm	100 g m^{-2}	
Guttation (over grass)	1.5 mm	100 g m^{-2}	2×10^{-2}

does. Large droplets are found only at the high relative humidity in or below clouds and are short-lived (hours) because of the need for updrafts to keep them suspended.

8.1.1 Cloud and rainwater

Small droplets in the atmosphere form around a nonvolatile core. It is these materials that contribute to the initial solutes present in atmospheric water, and they result from the way clouds form. As rising air cools and its relative humidity rises, and although condensation might be expected once the relative humidity exceeded 100%, very high levels of supersaturation with respect to water vapor are required before small droplets form.

This arises because a high degree of surface curvature enhances the vapor pressure of liquids. The idea that the vapor pressure over a curved surface of radius r (p_r) is greater than that over a plane surface (p_∞) was represented in the Kelvin equation, first derived by W. Thomson:

$$\ln(p_r/p_\infty) = 2\sigma M/\rho_L R T r$$

where σ is the surface tension, M is the molecular weight, ρ_L is the density of the liquid and R and T take their normal meanings. In the atmosphere condensation is aided by the presence of airborne particles acting as nuclei for condensation. In particular, hygroscopic particles can start to absorb water below 100% relative humidity, while nonhygroscopic particles (dusts) require some degree of supersaturation. The effectiveness of salts as condensation nuclei arises because water activity can be much less than unity in concentrated electrolyte solutions. Salts thus lower the water vapor pressure over small droplets and prevent them from evaporating. The effect of sodium chloride on the equilibrium relative humidity over water droplets is shown in Fig. 8.1. The Kelvin effect is significant when droplet sizes are less than a micrometer, so it becomes relevant in processes such as the formation of cloud droplets. However, these initial droplets have low volume and will grow rapidly, so the effect is generally

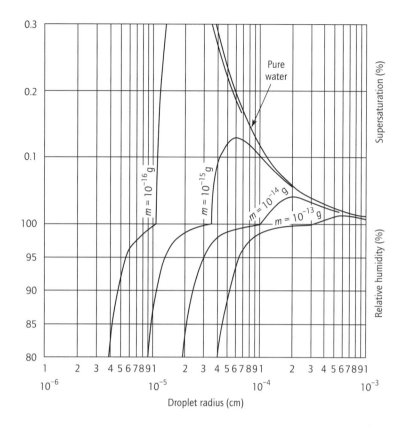

Fig. 8.1 The equilibrium relative humidity as a function of droplet radius for solutions containing various masses of sodium chloride. (From Mason 1975.)

neglected in the subsequent chemistry of larger droplets.

Sodium chloride and perhaps more generally sulfuric acid are typical components of cloud condensation nuclei. However, it is possible to postulate that a range of organic materials can act as cloud condensation nuclei. In the remote Arctic nucleation has been attributed to the oxidation products of the amino acid L-methionine (Leck & Bigg 1999). Dicarboxylic acids or humic acids may also affect the ability of aerosols to act as cloud condensation nuclei (Gelencser *et al.* 2000).

The presence of an electrolyte core to a cloud droplet is the major influence on its initial composition. Composition will obviously vary with size as the cloud droplets grow. Prupacher and Klett (1997) have offered a conceptual model developed

from Ogren (Fig. 8.2) to illustrate the change in concentration with radius. In the first stage, as a droplet grows between 1 and 10 μm the salt concentration, derived from the original cloud condensation nucleus, decreases. Then, up to 50 μm the concentration rises with increasing drop size. In this size range larger droplets grow by vapor diffusion more rapidly than the smaller ones and these larger drops have generally arisen from larger initial salt particles. Above 50 μm droplet growth occurs via coalescence, which is a process where large droplets accumulate smaller ones and thus become diluted. Beyond a few hundred micrometers in size cloud water is well mixed and concentration constant with size.

Further processes occur below the cloud as raindrops fall through subcloud air. Measurements of

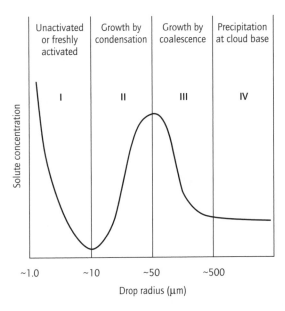

Fig. 8.2 A conceptual model of solute concentration as a function of droplet radius developed by Prupacher and Klett (1997) from Ogren. (Reproduced with kind permission from Kluwer Academic Publishers.)

monium bisulfate, and sulfuric acid. Sulfuric acid, being a liquid, does not crystalize at temperatures found in the lower atmosphere. The amount of water associated with aerosols is a function of humidity and the amount of saline material in the aqueous atmospheric aerosol. Concentrations can be exceedingly high, often supersaturated, which means that such systems depart from ideality, so their activity relations require quite complex descriptions. Pizter's formalism has been successfully adapted to aerosols (e.g. Clegg *et al.* 1998a, b).

Dew forms on surfaces as they cool by radiating to the night sky. Most measurements tend to be made on artificial surfaces, but these are not always good models of dew chemistry, which is so often associated with vegetation. This is different from that found on artificial surfaces, as we need to distinguish between dew that forms by condensation and guttation (Hughes & Brimblecombe 1994), and liquid exuded through the leaf stomata, which can exert a strong control on the composition of leaf wetness.

Ice can nucleate on hygroscopic particles, but solids such as the clays kaolinite and montmorillonite seem more effective (Prupacher & Klett 1997). Snow, graupel, and hail have ionic compositions that are similar to that of rainwater. However, where riming occurs (i.e. growth by collision with droplets that then freeze) the particles have a much higher scavenging efficiency (3–20 times) for solutes such as sulfate (Barrie 1991) and the nitrate ion, which may be about double the concentration in rainwater under similar conditions. Domine (1999) has shown that diffusive growth models of the incorporation of nitric and hydrochloric acid into snow crystals can describe their uptake in laboratory experiments, but Greenland snow appears to be somewhat undersaturated.

the composition of raindrops sampled at the base of stratus clouds have cation and anion concentrations that are independent of radius. Well below the cloud base the maximum concentration is usually found at the middle drop radii, and raindrops at ground level show a relationship between the chemical content and radius that has been gained from below-cloud processes (Bächmann *et al.* 1996).

8.1.2 *Other forms of atmospheric water*

Even in air under cloud-free conditions there is a small amount of liquid water associated with aerosol particles (Pilinis *et al.* 1989). The ability of salts to absorb water below equilibrium saturation vapor pressures means that the most substantial quantities will be as hygroscopic water. This is associated with the most common deliquescent salts in the atmosphere, typically sodium chloride, am-

8.2   NONVOLATILE SOLUTES

The most common inorganic anions in solution in the atmosphere are sulfates, chlorides, and nitrates. These are typically associated with the sodium, ammonium, and hydrogen ions and to a lesser extent the alkaline earths: calcium and mag-

Table 8.2 Characterization of remote coarse and fine aerosols (data from Yoshizumi & Asakuno 1986; Milne & Zika, 1993; Kawamura *et al.* 1996).

	Concentration ($\mu g\,m^{-3}$)	
	>2 μm	<2 μm
Sea salt	10.52	0.83
Soil	2.6	0.62
$NaNO_3$	0.69	0.07
NH_4NO_3	0.05	0.08
$H_2SO_4{}^{2-}$	0.04	1.9
NH^{4+}	0.01	0.54
$R(COO^-)_2$	0.01–0.1	
RNH_2COO^-	~1 nmol m^{-3}	

nesium. Hydrogen chloride and nitric acid are fairly volatile, with ammonium salts volatile under some conditions, so that they can also be lost from aerosols. In general most soluble salts and sulfuric acid are not particularly volatile, so there is partition into aerosols. Although dissolved salts provide a highly concentrated background electrolyte in aerosols, they are diluted in clouds or rain, with the concentrations typically below the millimolar range.

Water-soluble particles in the atmosphere have long been collected. Simple analysis of the ions only gives clues to their original chemical forms, but these are often taken to be sodium chloride, ammonium sulfates, bisulfates, and sulfuric acid. Examination of dry particles tends to confirm the importance of these compounds. In remote oceanic air the materials are generally identified as sea salt, ammonium sulfates, and sulfuric acid. Sea salt is typically predominant in the larger size range particles (>2 μm), while secondary sulfates from the oxidation of sulfur dioxide are found in smaller size ranges (<2 μm). Nitrates are a product of the oxidation of nitrogen oxides. From the point of view of aquatic chemistry in the atmosphere the original association between anions and cations in the initial salt is often not relevant.

Organic compounds in aerosols include a great many hydrocarbons derived from combustion processes and biological emissions. They fre-

quently contribute to a substantial fraction of the material in remote aerosols, but this material may not be very soluble in water. However, atmospheric transformations, most particularly oxidation, convert compounds from volatile low-solubility compounds into more soluble forms. Once again there are issues that limit their presence in solution. Many oxidation products (e.g. the alcohols, aldehydes, ketones, and acids) are still rather volatile if they have few carbon atoms, and can be lost to the gas phase. On the other hand, less volatile, larger oxidized organic molecules are frequently rather insoluble in water.

A fraction of the total organic matter present in aerosols dissolves in water, but there are still many uncertainties concerning organic materials in aerosols (see Huebert & Charlson 2000). In fogwater almost 80% of the water-soluble organic carbon can be identified as: (i) neutral/basic compounds; (ii) mono- and di-carboxylic acids; and (iii) polyacidic compounds (Descari *et al.* 2000). Some biological compounds, although not volatile, seem to find their way into the atmosphere. These are typified by sugars, amino acids, uric acid, urea, and humic substances (Cornell *et al.* 1998; Legrand *et al.* 1998; Leck & Bigg 1999).

Typical candidate oxidation products that have a low volatility but high solubility in water include bifunctional acids such as the dicarboxylic acids (oxalic, malonic, succinic, and hydroxy acids such as lactic acid) and some ring compounds represented by furancarboxylic acid (2-furoic acid). Saxena and Hildermann (1996) attempted a more rigorous exploration of likely water-soluble organic compounds based on estimates of their partition into aqueous particles. In addition to the acids noted above they recognized the potential for glyoxal, alkyldiols, and polyols, glycols, oxoalkanoic acids (e.g. pyruvic acid), and multifunctional acids (e.g. citric, lactic, tartaric acid) to be found in aerosol droplets. In addition to these oxygen-containing compounds, some nitrogen-containing organic compounds, such as nitrophenols, ethanolamine, and the amino acids, have physical properties that would suggest they should be found in the aerosol phase. Kames and Schurath (1992) have also shown that alkyl nitrates and other bifunctional nitrates should be soluble in aqueous atmospheric systems.

Some long chain hydrocarbons, although not soluble, may behave as surfactants on aerosol droplets. The most obvious surfactants would be the carboxylic acids and nitrates (Seidl 2000). Surface active compounds may also be associated with vegetation. Leaf wetness has an unusual composition because so much can be exuded as guttation through the stomata of plants. There are often high concentrations of dissolved carbon dioxide and plant derived ions, such as potassium (Hughes & Brimblecombe 1994). Once on the surface of vegetation "dew" can represent an important sink for soluble trace gases such as sulfur dioxide can be converted to sulfates. In regions such as Japan where high concentrations of yellow dust are deposited from the air in spring dew may become neutral or alkaline even where there is a high probability of acidification from sulfates and nitrates (Chung *et al.* 1999). Concentrations of ammonium and nitrite also tend to be higher in dew than in rain. However, as the leaf wetness dries nitrite and ammonium appears to be lost (Takenaka *et al.* 1999). The degradation of biological materials such leaf waxes (Fruekilde *et al.* 1998) has the potential to yield water-soluble organic compounds such as 4-oxopentanal in dew.

8.2.1 Gases

In general materials have to be nonvolatile or very soluble gases to partition into the liquid phase in the atmosphere. This means that typically we find salts, highly soluble gases (usually those that undergo extensive hydrolysis in solution), and some soluble organic compounds. The gases that are most typically transferred, in substantial fractions, into atmospheric water are carbon dioxide, sulfur dioxide, and ammonia, along with the stronger mineral acids. Oxygen, although only slightly partitioned into water in the atmosphere, is nevertheless important because its high partial pressure ensures that it is present in relatively high concentration in water (i.e $0.25\,mmol\,l^{-1}$ at 25°C).

8.2.2 Dissolution of gases

The dissolution of a gas in water can be written as an equilibrium between the gas ($X_{(g)}$) and aqueous phases $X_{(aq)}$, which can be represented as an equilibrium constant (the Henry's Law constant, K_H):

$$X_{(g)} \rightleftharpoons X_{(aq)}, K_H = \gamma X m X / p X, \text{ or } K_H = a X / p X \quad (8.1)$$

Where γX is the activity coefficient, mX is the concentration (in molar, $mol\,l^{-1}$ or molal mole $kg(H_2O)^{-1}$ units), pX is the pressure, and aX is the activity (where $aX = \gamma X m X$). Strictly speaking, pX should be fugacity rather than pressure, but these quantities are almost identical for most gases under ambient conditions. In rainwater the activity coefficient γX is fairly close to unity, but in aerosols there are very significant departures from ideality and, as mentioned above, these can be treated using the formalism of Kenneth Pitzer, which has become widely adopted for atmospheric chemistry (Clegg *et al.* 1998a, b).

We should also note that equation 8.1 represents the solubility of a gas that does not undergo extensive hydrolysis. Typically the constant K_H is given the units $mol\,l^{-1}\,atm^{-1}$ (or $mol\,kg^{-1}\,atm^{-1}$) but this offends SI conventions, so there are arguments for presenting the constant in dimensionless terms. Nevertheless, the dimensioned form is rather convenient for atmospheric chemists and typical values are given this way in Table 8.3. The Henry's Law constant is a sensitive function of temperature, typically becoming larger as the temperature falls, so there is the need to tabulate enthalpy-related parameters.

It is clear from Table 8.3 that hydrogen peroxide is a remarkably soluble gas. In fact it is so soluble that it partitions effectively into the liquid phase of clouds. This can be seen from a simple equation for equipartitioning of a trace gas between the air and suspended aqueous droplets. The number of moles in the gas phase is pX/RT, while the number of moles in the liquid phase will be the concentration times the volume of water, i.e. $K_H pXL$, where L is the volume of liquid water present in air. The situation where equal amounts are present in the gas and liquid phase can be written:

Table 8.3 Henry's Law constants $(K_H(T_o)$ at 298 K) and temperature dependencies for gases that do not undergo extensive hydrolysis. Data are from Sander (1999), but the very extensive tabulation lists a great range of values and substances. Values for other temperatures, T, may be calculated from

$$K_H(T) = K_H(T_o)\exp(-\mathrm{d}\ln K_H/\mathrm{d}(1/T)\{1/T - 1/T_o\}).$$

Gas	$K_H(T_o)$ (mol l^{-1} atm^{-1})	$-\mathrm{d}\ln K_H/\mathrm{d}(1/T)$ (K)
Acetone	32	5800
Chloroform	0.25	4500
Ethanol	0.019	6600
Benzene	0.16	4100
Hydrogen peroxide	8.3×10^4	7400
Methane	1.4×10^{-3}	1600
Methanol	0.022	5200
Methylchloride	0.094	3000
Nitrous oxide	0.024	2800
Nitric oxide	1.9×10^{-3}	1700
Nitrogen dioxide	1.2×10^{-2}	2500
Ozone	9.4×10^{-3}	2500

$$pX/RT = K_H pXL$$

This can be used to define a critical value of the Henry's Law constant where this equipartitioning will occur:

$$K_{H,crit} = RTL$$

If there is a gram of liquid water (i.e. 0.001 liter) in every cubic meter of air the critical Henry's Law constant amounts to 4×10^4 mol l^{-1} atm^{-1}. Thus in Table 8.3 only hydrogen peroxide appears soluble enough to partition into cloud water, as its K_H exceeds the critical equipartitioning value $(K_{H,crit})$.

However, this does not mean that hydrogen peroxide is the only gas to partition strongly into cloud water. Gases that undergo subsequent hydrolysis reactions can also partition into the liquid phase. These processes, and their associated equilibria, can be typified by the dissolution of formaldehyde, formic acid (a weak acid), and hydrochloric acid (a strong acid):

$$HCHO(g) \rightleftharpoons HCHO(aq)$$
$$H_2O + HCHO(aq) \rightleftharpoons H_2C(OH)_2(aq)$$

$$K_H = aHCHO/pHCHO$$
$$K_{Hy} = aH_2C(OH)_2/aHCHO$$

$$HCOOH(g) \rightleftharpoons HCOOH(aq)$$
$$HCOOH(aq) \rightleftharpoons H^+(aq) + HCOO^-(aq)$$

$$K_H = aHCOOH/pHCOOH$$
$$K' = aH^+ \, aHCOO^-/aHCOOH$$

$$HCl(g) = H^+(aq) + Cl^-(aq)$$

$$K_H = aH^+ \, aCl^-/pHCl$$

The equations above show that in the case of a strong acid the equilibrium between the gas phase and the undissociated aqueous species is often omitted. It is difficult to assign equilibrium constants to the dissociation of strong electrolytes, but this equilibrium step can be ignored without loss of generality, because the undissociated species in a strong electrolyte are low in relative concentration.

Where gases hydrolyze or undergo other equilibrium reactions, partition between air and droplets requires additional dissolved species to be considered. These can be derived for individual cases, but as an example take the dissolution of a weak acid, such as formic acid. We can write the total formic acid concentration in solution, $T_{HCOOH(aq)}$:

$$T_{HCOOH(aq)} = HCOOH(aq) + HCOO^-(aq)$$

$$= K_H pHCOOH + K_H K' pHCOOH/H^+$$

The psuedo-Henry's Law constant $(K^{\ddagger})$ to describe this dissolution in terms of the total amount of dissolved formate can be written:

$$K^{\ddagger} = K_H\left(1 + K'/H^+\right) = T_{HCOOH(aq)}/pHCOOH(g)$$

Aldehydes are especially notable in the atmosphere as they readily hydrolyze in water to form

Table 8.4 Henry's Law constants and hydrolysis constants for gases that undergo reaction in water (at 298 K). The values of K_{Hy} are hydrolysis constants and K' and K" are first and second dissociation constants. Henry's Law constants can be found in Sander (1999), but hydrolysis constants and information on aldehydes (Findlayson-Pitts & Pitts 2000) and organic acids (Clegg et al. 1996; Khan et al. 1996) can be found in other sources.

Simple hydrolysis	$K_H(T_o)$ (mol l^{-1} atm^{-1})	K_{Hy}	
Formaldehyde	1.3	2.3×10^{-3}	
Glyoxal	≥ 1.4	2.2×10^{-5}	
Methylglyoxal	1.4	2.7×10^{-3}	
Weak acid or base	$K_H(T_o)$ (mol l^{-1} atm^{-1})	K' (mol l^{-1})	K" (mol l^{-1})
Formic acid	5.53×10^3	1.77×10^{-4}	
Acetic acid	5.50×10^3	1.75×10^{-5}	
Pyruvic acid	3.11×10^5	3.4×10^{-3}	
Oxalic acid	7.2×10^8	5.29×10^{-2}	5.33×10^{-5}
Sulfur dioxide	1.23	1.3×10^{-2}	6.6×10^{-8}
Carbon dioxide	3.43×10^{-2}	4.3×10^{-7}	4.7×10^{-11}
Hydrogen sulfide	0.12	1.8×10^{-7}	
Ammonia	62	1.7×10^{-5}	
Strong acid	$K_H(T_o)$ (mol kg^{-2} atm^{-1})		
Hydrochloric acid	2.04×10^6		
Methanesulfonic acid	6.5×10^{13}		
Nitric acid	2.45×10^6		

glycols (Montonya & Mellado 1995), which significantly increases their solubility. In the case of formaldehyde this is by a factor of almost 2000. This hydrolysis is particularly significant for glyoxal (CHOCHO), which has an effective Henry's Law constant of about 3×10^5 mol l^{-1} atm^{-1}. This means that it largely partitions from the gas phase into rainwater.

Aldehydes undergo further reactions in solution, particularly with dissolved S(IV) anions:

$$HCHO + HSO_3^- \rightleftharpoons CH_2(OH)SO_3^-$$

The product here is the hydroxymethansulfonate ion. The formation constant is strongest with formaldehyde, but glyoxal and hydroxyacetaldehyde are also likely to have a substantial sulfonate formation in aqueous solutions in the atmosphere, which can be responsible for enhancing the dissolution of sulfur dioxide. The reaction of S(IV) with aldehydes tends to be relatively slow compared with oxidation to sulfuric acid by oxidants found in atmospheric water (most typically aqueous hydrogen peroxide). Thus it is only under relatively alkaline conditions >pH5 and where oxidation rates by H_2O_2 are low that the formation of these adducts with aldehydes becomes important. Such conditions can lead to as much as an order of magnitude increase in S(IV) concentrations in droplets over the value expected simply on the basis of hydration of dissolving SO_2.

Dissolution within droplets can also be affected by other types of equilibrium reactions, most notably the chelation with metal ions. This requires ligands that have large stability constants with metal ions, because concentrations will typically be low in atmospheric water. Oxo-ligands are the most likely and the polycarboxylic acids in particular seem to be strong acids that ensure they highly charged. The few studies of metal complexes in atmospheric water have focussed on oxalate complexes with iron and their potential for photochemistry (Zuo & Holgne 1992). Humic substances in rainwater probably form exceedingly strong metal complexes (Spokes *et al.* 1996) because they contain many oxy and hydroxy groups, and along with nitrogen atoms are potential electron donors.

Surface active compounds change the surface tension of water droplets in the atmosphere (Seidl 2000). There is also experimental evidence that the Henry's Law constants for organic compounds increased in the presence of surfactants via micelle formation (Vane & Giroux 2000). This is hardly surprising and it is a potential route for a solubility increase among compounds that are not highly polar. Commercial surfactants used in these kinds of experiments are often derived from strong organic acids, so they differ from the weak carboxylic acids most often discussed in atmospheric systems. There seems to be about $100 \, \text{pmol m}^{-3}$ of surfactant in continental air (Sukhapan & Brimblecombe 2002). There is the potential for surface activity on atmospheric aerosols to enhance the solubility of low-polarity organic compounds, and surfactants can also affect the rate of gas dissolution in aerosols and perhaps even global albedo (Facchini *et al.* 1999).

So far we have assumed that the gas and liquid phases are able to reach equilibrium relatively quickly. This seems reasonable, as most of the hydrolysis reactions we have discussed above proceed fairly rapidly. However, such an approach neglects the transfer of a trace gas to an atmospheric solution. This can be broken down into a number of steps: (i) diffusion of the gas to the surface of the liquid; (ii) transfer across the interface, which also involves moving that gas through a film; (iii) hydration or ionization; (iv) diffusion through the liquid phase; and ultimately (v) any other chemical reaction.

The speed of each of these processes is often described in terms of characteristic or relaxation times. For many processes the chemical reaction is slower than the transfer processes and hydration.

1 The time for gas-phase diffusion to achieve equilibrium with the surface of a liquid droplet depends on drop radius, but even with large droplets $100 \, \mu\text{m}$ in radius it is less than a millisecond.

2 The time to attain interfacial equilibrium is relatively rapid and is a function of the accommodation coefficient (i.e. the fraction of molecules that strike the surface which dissolve) and Henry's Law constant of the gas. Soluble gases have much longer characteristic times than less soluble ones. In the case of hydrogen peroxide this may be as long as a second.

3 Ionization equilibria are typically fast in the case of the hydration of aqueous SO_2 the hydration reaction proceeds with a rate constant of 3.4×10^6 s^{-1}, which allows the formation of the bisulfite anion to be exceeding rapid. On the other hand the hydrolysis of some of the aldehydes, such as glyoxal, is exceedingly slow and may take many hours.

4 The characteristic time for diffusion in a spherical droplet is given as $r^2/D\pi^2$, where r is the droplet radius and D the diffusion constant of the species in solution. As D is typically $1.8 \times 10^{-9} \, \text{m}^2 \, \text{s}^{-1}$ characteristic times for $100 \, \mu\text{m}$ radius droplets will be close to a second. This does not consider whether droplets are mixed by internal circulation as they fall through the air. This process can increase the rate of achieving a homogeneous distribution in droplets larger than $100 \, \mu\text{m}$ in radius.

Chemical reactions within the droplet are of such general wide importance that they deserve treatment in a section of their own. In general the slower chemical reactions require droplets with longer lifetimes to be significant. Long lifetimes require the droplet to be small so as to have a low fall velocity, which in turn means that the distances over which diffusion takes place are also small, so characteristic times can be relatively short.

8.3 REACTIONS AND PHOTOCHEMISTRY

Section 8.2 treated equilibrium processes in solution, while this section deals with reactions that are essentially irreversible. The liquid phase can promote reactions that may not occur in the gas phase. Ravishankara (1997) has argued that two rather unreactive gas-phase species can potentially react effectively in the aqueous aerosol. The argument goes that reactions between filled shell molecules are slow in the gas phase because of the high energy barriers, but can be faster in the liquid-phase reactions because of ionic reaction pathways.

In Section 8.2.2 hydrolysis emerged as an important process. These reactions were essentially reversible. A number of gases will hydrolyze rapidly and virtually irreversibly in water. The oxidation of HCFC-124 (CF_3CFClH) leads to trifluoroacetylfluoride in the atmosphere, which readily dissolves in water and rapidly hydrolyzes to trifluoroactic acid. This is a stable compound and may well accumulate in surface waters (Tromp *et al.* 1995).

$$CF_3CFO + H_2O \rightarrow CF_3COOH + HF$$

Phosgene is produced by OH-initiated oxidation; although a chlorinated hydrocarbon it is not very soluble in water, but once dissolved it hydrolyzes rapidly (Helas & Wilson 1992).

$$COCl_2 + H_2O \rightarrow 2HCl + CO_2$$

There is also the potential for aqueous-phase production of low-solubility gases such as carbonyl sulfide in rainwater (Belviso *et al.* 1987), which could ultimately degas. Photochemical production of acetone from citric acid (Abrahamson *et al.* 1994) also seems a possibility in iron-rich systems.

8.3.1   Saline droplets

Sea-salt particles are frequently found to have chloride concentrations much lower than expected from maritime ionic ratios. This can be easily explained in terms of a displacement by dissolved acids at high concentrations in aqueous marine aerosols:

$$HNO_3(g) + Cl^- \rightarrow HCl(g) + NO_3^-$$

$$H_2SO_4 + Cl^- \rightarrow HCl(g) + HSO_4^-$$

One would also expect fluoride to be even more strongly depleted in the marine aerosol as it has a lower Henry's Law constant than HCl (Clegg & Brimblecombe 1988).

Reactions can also occur directly with solid sea-salt particles. Various nitrogen oxides can react with sodium chloride (Findlayson-Pitts & Pitts 2000) to give ClNO:

$$2NO_2(g) + NaCl(s) \rightarrow NaNO_3(s) + ClNO(g)$$

with the potential for an analogous bromine chemistry.

8.3.2   Sulfur oxidations

The oxidation of aqueous sulfur dioxide has been studied for many decades as the importance of its extensive droplet chemistry was realized very early. Although chemical engineering and flash photolysis studies suggested radical chain reactions contributed to the autoxidation of sulfites the details are often neglected and the reactions were treated in a simpler overall manner. At typical atmospheric pH values (2–6) most of the S(IV) is present as bisulfite anions.

$$0.5O_2 + HSO_3^- \rightarrow H^+ + SO_4^{2-}$$

The oxidation of sulfur(IV) by molecular oxygen is very slow, although in polluted environments there is the potential for the reaction to become catalyzed by a range of transition metals, such as iron and manganese (Warneck 1999b). The mechanisms are seen as occurring through an electron transfer such that the metal is reduced in solution, so we might represent it:

$$M(III)(OH)_n + HSO_3^-$$
$$\rightarrow M(II)(OH)_{n-1} + SO_3^- + H_2O$$

Essentially initiating a radical chain:

$$SO_3^- + O_2 \rightarrow SO_5^-$$

$$SO_5^- + SO_3^{2-} \rightarrow SO_4^- + SO_4^{2-}$$

$$SO_4^- + SO_3^{2-} \rightarrow SO_3^- + SO_4^{2-}$$

This was a mechanism originally proposed by Bäckström in the 1930s, but has found much support through mechanistic studies (Hayon *et al.* 1972). The metal catalyzed pathways are likely to be more important at relatively modest acidities because the reactions slow down as pH decreases and the reaction generates acidity overall with the conversion of sulfurous to sulfuric acid. Organic materials such as terpenes may decrease the chain length and thus reduce the overall oxidation rate (Ziajka & Pasuik-Bronikowska 1999).

Penkett *et al.* (1979) showed that other oxidants, most particularly hydrogen peroxide and ozone, are capable of dissolving and oxidizing bisulfite to sulfate. The hydrogen peroxide route is a particularly significant one as the reaction is faster in acid solution. This means that it would not slow down as the system became more acidic with the production of sulfuric acid. The oxidation by ozone can readily be represented:

$$HSO_3^- + O_3 \rightarrow H^+ + SO_4^{2-} + O_2$$

However, the exact rearrangement during this oxidation can be complex (Hoffmann 1986). Hydrogen peroxide and organo-peroxides will react:

$$ROOH + HSO_3^- \rightleftharpoons ROOSO_2^- + H_2O$$

$$ROOSO_2^- \rightarrow ROSO_3^-$$

$$ROSO_3^- + H_2O \rightarrow ROH + SO_4^{2-} + H^+$$

Alternatively, one can imagine OH in atmospheric droplets initiating an opening step and subsequent reaction chain:

$$OH + HSO_3^- \rightarrow SO_3^- + H_2O$$

This brings us on to the general question of the overall fraction of oxidation taking place through various mechanisms. Warneck (1999b) has investigated the efficiency of various reactions contributing to the oxidation of sulfur dioxide and nitrogen dioxide in cloud water. Ozone and hydrogen peroxide are the most important oxidants, in the aqueous phase, but the reaction of peroxynitric

acid with the bisulfite anion can make a significant contribution (Warneck 1999a).

$$HOONO_2 + HSO_3^- \rightarrow 2H^+ + NO_3^- + SO_4^{2-}$$

In more polluted situations hydrogen peroxide could be overwhelmed by much larger amounts of sulfur dioxide in the air. Under these types of situations oxidation by hydrogen peroxide and ozone will be typically less important than oxidation by OH, Br_2^- and Cl_2^- (Herrmann *et al.* 2000).

8.3.3 *Nitrogen compounds*

The nitrogen chemistry of droplets has been less extensively studied than sulfur chemistry. Peroxynitric acid was mentioned in the section above. It is a very soluble product of atmospheric oxidation (Macleod *et al.* 1988). It is a moderately strong acid (pKa ≈ 5) and if the pH is low enough for most of the acid is in the molecular form it is relatively stable and so available to react with S(IV). As seen in the equation above, this also represents a source of nitric acid in droplets. However, most nitrate is formed from the oxidation of nitrogen dioxide to nitric acid in the gas phase by reaction with OH radicals. In addition to oxidizing S(IV), peroxynitric acid also contributes appreciably to nitrite (Warneck 1999b):

$$HOONO_2 \rightleftharpoons H^+ + NO_4^-$$

$$NO_4^- \rightarrow NO_2^- + O_2$$

Although we have not mentioned reactions in anything but the liquid phase it is worth noting observations that nitrite oxidation seems to proceed very quickly in freezing particles (Takenaka *et al.* 1998).

8.3.4 *Organic solutes*

Cloud and fog water processes are potentially important contributors to secondary organic aerosol formation (Blando & Turpin 2000). Organic vapors dissolve within suspended droplets and participate in aqueous-phase reactions. Typically aldehydes, ketones, alcohols, monocarboxylic acids, and organic peroxides can lead to carboxylic acids (especially polyfunctional ones), such as polyols, glyoxal, and esters. The organic compounds that dissolve in atmospheric water most likely react via photochemically induced oxidation reactions, which are covered in the section below.

Dimerization might well occur with dissolved compounds such as methacrylic acid (Khan *et al.* 1992), which forms from the oxidation of isoprene. However, the low concentrations of organic compounds expected in most atmospheric droplets limits the potential for polymerization reactions.

In systems with a rich biology there is an opportunity for biochemical reactions. Thus urea might be expected to degrade biologically:

$$NH_2CO\,NH_2 + H_2O \rightarrow 2NH_3 + CO_2$$

In leaf wetness enzymes such as the peroxidases are present. These could also catalyze the oxidative cross-linking and polymerization of organic compounds utilizing hydrogen peroxide and other organic peroxides that are present in guttation (Kerstetter *et al.* 1998). Legrand *et al.* (1998) showed that at coastal Antarctic sites, bacterial decomposition of uric acid is a source of ammonium, oxalate, and cations (such as potassium and calcium) in aerosols, with a subsequent large ammonia loss from ornithogenic soils to the atmosphere.

8.3.5   Radical and photochemical reactions

In the past few decades the photochemistry and radical chemistry of atmospheric water droplets has been seen as increasingly important (Herrmann *et al.* 1999). Photolysis frequencies in the aqueous phase are more rapid than might be expected, as the actinic flux inside cloud droplets is on average more than twice as large as compared to the interstitial air (Ruggaber *et al.* 1997). In addition, we can expect different types of photochemistry in aqueous systems compared with that which has become familiar in the gas phase (Faust 1994). Spectral shifts can stabilize some molecules, so, for example, HCHO is photochemically degraded in the gas phase, but when hydrolyzed to

$CH_2(OH)_2$ in droplets it is not sensitive to photolysis. There is also the potential for some reactions to be photosensitized, perhaps through the presence of the ferric ion. Despite the fact that the importance of these processes in the liquid phase was pointed out more than 20 years ago, the development of detailed mechanisms was at first rather slow and study of these reactions *in situ* is not easy. The understanding of the balance of reactions in the liquid phase remains limited compared with that in the gas phase.

8.3.6   Peroxides

The production and loss of hydrogen peroxide, and the related processes with the HO_2 and OH radical, lie at the heart of this liquid phase chemistry. Hydrogen peroxide partitions very effectively into the liquid phase and is an important oxidant. Although hydrogen peroxide can be photodissociated in solution to give OH this is less effective than direct transfer into aqueous solution.

Just as trace gases dissolve in droplets, radical species can also be absorbed into solution. Some of the most notable for droplet chemistry are OH, HO_2, NO_3, and CH_3O_2. This dissolution process represents the most important source of aqueous HO_2. The dissolved hydroperoxide ion is a moderately strong acid (pKa 4.88) and gives O_2^-.

$$HO_2 \rightleftharpoons H^+ + O_2^-$$

The hydroperoxide radical can also be produced in solution. This can be by reaction with OH:

$$H_2O_2 + OH \rightarrow HO_2 + H_2O$$

In addition to transfer into solution from the gas phase, hydrogen peroxide can be produced in solution via Fe(II)(aq) mediated photoproduction. This has been observed in simulated cloudwater experiments. Potential electron donors for these types of processes, such as oxalate, formate, or acetate, are commonly found in cloudwater (Siefert *et al.* 1994).

When H_2O_2 is abundant in solution nearly all of the OH produced comes from an iron(II)-HOOH photo-Fenton reaction mechanism

$$Fe(II) + HOOH \rightarrow Fe(III) + OH + OH^-$$

initiated by photoreduction of Fe(III) to Fe(II) in the presence of HOOH (Arakaki & Faust 1998).

The O_2^- from the peroxide system can react rapidly with dissolved ozone to give O_3^-:

$$O_2^- + O_3 \rightarrow O_2 + O_3^-$$

$$O_3^- \rightarrow O_2 + O^-$$

$$O^- + H^+ \rightarrow OH$$

Methyleneglycol ($CH_2(OH)_2$, hydrated formaldehyde) is an important sink of OH:

$$CH_2(OH)_2 + OH \rightarrow CH(OH)_2 + H_2O$$

$$CH(OH)_2 + O_2 \rightarrow HCOOH + HO_2$$

The formic acid produced can react further with OH to oxidize to carbon dioxide. Bicarbonate ions, although ubiquitous in atmospheric solutions, react only slowly with OH.

The alkylperoxy radical CH_3O_2 can also dissolve effectively, but it is probably converted to the peroxide

$$CH_3O_2 + HO_2 \rightarrow O_2 + CH_3OOH$$

and as CH_3OOH is not very soluble it can be readily lost from the droplet.

8.3.7 Nitrite radical chemistry

Dissolving NO_3 can react with chloride or the bisulfite anion which means that the presence of Cl tends to moderate the chemistry of NO_3 through the reversible reactions (Buxton *et al.* 1999b)

$$NO_3 + Cl^- \rightarrow NO_3^- + Cl$$

The concentration may be much modified by the presence of aldehydes, which have a high rate of reaction with the nitrate radical, converting them via hydrogen abstraction to the acids. There is the possibility of nitration of disolved phenols by NO_3 or NO_2^+ (Belloli *et al.* 1999) to form potentially carcinogenic and soluble nitrophenols.

8.3.8 Halogen chemistry

Saline droplets in the atmosphere, especially where they are concentrated, represent an important source of atomic bromine and chlorine that have a range of gas-phase reactions. In addition to the processes with nitrogen and sulfur radicals noted in the section above there is the potential for reactions via OH (e.g. Knipping *et al.* 2000).

$$OH + Cl^- \rightarrow HOCl^-$$

$$HOCl^- + H^+ \rightarrow Cl + H_2O$$

A further important sequence is:

$$O_3^- + H^+ + Br^- \rightarrow HOBr + O_2^-$$

$$HOBr + Cl^- \rightarrow BrCl + H_2O$$

This represents an important source of BrCl to the gas phase (Vogt *et al.* 1996; Disselkamp *et al.* 1999). Alternate drivers for the same types of processes arise from the dissolution of the hypochlorous and hypobromous acids

$$HOCl(g) \rightarrow HOCl(aq)$$

$$HOCl(aq) + Cl^- + H^+ \rightarrow Cl_2 + H_2O$$

$$Cl_2 + HO_2 \rightarrow Cl_2^- + H^+ + O_2$$

$$Cl_2^- \rightleftharpoons Cl^- + Cl$$

The equilibrium constant is of the order of $1/10^5$, which means that in typical cloud water Cl dominates over Cl_2^-. Reactions with SO_4^- represent a further source (Buxton *et al.* 1999a):

$$SO_4^- + Cl^- \rightarrow Cl + SO_4^{2-}$$

Chlorine atoms can be lost with the production of OH (Buxton *et al.* 2000):

$$Cl + H_2O \rightarrow Cl^- + H^+ + OH$$

These radicals undergo interconversion, being equilibrated by chloride, sulfate, and hydrogen ions such that no one radical acts as a sink. Thus, at high Cl^- concentrations, for example, the radical chemistry of Cl_2 and Cl will be important, while at high sulfate that of SO_4^- will predominate (Buxton et al. 1999a).

8.3.9 Organic chemistry

There is also an active organic chemistry in sunlit droplets. The aqueous-phase photolysis of biacetyl is an important source of organic acids and peroxides to aqueous aerosols, and fog and cloud drops. The half-life of aqueous-phase biacetyl with respect to photolysis is approximately 1.0–1.6 h (solar zenith angle 36°). Major products of aqueous biacetyl photolysis are acetic acid, peroxyacetic acid, and hydrogen peroxide, with pyruvic acid and methylhydroperoxide as minor photoproducts. Typical reductants in atmospheric waters are likely to be formate, formaldehyde, glyoxal, phenolic compounds, and carbohydrates (Faust et al. 1997).

Dew chemistry is likely to involve a novel range of organic and nitrogen compounds (Kerstetter et al. 1998) and typically has a higher pH than rainwater. The photolysis of aqueous-phase nitrous acid and nitrites could play a significant role in initiating oxidation reactions through the photoproduction of the OH radical (Arakaki et al. 1999).

Continental aerosol chemistry is likely to be a little different to that in marine clouds, as several processes are likely to consume more OH. Most notably these are:
1 Larger formic acid concentrations in the gas phase, which can dissolve and consume OH.
2 Transition metals can scavenge HO_2/O_2^- from the system.
3 Higher concentrations of SO_2 to reduce H_2O_2.

Aqueous phase chemistry has been detailed in the CAPRAM model (chemical aqueous-phase radical mechanism) (Herrmann et al. 1999, 2000).

8.4 CONCLUSIONS

Liquid water in the atmosphere represents an important part of the process of removing trace substances from the atmosphere. The residence time of water in the atmosphere is a matter of days (4–10) and the lifetime of raindrops and dewdrops is considerably shorter. However, the effectiveness of aqueous systems in removing gases from the atmosphere requires these gases to be very soluble in water or to undergo rapid reactions.

Chemistry within the droplets may serve to reprocess materials. This is particularly evident where reactions within the droplet produce gases that have a low solubility and become lost from the liquid phase.

The influence of droplet phases on atmospheric chemistry can be more subtle than this. Sequestration can alter the chemistry of the gas phase, for example, through the high solubility of peroxyradicals (Monod & Carlier 1999), which has an impact of ozone concentrations. Lelieveld and Crutzen (1990) suggested that the removal of HO_2 from the gas phase into cloudwater will prevent it from oxidizing NO to NO_2. This limits the gas-phase production of O_3 from photolysis of NO_2 and thus effectively leads to a loss of ozone from the troposphere:

$$OH + O_3 \rightarrow HO_2 + O_2$$

$$HO_2 \rightleftharpoons O_2^- + H^+$$

$$O_2^- + O_3 \rightarrow O_2 + O_3^-$$

$$O_3^- \rightarrow O_2 + O^-$$

$$O^- + H^+ \rightarrow OH$$

which sums to give

$$2O_3 \rightarrow 3O_2$$

The chemistry of the liquid phase has been more difficult to resolve, because although gas-phase chemistry is affected by the liquid phase it has been possible to treat the liquid phase simply as a sink as a first approximation. When considering the liquid phase the chemistry has often had to involve multiphase considerations from the outset. Nevertheless, the past decade has seen considerable advances in consideration of the radical chemistry and photochemistry of droplets in the atmosphere.

REFERENCES

Abrahamson, H.B., Rezvani, A.B. & Brushmiller, J.G. (1994) Photochemical and spectroscopic studies of complexes of iron(III) with citric-acid and other carboxylic-acid. *Inorganica Chimica Acta* **226**, 117–27.

Arakaki, T. & Faust, B.C. (1998) Production of H_2O_2 sources, sinks, and mechanisms of hydroxyl radical (OH)–O–photoproduction and consumption in authentic acidic continental cloud waters from Whiteface Mountain, New York: The role of the Fe(r) (r = II, III) photochemical cycle. *Journal of Geophysical Research (Atmospheres)* **103**, 3487–504.

Arakaki, T., Miyake, T., Hirakawa, T. & Sakugawa, H. (1999) H dependent photoformation of hydroxyl radical and absorbance of aqueous-phase N(III) (HNO$_2$ and NO$_2^-$). *Environmental Science and Technology* **33**, 2561–5.

Bächmann, K., Ebert, P., Haag, I. & Prokop, T. (1996) The chemical content of raindrops as a function of drop radius. 1. Field measurements at the cloud base and below the cloud. *Atmospheric Environment* **30**, 1019–25.

Barrie, L.A. (1991) Snow formation and processes in the atmosphere that influence its chemical composition. In Davies, T.D. Tranter, M. & Jones, H.G. (eds), *Seasonal Snowpacks*. Springer-Verlag, Berlin.

Barry, R.G. & Chorley, R.J. (1998) *Atmosphere, Weather and Climate*. Routledge, London.

Belloli, R., Barletta, B., Bolzacchini, E., Meinardi, S., Orlandi, M. & Rindone, B. (1999) Determination of toxic nitrophenols in the atmosphere by high-performance liquid chromatography. *Journal of Chromatography A* **846**, 277–81.

Belviso, S., Mihalopoulos, N. & Nguyen, B.C. (1987) The supersaturation of carbonyl sulfide (OCS) in rain waters. *Atmospheric Environment* **21**, 1363–7.

Blando, J.D. & Turpin, B.J. (2000) Secondary organic aerosol formation in cloud and fog droplets: a literature evaluation of plausibility. *Atmospheric Environment* **34**, 1623–32.

Brimblecombe, P. & Todd, I.J. (1977) Sodium and potassium in dew. *Atmospheric Environment* **11**, 649–50.

Buxton, G.V., Bydder, M. & Salmon, G.A. (1999a) The reactivity of chlorine atoms in aqueous solution. Part II. The equilibrium $SO_4^- + Cl^- = Cl + SO_4^{2-}$. *Physical Chemistry Chemical Physics* **1**, 269–73.

Buxton, G.V., Salmon, G.A. & Wang, J.Q. (1999b) The equilibrium $NO_3 + Cl^- = NO_3^- + Cl^-$: a laser flash photolysis and pulse radiolysis study of the reactivity of NO_3 center dot with chloride ion in aqueous solution. *Physical Chemistry Chemical Physics* **1**, 3589–93.

Buxton, G.V., Bydder, M., Salmon, G.A. & Williams, J.E. (2000) The reactivity of chlorine atoms in aqueous solution. Part III. The reactions of Cl with solutes. *Physical Chemistry Chemical Physics* **2**, 237–45.

Chung, Y.S., Kim, H.S. & Yoon, M.B. (1999) Observations of visibility and chemical compositions related to fog, mist and haze in south Korea. *Water Air and Soil Pollution* **111**, 139–57.

Clegg, S.L. & Brimblecombe, P. (1988) Hydrofluoric and hydrochloric-acid behavior in concentrated saline solutions. *Journal of the Chemical Society Dalton* 705–10.

Clegg, S.L., Brimblecombe, P. & Khan, I. (1996) The Henry's law constant of oxalic acid and its partitioning into the atmospheric aerosol. *Idojárás* **100**, 51–68.

Clegg, S.L., Brimblecombe, P. & Wexler, A.S. (1998a) A thermodynamic model of the system H–NH$_4$–Na–SO$_4$–NO$_3$–Cl–H$_2$O at 298.15 K. *Journal of Physical Chemistry* **102A**, 2155–71.

Clegg, S.L., Brimblecombe, P. & Wexler, A.S. (1998b) A thermodynamic model of the system H–NH$_4$–SO$_4$–NO$_3$–H$_2$O at tropospheric temperatures. *Journal of Physical Chemistry* **102A**, 2137–54.

Cornell, S.E., Jickells, T.D. & Thornton, C.A. (1998) Urea in rainwater and atmospheric aerosol. *Atmospheric Environment* **32**, 1903–10.

Descari, S., Facchini, M.C., Fuzzi, S. & Tagliavini, E. (2000) Characterization of water-soluble organic compounds in atmospheric aerosol: a new approach. *Journal of Geophysical Research (Atmospheres)* **105**, 1481–9.

Disselkamp, R.S., Chapman, E.G., Barchet, W.R., Colson, S.D. & Howd, C.D. (1999) BrCl production in NaBr/NaCl/HNO$_3$/O$_3$ solutions representative of sea-salt aerosols in the marine boundary layer. *Geophysical Research Letters* **26**, 2183–6.

Domine, F. (1999) Incorporation of trace gases into ice particles. In Borrell, P.M. & Borrell, B.P. (eds), *Transport and Chemical Transformation in the Troposphere, Volume 1*. WIT Press, Southampton.

Facchini, M.C., Mircea, M., Fuzzi, S. & Charlson, R.J. (1999) Cloud albedo enhancement by surface-active organic solutes in growing droplets. *Nature* **401**, 257–9.

Faust, B.C. (1994) Photochemistry of clouds, fogs, and aerosols. *Environmental Science & Technology* **28**, A217–22.

Faust, B.C., Powell, K., Rao., C.J. & Anastasio, C. (1997) Aqueous-phase photolysis of biacetyl (an alpha-dicarbonyl compound): A sink for biacetyl, and a source of acetic acid, peroxyacetic acid, hydrogen peroxide, and the highly oxidizing acetylperoxyl radical in aqueous aerosols, fogs, and clouds. *Atmospheric Environment* **31**, 497–510.

Findlayson-Pitts, B.J. & Pitts, J.N. (2000) *Chemistry of the Upper and Lower Atmosphere*. Academic Press, San Diego.

Fruekilde, P., Hjorth, J., Jensen, N.R., Kotzias, D. & Larsen, B. (1998) Ozonolysis at vegetation surfaces: a source of acetone, 4-oxopentanal, 6-methyl-5-hepten-2-one, and geranyl acetone in the troposphere. *Atmospheric Environment* **32**, 1893–902.

Gelencser, A., Sallai, M., Krivacsy, Z., Kiss, G. & Meszaros, E. (2000) Voltammetric evidence for the presence of humic-like substances in fog water. *Atmospheric Research* **54**, 157–65.

Hayon, E., Treinin, A. & Wilf, J. (1972) *Journal of the American Chemical Society* **94**, 47.

Heitnzenberg, J. (1998) Condensed water aerosols. In Harisson, R.M. & van Grieken, R. (eds), *Atmospheric Particles*. Wiley, Chichester.

Helas, G. & Wilson, S.R. (1992) On sources and sinks of phosgene in the troposphere. *Atmospheric Environment* **26**, 2975–82.

Herrmann, H., Ervens, B., Nowacki, P., Wolke, R. & Zellner, R. (1999) A chemical aqueous phase radical mechanism for tropospheric chemistry. *Chemosphere* **38**, 1223–32.

Herrmann, H., Ervens, B., Jacobi, H.W., Wolke, R., Nowacki, P. & Zellner, R. (2000) CAPRAM2.3: a chemical aqueous phase radical mechanism for tropospheric chemistry. *Journal of Atmospheric Chemistry* **36**, 231–84.

Hoffmann, M.R. (1986) On the kinetics and mechanisms of oxidation of aquated sulfur dioxide by ozone. *Atmospheric Environment* **20**, 1145–54.

Huebert, B.J. & Charlson, R.J. (2000) Uncertainties in data on organic aerosols. *Tellus* **52B**, 1249–55.

Hughes, R.N. & Brimblecombe, P. (1994) Dew and guttation: formation and environmental significance. *Agricultural and Forest Meteorology* **67**, 173–90.

Kames, J. & Schurath, U. (1992) Alkyl nitrates and bifunctional nitrates of atmospheric interest: Henry's law constants and their temperature dependencies. *Journal of Atmospheric Chemistry* **15**, 79–95.

Kawamura, K., Kasukabe, H. & Barrie, L.A. (1996) Source and reaction pathways of dicarboxylic acids, keto acids and dicarbonyls in Arctic aerosols: one year of observations. *Atmospheric Environment* **30**, 1709–22.

Kerstetter, R., Zepp, R.G. & Carreira, L. (1998) Peroxidases in grass dew derived from guttation: possible role in polymerization of soil organic matter. *Biogeochemistry* **42**, 311–23.

Khan, I., Brimblecombe, P. & Clegg, S.L. (1992) The Henry's law constants of pyruvic and methacrylic acids. *Environmental Technology* **13**, 587–93.

Khan, I., Brimblecombe, P. & Clegg, S.L. (1996) Solubilities of pyruvic acid and the lower (C1–C6) carboxylic acids. Experimental determination of equilibrium vapor pressures above pure aqueous and salt solutions. *Journal of Atmospheric Chemistry* **22**, 285–302.

Knipping, E.M., Lakin, M.J., Foster, K.L. *et al.* (2000) Experiments and simulations of ion-enhanced interfacial chemistry on aqueous NaCl aerosols. *Science* **288**, 301–6.

Leck, C. & Bigg, E.K. (1999) Aerosol production over remote marine areas—a new route. *Geophysical Research Letters* **26**, 3577–80.

Legrand, M., Ducroz, F., Wagenbach, D., Mulvaney, R. & Hall, J. (1998) Ammonium in coastal Antarctic aerosol and snow: Role of polar ocean and penguin emissions. *Journal of Geophysical Research-Atmospheres* **103**, 11043–56.

Lelieveld, J. & Crutzen, P.J. (1990) Influences of cloud photochemical processes on tropospheric ozone. *Nature* **343**, 227–33.

Macleod, H., Smith, G.P. & Golden, D.M. (1988) Photodissociation of pernitric acid (HO_2NO_2) at 248 nm. *Journal of Geophysical Research* **93**, 3813–23.

Mason, B. (1975) *Clouds and Rainmaking*. Cambridge University Press, Cambridge.

Milne, P.J. & Zika, R.G. (1993) Amino-acid nitrogen in atmospheric aerosols—occurrence, sources and photochemical modification. *Journal of Atmospheric Chemistry* **16**, 361–98.

Monod, A. & Carlier, P. (1999) Impact of clouds on the tropospheric ozone budget: direct effect of multiphase photochemistry of soluble organic compounds. *Atmospheric Environment* **33**, 4431–46.

Montonya, M.R. & Mellado, J.M.R. (1995) Hydration constants of carbonyl and dicarbonyl compounds. Comparison between electrochemical and no electrochemical technique. *Portugaliae Electrochimica Acta* **13**, 299–303.

Penkett, S.A., Jones, B.M.R., Brice, K.A. & Eggleton, A.E.J. (1979) The importance of atmospheric ozone and hydrogen peroxide in oxidizing sulfur dioxide in cloud and rain water. *Atmospheric Environment* **13**, 323–37.

Pilinis, C., Seinfeld, J.H. & Grosjean, D. (1989) Water content of atmospheric aerosols. *Atmospheric Environment* **23**, 1601–6.

Prupacher, H.R. & Klett, J.D. (1997) *Microphysics of Clouds and Precipitation*. Kluwer, Dordrecht.

Ravishankara, A.R. (1997) Heterogeneous and multiphase chemistry in the troposphere. *Science* **276**, 1058–65.

Ruggaber, A., Dlugi, R., Bott, A., Forkel, R., Herrmann, H. & Jacobi, H.W. (1997) Modelling of radiation quantities and photolysis frequencies in the aqueous phase in the troposphere. *Atmospheric Environment* **31**, 3135–48.

Sander, R. (1999) Compilation of Henry's law constants for inorganic and organic species of potential importance in environmental chemistry (http://www.mpch-mainz.mpg.de/~sander/res/henry.html).

Saxena, P. & Hildemann, L.M. (1996) Water-soluble organics in atmospheric particles: a critical review of the literature and application of thermodynamics to identify candidate compounds. *Journal of Atmospheric Chemistry* **24**, 57–109.

Seidl, W. (2000) Model for a surface film of fatty acids on rainwater and aerosol particles. *Atmospheric Environment* **34**, 4917–32.

Siefert, R.L., Pehkonen, S.O., Erel, Y. & Hoffmann, M.R. (1994) Iron photochemistry of aqueous suspensions of ambient aerosol with added organic-acids. *Geochimica et Cosmochimica Acta* **58**, 3271–9.

Spokes, L.M., Lucia, M., Campos, A.M. & Jickells, T.D. (1996) The role of organic matter in controlling copper speciation in precipitation. *Atmospheric Environment* **30**, 3959–66.

Sukhapan, J. & Brimblecombe, P. (2002) Ionic surface active compounds in atmospheric aerosols. *Scientific World Journal* **2**, 1138–46.

Takenaka, N., Daimon, T., Ueda, A. *et al.* (1998) Fast oxidation reaction of nitrite by dissolved oxygen in the freezing process in the tropospheric aqueous phase. *Journal of Atmospheric Chemistry* **29**, 135–50.

Takenaka, N., Suzue, T., Ohira, K., Morikawa, T., Bandow, H. & Maeda, Y. (1999) Natural denitrification in drying process of dew. *Environmental Science and Technology* **33**, 1444–7.

Tromp, T.K., Ko, M.K.W., Rodriguez, J.M. & Sze, N.D. (1995) Potential accumulation of CFC-replacement degradation product in seasonal wetlands. *Nature* **376**, 327–30.

Vane, L.M. & Giroux, E.L. (2000) Henry's law constants and micellar partitioning of volatile organic compounds in surfactant solutions. *Journal of Chemical Engineering Data* **45**, 38–47.

Vogt, R., Crutzen, P.J. & Sander, R. (1996) A mechanism for halogen release from sea-salt aerosol in the remote marine boundary layer. *Nature* **383**, 327–30.

Warneck, P. (1999a) *Chemistry of the Natural Atmosphere, Volume 71*, 2nd edn. Academic Press: San Diego.

Warneck, P. (1999b) The relative importance of various pathways for the oxidation of sulfur dioxide and nitrogen dioxide in sunlit continental fair weather clouds. *Physical Chemistry Chemical Physics* **1**, 5471–83.

Yoshizumi, K. & Asakuno, K. (1986) Characterization of atmospheric aerosols in Chichi of the Ogasawara (Bonin) Islands. *Atmospheric Environment* **20**, 151–5.

Ziajka, J. & Pasuik-Bronikowska, W. (1999) Effect of alpha-pinene and cis-verbenol on the rate of S(IV) oxidation catalyzed by Fe. In Borrell, P.M. & Borrell, B.P. (eds), *Transport and Chemical Transformation in the Troposphere, Volume 1*. WIT Press, Southampton.

Zuo, Y.G. & Holgne, J. (1992) Formation of hydrogen-peroxide and depletion of oxalic-acid in atmospheric water by photolysis of iron(III) oxalato complexes. *Environmental Science and Technology* **26**, 1014–22.

9 Atmospheric Particulate Matter

URS BALTENSPERGER, STEFAN NYEKI, AND
MARKUS KALBERER

9.1 INTRODUCTION

In the past several decades, atmospheric aerosols have increasingly been recognized as constituting one of the major uncertainties in the current understanding of climate change (IPCC 1996, 2001). The uncertainty is mainly due to the large variability in their physical and chemical properties, as well as their temporal and spatial distributions. Atmospheric aerosols may originate from either naturally occurring or anthropogenic processes. Major natural aerosol sources include emissions from the oceans such as sea spray, volcanoes, and mineral dust from arid regions, while major anthropogenic sources include emissions from industry and combustion processes. Further classification within each category into so-called primary and secondary sources may also be made. Aerosols directly emitted into the atmosphere constitute primary sources, while secondary sources arise from the gas-to-particle conversion of gaseous precursor compounds such as nitric oxide and nitrogen dioxide (collectively known as NO_x), sulfur dioxide (SO_2), and hydrocarbons.

Characterization of the life cycle of atmospheric aerosols is a complex and many-faceted issue. A schematic is illustrated in Fig. 9.1, which summarizes: (i) aerosol sources; (ii) transformation mechanisms in the atmosphere; and (iii) aerosol sink processes. As aerosol size is one of the most important parameters in describing aerosol properties and their interaction with the atmosphere, its determination and use is of fundamental impor-

tance. Aerosol size covers several decades in diameter and hence a variety of instruments are required for its determination. This necessitates several definitions of the diameter, the most common being the geometric diameter d, which is used here. Another commonly used aerosol diameter is the aerodynamic diameter, which normalizes for density and shape. A good discussion of these aspects is given in Reist (1993) and Hinds (1999). With reference to Fig. 9.1, the size fraction with diameter $d > 1$–$2\,\mu m$ is usually referred to as the coarse mode, and the fraction $d < 1$–$2\,\mu m$ is the fine mode. The latter mode can be further divided into the accumulation ($d \sim 0.1$–$1\,\mu m$), Aitken ($d \sim 0.01$–$0.1\,\mu m$) and nucleation ($d < 0.01\,\mu m$) modes. Due to the d^3 dependence of aerosol volume (and mass), the coarse mode is typified by a maximum volume concentration and, similarly, the accumulation mode by the surface area concentration and the Aitken and nucleation modes by the number concentration.

Aerosol formation arises from heterogeneous or homogeneous nucleation. The former refers to condensation growth on existing nuclei, and the latter to the formation of new nuclei through condensation. Heterogeneous nucleation preferentially occurs on existing nuclei. Condensation onto a host surface occurs at a critical supersaturation, which is substantially lower (<1–2%) than for homogeneous nucleation in the absence of impurities (>300%). Examples of gas-to-particle conversion are combustion processes and the ambient formation of nuclei from gaseous organic emis-

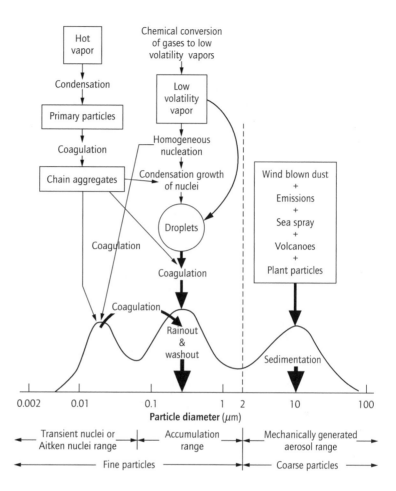

Fig. 9.1 Schematic of the aerosol surface area concentration illustrating the Aitken, accumulation and coarse modes. (Reprinted from *Atmospheric Environment* (Whitby & Sverdrup 1980). Copyright 1980, with permission from Elsevier Science.)

sions. The high initial number concentration of particles with $d < 0.1\,\mu m$ are rapidly reduced through coagulation, resulting in aerosol lifetimes of the order of minutes for these smallest particles.

Particles in the Aitken/accumulation mode typically arise from either: (i) the condensation of low-volatility vapors; or (ii) coagulation. Atmospheric aerosols may be composed of a range of chemical species. When particles are chemically distinct from one another, the aerosol is termed an external mixture. Alternatively, when particles have a similar composition they are known as an internal mixture. This may physically range from complete mixing of chemical components (e.g. hy-

groscopic components in a cloud droplet) to various other structures, such as a coated aerosol (e.g. sulfate coating on soot aerosols). Particles in the accumulation mode have a longer atmospheric lifetime than other modes, as there is a minimum efficiency in sink processes. Of these processes, wet deposition (in-cloud and below-cloud scavenging) is the major sink process.

Particles in the coarse mode are usually produced by weathering and wind erosion processes. Dry deposition (primarily sedimentation) is the dominant removal process. Chemically their composition reflects their sources, as demonstrated by mineral dust from arid regions and sea salt from oceans. Organic compounds such as biological (spores, pollens, and bacteria) and biogenic particles resulting from direct emission of hydrocarbons into the atmosphere may also be constituents of the coarse mode. As the sources and sinks of the coarse and fine modes are different, there is only a weak association of particles in both modes.

The above brief summary of atmospheric aerosols serves to illustrate the complex processes involved in modeling their behavior and assessing their influence on climate. It is mainly for the latter reason that interest in aerosols has grown of late and hence this chapter focuses on the climate effect of aerosols. Aerosol types and composition are considered first, followed by their interaction with radiation through the so-called direct and indirect aerosol effects. Further reading on aerosol fundamentals and atmospheric aerosols may be found in the following list of reprinted and new books: Levine (1991), Baron and Willeke (2001), Hobbs (1993), Reist (1993), Charlson and Heintzenberg (1995), Singh (1995), Pruppacher and Klett (1997), Seinfeld and Pandis (1998), Brasseur *et al.* (1999), Finlayson-Pitts and Pitts (1999), and Hinds (1999). A list of acronyms used throughout the following discussion may be found at the end of this chapter.

9.2 SIZE DISTRIBUTION, COMPOSITION, AND CONCENTRATION

Knowledge of the different moments of an aerosol size distribution and its composition are of primary importance, as most other parameters of interest may be deduced from this information. The atmospheric aerosol may be classified into several categories, according to source and geographic location.

Figure 9.2 illustrates typical size distributions for number, surface, and volume concentration for various aerosol types. A more recent compilation of size distributions, mainly from the Atlantic and European regions (Raes *et al.* 2000), exhibits similar features to those in Fig. 9.2. The number size distribution of the atmospheric aerosol may be approximated by an empirical power law equation for radii $r > 0.1\,\mu m$ (e.g. Jaenicke 1988, and references therein):

$$dN/d\log r = cr^{-\nu} \qquad (9.1)$$

where N is the aerosol number concentration, c is a constant and ν depends on the aerosol type. The large range in magnitude of N is demonstrated by values $<20\,cm^{-3}$ for the polar regions and $>10^5\,cm^{-3}$ for an urban aerosol. Not only is there a large geographic variation in aerosol concentrations, but the vertical extent also varies substantially. For instance, a background aerosol at $3\,km$ over the oceans or $5\,km$ over the continents is typified by $N \sim 150\,cm^{-3}$ and the stratospheric aerosol at $20\,km$ by $N \sim 10\,cm^{-3}$. Urban, remote continental, and remote marine aerosol models from Fig. 9.2 may be described as a first approximation by typical chemical compositions appearing in Table 9.1. Sulfate and organics are the major aerosol components of urban and remote continental regions and sodium chloride of remote marine regions. The aerosol chemical composition and geographic type are thus recognized as being fairly specific to an aerosol source and are discussed below in greater detail.

9.3 AEROSOL SOURCES

The different types of aerosol sources mentioned above, i.e. natural and anthropogenic, primary and secondary, are described in greater detail in this section. An estimate of annual atmospheric contributions to major aerosol sources is given in Table 9.2. Estimates suggest that anthropogenic emissions of SO_4^{2-} are greater than from natural sources. As a result of large spatial and temporal variations in aerosol sources and sinks, the range of values in Table 9.2 highlights the uncertainties involved in estimating emissions. Owing to the short atmospheric lifetime of aerosols it should be remembered that globally averaged figures in Table 9.2 are not necessarily indicative of local burdens. For instance, natural sources will dominate on a global scale due to their large areal emis-

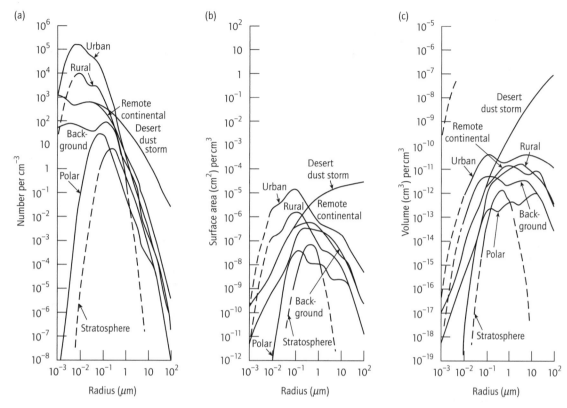

Fig. 9.2 (a) Number, (b) surface, and (c) volume size distributions (dN/dlog$r$, dS/dlog$r$, and dV/dlog$r$, respectively) for various atmospheric aerosols. Aerosol types are self-explanatory apart from background aerosol, which refers to the tropospheric aerosol 5 km above the continents and 3 km above the oceans. (From Jaenicke 1988.)

sion sources (e.g. deserts and oceans). In contrast, anthropogenic emissions from industrialized regions in Europe, the USA, and East Asia, which are relatively smaller, are likely to exceed the contributions from natural sources.

9.3.1　*Natural sources*

Primary emissions

Sea-salt aerosol. Sea-salt aerosol results from the bursting of bubbles, formed by wave and wind action at the ocean surface (e.g. O'Dowd *et al.* 1997). As a result, sea-spray droplets are ejected and either return to the water surface or evaporate to form inorganic/organic aerosols, which may then

be entrained into the marine boundary layer (MBL) by wind turbulence. The wind speed is a major factor in controlling the sea-salt aerosol concentration and roughly exhibits a linear dependence. The fairly global uniformity of sea water composition is reflected in the composition of sea-salt aerosol, although enrichment of particular species during formation or subsequent chemical transformation in the atmosphere may occur. The composition of sea-salt aerosol is given in Table 9.1, and is mainly the salts NaCl, KCl, $CaSO_4$, and Na_2SO_4. Soluble and insoluble organic compounds may also be an important component, the fraction depending on a number of parameters, such as location and time of season. As a result of the uniform trace aerosol

Table 9.1 Typical mass composition ($\mu g\,m^{-3}$) of various chemical species in urban, remote continental, and remote marine aerosol types (adapted from Pueschel 1995).

Element or compound	Urban aerosol – photochemical smog	Remote continental aerosol	Remote marine aerosol
SO_4^{2-}	16.5	0.5–5	2.6
NO_3^-	10	0.4–1.4	0.05
Cl^-	0.7	0.08–0.14	4.6
Br^-	0.5	–	0.02
NH_4^+	6.9	0.4–2.0	0.16
Na^+	3.1	0.02–0.08	2.9
K^+	0.9	0.03–0.01	0.1
Ca^{2+}	1.9	0.04–0.3	0.2
Mg^{2+}	1.4	–	0.4
Al_2O_3	6.4	0.08–0.4	–
SiO_2	21.1	0.2–1.3	–
Fe_2O_3	3.8	0.04–0.4	0.07
CaO	–	0.06–0.18	–
Organics	30.4	1.1	0.9
EC	9.3	0.04	0.04
Total	112.9	2.99–12.44	12.04
Selected mass fractions and molar ratios			
SO_4^{2-} (%)	15.9	30.2–45.7	22.6
NO_3^- (%)	9.6	13.3–22.7	0.44
NH_4^+/SO_4^{2-}	2.2	2.1–3.4	0.47

chemical composition over the oceans, source regions cannot be easily identified.

The mass median diameter of sea-salt aerosol near the sea surface is about 8 µm, which results in a short lifetime and inefficient light-scattering properties. However, smaller sea-salt aerosols may be transported over the ocean surface and a global annual emission of 1300 Tg yr^{-1} (Table 9.2) is considered as representative of the fraction present in the lower MBL.

Mineral dust. Wind-blown mineral dust from desert and semiarid regions is an important source of tropospheric aerosols (Duce 1995) and of particular interest in paleoclimatological studies due to its inert properties (Leinen & Sarnthein 1989). Mineral dust arises from the physical and chemical weathering of rock and soils. The wind speed is the main controlling factor in entraining particles into the atmosphere, followed by other factors, such as soil moisture and surface composition. Atmospheric size distributions in the vicinity of soil sources are generally bimodal, in which the range $d \sim 10$–$200\,\mu m$ consists mainly of quartz grains, and the range $d < 10\,\mu m$ of clay particles. Quartz grains will preferentially sediment close to their source, resulting in a fractionation process from a quartz/clay to a clay aerosol with increasing distance. The size distribution will also change with downwind distance as a result of sedimentation. Measurements have indicated that the modal volume diameter changes from $d \sim 60$–$100\,\mu m$ to $d \sim 2\,\mu m$ at a distance of 5000 km, where it appears to stabilize.

The principal elemental constituents of mineral dust (oxides and carbonates of Si, Al, Ca, Fe)

Table 9.2 Global emission source strengths for atmospheric aerosols ($Tg\,yr^{-1}$).

Aerosol component	d'Almeida *et al.* (1991)	Pueschel (1995)	IPCC (1995)*	Aerosol size mode
Natural				
Primary				
Sea salt	1,000–10,000	300–2,000	1,300	Coarse
Mineral dust	500–2,000	100–500	1,500	Mainly coarse
Primary organic aerosols/biological debris	80	3–150	50	Coarse
Volcanic ash	25–250	25–300	33	Coarse
Secondary	345–1,100			
Sulfate from biogenic gases	121–452	90	Fine	
Sulfate from volcanic SO_2	9	12	Fine	
Nitrate from NO_x	75–700	22	Fine/coarse†	
Organics from biogenic VOC	15–200	55	Fine	
Natural total	1,950–13,430	648–4,311	3,062	
Anthropogenic				
Primary				
Industrial dust	10–90	167	100	Coarse/fine
Biomass burning	3–150	29–72	80	Fine
Soot (all sources)	24	10	Mainly fine	
Secondary	175–325			
Sulfate from SO_2	70–220	140	Fine	
Nitrate from NO_x	23–40	40	Fine/coarse	
Ammonium from NH_3	269			
Organics from VOC	15–90	10	Fine	
Anthropogenic total	188–565	597–882	380	
Overall total	2,138–13,995	1,245–5,193	3,442	

* "Best" estimate.
† Relative fractions uncertain.
Numbers have changed slightly in IPCC (2001), which appeared after finalization of this chapter.

bear a close resemblance to the average crustal composition and may be used to identify source regions. Due to the inert nature of mineral dust, chemical transformation processes in the atmosphere are thus considered minor, although surface chemical reactions may be important, e.g. reactions with gaseous HNO_3 and SO_2, hence enhancing coarse mode sulfate and nitrate, respectively.

Mineral dust is estimated to contribute 1500 Tg yr^{-1} to global atmospheric emissions (Table 9.2), which originates from areas totalling about 10% of the Earth's surface. This compares to a similar sea-salt emission from oceans covering an area larger than 70% of the Earth's surface. Principal source regions of mineral dust cover about one-third of the land surface and include the Saudi Arabian peninsula, the US Southwest, and the Sahara and Gobi deserts. Only the latter two regions are significant sources of long-range transported dust, occurring mainly westward over the tropical North Atlantic and eastward over the North Pacific, respectively. Although mineral dust is considered a natural emission, recent work has suggested that between 20 and 50% of the current atmospheric burden may arise from disturbed soils through human activity (Sokolik & Toon 1996; Tegen *et al.* 1996).

Primary organic aerosols/biological debris. These categories of natural emissions from the biosphere have only recently been investigated in any detail. Continental sources mainly arise from vegetation (plant waxes and fragments, pollen, spores, fungi, and decaying material), while marine sources consist of organic surfactants formed via bubble bursting. Typical size distributions are dominated by the coarse mode. Continental and marine source burdens are similar, with a total global emission of $50\,\mathrm{Tg\,yr^{-1}}$, although the scarcity of data makes such an estimate very uncertain.

Volcanic emissions. Although volcanic activity occurs on a sporadic basis and is mainly located in the northern hemisphere, the more recent outbreaks of Mount St Helens (USA, 1980), El Chichón (Mexico, 1982), and Mount Pinatubo (Philippines, 1991) have highlighted the importance of volcanic emissions to the atmosphere. These emissions are composed of ash (principally SiO_2, Al_2O_3, and Fe_2O_3), gases (SO_2, H_2S, CO_2, HCl, HF), and water vapor. Mount Pinatubo was estimated to have emitted $9\,\mathrm{Tg}$ sulfur (S) compared to $3.5\,\mathrm{Tg(S)}$ for El Chichón, of which a large fraction in the form of SO_2 was buoyantly injected into the stratosphere. Table 9.2 gives an estimated emission of $33\,\mathrm{Tg\,yr^{-1}}$ for ash particles in the coarse mode, which mainly limit their impact to the regional scale and to the troposphere. Of greater long-term importance is the emission of SO_2 into the stratosphere and the subsequent formation of H_2SO_4 aerosol droplets, which generally exhibit a bimodal structure in the Aitken accumulation mode size range. Whereas aerosols in the troposphere have an average lifetime of 5–7 days, due mainly to wet scavenging by clouds, stratospheric aerosols have a 6–9-month lifetime, due to the thermal stratification of the stratosphere and the absence of wet scavenging. Stratospheric aerosol removal mechanisms are primarily sedimentation, subsidence, and exchange with the upper troposphere. Although Table 9.2 indicates that explosive volcanic emissions may only contribute up to 10–20% of the total natural sulfur emission to the atmosphere, the radiative impact of stratospheric aerosols may be quite significant. Figure 9.3 illustrates the effect that El Chichón and Mount Pinatubo had on stratospheric optical properties. Peak values of the integrated aerosol backscatter, as measured with a ground-based lidar (Osborn *et al.* 1995), occurred about 6 months after each eruption. During this time, H_2SO_4 droplets of a sufficient size to interact with radiation were formed. A nonvolcanic background aerosol is also evident in Fig. 9.3 and is attributed to the formation of H_2SO_4 from the upward flux of COS (carbonyl sulfide), emitted from the oceans. The climatic effect of stratospheric aerosols is further described in Section 9.5.2, while a good review of their formation, properties, and effects is given by Pueschel (1996).

Secondary emissions

Secondary natural aerosols may be formed from a number of natural precursor gas sources, containing sulfur, nitrogen, and hydrocarbons. The main natural source is the release of gaseous dimethyl sulfide, or DMS (CH_3SCH_3), from the oceans. DMS is formed from the biological activity of phytoplankton and eventually forms aerosol sulfate via the photooxidation to methanesulfonic acid and SO_2. The contribution to sulfate from DMS and other sources, except sea water, is known as nss (no sea salt) sulfate to differentiate it from sea water as a source. Emissions of nss sulfate are estimated at $90\,\mathrm{Tg\,yr^{-1}}$, toward which other marine sulfur sources, such as H_2S, CS_2, and COS contribute less than 10% of total sulfur emissions. The seasonal variation of DMS follows the ocean productivity cycle and may be a magnitude higher in the summer than in the winter season. Concentrations of nss sulfate in the remote MBL generally range from 20–$800\,\mathrm{ng\,m^{-3}}$ for the southern hemisphere oceans to 400–$3000\,\mathrm{ng\,m^{-3}}$ for the northern Atlantic and illustrate the enhanced contribution from anthropogenic sources in the northern hemisphere (Heintzenberg *et al.* 2000).

Typical global concentrations above the MBL rapidly decline with altitude to several $\mathrm{ng\,m^{-3}}$ in

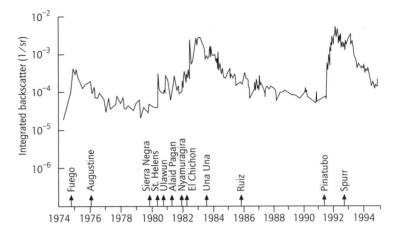

Fig. 9.3 The integrated aerosol backscatter (a measure of the total stratospheric column of aerosol) using a ground-based lidar at $\lambda = 0.694\,\mu m$. Injections of aerosols associated with El Chichón and Mount Pinatubo are seen to be superimposed on a natural background. (From Osborn *et al.* 1995.)

the troposphere. In comparison, typical values over the continents are $<100\,ng\,m^{-3}$. These concentrations may be put into context by considering that $<50\%$ of sulfate in the MBL is of marine origin, while the rest may be attributed to soil dust and anthropogenic sulfate.

The formation of nitrate aerosols from nitrogen precursor gases has two main natural sources: (i) NO_x from lightning and soils; and (ii) N_2O from bacterial activity in soils and the oceans. An emission rate of $22\,Tg\,yr^{-1}$ is estimated in Table 9.2.

Secondary organic aerosols or SOAs are formed by oxidation products of volatile organic compounds (VOCs). These reactions are mainly initiated by reactions with ozone and OH and NO_3 radicals. Some products of these oxidation reactions have a low enough vapor pressure for partitioning between the gas phase and the aerosol phase to become significant. The equilibrium constant K, describing the partitioning of a compound between the gas and the particle phase, is mainly a function of its vapor pressure and of its activity in the particulate phase.

$$K = R_G T / \gamma p M_{OM} \qquad (9.2)$$

where R_G is the gas constant, T is the temperature, M_{OM} is the mean molecular weight of the organic

particulate phase into which the compound partitions, γ is the activity coefficient of the compound in the particulate phase, and p is its vapor pressure. Thus the formation of SOA is described by K for all compounds that are found in the organic aerosol phase. This, however, is a challenging task, since generally only a small fraction of the organic aerosol mass can be analyzed on a molecular level. However, laboratory experiments have shown that for most cases the overall aerosol yield of a compound can be parameterized assuming two hypothetical compounds with different K. This allows the SOA mass in modeling studies to be estimated without knowing the actual molecular composition of the organic particulate phase. The aerosol yield Y of a single compound is defined as:

$$Y = \Delta M_0 / \Delta HC = M_0 \sum_i \frac{a_j K_i}{1 + K_i} \qquad (9.3)$$

where ΔM_0 is the organic mass formed, ΔHC is the fraction of hydrocarbon reacted, and α_i is a stochiometric coefficient. The aerosol yield of an organic compound was shown to depend on its structure (e.g. number and location of double bonds) and on the gaseous reaction partners available (Griffin *et al.* 1999a). In general, compounds with a larger number of carbon atoms have a larger aerosol yield. However, compounds with fewer than six carbon atoms are mostly unable to form SOAs, because their oxidation products are too small and thus too volatile to partition into the particle phase.

Terpenes emitted from plants are considered to be the largest class of natural VOCs able to form SOAs. Guenther *et al.* (1995) estimated the global emission of terpenes and other reactive organic compounds at 127 Tg yr^{-1}, which is equally distributed between the northern and southern hemispheres. From emission data and laboratory measurements of aerosol formation yields (Griffin *et al.* 1999a), the global formation rate of SOA has been estimated at 13–24 Tg yr^{-1} from this source (Griffin *et al.* 1999b).

Information on the molecular composition of biogenic SOA is still largely unavailable. In general, the small amounts of sample available for analysis and the highly oxidized compounds make chemical analysis difficult. Studies concerning the composition of laboratory-generated aerosols at the molecular level have only recently begun to resolve an appreciable amount of the organic particle mass. In ambient aerosol samples only 10–20% of the organic aerosol mass can usually be analyzed on a molecular level. In ambient samples it is difficult to distinguish between primary and secondary organic mass, as no simple experimental methods exist to separate these two fractions.

It is assumed that SOA mass grows mainly on existing aerosol particles (such as salt aerosols or sulfuric acid nuclei). However, some events have been observed in field measurements where nucleation of new particles was attributed to low-volatility organic compounds (Marti *et al.* 1997). At present it is unclear how important this process is.

9.3.2 *Anthropogenic sources*

Primary emissions

Biomass burning. Natural wild fires and anthropogenic fires (e.g for agricultural clearing) are commonly termed biomass burning (Levine 1991). The latter source has grown so quickly in the past two decades that it is estimated to account for 95% of all biomass emissions. Table 9.2 gives an anthropogenic component of 80 Tg yr^{-1}, which consists of soot, sulfate, nitrate, and incomplete combustion products containing carbonaceous compounds. Release of biomass products to the atmosphere occurs mainly in the Tropics during the dry seasons, i.e. December to March in the northern hemisphere and June to September in the southern hemisphere. Although Table 9.2 only considers particle emissions, gaseous emissions of CO_2, CO, CH_4, and VOCs are also important, and the latter may result in SOA formation.

Industrial aerosols. Aerosol emissions from industrial processes, estimated at 100 Tg yr^{-1}, have a diverse number of sources. Major sources in industrialized countries include: coal and mineral dust from mining, aerosols formed from incombustible inorganic compounds in oil and coal fuels, stone-crushing, cement manufacture, metal foundries, and grain elevators. However, decreasing emissions in industrialized nations are occurring against rapidly rising emissions in emerging nations, where the implementation of modern technologies is not keeping pace with rapid economic and industrial development.

Soot aerosols. Soot is a ubiquitous component of the atmospheric aerosol. As a result of it being chemically relatively inert and having poor hygroscopic properties, it may be used as an anthropogenic tracer. For instance, a recent study of soot in an alpine ice-core exhibited enhanced soot concentrations since the beginning of industrialization in the mid-nineteenth century and has been attributed to anthropogenic activity in Europe (Lavanchy *et al.* 1999). Soot aerosols are generated in incomplete combustion processes and consist mainly of an organic fraction and an inorganic graphite-like carbon fraction. These two fractions are often not clearly distinguished in the literature, which is in large part due to technical difficulties in distinguishing them. The graphite-like carbon fraction is also often known as black carbon (BC) or elemental carbon (EC). Depending on the source and burning conditions, the amount of soot is highly variable. For instance, biomass soot has a high organic carbon content, in contrast to diesel soot with a high EC content. Emissions of anthropogenic soot aerosols from fossil fuel combustion are estimated at ~10 Tg yr^{-1}, which include contributions from biomass sources.

Secondary emissions

Sulfate and nitrate aerosols. The main atmospheric source of secondary particles is the oxidation of SO_2 and NO_x. It is estimated that about 50% of SO_2 and NO_x is oxidized before being deposited (Langner & Rodhe 1991), where oxidation may occur in either the gas or condensed phases. Gasphase oxidation to both H_2SO_4 and HNO_3 is dominated by OH. For the condensed phase, about 50% of NO_x and ≥80% of SO_2 is oxidized to HNO_3 and H_2SO_4, respectively, by heterogeneous reactions. Oxidation of SO_2 always results in the formation of aerosol mass, due to the low H_2SO_4 vapor pressure, and is in contrast to HNO_3, which is distributed between the gas and aerosol phases. The chemical transformation of gases into particles depends on many factors, including chemical reaction kinetics and physical factors such as plume mixing and dispersion, oxidant concentration, sunlight, and catalytic aerosol surfaces. Despite this, the conversion rates of SO_2 are generally around 1–2% per hour and somewhat higher for nitrate. Table 9.2 indicates that current sulfate emissions from anthropogenic sources exceed natural sources, while different estimates exist in the case of nitrate.

The largest source of secondary anthropogenic aerosol comes from fossil fuel emissions of SO_2 and subsequent conversion to H_2SO_4. Over continental surfaces in the PBL or above, where gaseous ammonia is present, H_2SO_4 forms NH_4HSO_4 and $(NH_4)_2SO_4$. These components may exist simultaneously and are illustrated by varying molar ratios of NH_4^+/SO_4^{2-} according to the atmospheric aerosol type, as in Table 9.1. In regions of the stratosphere and upper troposphere, H_2SO_4 is found to be the major aerosol component.

Atmospheric emissions from fossil fuel combustion have been increasing since the beginning of industrialization in about 1850. The current industrial SO_2 emission of about 70–90 Tg S yr^{-1} accounts for about 80–85% of the total SO_2 annual flux in the northern and 30% in the southern hemisphere (see Berresheim *et al.* 1995; Möller 1995). Ninety percent of these emissions arise in industrialized regions of the northern hemisphere. Little mixing occurs into the southern hemisphere due to a long inter-hemispheric mixing time, of ~1 year, which compare to aerosol lifetimes of ~1 week.

Current estimates of NO_x emissions from biomass burning and NH_3 oxidation (~17 Tg N yr^{-1}) are lower than the anthropogenic value of ~32 Tg N yr^{-1} from fossil fuel combustion. The formation of HNO_3 from NO_x is a major removal mechanism for tropospheric NO_x as most HNO_3 is subsequently lost through wet and dry deposition.

Sulfate aerosol concentrations are more stable to fluctuations in H_2SO_4 concentration, temperature and humidity conditions, in contrast to ammonium nitrate and chloride aerosols. For these aerosols, the reversible reactions shown below occur to form the parent gaseous components under conditions of low atmospheric ammonia concentration, high temperature and low humidity:

$$NH_4NO_3(s) \leftrightarrow NH_3(g) + HNO_3(g)$$

$$NH_4Cl(s) \leftrightarrow NH_3(g) + HCl(g)$$

where (g) and (s) denote the gaseous and solid phases, respectively. The main source of atmospheric HCl is from refuse incineration and coal combustion.

Ammonia plays an important role in the neutralization of acid species, as it is the most common atmospheric alkaline gas. Conversion to ammonium salts is a function of not only altitude, but also of temperature and humidity. Major natural and anthropogenic sources respectively include: (i) soils and organic decomposition, (ii) fertilizers and animal farming, and (iii) catalyzed vehicle. The annual emission of ammonia is estimated at 269 Tg yr^{-1}.

The aqueous-phase production of aerosol material on cloud droplets is an important mechanism in nonprecipitating clouds. Cloud droplets, which form on CCN, may undergo on average 10 evaporation/condensation cycles before precipitable droplets are formed. During this process, gaseous species are scavenged and undergo chemical transformation, while aerosols and other droplets are scavenged by coagulation and phoresis mechanisms. As a result, the aerosol mass and hygroscopicity increases, in turn increasing the CCN activity. The conversion of SO_2 and NH_3, which are dissolved in droplets, appears to be an efficient process for the production of NH_4HSO_4 and $(NH_4)_2SO_4$. Such processes have been postulated as responsible for the rather uniform composition of the background tropospheric aerosol (e.g. Raes *et al.* 2000).

Secondary organic aerosol. Fossil fuel combustion and biomass burning, caused by human activities, are the main sources of anthropogenic VOCs that can lead to SOAs. VOC emissions due to fossil fuel combustion are estimated at ~65 Tg yr^{-1}, i.e. 60% of total anthropogenic emissions (Piccot *et al.* 1992). The same principles apply to anthropogenic SOA formation as for SOA formed from biogenic precursors compounds.

Aromatic compounds are the main class of anthropogenic VOCs that lead to significant aerosol formation. Pandis *et al.* (1992) estimated that aromatics of anthropogenic emissions can be responsible for about two-thirds of the total SOA formation in an urban atmosphere. Aerosol yields of different single aromatic compounds, when compared to the SOA yield of whole gasoline vapor, show that aromatics are mainly responsible for SOA from fossil fuel VOC emissions (Odum *et al.* 1997). As in the case of biogenic aerosol particles, little is known about the molecular composition of anthropogenic SOA.

9.4   HETEROGENEOUS CHEMISTRY

In a general sense, heterogeneous reactions are reactions involving more than one phase, such as reactions of gaseous molecules with compounds on a solid or liquid surface or in bulk liquids. Some reac-

tions that are unfavorable in the gas phase are able to occur on the surfaces or in the bulk of atmospheric aerosol particles. Many authors (e.g. Ravishankara 1997) further split these heterogeneous reactions into heterogeneous reactions in a narrower sense, i.e. on solid surfaces, and multiphase reactions in liquids. In the following, the term "heterogeneous reactions" is used in the broad sense.

There are two different implications when considering such reactions. First, they modify the aerosol composition; second, they influence gas phase chemistry. Although the importance of heterogeneous reactions in tropospheric and especially in stratospheric chemistry (e.g. ozone depletion in polar regions) has been realized, only few experimental data exist compared to gaseous reactions, which are relatively well characterized. One of the major difficulties in investigating heterogeneous reactions in laboratory experiments is to accurately simulate relevant aerosols, as ambient particles are often complex mixtures of organic and inorganic compounds, and water.

The most basic example of a heterogeneous reaction, the adsorption of a gaseous molecule onto a solid particle surface, can be described with the following equation:

$$A(g) + \{B\} \rightarrow \{AB\}$$

where species in { } denote surface-bound compounds. The rate with which gaseous molecules adsorb to aerosol particles is described by:

$$J = N_{mol} \sum_i N_i(d) B(d) \qquad (9.4)$$

where J is the flux, N_{mol} is the number concentration of the molecule of interest, d is the aerosol diameter, $N_i(d)$ is the particle number concentration with particle diameter d and $B(d)$ is the attachment coefficient. A variety of different equations for $B(d)$ exist in the literature, among which a simplified formula gives $B(d)$ as:

$$B(d) = \frac{2\pi D d}{\dfrac{8D}{cd\gamma} + \dfrac{1}{1 + 2\lambda_{MFP}/d}} \qquad (9.5)$$

where D, c, and λ_{MFP} are the diffusion coefficient, the mean thermal velocity, and the mean free path of the diffusing molecule, respectively. The parameter γ is the dimensionless sticking coefficient, defined as the ratio of reactive gas molecule-particle collisions to total gas molecule-particle collisions, i.e. an indication of adsorption efficiency.

When $\gamma = 1$, eqn 9.5 illustrates that for small particles (i.e. $d \ll \gamma_{MFP}$) $B(d) \propto d^2$, while for large particles (i.e. $d \gg \lambda_{MFP}$) $B(d) \propto d$. However, for $\gamma \leq 0.01$, eqn 9.5 may be simplified to the following for all particle diameters of interest ($d \leq 10\,\mu\text{m}$):

$$B(d) = \frac{1}{4}\pi c d^2 \gamma \qquad (9.6)$$

Equations 9.4 to 9.6 describe an overall adsorption, but do not take the chemical nature of this process into account. Several different categories may be distinguished, which is sometimes a reason for confusion in the literature. A gaseous molecule can be physically (or reversibly) adsorbed, i.e. no covalent bonding is formed between the adsorbent and the adsorbate. In many cases this is the first step in a heterogeneous reaction. The reversibly adsorbed molecule may desorb again from its physisorbed state, hence resulting in equilibrium between the gaseous and the particulate phases. However, the adsorption process can also result in covalent bonding of the gas molecule and a particle surface compound (chemisorption). If the compound remains irreversibly adsorbed on the particle (i.e. the residence time on the particle is longer than the lifetime of the aerosol in the atmosphere), adsorption then represents a permanent sink for the gaseous compound in its atmospheric life cycle.

The adsorbed molecule may also undergo a chemical reaction on/in the aerosol, resulting in either an altered aerosol composition or a release of a reaction product into the gas phase, or both. Depending on the aerosol chemical and physical (i.e. solid or liquid) composition, heterogeneous reactions may be quite diverse, and are governed by many factors.

9.4.1 Dry aerosols

Reactions of gaseous compounds with dry, solid aerosols have remained relatively uncharacterized. It has been known for quite a while that the reaction between sea-salt aerosols and gaseous nitric acid results in the release of HCl into the gas phase:

$$HNO_3(g) + NaCl(s) \rightarrow HCl(g) + NaNO_3(s)$$

Another example is the reduction NO_2 undergoes on the surface of soot aerosol particles, where nitrous acid (HONO) is the main gaseous reaction product.

$$NO_2(g) + \{C_{red}\}(s) \rightarrow HONO(g) + \{C_{ox}\}(s)$$

The products $\{C_{red}\}(s)$ and $\{C_{ox}\}(s)$ denote a reduced and an oxidized compound on the soot particle surface, respectively (Ammann *et al.* 1998). Nitrous acid is an important precursor of the most important oxidant in the lower troposphere, the OH radical, and is easily photolyzed in the atmosphere. The reaction efficiency appears to largely depend on the chemical characteristics of soot, rendering the applicability to ambient atmospheric conditions difficult. However, on other chemically inert surfaces, NO_2 appears to react with absorbed water to nitrous and nitric acids.

$$2NO_2(g) + H_2O(ads) \rightarrow HONO(g) + HNO_3(ads)$$

where HONO is released into the gas phase and HNO_3 remains predominantly on the surface. Several studies on the interaction of gaseous HNO_3 with crustal aerosols exist (e.g. Phadnis & Carmichael 2000), while other reactions such as the interaction of OH radicals are mostly speculative (Saylor 1997).

9.4.2 Liquid aerosols

When considering liquid aerosols (i.e. cloud, fog, or haze droplet), eqns 9.4 to 9.6 are also applicable, where γ is then referred to as the uptake coefficient. After accommodation, the molecule dissolves into the bulk solution of a droplet. Chemical reactions occur mainly in the bulk, although it has been found that reactions may occur on the surface of a liquid aerosol in some systems. One can envisage the overall interaction of a gas molecule with a liquid aerosol droplet using an electrical circuit analogy (Davidovits *et al.* 1995). In this model the overall measured uptake of a gas molecule into a liquid aerosol (γ_{meas}) can be described as the sum of dimensionless coefficients:

$$\frac{1}{\gamma_{meas}} = \frac{1}{\gamma_{diff}} + \frac{1}{\alpha} + \frac{1}{\gamma_m + \gamma_{sol}} \qquad (9.7)$$

The first two terms represent processes in the gas phase (diffusion to the droplet, γ_{diff}) and at the gas/liquid interface (accommodation on the droplet surface, α), respectively. The third term describes processes in the droplet, i.e. the chemical reaction (γ_{rn}) and the solubility of the gas molecule in the liquid (γ_{sol}). Chemical reactions in the liquid depend heavily on the chemical composition of the liquid aerosol, i.e. parameters such as the ionic strength and redox potential play an important role. Thus different reactions occur in cloud or haze droplets, as a result of differing water content.

An important example of heterogeneous reactions in cloud droplets is the aqueous oxidation of SO_2. The main oxidation paths are reactions with aqueous H_2O_2 and ozone:

$SO_2(g) \rightarrow SO_2(aq)$ accommodation, solution

$SO_2(aq) + H_2O_2(aq) \rightarrow H_2SO_4$ in liquid

$SO_2(aq) + O_3(aq) + H_2O \rightarrow H_2SO_4 + O_2$ in liquid

Another important example of a heterogeneous reaction scheme in liquid aerosols is the nighttime formation of nitric acid:

$NO_2 + O_3 \rightarrow O_2 + NO_3$ in gas phase

$NO_3 + NO_2 + M \rightarrow N_2O_5 + M$ in gas phase

$N_2O_5(aq) + H_2O \rightarrow 2\,HNO_3$ in liquid

During the day, these reactions are of minor importance due to the rapid photolysis of NO_3 radicals. Model calculations have shown that with these reactions, the yearly average global NO_x burden decreases by 50%, due to a decreased residence time in the atmosphere (Dentener & Crutzen 1993). Observed nitrate wet deposition patterns in North America and Europe are better simulated if these aerosol reactions are included.

In conclusion, heterogeneous reactions are of potential importance in atmospheric chemistry, but experimental studies remain sparse.

9.5   CLIMATE FORCING

Aerosols are now recognized as being a major uncertainty in studies of global climate change (IPCC 2001). Emissions of anthropogenic aerosols to the atmosphere may explain the lower observed temperature increase than is otherwise predicted for greenhouse gas emissions. Aerosols are considered to be responsible for a negative forcing or cooling of the Earth-atmosphere system, in contrast to a positive forcing, i.e. "warming," from greenhouse gases. In this definition, a forcing refers to a natural or anthropogenic perturbation in the radiative energy budget of the Earth's climate system.

Aerosols may influence the atmosphere in two important ways, by direct and indirect effects (Charlson *et al.* 1987, 1992; Charlson & Heintzenberg 1995; Andreae & Crutzen 1997; Baker 1997). Direct effects refer to the scattering and absorption of shortwave radiation and the subsequent influence on the climate system and planetary albedo (see Schwartz 1996). Indirect effects refer to a complex positive feedback system, whereby an increase in CCN arises from an increase in anthropogenic aerosol concentration. As a consequence, cloud droplets become smaller for a given cloud liquid water content, thereby increasing cloud albedo and resulting in a negative forcing. The principal aerosol–cloud interactions are schematically summarized in Fig. 9.5, which illustrates the life cycle in marine and continental air respectively. Many of the physical and chemi-

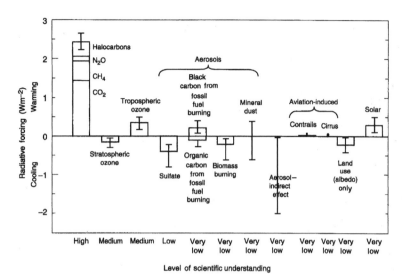

Fig. 9.4 Estimates of globally and annually averaged anthropogenic radiative forcing from pre-industrial times to the present. Confidence levels (from high to very low) in each estimate are reflected in the uncertainty range of the error bars. Due to the episodic and sporadic nature of volcanoes, they are omitted from the above estimates, although their effects on time scales of the order of decades may be significant. (From IPCC 2001.)

cal aspects of the life cycle have already been considered in Sections 9.1 to 9.4.

The climate effects of aerosols are still poorly quantified and it is at present still unclear what magnitudes are involved. Figure 9.4 illustrates recent estimates of the mean annual radiative forcing for various climate change mechanisms, averaged globally for the period 1750–2000 (IPCC 2001). Current estimates suggest that positive greenhouse forcing is dominated by anthropogenic emissions of CO_2, CH_4, N_2O, and halocarbons. Additional mechanisms, such as the depletion of stratospheric ozone, the formation of tropospheric ozone by photochemical smog, and variation in the solar irradiance, are believed to contribute smaller forcings. When viewed on a global scale, shortwave radiative forcing from anthropogenic aerosols is considered to offset part of the long-wave radiative forcing due to greenhouse gases. However, on a more regional scale, the large uncertainty in aerosol radiative forcing compared to that for greenhouse gases does not allow a meaningful net value from all mechanisms to be defined.

As already mentioned, aerosols may be responsible for the lower than predicted temperature in-

crease in the northern hemisphere. While this may apply on a global basis, the same argument cannot be applied on a regional basis for a number of reasons. First, greenhouse gases have lifetimes measured in decades to centuries and are globally well mixed, in contrast to aerosols, which have lifetimes of less than one week on average and are geographically variable in extent. Second, aerosol forcing responds more rapidly to changes in aerosol emissions than greenhouse forcing, which is still influenced by accumulated past emissions. Third, aerosol radiative forcing is more restricted to source and downwind regions, which may then experience an overall negative forcing. Lastly, greenhouse gas forcing is not as diurnally and seasonally variable as aerosol forcing, which has a greater influence during: (i) daylight, (ii) cloudless conditions, and (iii) summer. Such difficulties greatly hinder the modeling of aerosol direct and indirect effects.

It is important to note that in the above definition of climate forcing, complex feedback systems, such as the hydrological cycle, are not considered, as they are in the overall discussion of climate change. Although this simplifies the considera-

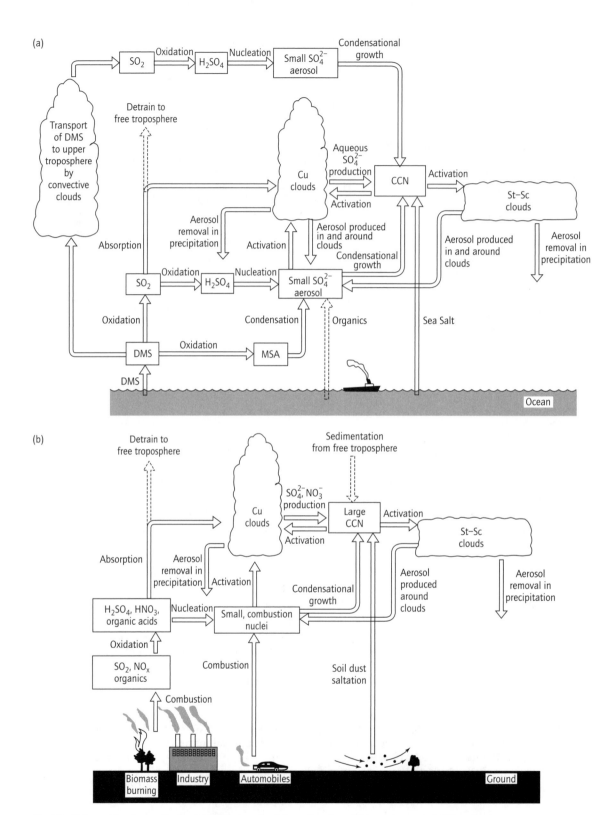

Fig. 9.5 Schematic of aerosol–cloud interactions for (a) marine air and (b) continental air. Cloud types: Cu, cumulonimbus; St, stratus; Sc, stratocumulus. (From Hobbs 1993.)

tion of aerosols, it does not fully consider their overall impact. In order to assess the direct aerosol effect on climate change, the following aspects of tropospheric aerosols are further considered: (i) the radiative properties of aerosols; and (ii) estimates of aerosol direct radiative forcing.

9.5.1 Direct aerosol effects: radiative properties of aerosols

The interaction of aerosols with radiation depends primarily on the aerosol diameter and the wavelength γ of radiation. The largest interaction occurs when the aerosol size parameter $\pi d/\lambda \sim 1$. As a consequence, the longer-lived accumulation mode interacts with the shortwave solar radiation spectrum to a greater extent than with the longwave infrared spectrum emitted by the Earth's atmosphere and surface, which peaks at $\lambda \sim 7\,\mu m$.

Optical properties of aerosols may be described using a number of parameters. Scattered or absorbed radiation from an incident beam is defined by the total scattering (σ_{SP}) and absorption coefficients (σ_{AP}), which are a measure of the fractional change in beam intensity per meter. Radiation scattered into the backward hemisphere is correspondingly described by the hemispheric backscattering coefficient (σ_{BSP}). The sum of σ_{SP} and σ_{AP} is the extinction coefficient σ_{EXT} and when integrated over the beam path length dl gives the aerosol optical depth (AOD: Ω).

$$\sigma_{SP} + \sigma_{AP} = \sigma_{EXT} \qquad (9.8)$$

$$\Omega = \int \sigma_{EXT} dl \qquad (9.9)$$

An important parameter in global aerosol models is the ratio of scattering to extinction, otherwise known as the single scattering albedo ω_o:

$$\omega_o = \sigma_{SP}/\sigma_{EXT} \qquad (9.10)$$

The extinction coefficient and its components are often approximated as being proportional to $\lambda^{-\mathring{a}}$ where $\mathring{a}$ is the Ångström exponent and K is a con-

stant (eqn. 9.11). Furthermore, $\mathring{a}$ is related to the Junge parameter v by eqn 9.12.

$$\sigma_{EXT} = K\lambda^{-\mathring{a}} \qquad (9.11)$$

$$\mathring{a} = v - 2 \qquad (9.12)$$

Typical values of $\mathring{a}$ range from ~4 for gases, ~2 for urban aerosols, and ~1–2 for rural haze, to ~0 for coarse aerosols. Hence if $\mathring{a}$ can be measured then information on the number size distribution is gained and vice versa.

The asymmetry factor g is defined as the cosine weighted mean of the angular scattering phase function $\beta(\phi)$, where $\beta(\phi)$ describes the amount of light scattered through an angle ϕ:

$$g = \frac{\int \beta(\phi)\cos\phi\, d(\cos\phi)}{\int \beta(\phi) d(\cos\phi)} \qquad (9.13)$$

The value of g ranges from -1 for complete back scattering to $+1$ for complete forward scattering. The asymmetry factor is important in radiative models of the atmosphere and takes the angular scattering of radiation into account. As direct measurement is not possible, it may be parameterized by the hemispheric backscattering ratio σ_{BSP}/σ_{SP}.

The radiative effect of aerosols on global climate may be predicted using two computer model types: (i) a general circulation model (GCM) providing information on atmospheric state parameters (wind and temperature fields, etc.) and radiative properties; and (ii) a chemical transport model (CTM) to provide information on aerosol emission, transport, transformation, and removal. Early models used prescribed databases of measured aerosol parameters (e.g. d'Almeida *et al.* 1991; Koepke *et al.* 1997), while recent progress has focused on developing algorithms to dynamically model many complex aspects of atmospheric aerosols, such as the interaction with radiation or clouds. Global distributions of sulfate aerosol have been widely modeled. A good review and comparison of currently used sulfate GCM/CTM models is given by Barrie *et al.* (2001) and IPCC (2001). More

Table 9.3 Specific scattering (α_{SP}) and absorption (α_{AP}) coefficients for the indicated aerosol species. Also indicated is $f(RH = 80\%)$, the fractional increase in σ_{SP} at RH = 80%. All measurements are for $\lambda = 0.525\,\mu m$. Figures in parentheses represent approximate standard deviations (adapted from IPCC 1995).

Optical property	Sulfate	Organic carbon	Mineral dust	Soot	Fine aerosol
$\alpha_{SP}\,(m^2\,g^{-1})$	5 (3.6–7)	5 (3.0–7)		3	3 (2–4)
$\alpha_{AP}\,(m^2\,g^{-1})$	0	0	$\alpha_{EXT} \sim 0.7$	10 (8–12)	(0–10)
$f(RH = 80\%)$	1.7 (1.4–4)	1.7 (1.4–4)	<0.05	(0–1.7)	1.7 (0–4)

recent model studies have been widened to include global distributions of soot (e.g. Cooke *et al.* 1999), mineral dust (e.g. Miller & Tegen 1998), and sea salt (e.g. Gong *et al.* 1997).

GCM/CTM models approach global direct radiative forcing by considering column-integrated and local aerosol optical properties. Both approaches are necessary to determine the vertical structure of the atmosphere, and to validate ground-based against aircraft and satellite observations. As a number of assumptions are inherent in such calculations, measurements of column-integrated and local properties are ideally conducted simultaneously (Quinn *et al.* 1996). Much progress has been made in this respect through recent column-closure experiments in internationally coordinated field campaigns such as TARFOX and ACE II (Russell *et al.* 1999; Russell & Heintzenberg 2000).

Column-integrated properties refer to aerosol properties averaged throughout the atmospheric column. Such measurements have been commonly conducted using Sun photometers by subtracting the attenuation from atmospheric gases and water vapor to give the AOD. Despite numerous *in situ* Sun photometer measurements, satellites are gaining in importance, due mainly to their large spatial coverage of the Earth's surface. These aspects are discussed in Section 9.6.

Local measurements involve directly measuring the above-defined aerosol parameters. Aerosol optical properties may be derived from *in situ* measurement or by calculating their properties through knowledge of the aerosol chemical composition, size distribution, and refractive index. As

many CTMs calculate global aerosol mass concentration distributions, the ability to simultaneously derive distributions of scattering and absorption by aerosols is advantageous. Parameters that relate aerosol scattering and absorption coefficients to measurements of the aerosol concentration and composition are the specific scattering α_{SP} and specific absorption α_{AP} efficiencies.

$$\alpha_{SP} = \sigma_{SP}/M \qquad (9.14)$$

$$\alpha_{AP} = \sigma_{AP}/M \qquad (9.15)$$

These parameters are given for either a particular chemical species or a mass concentration M for a particular aerosol mode. Typical values are given in Table 9.3 and illustrate a mean value of $\alpha_{SP} \sim 5\,m^2\,g^{-1}$ for sulfate and $\alpha_{AP} \sim 10\,m^2\,g^{-1}$ for soot. The latter high value emphasizes the important role of soot aerosols in atmospheric extinction despite the relatively small atmospheric burden. Specific efficiencies are dependent on a number of parameters: the size distribution, the wavelength of incident light, and the ambient relative humidity (RH). The latter is of importance due to the uptake of water, resulting in an increase in d, and hence in σ_{SP}. For a pure component aerosol (e.g. inorganic salts), growth into an aqueous droplet will abruptly occur at the deliquescence point. The phase transition is characteristic of the chemical species (e.g. $(NH_4)_2SO_4$, NaCl, and NH_4HSO_4 deliquesce at RH = 80, 75, and 39%, respectively), but becomes poorly defined for mixed aerosols. Upon reduction of RH, recrystallization (or efflorescence) of pure salts occurs at a lower RH than the deliquescence

Table 9.4 Representative values of observed aerosol optical properties in the lower troposphere for $\lambda = 0.5$–$0.55\,\mu m$ and RH < 60% (adapted from IPCC 1995).

Parameter	Polluted continental	Clean continental	Clean marine
Optical depth (Ω)	0.2–0.8	0.02–0.1	0.05–0.1
Single scattering albedo (ω_o)	0.8–0.95	0.9–0.95	~1
Back/total scattering ratio (R)	0.1–0.2	0.13–0.21	0.15
Total scattering coefficient (σ_{SP}; m^{-1})	50–300×10^{-6}	5–30×10^{-6}	5–20×10^{-6}
Absorption coefficient (σ_{AP}; m^{-1})	5–50×10^{-6}	1–10×10^{-6}	$<0.05 \times 10^{-6}$
Fine mass concentration ($\mu g\,m^{-3}$)	5–50	1–10	1–5
CN number concentration (cm^{-3})	10^3–10^5	10^2–10^3	$<10^2$
CCN number concentration (cm^{-3}; 0.7–1% supersaturation)	1000–5000	100–1000	10–200
Ångström exponent ($\mathring{a}$)	1–2	1–2	1.5–2.1

point, due to the droplet remaining in a supersaturated state. For atmospheric aerosols, no effect is generally observed under RH = 40%. Table 9.3 also demonstrates the fractional increase in σ_{SP} at RH = 80% with respect to a "dry" value at RH < 30%, referred to as f(RH = 80%). Values indicate a factor 1.7 increase in most cases, although the standard deviations in parentheses demonstrate the dependence on chemical composition. Soot does not always exhibit an increase below RH = 80% and a value is therefore only presented as a standard deviation. Apart from the large increase in σ_{EXT}, enhanced absorption in the polluted marine and urban models results from increased absorption in an internal mixture containing soot. These results demonstrate the significance of water uptake on optical properties, especially as such measurements are still sparse for ground locations and almost nonexistent for the free troposphere. When the RH increases beyond RH = 100%, aerosol–cloud interactions become important, and these are considered in Section 9.5.3. Table 9.4 summarizes the range of optical properties representative of polluted continental, clean continental, and clean marine aerosol types. Polluted continental aerosol concentrations are nearly a factor 10 larger than for remote regions, while a lower value of ω_o indicates a higher proportion of absorbing species. Large efforts are presently under way to obtain detailed aerosol databases for various representative locations around the globe. The Global

Atmosphere Watch (GAW) program of the WMO has, for instance, been established to collect data over decadal time scales to monitor long-term changes in atmospheric composition.

9.5.2 *Direct aerosol effects: estimates of direct forcing*

The application of GCMs/CTMs in assessing the influence of atmospheric aerosols is made difficult by the large spatial and temporal variation in their properties. This is illustrated by the fact that present assessments of the role of aerosols have largely come from models as opposed to observations. Most models predict a regional offset of greenhouse forcing by sulfate aerosols in the industrialized regions of the eastern USA, central Europe and eastern China. However, as mentioned above, the regional forcing is not expected to be indicative of regional climate response, as atmospheric circulation may result in a nonlocal response to local forcing.

Direct radiative forcing is estimated to be $-0.4\,W\,m^{-2}$ for sulfate, $-0.2\,W\,m^{-2}$ for biomass burning aerosols, $-0.1\,W\,m^{-2}$ for fossil fuel organic carbon, and $+0.2\,W\,m^{-2}$ for fossil fuel black carbon aerosols (IPCC 2001). Uncertainties are given as a factor of 2 for sulfate and fossil fuel black carbon aerosols, and as a factor of 3 for biomass burning aerosols and fossil fuel organic carbon. For mineral dust aerosols, a range of -0.6 to $+0.4\,W\,m^{-2}$ is indicated. The northern hemisphere to southern hemisphere is forcing ratio ≫1 except for biomass burning aerosols (<1), while no estimate is given for mineral dust aerosols (IPCC 2001). In general, the variation in the above estimates is mainly due to the differing sophistication of GCMs/CTMs, while the uncertainty depends on assumed optical properties and the global modeled distribution of aerosol species. Sulfate and biomass aerosols have been most widely modeled to date, but additional aerosol types, such as volcanic emission, mineral dust, and carbonaceous aerosols, are being increasingly included in GCM/CTM simulations.

Volcanic emissions to the stratosphere influence the climate by warming the lower stratosphere through aerosol absorption and by reducing the net radiation transmitted to the troposphere and surface. However, as stratospheric aerosol residence times are of the order of 1 year, the radiative influence is also restricted to similar time scales. The El Chichón (1982) and Mount Pinatubo (1991) eruptions allowed the climate effect of a large transient forcing to be studied. GCM temperature predictions over short time scales were found to be in reasonable agreement with observations. Mount Pinatubo is estimated to have contributed a maximum forcing of $-4\,W\,m^{-2}$ and about $-1\,W\,m^{-2}$ up to 2 years later (Hansen *et al.* 1992; McCormick *et al.* 1995), which illustrates the large transient cooling effect when compared to other radiative forcing mechanisms.

The organic aerosol fraction has only recently been considered. Field studies have observed similar values of α_{SP} for organics and sulfate (e.g. Hegg *et al.* 1997). In contrast, the EC fraction of soot aerosols absorbs light and may contribute to a positive radiative forcing (Penner 1995; Schult *et al.* 1997) when ω_o is less than ~0.85. Current estimates for various aerosol types (fossil fuel EC, biomass EC) and various mixing states (internal, external, core treatment) place the globally averaged forcing in the range +0.2 to $+0.5\,W\,m^{-2}$ (see Jacobson 2000, 2001, and references therein). Mineral dust is also receiving more attention (Sokolik & Toon 1996; Tegen *et al.* 1996; Miller & Tegen 1998; IPCC 2001). The influence of sea salt on aerosol radiative properties in the MBL has until recently been considered negligible. However, new findings indicate that the contribution of sea salt to a negative radiative aerosol forcing may be larger than previously considered (Murphy *et al.* 1998; Haywood *et al.* 1999).

9.5.3 *Indirect aerosol effects: aerosol effects on clouds*

The formation of nss sulfate via the emission of DMS has been proposed as a cloud–climate feedback mechanism. The present greenhouse warming of the Earth's surface/atmosphere is considered to warm the ocean surface waters, which in turn lead to an increase in phytoplankton activity and hence DMS emissions. As a result of increased cloud condensation nuclei (CCN) concentration due to aerosol formation from DMS, the albedo of marine stratiform clouds may increase (Twomey 1977), thereby offsetting a global temperature rise (Charlson *et al.* 1987). While the mechanism and different pathways are complex, the large areal extent of stratiform clouds, covering 25% of the oceans, renders such a feedback mechanism of potential importance. Cloud lifetimes and precipitation frequencies are also thought to be affected. However, the aerosol–climate–DMS feedback theory has not been conclusively proven to date. In order to assess the indirect effects of aerosols cloud properties are considered further.

The presence of CCN in the atmosphere allows cloud droplet formation to occur at supersaturations below 1–2% (i.e. RH = 101–2%). Without their presence, supersaturations of several hundred percent would be required. The physical mechanisms of these processes have long been investigated (Pruppacher & Klett 1997), but atmospheric observations and secular studies remain sparse due to the overall complexity of the aerosol–cloud effect on climate. Not only are aerosol–cloud interactions difficult to model, but cloud parameterizations themselves are still an outstanding issue.

The critical supersaturation at which CCN become activated depends on aerosol composition, size, and age, and may be described by the Köhler theory (e.g. Pruppacher & Klett, 1997). Growth of aerosols to RH = 100% was described in Section 9.5.1. As RH increases beyond 100% the aerosol droplets continue to grow until a critical supersaturation S_C is reached, corresponding to a critical diameter d_{CRIT} at which the aerosol becomes "activated." At this point droplets are in an unstable equilibrium with their environment and may grow uncontrollably or return to a stable equilibrium, where they will exist as unactivated droplets or haze.

The value of S_C depends on the solubility of chemical components and dry diameter of the aerosol. The larger both parameters are, the lower the critical supersaturation required to activate the aerosol. As a result the larger, hygroscopic aerosols tend to form CCN first. Typical supersaturations in marine stratus clouds, estimated at 0.1%, imply that dry aerosol diameters with $d >$ 0.1 μm will be activated to produce cloud droplets, whereas for $d < 0.1$ μm the aerosol will remain interstitial to cloud droplets.

Effective CCN sources are, in general, secondary aerosols of either natural or anthropogenic origin. Non-sea-salt sulfate is considered to be the major natural source, while biomass/organic and sulfate aerosols are major anthropogenic sources. The ability of pure inorganic salts and acids such as nitrates, sulfates, or sulfuric acid to act as CCN is relatively well investigated in experimental and modeling work. The role of organic aerosols or the mixture of organic and inorganic particles, however, is less characterized and largely dependent on the hygroscopic behavior of the organic compounds. Studies have shown that smoke particles from biomass burning exhibit a CCN activation of up to 100%, in contrast to an activation of less than several percent for fresh soot from petroleum fuel. Other studies suggest that organics mixed with inorganic salt aerosols alter their hygroscopic behavior. Organics were found to reduce the critical supersaturation needed to activate inorganic aerosols (Shulman *et al.* 1996). Others found that even large amounts of hydrophobic organics did not affect the hygroscopic growth of inorganic particles below 100% humidity. Soot aerosols from biomass burning containing large parts of organics have been shown to act as CCN. Field measurements, e.g. in the Amazon region, found significant influence of biomass burning events and CCN concentrations (Kaufman *et al.* 1998). The role of organics seems to be an important but as yet unresolved aspect in the activation of ambient aerosols to cloud droplets.

Cloud physical properties over continental regions differ from those over oceans. Over land, larger CCN concentrations result in increased cloud droplet concentrations (N_{CLOUD}), and since the LWC of both cloud types are similar, continental clouds have a smaller average droplet size than marine clouds. Observations indicate that marine cumulus clouds have a median value $N_{CLOUD} \sim 45$ cm^{-3} and a broad droplet size spectrum with median at $d \sim 30$ μm, while continental cumuli have a median value $N_{CLOUD} \sim 230$ cm^{-3} and a narrower size spectrum with median ~10 μm. Parameterizations of clouds use the cloud optical thickness δ, which is defined as:

$$\delta \propto L/r_e \qquad (9.16)$$

where L is the cloud LWC and r_e is the effective radius of cloud droplets. L may be approximated from eqn 9.16 and, when solving for r_e and differentiating, eqn 9.18 results.

$$L \propto \frac{4}{3} \pi r_e^3 N_{CLOUD} \qquad (9.17)$$

$$\frac{\Delta \delta}{\delta} = \frac{1}{3} \frac{\Delta N}{N} \qquad (9.18)$$

Hence for a constant LWC, an increase in N as a result of anthropogenic sources will result in a linear increase in δ. This simple approach is, however, complicated by many other factors apart from the inherent assumptions in the above derivation. A further cloud parameter of importance is the albedo A_o, which may be approximated by (Hobbs 1993):

$$A_o = \frac{(1-g)\delta}{1+(1-g)\delta} \qquad (9.19)$$

For scattering of solar radiation by clouds, $g \sim 0.85$ such that:

$$A_o \approx \frac{\delta}{\delta + 6.7} \qquad (9.20)$$

Hence a greater change in albedo occurs when δ is low. Using the above equations, and assuming that the cloud LWC and depth remain constant then a parameter known as the susceptibility may be derived:

$$\frac{\Delta A_o}{\Delta N} = \frac{A_o(1-A_o)}{3N} \qquad (9.21)$$

The susceptibility $\Delta A_o/\Delta N$ is most sensitive to change when N is small and when A_o lies between ~ 0.25 and 0.75, conditions that are both typical of marine clouds. Hence, small increases in anthropogenic CCN concentrations are likely to have a greater influence in marine than continental regions, i.e. in the southern hemisphere. As clouds are already optically thick to longwave radiation, only shortwave radiation is influenced by cloud properties. As a consequence, enhancement of the shortwave albedo is considered to result in increased reflection of solar radiation back to space and a cooling of the Earth's surface.

The susceptibility of marine stratiform clouds to increased CCN concentrations has been observed in satellite images where ship stack exhausts have resulted in an increase in cloud albedo. The increase in droplet concentration and reduction in size have been confirmed simultaneously by *in situ* and remote sensing measurements. Observational evidence suggests that for a factor 10 increase in the aerosol size range $d \sim 0.1$–$0.3\,\mu m$, a 2–5-fold increase in droplet concentration results. An additional important effect, as a result of the above processes, is a reduction in precipitation efficiency and increased cloud lifetime (Albrecht 1989). Such observations are at present difficult to quantify and the complex

processes involved are as yet still very poorly understood.

9.5.4   Indirect aerosol effects: estimates of indirect forcing

As a consequence of the large uncertainties in the interaction of aerosols with clouds, an estimate of the sign and magnitude of indirect forcing is fraught with assumptions. Present estimates of aerosol indirect forcing in Fig. 9.4 give a range of uncertainty from 0 to $-2.0\,W\,m^{-2}$, which indicate a similar sign and magnitude to direct forcing (IPCC 2001). This report summarizes the rapid advances that have been made in understanding and assessing the indirect aerosol effect since the last report was published in 1996. Among the numerous new uncertainties and findings that have been highlighted in recent studies, several investigations have provided evidence that the incorporation of soot aerosols from biomass burning may be influencing cloud albedo. A satellite study of cumulus and stratocumulus clouds over the Amazon basin during the burning season showed that the average droplet diameter decreased from 14 to $9\,\mu m$, with a corresponding increase in cloud reflectance from 0.35 to 0.45 when smoke aerosols were present (Kaufman & Fraser 1997).

9.6   TROPOSPHERIC AND STRATOSPHERIC AEROSOLS: REMOTE SENSING

In order to reduce the uncertainty in our understanding of climate change, satellite measurements of atmospheric gases and aerosols are rapidly gaining in importance (Browell *et al.* 1998; King *et al.* 1999). Due to the spatial and temporal variability of aerosol sources and sinks, obtaining global distributions of tropospheric aerosols from point measurements is an arduous task. Even aircraft studies are limited by a necessarily large and complex infrastructure, which is especially true of stratospheric measurements. The main advantage of satellites therefore lies in their ability to measure global distributions of atmospheric constituents. In addition, subtle changes in atmospheric composition may be monitored using accurate calibration standards, and is necessary if anthropogenic-induced climate change is to be detected. As a multitude of satellite sensors with varying capabilities exist and are planned for the future, the discussion below can give only a brief overview of some highlights. Further details in good review articles can be found in both of the above-mentioned references.

Although satellites with instrument packages dedicated to aerosol measurements have only been launched in recent years, the retrieval of AOD has been possible since the late 1970s from two nadir-viewing satellites. The advanced very high resolution radiometer (AVHRR) is a five-band sensor, aboard the National Oceanic and Atmospheric Administration's series of satellites, and is dedicated to weather and ocean observations. The total ozone mapping spectrometer (TOMS) is an ultraviolet scanning monochromator, aboard various satellites, dedicated to measuring the total ozone column depth. Retrieval of the AOD over clear-sky regions of the ocean has proven to be accurate (e.g. Husar *et al.* 1997), due to the low and well defined reflectance of the ocean surface at solar wavelengths. In comparison, retrieval over land surfaces has remained difficult, and this stems from several problems: the complex variation in surface reflectance with viewing angle, wavelength, season of year (summer/winter biological cover), and surface features (agricultural land, mountainous regions, deserts, etc.). Nevertheless, AOD has been retrieved over land surfaces (e.g. AVHRR and TOMS data), but requires a number of assumptions, more specifically on the aerosol size distribution, surface reflectance, and ω_0. Progress has been made in several areas by using novel retrieval algorithms (Kaufman *et al.* 1997; Herman *et al.* 1997) and dedicated aerosol sensors (see King *et al.* 1999), although a comprehensive network of ground-based Sun photometers and sky radiometers are still presently required to provide ground-truth verification. Several of the above-mentioned aerosol parameters can now be retrieved with less ambiguity using several techniques aboard recent satellite sensors: (i) at multiple wavelengths and angles (moderate resolution imaging spectrometer (MODIS) and multiangle imaging spectroradiometer (MISR) sensors aboard the Terra satellite); and (ii) using polarized radiance (the POLDER (polarization and directionality of the Earth's reflectances) sensor aboard the ADEOS I and II satellites).

An example of POLDER data is shown in Plate 9.1a, b (facing p. 180), which exhibits the global radiative forcing due to aerosols in winter and summer (Boucher & Tanré 2000). A negative forcing is clearly seen over regions of enhanced aerosol concentrations, such as over the US east coast (industrial aerosols), the Mediterranean, and West Africa (Saharan dust plume). A global mean clear-sky shortwave perturbation of -5 to $-6\,\mathrm{W\,m^{-2}}$ was estimated from these measurements.

Despite these advances, aerosol properties as a function of altitude are also required for GCM/CTM models. Aerosol forcing depends not only on the vertical distribution of aerosols but also on their scattering and absorbing properties. Several remote sensing systems have allowed vertically resolved aerosol information to be retrieved, such as the Stratospheric Aerosol and Gas Experiment (SAGE) satellite and the Lidar in Space Technology Experiment (LITE) aboard the space shuttle. The SAGE II was launched in 1984 and has allowed the observation of aerosols and gases in the stratosphere and upper troposphere (i.e. above 6 km). The sensor uses the solar occultation technique to obtain 15 sunrise and sunset observations per day (limb-viewing) with 1 km vertical and 200 km horizontal resolutions. As a result of interference from clouds, the limb-viewing technique is generally restricted to observations above 6 km. SAGE II has a high sensitivity and is therefore able to measure aerosols in the stratosphere and upper troposphere, where concentrations are much lower than in the lower troposphere and PBL. For instance, typical AOD values ($\lambda = 1.0\,\mu\mathrm{m}$) are 0.02–0.5 in the troposphere, and may be compared with a 2×10^{-4} to 2×10^{-3} range for the unenhanced upper troposphere.

Two global maps of the aerosol extinction ($\lambda = 1.02\,\mu\mathrm{m}$) at altitudes of 6.5 and 12.5 km are shown in Plate 9.2a, b (facing p. 180), respectively, and have been averaged over several years for the

September to November period (Kent *et al.* 1998a). The main features in Plate 9.2a are the high extinction levels at high northern latitudes, attributed mainly to anthropogenic emissions, and distinct bands of rather uniform zonal distribution. Enhanced extinction over Brazil was attributed to biomass burning, which peaks at this time of year. Plate 9.2b exhibits similar features, but the elevated extinction at high latitudes is due to the observation of stratospheric aerosols as a consequence of the tropopause lying below 12.5 km. An aerosol climatology of SAGE observations has indicated a higher aerosol concentration in the northern than in the southern hemisphere and may be attributed to the larger number of land and anthropogenic sources (Kent *et al.* 1998b). An enhancement in springtime concentrations above 20° either side of the Equator and in each hemisphere has also been observed.

SAGE II has gathered a comprehensive database of globally averaged aerosol maps, which has helped the understanding of atmospheric dynamics. In addition, it has been possible to conduct a robust comparison of *in situ* and remote measurements due to the long lifetime of stratospheric aerosols. The aerosol direct effect in the upper atmosphere is minimal during volcanic quiescent years; the utility of SAGE II in providing information to assess the indirect aerosol effect is perhaps of greater significance. Properties of cirrus clouds and their extent, whether modified by natural or anthropogenic influences (see IPCC 1999) are a further major uncertainty in climate change studies.

LITE was a nadir-pointing backscatter lidar flown aboard the *Discovery* space shuttle mission in September 1994. The objectives of LITE were to assess lidar space technology for use in future satellites, and to demonstrate how aerosol distributions and optical properties could be monitored in the troposphere and stratosphere (e.g. Kent *et al.* 1998a; Osborn *et al.* 1998). One of the main advantages demonstrated by LITE was the ability to measure vertical profiles down to the Earth's surface through openings in the cloud-cover, hence allowing a larger database to be gathered than would be possible with a limb-viewing system. An example of stratospheric aerosol distributions is shown in Plate 9.3 (facing p. 180), where the aerosol scattering ratio ($\lambda = 0.532\,\mu m$) on September 17, 1994 is shown (Osborn *et al.* 1998). The white line indicates the tropopause height, which forms the lower limit to the stratospheric aerosol layer. Stratospheric aerosols, mainly composed of sulfuric acid, appear to form a stable and globally distributed layer above the tropopause. This so-called Junge layer is long-lived due to the absence of wet-removal processes and the strong temperature stratification of the stratosphere.

Although LITE was only flown for a 10-day period, the potential to retrieve temporally and spatially resolved aerosol parameters in three dimensions proved to be very promising. Looking into the future, the satellite remote sensing capability will be strengthened and widened with the launch of several new additional sensors dedicated to aerosol measurements. Major advances in the understanding of climate change are expected by utilizing better calibration techniques and advanced computer analysis of multiple, cross-referenced satellite databases. Further progress is also foreseen in the refinement of column-closure experiments (ground-based and *in situ* aircraft measurements), which is essential if satellite measurements are to be verified.

APPENDIX: NOMENCLATURE

$\mathring{a}$	Ångström exponent
A_o	cloud albedo
α	accommodation coefficient
α_i	stochiometric coefficient
α_{AP}	aerosol specific absorption efficiency
α_{SP}	aerosol specific scattering efficiency
$B(d)$	attachment coefficient
$\beta(\phi)$	angular scattering phase function
c	the mean thermal velocity of a diffusing molecule
d	aerosol geometric diameter

d_{CRIT}	aerosol critical diameter at which activation occurs
D	diffusion coefficient of a diffusing molecule
δ	cloud optical thickness
$f(\mathrm{RH} = 80\%)$	fractional increase in σ_{SP} at RH = 80% with respect to RH = 30%
g	asymmetry factor of a scattering aerosol
γ	dimensionless sticking coefficient
γ_{meas}	measured uptake coefficient
γ_{diff}	gas transport coefficient
γ_{rn}	reaction coefficient
γ_{sol}	solubility coefficient
HC	mass of hydrocarbon
η	viscosity of air
J	flux of gaseous molecules sticking to aerosol particles
K	equilibrium constant for partitioning of a compound between the organic and gaseous phase
L	cloud liquid water content
λ_{MFP}	mean free path of a diffusing molecule
λ	wavelength of light
M	fine mode aerosol mass concentration
M_0	organic aerosol mass formed
M_{OM}	mean molecular weight of organic matter
N	aerosol number concentration
N_{CLOUD}	cloud droplet concentration
$N_i(d)$	the particle number concentration with particle diameter d
N_{mol}	the molecular number concentration
ω_o	aerosol single scattering albedo; σ_{SP}/σ_{EXT}
Ω	aerosol optical depth; the vertical integral of σ_{EXT} over the light beam path length dl
p	vapor pressure
v	Junge parameter

R	aerosol back/total scattering ratio; σ_{BSP}/σ_{SP}
R_G	gas constant
r	aerosol geometric radius
r_e	effective cloud droplet radius
S_C	critical supersaturation at which aerosol activation into a cloud droplet occurs
σ_{AP}	aerosol absorption coefficient
σ_{BSP}	aerosol hemispheric backscatteringing coefficient
σ_{EXT}	aerosol extinction coefficient; $\sigma_{SP} + \sigma_{AP}$
σ_{SP}	aerosol total scattering coefficient
T	temperature
Y	aerosol yield from a single compound

REFERENCES

Albrecht, B.A. (1989) Aerosols, cloud microphysics, and fractional cloudiness. *Science* **245**, 1227–30.

Andreae, M.O. & Crutzen, P.J. (1997) Atmospheric aerosols: biogeochemical sources and role in atmospheric chemistry. *Science* **276**, 1052–8.

Ammann, M., Kalberer, M., Jost, D.T. *et al.* (1998) Heterogeneous production of nitrous acid on soot in polluted air masses. *Nature* **395**, 157–60.

Baker, M.B. (1997) Cloud microphysics and climate. *Science* **276**, 1072–8.

Baron, P.A. & Willeke, K. (eds) (2001) *Aerosol Measurement*. John Wiley, New York.

Barrie, L.A., Yi, Y., Leaitch, W.R. *et al.* (2001) A comparison of large scale atmospheric sulfate aerosol models (COSAM): overview and highlights. *Tellus, Series B* **53**, 615–45.

Berresheim, H., Wine, P.H. & Davis, D.D. (1995) Sulfur in the atmosphere. In Singh, H.B. (ed.), *Composition, Chemistry and Climate of the Atmosphere*. Van Nostrand Reinhold, New York.

Boucher, O. & Tanré, D. (2000) Estimation of the aerosol perturbation to the Earth's radiative budget over oceans using POLDER satellite aerosol retrievals. *Geophysics Research Letters* **27**, 1103–6.

Brasseur, G.P., Orlando, J.J. & Tyndall, G.S. (eds) (1999) *Atmospheric Chemistry and Global Change*. Oxford University Press, Oxford.

Browell, E.V., Ismail, S. & Grant, W.B. (1998) Differential absorption lidar (DIAL) measurements from air and space. *Applied Physics B* **67**, 399–410.

Charlson, R.J. & Heintzenberg, J. (eds) (1995) *Aerosol Forcing of Climate*. Wiley, Chichester.

Charlson, R.J., Lovelock, J.E., Andreae, M.O. & Warren, S.G. (1987) Oceanic phytoplankton, atmospheric sulfur, cloud albedo and climate. *Nature* **326**, 655–61.

Charlson, R.J., Schwartz, S.E., Hales, J.M. *et al.* (1992) Climate forcing by anthropogenic aerosols. *Science* **255**, 423–30.

Cooke, W.F., Liousse, C., Cachier, H. & Feichter, J. (1999) Construction of a 1° × 1° degree fossil fuel emission data set for carbonaceous aerosol and implementation and radiative impact in the ECHAM4 model. *Journal of Geophysical Research* **104**, 22, 137–62.

d'Almeida, G.A., Koepke, P. & Shettle, E.P. (eds) (1991) *Atmospheric Aerosols: Global Climatology and Radiative Characteristics*. Deepak, Hampton.

Davidovits, P., Hu, Jh., Worsnop, D.R., Zahniser, M.S. & Kolb, C.E. (1995) Entry of gas molecules into liquids. *Faraday Discussions* **100**, 65–82.

Dentener, F.J. & Crutzen, P.J. (1993) Reaction of N_2O_5 on tropospheric aerosols: Impact on the global distributions of NO_x, O_3, and OH. *Journal of Geophysical Research* **98**, 7149–63.

Duce, R.A. (1995) Sources, distributions and fluxes of mineral aerosols and their relationship to climate. In Charlson, R.J. & Heintzenberg, J. (eds), *Aerosol Forcing of Climate*. Wiley, Chichester.

Finlayson-Pitts, B. & Pitts, J. (1999) *Chemistry of the Upper and Lower Atmosphere*. Acadamic Press, New York.

Gong, S.L., Barrie, L.A. & Blanchet, J.P. (1997) Modeling sea-salt aerosols in the atmosphere. Part 1: Model development. *Journal of Geophysical Research* **102**, 3805–18.

Griffin, R.J., Cocker, D.R., Flagan, R.C. & Seinfeld, J.H. (1999a) Organic aerosol formation from the oxidation of biogenic hydrocarbons. *Journal of Geophysical Research* **104**, 3555–67.

Griffin, R.J., Cocker, D.R., Seinfeld, J.H. & Dabdub, D. (1999b) Estimate of global atmospheric organic aerosol from oxidation of biogenic hydrocarbons. *Geophysical Research Letters* **26**, 2721–4.

Guenther, A., Hewitt, C., Erickson, D. *et al.* (1995) A model of natural volatile organic compound emissions. *Journal of Geophysical Research* **100**, 8873–92.

Hansen, J.E., Lacis, A., Ruedy, R. & Sato, M. (1992) Poten-tial climate impact of Mount Pinatubo eruption. *Geophysical Research Letters* **19**, 215–218.

Haywood, J.M., Ramaswamy, V. & Soden, B.J. (1999) Tropospheric aerosol climate forcing in clear-sky satellite observations over the oceans. *Science* **283**, 1299–303.

Hegg, D.A., Livingston, J., Hobbs, P.V., Novakov, T. & Russell, P. (1997) Chemical apportionment and aerosol column optical depth off the mid-Atlantic coast of the United States. *Journal of Geophysical Research* **102**, 25,293–303

Heintzenberg, J., Covert, D.C. & Van Dingenen, R. (2000) Size distribution and chemical composition of marine aerosols: a compilation and review. *Tellus* **52B**, 1104–22.

Herman, M., Deuzé, J.L., Goloub, P., Bréon, F.M. & Tanré, D. (1997) Remote sensing of aerosols over land surfaces including polarization measurements and application to POLDER measurements. *Journal of Geophysical Research* **102**, 17,039–49.

Hinds, W.C. (1999) *Aerosol Technology: Properties, Behavior, and Measurements of Airborne Particles*. Wiley, New York.

Hobbs, P.V. (ed.) (1993) *Aerosol–Cloud–Climate Interactions*. Academic Press, San Diego.

Husar, R.B., Prospero, J.M. & Stowe, L.L. (1997) Characterization of tropospheric aerosols over the oceans with the NOAA Advanced Very High Resolution Radiometer optical thickness operational product. *Journal of Geophysical Research* **102**, 16,889–909.

IPCC (1996) *Climate Change 1995. The Science of Climate Change*. Cambridge University Press, Cambridge.

IPCC (1999) *Aviation and the Global Atmosphere*. Cambridge University Press, Cambridge.

IPCC (2001) *Climate Change 2001: The Scientific Basis*. Cambridge University Press, New York.

Jacobson, M.Z. (2000) A physically-based treatment of elemental carbon optics: implications for global direct forcing of aerosols. *Geophysical Research Letters* **27**, 217–20.

Jacobson, M.Z. (2001) Strong radiative heating due to the mixing state of black carbon in atmospheric aerosols. *Nature* **409**, 695–7.

Jaenicke, R. (1988) Atmospheric physics and chemistry. In Fischer, G. (ed.), *Meteorology: Physical and Chemical Properties of Air*. Springer-Verlag, Berlin.

Kaufman, Y.J. & Fraser, R.S. (1997) The effect of smoke particles on clouds and climate forcing. *Science* **277**, 1636–9.

Kaufman, Y.J., Tanré, D., Gordon, H.R. *et al.* (1997) Passive remote sensing of tropospheric aerosol and atmospheric correction for the aerosol effect. *Journal of Geophysical Research* 102, 16,815–30.

Kaufman, Y.J., Hobbs, P.V., Kirchhoff, V.W.J.H. *et al.* (1998) Smoke, Clouds, and Radiation—Brazil (SCAR-B) experiment. *Journal of Geophysical Research* 103, 31,783–808.

Kent, G.S., Trepte, C.R., Skeens, K.M. & Winker, D.M. (1998a) LITE and SAGE II measurements of aerosols in the southern hemisphere upper troposphere. *Journal of Geophysical Research* 103, 19,111–27.

Kent, G.S., Trepte, C.R. & Lucker, P.L. (1998b) Long-term Stratospheric Aerosol and Gas Experiment I and II measurements of upper tropospheric aerosol extinction. *Journal of Geophysical Research* 103, 28,863–74.

King, M.D., Kaufman, Y.J., Tanré, D. & Nakajima, T. (1999) Remote sensing of tropospheric aerosols from space: past, present and future. *Bulletin of the American Meteorological Society* 80, 2229–59.

Koepke, P., Hess, M., Schult, I. & Shettle, E.P. (1997) *Global Aerosol Data Set*. Report 243, Max Planck Institute for Meteorology, Hamburg.

Langner, J. & Rodhe, H. (1991) A global three-dimensional model of the tropospheric sulfur cycle. *Journal of Atmospheric Chemistry* 13, 255–63.

Lavanchy, V.M.H., Gäggeler, H.W., Schotterer, U., Schwikowski, M. & Baltensperger, U. (1999) Historical record of carbonaceous particle concentrations from a European high-alpine glacier (Colle Gnifetti, Switzerland). *Journal of Geophysical Research* 104, 21,227–36.

Leinen, M. & Sarnthein, M. (eds) (1989) *Paleoclimatology and Paleometeorology: Modern and Past Patterns of Global Atmospheric Transport*. Kluwer Academic, Dordrecht.

Levine, J.S. (ed.) (1991) *Global Biomass Burning: Atmospheric, Climatic and Biospheric Implications*. MIT Press, Cambridge, MA.

McCormick, M.P., Thomason, L.W. & Trepte, C.R. (1995) Atmospheric effects of the Mount Pinatubo eruption. *Nature* 373, 399–403.

Marti, J.J., Weber, R.J., McMurry, P.H., Eisele, F., Tanner, D. & Jefferson, A. (1997) New particle formation at a remote continental site: assessing the contributions of SO_2 and organic precursors. *Journal of Geophysical Research* 102, 6331–9.

Miller, R. & Tegen, I. (1998) Climate response to soil dust aerosols. *Journal of Climate* 11, 3247–67.

Möller, D. (1995) Sulfate aerosols and their atmospheric precursors. In Charlson, R.J. & Heintzenberg, J. (eds), *Aerosol Forcing of Climate*. John Wiley, Chichester.

Murphy, D.M., Anderson, J.R., Quinn, P.K. *et al.* (1998) Influence of sea salt on aerosol radiative properties in the southern Ocean marine boundary layer. *Nature* 392, 62–5.

O'Dowd, C., Smith, M.H., Consterdine, I.E. & Lowe, J.A. (1997) Marine aerosol, sea-salt, and the marine sulfur cycle: a short review. *Atmospheric Environment* 31, 73–80.

Odum, J.R., Jungkamp, T.P.W., Griffin, R.J., Flagan, R.C. & Seinfeld, J.H. (1997) The atmospheric aerosol-forming potential of whole gasoline vapor. *Science* 276, 96–9.

Osborn, M.T., DeCoursey, R.J., Trepte, C.R., Winker, D.M. & Woods, D.C. (1995) Evolution of the Pinatubo volcanic cloud over Hampton, Virginia. *Geophysical Research Letters* 22, 1101–4.

Osborn, M.T., Kent, G.S. & Trepte, C.R. (1998) Stratospheric aerosol measurements by the Lidar In-Space Technology Experiment. *Journal of Geophysical Research* 103, 11,447–53.

Pandis, S.N., Harley, R.A., Cass, G.R. & Seinfeld, J.H. (1992) Secondary organic aerosol formation and transport. *Atmospheric Environment* 26, 2269–82.

Penner, J.E. (1995) Carbonaceous aerosols influencing atmospheric radiation: black and organic carbon. In Charlson, R.J. & Heintzenberg, J. (eds), *Aerosol Forcing of Climate*. Wiley, Chichester.

Phadnis, M.J. & Carmichael, G.R. (2000) Numerical investigation of the influence of mineral dust on the tropospheric chemistry of East Asia. *Journal of Atmospheric Chemistry* 36, 285–323.

Piccot, S.D., Watson, J.J. & Jones, J.W. (1992) A global inventory of volatile organic compound emissions from anthropogenic sources. *Journal of Geophysical Research* 97, 9897–912.

Pruppacher, H.R. & Klett, J.D. (1997) *Microphysics of Clouds and Precipitation*, 2nd edn. Reidel, Dordrecht.

Pueschel, R.F. (1995) Atmospheric aerosols. In Singh, H.B. (ed.), *Composition, Chemistry and Climate of the Atmosphere*. Van Nostrand Reinhold, New York.

Pueschel, R.F. (1996) Stratospheric aerosols: Formation, properties, effects. *Journal of Aerosol Science* 27, 383–402.

Quinn, P.K., Anderson, T.L., Bates, T.S. *et al.* (1996) Closure in tropospheric aerosol–climate research: a review and future needs for addressing aerosol direct short-wave radiative forcing. *Contributions in Atmospheric Physics* 69, 547–77.

Raes, F., Van Dingenen, R., Vignatti, E. *et al.* (2000) Formation and cycling of aerosols in the global tropo-sphere. *Atmospheric Environment* **34**, 4215–40.

Ravishankara, A.R. (1997) Heterogeneous and multi-phase chemistry in the troposphere. *Science* **276**, 1058–65.

Reist, P.C. (1993) *Aerosol Science and Technology*, 2nd edn McGraw-Hill, New York.

Russell, P.B. & Heintzenberg, J. (2000) An overview of the ACE-2 clear sky column closure experiment (CLEARCOLUMN). *Tellus* **52B**, 463–83.

Russell, P.B., Hobbs, P.V. & Stowe, L.L. (1999) Aerosol properties and radiative effects in the United States East Coast haze plume: an overview of the Tropos-pheric Aerosol Radiative Forcing Observational Exper-iment (TARFOX). *Journal of Geophysical Research* **104**, 2213–22.

Saylor, R.D. (1997) An estimate of the potential signifi-cance of heterogeneous loss to aerosols as an addi-tional sink for hydroperoxy radicals in the troposphere. *Atmospheric Environment* **31**, 3653–8.

Schult, I., Feichter, J. & Cooke, W.F. (1997) Effect of black carbon and sulfate aerosols on the global radia-tion budget. *Journal of Geophysical Research* **102**, 30,107–17.

Schwartz, S.E. (1996) The Whitehouse Effect — shortwave radiative forcing of climate by anthro-pogenic aerosols: an overview. *Journal of Aerosol Science* **27**, 359–82.

Shulman, M.L., Jacobson, M.C., Carlson, R.J., Synovec, R.E. & Young, T.E. (1996) Dissolution behavior and surface tension effects of organic compounds in nucle-ating cloud droplets. *Journal of Geophysical Research* **23**, 277–80.

Seinfeld, J.H. & Pandis, S.N. (1998) *Atmospheric Chem-istry and Physics: From Air Pollution to Climate Change.* John Wiley, New York.

Singh, H.B. (ed.) (1995) *Composition, Chemistry and Cli-mate of the Atmosphere.* Van Nostrand Reinhold, New York.

Sokolik, I. & Toon, O.B. (1996) Direct radiative forcing by anthropogenic airborne mineral aerosols. *Nature* **381**, 681–3.

Tegen, I., Lacis, A.A. & Fung, I. (1996) The influence on climate forcing of mineral aerosols from disturbed soils. *Nature* **380**, 419–22.

Twomey, S.A. (1977) The influence of pollution on the short-wave albedo of clouds. *Journal of Atmospheric Science* **34**, 1149–52.

Whitby, K.T. & Sverdrup, G.M. (1980) California aerosols: their physical and chemical characteristics. *Advances in Environmental Science and Technology* **9**, 477–517.

10 Atmospheric Dispersion and Air Pollution Meteorology

DAVID CARRUTHERS

10.1 INTRODUCTION

Gases and particulate pollutants emitted into the atmosphere are dispersed by movements of air that carry them from the source and diffuse them into larger volumes of air by turbulent eddies. This dispersion and dilution is affected by density differences between the pollutants and atmosphere, by deposition, and also by many complex effects including those due to complex terrain and buildings.

In this chapter we consider dispersion within a distance of about 100 km from the source, and focus on the lowest part of the atmosphere or atmospheric boundary layer where the air is turbulent. We first describe the essential features of the atmospheric boundary layer in idealized flat terrain conditions. We then consider complicating effects such as complex terrain, surface roughness, and differential surface heating. Dispersion of pollutants under a range of different conditions is then covered, the emphasis being on the physical descriptions rather than on models; the latter are considered in Part 2. A pictorial of some of the features and processes influencing dispersion and air pollution meteorology is shown in Fig. 10.1.

10.2 THE ATMOSPHERIC BOUNDARY LAYER

A satisfactory understanding of the mean and turbulent motions and temperature distributions within the atmospheric boundary layer is essential before the dispersion of pollutants may be understood or described in detail, and this forms the subject matter of this section. For a fuller discussion the reader is referred, for example, to Stull (1988).

10.2.1 Factors determining boundary layer structure

The motions and the temperature distribution in the atmospheric boundary layer are broadly governed by three groups of factors, which are as follows:

1 The heat flux into the atmosphere at the surface. This is determined by such factors as the solar radiation reaching the surface, the absorption and release of latent heat by water vapor near the surface, especially around vegetation and over water sources, and heat either absorbed by or released from the surface. The temporal and spatial variations in the heat flux are also important in determining the structure of the boundary layer. Examples of the atmosphere affected by spatial gradients in heat fluxes are urban heat islands and sea breezes. An example of the temporal variation in the heat flux is that caused by the regular diurnal changes in solar radiation.

2 The roughness and changes in elevation of the Earth's surface near and for some distance upwind of the region of interest. The roughness effect is caused by many small surface obstacles or roughness elements obstructing and hence decelerating

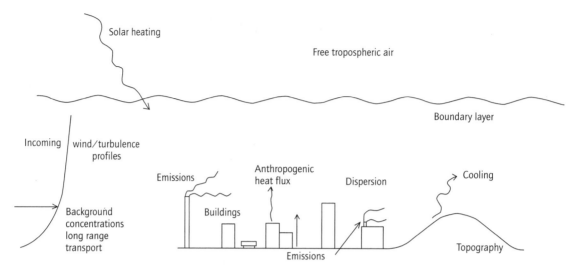

Fig. 10.1 Features of atmospheric dispersion and air pollution meteorology.

the airflow; these may range in scale from grass to buildings and even ranges of hills. A roughness length can be used to characterize the effect of these elements on the flow above them; this length is typically about 1/30 of the height of the obstacles. Thus, for the examples given, the roughness length ranges from 0.01 m to 1 m to over 10 m. The effects of changes in elevation on the airflow are characterized by both the heights and the length scales of the terrain features. Effects include acceleration and deceleration of the flow, streamline convergence or divergence, streamline deflection, blocking, and flow separation.

3 The airflow at the top of the boundary layer, free from the influence of ground roughness, is usually referred to as the geostrophic wind, and is determined by large-scale pressure gradients. Temporal and horizontal gradients in these gradients are important and, in general, lead to convergence and divergence of the horizontal flow and thence to upward/downward motions.

10.2.2 *Idealized boundary layers over flat terrain*

It is now well verified that the atmospheric boundary layer over flat terrain has an approximately

similar structure over defined ranges of meteorological conditions. This means that the physical mechanisms determining the structure are similar for similar types of boundary layer and that the vertical profiles of the mean wind speed, turbulence, temperature, and other quantities can be described in terms of the height above the ground z, when each of these quantities is divided by a value at a particular height (e.g. 10 m or a surface value) and the height is divided by a particular length scale (e.g. the roughness length z_0 or the boundary-layer depth h). The advantage of similarity is that the structure of the boundary layer at all heights can be described in terms of a small number of boundary layer parameters.

Within the boundary layer there are three broad types of structure corresponding to: (i) unstable/convective boundary layers, (ii) neutral boundary layers, and (iii) stable boundary layers. Typical conditions characterizing these structures are: light winds and a high surface heat flux; high winds and/or a small positive or negative surface heat flux; and nighttime conditions of light winds and negative surface heat flux, respectively. The parameters defining the structure of these boundary layers are quite different for each case.

Convective boundary layer

Figure 10.2 shows typical profiles of mean wind, turbulence, represented by the standard deviation of the wind in the transverse (across wind) and vertical direction, and temperature of the convective boundary layer, together with measured data. In the case of the mean wind there is a sharp increase near the surface, however throughout most of the boundary layer the mean flow is fairly uniform. The transverse component of turbulence σ_v is fairly uniform, while the vertical component

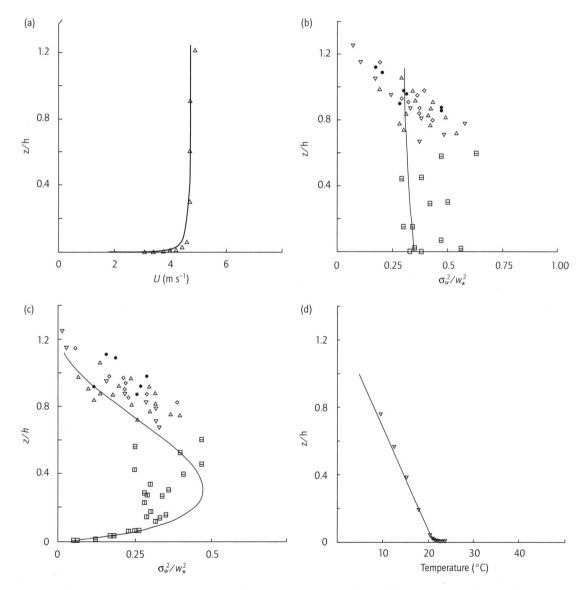

Fig. 10.2   Typical values in the convective boundary layer superimposed over field data from Caughey and Plamer (1979). (a) Mean wind. (b) Transverse component of turbulence. (c) Vertical component of turbulence. (d) Temperature.

σ_w increases above the surface and subsequently decreases towards the top of the boundary layer. The temperature decreases sharply with height near the surface and then decreases at a rate close to the adiabatic lapse rate.

In thermal convection the turbulence is mainly determined by the heat flux at the surface (F_{θ_0}), which produces hot thermals of air that rise from the surface. Since the thermals gain in scale and speed as they rise, the depth h of the boundary layer that limits their scale is also important. By dimensional analysis (Deardorff 1985) it follows that the characteristic velocity of these thermals is

$$w_\star = \left[gF_{\theta_0} h/(\rho c_p T_0)\right]^{\frac{1}{3}} \qquad (10.1)$$

where T_0 is the temperature at the surface, g is the acceleration due to gravity, ρ is the density and c_p is the specific heat at constant pressure. For typical values in the convective boundary layer for F_{θ_0}, h, and T_0 respectively of $100\,\mathrm{W\,m^{-2}}$, $1000\,\mathrm{m}$ and $300\,\mathrm{K}$, $w_\star \approx 1.4\,\mathrm{m\,s^{-1}}$, and this is a typical scale for the convective turbulence.

Near the surface—that is, up to about one-tenth of the boundary layer height—the airflow is strongly affected by the turbulence generated by the wind shear, since the mean wind decreases to zero at the surface, and the similarity based simply on the heat flux no longer applies. The wind shear gives rise to a shear stress τ which, at the surface, is usually expressed in terms of a friction velocity $u_\star$ as $\rho u_\star^2$. Very close to the surface the mean wind speed is solely determined by $u_\star$, the roughness length and the distance above the surface, i.e.

$$U(z) = \frac{u_\star}{\kappa} \ln\left(\frac{z}{z_0}\right) \qquad (10.2)$$

where κ is a constant, known as von Karman's constant, whose value is about 0.4. Near the surface the vertical turbulence is given by

$$\sigma_w \approx 1.3\,u_\star \qquad (10.3)$$

An important parameter for determining the structure of the boundary layer is the Monin–Obukhov (L_{MO}) length, which depends on

the ratio of mechanical (wind shear) to convective generation of turbulence as follows:

$$L_{MO} = \frac{-u_\star^3}{\kappa g F_{\theta_0}/(\rho c_p T_0)} \qquad (10.4)$$

L_{MO} is negative in convective conditions but it is the magnitude of L_{MO} that determines the structure. Thus for $z > |L_{MO}|$ (or at heights greater than $|L_{MO}|$) convective motions dominate the boundary layer turbulence, while for $z < |L_{MO}|$, near the surface, production of turbulence by shear stresses dominates.

Neutral boundary layer

Typical vertical profiles of wind speed and the σ_w in the neutral boundary layer are shown in Fig. 10.3. The mean wind velocity profile increases continuously to its maximum value, the geostrophic wind at the top of the layer. Over the lowest $100\,\mathrm{m}$ the profile is logarithmic and depends on roughness length (eqn 10.2) (Panofsky 1974), while in the upper part of the boundary layer, the profile depends on (z/h), and the power-law profile

$$U/U_G = (z/h)^p \qquad (10.5)$$

is a useful approximation; p depends on the roughness length.

There is a systematic turning of the wind with height due to the Coriolis acceleration (to the left for an observer facing away from the wind in the northern hemisphere, but in the opposite direction in the southern hemisphere), ranging from about 5 to 20°. The direction and magnitude of this turning is quite sensitive to changes in surface roughness and upper-layer conditions. In marked contrast to the convective boundary layer, turbulence in the neutral boundary layer is largely controlled by the surface shear stress and $h/|L_{MO}| \leq 0.1$. Since most of the turbulence is generated close to the surface the magnitude of the vertical fluctuations decreases with height (Fig. 10.3b).

The depth h of the neutral boundary layer may be determined by a balance of frictional forces and the Coriolis force. However, it may be limited by

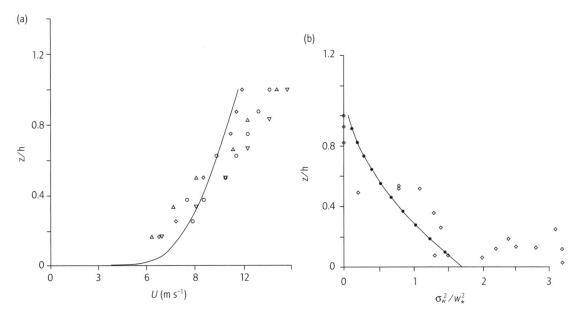

Fig. 10.3 Typical vertical profiles in a neutral boundary layer superimposed over field data from Lenschow *et al.* (1988). (a) Mean wind. (b) Vertical component of turbulence.

how much the mechanically generated turbulence can entrain the stable air above the boundary layer. In the former case $h \sim 0.3u_*/f$, where f is the Coriolis parameter ($\sim 10^{-4} s^{-1}$ for mid-latitudes), but in the latter case no simple formula can be applied. Typically a neutral boundary layer is some hundreds of meters deep.

Stable boundary layer

In steady or slow-varying conditions the stable boundary layer over land usually exists when the surface heat flux, F_{θ_0}, is negative. Over land this usually occurs between the hours of sunset and sunrise when the surface is cooling. A stable boundary layer can also occur when warm air is advected over a cool surface; for example, in winter when warm air from over the sea is advected over cold ground. The stable boundary layer is the most variable of the three kinds of boundary layers and is usually in a continuous state of evolution or even slow oscillation. Unlike the convective and neutral boundary layers, it is very sensitive to the slope

and undulations of the local terrain and to the properties of the airflow above the boundary layer. Typical vertical profiles in the stable boundary layer are shown in Fig. 10.4.

In the stable boundary layer the stable density gradient (formed by the cooler air near the ground) causes damping of the large scales of turbulence, leading to lower turbulent intensities and smaller length scales than occur in convective or neutral boundary layers. As in the neutral boundary layer, the main source of turbulence is the airflow over the ground, but this diffuses up so slowly that the upper part of the stable boundary layer is often dominated by locally generated turbulence.

The boundary layer begins to change from its neutral form when the stabilizing buoyancy forces are comparable with shear stresses. This occurs when

$$h/L_{MO} \geq 0.3 \qquad (10.6)$$

The depth of the stable boundary can approximately be represented by

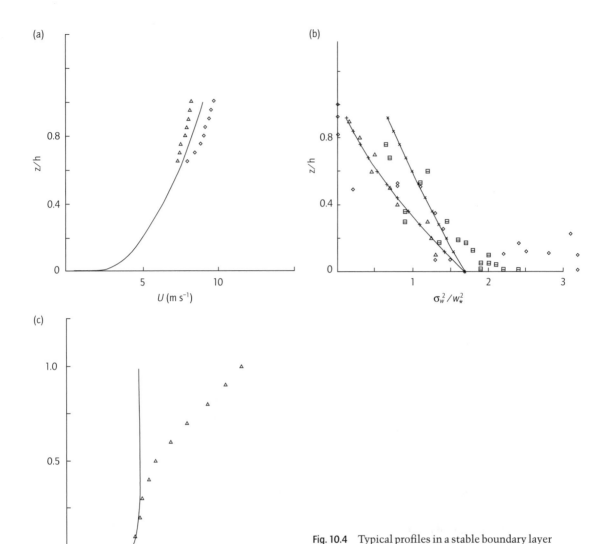

Fig. 10.4   Typical profiles in a stable boundary layer superimposed over data from Lenschow *et al.* (1988) and Caughey *et al.* (1979). (a) Mean wind. (b) Vertical component of turbulence. (c) Temperature profile.

$$h = \frac{0.3 u_\star / |f|}{1 + 1.9\, h/L_{MO}} \qquad (10.7)$$

which is similar to the expression for h in neutral conditions as h/L_{MO} tends to zero (Nieuwstadt 1981). In very light winds the depth of the stable boundary layer can be as small as a few tens of meters.

The intensity of the stratification in the boundary layer is frequently represented by the buoyancy frequency given by

$$N = \left(\frac{g}{\theta} \frac{d\theta}{dz} \right)^{\frac{1}{2}} \qquad (10.8)$$

where θ is the potential temperature. This parameter is important in determining how the flow pattern is influenced by complex terrain. Typical values of N in a stable boundary layer are $0.02\,\text{s}^{-1}$ so that the typical period of vertical oscillation in the flow is hundreds of seconds or a few minutes.

Because of the small levels of turbulence or mixing, the stable boundary layer is more responsive to Coriolis accelerations. This leads to the direction of the wind changing over the depth of the layer, typically by $25-40°$. Van Ulden and Holtslag (1985) showed that the mean deviation between 20 and 200 m increased from $12 \pm 12°$ to $40 \pm 20°$ as the stability increased from neutral to stable ($h/L_{MO} = 10$). This change in angle of the mean wind is important for estimating lateral dispersion.

10.2.3 h/L_{MO} as a measure of boundary layer stability

In the previous sections frequent reference has been made both to the Monin–Obukhov length and to the boundary layer height. By substituting the value for $gF_{\theta_0}/(\rho c_p T_0)$ from eqn 10.1 into eqn 10.4, it follows that

$$h/L_{MO} = -\kappa\frac{W_*^3}{u_*^3} \qquad (10.9)$$

Therefore the ratio h/L_{MO} also characterizes the relative importance throughout the convective boundary layer of the buoyancy-generated turbulence to the surface shear-generated turbulence. In fact h/L_{MO} has been found to be the most appropriate ratio for defining the state of the boundary layer for dispersion calculations for all types of boundary layer (Holtslag & Nieuwstadt 1986). This is illustrated in Fig. 10.5, which describes these different regions of the boundary layer. Also marked on the figure is the Pasquill Stability Category (Pasquill & Smith 1983), which has frequently been used to classify boundary layer structure, but which has largely been superseded by classification according to h/L_{MO}. Pasquill categories range from A (very convective), to D (neutral), through to G (very stable).

In the surface layer vertical profiles can be ex-pressed as general functions of z/L_{MO}, while for $z > |L_{MO}|$ the height of the boundary layer is important. Examples of how these parameters can be used to describe the vertical profiles are given for the mean wind U (all stabilities) and vertical component of turbulence (convective conditions) and are as follows:

$$U = \frac{u_*}{k}\left\{\ln\frac{(z+z_0)}{z_0} + \psi\left(\frac{(z+z_0)}{L_{MO}}\right) - \psi\left(\frac{z_0}{L_{MO}}\right)\right\}$$

$$(10.10)$$

where ψ depends on whether L_{MO} is greater or less than 0 (Dyer & Hicks 1970; Holtslag & de Bruin 1988) and

$$\sigma_w^2 = \left(1-0.8\frac{z}{h}\right)^2\left[0.4\left(2.1w_*\left(\frac{z}{h}\right)^{1/3}\right)^2 + \left(1.3u_*\right)^2\right]$$

$$(10.11)$$

10.2.4 Boundary layers in changing conditions

It is stated in Section 10.2.1 that the boundary layer is affected by the spatial and temporal changes both at the surface and in the airflow above the top of the layer. In this section we describe these changes and also when it is possible to ignore such changes in classifying, describing, or modeling the boundary layer, this justifying use of similarity profiles.

Diurnal changes

Figure 10.6 shows the variation of the height of the boundary layer in convective conditions, caused by the erosion of stable layer aloft by turbulent eddies. The figure shows that the depth can double in a period T_h of the order of $10^4\,\text{s}$ (~3 hours). Because this time is larger than a typical time scale for the adjustment of eddies (T_L) there is local adjustment and the similarity profiles represent local structure. However, at the end of the day when the surface heating reduces and the surface heat flux reverses, the structure of the boundary layer is no longer described by any of the ideal cases of the

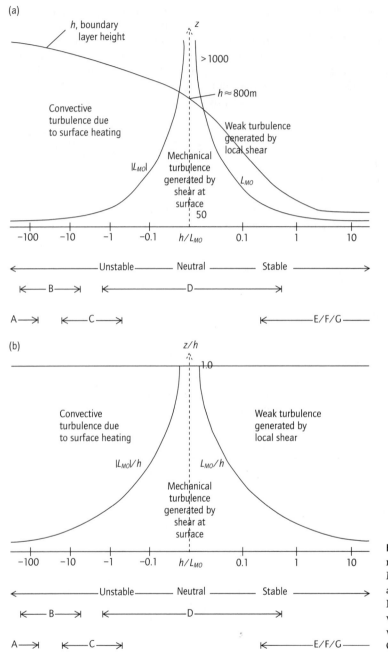

(a)

h, boundary layer height

z

> 1000

$h \approx 800$m

Convective turbulence due to surface heating

Weak turbulence generated by local shear

$|L_{MO}|$

Mechanical turbulence generated by shear at surface

L_{MO}

50

| -100 | -10 | -1 | -0.1 | h/L_{MO} | 0.1 | 1 | 10 |

←——————Unstable———— Neutral ———— Stable ——————→

|←— B —→| |←———————— D ————————→|

A—→| |←— C —→| |←———— E/F/G ————

(b)

z/h

1.0

Convective turbulence due to surface heating

Weak turbulence generated by local shear

$|L_{MO}|/h$

L_{MO}/h

Mechanical turbulence generated by shear at surface

| -100 | -10 | -1 | -0.1 | h/L_{MO} | 0.1 | 1 | 10 |

←——————Unstable———— Neutral ———— Stable ——————→

|←— B —→| |←———————— D ————————→|

A—→| |←— C —→| |←———— E/F/G ————

Fig. 10.5 (a) Dimensional representation of variation of Monin–Obukhow length with atmospheric stability. (b) Nondimensional representation of variation of Monin–Obukhov length with atmospheric stability. Based on Golder (1972).

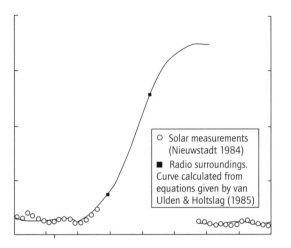

Fig. 10.6 Diurnal variation of boundary layer depth at Cabauw (the Netherlands), May 31.

○ Solar measurements
(Nieuwstadt 1984)

■ Radio surroundings. Curve calculated from equations given by van Ulden & Holtslag (1985)

boundary layer we have described. In this case, the surface layer is stable, while the upper regions are slightly unstable; this leads to plumes "lofting" or being trapped above the ground. Conversely, in the morning when the heat flux changes from negative to positive, the surface layer is typically unstable, while the upper layers are stable.

Complex surfaces

In this section we describe the essential processes governing the changes in airflow over complex surfaces (changes in surface elevation, roughness, and temperature). This is an essential step in deciding when it is necessary for dispersion models to take complex flows into account.

Changes in surface elevation. Changes in the elevation or slope of the ground affect the boundary layer flow in different ways depending on the relative importance of the inertia of the oncoming flows to the buoyancy forces (which we denote by a parameter F). (Hunt *et al.* 1988b; Carruthers & Hunt 1990). In the following sections we describe how the flow changes when the boundary layer changes progressively from neutral through to very stable (Fig. 10.7). The general patterns in con-

vective flows are similar to neutral flows except when local heating effects dominate the flow.

(i) Near neutral flows, $0 \leq h/L_{MO} \leq 1$, Pasquill Class D/E. Consider flow over an isolated hill of height H and length L. L is defined, for convenience, as the horizontal distance from the summit to where the surface elevation is approximately $H/2$. It has been found (Jackson & Hunt 1975) that H, L, and the ratio H/L are important in determining how the hill affects the flow. For many hills, the ratio of H/L is less than about 1/3, although of course there are steep hills, mountains and cliffs where this is not true. When $H/L \leq 1/3$, the airflow passes over and round the hill and does not reverse its direction. The streamlines are deflected vertically by the slope of the hill up to a height h_m, which is of the order of L, and the change in the horizontal velocity increases quite sharply near the surface, reaching a maximum at a height h_m of $\Delta U \sim (H/L)\, U(h_m)$, with a speed-up $\Delta U/U \approx H/L$. Because the approach wind speed $U(z)$ increases with height, the vertical velocity over the hill also increases with height up to h_m. The same pattern of changes occurs over hills where H/L is significantly greater than 1/3. The main effect of steep slopes is to induce recirculating flow on either the upwind or downwind slopes.

When the atmosphere is slightly stable, the turbulence is reduced and the shear in the mean velocity profile is increased when compared with the neutral boundary layer. Since the mean velocity now increases more rapidly with height, deflecting the airflow at a height h_m well above the hill has a relatively greater effect on the surface wind. Therefore the relative increase in wind speed is greater and the vertical and horizontal deflection of streamlines is greater. The stratification is now great enough to affect the velocity profile approaching the hill, but not great enough for buoyancy forces to be comparable with the inertia associated with the mean flow over the hills. Consequently, the changes in U and streamwise deflection have a similar distribution over the hills as for neutral flows.

So far we have only considered isolated hills, but hills usually/often occur in groups. Field, laboratory, and numerical studies of airflow over

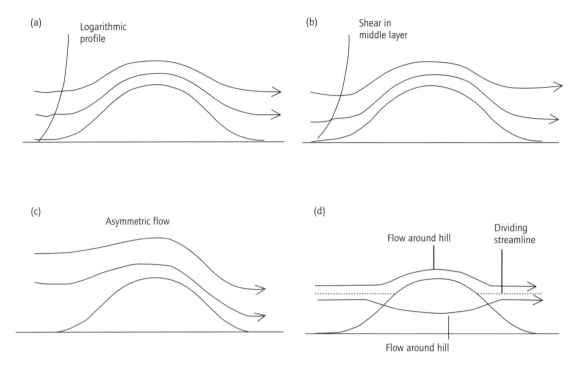

Fig. 10.7 Flow patterns over a three-dimensional hill for (a) neutral flow, (b) weak stratification, (c) moderate stratification, and (d) strong stratification.

groups of hills and valleys have shown that similar speed-ups and streamline deflections occur over the hill tops, but that the wakes are changed by the presence of downwind hills; in particular, the extent of the wakes and recirculating regions is reduced. When the wind is at an angle to a valley with sleep slopes, significant helical motions occur, which may determine the large-scale turbulence associated with that valley, which in turn may have a significant impact on dispersion.

(ii) Moderately stable stratification.

$$h/L_{MO} \geq 1, \; F_H = (U/NH) > 1,$$
$$\text{Pasquill Class E/F}$$

In these conditions the stratification or stable density gradient is large enough, and the wind speed is low enough, for the buoyancy forces to be significant compared with inertial forces of the airflow

over the hills. The whole distribution of the flow then begins to change. For a uniform temperature gradient, waves appear in the flow, especially lee waves, and downwind of a hill the mean wind speed near the surface alternately increases and decreases with downwind distance. The wind speed at the crest and upwind decreases but it increases on the lee slopes. If the waves are strong enough, stagnant or recirculating regions may form on the lee slopes. If the airflow is neutral and there is a strong inversion, the wind speed may reach a maximum value at the hill top.

(iii) Strong stable stratification.

$$h/L_{MO} \geq 3, \; F_H < 1, \; \text{Pasquill Class F/G}$$

When the stratification is even stronger, the hills are higher, and the wind speed is low enough, the buoyancy forces are so strong relative to the inertia of the oncoming flow that all the air cannot flow over the top of the hills (Fig. 10.9). An approximate criterion for this condition is

$$F_H = \frac{U(H)}{N(H)H} \le 1 \qquad (10.12)$$

where the relevant buoyancy frequency is defined by the temperature gradient at the hill top. Thus if the wind speed at hill height is $2\,\mathrm{m\,s^{-1}}$ and the buoyancy frequency is $0.01\,\mathrm{s^{-1}}$, there is some flow around the hill if the hill height H is greater than $200\,\mathrm{m}$, with the depth of the layer flowing around the hill increasing as the hill height increases. Thus in this case the air stream, above a critical height H_c, has sufficient kinetic energy to pass over the crest, whereas air below H_c flows around the hill or is blocked. Laboratory, field, and numerical studies indicate that this critical height $H_c \approx H(1 - F_H)$ (Sheppard 1956).

The importance of this situation for atmospheric dispersion is that plumes released from sources below the critical height, H_c, move approximately horizontally; they impact onto the hill and either "split" as they pass each side of the hill, or move around one side or the other. Upwind of a ridge the plumes below H_c can slowly drift onto the surface. Plumes released above H_c pass over the hill as if the ground level has been raised to H_c.

An important feature of these highly stable air flows is that they often contain low-frequency horizontal fluctuations that can lead to streamlines moving from one direction to another around the hill. There may also be slow fluctuations, over a period of less than one hour, of the temperature gradient, which means that the central height H_c can fluctuate; thus plumes can pass first over and then around a hill.

(iv) Strong stratification with drainage winds. So far we have considered the boundary layer flow to be driven by the large-scale synoptic gradients. It has been assumed that the wind speed well above the hills is the relevant speed for defining the flow regime. However, there are situations where this is not the case; in particular, when the wind speed is low enough, the stability is strong enough, and the height and slopes of the hills are large enough, the airflow over the hills can be driven by buoyancy

forces produced by temperature differences between the surface of the hills and the air above. These are called drainage, downslope, or katabatic winds when the slopes are cool, and upslope or anabatic winds when the slopes are heated.

Changes in roughness. The change of roughness of terrain affects the airflow by increasing or decreasing the resistance to the flow near the surface. The quantity that defines the effect on the airflow of the roughness "elements" (buildings, trees, grass, water waves, etc.) is the roughness length, z_0, as explained in Section 10.2.1. Since the mean velocity $U(z)$ in the boundary layer just above the roughness elements has the form (given in eqn 10.2)

$$U(z) = \frac{u_*}{k} \ln\left(\frac{z}{z_0}\right) \qquad (10.13)$$

it follows that the effect of a change of roughness varies approximately in proportion to the log of the roughness length $\ln(z_0)$. Therefore, a measure of the effect of a change in roughness from terrain (1) to terrain (2) is the parameter $\ln(z_{02}/z_{01})$. This ratio is positive when the air flows toward the rougher terrain. Also as the air flows over such a roughness change, the surface shear stress u_* increases. As an example, in near neutral conditions, the effect of a change in roughness length by a factor of 50 from a coastal site ($z_0 = 0.01\,\mathrm{m}$) to a suburban site ($z_0 \approx 0.5\,\mathrm{m}$) leads typically to a decrease in surface wind speed at $10\,\mathrm{m}$ of about 30%, but at the same time, the surface shear stress u_* increases by about 30%. Therefore the turbulence intensities $\sigma_w/U(z)$, $\sigma_v/U(z)$, which are proportional to u_*/U, increase by a factor of about two.

Spatial changes in temperature. Significant horizontal gradients in the temperature (T) give rise to vertical gradients of mean velocity above and within the boundary layer and they also lead to significant changes in the wind direction with height. These changes may be caused by sloping ground, by large-scale weather systems, by land–sea temperature changes, or by temperature contrasts caused by urban heating effects. An estimate of the magnitude of this effect can be obtained from the thermal wind equation

$$f\frac{\partial U}{\partial z} = -\frac{g}{T_0}\frac{\partial T}{\partial y} \qquad (10.14)$$

so that the change in the horizontal wind ΔU can be approximated by

$$\Delta U \sim gh\Delta\theta/(fLT_0) \qquad (10.15)$$

which applies for a height of order h, and if T changes by ΔT over a lateral distance L. Thus for a temperature change of $1°C$ over $100\,\text{km}$, the variation in wind speed over the depth of the boundary layer is approximately $3\,\text{m s}^{-1}$. The importance of such synoptic changes on the dispersion of pollution is that they affect the depth of the boundary layer within which pollutants are confined. However, the rates of change of h are usually sufficiently slow for the boundary layer to adjust so that "similar" profiles can persist during such changes.

Land–sea temperature contrasts result in the development of both sea breeze circulations (Fig. 10.8) and internal thermal boundary layers (Fig. 10.9). A sea breeze occurs when a temperature

difference develops at the coast, resulting in a pressure gradient, which drives the cool air overlying the sea inland. Urban heat islands are caused by anthropogenic heating and can result both in changes in the mean flow and in the development of internal thermal boundary layers (Oke 1987). In this case the urban circulation can advect pollution into the center of a large urban area.

Rapid spatial and temporal changes associated with weather phenomena (e.g. fronts, clouds, thunderstorms) can also change the structure of the boundary layer. The vertical profiles of mean velocity and direction, temperature, and turbulence then change rapidly and are quite different to the profiles in slowly changing conditions.

10.3 ATMOSPHERIC DISPERSION

10.3.1 Motions of gases or particles near a source

In this section we describe how airflow over and around a source transports and disperses material that is released into the ambient flow. Such a release may consist of one or more polluting gases or particulates mixed within harmless products such as carbon dioxide and water vapor. These releases are frequently emitted with both buoyancy (positive and negative) and momentum. Neither negatively buoyant releases nor the specific features of particulates are discussed herein.

When material is released into the atmosphere it does not immediately travel with the velocity of the airflow, and so initially there are differences between the velocities of the released gases and the airflow. These are confined within a thin layer around the release gases. The small-scale eddying motions induced by the velocity gradients cause

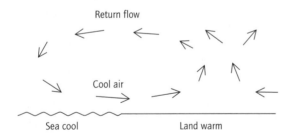

Fig. 10.8   Idealized sea-breeze circulation.

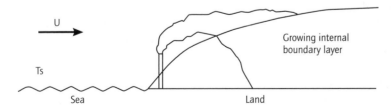

Fig. 10.9   Effect of coastline on a dispersing plume.

this layer to thicken and deform and the material starts to move in the direction of the airflow. Eventually the mixing is completed and the material follows the motions of the airflow even at the smallest scales.

The time and advection distance required for this total mixing and the distribution of the release gases when it is completed depend on a number of factors. These include principally the initial buoyancy and momentum of the pollutant at its release point, together with the height and direction of the release; and also ambient airflow around the source. The initial rise of release gases due to their buoyancy and momentum is known as plume rise. For a review of many of the factors that affect the plume rise, see Briggs (1984); the main processes are shown in Fig. 10.10.

In general the greater the momentum of the release gases either at the source or produced by buoyancy forces associated with differences in its density with the surroundings, or the less the environmental turbulence, the longer it takes for mixing to occur. For large plumes from tall chimneys

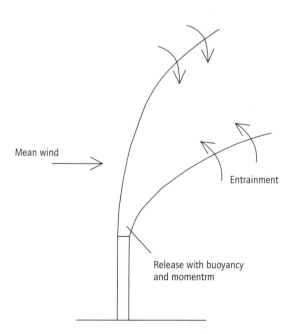

Mean wind

Entrainment

Release with buoyancy and momentrm

Fig. 10.10 Processes near the source of a buoyant release.

(e.g. power station stacks), it may take several kilometers from the source; for weakly buoyant releases near the ground, the distance may be less than 100 m.

10.3.2 Dispersion processes

Consider now an ideal or effective source that releases Q units of pollutant per second into an airflow having mean velocity U, such that initially the cross-sectional area of the dispersing material or plume is A. It follows that the initial concentration (C) of the pollutant is given by

$$C = Q/(UA) \qquad (10.16)$$

and thus the initial concentration is proportional to the source strength but inversely proportional to the wind speed.

For a laminar airflow only molecular diffusion would occur downstream of the source, and the concentration, C, would tend to the form given by

$$C(x, y, z) = \frac{Q}{2\pi U \sigma_y \sigma_z}$$
$$\times \exp\left\{-\frac{1}{2}\left(\frac{(y-y_s)^2}{\sigma_y^2} + \frac{(z-z_s)^2}{\sigma_z^2}\right)\right\}$$
$$(10.17)$$

where $\sigma_y^2 = \sigma_z^2 = 2D(x-x_s)/U$, D_m is the coefficient of molecular diffusion, x_s is the source location and the flow is in the x direction. σ_y and σ_z are the lateral and vertical standard deviations of the concentration distribution and are therefore a measure of its spread or cross-section. A schematic of the concentration distribution is shown in Fig. 10.11.

An identical expression for the concentration distribution in a turbulent flow can be obtained if it is assumed that turbulent diffusion can be represented by a constant turbulent diffusivity analogous but much greater than D_m, and if C represents the mean concentration; that is, the concentration averaged over times greater than the turbulent time scales but less than that associated with changing meteorology. For shorter averaging times

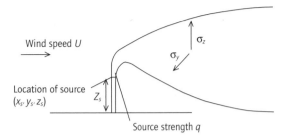

Fig. 10.11   Pictorial of a Gaussian plume.

the concentration will vary significantly from one averaging period to the next because of the random movement of the plume by turbulence and because of in-plume structure, which occurs as the plume mixes with ambient air. Expressions similar to eqn 10.17 have formed the basis for both discussions of dispersion and many dispersion models. Referring to eqn 10.17, we see that it describes the mean concentration in terms of the vertical and horizontal plume spreads, the mean wind speed, and a distribution about a plume centerline (y_s, z_s) at downstream distance x. Since the plume spread and the location of the plume centerline depend on the mean and turbulent wind field, the expression provides a mechanism by which concentrations can be approximately quantified in terms of the boundary layer structure. How this structure varies with h/L_{MO} over homogeneous terrain and in changing conditions was discussed in Section 2. In the following sections we describe how σ_y, σ_z vary with distance downstream of the source. In subsequent sections we consider features that affect the streamline height.

Turbulent diffusion near the source

If the external flow is turbulent, particles of pollutant from the source are randomly displaced downwind of the source. Close to the source even if the turbulence is quite inhomogeneous, the root mean square amplitudes of these displacements in the horizontal and vertical direction, $\left(\overline{Y^2}\right)^{1/2}$, $\left(\overline{Z^2}\right)^{1/2}$, are proportional to the root mean square of the turbulent velocity fluctuations σ_v, σ_w. This leads

to the result that, after a time of travel t from the source,

$$\left(\overline{Y^2}\right)^{1/2} = \sigma_v t$$
$$\left(\overline{Z^2}\right)^{1/2} = \sigma_w t \qquad (10.18)$$

where the averaging time is greater than the time scale of the turbulence, but smaller than the time scale for changes in meteorology.

If the turbulence is weak compared with the mean velocity at the source $(\sigma_v, \sigma_w \ll U)$, then the travel time $t = (x - x_s)/U$ and

$$\left(\overline{Y^2}\right)^{1/2} = \frac{\sigma_v}{U}(x - x_s)$$
$$\left(\overline{Z^2}\right)^{1/2} = \frac{\sigma_w}{U}(x - x_s) \qquad (10.19)$$

then near the source the mean concentration is given by eqn 10.17 for neutral and stable conditions with $\sigma_y = \left(\overline{Y^2}\right)^{1/2}$ and $\sigma_z = \left(\overline{Z^2}\right)^{1/2}$. If the turbulence is strong relative to the mean velocity, $(\sigma_v, \sigma_w \gg U)$, then the mean concentration may or may not reach a steady state.

In a convective boundary layer $(h/L_{MO} < -0.3)$ the vertical component of turbulence is significantly non-Gaussian, with a high probability of weak downflow and a low probability of strong upflow (Hunt *et al.* 1988a). Consequently, the height (z_{max}) at which there is the maximum concentration decreases with distance downstream, although the height $(\overline{z})$ of the mean concentration increases with distance downstream. This is inconsistent with a Gaussian profile for which $z_{max} = \overline{z}$ near the source, and thus a modified non-Gaussian concentration profile is required to describe the vertical concentration distribution in this case.

Diffusion far from the source in unstratified turbulence

Since turbulence consists of many scales of eddies that are weakly correlated with each other, the different parts of the pollutant released into the turbulent airflow move in different directions. Because of this the root-mean-square displacements $\left(\overline{Y^2}\right)^{1/2}$, $\left(\overline{Z^2}\right)^{1/2}$ do not continue to increase linearly with time (as in eqn 10.18). In a homogeneous turbulent flow (such as a wind tunnel) with a particular scale of turbulence L_x, (and a turbulent time scale $T_L \approx L_x/\sigma_w$) it is found that when the travel time $t > 3T_L$,

$$\sigma_y = \left(\overline{Y^2}\right)^{1/2} \propto \left(tT_L^{(v)}\sigma_v^2\right)^{1/2}$$
$$\sigma_z = \left(\overline{Z^2}\right)^{1/2} \propto \left(tT_L^{(w)}\sigma_w^2\right)^{1/2} \qquad (10.20)$$

Equation 10.20 shows that σ_y, σ_z increase with the square root of travel time $(t^{1/2})$, and therefore at a slower rate than that given in eqn 10.18. The transition between these two limits depends on the spectrum of the atmospheric turbulence. However, it is now recognized that in practice the limit defined by eqn 10.20 is seldom found in the atmosphere for the reasons discussed below. An example using the dispersion model ADMS (Carruthers *et al.* 1997) of how σ_z and σ_y vary with distance downstream of an elevated source in different boundary layers is shown in Fig. 10.12. The plume spread is much more rapid in convective conditions.

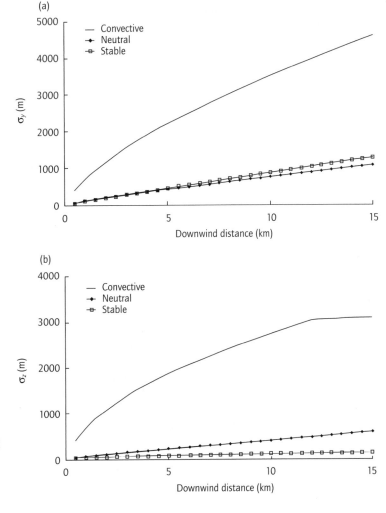

Fig. 10.12 Variation of plume spread with distance for an elevated source (a) Transverse component σ_y. (b) Vertical component σ_z.

Unsteadiness of the dispersing plumes. The ide-alized flow of a turbulent layer in a statistical steady state seldom corresponds to the actual state of the atmospheric boundary layer. Therefore lateral spreading of a plume does not follow the form of eqns 10.16 or 10.18 primarily because of large mesoscale eddies, fronts, and synoptic-scale variations. In order to calculate the magnitude of this effect it is necessary to use either measurements of the variation of the wind direction over significant periods (>30 min), or local climatological estimates.

Effects of vertical structure in the boundary layer. As was explained in Section 2.1, the turbulence in the boundary layer varies in intensity and scale across it. In neutral and convective layers the eddy scales are of the same order as the height above the surface. Thus if a pocket of gas is dispersed downwards from an elevated source it is mixed into small-scale eddies with a low diffusivity, but if it is dispersed upwards it is diffused further by large eddies. This inhomogeneous dispersion affects the growth of $\overline{Z^2}$ in a way not described by eqn 10.18; it can lead to

$$\left(\overline{Z^2}\right)^{1/2} \propto t \qquad (10.21)$$

Vertical diffusion is also prevented by an elevated inversion layer at the top of the boundary layer, where both the turbulence scale and vertical mixing reduce significantly.

Effects of wind shear. In the neutral and stable boundary layers there is a significant vertical variation of the mean horizontal velocity components U and V. The effect of this shear on the distribution of mean concentration is illustrated in Fig. 10.13, which shows the spreading sequence of puffs emitted from the source. As they diffuse vertically into the layers of faster and slower moving fluid, they are bent over; in the faster stream their ends are separated, but in the slower stream they hardly move apart. Therefore, at a given value of distance downstream of the release, the concentration below the source is greater than above it and so the position of maxi-

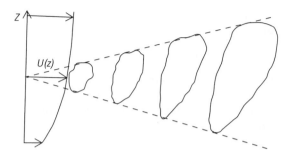

Fig. 10.13   Effect of wind shear on puffs of pollutant.

mum concentration and the mean height of the plume both decrease.

Averaging times

The expressions in eqn 10.20 describe how the mean widths of σ_y, σ_z of a plume vary with distance or travel time from the source when the concentration is averaged over a time greater than that of the largest turbulent eddies in a boundary layer $(T_L - h/\sigma_w \sim 10^3\,\text{s})$. On the other hand, they say nothing about the mean widths $\hat{\sigma}_y$, $\hat{\sigma}_z$ at a particular moment in time, which are important in understanding fluctuations in concentration over short time scales. Near the source, say at $x < UT_L$, where the plume has a thin meandering form, $\hat{\sigma}_y$, $\hat{\sigma}_z \ll \sigma_y$, σ_z, but further downstream $\hat{\sigma}_y$, $\hat{\sigma}_z$ and σ_y, σ_z are comparable.

For averaging times less than T_L there are large fluctuations in concentration with time at a point, and these increase as the averaging times decreases. In this case a model based on eqn 10.17 gives only a poor indication of the concentration likely to be observed at any particular time. For these averaging times it is more useful to describe the concentration probability distribution. This depends on both in-plume structure and the meandering of the plume by turbulence (see, for example, Thomson 1990; Mylne & Mason 1991).

10.4   MEAN CONCENTRATIONS

The preceding sections have discussed the range of

atmospheric conditions that occur in the boundary layer (Section 10.2) and how these impact on dispersion (Section 10.3). In this section we describe in qualitative terms how this impacts on the mean concentration of pollutants.

10.4.1 Flat terrain

In Figs 10.14 and 10.15 a dispersion model based on many of the principles described in this chapter, ADMS 3 (Carruthers *et al.* 1997), has been used to illustrate how in general terms the ground level concentration (g.l.c.) of a pollutant emitted from a point source varies with distance downwind from the source over flat terrain. Figure 10.14 is for a ground level source, while Fig. 10.15 is for an elevated source. The concentration is for an averaging time of one hour, which is generally smaller than the time scale of typical changes in meteorology, but larger than the time scales of turbulence fluctuations in concentration. The weather conditions range from Case 1 (very unstable), through to Case 2 (neutral), through to Case 3 (very stable).

For the ground level source the maximum g.l.c. occurs at the release point and then decays away from the source, with the rate of decay being highest for the most unstable conditions, because of rapid turbulent mixing, and lowest for the most stable case, because of very weak turbulence. Thus ground level concentrations remain quite high

well downwind of sources in stable conditions. If there are many sources—for instance, roads in an urban area—this can lead to significant build-up in concentrations.

For the elevated source (Fig. 10.15), the position of maximum g.l.c. is downstream of the source, since the pollutant has to be mixed down to the surface. The highest g.l.c.s occur in the most unstable conditions because of rapid vertical mixing and also because the skewed nature of the vertical component of turbulence brings the plume centerline down toward the surface (Section 10.3.2). This also results in the position of maximum g.l.c. being closest to the source for unstable conditions. As conditions become less convective and finally stable the maximum g.l.c. decreases, while the downwind distance of greater impact at ground level increases significantly.

10.4.2 Complex effects

The ways in which mean and turbulent airflow are affected by complex effects have been described in Section 10.2.4. Here we describe how these changes in flow impact on concentrations of dispersion pollutant (for a more comprehensive review see Hunt 1985). In fact, depending on the nature of the source and its location relative to the region of complex surface conditions and the receptor point, quite different aspects of the mean

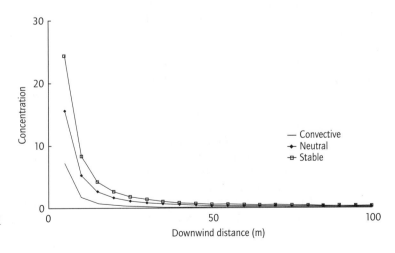

Fig. 10.14 Variations of ground level concentration with downwind distance for a ground-level source.

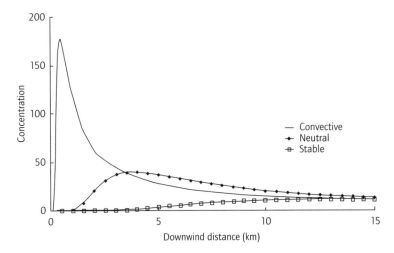

Fig. 10.15 Variation of ground level concentration with distance for an elevated source.

flow and turbulence are important. Some of the main influences on the ground level concentrations are described below (and see also Fig. 10.16).

1 The height of the plume centerline above the ground has a very large influence on ground-level concentration, since it determines how close the dispersing plume is to the ground.

2 The variations of the mean flow either side of and above and below the mean streamline. Wherever the mean flow changes, the streamlines converge and diverge, thus narrowing or widening the plume from a source. For plumes impinging onto hills, they diverge in both directions, but over hill tops the narrowing in the vertical direction of a plume is associated with divergence in the transverse direction.

3 The vertical gradient of the mean velocity or shear. Both over hills, especially in their wakes, and over roughness changes, the vertical shear of the mean wind is increased. This tends to lower the height of the position of maximum concentration in the plume. Downwind of structures or hills with large slopes, the mean flow can change direction and recirculate. Then the shear is very strong and has a correspondingly marked effect on the dispersion, especially in the recirulating region.

4 The turbulence on and near the mean streamline. For an elevated source it is particularly important to know the structure of the turbulence

between the plume centerline and the ground for estimating surface concentration. In some cases the mean streamline approaches the surface but the turbulence intensity is also reduced by the accelerating flow or increasing stable stratification. In that case the surface concentration may not be changed very much by the reduction in height of the mean streamline. On the other hand, enhanced thermal convection on the slopes of a hill can combine with a decrease in the mean streamline height to increase the concentration at the surface.

10.5 CONCLUSIONS

In this chapter we have seen that the structure of flow and turbulence within the atmospheric boundary layer determine how releases of pollutant into the atmosphere are transported and dispersed. Conveniently, as a result of similarity, in idealized conditions the structure can be described by a number of key parameters, including the surface roughness, Monin–Obukhov length, and boundary layer height. The ratio h/L_{MO} can be used to classify the boundary layer type.

The surface concentrations of pollutants are determined by the vertical and cross wind spreads of the dispersing cloud or plume, which depend on turbulence intensities and mean flow conver-

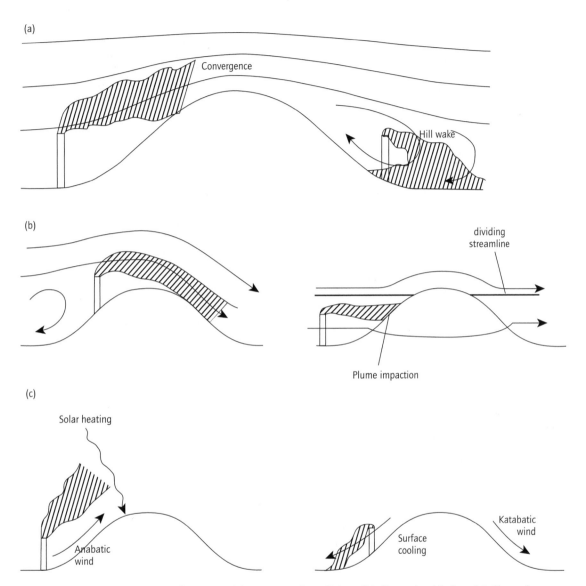

Fig. 10.16 Dispersion over a complex terrain: (a) near-neutral conditions; (b) effects of stable flow; (c) effects of thermal winds.

gence/divergence, and by the height of the plume, which is determined by the mean streamline through the plume and plume rise. We have shown how these quantities are affected by boundary layer structure, and we have also considered the influence of some complex effects, such as complex terrain, roughness changes, and mean thermal gradients. Inevitably, some other effects have been omitted. These include the influence of buildings and dispersion in calm conditions. However, the principles discussed in this chapter also apply to these cases.

ACKNOWLEDGMENTS

The material presented in this chapter draws on input from many colleagues. These include Julian Hunt, Alan Robins, David Thomson, and Christine McHugh.

REFERENCES

Briggs, G.A. (1984) Plume rise and buoyancy effects. In Randerson, D. (ed.), *Atmospheric Science and Power Production*. Report DOE/TIC-27601, Technological Information Center, Office of Science and Technology Information, US Department of Energy, Washington, DC.

Carruthers, D.J. & Hunt, J.C.R. (1990) Fluid mechanics of airflow over hills: turbulence, fluxes and waves in the boundary layer. In *Atmospheric Processes over Complex Terrain*. American Meteorological Society, Washington, DC.

Carruthers, D.J., Edmunds, H.A., Bennet, M. *et al.* (1997) Validation of the ADMS dispersion model and assessment of its performance relative to R-91 and ISC using archived LIDAR data. *International Journal of Environment and Pollution* **8**, 264–78.

Caughey, S.J. & Palmer, S.G. (1979) Some aspects of turbulence structure through the depth of the convective boundary layer. *Quarterly Journal of the Royal Meteorological Society* **105**, 811–27.

Caughey, S.J., Wyngaard, J.C & Kaimal, J.C. (1979) Turbulence in the evolving stable boundary layer. *Journal of Atmospheric Science* **6**, 1041–52.

Deardorff, J.W. (1985) Laboratory experiments on diffusion: the use of convective mixed layer scaling. *Journal of Climate and Applied Meteorology* **29**, 91–115.

Dyer, A.J. & Hicks, B.B. (1970) Flux-gradient relationships in the constant flux layer. *Quarterly Journal of the Royal Meteorological Society* **96**, 715–21.

Golder, D.G. (1972) Relations among stability parameters in the surface layer. *Boundary Layer Meteorology* **41**, 47–58.

Holtslag, A.M.M. & de Bruin, H.A.R. (1988) Applied modeling of the nighttime surface energy balance over land. *Journal of Applied Meteorology* **27**, 689–704.

Holtslag, A.M.M. & Nieuwstadt, F.T.M. (1986) Scaling the atmospheric boundary layer. *Boundary Layer Meteorology* **36**, 201–9.

Hunt, J.C.R. (1985) Turbulent diffusion from sources in complex flows. *Annual Review of Fluid Mechanics* **17**, 447–85.

Hunt, J.C.R., Kaimal, J.C. & Gaynor, E. (1988a) Eddy structure in the convective boundary layer—new measurements and new concepts. *Quarterly Journal of the Royal Meteorological Society* **114**, 827–58.

Hunt, J.C.R., Richards, K.J. & Brighton, P.W.M. (1988b) Stably stratified flow over low hills. *Quarterly Journal of the Royal Meteorological Society* **114**, 859–86.

Jackson, P.S. & Hunt, J.C.H. (1975) Turbulent wind flow over a hill. *Quarterly Journal of the Royal Meteorological Society* **101**, 919–55.

Lenschow, D.H. *et al.* (1988) Aircraft measurements, over Oklahoma, of the nocturnal boundary layer, obtained in the SESAME experiment of 1979. *Boundary Layer Meteorology* **42**, 95–121.

Mylne, K.R. & Mason, P.J. (1991) Concentration fluctuation measurements in a dispersing plume at a range of up to 1000 m. *Quarterly Journal of the Royal Meteorological Society* **117**, 177–206.

Nichols, S. & Readings, C.J. (1979) Aircraft observations of the structure of the lower boundary layer over the sea. Flights around the UK. *Quarterly Journal of the Royal Meteorological Society* **105**, 785–802.

Nieuwstadt, F.T.M. (1981) The steady-state height and resistance laws of the nocturnal boundary layer: theory compared with Cabauw observations. *Boundary Layer Meteorology* **20**, 3–17.

Nieuwstadt, F.T.M. (1984) The turbulent structure of the stable nocturnal boundary layer. *Journal of Atmospheric Sccience* **41**, 2202–16.

Oke, T.R. (1987) *Boundary Layer Climates*, 2nd edn. Methuen, London.

Panofsky, H.A. (1974) The atmospheric boundary layer below 150 m. *Annual Review of Fluid Mechanics* **6**, 147–77.

Pasquill, F. & Smith, F.B. (1983) *Atmospheric Diffusion*, 3rd edn. Ellis Horwood, New York.

Sheppard, P.A. (1956) Air flow over mountains. *Quarterly Journal of the Royal Meteorological Society* **82**, 528–9.

Stull, R.B. (1988) *An Introduction to Boundary Layer Meteorology*. Kluwer Academic Publishers, London.

Thomson, D.J. (1990) A stochastic model for the motion of particle pairs in isotropic high-Reynolds-number turbulence, and its application to the problem of concentration varience. *Journal of Fluid Mechanics* **210**, 113–53.

Van Ulden, A.P. & Holtslag, A.M.M. (1985) Estimation of atmospheric boundary layer parameters for diffusion applications. *Journal of Climate and Applied Meteorology* **24**, 1196–207.

11 Synoptic-Scale Meteorology

DOUGLAS J. PARKER

11.1 INTRODUCTION

The adjective "synoptic" literally means the consideration of simultaneous observations: in considering weather patterns, which vary in three physical dimensions and time, it was the development of synoptic charts in the nineteenth century that led to systematic and scientific study of meteorology, alongside considerable improvements in forecasting skill. The standard observations ("synops") that go into the production of "synoptic charts" are obtained from "synoptic stations" at the "synoptic hours" of 0000, 0600, 1200, and 1800 UTC. The typical spatial coverage of these stations means that they are well suited to study systems on scales of several hundred kilometers and upwards, with temporal scales on the order of days: for this reason, the term "synoptic," which literally means "simultaneous," has been transposed into a specification of horizontal scales of a few hundred to several thousand kilometers. In the mid-latitudes this corresponds to the scales of frontal cyclones and blocking highs, for example. In point of completeness, however, it should not be forgotten that the strict definition of the term "synoptic" allows us to discuss global synoptic analyses as well as meso- or microscale synoptic analysis—of an urban region, for instance.

The origins of synoptic analysis being in the study of instantaneous datasets, this kind of analysis is centered on **diagnostics** derived from the synoptic fields. For example, a key aspect of synoptic analysis is to diagnose areas of vertical motion, a field that is not directly measured, from the fields of horizontal wind and temperature. The standard models used for such analysis are based on balanced theories, in which the wind and temperature are assumed to be close to a state in which rotational and pressure gradient forces are in balance. In times when numerical weather prediction far outperforms human forecasters, such diagnostic methods are the ways in which we can interpret complicated numerically computed analyses, and even modify numerical forecasts to forecasting advantage. This chapter describes such models and the way they are applied in real contexts.

11.2 BASIC PHYSICAL DESCRIPTIONS AND MODELS

This section deals with an overview of the balanced models that govern synoptic meteorology. Various textbooks (Carlson 1991; Holton 1992) give detailed derivations of the basic equation sets, and here they are introduced only briefly. In particular, much of the following discussion is concerned with the "dry" evolution of synoptic systems.

11.2.1 Statics, stability, and parcel methods

To a good approximation the air is an ideal gas, with its temperature, T, pressure, p, and density, ρ, related by

$$p = \rho RT \qquad (11.1)$$

in which R is the gas constant ($R = 287 \, \mathrm{J \, kg^{-1} \, K^{-1}}$) for dry air. When the air contains water vapor the ideal gas law may be expressed as

$$p = \rho RT \left(\frac{1 + r/\varepsilon}{1 + r} \right) \qquad (11.2)$$

where r is the mixing ratio, defined as the mass of water vapor per unit mass of dry air, and $\varepsilon = R/R_v = 0.622$ is the ratio of gas constants for dry air and water vapor. In practice this is accommodated by using the virtual temperature

$$T_v = T \left(\frac{1 + r/\varepsilon}{1 + r} \right) \qquad (11.3)$$

T_v is the quantity that may be related to density and thereby buoyancy in the equations of motion for the air. Since r is a positive definite quantity and $0 < \varepsilon < 1$, air with a higher moisture content is less dense and has a higher value of T_v. At higher temperatures, in the tropical boundary layer, for example, where the saturation mixing ratio can be 4% or higher, T_v may differ from T by several Kelvin in moist air. However, at cooler mid-latitude and upper air temperatures, the difference between T_v and T is usually small, so in the coming discussion T will be used.

To a good approximation on synoptic scales, the atmosphere is hydrostatic. This is to say that the pressure distribution in the air is dominated by the weight of the air column above any given level, and that the dynamic pressure associated with the acceleration of air parcels is insignificant on these scales. Formally this leads to the hydrostatic relation:

$$\frac{\partial p}{\partial z} = -\rho g \qquad (11.4)$$

There are some immediate and important consequences of this relation:

1 Since density is a positive definite quantity the pressure always decreases upwards under hydrostatic conditions. It is this property that makes pressure a useful surrogate for physical height (for example, in altimeters). Since pressure decreases monotonically with height, synoptic meteorology is often described in "pressure coordinates"—for example, (x, y, p) replacing the Cartesian (x, y, z) for flow in rectangular geometry, or (χ, λ, p) replacing the spherical polars (χ, λ, z). Using pressure as a height scale means that we have a coordinate which is also a state variable for the air.

2 Integrating the hydrostatic relation from a given level upwards shows that the pressure at any level in a hydrostatic atmosphere corresponds to the weight of the column of air above that level, per unit area.

3 We can deduce an approximate **scale height** for the rate of decrease of pressure with height: if it is taken that T variations are small and T may be treated as a constant, eqn 11.4 can be integrated over height (using the ideal gas law) to give

$$p = p_0 \exp \left\{ -\frac{gz}{RT} \right\} \qquad (11.5)$$

where p_0 is the pressure at $z = 0$, implying a scale height z_s of

$$z_s = \frac{RT}{g} \approx 8 \times 10^3 \, \mathrm{m} \qquad (11.6)$$

Since pressure corresponds to the weight of the air column, this shows that the bulk of the atmosphere's mass lies in the troposphere.

4 The hydrostatic relation also leads directly to a useful relation describing the **thickness**, $z_1 - z_0$, of the air column between two pressure levels, p_0 and p_1. Integrating eqn 11.4, using the ideal gas law to eliminate density, the thickness is be found to be

$$z_1 - z_0 = \frac{R\overline{T}}{g} \ln \frac{p_0}{p_1} \qquad (11.7)$$

in which $\overline{T}$ is a mean temperature of the layer. The thickness is an extremely useful quantity in relating mean-state pressure and temperature anomalies. The 1000–500 hPa thickness is often plotted on surface charts, as an aid to estimating mean temperature of the lower troposphere. For forecasting purposes, it is taken that snow is the likely form of any precipitation which occurs in air whose 1000–500 hPa thickness is at most 528 dm (in the UK) or 540 dm (in the USA; a higher value because of differing climatological stability over the North American continent), because these values correspond to surface temperatures of less than about 0°C. Thickness is also useful in allowing us to make heuristic predictions of the atmospheric pressure response to thermal changes: warming a layer of the atmosphere increases the thickness and forces the pressure levels apart. This tends to create a high-pressure anomaly above the heating and low pressure below the heating (Fig. 11.1). Examples where this can be seen to apply are the warm cores of tropical cyclones, the formation of heat lows over the Iberian peninsula

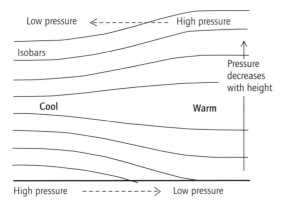

Fig. 11.1 A schematic of the way in which the concept of thickness allows us to associate thermal anomalies with pressure anomalies. On this vertical section the warm region is associated with a greater thickness, where the isobars are pushed apart. There is a tendency for the resulting pattern to give a low at the surface under the warm air and a high aloft, relative to a horizontal surface (dashed arrows indicate the direction of the pressure gradient force). Such pressure patterns help to explain the origins of the sea breeze, flowing down the pressure gradient from the cool maritime air to the warm land air, as well as the circulation patterns in warm core cyclones such as hurricanes, if the thermal wind relation (eqn 11.24) is invoked.

or the Australian desert, and the generation of land–coast pressure differences leading to the sea breeze circulation.

Although this relationship between atmospheric temperature and pressure is useful for deducing the pressure response to thermal changes, it should not be forgotten that the full vertical profile must be considered, and that warm air does not automatically imply a low at the surface. Indeed, cold core "cut-off" lows are common over the Atlantic sector, and a blocking high may exhibit a warm anomaly at the surface: in these cases the vortex intensity, and pressure anomaly, simply increases with height.

Invoking the hydrostatic relation in cases where the atmosphere has been externally forced—by heating, for example—implies that some adjustment has taken place, after which the atmosphere has gained a new, slowly varying state. For instance, in the generation of a heat low, the hydrostatic condition implies that mass has been removed from the column in order to generate a low at the surface, and this process must have involved air motion. There remains debate as to the exact nature of this process, but in synoptic analysis it is not thought to be of primary importance. In invoking the hydrostatic assumption it is said that we have "filtered" out the more rapid motions.

Any state property of air may be determined by two other state variables; for instance $T = T(p, \rho)$ according to the ideal gas law. In order to determine the evolution of two state variables when one other varies (the common example in the atmosphere being variation of pressure as air ascends or descends), another condition is required. In practice, this condition is the first law of thermodynamics, applied under the observation that heat exchange in the atmosphere is extremely weak on synoptic time scales (air is a poor conductor of heat), except in special instances such as the latent heat release in clouds and the turbulent exchanges in the boundary layer. Under such circumstances of zero heat exchange the first law may be written as

$$0 = c_p dT - \frac{1}{\rho} dp \qquad (11.8)$$

where c_p is the specific heat capacity at constant pressure, from which we integrate to obtain

$$T = \theta \left(\frac{p}{p_0} \right)^{\frac{R}{c_p}} \qquad (11.9)$$

(where for dry air $R/c_p = 2/7$). This is to say that for any given parcel of air, the temperature and pressure are related by eqn 11.9, with θ a constant for that parcel. We can rearrange this and label each parcel of air according to its value of θ: θ may vary in space, but if the air is disturbed materially, each parcel carries its θ value with it. θ is the **potential temperature** of the air and is conserved in adiabatic motion (for which $Q = 0$). θ is useful not only because it allows us to compute the thermodynamic properties of air as it moves between pressure levels, but also because it allows us to compare air parcels. Returning to eqn 11.9, it can be seen that on a given pressure level, air of higher temperature also has higher θ: since temperature is inversely proportional to density on this pressure level (from the ideal gas law), θ differences relate directly to density differences on a pressure level. It is the density differences that provide buoyancy and determine the principal driving forces of atmospheric flow. If the basic state θ profile increases with height, air that is lifted to a level of lower pressure will find itself surrounded by air with a higher potential temperature: the displaced air will be cooler and denser than its surroundings and tend to sink, and the atmosphere is stable. If the θ profile decreases with height, upwardly displaced air parcels are warmer and less dense than their environment, and continue to rise. These simple arguments lead to the conclusion that the static stability of an air profile depends on the sign of $\partial\theta/\partial z$:

$\partial\theta/\partial z > 0$: stable
$\partial\theta/\partial z = 0$: neutral
$\partial\theta/\partial z < 0$: unstable

Each of these conditions occurs in the atmosphere and characterizes particular flows: for example, unstable conditions generally occur in a boundary layer that is being forced by solar heating at the ground, and correspond to well developed turbulent convection. High stability is characteristic of the stratosphere, frontal zones, and a stable "inversion," capping the boundary layer. A common measure of static stability is the Brunt–Väisälä frequency, N, defined by

$$N^2 = \frac{g}{\theta_0} \frac{\partial\theta}{\partial z} \qquad (11.10)$$

where θ_0 is a constant reference value. N represents a characteristic frequency of oscillations of displaced air parcels in an atmosphere that is stable. Typically, in the troposphere, N has a value of around $10^{-2}\,\text{s}^{-1}$.

When moist effects are introduced, the question of stability becomes more complex. However, despite the increased complexity, the different possibilities for stability of a given profile can be related usefully to different cloud and weather phenomena, so it is worth giving some thought to this topic (a very detailed discussion of moist atmospheric thermodynamics and stability is given in Emanuel 1994).

When unsaturated air is lifted adiabatically to lower pressures, its dew point temperature falls less rapidly than its temperature, so that at some altitude, known as the **lifting condensation level** (LCL), these values become equal and the air becomes saturated. For saturated air, the state variables are linked by the conditions for saturation: the condition that the air remains saturated as it rises beyond the LCL determines the evolution of temperature and pressure for a given air parcel. This is to say that just as the condition of adiabatic motion for unsaturated air determines a relationship between T and p for a given air parcel, the condition of saturated motion determines an equivalent relationship for saturated parcels, by fixing the water vapor content to a known function of temperature and pressure. In this way it is possible to construct (given one or two approximations) "equivalent potential temperature," θ_e, with the following properties:

• θ_e is conserved in both saturated and unsaturated motion;
• θ_e is a function of (T, p) in saturated air, and increases monotonically with T.

There are different assumptions that may be made when evaluating θ_e, regarding the treatment of the different phases of water (for instance, it may be assumed that all condensed water is rained out instantly and no longer contributes to the air parcel thermodynamics). Accurate empirical formulae for calculating pseudoadiabatic θ_e have been derived by Bolton (1980):

$$\theta_e = T\left(\frac{1000}{p}\right)^{0.2854(1-0.28 \times r)}$$
$$\times \exp\left[\left(\frac{3376}{T_L} - 2.54\right) \times r(1+0.81r)\right] \quad (11.11)$$

where p is in hPa and the temperature at the condensation level can be obtained from

$$T_L = \frac{2840}{3.5 \ln T - \ln e - 4.805} + 55 \quad (11.12)$$

in which e is the vapor pressure in hPa.

Note that θ_e is homeomorphic to the "wet-bulb potential temperature," θ_w, and for the purposes of understanding, the two are interchangeable (though the exact functional values differ). Lines of constant θ_e are also lines of constant θ_w and they are termed pseudoadiabats. Note also that although θ_e is an exact thermodynamic variable in saturated flow, meaning that it is a function of exactly two other variables such as (p, T), and can be linked directly to buoyancy, in unsaturated air a third quantity, the air's moisture content, must be known in order to infer buoyancy from θ_e and pressure, say. For example, two unsaturated parcels of air, of the same pressure and temperature, may have differing water vapor contents, and thereby have different values of θ_e (implying that one reaches saturation more rapidly than the other). This means that although θ_e is an attractive variable in that it is universally conserved (under some reasonable approximations), the moisture content of the air must generally still be carried as a variable in order to determine atmospheric evolution.

When unsaturated air is lifted, it cools adiabatically, from eqn 11.9. When saturated air is lifted, it gains energy from the latent heat release of condensation, so that it cools at a slower rate than an unsaturated air parcel. Seen another way, the unsaturated air parcel conserves θ but the saturated air experiences an increase in θ due to the latent heat release as it rises. This means that an air profile which is stable to dry ascent may be unstable to moist ascent. Making the initial assumption of dry stability, there are two principal possibilities, depending on the state of subsaturation of the air profile, and there is some common confusion in defining the conditions for stability; they are summarized here:

First, the θ_e profile of the air determines the **convective stability** of the air (sometimes termed "potential instability"). This means that if the air becomes saturated, parcels will move along lines of constant θ_e, and arguments entirely analogous to the ones determining unsaturated static stability determine moist convective stability criteria depending on $\partial \theta_e / \partial z$. This condition relates to the stability of a layer of the atmosphere: if it is lifted and cooled to saturation, how will small perturbations behave in the saturated layer? Convective instability is manifest in cloud forms such as altocumulus, where a layer of ascending air becomes saturated and then evolves into convective cells.

Second, in many instances it is more useful to consider the stability according to the ascent of an isolated parcel of air over a significant vertical distance. In this case, the parcel reaches its lifting condensation level and subsequently rises along a pseudoadiabatic trajectory (conserving θ_e). At some stage, the parcel may then cross the environmental profile and become positively buoyant (region "PA" in Fig. 11.2). In this instance the air profile is "**conditionally unstable**," meaning that small perturbations are stable but significant displacements of a parcel may lead to free convection.

Conditional stability can be inferred from the slope of the environmental sounding with respect to dry and moist (pseudo-) adiabats on a thermodynamic diagram. If the sounding lies between the adiabats and pseudoadiabats, then the air is stable to unsaturated motion but there is a possibility of saturated parcels of air following a pseudoadiabat

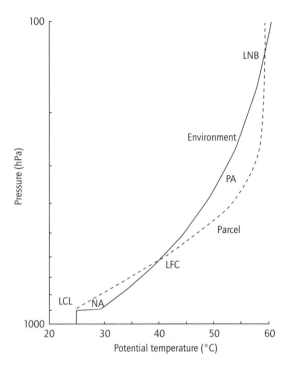

Fig. 11.2 A simple profile of air that is conditionally unstable (the "Environment" curve). A parcel of air that is lifted from the surface is initially cooler than its surroundings but, following condensation (at the lifting condensation level, LCL), ascends along a pseudoadiabat and can become buoyant (at the level of free convection, LFC). This kind of behavior leads to cumulus and cumulonimbus convection. The parcel finally regains the environmental curve and loses its buoyancy at the level of neutral buoyancy (LNB). For deep convection this can be the tropopause.

that crosses the sounding, leading to conditional instability. In practice, conditional instability occurs when there is a relatively cool mid-to-lower troposphere and a moist, humid surface layer: these conditions will tend to imply convective instability also. However, conditional instability does not depend on the moisture content of the air above the initial level of the air parcel: just the θ_e of the ascending parcel.

Conditional instability is characteristic of deep cumulus and cumulonimbus clouds. Usually, advection of cool air over a warm and moist boundary

layer sets up this kind of profile, and various **triggering** mechanisms act to lift air from the surface until it becomes unstable and rises rapidly, gaining potential energy as a result of its buoyancy. Adiabatic descent in response to the cloud acts to reduce the instability by warming the mid-troposphere, as do precipitation-driven downdraughts, which cool the boundary layer.

There are a number of measures of conditional instability that are used for forecasting. The Convective Available Potential Energy (CAPE) is a measure of the energy available to a lifted parcel of air rising from a given level up to cloud top, and is computed as

$$CAPE = -R\int_{p_i}^{p_n}\left(T_p - T_{env}\right)d\ln p \quad (11.13)$$

where T_p denotes the temperature of the rising parcel of air and T_{env} is the temperature of the environment. The integral is taken from the initial parcel pressure, p_i, up to the level of neutral buoyancy, p_n, at which the buoyant, rising parcel returns to the environmental profile (Fig. 11.2). Note that the CAPE is a function of the initial choice of ascending air parcel, a choice that influences p_i, p_n, and $T_p(p)$, but does not influence T_{env}. Higher values of CAPE indicate a likelihood of more intense convection. Other, simpler diagnostics of conditional instability exist: typically they may employ a measure of the difference between θ_e (or a surrogate, such as dew point depression) at two levels, chosen to represent the lower and middle troposphere, in order to estimate the buoyancy of lifted parcels.

Air profiles, as obtained from radiosonde ascents or from model fields, are generally plotted on thermodynamic diagrams of some kind, for which the axes of the diagram are state variables and, for obvious convenience, one of the axes is a function of p, so that height is upwards on the diagram. While the specifics of such diagrams differ according to convention and purpose, the generalities are the same:

1 Data are plotted as curves of (T, p) (solid line, by convention) and (T_d, p) (dashed line), where T_d is the dew point.

2 Adiabats are marked on the diagram: these are the routes that unsaturated air parcels will follow in ascent or descent.

3 Pseudoadiabats are also marked, to indicate the trajectories of saturated parcels.

4 Lines of constant humidity mixing ratio indicate the evolution of dew point for unsaturated air.

Given any initial air state of (p, T, T_d), causing this air to ascend to a lower pressure implies that the temperature will evolve by following an adiabat and the dew point by following a line of humidity mixing ratio, until these meet, at which point the air is saturated. Subsequent ascent will follow a pseudoadiabat. An example of the most commonly used thermodynamic diagram in the UK, the "tephigram," is shown in Fig. 11.3.

11.2.2 Dynamics

The discussion has so far considered a static atmosphere, and the conditions for instability of such an atmosphere when parcels are displaced. The study of the full dynamic evolution of weather systems requires a more precise description of the fluid flow. The basic equation set used to describe most atmospheric flows is the "primitive equations," which are derived as an inviscid, shallow-atmosphere approximation to the Navier–Stokes equations in spherical geometry (see Holton 1992). Molecular diffusion and viscosity terms are negligible on synoptic scales, due to the extremely high Reynolds numbers of macroscopic atmospheric phenomena. However, it is usually necessary to include turbulent dissipation terms on the

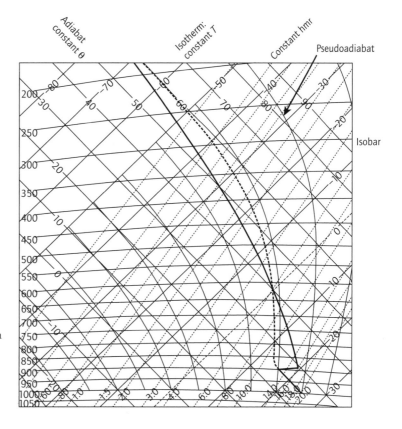

Fig. 11.3 The curves of Fig. 11.2 represented on a tephigram, as an example of one of a number of thermodynamic diagrams in common use. By convention, the dew point curve would be plotted on the (T, p) axes as a dashed line (not shown here). The dew point of an unsaturated air parcel follows a line of constant humidity mixing ratio as its pressure changes.

right-hand side of the momentum and thermodynamic equations.

It is the primitive equations that form the basis for numerical weather prediction (NWP) models and from which simpler sets are derived. For synoptic-scale flows on horizontal scales less than the planetary scale, these equations are commonly approximated to an "f-plane" or a "β-plane," in which cartesian coordinates are used and the Coriolis parameter, $f = 2\Omega \sin \chi$ (where χ is the latitude and Ω is the planetary angular rotation rate), is approximated by a constant (f-plane; $f = f_0$) or a linear function of the cartesian latitude (β-plane; $f = f_0 + \beta y$). In general, in the following discussions the northern hemisphere is assumed, so that $f_0 > 0$. The hydrostatic approximation is also used as a standard for synoptic flows.

Often, the equations are converted to a pressure coordinate system, (x, y, p) (see Carlson 1991). Pressure coordinates have the advantages that incompressibility is exact in the continuity equation, and that the Boussinesq approximation is not needed in order to simplify the pressure terms, but these coordinates yield a more complicated lower boundary condition and the necessity of describing vertical motion in terms of a less intuitive pressure tendency. Hoskins and Bretherton (1972) outlined a "pseudo height" vertical coordinate, which is a monotonically decreasing function of pressure and approximates quite closely to physical height: for simplicity of interpretation, this coordinate is used here. The basic working equation set is now:

$$\frac{Du}{Dt} - fv = \frac{\partial \phi}{\partial x} \qquad (11.14)$$

$$\frac{Du}{Dt} - fu = \frac{\partial \phi}{\partial y} \qquad (11.15)$$

$$\frac{\partial \phi}{\partial z} = \frac{g\theta}{\theta_0} \qquad (11.16)$$

$$\nabla \cdot (\rho_r \mathbf{u}) = 0 \qquad (11.17)$$

$$\frac{D\theta}{Dt} = S \qquad (11.18)$$

where (u, v, w) are the components of zonal, meridional, and vertical velocity, ϕ is the geopotential, $\rho_r(z)$ is a reference density, and S is the diabatic source term. The normal cartesian total derivative is used. The "vertical velocity," w, in these coordinates is a simple function of the pressure tendency, ω, and density; $w = -\rho g \omega$ (meaning that imposition of a boundary condition related to vertical velocity on w remains an approximation). The equations are identical to a Boussinesq, anelastic set in true height coordinates, in which case ϕ represents p/ρ_r.

QG theory

The classic simplification of eqns 11.14–11.18, which applies to synoptic-scale systems and has, in many respects, defined the subject of synoptic meteorology, is to nondimensionalize and then expand in the Rossby number

$$Ro = \frac{U}{f_0 L} \qquad (11.19)$$

where U is a velocity scale and L a length scale. At first order the expansion yields

$$u_g = -\frac{1}{f_0} \frac{\partial \phi}{\partial y} \qquad (11.20)$$

$$v_g = \frac{1}{f_0} \frac{\partial \phi}{\partial x} \qquad (11.21)$$

where $\mathbf{v}_g = (u_g, v_g)$ is the well known geostrophic wind, which is the horizontal wind for which the Coriolis force balances the pressure gradient force. A consequence of the force balance is that the wind is directed along the isobars, causing a cyclonic circulation (same sense as the planetary rotation) around a low-pressure system and an anticyclonic circulation around a high. In fact, the geopotential, ϕ (or in true height coordinates, the dynamic pressure, p/ρ), represents a streamfunction for the geostrophic wind. The geostrophic wind may usefully be employed to estimate wind speeds from charts of isobars. However, for synoptic analysis and forecasting another, deeper relationship, the

thermal wind balance, exists. The thermal wind balance comes from combining the hydrostatic eqn 11.16 with the equations for the geostrophic wind (eqns 11.20 and 11.21) to obtain

$$\frac{\partial u_g}{\partial z} = -\frac{g}{f_0\theta_0}\frac{\partial\theta}{\partial y} \tag{11.22}$$

$$\frac{\partial v_g}{\partial z} = \frac{g}{f_0\theta_0}\frac{\partial\theta}{\partial x} \tag{11.23}$$

or, in vector form

$$\frac{\partial \mathbf{u}_g}{\partial z} = \frac{g}{f_0\theta_0}\mathbf{k}\times\nabla\theta \tag{11.24}$$

where **k** is the unit vector in the z direction. The thermal wind relation is one of the fundamental rules, which is useful both for everyday weather prediction and for deeper, theoretical analysis. For everyday purposes a useful corollary of eqn 11.24 (obtained by considering the sign of the vector product of *u* with $\mathbf{u}_z$) is that if the wind backs with height (turns cyclonically) then there is cold advection, and if the wind veers with height (turns anticyclonically) there is warm advection (Fig. 11.4). In terms of understanding weather systems, the thermal wind relation explains the westerly shear of the mean wind with height in midlatitudes, in balance with the mean equatorward temperature gradient, and the tendency for the circulation in warm core systems such as tropical cyclones to become more anticyclonic with height. The vertical structure of cyclones and anticyclones is sometimes understood more easily in terms of the QG relative vorticity

$$\xi_g = \frac{\partial v_g}{\partial x} - \frac{\partial u_g}{\partial y} = \frac{1}{f_0}\nabla_h^2\phi \tag{11.25}$$

where ∇_h includes only the horizontal derivatives. This function has the same sign as f for cyclones (i.e. positive in the northern hemisphere) and the opposite sign for anticyclones. Using the thermal wind balance (eqns 11.22 and 11.23), the QG relative vorticity varies in height according to

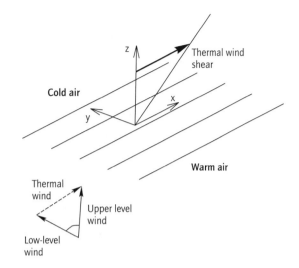

Fig. 11.4 A schematic of the relationship between the thermal wind (shear) and thermal advection. In this example, southerly flow at low levels, implying warm advection, leads to a clockwise rotation of the wind vector with height (veering).

$$\frac{\partial\xi_g}{\partial z} = \frac{g}{f_0\theta_0}\nabla_h^2\theta \tag{11.26}$$

For localized anomalies of θ, we can crudely estimate the Laplacian term in order to obtain an estimate of the right-hand side of eqn 11.26 as

$$\nabla_h^2\theta \sim -\frac{1}{L^2}\frac{g}{f_0\theta_0}\theta \tag{11.27}$$

from which we can associate a warm anomaly with a decrease in vorticity with height (e.g. tropical cyclone), and a cold anomaly with an increase in vorticity with height (e.g. cut-off low).

The geostrophic wind is degenerate, in that there is redundancy in the continuity and thermodynamic equations. In effect this means that advection of the geostrophic quantities, u_g, v_g, and θ, by the geostrophic winds alone, will tend to take the system out of balance. For example, a wind flow that is sheared with height may act to advect warm air over a surface cyclone, thereby tending to increase the vorticity at the surface and decrease it aloft (thermal wind relation). Such changes in vorticity can only occur through ageostrophic motion, such as vertical velocity stretching the surface vorticity. Hence in order to describe the time-evolution of the geostrophic winds, it is necessary to consider the Ro expansion further (i.e. consider an expansion of the form

$$u = u_g + Ro \times u_{ag} + O(Ro^2) \qquad (11.28)$$

where u_{ag} is the "ageostrophic" zonal wind, and take terms in $O(Ro)$. To next order in Ro we obtain the quasigeostrophic (QG) equations ("quasi" because they include the effect of an ageostrophic wind), which are the basis for an understanding of synoptic weather systems:

$$\frac{D_g u_g}{Dt} - f_0 v_{ag} - \beta y v_g = 0 \qquad (11.29)$$

$$\frac{D_g v_g}{Dt} + f_0 u_{ag} + \beta y u_g = 0 \qquad (11.30)$$

$$\frac{\partial \mathbf{u}_g}{\partial z} = \frac{g}{f\theta_0} \mathbf{k} \times \nabla \theta \qquad (11.31)$$

$$\rho_r \left(\frac{\partial u_{ag}}{\partial x} + \frac{\partial v_{ag}}{\partial y} \right) + \frac{\partial(\rho_r w)}{\partial z} = 0 \qquad (11.32)$$

$$\frac{D_g \theta}{Dt} = -w \frac{\theta_0}{g} N^2 + S \qquad (11.33)$$

where the geostrophic Lagrangian derivative is

$$\frac{D_g}{Dt} = \frac{\partial}{\partial t} + u_g \frac{\partial}{\partial x} + v_g \frac{\partial}{\partial y} \qquad (11.34)$$

$N^2(z)$ is a function of the basic state, reference temperature profile, and small products of $\mathbf{u}_{ag}$ and β have been neglected. It is also useful to construct a QG vorticity equation for the vertical component of the geostrophic vorticity, $\zeta_g = f + \xi_g$,

$$\frac{D_g}{Dt}(\zeta_g) = \frac{f_0}{\rho_r} \frac{\partial}{\partial z}(\rho_r w) \qquad (11.35)$$

From eqn 11.35 it can be seen that the principal tendency in geostrophic absolute vorticity is due to convergence or divergence leading to vortex stretching in the vertical. This highlights the importance of the ageostrophic wind in QG dynamics: the only way to derive a tendency in the geostrophic fields is through ageostrophic motion.

There are two useful ways of simplifying the QG equation set. The first and perhaps more obvious is to combine these equations in a form that eliminates time derivatives and leaves ageostrophic terms on the left-hand side, as flows forced by geostrophic terms on the right: this leads to the well known "omega equation" (or, for two-dimensional flows, the equivalent Sawyer–Eliassen equation). The second route is equally useful and involves eliminating ageostrophic terms altogether, to obtain evolution equations for a geostrophic variable, the QG potential vorticity (PV), which has a profound role in synoptic meteorology.

Diagnosing the ageostrophic wind: the omega equation. Combining eqn 11.35 with eqn 11.33, using the thermal wind balance (eqn 11.26) to eliminate time derivatives, gives a single equation for ageostrophic vertical velocity, w:

$$N^2 \nabla_h^2 w + f_0^2 \frac{\partial}{\partial z} \left(\frac{1}{\rho_r} \frac{\partial}{\partial z}(\rho_r w) \right) = f_0 \frac{\partial}{\partial z} \left(\mathbf{u}_g . \nabla \zeta_g \right)$$

$$- \frac{g}{\theta_0} \nabla_h^2 \left(\mathbf{u}_g . \nabla \theta \right) + f_0 \beta \frac{\partial v_g}{\partial z} + \frac{g}{\theta_0} \nabla_h^2 S$$

$$(11.36)$$

This equation is generally known as the **omega equation**, since in true pressure coordinate form it would be used to solve for the pressure tendency, ω. It has the form of:

$$L[w] = F[\Phi] \qquad (11.37)$$

where L is a linear differential operator and F is a forcing operator that is dependent on geostrophic streamfunction $\Phi = \phi/f_0$. The boundary condition on w is the kinematic boundary condition of zero normal flow at the lower surface, or a pressure tendency for ω, which, for flat terrain, is often regarded to be small. From eqn 11.37, it is apparent that the ageostrophic wind is *determined by* the geostrophic fields, and it is important to remain aware that the ageostrophic wind represents only the next order of expansion in Rossby number, not the full departure of the wind field from the geostrophic wind. In particular, gravity waves remain unrepresented in QG dynamics. The ageostrophic wind exists in order to maintain geostrophic balance in a time-dependent solution. However, from a perspective of forecasting and understanding, diagnosis of vertical motion through forms such as eqn 11.37 is critical to questions of cyclonic development and cloud formation.

The solution to eqn 11.37 depends on solution of a second-order partial differential equation according to particular forcing, F, and boundary conditions. In order to gain some insight into the physical meaning of this equation, it is useful to consider different partitions of $F[\phi]$ and some rule-of-thumb inversions of the operator, L. In particular, for constant N and ρ_r the height coordinate can be rescaled to $z^* = (N/f)z$, in which case L is the Laplacian operator. This means that we can gain some intuition into the solution according to various forcings, from our understanding of this well known operator. Then, for many functions (e.g. sinusoidal), we can make the approximation that

$$L[w] \sim -\frac{N^2}{L^2} w \qquad (11.38)$$

where L is a length scale for the system.

In the following analysis we will neglect the term involving β in the forcing of the omega equation (eqn 11.36). The traditional partition of $F[\Phi]$ given above is into a term involving vorticity advection and another involving thermal advection, plus the diabatic source term (see Carlson (1991) for discussion of various other expressions of $F[\Phi]$). The forcing may be rewritten

$$F[\phi] = -f_0 \frac{\partial}{\partial z}(\text{VA}) + \frac{g}{\theta_0}\nabla_h^2(\text{TA}) + \frac{g}{\theta_0}\nabla_h^2 S \qquad (11.39)$$

where the new terms are the **vorticity advection**

$$\text{VA} = -\mathbf{u}_g.\nabla\zeta_g \qquad (11.40)$$

and the **thermal advection**

$$\text{TA} = -\mathbf{u}_g.\nabla\theta \qquad (11.41)$$

These terms can be computed exactly from model data, but may also be estimated by eye, as in eqn 11.38:

$$w \sim -\frac{L^2}{N^2} L[w] = -\frac{L^2}{N^2} F[\Phi]$$

$$\sim \frac{f_0 L^2}{N^2} \frac{\partial}{\partial z}(\text{VA}) + \frac{g}{\theta_0 N^2}(\text{TA}) + \frac{g}{\theta_0 N^2} S \qquad (11.42)$$

This approximate relationship relates upward motion to:

1 A positive vertical gradient of vorticity advection (VA); for instance, the advection of an upper-level vortex tends to produce upward motion ahead of the vortex.

2 Positive thermal advection (TA) or positive diabatic heating (S). These latter two are in some way intuitive, in that warming tends to cause upward motion, but notice that it is not thermal anomalies which lead to upward motion in the QG system, but the **thermal tendency**.

These vorticity and thermal advection terms are the ones traditionally used by forecasters. However, use of two separate forcings based on geostrophic diagnostics that are essentially linked through thermal wind balance is rather unsatisfactory, especially in instances where the signs of these forcings may be opposite. A refinement of the description of $F[\phi]$ has been through the construction of **Q**-vectors (see Sanders & Hoskins 1990). The **Q**-vector is defined to be

$$\mathbf{Q} = -\frac{g}{\theta_0} \begin{pmatrix} \dfrac{\partial u_g}{\partial x}\dfrac{\partial\theta}{\partial x} + \dfrac{\partial v_g}{\partial x}\dfrac{\partial\theta}{\partial y} \\[2mm] \dfrac{\partial u_g}{\partial y}\dfrac{\partial\theta}{\partial x} + \dfrac{\partial v_g}{\partial y}\dfrac{\partial\theta}{\partial y} \end{pmatrix} \qquad (11.43)$$

in which case we have

$$F[\Phi] = 2\nabla.\mathbf{Q} \qquad (11.44)$$

The **Q**-vector field is often computed numerically, as a model diagnostic, but also may be inferred by eye from thermal and wind fields. It is possible to rewrite the **Q**-vector in the form

$$\mathbf{Q} = -\frac{g}{\theta_0}|\nabla\theta|\mathbf{k} \times \frac{\partial\mathbf{u}_g}{\partial s} \qquad (11.45)$$

where s is a coordinate following the θ contours in the direction of the thermal wind. Interpreting this geometrically, take the vector change in $\mathbf{u}_g$ in the direction of the thermal wind, and rotate this vector 90° anticyclonically to obtain the direction of the **Q**-vector. In order to interpret **Q**-vectors in terms of the ageostrophic response, once again think of the simplistic inversion of the Laplacian (eqn 11.38), in which case descent is associated with divergence of **Q** and ascent is associated with convergence of **Q**. Considering the continuity equation (11.17) means that low-level **Q**-vectors often point in the direction of the low-level ageostrophic wind.

As an example of using **Q**-vectors, consider a local jet maximum, where the wind is roughly along the direction of the thermal wind (Fig. 11.5). Then the **Q**-vector forcing implies ascent to the left side of the jet exit and to the right of the jet entrance.

Q-vectors also have a useful role in diagnosing regions of frontogenesis. If we consider the rate of change of the square of the local thermal gradient, then it is relatively straightforward to show that at low levels it is proportional to the scalar product of the gradient and the **Q**-vector:

$$\frac{D_g}{Dt}|\nabla\theta|^2 = \frac{D_g}{Dt}\left[\left(\frac{\partial\theta}{\partial x}\right)^2 + \left(\frac{\partial\theta}{\partial y}\right)^2\right] = 2\frac{g}{\theta_0}\mathbf{Q}\cdot\nabla\theta$$

$$(11.46)$$

(in which we have used the observation that at low levels $w \to 0$). This indicates that the low-level thermal gradient increases where the **Q**-vectors and thermal gradient vectors are aligned.

Overall, the choice of partition of the omega equation forcing is rather subjective. Thermal

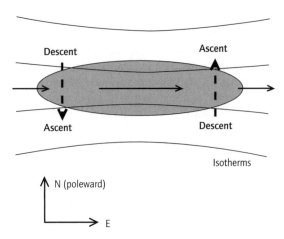

Fig. 11.5 One example of the use of **Q**-vectors in diagnosing vertical motion in synoptic scale flows: the vertical circulation associated with an upper level jet. The jet maximum is shaded, the geostrophic wind is indicated by continuous vectors, and the continuous contours are isotherms. Along the jet axis, the vector change in the geostrophic wind is directed along the wind direction, so that the **Q**-vectors are as indicated by the dashed vectors. Ascent occurs at the "left exit" and "right entrance" to the jet.

advection and vorticity advection are intuitive processes that may be rather easier to interpret, but the **Q**-vector form is more concise and is unambiguous in cases where the thermal and vorticity advection terms may cancel. In addition, the **Q**-vector form has some advantages in further diagnostics such as those relating to frontogenesis.

QG potential vorticity. The second useful derivation from the QG equations is to eliminate the ageostrophic terms between eqns 11.35 and 11.33, to arrive at a time-dependent equation involving purely geostrophic quantities. This equation may be reorganized to the remarkably simple form

$$\frac{D_gP}{Dt} = \frac{f_0}{\rho_r}\frac{\partial}{\partial z}\left(\frac{g\rho_r}{\theta_0 N^2}S\right) \qquad (11.47)$$

where P is the QG potential vorticity (PV), defined by

$$P = f + \nabla^2 \Phi + \frac{f_0^2}{\rho_r} \frac{\partial}{\partial z} \left(\frac{\rho_r}{N^2} \frac{\partial \Phi}{\partial z} \right) \qquad (11.48)$$

in which Φ is the geostrophic streamfunction, $\Phi = \phi/f_0$. Thus the QG PV is *conserved* in adiabatic motion. The QG PV has a profound role in atmospheric and oceanic dynamics, and some general discussion of PV is given in Section 11.2.4. In QG systems, eqn 11.48 can be solved ("inverted"), with suitable boundary conditions such as the Neumann boundary condition of θ on the horizontal lower surface, to give the geostrophic streamfunction, Φ. Knowing the geostrophic wind field then allows the PV field to be advected, through eqn 11.47. Thus the PV evolution offers an alternative route to solving and understanding the evolution of the QG system. PV is also attractive in that it unites the geostrophic variables in a single scalar field: in QG dynamics, vortices are inseparably related to thermal anomalies through thermal wind balance, and the PV encapsulates this (consider writing the QG PV as

$$P = f + \xi_g + \frac{f_0 g}{\rho_r \theta_0} \frac{\partial}{\partial z} \left(\frac{\rho_r}{N^2} \theta \right) \qquad (11.49)$$

which expresses it as a linear function of vorticity and potential temperature). Many meteorologists prefer to identify an atmospheric feature as a PV feature for this reason.

Inversion of the PV equation (11.48) again brings forward the natural height scale z^*. The definition of z^* implies a natural ratio of scales for the system, of $L/H \sim N/f$, where H is a height scale. This ratio of scales typifies many synoptic systems. It also indicates a constraint on modeling gridbox sizes, if synoptic weather systems are to be represented isotropically; the ratio of the horizontal to the vertical resolution of a model should ideally correspond to N/f. From consideration of z^*, it is useful to construct a Rossby radius

$$L_R = NH/f \qquad (11.50)$$

and a Rossby height scale

$$H_R = fL/N \qquad (11.51)$$

to characterize the scales of influence of a given PV or omega-forcing anomaly. These are the scales over which balanced flows tend to evolve. For instance, a mesoscale convective system may produce a localized PV anomaly and some localized forcing of ageostrophic motion: the geostrophic and ageostrophic flows that develop as a response to this convective system may be expected to extend over a horizontal region determined by the Rossby radius, L_R.

SG theory

Hoskins and Bretherton (1972) introduced semigeostrophic (SG) theory, as a refinement of QG theory, to allow for the short length scales that occur across fronts. The basis of SG dynamics is to define different Rossby numbers according to the across- and along-front wind and length scales: terms scaling with the weak across-front wind and the large along-front length scale tend to yield low Rossby number and are negligible to first order. The result of this analysis is an equation set resembling the QG set, but with ageostrophic winds explicitly in the Lagrangian derivative. Since this theory is primarily applied to frontal zones it is here presented in two dimensions, and for an f-plane: extension to three dimensions is documented by Hoskins (1975). Now the basic equations are

$$v_g = \frac{1}{f_0} \frac{\partial \phi}{\partial x} \qquad (11.52)$$

$$\frac{g}{\theta_0} \theta = \frac{\partial \phi}{\partial z} \qquad (11.53)$$

$$\frac{Dv_g}{Dt} + f_0 u_{ag} = 0 \qquad (11.54)$$

$$\rho_r \frac{\partial u_{ag}}{\partial x} + \frac{\partial(\rho_r w)}{\partial z} = 0 \qquad (11.55)$$

$$\frac{D\theta}{Dt} = S \qquad (11.56)$$

where x is the cross-frontal direction and the

Lagrangian derivative includes advection due to the across-front ageostrophic flow (u_{ag}, w)

$$\frac{D}{Dt} = \frac{\partial}{\partial t} + (u_g + u_{ag})\frac{\partial}{\partial x} + v_g\frac{\partial}{\partial y} + w\frac{\partial}{\partial z} \quad (11.57)$$

This system is simplified by transformation to geostrophic coordinates (as distinct from the usual physical coordinates, (x, z))

$$(X, Z, T) = \left(x + v_g/f, z, t\right) \quad (11.58)$$

so that the total derivative reduces to

$$\frac{D}{Dt} = \frac{\partial}{\partial T} + u_g\frac{\partial}{\partial X} + v_g\frac{\partial}{\partial Y} + w\frac{\partial}{\partial Z} \quad (11.59)$$

Then, after manipulation as outlined in the QG derivation above, a Sawyer–Eliassen equation (related to an omega equation) can be derived for the ageostrophic streamfunction.

More details of the mathematical formulation in geostrophic coordinates are given by Hoskins (1975). An equation for PV conservation is obtained after some quadratic terms have been neglected:

$$\frac{Dq}{Dt} = \frac{\zeta}{\rho}\frac{\partial S}{\partial Z} \quad (11.60)$$

with the SG PV given by

$$q = \frac{1}{\rho}\zeta\frac{\partial\theta}{\partial Z} \quad (11.61)$$

The SG equations in geostrophic (X, Y, Z) are structurally very similar to the QG equations in physical coordinates (x, y, z) and in consequence the interpretation of the solutions follows the ideas outlined for QG dynamics. However, in interpreting the SG results, they must be transformed back to physical (x, y, z) space. It is this nonlinear transformation that accounts for the influence of ageostrophic advection—the key process differentiating QG and SG dynamics. Notably, the Jacobian of the transformation turns out to be

$$J \equiv \frac{\partial(X, Y, Z)}{\partial(x, y, z)} = \zeta/f = \frac{f}{1 - \frac{1}{f}\frac{\partial v_g}{\partial X}} \quad (11.62)$$

so that transforming back to physical space effectively "contracts" the results in zones of high vorticity. It is this process that allows the SG equations to be useful for describing fronts, which are amplified in regions of high vorticity. In practical terms, solving numerically in geostrophic space on a regular grid and transforming back to physical space automatically contracts the grid in physical space, in regions of high vorticity. Singular fronts, at which there are discontinuities in the wind and temperature, evolve out of smooth initial synoptic systems at the point where the vorticity becomes singular, since in geostrophic space vorticity is a nonlinear function. There are two canonical large-scale flows that lead to SG frontogenesis: deformation and horizontal shear. Deformation acts to contract any temperature field along its "diffluent" axis (Fig. 11.6a), while shear, as occurs in a growing baroclinic wave, for example, will produce strong gradients along a line of cyclonic vorticity (Fig. 11.6b). Both mechanisms are observed, and the characteristics of each kind of front are different: the deformation front is more "stationary" in the flow, acting more like a material boundary than the shear front, which generally exists as a component in a propagating wave, with significant across-front vertical shear of the wind.

Formally, three-dimensional SG dynamics involves the same level of approximation as QG dynamics: there is no formal reason to prefer one to the other. However, the SG equations are in two dimensions able to represent frontal zones to very good accuracy, and for this reason they are generally favored for the consideration of systems such as the frontal cyclone, where the synoptic cyclone develops fine-scale quasi two-dimensional fronts. SG dynamics also leads to some other improvements over QG, such as a physical representation of the observed asymmetry between the intensity of highs and lows in a developing baroclinic wave.

(a)

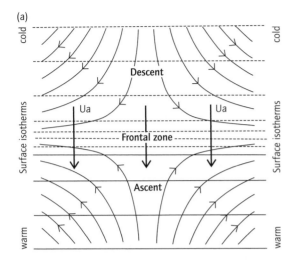

(b)

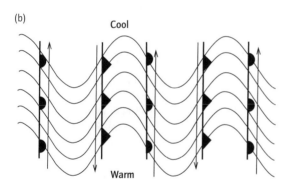

Fig. 11.6 The canonical large-scale flows that lead to frontogenesis, viewed in horiozontal sections. (a) The deformation flow intensifies the component of the temperature gradient lying along its confluent axis. Also indicated is the ageostrophic cross-frontal circulation, which through SG theory intensifies the frontogenesis process in a nonlinear way. At low levels thermal wind balance leads to an along-front geostrophic flow with the cold air on its right (not shown). (b) Horizontal shear leads to locally intense temperature gradients which may grow to a singularity in zones of cyclonic vorticity. Note the relationship of this system to the Rossby wave of fig. 11.11.

11.2.3 Diabatic effects

Stress terms on the right-hand side of the momentum equations arise from unresolved momentum flux convergence. In the troposphere and stratosphere these terms appear as a result of mixing by convection and at significant dynamical features such as fronts or jetstreaks, or as a result of momentum flux convergence by inertia-gravity waves. The inertia-gravity waves are oscillations on relatively high frequencies, from a synoptic perspective, and are effectively filtered out of the balanced equation sets: they are generated at orography, as well as at weather features such as fronts and cumulonimbus systems. None of these means of imposing stresses on airflow is particularly well understood, although some attempts to represent them in numerical models are being made. However, the stresses occurring at the Earth's surface and communicated by turbulence into the boundary layer are better understood and can be discussed in relatively simple terms.

As a first approximation, if the drag force on an air parcel in the boundary layer exactly opposes the direction of the parcel's velocity, a simple force diagram (Fig. 11.7) indicates that the *velocity* will be caused to deviate from the geostrophic wind *toward low pressure*. A particularly well known and useful paradigm that demonstrates this more quantitatively is that of the "Ekman layer." In the first instance, if the boundary layer flow is approximately independent of horizontal x and y, and the turbulent diffusion approximates to a high viscosity, K, for steady flow the horizontal momentum equations (11.14 and 11.15) become

$$-f(v - v_g) = K \frac{\partial^2 u}{\partial z^2} \qquad (11.63)$$

$$f(u - u_g) = K \frac{\partial^2 v}{\partial z^2} \qquad (11.64)$$

Using boundary conditions of $(u, v) = 0$ at $z = 0$ and $(u, v) \to (u_g, v_g)$ as $z \to \infty$, the solution is obtained as

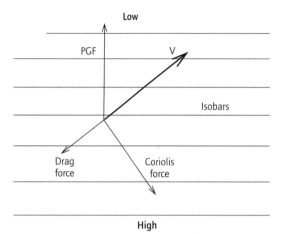

Fig. 11.7 A schematic of the force balance when drag is opposed to the wind direction, leading to a turning of the wind vector towards low pressure.

$$u = u_g\left(1 - \exp\left(-\frac{z}{\delta}\right)\cos\frac{z}{\delta}\right) - v_g \exp\left(-\frac{z}{\delta}\right)\sin\frac{z}{\delta}$$

$$(11.65)$$

$$v = v_g\left(1 - \exp\left(-\frac{z}{\delta}\right)\cos\frac{z}{\delta}\right) + u_g \exp\left(-\frac{z}{\delta}\right)\sin\frac{z}{\delta}$$

$$(11.66)$$

where the characteristic boundary layer depth

$$\delta = \sqrt{\frac{2K}{f}} \qquad (11.67)$$

A typical value of δ might be 500 m. This solution represents the "Ekman spiral"; for instance, if the geostrophic wind is a constant westerly wind, $(u_g, 0)$, the limiting surface wind direction is southwesterly (northern hemisphere; $f > 0$), and the wind vector rotates anticyclonically with height round toward the geostrophic wind. Effectively this confirms that the flow under the influence of turbulent drag deviates toward low pressure.

If the conditions are relaxed to allow for weak horizontal gradients, the Ekman solution may be used to deduce a weak vertical motion, w_E, at the top of the Ekman layer: from continuity (eqn 11.17), assuming constant density in the relatively shallow boundary layer and integrating with height

$$w_E = \frac{\delta}{2}\xi_g \qquad (11.68)$$

This "Ekman pumping" implies that in regions of positive relative vorticity there is upward motion forced at low levels. The upward motion is compensated by a vertical "squashing" of the air in the troposphere, leading to a reduction in the relative vorticity (from eqn 11.35). In effect, the boundary layer stresses on the airflow extract energy from a cyclone through this Ekman pumping mechanism, and this is a way in which vortices (positive or negative relative vorticity) spin down under the influence of boundary layer viscous effects.

The turbulence in the boundary layer can be a strong function of the solar heating at the surface, in which case K can fall suddenly at nightfall to a small value. Then, the Coriolis force acting on the air in the boundary layer is no longer balanced by turbulent dissipation terms and leads to an acceleration

$$\frac{du}{dt} - f(v - v_g) = 0 \qquad (11.69)$$

$$\frac{dv}{dt} + f(u - u_g) = 0 \qquad (11.70)$$

the solution of which is an "inertial oscillation" of frequency f (Fig. 11.8). The effect of this oscillation is to cause the wind direction to fluctuate during the night, at the altitudes where there had been a significant departure of the horizontal wind from the geostrophic value, usually in the lower hundreds of meters above the ground. Notably the amplitude of the wind increases with time in the early stages of the oscillation and at some point in the cycle it will exceed the geostrophic wind. This acceleration of the wind speed at sunset is referred to as the "nocturnal jet."

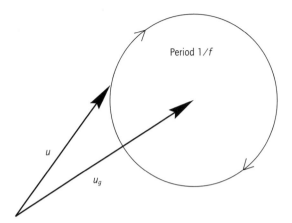

Fig. 11.8 The inertial oscillation: when drag retards the wind it tends to turn toward the left (NH). If this drag subsequently diminishes, the wind vector describes an inertial oscillation about the geostrophic wind vector. At some point in the cycle, this leads to a wind magnitude exceeding that of the geostrophic wind.

From the omega equation (11.36) it can be seen that a positive diabatic heating, S, will tend to force positive vertical motion (from a naive inversion of the Laplacian operators, as in eqn 11.38), as would be expected. The influence on the PV is a positive PV source below the level of maximum heating and a negative PV source above (in some senses there is no real PV source (Haynes & McIntyre 1987, 1990), merely a PV redistribution, and the volume-integrated PV remains constant unless there are boundary sources). The net effect on the air passing through the region of latent heating depends crucially on the airflow structure: if the air simply flows upwards in the region of heating, it gains PV at lower levels and loses it aloft, leading to a PV maximum at midlevels (Fig. 11.9a). If, in contrast, the heating is short-lived, or the air is passing horizontally through the region of heating, low-level air gains positive PV and upper-level air gains negative PV, as in the squall line models of Hertenstein and Schubert (1991) or Parker and Thorpe (1995) (Fig. 11.9b). The situation as sketched in Fig. 11.9b is really rather complicated: individual parcels of air may descend in the unsaturated environment of convective cells,

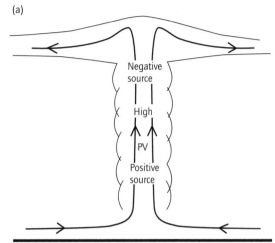

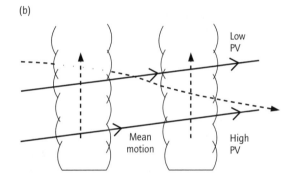

Fig. 11.9 Schematics of the relationship between propagation and PV sources for diabatic heating in a convective system. In (a) the source is stationary; as air ascends in the convective clouds it gains PV at low levels to generate a positive anomaly. Above the level of maximum heating the PV source becomes negative and the resulting PV structure is a single positive anomaly at mid-levels. In (b) the source is moving, and air parcels pass though the PV sources from left to right (dashed lines represent air parcel trajectories and continuous lines show the mean air motion). Low-level air gains positive PV and upper-level air gains negative PV on the large scale. The resulting PV structure is a dipole in the vertical.

ascend rapidly within the cells themselves, or even descend rapidly in the downdraughts of storms. The response of the large-scale, averaged state to this convection is not well understood, but would typically correspond to a large-scale mean heating with a structure related to that within an individual cumulonimbus.

Closing the system by coupling the cloud diabatic sources to the synoptic dynamics is one of the poorly understood meteorological problems. Some very simple models have used a modified "CISK" parametrization, to specify the heating as a function of boundary layer convergence (calculated from the ageostrophic winds, which are themselves forced by the latent heating, or from the boundary-layer convergence), in a representation of "triggered" convection. However, the atmospheric horizontal scales over which triggering is the controlling process are thought to be restricted to the mesoscale; perhaps pertaining to frontal convection on tens of kilometers but not to widespread convection over hundreds of kilometers in the cold sector of a frontal cyclone, for example.

Other simple parametrizations have employed an assumption of near-neutral stability to convection, in which case there is a heat source proportional to the local value of (upward) vertical motion, which leads to a reduced (but still positive) effective static stability. If a negative effective stability occurs, the various inversion operators for the balanced systems (e.g. $L[w]$ for the omega equation) change their properties significantly, by ceasing to be elliptic. In particular, it is possible then for the solutions to lose uniqueness, in which case "free" solutions may be obtained, and the results are not in general well determined.

Neither of the above approaches to representing cloud diabatic processes is ideal, and there is some evidence that more than one parametrization may need to be combined if realistic representations are to be achieved (Lagouvardos *et al.* 1993). Operational NWP models split clouds into large-scale, resolved clouds and subgrid convective clouds. This split reflects physical differences in cloud form and origin (stratiform versus convective), as well as being a practical necessity for modeling

with a finite horizontal resolution. The operational convective parametrization may be dealt with in a variety of ways, and its implementation has a strong bearing on the results of global numerical models, largely because of the impact of moist convection on tropical dynamics. Increasing computer power has meant progressively finer numerical grids, and we are now at the stage where the resolution of models is close to the scales of convective systems, and such systems are being neither resolved nor parametrized.

11.2.4 More on PV

Study of dynamical meteorology has in recent years been suffused with the concept of potential vorticity, yet many meteorologists, in particular some practicing forecasters, remain skeptical as to the usefulness of PV in real analysis. In this light, the important properties of PV are reviewed here. PV is often regarded as a rather mysterious quantity, shrouded in subtle mathematics: while it is true that the mathematics behind PV dynamics can be deep, the fundamentals of PV behavior are quite intuitive, given some basic knowledge of the dynamics.

Hoskins *et al.* (1985) (HMR) highlighted the two aspects of PV analysis that make it useful: its conservation and its invertibility. Conservation is exact in adiabatic, inviscid flows, even for the full inviscid Navier–Stokes equations, in which case

$$P = \frac{1}{\rho}\zeta.\nabla\theta \qquad (11.71)$$

where ζ is the absolute vorticity vector. If the atmosphere approximates adiabatic conditions (typically over times of many hours or a few days) the PV is a dynamical tracer for the air. It is the dynamical part of this that particularly fascinates theoreticians; the fact that there is a tracer which reflects properties of the wind field. For practical purposes the basic state structure of PV in the atmosphere makes conservation particularly useful: the PV becomes large in the stratosphere due to the high static stability above the tropopause, and therefore anomalies in PV at upper levels (and even descending to the surface in tropopause folds) reflect material motion of air of stratospheric and tropospheric origin. When correlated with trace gases such as water vapor or ozone, the PV can demonstrate deep links between the chemistry and the dynamics of the atmospheric flow. In midlatitudes the dynamical tropopause, often defined to lie at PV = 2 PVU (1.0 or 1.5 PVU are also commonly used, where the PV unit is defined as 1 PVU $= 10^{-6}\,m^2\,K\,kg^{-1}\,s^{-1}$; see HMR) has become widely used in diagnostic studies, as this is approximately a material surface whose evolution can be followed through time.

Plotting the PV on an isentropic surface is often preferred to plotting PV on height or pressure surfaces, since the isentropic surface is approximately a material surface. This means that PV anomalies can credibly be tracked in two dimensions without concern for motion out of the plane (except in regions of significant diabatic processes).

Invertibility implies that the flow field can be reconstructed from a knowledge of the PV (with suitable boundary conditions), and it is this property that gives PV real practical usefulness in understanding the evolution of weather systems. In order for the PV to be invertible, there must be some balance condition on the flow fields (HMR), and such a condition (thermal wind balance) is essential to the QG and SG theories underpinning synoptic analysis. Heuristically the PV may be inverted "by eye," especially in cases of large horizontal scale where the vertical component of the thermal gradient is dominant: in such cases, PV approximates by

$$P = \frac{1}{\rho} \zeta \frac{\partial \theta}{\partial z} \qquad (11.72)$$

(compare with the SG PV in eqn 11.61). Cartoons of the thermal and wind fields associated with such a PV element have been produced by Thorpe (in HMR), for example: a positive anomaly of PV corresponds to a combination of positive stability and positive vorticity, while a negative anomaly of PV corresponds to a combination of negative stability and negative vorticity (Fig. 11.10). Similarly, in regions where there is no PV anomaly, negative

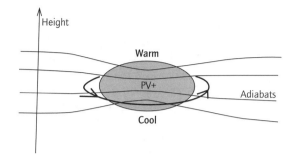

Fig. 11.10 Schematic of the flow and thermal pattern associated with a positive PV anomaly.

anomalies of static stability tend to correspond to positive vorticity, a result that is entirely in line with the concept of vortex stretching in a fluid.

The ability to invert PV heuristically is central to its usefulness in synoptic dynamics, since this subject is often concerned with interpretation of observed and modeled weather systems. A number of authors have used "piecewise PV inversion" to discuss the influence of individual PV features (such as tropopause folds) on the evolution of weather system. This technique involves identifying a particular PV structure, subtracting it from some chosen "basic state," and inverting both new and old PV fields to find the magnitude of the change in the wind and temperature fields associated with the PV change. For instance, it is often important to understand how an upper-level PV feature associated with a lowered tropopause (or the error in the specification of such a feature in the initial state) can be attributed to wind and temperature fields near the surface. One critical facet of the attribution of the flow to the PV structure is the role of the boundary conditions for the PV inversion. It is quite possible to have an atmosphere of constant PV, with a complex wind and thermal structure in the interior of the domain, and this internal structure is associated with (attributed to) the thermal fields on the domain boundaries. For instance, the thermal wind balance indicates that a low-level warm anomaly will be associated with a cyclonic vortex at the surface, in the absence of significant PV structure.

One way of relating the thermal boundary conditions to "PV thinking" is through the "Bretherton analogy" (Bretherton 1966), whereby a warm anomaly on the lower boundary is equivalent to an infinitesimal layer of positive PV anomaly and a warm anomaly on an upper boundary is equivalent to an infinitesimally thin layer of negative PV anomaly. In consequence, warm anomalies near the ground tend to act like positive PV, and are usually associated with positive vorticity: this is an observation that could equally well be made through direct consideration of the thermal wind relation. However, seeing boundary thermal anomalies in PV terms allows us to understand Rossby waves on the surface in terms of simple theory, as will be seen in Section 11.2.5.

An example of the power of considering such balanced flows in PV terms is that of an elevated zone of latent heating. It is known that such heating leads to a PV source within the heating zone that is proportional to the vertical gradient of the heating: a positive source below the heating maximum and a negative source above it. In consequence, in the absence of strong vertical motion, the response to propagating heating is a PV dipole; positive below negative (Fig. 11.9b). The heuristic inversion of this PV structure, as outlined above, implies a thermal structure that represents a warm anomaly between the PV anomalies (in the region of maximum heating, as required). Also, each element in the dipole may be associated with a vortex at that level. This latter observation demonstrates how the PV analysis encapsulates more than a simple thermal analysis, because it contains the thermal wind balance condition within its definition. From a knowledge of heating, a source term in the thermodynamic equation, we have deduced PV source terms, and thereby a wind structure that is guaranteed to be in balance with the new thermal field. This process could have been deduced by invoking the thermal wind balance directly from the thermal anomaly, but the PV inversion method deals with balance implicitly.

In overview, provided that we know a balance condition, the properties of conservation and invertibility together mean that the atmospheric evolution can be described purely in PV terms: a knowledge of the PV at a given instant allows us to invert to get the flow fields. These flow fields may then be applied to advect the conserved PV to produce an evolving atmospheric state. In practice, many balanced (QG or SG) models do follow this program of PV conservation and inversion.

Beyond its fundamental role as a conserved dynamical quantity that encapsulates a balance condition, PV is also useful as a diagnostic of atmospheric stability. Thinking intuitively once again, with PV being simply the product of stability and vertical vorticity, it can be seen that negative PV generally corresponds to negative static stability (static instability) or negative vorticity (implying inertial instability). As shown by Hoskins (1974), this association of negative PV with instability is a necessary condition for **symmetric instability**, which may occur in practice in regions where the flow is neither statically unstable nor inertially unstable, in their pure forms. Parcel theory has shown that the relative slopes of isentropes and surfaces of constant angular momentum, $M = v_g + fx$, determines an equivalent condition for symmetric instability: the instability may be manifest where the isentropes are steeper than the momentum surfaces.

The derivation of the conservation relation for PV requires that there is a thermodynamic quantity, θ, which is a single-valued function of p and ρ. Consequently there is an infinite number of other potential vorticities which could be constructed: the use of θ is chosen to make invertibility a useful and intuitive process. In consideration of flows where latent heating or cooling is active, the wet-bulb potential temperature (θ_w, homeomorphic to θ_e) is often used as a more exactly conserved field than θ. Because of this, a wet-bulb potential vorticity, PV_w, is often invoked, and this is widely used as a diagnostic of moist symmetric instability. However, it must not be forgotten that θ_w used to calculate PV_w is not a thermodynamic variable, depending on the three variables, p, ρ, and humidity, so the PV_w field is neither conserved nor invertible (θ_w does not determine buoyancy unless the air is saturated). Examples of the nonconservation of PV_w are given by Cao and Cho (1995) and Parker (1999).

11.2.5 Rossby waves

Having developed an understanding of balanced flows and in particular the role of PV in QG dynamics, it is now possible to discuss the most fundamental mode of variability of the synoptic-scale atmosphere; the Rossby wave (see Gill 1982 for a more complete discussion). Rossby waves occur where there is a gradient in PV, or a gradient in temperature at the ground (which may be seen as a PV gradient, using the Bretherton analogy of Section 11.2.4). For the sake of argument we may consider a discrete boundary between zones of differing PV, on which is superposed a wavelike disturbance (Fig. 11.11). Where there is an incursion of high PV into the lower PV area, we anticipate (from a heuristic PV inversion) a cyclonic anomaly of circulation, and where the lower PV intrudes into the higher PV, anticyclonic circulation occurs. The result of these circulations is a wavelike flow across the PV gradient that is out of phase with the PV anomalies. By eye, it can be seen that the flow across the gradient will tend to cause the PV wave to propagate, and the resulting propagating wave is the essence of a Rossby wave. For the climatological northward PV gradient in mid-latitudes, due to the meridional gradient of the Coriolis parameter, Rossby waves propagate westwards relative to the flow: in the westerly jets the net result is eastward propagation of Rossby waves at upper levels. In contrast, the equatorward temperature gradient in mid-latitudes means that Rossby waves at the surface tend to propagate eastwards. An interaction between upper- and lower-level Rossby waves

leads to a positive feedback, with the circulation of the upper and lower waves reinforcing each other, and this mechanism is the basis behind baroclinic instability (see HMR).

For the purposes of conceptual understanding, a relatively simple barotropic Rossby wave can be considered. Consider two-dimensional QG flow, independent of latitude, y, on a β-plane, under barotropic conditions of zero horizontal temperature gradient. Since the **Q**-vector is identically zero, there is no vertical motion, and the vorticity equation (11.35) reduces to

$$\frac{\partial}{\partial t}\frac{\partial^2 \Phi}{\partial x^2} + U_0 \frac{\partial}{\partial x}\frac{\partial^2 \Phi}{\partial x^2} + \beta \frac{\partial \Phi}{\partial x} = 0 \quad (11.73)$$

where U_0 is a uniform zonal wind. Writing the geostrophic streamfunction in modal form, with k a wavenumber and ω a frequency, leads to the dispersion relation

$$\omega = U_0 k - \frac{\beta}{k} \quad (11.74)$$

From consideration of the phase speed

$$c = \frac{\omega}{k} = U_0 - \frac{\beta}{k^2} \quad (11.75)$$

it can be seen that such Rossby waves on the planetary vorticity gradient, β, propagate westwards relative to the mean flow. However, as noted above, given that the mid-latitude flow is dominated by the westerly jets, upper-level Rossby waves tend to propagate eastwards relative to the surface. There is also an important possibility of stationary waves.

In other contexts, the relevant (potential) vorticity gradient may be reversed from the gradient in Coriolis parameter. For instance, the mean pole-to-Equator temperature gradient at the surface implies an equivalent negative PV gradient, through thermal wind balance or the Bretherton analogy. Easterly jets, such as the African Easterly Jet, can also exhibit reversed PV gradients. Work by Snyder and Lindzen (1991) and Parker and Thorpe (1995) has discussed the role of sources of PV due to latent

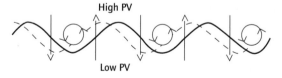

High PV

Low PV

Fig. 11.11 Schematic of a Rossby wave on a boundary between two regions of differing PV. The thick contour denotes the interface between the regions and the dashed contour is the subsequent position of this interface. Other arrows indicate the circulation induced by the waves on the interface.

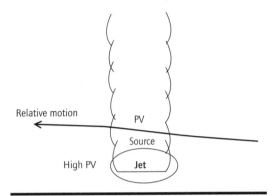

Fig. 11.12 Sketch of the contribution of a diabatic PV source to propagation of a Rossby wave. The low-level jet lies to the east of a low-level PV anomaly (as in Fig. 11.11) and advects warm air, leading to positive equivalent PV. It also encourages convective latent heating, so positive PV anomalies above the jet are also formed.

heating, providing an equivalent PV gradient that is of opposite sign to that of the Coriolis parameter. In this way latent heating may act to reinforce a low-level Rossby wave propagating on the pole-to-Equator temperature contrast (Fig. 11.12).

11.2.6 Tropical considerations

The Tropics account for a large proportion of the Earth's surface but it is generally acknowledged that understanding and prediction of tropical weather systems is relatively poor. Synoptic methods have typically been developed by scientists in the mid-latitudes, for application to mid-latitude systems, and do not always translate well to a tropical context. In terms of the dynamics of atmospheric flows, certain considerations characterize the tropical regime; in particular, convection and low Coriolis parameter.

Convection

The Tropics are (defined as being) a region of high insolation, and this leads to significant development of dry and moist convection; overall,

diabatic processes are relatively strong components of the thermodynamic balance over synoptic time scales. In particular, deep cumulonimbus convection characterizes the Tropics, as the principal rain-bringing cloud form, and as an agent of thermal and chemical transport. The tropopause is typically high and some cumulonimbus systems can extend beyond 15 km in height. These cumulonimbus clouds are observed to have complex morphology, organizing into mesoscale convective systems (MCSs), and being controlled by an interaction between their own internal dynamics and the environmental state (such as CAPE and wind shear). Although they are traditionally regarded as "subgrid" phenomena from the perspective of NWP modeling, it is now recognized that the scales of a well developed MCS impinge on the resolution of forecast models (of order 100 km) and the systems are being neither resolved nor parameterized.

In mid-latitudes, a major goal of synoptic analysis is to identify regions of significant upward motion, as being associated with precipitation, vertical transport, and surface pressure development. In contrast, in the Tropics it is the mesoscale cumulonimbus convection that produces rainfall and rapid vertical transport, and this convection is not *necessarily* coincident with larger-scale mean ascent (although upward motion is one process that enhances moist convection).

There has been some debate in recent years over the correct way in which to interpret and parameterize cumulonimbus convection; more recently some consensus has been achieved. On the mesoscale (the physical and temporal scale of the convective systems themselves), the occurrence of convection requires a combination of instability (CAPE) and a triggering mechanism (essentially an anomaly of upward motion, to release the conditional instability). Short-term forecasts need to identify areas of instability as well as triggering agents, such as orography, surface wetness anomalies, or the remnants of the cold pool of a previous convective system. In predicting the occurrence of convection on these scales, an operational forecaster needs to rely on empirical methods of prediction, since NWP model predictions of moist convection are generally unreliable. Typically the forecaster may use the model predictions of winds and thermodynamic fields to make a prediction of convection based on local knowledge and the recent history of convective storms.

On larger temporal and spatial scales, the process of triggering may be relatively rapid, and convection may be seen as being in a sustained equilibrium with the radiative forcing (primarily shortwave heating at the surface and longwave cooling aloft). The convective systems act to warm the middle to upper troposphere and cool the boundary layer (reduce its θ_e) through material transport in downdraughts. On these larger scales the problem in understanding and predicting the convective cloud distribution is to quantify the relationship between the energy budgets in the equilibrium state. For instance, the low-level θ_e may determine the thermal profile aloft, through a convective equilibrium condition (such as a requirement of zero CAPE). The low-level θ_e is then determined by a balance between surface fluxes (determined by low-level winds and sea surface temperatures over the ocean; very much more complex over land), boundary layer entrainment, and convective downdraughts (themselves related to the complex morphology of the convective systems). The wind fields that partially control these fluxes are dependent on the thermal and pressure structures set up by the radiative–convective equilibrium.

In all, the discussion in terms of the radiative–convective equilibrium has proposed that for large-scale convective fields, triggering by large-scale upward motion is not the primary control on the clouds. In general, the large-scale vertical motion is relatively weak, and the rapid updraughts in storms are to a great extent compensated by subsidence in the clear air. Individual convective systems respond to their local environmental conditions: instability, wind shear, and triggering mechanisms. The convective storms then act to modify and control the large-scale thermal state and it is this large-scale thermal field that, through associated pressure gradients, leads to large-scale horizontal and vertical motions. To understand how the large-scale wind fields feed

back on the convective field, we need to take into account the feedbacks on the energy balance between the troposphere and the boundary layer. On the large scale, then, the role of upward motion is not to trigger convection but to modify the thermodynamic profile. Whereas in mid-latitudes a major goal of synoptic analysis is to diagnose areas of ascent, which are likely to lead to precipitation, in the Tropics large-scale ascent is just one factor controlling the stability of the profile and the occurrence of rainfall.

As remarked above, the line between resolved and subgrid convection is not clear. Although equilibrium assumptions are favored for many large-scale models, they may not be appropriate to many synoptic circumstances, in which the convection operates close to the scale of the synoptic variability. For instance, in the monsoon over sub-Saharan West Africa, high CAPE is common, and an equilibrium state seems inappropriate: the convection seems to occur intermittently, every few days, as an interaction between the synoptic structure of CIN and triggering by orography or surface wetness. Once formed, convective systems over West Africa may propagate as coherent storms for many hours (comparable with the time scale of synoptic variability) and may significantly modify their synoptic environment.

Coriolis force

As we approach the Equator, the Coriolis parameter approaches zero and the Coriolis force becomes negligible. At the Equator there is no possibility of geostrophic balance, so any horizontal pressure gradient must either be balanced by frictional drag or lead to an acceleration of the air.

Much of the tropical zone is dominated by the low-level trade winds, which flow down the pressure gradient toward the climatological low pressure associated with the ITCZ, and are turned by the Coriolis force to give them an easterly component. When the ITCZ departs from the Equator and one branch of the trades is moving poleward, a westerly component develops. Climatologically, easterlies are observed at low levels near the Equator, but the zonal mean picture masks considerable variations around the globe and westerlies are not uncommon.

As a consequence of the low Coriolis force, strong horizontal temperature gradients are rare near the Equator. Moving a few degrees poleward, however, the strength of the insolation means that significant temperature gradients can be sustained: a notable example is the monsoon over West Africa in the summer months, in which there is a marked transition between high temperatures over the Sahara and lower temperatures over the tropical Atlantic Ocean. This thermal transition is accompanied by a wind shear which is qualitatively in thermal wind balance, and leads to the African Easterly Jet at about 600–700 hPa and a latitude of 10–20° N. Tropical cyclones are also sensitive to the Coriolis parameter, and are therefore not formed within around 5° of the Equator. Consequently, in the Tropics there is no simple, immediate separation of the dynamics according to geostrophically balanced flow. It has been said that all tropical flows are mesoscale phenomena, since no terms in the equations can automatically be negelected.

On the Equator, the β-plane approximation ($f = \beta y$) is a relatively good one. A common treatment is to linearize the equations of motion on the β-plane and separate variables into a vertical structure function and a horizontal structure satisfying the equations for a shallow homogeneous layer (Gill 1982, chapter 6). Within this framework, the existence of the gradient in planetary vorticity expressed through β leads to the possibility of westward-propagating Rossby waves (Gill 1982, chapter 11). Another possibility of wave dynamics at the Equator is through eastward propagation of Kelvin waves. The Kelvin wave solution is one for which the meridional wind, v, is identically zero: in this case by considering the meridional momentum equation we have on the Equator

$$u = -\frac{1}{\beta}\frac{\partial^2 \phi}{\partial y^2} \qquad (11.76)$$

The result is that, for the Kelvin wave, high pressure on the Equator is associated with westerlies and low pressure with easterlies. The zonal mo-

mentum equation then implies that the westerlies will accelerate on the eastern side of a high-pressure anomaly, leading to eastward propagation of a disturbance. A final wave solution near the Equator (apart from inertia-gravity modes) is the Rossby-gravity wave (see Gill 1982, p. 440): this wave has the behavior of a gravity wave for high positive wave number and a Rossby wave for high negative wave number, while its group speed is always eastward. This mode exhibits the possibility of a stationary solution (with eastward group propagation).

A key result in consideration of forced solutions near the Equator is obtained from the steady, linearized vorticity equation (see Gill 1982, p. 465), and states that vertical stretching is associated with poleward motion,

$$\beta v = f\frac{\partial w}{\partial z} \qquad (11.77)$$

a relationship that is connected with "Sverdrup balance" for the ocean (and note the relationship to the β term in the omega equation (11.36). These ideas now allow us to infer the response of the tropical atmosphere to localized heating: the composition of the westward-propagating Rossby wave response and eastward-propagating Kelvin wave response has led to a well known "Matsuno–Gill" model of the effects of tropospheric heating close to the Equator (say as a result of changes in the convective field, due to boundary layer θ_e changes). The form of this response for symmetric heating is a pair of low-level cyclonic circulations to the west, straddling the Equator, and easterlies to the east (as shown schematically in Fig. 11.13). To the east, this response is due to Kelvin wave propagation with zero meridional flow and easterly winds at low levels. Around the longitude of the heating, the lower tropospheric vertical vorticity is stretched and the low-level air moves polewards, in accordance with eqn 11.77. Some compensating equatorward flow occurs to the west, as a result of Rossby wave propagation. Similar results for heating that is located away from the Equator are discussed by Gill (1982).

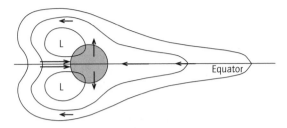

Fig. 11.13 A schematic of the low-level response to heating located symmetrically across the Equator. The heating region is shaded and upward motion approximately matches this zone. The response to the west is a pair of low-pressure centers with corresponding cyclonic circulation, while easterlies occur to the east. The meridional scale of this response is controlled by an equatorial Rossby radius related to β and stability, typically of the order of 1000 km.

11.3 APPLICATIONS TO WEATHER SYSTEMS

Synoptic meteorology had its origins and very definition in the resolution of observations that were available to the analyst. In recent times, the use of those observations has passed to the electronic computer, which assimilates the data into a previous forecast field to generate an analysis of greater horizontal resolution than the observation set. In addition to this, the computer model is now diagnosing weather patterns (precipitation, turbulence, etc.), at its given resolution, explicitly. This means that the boundary between synoptic meteorology and mesoscale meteorology is in practice becoming blurred: synoptic meteorology may still be defined as that involving scales well represented by balanced theories, but the practicing forecaster is routinely analyzing weather patterns on much finer scales than the synoptic observing network.

11.3.1 Frontal cyclones

The most deeply studied atmospheric weather system is undoubtedly the frontal cyclone, and the basic model of this object is centered on balanced dynamics. It was constructed by the Bergen school

of V. Bjerknes and co-workers around the end of the First World War and the mid-latitude frontal cyclone (Fig. 11.14a) remains the most well known and well studied of synoptic phenomena, largely because of its importance to the developed nations in Europe and North America, but also because of its significance as a very coherent structure in the complexity of nonlinear atmospheric flow. Explanations of the dynamics and evolution of such cyclones have evolved to some extent, and it is now appreciated that the Norwegian model on its own is inadequate to describe some complex aspects of the cyclone, such as its transport properties. Some refinements in the Norwegian model are indicated in Fig. 11.14.

The Bergen school originally identified frontal cyclones as unstable growing waves on a "polar front." Although a coherent polar front is not identified in global analyses, it remains true that these cyclones represent baroclinic instabilities of the basic pole-to-Equator temperature contrast, extracting potential energy from the basic state and converting it to potential and kinetic energy of the cyclone. The baroclinic zone on which the cyclones grow is typically marked by westerly wind shear with height (through the thermal wind relation), leading to an upper tropospheric jet associated with a change in tropopause height (e.g Fig. 11.15). Thus the location of the jetstream is an important indicator of the baroclinic zone, apart from importance in its own right as a wind feature.

The cyclogenesis process was first accurately understood in terms of linear QG instability models (Charney 1947; Eady 1949; discussed by Gill 1982 and HMR), which identified the cyclones as components of an unstable baroclinic wave. The idealized Eady wave in particular may be seen as a periodic sequence of highs and lows as a result of Rossby waves at upper and lower levels. For suitably long wavelengths, the upper and lower waves are able to couple and lead to an exponentially growing wave: the lows of the upper wave augment the cyclonic flow at lower levels and vice versa. The tilt of the cyclone with height turns out to be westwards for unstable waves, which is a useful guide for interpreting upper- and lower-level

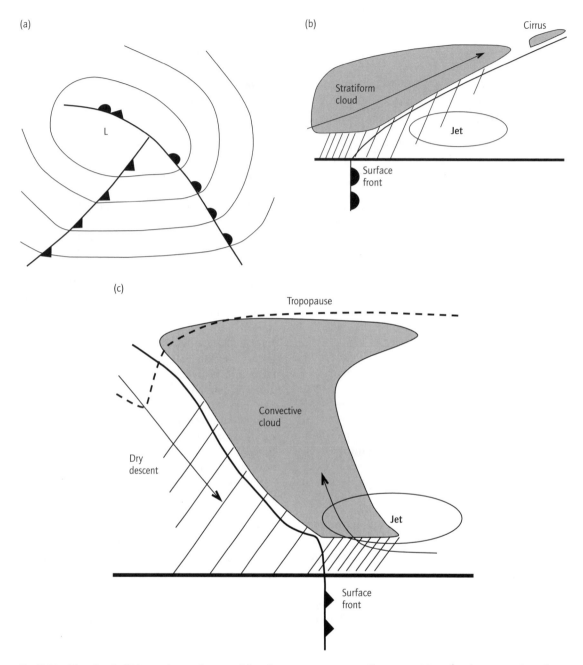

Fig. 11.14 The classical Norwegian cyclone model and some more recent refinements. Note that fronts are plotted on the warm air side of each frontal zone. (a) Isobars and fronts in a typical alignment; a "kink" in the isobars at each front is related to locally increased low-level vorticity. (b) The warm front is characterized by stratiform cloud, decreasing in depth ahead of the front. The jet (into the section) in the cold air ahead of the front is related to the cold conveyor belt and decreases with height according to thermal wind balance. (c) A classical cold front is characterized by convective clouds and a narrower, more intense band of precipitation than anticipated at a warm front. The low-level jet ahead of the front (into the section) is related to high low-level PV due to latent heating and/or a band of warm air at low levels ahead of the cold front; the jet is linked to the warm conveyor belt. There is commonly a stronger jet along the front at the tropopause level, with a significant change in tropopause height at the front (the tropopause may often form a "fold" in the descending zone behind the front as indicated here). (*Continued*)

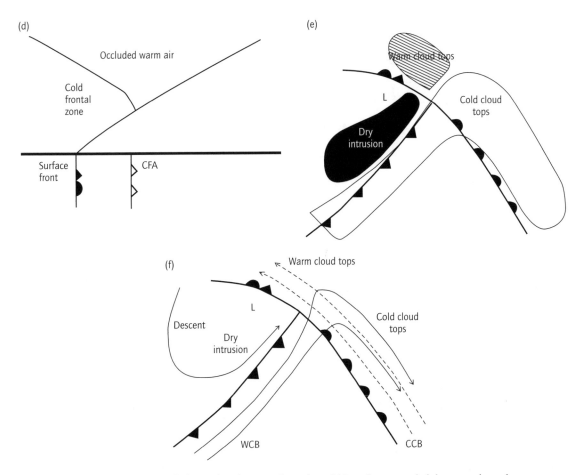

Fig. 11.14 (*Continued*) (d) A simple form of occlusion, where the cold front has ascended the warm frontal zone, leading to a "cold front aloft" (CFA). As may be anticipated, in a region of the cyclone with high curvature, strong gradients, and significant cloud processes, various complex morphologies are possible. (e) A simple translation of the typical cloud forms onto the surface frontal analysis of (a). The details of these patterns vary significantly, according to the morphology of the cyclone and the fronts. (f) A schematic of the airflows in the surface frontal analysis of (a). The warm and cold conveyor belts (WCB and CCB) bring about warm advection and contribute to the cloud and frontal dynamics. The interaction of the WCB with the dry intrusion leads to the form of the cold front (classical or split) according to their relative motion.

charts in terms of cyclone development. In practice, finite amplitude interactions may be observed instead of periodic waves at upper and lower levels, but the instability mechanisms are essentially the same. Nonlinear integrations of the modal wave solutions generate fields with remarkable similarity to observed frontal cyclones.

In addition to the understanding of baroclinic instability in cyclogenesis, it has been found that the developing cyclones can be significantly conditioned by a mean barotropic shear in the basic state (a zonal flow that is sheared in the meridional direction, and has no vertical shear, so does not involve a thermal wind component). Thorncroft *et*

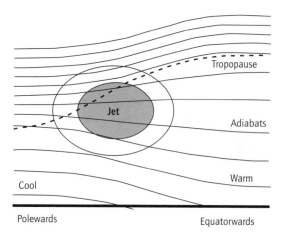

Fig. 11.15 A meridional cross-section through the midlatitude westerly jet, indicating its relationship with the thermal pattern (through thermal wind balance) and with the tropopause. The location of the jetstream is then coincident with the zone of thermal and PV gradients, which give rise to instability in the form of frontal cyclones.

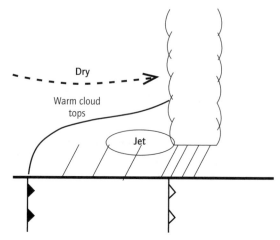

Fig. 11.16 Schematic of a split cold front, in which dry air overruns the surface cold front, possibly associated with a dry intrusion.

al. (1993) have studied idealized life cycles of such developing cyclones and discussed two paradigms, LC1 and LC2, which have enjoyed some significant attention. Very superficially, the two life cycles are distinguished by an enhanced cyclonic shear in the basic state of LC2, which leads to a cyclonic wrapping-up of the upper-level trough to produce a significant upper-level cut-off cyclone at high latitudes. Hoskins and West (1979) and Davies *et al.* (1991) have discussed the influence of such barotropic shear on the frontal structure: large-scale shear of the zonal wind with latitude acts to control the relative intensities of the cold and warm fronts (cyclonic shear, as in LC2, enhances warm frontogenesis), as well as some fundamental aspects of the airflow patterns.

The basic observational picture of fronts constructed as part of the Bergen cyclone model remains relatively unchanged: the cold front follows the warm front, and they bound the warm sector. Where these fronts merge, one may undercut the other, leading to an occlusion. The classical warm and cold fronts are sketched in Fig. 11.14b and c: the warm front is characterized by more stratiform

types of cloud and precipitation, while the cold front tends to be steeper and exhibit cumulonimbus convection. Such a cold front is now termed a "classical" cold front, to distinguish it from a less intuitive, but relatively common, form of "split cold front." At a split cold front (Fig. 11.16), there is significant advection *through* the frontal zone, leading to the advance of a layer of dry air above the surface cold front. At the nose of this dry air is an upper cold front, at which cumulonimbus convection may occur, and the whole zone between the upper and lower cold fronts is characterized by conditional instability. Consideration of the split front demonstrates that it is impossible to regard the fronts as airmass boundaries: these phenomena act as steep waves in the airflow, which may or may not be stationary relative to the air locally. There is significant along-front thermal advection (naturally, since the frontal cyclone as a whole is composed of propagating Rossby waves), which may contribute to the frontal propagation. The results of Parker (1999) indicate that the occurrence of the split cold front may be favored in systems with a relatively low steering level, for which there is stronger upper-level cross-frontal advection, while classical cold fronts occur in baroclinic models with stronger upper-level control, a higher steering level, and consequently weaker cross-frontal flow in the mid-troposphere.

The upper fronts associated with a split cold front may on occasion propagate right across the warm frontal zone, leading to a propagating rain-band at the normally stratiform warm front (e.g. Browning *et al.* 1995). Another form of upper cold front is the "cold front aloft," which occurs at a warm occlusion, where the cold front is rising over the warm frontal zone, but may remain coherent at upper levels (Fig. 11.14d). Locatelli *et al.* (1994) have related such upper-level features to significant weather patterns at the surface. In general, the nature of the occlusion is a matter of some disagreement and certainly exhibits considerable variability. Some authors have argued in terms of a "bent-back warm front" to describe the evolving surface thermal pattern in certain systems. "Occlusion" refers to the lifting of the warm air between the converging cold and warm fronts, while the term "seclusion" refers to the cutting off of a pool of warm air, polewards of the main warm sector.

Advection of dry air above the surface cold front is normally associated with a dry intrusion of air descending from the poleward direction, behind and then above the cold front. Such dry intrusions tend to be very clear in WV imagery and may be associated with stratospheric air, of high PV and ozone concentrations, being brought almost down to the surface (Fig. 11.14e and f). At the cold front, the destabilizing effect of the advection of dry air at low- to mid-levels in the dry intrusion is enhanced by warm advection from the equatorward direction ahead of the cold front in a "warm conveyor belt" (Fig. 11.14f), which is related to the low-level jet observed ahead of cold fronts (Fig. 11.14c). From thermal wind balance at a simple cold front (as constructed in simple dry models such as the "deformation front"), one would expect low-level along-front flow with the cold air on its right, in contrast to the direction of the observed jet ahead of the cold front. This jet is observed to be associated with a low-level warm band ahead of the front and/or with PV anomalies at low levels generated by latent heating in the frontal clouds. The warm advection in the warm conveyor belt may be important in the frontal propagation (Browning & Pardoe 1973), as it is associated with the surface Rossby wave, and certainly acts to increase the conditional and convective instability that leads to the cumulonimbus convection at the cold front. The term "conveyor belt" was coined in the late 1960s and tends to imply a material flux of air from the equatorward direction, being carried poleward and ascending as it approaches the center of the cyclone. Theoretically speaking, identification of the cyclone as a propagating wave tends to suggest that it may be more reasonable to see the conveyor belt as an Eulerian rather than a Lagrangian feature, with air parcels entering the warm conveyor belt from the east before ascending above the cold front. However, case study analyses of air parcel trajectories in mid-latitude cyclones (Wernli 1997) have indicated that in the active part of the cyclone the conveyor belts may indeed act as coherent Lagrangian features, with air parcels remaining within the conveyor belt throughout their passage along the frontal zone.

A "cold conveyor belt" has also been identified (Fig. 11.14f) in the cold air ahead of the warm front. Both the warm and the cold conveyor belts, through bringing about warm advection, may naturally be linked to the high vorticity that must occur in a frontal zone (from SG theory, or the Bretherton analogy).

Another important consequence of the convection that commonly occurs at the cold front is the possibility of boundary layer uplift. In such cases the cold front may resemble a squall line at which the surface front behaves like a gust front, or a laboratory density current (see review by Smith & Reeder 1988). The low-level cold air at the front lifts the warm sector boundary layer air off the ground, and undercuts it with cleaner, colder air produced in convective downdraughts (Browning & Harrold 1970). The combination of dry intrusions of stratospheric air into the middle and lower tropopause with rapid lifting of warm sector boundary layer air means that cold fronts make a significant contribution to larger-scale chemical transport.

In the Bergen cyclone model, the distinct

airmasses separated by the fronts have distinct weather characteristics. More recently these weather characteristics of the different sectors have been identified with the evolution of stability properties due to differential advection in different sectors of the wave, rather than to the history of distinct masses of air. In particular, Parker (1999) has discussed the evolution of θ_e in simple baroclinic waves and shown that this can account for some of the observed weather characteristics. Basic state profiles of θ_e tend to have less vertical variation than the θ profiles, because of the capacity of the warmer lower-level air to hold more water vapor. This means that in areas of descending air from the poleward direction ("downgliding" air; HMR) the convective stability tends to be reduced: a mid-level maximum in descending, equatorward flow leads to significant mid-level reduction in θ_e, while θ is compensated through adiabatic warming. Such behavior is characteristic of the air behind the cold front, regardless of the surface over which it passes.

Frontal cyclones are known to grow as instabilities of the mid-latitude baroclinic zones, and naturally the cold and warm fronts within these cyclones may also be unstable, to secondary cyclones (see review by Parker 1998). Such cyclones are generally of smaller scale (say 500 km) and may grow very rapidly (occasionally into severe weather systems), making them extremely important yet difficult to forecast. A great deal of recent research effort has been directed at understanding these secondary systems, and some basic conclusions are now reasonably well accepted:

1 Like the parent cyclones, the secondaries may develop as baroclinic waves. However, there is also a strong or even dominant component of barotropic instability (feeding off the basic state kinetic energy) at these scales. The relative importance of barotropic and baroclinic instability depends on the geometrical structure of the PV anomalies at the front.

2 Most fronts are unstable to secondary cyclone growth, but few exhibit intensely growing cyclones. This may be due to the rapid saturation of unstable linear modes in the nonlinear regime, or to the effect of frontogenesis processes (the large-

scale flows in the parent cyclone leading to an unsteady, time-dependent front) in suppressing instability in many cases.

3 Moist processes become more important on shorter horizontal length scales, and there is evidence that secondary cyclones may receive significant energy from cloud feedbacks.

11.3.2 Mountains

Flow influenced by hills and mountains is of great importance in global climatology and in the determination of local weather patterns. The horizontal scales of surface orography range from the continental scales downwards, and the way in which the atmosphere responds to such features depends on the relevant scales of wind speed, height, length, and Coriolis parameter. The principal controlling parameters are the Rossby number (eqn 11.19) and the Froude number

$$Fr = \frac{U}{NH} \qquad (11.78)$$

where H is the mountain height scale.

Low Rossby number flow

Here the nature of the flow is interpreted in terms of balanced (e.g. QG) dynamics. The basic response of the atmospheric column as it passes over the mountains is to be "squashed" in the vertical and thereby, from the QG vorticity equation, to form a relative anticyclone. Less obviously, the air column may tend to stretch before it reaches the hill and in the lee of the hill, to generate a low upstream and downstream (Fig. 11.17). Lee lows in particular are common features of various parts of the world, such as northern Italy in the lee of the Alps. The superposition of the anticyclone and cyclones means more intense anticyclonic winds associated with the hill. Upstream, these winds are related to **barrier winds**: close to a steep hill where there is a pressure gradient along the contours or orography, there can be little Coriolis force to balance the pressure gradient (the horizontal flow up the hill is restrained by stability) and the flow tends to follow

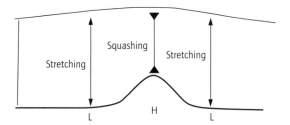

Fig. 11.17 The stretching and squashing of air columns as they approach and pass over a ridge leads to characteristic patterns of highs and lows.

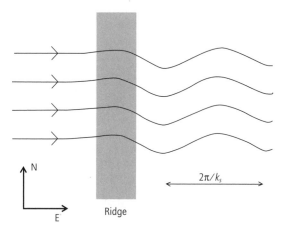

Fig. 11.18 Rossby waves form downstream of a N–S oriented ridge, and are significant in conditioning the synoptic activity in the lee.

the pressure gradient on the upstream side. A classical result for large mountain systems in this regime of low Ro is that there is a possibility of generating a stationary Rossby wave train downstream of the mountains. For instance, consideration of the dispersion relation for barotropic Rossby waves (eqn 11.74) shows that stationary waves may occur in westerly flow ($U_0 > 0$) for a value of wave number

$$k_s = \sqrt{\frac{\beta}{U_0}} \qquad (11.79)$$

in which case the group speed implies that the wavetrain develops downwind, to the east of the mountain range (Fig. 11.18).

For a distinct cyclone approaching orography on these scales, there is a tendency to move polewards as the cyclone crosses the orography. This may be seen as the circulation around the low tending to lead to cyclonic development to the poleward side as air descends the terrain in this region (see Carlson 1991).

In many cases the influence of hills on a cyclone has been described as the decay of the cyclone on the upstream side and the formation of a new lee low on the downstream side. A number of authors have pointed out, however, that this behavior can be interpreted more simply as the propagation of a cyclone over the hill, being masked by the superposition of the hill's anticyclonic circulation. If upper-level conserved features are considered, this interpretation is often found to be appropriate: a

coherent PV anomaly propagates over the orography and, while the surface cyclone may temporarily be masked over the hills, the PV anomaly remains coherent.

High Ro

High Fr. In this regime, the kinetic energy of upstream parcels of air is high compared with the potential energy required to reach the summit of the hill. The flow is principally *over* the orography and to a good approximation linear theory may be applied. In the linear theory, gravity wave modes may appear in the stable atmosphere above the hill: the dispersion characteristics of these waves depend sensitively on the basic state profiles of stability and wind shear (see Holton 1992). There are two principal regimes:

1 Trapping occurs at some layer above the hills: energy is contained below this layer and a wave train may occur in the lee of the hills. This state corresponds to relatively high drag.

2 Wave energy propagates upwards until a critical layer is reached, at which the wave amplitude becomes large and the waves break, depositing energy and momentum in that layer. The waves act to cause a drag force between the hill and the

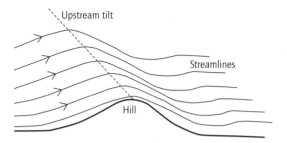

Fig. 11.19 The upstream tilt of gravity waves with height occurs in order to balance momentum fluxes at the ground: its result is a tendency for higher winds in the lee. Downslope winds are also related to nonlinear effects in the flow.

atmosphere, through a vertical flux of horizontal momentum, and in order that the drag be of the right sign (tending to decelerate the airflow) the wave crests tilt upstream with height. This tilt of the waves tends to lead to a confluence of the streamlines on the downwind side of the hill (Fig. 11.19), which implies a speed-up of the flow in the lee. This effect is also forced by nonlinear effects in the airflow, and downslope windstorms are relatively common hazards of hilly areas.

Low Fr. For typical values of N and U, this regime is somewhat extreme for the atmosphere, but in light winds and conditions of high stability, low-*Fr* conditions may be observed at high mountains. The basic balance is that the kinetic energy of air upstream is significantly less than the potential energy required to reach the summit: in this case there must be flow splitting and a stagnation point. For the case of long ridges there may be damming of the low-level air, which is unable to ascend the ridge, and strong barrier winds may occur on the upslope side.

11.3.3   Tropical cyclones

Tropical cyclones are among the most severe and hazardous of weather patterns, with the potential to devastate shipping and coastal land areas, through a combination of high winds, extreme sea state, and heavy precipitation. They occur over the tropical oceans and although they may pass over land, they cannot persist without the latent heat fluxes from the ocean. In the Atlantic, intense tropical cyclones are known as "hurricanes," and in the Pacific they are called "typhoons"; weaker cyclones are termed "tropical depression," increasing in intensity to "tropical storm," and cyclones are named once they reach tropical storm strength. Tropical cyclones are routinely observed and their progress is predicted with a variety of modeling techniques. However, forecasting of the movement of the storms remains imperfect and the storm dynamics is an area of active research.

Initiation of tropical cyclones is known to depend on several criteria:
1 A warm core precursor vortex is needed.
2 The sea surface temperature must be greater than 26°C.
3 The tropospheric wind shear must be relatively weak.
4 There must be a decrease of θ_e with height in the lower troposphere, and a humid mid-troposphere.
5 The latitude must correspond to significant Coriolis parameter (so these storms do not form within 5° of the Equator).

Once formed, tropical cyclones are maintained through the strong sensible and latent heat fluxes from the ocean, which are amplified by the cooling of air as it moves in toward the very low pressure at the center. The cyclones exhibit a warm core, corresponding, through eqn 11.26, to a decrease in the amplitude of vorticity with height and an anticyclone at the tropopause. This warm core is maintained by latent heat release in the convective clouds.

It is a common mistake to observe hurricane force winds (as defined on the Beaufort Scale) in a mid-latitude cyclone, and consequently describe the cyclone as a hurricane. Mid-latitude cyclones derive their energy from baroclinic mechanisms (involving horizontal temperature gradients and associated thermal wind shear) and barotropic mechanisms (extracting energy from horizontal wind shear), which lead to particular structures and weather patterns, such as strong axial asymmetry, and fronts. Hurricanes, in contrast, extract energy from the warm sea surface in a quasi-axisymmetric manner. While many hurricanes may "recurve" — for example, northward along the eastern seaboard of North America — they inevitably interact with mid-latitude baroclinicity and westerly wind shear, to become essentially baroclinic storms. One remote cousin of the hurricane that may affect the mid-latitude oceans, however, is the mesoscale "polar low," which in a significant proportion of cases may derive its energy from relatively high fluxes from the sea surface into extremely cold polar airflows.

11.4   ON AIRMASSES

A widely used and abused concept in atmospheric science is that of the "airmass"; a term that may variously refer to wide bodies of the atmosphere separated by fronts, or to small volumes of air with particular trace chemical contents. Many dynamical meteorologists dislike the use of this term: here an attempt is made to discuss the applicability of the idea of an airmass, and the more precise terminology that may be applied.

Within general fluid flows, it is known that relatively simple large-scale flow fields may lead to distortion of a tracer field to extremely short scales and complex structures. For example, a volume of air that is 100 m deep, in a shear of 5 m s^{-1} km^{-1}, which is typical for the mid-latitude troposphere, will after 1 day be sheared over 50 km. In turbulent flow fields, the tracer dispersion is more rapid and comprehensive, and in fully developed turbulence, any two adjacent elements of the fluid will ultimately become separated by an arbitrary distance in the flow. Reference is often made to a "parcel" of fluid, which is a small volume of fluid that moves as a coherent fluid element: this is a construct that enables us to imagine the complex processes involved in real fluid flows. For example, we consider the forces on a fluid "parcel" in deriving the Navier–Stokes equations, despite knowing that such parcels could not be experimentally constructed or observed. Similarly, one might consider the chemical reactions occurring in a parcel of air as it moves through the atmosphere: in reality volumes of air are being stretched and

sheared until they are dispersed widely, but provided that our mentally constructed parcel of air is small enough, it will remain coherent over the time we wish to consider. It might be said that the parcel represents an infinitesimal volume of fluid, which may still describe physical evolution. The danger of employing the term "airmass" in this context is that it may be taken to imply a relatively large volume of air (as is often the case in discussion of synoptic systems), which in practice will not behave as a coherent mass of air except over sufficiently short time scales. The term "airmass" should here be used with the same reservations and caution employed in discussing a "parcel" of fluid.

"Airmasses" were invoked in the construction of the Norwegian model, and for many years fronts were still regarded as boundaries between large zones of air of differing properties. More recently it has been realized that the fluid dynamics of the atmosphere is generally not well described by such discrete models. At a given instant a front will very often separate regions in which the air has characteristically different properties, but the air in these regions is not static and is generally evolving under considerable vertical shears and horizontal strains. This is to say that the properties of a column of air within an "airmass" are determined by the relative advection at each level within that column. It is important to accept that the airmass model only works well as an instantaneous one, and is not good for describing the evolution or transport of dynamical or tracer fields within those airmasses. There remain some useful empirical descriptions of weather types in terms of airmasses, but for the purposes of describing atmospheric evolution and transport it would be better to work in terms of stability theory: in considering a horizontally homogeneous airmass we are really considering a one-dimensional column of the atmosphere. Transposing airmass arguments into arguments relating to an atmospheric column means that we can reintroduce the important processes of vertical shear of horizontal advection into the model.

A significant example of the failure of airmass concepts to describe synoptic evolution is in the case of a cold front for which cross-frontal flow is significant: an airmass model of a front cannot admit this behavior. Reeder and Smith (1988) have described observations of such fronts over Australia, and related them usefully to the SG frontal model. Parker (1999) has used such SG models to describe the evolution of tracer fields in a frontal zone and highlighted the importance of cross-frontal advection.

11.5 PRACTICALITIES: HOW TO PERFORM A SYNOPTIC ANALYSIS

In many instances it is useful to perform a synoptic analysis; for instance, when considering a case study of a pollution event or a case of severe weather. Performing such an analysis is a matter involving a certain amount of experience and is sometimes regarded as a black art by those not regularly experienced in looking at synoptic data. Certainly it is true that there is an almost limitless amount of general information and intuition that is built up by bench forecasters over their careers. However, it is also true that much of the information required to describe the synoptic situation succinctly can be obtained by routine methods. This is particularly true in the age of information technology, when recent observations as well as the results of explicit NWP models are available to most of us through the internet. In particular, the NWP products offer detailed maps of fields such as precipitation, which were formerly inferred by the forecaster on the basis of experience and diagnostic rules.

Practice and experience at looking at synoptic data can be gained through use of the internet and consultation of textbooks (such as Bader *et al.* 1995). What follows is not a definitive guide to analyzing a synoptic situation but a checklist of topics that may bear on the analysis and should be considered.

11.5.1 Synoptic charts

The traditional surface chart showing isobars, lows, highs, fronts, troughs, and perhaps thickness contours (Section 11.2.1) is usually the starting point for synoptic discussion. Application of the standard paradigms of cold and warm fronts, for example, and the overall cyclone structure indicates the form of the weather patterns. The geostrophic wind may be inferred from the isobars and, in combination with the thickness pattern, this indicates the sense of the thermal advection. Interpretation beyond this level of detail becomes more difficult for the nonpractitioner: the weather associated with fronts in winter may be significantly different to that in summer, for example, and varies according to location. Confirmation of the details of the weather patterns should be inferred from other corroborative sources.

Upper air plots, apart from their immediate importance for upper-level applications such as aircraft forecasting, are useful in determining stability and cyclone structure. It is common to use charts of geopotential height of a pressure surface (say the 200 hPa height): a low-pressure anomaly on a height level corresponds, conveniently, to low height of a pressure surface. Such upper-level charts are useful in indicating the vertical tilt of cyclones, as an indication of their potential for baroclinic growth, as well as the location of upper-level jetstreams, associated with baroclinicity and a forcing of low-level cyclonic development (Bader *et al.* 1995).

11.5.2 Imagery

Use of satellite imagery improves with practice, but with some thought the basic IR and VIS channels can immediately give information about the synoptic situation.

1 The VIS (visible) image gives quite intuitive information about the nature of the reflecting surface for solar radiation: white features may be cloud, fog, or snow, while the sea surface and various land surfaces are relatively dark. Thin clouds such as cirrus may sometimes appear opaque.

2 The IR image indicates the temperature of the upper surface of the emitting layer: the tops of clouds appear cold, according to altitude, while the land surface appears warm and cold according to the diurnal cycle. The convention is to shade this

image with cold features white, so that high cloud tops are white, low clouds are grey, and the land/sea surface is usually dark.

3 The water vapor (WV) image of geostationary satellites is slightly more difficult to interpret, but roughly speaking this indicates the height above which the atmosphere becomes relatively dry. When there has been significant descent at upper levels, this leads to a dry upper troposphere and the WV image is shaded dark. In regions of high upper tropospheric water vapor the image is shaded white.

4 The so-called "channel 3" image (polar-orbiting satellites) detects radiation in a band consisting of both shortwave solar radiation scattered back from Earth (which dominates the signal during the day) and longwave radiation emitted from terrestrial sources (which dominates at night). The change in the source by day or night leads to a switch-over in the appearance of sources: cold cloud tops, for example, emit relatively little radiation but may scatter solar radiation strongly by day. For this reason, the negative is commonly produced for daytime images, and labeled with a minus sign: channel 3–. Some sophisticated use of channel 3 is outlined by Bader *et al.* (1995), but the principal use of this channel for qualitative analysis is in diagnosis of cloud droplet sizes by day. In contrast to the visible image, the cloud albedo in channel 3– is dominated by droplet size, with large drops absorbing in this band. Thus, clouds composed of relatively large drops appear dark in channel 3–, in contrast to a bright visible image.

Satellite imagery fields can give an indication of the intensity (size and depth of cells) of cumulus/cumulonimbus in the cold sector, the nature of a cold front (split/classical), and the relative intensity of cold and warm fronts. The IR channel and the WV channel in particular give striking views of a dry intrusion. The imagery is also useful for confirming the nature of rather poorly defined features such as troughs, occlusions, and frontal waves.

Radar imagery is useful in locating precipitation to high spatial and temporal accuracy, especially, for example, in cases of cumulonimbus convection. Care must be taken in inferring sur-

face precipitation rates from radar estimates, which are made in an elevated section that ascends away from each radar (see Bader *et al.* 1995). For validation it is possible to use surface station rainfall data.

11.5.3 Synoptic observations

The raw observations, SYNOPS from surface stations and TEMPS from upper air stations, are the bread and butter of the bench forecaster's work, and are considered essential for determining the quality and details of a synoptic forecast. For specific applications such as a case study, the local conditions from the nearest reporting synoptic station are of considerable importance. More significant is likely to be the nearest radiosonde ascent, which gives far more vertical resolution than one or two horizontal charts on different levels. Quantities that can be obtained from the sounding include conditional or convective instability, boundary layer characteristics, thermal advection (from the turning of the wind; eqn 11.24 and ensuing comments), local tropopause height, cloud base, cloud top, and freezing level.

11.5.4 Model products

Now that model analyses are becoming more easily obtained, there are large, almost bewildering, amounts of diagnostics available. The availability of explicit fields of convective rainfall, for example, can answer what were hitherto difficult questions of interpretation without any discussion. However, it remains important to understand and interpret the situation in terms of traditional (e.g. QG) paradigms; the interaction of highs and lows and thermal advection. A key advantage of model diagnostics is that PV is often provided explicitly, giving a useful conserved field with the key dynamical role in synoptic dynamics that is discussed above. All this said, the explicit nature of model output means that presentation of such diagnostics rarely includes coherent structures such as fronts (Hewson 1998), and it is left to the user to infer these on the basis of the model fields, the imagery, and a large dash of experience.

It is here that a prior look at the bench forecaster's synoptic analysis is invaluable.

1 As on a traditional synoptic chart, the starting point is probably the **mean sea level pressure** (MSLP) field from which geostrophic wind can be inferred. This field represents a pressure that has been extrapolated down to a constant geopotential height, so that horizontal gradients leading to a geostrophic wind, for example, may be inferred sensibly. The field of surface pressure is used internally in models but over land this field generally reflects the orography more strongly than weather systems, and is not particularly useful for synoptic analysis. As noted above, fronts are often not explicitly marked on presentations of this data.

2 For the analysis of weather systems, low-level dynamical fields, such as $\mathbf{u}$, θ and θ_e or θ_w, are usually discussed on some pressure surface away from the ground, say at 850 or 925 hPa, where surface processes and the diurnal cycle are less significant and these fields are more accurately conserved (of course, for discussion of surface conditions the surface fields themselves should be used). Fronts may be inferred from these low-level thermal fields, as can the important thermal structure of the cyclone and, in conjunction with the winds, the thermal advection.

3 The low-level circulation is commonly discussed in terms of vorticity or PV at a similar level, say 850 hPa. The useful property of PV conservation is less applicable at low levels, due to boundary layer mixing processes, so it is not particularly preferred as a diagnostic. Consideration of vorticity at low levels is a useful indicator of cyclones where the pressure anomaly is masked by a strong larger-scale gradient of pressure (implying a rapidly moving system; Fig. 11.20), so this field is particularly important for smaller-scale cyclones.

4 Upper-level structure may be inferred in a number of complementary ways, and since a three-dimensional field is being considered, it is often necessary to think of several fields in order to get a complete picture. Useful diagnostics are:

(i) Height of the PV = 2 PVU surface is a common indication of the tropopause level. Although the height is not a conserved field (unlike PV on a θ surface, for example), this is a useful measure of interaction, since the interaction depends on the vertical spacing of PV features relative to H_R (eqn 11.51). Height of the 1 PVU surface is often used to diagnose tropopause folds, which can be weak in height of 2 PVU. The region of strong gradients in this field occurs at the location of the upper-level jet, where the tropopause has a steep gradient.

(ii) PV or θ on a pressure surface, say 200 hPa, are useful indicators of the position of the jetstream, the baroclinic zone at upper levels, and upper-level vortices.

(iii) PV on a θ surface with vector winds is a chart liked by purists, since the anomalies are approximately conserved on synoptic time scales. Again, this chart indicates the baroclinic zone and the jetstream, and features can be tracked convincingly through the evolution of a system. The choice of the θ surface can be important, since the height of this surface changes significantly in the baroclinic region.

5 Vertical velocity is a useful diagnostic, which indicates regions that are likely to be cloudy (although convective rain may occur in the absence of significant large-scale forced ascent). It is best considered on some mid-tropospheric level, such as 700 hPa, unless specific applications are being considered.

6 Rainfall is now also given explicitly: the numerical model usually makes a split between convective (parametrized, e.g. cumulonimbus convection) and large-scale (explicitly modeled, e.g. stratiform frontal rain) precipitation. Apart from the practical numerical requirement for this split, it also usefully differentiates between qualitatively different weather patterns. Such fields should always be corroborated by considering other data, such as the satellite and radar imagery: representation of cloud processes in NWP models has certain inaccuracies that are hard to quantify.

7 Measures of conditional instability, such as the CAPE (eqn 11.13), are often plotted as diagnostics for prediction of cumulonimbus convection.

(a)

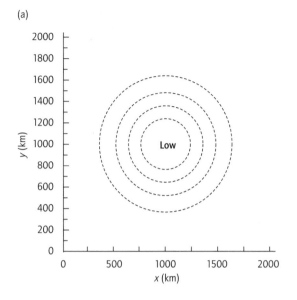

(b)

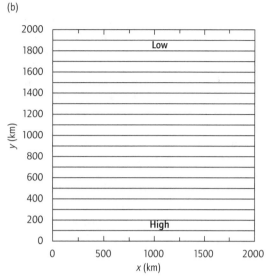

(c)

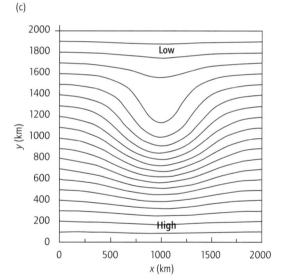

Fig. 11.20 The superposition of (a) a cyclonic pressure pattern with (b) a strong meridional pressure gradient (implying a rapidly moving cyclone) can act to mask the cyclone's pressure signal (in (c)), despite the fact that its thermal or PV pattern may be identical to a stationary system.

11.6 CONCLUSIONS

Synoptic meteorology is in many ways in advance of many other areas of science, in that numerical methods have here taken a central role in the practical subject, largely replacing human analysis in the generation of the primary weather forecast.

While we may realistically face a future in which there is no longer any human intervention in the commercial production of a weather forecast, synoptic meteorology remains a vibrant area of study and scientific progress. This is in part due to the real importance of synoptic weather systems in the rapidly developing area of atmospheric chemistry, and in part due to the expansion of the bench forecaster's role into prediction at ever smaller system scales. Fronts, cumulonimbus convection, and the planetary boundary layer are all phenomena that involve the synoptic meteorologist directly, and the ways in which each modifies weather systems and the transport of trace chemicals are still very poorly understood. Interpretation of local events, on scales far smaller than the computer may resolve, will always require understanding and interpretation of synoptic processes. At a more profound level, even if we are to reach a stage where numerical weather prediction no longer requires human intervention, most of us would agree that it will be necessary for humans to understand the products of those predictions at some real level of insight such as that provided by balanced dynamics.

ACKNOWLEDGMENTS

In preparing this chapter I have made use of notes taken from lectures by Alan Thorpe, as well as calling on many helpful discussions with colleagues at the JCMM in Reading.

REFERENCES

Bader, M.J., Forbes, G.S., Grant, J.R., Lilley, R.B.E. & Waters, A.J. (1995) *Images in Weather Forecasting: A Practical Guide for Interpreting Satellite and Radar Imagery*. Cambridge University Press, Cambridge.

Bolton, D. (1980) The computation of equivalent potential temperature. *Monthly Weather Review* **108**, 1046–53.

Bretherton, F.P. (1966) Critical layer instability in baroclinic flows. *Quarterly Journal of the Royal Meteorological Society* **92**, 325–34.

Browning, K.A. & Harrold, T.W. (1970) Air motion and precipitation growth at a cold front. *Quarterly Journal of the Royal Meteorological Society* **96**, 369–89.

Browning, K.A. & Pardoe, C.W. (1973) Structure of low-level jet streams ahead of mid-latitude cold fronts. *Quarterly Journal of the Royal Meteorological Society* **99**, 619–38.

Browning, K.A., Clough, S.A., Davitt, C.S.A., Roberts, N.M. & Hewson, T.D. (1995) Observations of the mesoscale sub-structure in the cold air of a developing frontal cyclone. *Quarterly Journal of the Royal Meteorological Society* **121**, 1229–54.

Cao, Z. & Cho, H.-R. (1995) Generation of moist potential vorticity in extratropical cyclones. *Journal of Atmospheric Science* **52**, 3263–81.

Carlson, T.N. (1991) *Mid-latitude Weather Systems*. HarperCollins, London.

Charney, J.G. (1947) The dynamics of long waves in a baroclinic westerly current. *Journal of Meteorology* **4**, 135–63.

Davies, H.C., Schär, C. & Wernli, H. (1991) The palette of fronts and cyclones within a baroclinic wave development. *Journal of Atmospheric Science* **48**, 1666–89.

Eady, E.T. (1949) Long waves and cyclone waves. *Tellus* **1**, 33–52.

Emanuel, K.A. (1994) *Atmospheric Convection*. Oxford University Press, Oxford.

Gill, A.E. (1982) *Atmosphere–Ocean Dynamics*. Academic Press, New York.

Haynes, P.H. & McIntyre, M.E. (1987) On the evolution of vorticity and potential vorticity in the presence of diabatic heating and frictional or other forces. *Journal of Atmospheric Science* **44**, 828–41.

Haynes, P.H. & McIntyre, M.E. (1990) On the conservation and impermeability theorems for potential vorticity. *Journal of Atmospheric Science* **47**, 2021–31.

Hertenstein, R.A. & Schubert, W.H. (1991) Potential vorticity anomalies associated with squall lines. *Monthly Weather Review* **119**, 1663–72.

Hewson, T.D. (1998) Objective fronts. *Meteorological Applications* **5**, 37–65.

Holton, J.R. (1992) *An Introduction to Dynamic Meteorology*. Harcourt Publishers, London.

Hoskins, B.J. (1974) The role of potential vorticity in symmetric stability and instability. *Quarterly Journal of the Royal Meteorological Society* **100**, 480–2.

Hoskins, B.J. (1975) The geostrophic momentum approximation and the semi-geostrophic equations. *Journal of Atmospheric Science* **32**, 233–42.

Hoskins, B.J. & Bretherton, F.P. (1972) Atmospheric frontogenesis models: mathematical formulation and solution', *Journal of Atmospheric Science* **29**, 11–37.

Hoskins, B.J. & West, N. (1979) Baroclinic waves and frontogenesis. Part II: uniform potential vorticity jet flows—cold and warm fronts. *Journal of Atmospheric Science* **36**, 1663–80.

Hoskins, B.J., McIntyre, M.E. & Robertson, R.W. (1985)

On the use and significance of isentropic potential vorticity maps. *Quarterly Journal of the Royal Meteorological Society* **111**, 877–946.

Lagouvardos, K., Lemaitre, Y. & Scialom, G. (1993) Importance of diabatic processes on ageostrophic circulations observed during the FRONTS 87 experiment. *Quarterly Journal of the Royal Meteorological Society* **119**, 1321–45.

Locatelli, J.D., Martin, J.E. & Hobbs, P.V. (1994) A wide cold frontal rainband and its relationship to frontal topography. *Quarterly Journal of the Royal Meteorological Society* **120**, 259–75.

Parker, D.J. (1998) Secondary frontal waves in the North Atlantic region: A dynamical perspective of current ideas. *Quarterly Journal of the Royal Meteorological Society* **124**, 829–56.

Parker, D.J. (1999) Passage of a tracer through frontal zones: A model for the formation of forward-sloping cold fronts. *Quarterly Journal of the Royal Meteorological Society* **125**, 1785–800.

Parker, D.J. & Thorpe, A.J. (1995) Conditional convective heating in a baroclinic atmosphere: a model of convective frontogenesis. *Journal of Atmospheric Science* **52**, 1699–711.

Reeder, M.J. & Smith, R.K. (1988) On air motion trajectories in cold fronts. *Journal of Atmospheric Science* **45**, 4005–7.

Sanders, F. & Hoskins, B.J. (1990) An easy method for estimation of Q vectors from weather maps. *Weather and Forecasting* **5**, 346–53.

Smith, R.K. & Reeder, M.J. (1988) On the movement and low-level structure of cold fronts. *Monthly Weather Review* **116**, 1927–44.

Snyder, C. & Lindzen, R.S. (1991) Quasi-geostrophic wave-CISK in an unbounded baroclinic shear. *Journal of Atmospheric Science* **48**, 78–86.

Thorncroft, C.D., Hoskins, B.J. & McIntyre, M.E. (1993) Two paradigms of baroclinic-wave life-cycle behavior. *Quarterly Journal of the Royal Meteorological Society* **119**, 17–55.

Wernli, H. (1997) A Lagrangian-based analysis of extratropical cyclones. II: a detailed case-study. *Quarterly Journal of the Royal Meteorological Society* **123**, 1677–706.

12 Atmospheric Removal Processes

BRAD D. HALL

12.1 INTRODUCTION

The Earth's surface is the ultimate sink for trace gases and particles emitted at the surface or formed in the atmosphere. The transfer of trace gases and particles to the surface occurs through two pathways: wet and dry deposition. Wet deposition involves precipitation. Dry deposition does not. The relative importance of one pathway over the other depends on the nature of the substance (phase, reactivity, solubility), the occurrence of precipitation, and the nature of the surface cover.

In this chapter, the basic principles of wet and dry deposition are discussed. The chapter is divided into three sections: dry deposition of gases, dry deposition of particles, and wet deposition. The bulk resistance model is the primary tool used for predicting dry deposition. This model and variations of it are discussed in detail. More complex models are also briefly discussed to highlight the advantages and disadvantages of the bulk resistance approach. Various methods used to measure and estimate dry deposition rates are presented, along with some of the associated assumptions and instrument requirements. The treatment of particle dry deposition is similar to that of gases, except that gravitational settling must be taken into account. For this reason, particle dry deposition is presented separately. Finally, the general concepts of wet deposition are presented. The nucleation and growth of aerosol particles from condensation nuclei are the first steps toward wet removal of both particles and gases. The complex processes associated with nucleation and growth are beyond the scope of this chapter. The treatment of wet removal is limited to the collection and removal of particles and gases by existing hydrometeors.

12.2 DRY DEPOSITION OF GASES

Dry deposition occurs as trace gases and particles are adsorbed or react on objects (plants, soil, water, buildings, etc.) at the Earth's surface. Factors that govern the dry deposition of gases include the level of atmospheric turbulence, the nature of the gas (solubility, reactivity, molecular diffusivity), and the nature of the surface itself. Obviously, a chemical or physical sink must exist for uptake to occur. Likewise, without turbulence even a perfectly adsorbing surface would be ineffective at removing gaseous material from the atmosphere because dry deposition would be limited by molecular diffusion. Vegetation represents an important sink for many trace gases (O_3, SO_2, HNO_3, H_2O_2, etc.). This is because vegetated surfaces often provide both an effective sink (leaf pores, or stomata) and the turbulence necessary for efficient uptake.

Because of the complex nature of turbulent transport, it is impractical to treat deposition explicitly in chemical transport models. The simulation of transport from the bulk atmosphere to the surfaces of individual surface elements is too complex for all but the most limited of models. Instead, the concept of a deposition velocity is used to relate the flux, F (moles $m^{-2} s^{-1}$), to the concentration at some reference height above the surface,

$$F = -V_d C \qquad (12.1)$$

where V_d is the deposition velocity ($m\,s^{-1}$) and C is the concentration (moles m^{-3}) of the depositing species at a reference height. For a depositing species (downward flux), the flux is negative and V_d is positive. Because C will be a function of height for a depositing species, V_d will also be a function of height. Therefore, V_d is normally associated with a reference height close to the surface, such as 10 m.

Equation 12.1 has the form of a concentration difference ($C_{bulk} - C_{surface}$) times a transfer coefficient under the condition that $C_{surface} \ll C_{bulk}$. The advantage of the deposition velocity concept is that it can be easily represented in atmospheric models, often as a lower boundary condition. The disadvantage is that all of the complexities of turbulent transfer and uptake must be represented by a single term. Thus, it may be difficult to specify V_d under a variety of conditions. Nevertheless, use of eqn 12.1 is common and a great deal of effort has been made to develop methods for specifying V_d.

12.3 BULK RESISTANCE ("BIG LEAF") MODEL

Dry deposition is normally represented as a combination of three steps: (i) turbulent transport from the bulk atmosphere to a thin layer of stagnant air adjacent to the surface; (ii) molecular diffusion across this thin layer to the surface; and (iii) uptake at the surface. Each of these processes can be modeled in terms of a concentration difference and a resistance to transport.

If the steady-state flux is only in the vertical direction, then the flux associated with each process will be equal. The flux will also be equal to the overall concentration difference divided by the overall resistance (R_t) (Seinfeld & Pandis 1998),

$$F = \frac{-(C_3 - C_2)}{R_a} = \frac{-(C_2 - C_1)}{R_b}$$
$$= \frac{-(C_1 - C_0)}{R_c} = \frac{-(C_3 - C_0)}{R_t} \qquad (12.2)$$

where the concentrations and resistances are as in Fig. 12.1. The total resistance can be expressed in terms of the three resistances in series,

$$R_t = R_a + R_b + R_c = V_d^{-1} \qquad (12.3)$$

where R_a is the aerodynamic resistance, R_b is the quasi-laminar resistance associated with molecular diffusion, and R_c is the surface resistance.

Before the individual resistances are discussed, it is worth contrasting the bulk model with a more realistic multi-layer approach. The vertical transport through an absorbing layer of height h (neglecting transport at the base) can also be described as

$$V_d(h) = \frac{1}{C(h)} \int_0^h v_d \rho_p C\,dz \qquad (12.4)$$

where v_d is the local deposition velocity ($r_p = 1/v_d$ would be the local effective resistance), and ρ_p is the foliage density (foliage area per unit volume). Here we equate the flux at height h with the integral of local fluxes throughout the canopy. There is some question as to whether the right-hand side of eqn 12.4 is equivalent to a sum of three resistances representing three distinct processes. Bache (1986) compared the results of a numerical solution to a multilayer problem to those determined from the bulk model, under conditions of high and low surface resistance. He found that it is not always possible to conveniently split the sum $R_b + R_c$ into purely aerodynamic and surface components. He found that in the low surface resistance limit, the term R_c might contain an aerodynamic component depending on the definition of R_b. Nevertheless, the bulk model approach is commonly used to estimate V_d, and experimental studies are often designed to measure one or more of the resistances based on commonly used definitions.

12.3.1 Aerodynamic resistance

Material is brought from the bulk atmosphere to the vicinity of the sink by turbulent transport. The turbulence intensity is primarily a function of atmospheric stability and surface roughness. During the day, the turbulence intensity is high throughout a deep layer (known as the mixed layer) of the atmosphere, bringing a large reservoir of material in contact with the surface. The formation of a shallow boundary layer at night limits the amount of material in contact with the surface. Although the turbulence intensity is much lower beneath the nocturnal boundary layer, deposition can still occur. Furthermore, because the shallow nocturnal boundary layer is isolated from the well mixed air above, even low deposition rates can result in large reductions in concentration within the nocturnal boundary layer.

The aerodynamic resistance is a function of turbulence intensity and is independent of species for gases (for large particles gravitational settling must be taken into account). The aerodynamic resistance is normally based on gradient transport theory (K-theory) (Stull 1988) and Monin–Obukov similarity (Kaimal & Finnigan 1994). The gradient transport theory relates a constant vertical turbulent flux to the local concentration gradient

$$F = -K \frac{dC}{dz} \qquad (12.5)$$

where K ($m^2 s^{-1}$) is the eddy diffusivity. Applying eqn 12.5 to the aerodynamic layer and substituting the differential ($C_3 - C_2$) for dC yields,

$$R_a = -\frac{(C_3 - C_2)}{F} = \int_{z2}^{z3} K^{-1} dz \qquad (12.6)$$

Similarity theory suggests that the eddy diffusivity should be proportional to the friction velocity, $u_\star$ ($m s^{-1}$), and the height above the ground,

$$K = ku_\star z \qquad (12.7)$$

where the constant of proportionality, k, is the von Karman constant (0.4). Under diabatic (nonneutral) conditions, the proportionality is modified from its neutral form by functions that depend on the dimensionless height scale, z/L, where L is the Monin–Obukhov length (L is the length scale appropriate for the surface layer),

$$K = \frac{ku_\star z}{\Phi\left(\dfrac{z}{L}\right)} \qquad (12.8)$$

where Φ is an empirically derived diabatic profile function. The general form of the aerodynamic resistance follows.

$$R_a = \int_{z_2}^{z_1} \frac{\Phi(z/L)}{ku_\star z} dz \qquad (12.9)$$

In the idealized model (Fig. 12.1), z_2 is defined to be the edge of the quasi-laminar boundary layer. Because the absorbing layer often consists of a vertical distribution of irregularly shaped objects, such as leaves, the lower integration limit must represent the extent of influence of the aerodynamic processes, averaged throughout the absorbing

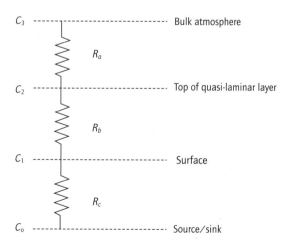

Fig. 12.1   Schematic of the bulk resistance model.

layer. This is normally taken as the roughness length, z_o.

Forms for the dimensionless stability function (Φ) vary in the literature (see Horvath *et al.* 1998), but those of Businger *et al.* (1971) associated with heat transfer are most commonly used (Kaimal & Finnigan 1994; Seinfeld & Pandis 1998).

$$\Phi = \begin{cases} (1-16z/L)^{-0.25} & -2 \leq z/L < 0 \quad \text{(unstable)} \\ 1 & z/L = 0 \quad \text{(neutral)} \\ (1+5z/L) & 0 < z/L \leq 1 \quad \text{(stable)} \end{cases} \quad (12.10)$$

Integration of eqn 12.10 leads to

$$R_a = \frac{1}{k\,u_*}[\ln(z/z_o) - \Psi] \quad (12.11)$$

where $\psi(z/L)$ is the integral form of the stability function. The corresponding aerodynamic resistances are as follows (Seinfeld & Pandis 1998):

$$R_a = \begin{cases} \dfrac{1}{ku_*}[\ln(z/z_o)+5(\xi-\xi_o)] & \text{(stable)} \\[2mm] \dfrac{1}{ku_*}[\ln(z/z_o)] & \text{(neutral)} \\[2mm] \dfrac{1}{ku_*}\left[\ln(z/z_o)+\ln\left[\dfrac{(\eta_o^2+1)(\eta_o+1)^2}{(\eta^2+1)(\eta+1)^2}\right]\right. \\[2mm] \left. +2(\tan^{-1}\eta-\tan^{-1}\eta_o)\right] & \text{(unstable)} \end{cases}$$
$$(12.12)$$

where $\xi = z/L$, $\xi_o = z_o/L$, $\eta_o = (1-16\xi_o)^{0.25}$, $\eta = (1-16\xi)^{0.25}$, valid for $-1 < \xi < 1$.

Within plant canopies, momentum is absorbed not only at the ground, but throughout the canopy. The mean level of momentum absorption is displaced to a height $z = d$, where d is the displacement distance. The displacement distance for momentum is normally about 75% of the mean canopy height. The roughness length and displacement distance can be estimated from measurements of friction velocity and wind speed at different heights. Alternatively, both quantities

can be determined as fit parameters to the logarithmic wind profile equation.

$$u(z) = \frac{u_*}{k}\ln\left(\frac{z-d}{z_o}\right) \quad (12.13)$$

For flow over plant canopies, the quantity $(z-d)$ should be substituted for (z) in eqns 12.8–12.12.

12.3.2 *Quasi-laminar resistance*

A laminar boundary layer may not actually exist adjacent to the surface elements. At best, it probably exists intermittently as leaves move in the wind. It is considered to exist in terms of the theory, however, and is used to represent the overall effective surface seen by the atmosphere. The flux across this boundary layer occurs by molecular diffusion for gases and by Brownian motion for particles. These processes occur independent of direction (diffusion can occur to both sides of a leaf). The flux is described in terms of a transfer coefficient, B, dimensionalized by the friction velocity (Wesely & Hicks 1977; Seinfeld & Pandis 1998):

$$F = -Bu_*(C_2 - C_1) \quad (12.14)$$

The quantity B^{-1} is sometimes called the Stanton number, and is often combined with the von Karman constant and expressed as kB^{-1}. Expressions for kB^{-1} have been derived from wind tunnel studies and from experiments on heat transfer to flat plates (see Wesely & Hicks 1977, and references therein). Field experiments have confirmed that kB^{-1} for SO_2, O_3, and water vapor can be described in terms of the Schmidt number ($Sc = \nu/D_i$) and the Prandtl number ($Pr = \nu/\alpha$), where ν is the kinematic viscosity of air, D_i is the molecular diffusivity of the depositing species, and α is the thermal diffusivity of air:

$$kB^{-1} = 2\left(\frac{Sc}{Pr}\right)^{2/3} = 2\left(\frac{\alpha}{D_i}\right)^{2/3} \quad (12.15)$$

In terms of eqn 12.15, R_b is given by

$$R_b = \frac{2}{ku_*}\left(\frac{\alpha}{D_i}\right)^{2/3} \qquad (12.16)$$

Early studies on heat, water vapor, and CO_2 transport (Thom 1972) suggest that kB^{-1} does not vary substantially between species and vegetation type. Thus, eqn 12.16 or a close analog is commonly used in dry deposition modeling. The common use of eqn 12.16 may be due more to the fact that the dry deposition of species such as O_3 and SO_2 is often dominated by the surface resistance (such that the exact form of R_b becomes relatively unimportant) than to the universal nature of the theory. Early studies showed substantially different behavior of kB^{-1} over rigid objects and agricultural crops planted in widely spaced rows (such as grape vines) compared to natural vegetation (forest, grass, bean crops). For species that deposit readily to surfaces, such as HNO_3 and H_2O_2 (for which R_b is relatively important), or for deposition to rigid artificial surfaces, such as buildings, alternatives to eqn 12.16 may be required (see Thom 1972; Garratt & Hicks 1973).

12.3.3   Relative magnitudes of R_a and R_b

The magnitudes of R_a and R_b are roughly similar over vegetated surfaces. Figure 12.2 shows examples of R_a and R_b associated with H_2O_2 deposition to a rough pine forest (Hall et al. 1999).

Both resistances are smallest during midday, when turbulence intensity is highest. Nighttime resistances are at least an order of magnitude larger than daytime resistances. In this example, the diurnal trend in R_b is due primarily to its u_*^{-1} dependence.

12.3.4   Surface resistance

The surface resistance is often the controlling factor in gas-phase dry deposition. Uptake at plant surfaces is a function of how readily the gas can reach reactive surfaces on or inside the plant cells where reaction takes place. Because deposition to vegetation is a major removal pathway for many species, parameterizations for surface resistance

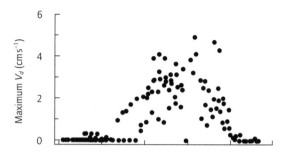

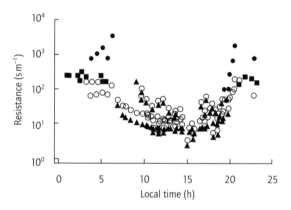

Fig. 12.2   Calculated resistances R_a and R_b (lower panel) and maximum deposition velocities computed as $V_d = (R_a + R_b)$ (upper panel) associated with H_2O_2 deposition to a pine forest in Saskatchewan, Canada, in July 1994. R_a is shown for stable (filled circles), neutral (filled squares), and unstable (filled triangles) conditions, while open circles denote R_b.

are often presented with specific reference to uptake by plants.

The primary areas involved in uptake at plant surfaces include the cuticular area (waxy skin of some leaves), leaf pores (stomata), the mesophyll (plant tissue within the stomatal cavity), and other wetted surfaces. Reaction at the ground or water surface offers a parallel pathway. The surface resistance is expressed as the parallel sum of the foliar resistance, R_{cf}, and the ground resistance, R_{cg}.

$$R_c = \frac{1}{R_{cf}} + \frac{1}{R_{cg}} \qquad (12.17)$$

The foliar resistance can be divided among the cuticular (R_{cut}), stomatal pore (R_{sp}), and mesophyllic resistances (R_m):

$$R_{cf} = \left[\frac{1}{R_{cut}} + \frac{1}{R_{sp} + R_m} \right]^{-1} (\text{LAI})^{-1} \qquad (12.18)$$

where the leaf area index (LAI) is used to provide the proper weighting of the foliar resistance relative to the ground resistance. The relative magnitudes of R_{cut}, R_{sp}, and R_m depend on the solubility of the depositing species and whether the stomata are open or closed. Daytime stomatal resistances for SO_2 and O_3 range from 30 to 1000 s m^{-1} for a variety of plant species (Baldocchi et al. 1987) with R_{sp} increasing with decreasing light level. Baldocchi et al. (1987) also showed an inverse relation between R_{sp} and LAI for O_3 (higher R_{sp} at lower LAI). At night or during periods of water stress, closed stomata lead to very high R_{sp} (reflective of no transfer via stomata). The mesophyll resistance depends predominately on solubility and reactivity. Highly soluble species, such as HNO_3, have essentially no resistance at the mesophyll. Reactive species, such as O_3, also have very small R_m, even though O_3 is much less soluble than HNO_3.

A more detailed model of R_c (Wesely 1989) has been included in the Regional Acid Deposition Model (RADM) (Wesely 1988). This model is described below because it provides a general, advanced method for estimating surface resistances for a variety of atmospheric species over many different surface types and conditions. The surface resistance employed in RADM is an extension of the basic idea conveyed in eqn 12.18, incorporating processes at different levels within a canopy (Fig. 12.3).

The surface resistance is a combination of parallel and series resistances associated with transport to various elements in the canopy:

$$R_c = \left[\frac{1}{R_{sp} + R_m} + \frac{1}{R_{lu}^i} + \frac{1}{R_{dc} + R_{cl}} + \frac{1}{R_{ac} + R_{gs}} \right]^{-1}$$
$$(12.19)$$

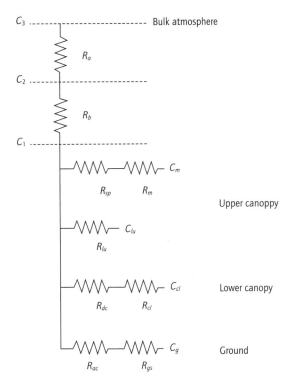

Fig. 12.3 Schematic of parallel resistances in the Wesely model.

where R_{sp} is the bulk stomatal pore resistance, R_m is the mesophyll resistance, R_{lu} is the resistance to uptake at the leaf cuticle and other surfaces in the upper canopy, R_{dc} is associated with buoyant convection in canopies, R_{cl} is associated with leaves, bark, and other surfaces in the lower canopy, R_{ac} is associated with transfer that depends on canopy height, and R_{gs} is the resistance at the ground. This model attempts to describe transport in a multilayer canopy using bulk properties. It is not designed to accurately represent each resistance in the context of a multilayer environment (values of a single resistance may not be comparable to those obtained by direct measurement). According to Wesely (1989), the resistances in eqn 12.19 are meant to represent bulk properties inferred from measurements of net vertical fluxes.

The stomatal resistance is estimated as the minimum stomatal resistance for water vapor, R_{sm}, scaled according to molecular diffusivity, times a factor that accounts for light level and temperature (Baldocchi et al. 1987):

$$R_{sp} = R_{sm} \frac{D_{H2O}}{D_i} \left[1 + \left(200(G + 0.1)^{-1}\right)^2\right]$$
$$\times \left[400(T_s(40 - T_s))^{-1}\right] \qquad (12.20)$$

where D_{H_2O} is the molecular diffusivity for water vapor, D_i is the molecular diffusivity for the depositing species, G is the solar radiation (W m^{-2}), and T_s is the surface temperature (°C). At night or when T_s is outside the range 0–40°C, R_{sp} is set to a large value to represent closed stomata. Values for minimum stomatal resistance for water vapor are tabulated as a function of season and land use in Table 12.1. The mesophyll resistance is determined from

$$R_m = \left[\frac{H^*}{3000} + 100 f_0\right]^{-1} \qquad (12.21)$$

where H^* is the effective Henry's Law coefficient (M atm^{-1}) and f_0 is a reactivity factor based on the electron activity for the half redox reaction in neutral aqueous solution and the second order rate coefficient for oxidation with S(IV) (see Table 12.2). The reactivity factor takes on values of 1 for high reactivity, 0.1 for moderate reactivity, and 0 for low reactivity. The upper canopy cuticular resistance for a particular species, R_{lu}^i, is given by,

$$R_{lu}^i = R_{lu} \left[\frac{1}{10^{-5} H^* + f_0}\right] \qquad (12.22)$$

where R_{lu} is tabulated in Table 12.1. This resistance will be important for soluble species such as SO$_2$, O$_3$, and H$_2$O$_2$. This resistance will also be important for species that are soluble in the waxy cuticle, although the use of H^* in eqn 12.22 does not account for this effect. The factor 10^{-5} in eqn 12.22 seems to be appropriate for O$_3$ and SO$_2$, but

may be too small for H$_2$O$_2$. Measurements (Hall et al. 1999) and chemical transport modeling (Sillman et al. 1998) suggest that the Wesely model underpredicts the deposition velocity of peroxides over forests. A factor of 10^{-3} might be more appropriate for peroxides.

The lower canopy resistance, R_{dc}, accounts for buoyant convection due to solar heating of the ground and by penetration of wind into canopies on sloped ground:

$$R_{dc} = 100\left(1 + 1000(G + 10)^{-1}\right)\left(1 + 1000\theta\right) \qquad (12.23)$$

where θ is the slope of the ground (radians). The resistance associated with lower canopy surfaces is given by

$$R_{cl} = \left[\frac{10^{-5} H^*}{R_{clS}} + \frac{f_0}{R_{clO}}\right] \qquad (12.24)$$

where R_{clS} and R_{clO} are associated with the uptake of SO$_2$ and O$_3$, respectively (Table 12.1). A similar expression is used for surfaces at the ground:

$$R_{gs} = \left[\frac{10^{-5} H^*}{R_{gsS}} + \frac{f_0}{R_{gsO}}\right] \qquad (12.25)$$

Dew and rain can change the surface resistance by preventing uptake via stomata, and by increasing uptake of highly soluble species. When the canopy surfaces are wet from rain or dew, Wesely (1989) suggests a modification of the upper canopy resistance for SO$_2$, O$_3$, and other species.

$$R_{lu}^{O_3} = \left[\frac{1}{3000} + \frac{1}{2R_{lu}}\right]^{-1} \qquad (12.26)$$

$$R_{lu}^{SO_2} = 100 \, (\text{s m}^{-1}) \qquad (12.27)$$

$$R_{lu}^i = \left[\frac{1}{3R_{lu}} + 10^{-7} H^* + \frac{f_0}{R_{luO}}\right]^{-1} \qquad (12.28)$$

Table 12.1 Resistances (s m^{-1}) used to compute R_c in the Wesely model (entries of 9999 indicate that there is no air–surface exchange via that pathway). (Reprinted from *Atmospheric Environment* **23**, M.L. Wesely, Parametrization of surface resistances to gaseous dry deposition in regional-scale numerical models, pp. 1293–1304, Copyright 1989, with permission from Elsevier Science.)

Resistance	Land-use type*										
	1	2	3	4	5	6	7	8	9	10	11
Seasonal category 1: midsummer with lush vegetation											
R_{sm}	9999	60	120	70	130	100	9999	9999	80	100	150
R_{lu}	9999	2000	2000	2000	2000	2000	9999	9999	2500	2000	4000
R_{ac}	100	200	100	2000	2000	2000	0	0	300	150	200
R_{gsS}	400	150	350	500	500	100	0	1000	0	220	400
R_{gsO}	300	150	200	200	200	300	2000	400	1000	180	200
R_{clS}	9999	2000	2000	2000	2000	2000	9999	9999	2500	2000	4000
R_{clO}	9999	1000	1000	1000	1000	1000	9999	9999	1000	1000	1000
Seasonal category 2: autumn with unharvested cropland											
R_{sm}	9999	9999	9999	9999	250	500	9999	9999	9999	9999	9999
R_{lu}	9999	9000	9000	9000	4000	8000	9999	9999	9000	9000	9000
R_{ac}	100	150	100	1500	2000	1700	0	0	200	120	140
R_{gsS}	400	200	350	500	500	100	0	1000	0	300	400
R_{gsO}	300	150	200	200	200	300	2000	400	800	180	200
R_{clS}	9999	9000	9000	9000	2000	4000	9999	9999	9000	9000	9000
R_{clO}	9999	400	400	400	1000	600	9999	9999	400	400	400
Seasonal category 3: late autumn after frost, no snow											
R_{sm}	9999	9999	9999	9999	250	500	9999	9999	9999	9999	9999
R_{lu}	9999	9999	9000	9000	4000	8000	9999	9999	9000	9000	9000
R_{ac}	100	10	100	1000	2000	1500	0	0	100	50	120
R_{gsS}	400	150	350	500	500	200	0	1000	0	200	400
R_{gsO}	300	150	200	200	200	300	2000	400	1000	180	200
R_{clS}	9999	9999	9000	9000	3000	6000	9999	9999	9000	9000	9000
R_{clO}	9999	1000	400	400	1000	600	9999	9999	800	600	600
Seasonal category 4: winter, snow on ground and subfreezing											
R_{sm}	9999	9999	9999	9999	400	800	9999	9999	9999	9999	9999
R_{lu}	9999	9999	9999	9999	6000	9000	9999	9999	9000	9000	9000
R_{ac}	100	10	10	1000	2000	1500	0	0	50	10	50
R_{gsS}	100	100	100	100	100	100	0	1000	100	100	50
R_{gsO}	600	3500	3500	3500	3500	3500	2000	400	3500	3500	3500
R_{clS}	9999	9999	9999	9000	200	400	9999	9999	9000	9999	9999
R_{clO}	9999	1000	1000	400	1500	600	9999	9999	800	1000	800
Seasonal category 5: transitional spring with partially green short annuals											
R_{sm}	9999	120	240	140	250	190	9999	9999	160	200	300
R_{lu}	9999	4000	4000	4000	2000	3000	9999	9999	4000	4000	8000
R_{ac}	100	50	80	1200	2000	1500	0	0	200	60	120
R_{gsS}	500	150	350	500	500	200	0	1000	0	250	400
R_{gsO}	300	150	200	200	200	300	2000	400	1000	180	200
R_{clS}	9999	4000	4000	4000	2000	3000	9999	9999	4000	4000	8000
R_{clO}	9999	1000	500	500	1500	700	9999	9999	600	800	800

* 1, Urban land; 2, agricultural land; 3, range land; 4, deciduous forest; 5, coniferous forest; 6, mixed forest; 7, water; 8, barren land, mostly desert; 9, unforested wetland; 10, mixed agricultural and range land; 11, rocky open areas with low-growing shrubs.

Table 12.2 Properties relevant to estimating resistances to dry deposition (data from Wesely 1989, Seinfeld & Pandis 1998, and Lind & Kok 1986).

Species	D_{H_2O}/D_I	H^* (M atm^{-1})*	Exponent (A)†	Reactivity (f_0)
Sulfur dioxide	1.89	1×10^5	−3020	0
Ozone	1.63	1×10^{-2}	2300	1
Nitrogen dioxide	1.6	1×10^{-2}	−2500	0.1
Nitrogen oxide	1.29	2×10^{-3}	−1480	0
Nitric acid	1.87	1×10^{14}	−8650	0
Hydrogen peroxide	1.37	1×10^5	−6800	1
Acetaldehyde	1.56	15	−6500	0
Formaldehyde	1.29	6×10^3	−6500	0
Methyl hydroperoxide	1.6	220	−5600	0.3
Peroxyacetic acid	2.0	540	−6170	0.1
Formic acid	1.6	4×10^6	−5740	0
Acetic acid	1.83	4×10^6	−5740	0
Ammonia	0.97	2×10^4	−3400	0
Peroxyacetyl nitrate	2.59	3.6	−5910	0.1
Nitrous acid	1.62	1×10^5	−4800	0.1
Pernitric acid	2.09	2×10^4	−1500	0
Hydrochloric acid	1.42	2.05×10^6	−2020	0

* Effective Henry's Law constant at a pH of about 6.5.
† H^* at the surface temperature is computed using $H^*(T) = H^* \exp[A(1/298 - 1/T)]$.

The three terms in eqn 12.28 account for a reduction in dry area, increased availability of water, and chemical reactivity, respectively. For deposition to open water, Seinfeld and Pandis (1998) suggest an alternative R_c proposed by Sehmel (1980):

$$R_c^i = \frac{25,400}{H^* T u_*} \quad (12.29)$$

where T is the surface temperature (Kelvin).

The Wesely surface resistance model includes data for 11 land-use types and five seasons. The model is designed to provide reasonable estimates of surface resistance to unstressed vegetation over time scales of a few weeks, and spatial scales consistent with the land-use types. To account for water stress, the stomatal resistance should be increased by a factor of 10. Wesley also suggests a minimum surface resistance of $10 \, s \, m^{-1}$ for species that exhibit an R_c near zero to avoid unrealistically high V_d over very rough surfaces.

12.3.5   General comments on the bulk model

There are two aspects of the bulk model approach that might lead to discrepancies when comparing model results to observations. First, the aerodynamic resistance is commonly derived from gradient transport theory (eqn 12.5), a theory that does not often apply within plant canopies. It is now well known that the local connection between flux and gradient is often lost within plant canopies, sometimes resulting in co-gradient transport (Raupach 1979; Cellier & Brunet 1992). Thus, expressions for the aerodynamic resistance, such as eqn 12.12, may not reliably describe the transport of material from the bulk atmosphere to the quasi-laminar layer. Second, the bulk model assumes that the vertical distribution of sources/sinks, such as leaves, can be described in a single term (R_c), which is often expressed as a parallel (i.e. vertical) sum of resistances. This essentially gives equal weight to the vertical distribution of resistances, implying that the concentration profile is constant with height (Bache 1986). If a source/sink exists, this cannot be true. Despite these shortcomings, the bulk model often performs well compared to observations. This is due in part to the fact that R_c is often the dominant resistance, and expressions for R_c have been based largely on observations of the residual term ($R_c = V_d^{-1} - R_a - R_b$). Thus, shortcomings in the theory are effectively "calibrated" into the R_c term.

When dry deposition rates are controlled largely by the stomatal resistance, measurements of latent heat flux may be useful for determining the surface resistance. Pleim *et al.* (1999) used latent heat fluxes to estimate the deposition velocity of O_3 to soybeans. Using the stomatal conductance ($1/R_s$) from latent heat measurements, along with a parallel resistance to account for deposition at the ground, their results were in good agreement with O_3 V_d measured by eddy covariance.

Comparisons between observed O_3 deposition velocities and those predicted by models such as that described above have shown that bulk models tend to predict the general diurnal trends of V_d (Padro *et al.* 1991, 1992; Zhang *et al.* 1996). Padro *et al.* (1991, 1992) noted that the bulk model does not explain variance on shorter time scales (hourly), which could be associated with passing clouds, abrupt changes in wind direction, and other small-scale processes. Large discrepancies may also be expected during stable periods. Long time scales (hours instead of minutes) and intermittent turbulence associated with stable conditions make it difficult to determine the aerodynamic resistance. It can also be difficult to measure small deposition velocities during stable conditions. The bulk model is also less useful during transition periods (around sunrise and sunset), when the assumption of steady state often breaks down.

While the bulk models are useful for predicting the general characteristics of dry deposition, more complex models have been used to study dry deposition in more detail (Meyers & Paw U 1987; Baldocchi 1988, 1992; Meyers *et al.* 1989, 1998). Multilayer models have been particularly useful

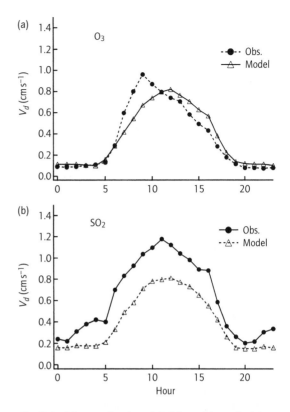

Fig. 12.4 Measured and modeled deposition velocities of O_3 and SO_2 above a deciduous forest (adapted from Finkelstein *et al.* 2000). Deposition velocities were measured by eddy covariance during summer.

for exploring factors that influence the surface resistance, such as leaf area index and canopy architecture, leading to improvements in bulk parameterizations of R_c.

Figure 12.4 shows deposition velocities of O_3 and SO_2 observed over a deciduous forest and corresponding V_d predicted using a multi-layer model (Finkelstein *et al.* 2000). Differences in the diurnal patterns of O_3 and SO_2 deposition velocity reflect differences in controlling factors. O_3 uptake is controlled largely by stomatal function. The sharp rise in V_d in early morning may occur as stomata open in the early morning. The gradual decline through midday may be caused by stomata closure, resulting from vapor pressure deficit. In contrast, SO_2 uptake follows the diurnal pattern of atmospheric resistance, indicating both turbulence and stomata control.

The model predicts the diurnal features of both O_3 and SO_2 deposition fairly well, but misses the peak O_3 V_d. The model underpredicts SO_2 V_d, which may indicate that SO_2 is subject to uptake along nonleaf surfaces. For comparison, the Wesely model predicts midday deposition velocities of about 0.7 and 0.9 cm s^{-1} for O_3 and SO_2 respectively, under conditions expected over a deciduous forest in summer.

12.3.6 Bidirectional fluxes

For some species, such as NH_3, both deposition and emission can occur. Ammonia is readily deposited to plants and wet surfaces because of its high solubility in water. Ammonia emissions have also been observed over fertilized agricultural land and, to a lesser degree, seminatural land (Sutton et al. 1998, and references therein).

The bulk resistance model can be used to describe ammonia deposition as outlined above. However, eqn 12.1 fails to describe bidirectional fluxes because implicit in it is the assumption of zero concentration at the surface. Including C_0 in eqn 12.2 leads to a modified resistance model with a nonzero surface concentration:

$$F = \frac{-(C_3 - C_0)}{R_a + R_b + R_c} \qquad (12.30)$$

The surface concentration is sometimes referred to as a compensation point. This simple model is able to describe bidirectional NH_3 fluxes as long as the exchange takes place through stomata (described by R_c).

Sutton et al. (1998) describe two forms of a resistance model designed to predict deposition and emissions that may occur within different regions of the canopy. Parallel fluxes to the cuticle and stomata are described in terms of a cuticle resistance (R_{cut}) and a stomatal resistance (R_s), as shown

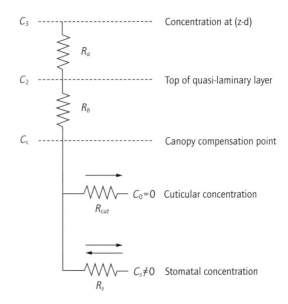

Fig. 12.5 Schematic of the bulk resistance model showing bidirectional fluxes.

in Fig. 12.5. Treating parallel fluxes in this way improved model performance for predicting NH_3 V_d to a wheat field. However, Sutton et al. (1998) point out that NH_3 may also be reemitted from the cuticle following deposition. To describe time-dependent deposition/emission cycles they replaced R_{cut} with a resistor–capacitor network in the framework of the electrical analog. This adds additional complexity to the model and is beyond the scope of this chapter. It is mentioned merely to demonstrate how the electrical analog might be modified to better describe unusual fluxes.

12.4   DRY DEPOSITION OF PARTICLES

The processes involved in the dry deposition of particles are similar to those associated with gases, except that particles are also subject to gravitational settling. Both particles and gases are transported similarly through the aerodynamic layer and are, therefore, described equally by the resist-

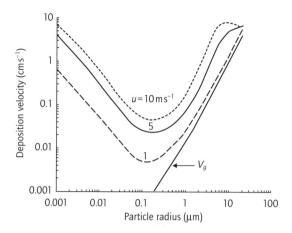

Fig. 12.6 Dry deposition velocity of unit density particles versus particle diameter for three wind speeds (10, 5, 1 m s^{-1}). (Reprinted from *Atmospheric Environment* (Slinn 1982). Copyright 1982, with permission from Elsevier Science.)

ance R_a. Diffusion in the laminar sublayer occurs through Brownian motion, analogous to molecular diffusion for gases. Particles can interact with the surface through Brownian motion, impaction, interception, and gravitational settling. The relative importance of these processes is governed largely by the particle size. Subsequently, particle size is a major controlling factor in the deposition of particles. The "stickiness" of the particles is also important, but it is not as critical as the surface properties that influence the surface resistance for gases (solubility and reactivity).

In general, the deposition of large particles ($D_p >$ 10 μm) is dominated by gravitational settling. Large particles can have short lifetimes in the atmosphere, on the order of several hours to a day (Fig. 12.6). Small particles ($D_p < 0.05$ μm) settle slowly, but are transported efficiently by Brownian motion and can also have short lifetimes. The relative importance of Brownian motion and gravitational settling can be also demonstrated by comparing the distance a particle travels in one second due to each process (Seinfeld & Pandis 1998). A 2 μm diameter particle diffuses a distance of about 4 μm and falls 200 μm due to gravity. A

0.02 μm diameter particle diffuses nearly 1000 times further than it falls. For particles in the accumulation region (0.05–2 μm), neither Brownian motion nor gravitational settling are effective. Therefore, these particles can have long lifetimes in the atmosphere (several days to weeks) unless removed by wet processes.

Phoretic processes, such as those associated with temperature gradients, concentration gradients, and electric fields, can also be important in particle deposition. For example, a temperature gradient can influence Brownian motion through differences in momentum associated with gases emerging from different directions. Phoretic effects are not considered here, except by default in the cases where experimental data have been used to provide estimates for model parameters. The reader is referred to Seinfeld and Pandis (1998) for a general discussion of thermophoresis and diffusophoresis, and to Opiolka *et al.* (1994) for a discussion of electrophoresis.

A bulk model for particle dry deposition is derived in the same manner as that for gas dry deposition. It is assumed that the particle concentration at the surface is zero. It is also assumed that the surface resistance can be ignored (i.e. particles adhere upon contact). Gravitational settling is often accounted for by splitting the flux through each layer into two parallel components ($F = F_d + F_s$), where F_d is the turbulent eddy flux described by R_a and R_b, and F_s is the flux associated with gravitational settling. The overall deposition velocity is described by

$$V_d = \frac{1}{R_a + R_b + R_a R_b V_g} + V_g \qquad (12.31)$$

where V_g is the gravitational settling velocity. The resistance $R_a R_b V_s$ is not a physical resistance. It is a virtual resistance that arises from adding F_s in parallel (Seinfeld & Pandis 1998).

The laminar sublayer resistance is given by

$$R_b = \frac{1}{u_* \left(Sc^{-2/3} + 10^{-3/St}\right)} \qquad (12.32)$$

where Brownian motion is described by the Schmidt number, and impaction is described by the Stokes number, St. The particle diffusivity is given by

$$D = \frac{k_b T C_c}{3\pi\mu D_p} \qquad (12.33)$$

where k_b is the Boltzmann constant $(1.38066 \times 10^{-23}\,\mathrm{J\,K^{-1}})$, C_c is the Cunningham correction factor, and μ is the viscosity of air. The Stokes number can be computed as

$$St = \frac{V_g u_\star^2}{g\nu} \qquad (12.34)$$

The settling velocity is derived from Stokes' Law:

$$V_g = \frac{\rho_p D_p^2 g C_c}{18\mu} \qquad (12.35)$$

where ρ_p is the particle density. The Cunningham correction factor is essentially unity for large particles $(D_p > 10\,\mu\mathrm{m})$ and increases as the particle diameter approaches the mean free path of the gas, λ (Seinfeld & Pandis 1998)

$$C_c = 1 + \frac{2\lambda}{D_p}\left[1.257 + 0.4\exp\left(-\frac{1.1 D_p}{2\lambda}\right)\right] \qquad (12.36)$$

The mean free path of air molecules can be computed as

$$\lambda = \frac{2\mu}{P\sqrt{\dfrac{8M}{\pi R T}}} \qquad (12.37)$$

where M is the molecular weight of air, and P is the pressure. The mean free path of air at 298 K is approximately 0.0651 μm (Seinfeld & Pandis 1998).

There is still considerable uncertainty associated with particle dry deposition (Nicholson 1998), particularly in the size range 0.05–2 μm. Some measurements have suggested that V_d in

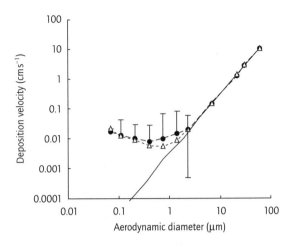

Fig. 12.7 Dry deposition velocity of particles versus particle size over open water (model, triangles; observations, circles; gravitational settling velocity, continuous line). (Reprinted with permission from Caffrey *et al.* 1998, Copyright 1998 American Chemical Society.)

this size range are larger than predicted by models (Nicholson 1998). However, there are significant uncertainties associated with measurement techniques. Order of magnitude differences between models and measurements are not uncommon. Models and measurements tend to show better agreement over smooth surfaces. Figure 12.7 shows recent measurements of particle dry deposition to Lake Michigan (Caffrey *et al.* 1998) along with predictions using the Williams model (Williams 1982). The Williams model is similar to the bulk deposition model described above, but contains improvements that account for the effects of wave breaking and spray formation in high winds, as well as the potential for particle growth in the region of high humidity close to the surface.

For particle deposition to plant canopies, the impaction term in eqn 12.32 causes difficulties. The term $10^{-3/St}$ results in a strong dependence on the Stokes number for $D_p \approx 0.1$–5 μm. Because there is a distribution of wind speeds and surface length scales in a canopy, the limit of inertial capture will not be as definite as suggested by the

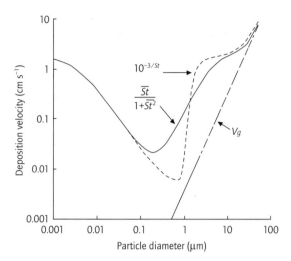

Fig. 12.8 Dry deposition velocity of particles calculated using two different impaction terms (with $u^* = 0.2\,\mathrm{m\,s^{-1}}$, $A = 0.01\,\mathrm{m}$). Also shown is the sedimentation velocity, V_g.

$10^{-3/St}$ term. Slinn (1982) suggested an alternative based on the characteristic radius of the surface collectors. Slinn (1982) refers to this impaction term as E_{IM}:

$$E_{IM} = \frac{\overline{St}}{1 + \left(\overline{St}\right)^2} \qquad (12.38)$$

where $\overline{St}$ is defined as

$$\overline{St} = \frac{V_g u_*}{cgA} \qquad (12.39)$$

where c is an unknown constant, expected to be near unity, and A is the radius of large collectors (leaves, blades of grass, etc.). Deposition velocities calculated as a function of particle size using eqn 12.32 with both forms of the impaction term are shown in Fig. 12.8.

Figure 12.8 provides an indication of the level of uncertainty associated with modeling particle deposition velocities in this size range over rough surfaces. Aerosol dry deposition remains an active

area of research and improvements in size-specific measurement techniques are required to improve the models, particularly over rough surfaces.

12.5 MEASURING DRY DEPOSITION

A number of techniques to measure dry deposition rates have been developed. Summaries of several techniques and their requirements have been reported by Baldocchi *et al.* (1988), Fowler & Duyzer (1989), and Dabberdt *et al.* (1993). The choice of technique depends largely on the type of instruments available to measure the species of interest and the assumptions that must be made in order to analyze the data. Deposition rates can be measured directly, by collecting material deposited on a surface, or by measuring the flux of material through a horizontal plane above a surface. Deposition rates can also be inferred indirectly through the measurement of a secondary quantity, such as a concentration gradient or variance. Indirect techniques are relatively easier to implement, but involve more assumptions then direct methods.

Collection. The analysis of material collected on a surface can be used to infer deposition rates to both natural and artificial surfaces. Surface extraction techniques (i.e. leaf washing and snow analysis) are more useful for particles than for gases because gases tend to bind to the surface. Artificial surfaces are useful primarily when the quasi-laminar and surface resistances are small. Otherwise, it is unlikely that the surface will mimic a natural surface (Seinfeld & Pandis 1998).

Chamber methods. Gaseous uptake can be measured by passing gas through a chamber containing an absorbing material (such as plants). Deposition rates are determined by measuring the concentration of the depositing species at the inlet and outlet of the chamber over time. Chamber techniques are used in order to study the factors that influence deposition.

Eddy covariance. The eddy covariance method (also known as eddy correlation) is used to obtain a direct measure of the flux though a horizontal plane above a surface. It is the most direct method available for dry deposition measurement and is the method of choice when adequate sensors are available. However, eddy covariance has been applied to only a handful of species because of stringent instrumentation requirements.

The technique involves the correlation of vertical velocity fluctuations (w') with fluctuations in concentration (c'), to measure the time-averaged vertical turbulent flux, $\overline{w'c'}$. Eddy covariance measurements require fast-response sensors to measure w' and c'. Typically, a sonic anemometer and a chemical sensor with response times of at least 0.2 s are employed. The basic requirement is that the instrument response should be fast enough to capture all time scales of turbulence that contribute significantly to the flux. Somewhat slower sensors can be employed when a similar flux, such as sensible heat, is measured simultaneously and used to correct the flux measured by the slower instrument. It is rare that sensors with response times longer than 1 s are employed, however.

Eddy accumulation. In the eddy accumulation method, two reservoirs (containers, filters, etc.) are used to store air associated with upward-moving and downward-moving eddies. This allows for the use of slower sensors to infer the flux from the concentration difference between the reservoirs. The traditional eddy accumulation method (Desjardins 1977) requires the use of fast proportional valves to sample air proportional to the vertical wind velocity. This difficult requirement has limited the success of the method. Bussinger and Oncley (1990) developed a modified version of the accumulation method that relaxed the need for proportional valves. This method (relaxed eddy accumulation) has been used to measure deposition and emission rates of a variety of compounds for which fast chemical sensors are not yet common, such as pesticides (Majewski *et al.* 1993), isoprene (Guenther *et al.* 1996), peroxides (Hall 1997), and monoterpenes (Beverland *et al.* 1996). Relaxed eddy accumulation is a promising technique likely to be used extensively in coming years until fast chemical sensors are available.

Gradient technique. The flux of a species to/from the surface can be estimated using eqn 12.5. If the eddy diffusivity is calculated as in eqn 12.8, the method is known as the aerodynamic gradient method. If the eddy diffusivity for a different species is measured directly (by measuring the flux and gradient simultaneously) and it is assumed that this diffusivity applies to the species of interest, it is known as the modified Bowen ratio method. Either way, it is usually assumed that the eddy diffusivity of the species of interest is equal to that of some other species, such as heat. For this to be true, the distribution of sources and sinks should be roughly similar.

Uncertainties and limitations. All of the micrometeorological techniques are subject to similar criteria to ensure that the measured flux truly represents the flux to the surface. The main criterion is that the surface exchange be steady-steady and one-dimensional (nondivergent). These criteria are discussed in more detail in Baldocchi *et al.* (1988), Fowler and Duyzer (1989), and Dabberdt *et al.* (1993). Even under the best of conditions, spatial variability limits the accuracy of turbulent flux measurements to about 10–20% (Baldocchi *et al.* 1988). Uncertainties associated with measurements of eddy diffusivities are approximately 30% (Meyers *et al.* 1996; Hall *et al.* 1999).

Flux divergence results from a boundary layer that is not fully developed. For this reason, a fetch to height ratio of 100 is typically sought (i.e. flat homogeneous terrain should exist for a distance of 100 times the effective measurement height). Measurements become more difficult in sloping or complex terrain. Measurements can still be carried out (McMillen 1988), but one should be careful to ensure that the measured flux is representative of the surface (i.e. more than one measurement site may be required).

The constant flux layer normally exists above the canopy to a height several times the length scale of the surface elements. The region just above the surface is called the roughness sublayer. The turbulence in this region is influenced by the individual surface elements, and does not follow surface-layer similarity. This layer often extends from the canopy top to 1.5 times the width of surface elements. For widely spaced elements the roughness sublayer may extend 3–4 times the height of the elements. The aerodynamic gradient method should be avoided in the roughness sublayer.

Fluxes of heat and water vapor can also influence turbulent fluxes of scalars when it is the density of the scalar that is measured. This is because fluxes of water vapor and heat can cause fluctuations in air density (Webb *et al.* 1980). The error introduced is small for species with deposition velocities greater than about $1\,\mathrm{cm\,s^{-1}}$ (Fowler & Duyzer 1989).

The steady-state assumption can lead to errors when the concentration of the species of interest changes rapidly with time. A rough estimate of this effect can be determined from,

$$\frac{\Delta C}{C} = V_d \frac{\Delta F}{F} \frac{\Delta t}{\Delta z} \qquad (12.40)$$

where $\Delta z = (z - d)$ and Δt is the measurement time interval (Fowler & Duyzer 1989). For example, if we require that nonsteady-state errors contribute no more that 10% to the error in the flux ($\Delta F/F = 0.1$), the deposition velocity is $0.01\,\mathrm{m\,s^{-1}}$, Δz is $10\,\mathrm{m}$, and Δt is 30 min, the maximum allowable change in concentration is 18%. In most cases this is easily achieved. However, problems can arise during transition periods around sunrise and sunset as the boundary layer and surface characteristics undergo rapid change.

Finally, nonsteady state can occur if there are chemical reactions taking place on time scales comparable to the time scale of turbulence. This can be an issue for O_3–NO–NO_2 measurements because the photochemistry can be rapid (Gao *et al.* 1991), as well as for very reactive hydrocarbon species. Micrometeorological techniques will often be applicable to the family total (i.e. $NO_x = NO + NO_2$) but not to the individual species.

12.6 WET DEPOSITION

The processes by which gases and particles are incorporated into precipitation and subsequently removed from the atmosphere involve a large range of spacial and temporal scales: from the molecular scale to those associated with thunderstorms and fronts. The problem is complicated by the existence of multiple phases (gases, liquid water, ice), a size distributions of both aerosol pollutants and scavengers, and chemical reactions that can occur within the scavengers. The complex nature of wet deposition makes explicit treatment in atmospheric models virtually impossible.

Although the terminology for various scavenging processes varies in the literature, the incorporation of species into cloud water is usually referred to as "rainout," while "washout" refers to the scavenging of species below clouds due to falling precipitation (Hanna *et al.* 1982). Below-cloud scavenging is usually represented as a first-order process, such that the rate of transfer of a species to a droplet is proportional to the concentration of that species. For the simple case of a homogeneous atmosphere with no chemical production or loss, the time rate of change of concentration of a gaseous species is related to the scavenging coefficient as

$$\frac{\mathrm{d}C}{\mathrm{d}t} = -\Lambda C \qquad (12.41)$$

where Λ ($\mathrm{s^{-1}}$) is the scavenging coefficient. A similar expression is also used for particle scavenging. The inverse of the scavenging coefficient is the characteristic scavenging time (τ), also referred to as the rainout lifetime. Scavenging coefficients are a function of time, space, rainfall characteristics, and the size distribution of scavenged aerosols. If the scavenging coefficient is constant with time, then the concentration can be expressed as

$$C(t) = C_0 e^{-\Lambda t} \qquad (12.42)$$

where C_o is the initial concentration.

The use of eqn 12.41 is limited to irreversible scavenging that is not dependent on previous scavenging events. Because precipitation events are naturally episodic, efforts to extend eqn 12.41 to cases in which scavenging is dependent on previous events (Giorgi & Chameides 1985) and to time-dependent scavenging (Xing & Chameides

1990) have been made. In this chapter, we consider only simple cases of independent scavenging events.

Another parameter used to describe wet deposition is the washout ratio, which is based on the ground-level concentration of a species in the precipitation and in air:

$$W_r = \frac{C_{precipitation}}{C_{air}} \qquad (12.43)$$

If the washout ratio is known, the concentration in the precipitation can be calculated and the flux of the species to the ground can be calculated in terms of the precipitation rate, p_o (mm h^{-1}):

$$F = p_o C_{precipitation} \qquad (12.44)$$

12.6.1 Scavenging of gases

The scavenging coefficient for soluble gases is a function of the solubility of the scavenged species, and may be affected by mass transfer limitations. Scavenging occurs within clouds as liquid water droplets scavenge soluble gases from the interstitial air. Droplets that grow large enough to fall may scavenge soluble gases as they fall. If droplets evaporate, the gases may be released back into the air. If a droplet with a large initial concentration of a species falls through a layer depleted in that species, mass transfer may occur from the droplet to the bulk atmosphere. Finally, chemical reactions taking place in the droplet can dramatically enhance the scavenging ability of the droplet. For treatment of reactive gases see Giorgi and Chameides (1985) and Seinfeld and Pandis (1998).

As a starting point, we consider a soluble gas initially at concentration C_o that establishes equilibrium with liquid cloud water. The concentration of the soluble species in the liquid phase, C_{aq} (mole l^{-1}), will be related to the equilibrium concentration in the gas phase, C^{eq}:

$$C^{eq} = \frac{C_{aq}}{H^*} \qquad (12.45)$$

where H^* is the effective Henry's Law coefficient (mol l^{-1} atm^{-1}). Note that a unit conversion may be necessary when using eqn 12.45 depending on the units associated with H^*. The right side of eqn 12.45 yields atmospheres, while gas-phase concentrations are normally given in moles per unit volume. For simplicity, the required unit conversion factor will not always be included in subsequent equations involving H^*.

By using eqn 12.45 in a mole balance on the soluble species, the fraction of a soluble species remaining in the gas phase after dissolution can be expressed as

$$\frac{C}{C_o} = \frac{1}{1 + 10^{-6} H^* RT \dfrac{L}{\rho_w}} \qquad (12.46)$$

where L (g m^{-3}) is the liquid water content of the cloud, 10^{-6} and RT are used for unit conversion (R is the gas constant (l atm mole^{-1} K^{-1})).

Figure 12.9 shows the fraction remaining in the gas phase as a function of H^* for three typical values of L. At equilibrium, highly soluble gases ($H^* > 10^6$) are almost completely removed from the gas phase. Species such as H_2O_2 and HNO_3 are scavenged very efficiently under equilibrium con-

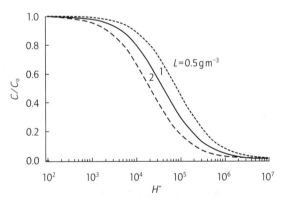

Fig. 12.9 Fraction remaining in the gas phase at equilibrium after dissolution into cloud droplets at three typical values of liquid water content ($T = 298$ K, $\rho_w = 1$ g cm^{-3}).

ditions. As cloud droplets that have scavenged nearly all of the soluble species from interstitial air grow large and fall, they have the opportunity to scavenge even more soluble gas as they pass through air still containing the soluble species. How much more they can scavenge depends largely on the size of falling droplets.

12.6.2 Below-cloud removal of gases

The rate of transfer (flux) of a soluble gas to a falling drop can be expressed in terms of a mass transfer coefficient, K_c $(m\,s^{-1})$, and the concentration difference between the bulk atmosphere and the air–liquid interface (Seinfeld & Pandis 1998):

$$F = K_c\left(C - C_{eq}\right) \qquad (12.47)$$

Applying Henry's Law,

$$F = K_c\left(C - \frac{C_{aq}}{H^\star}\right) \qquad (12.48)$$

The mass transfer coefficient is given by the empirical correlation (Frossling 1938):

$$K_c = \frac{D_g}{D_p}\left[2 + 0.6\left(\frac{D_p U_t}{\nu}\right)^{1/2}\left(\frac{\nu}{D_g}\right)^{1/3}\right] \qquad (12.49)$$

where D_g is the diffusivity of the species in air, D_p is the droplet diameter, and U_t is the terminal velocity of the droplet.

To arrive at an expression for the scavenging coefficient, Λ, we follow the method of Seinfeld and Pandis (1998) and calculate the rate of transfer of the gas into a single droplet as it falls. Assuming that the droplet does not evaporate during flight, the change in concentration with fall distance is given by

$$\frac{dC_{aq}}{dz} = \frac{6K_c}{U_t D_p}\left(C_g - \frac{C_{aq}}{H^\star}\right) \qquad (12.50)$$

The driving force for scavenging is the difference between the concentration of the species in the

bulk atmosphere and the gas-phase concentration near the surface of the droplet in equilibrium with the droplet concentration (as shown in eqns 12.47 and 12.50). If the concentration in the bulk atmosphere is much greater than C_{aq}/H, this implies irreversible scavenging and we can simplify eqn 12.50 to

$$\frac{dC_{aq}}{dz} = \frac{6K_c}{U_t D_p} C \qquad (12.51)$$

This approximation will hold for very soluble species $(H^\star > 10^5)$ such as H_2O_2 and HNO_3. If the droplet diameter and C are constant with height, eqn 12.51 can be integrated over the fall distance of the droplet.

$$C_{aq} = C_{aq}^o + \frac{6K_c}{U_t D_p} C\,z \qquad (12.52)$$

We also need to determine an expression for the below-cloud scavenging coefficient since this is what is often used in atmospheric modeling. The flux due to below-cloud scavenging is the integral of $C\Lambda$ from the ground to the cloud base. Assuming a homogeneous atmosphere,

$$F_{bc} = C\int_0^h \Lambda dz = \Lambda h C \qquad (12.53)$$

The below-cloud scavenging coefficient is calculated by equating eqn 12.53 with the flux determined by summing the change in concentration $(C_{aq} - C_{aq}^0)$ over all droplets (Giorgi & Chameides 1985; Seinfeld & Pandis 1998):

$$\Lambda = \frac{6 \times 10^{-3} p_o K_c}{U_t D_p} \qquad (12.54)$$

The scavenging coefficient is proportional to the rainfall rate, p_o. It is also a strong function of droplet size. Small droplets scavenge much more effectively than large droplets. This is because they fall more slowly than large droplets, and because they have larger mass transfer coefficients. For a rainfall rate of $1\,mm\,h^{-1}$ the gas-phase scavenging coefficient for a 0.001 cm droplet is approximately 10^8 times larger than that of a 1 cm droplet ($4.4 \times 10^5\,h^{-1}$ versus $3.6 \times 10^{-3}\,h^{-1}$) (Seinfeld & Pandis 1998).

The below-cloud uptake of a gas, W_{bc} (moles $m^{-3}\,s^{-1}$) can also be expressed in terms of the per-droplet flux (eqn 12.47) and the droplet number density, $n\,(m^{-3})$.

$$W_{bc} = \Lambda C = \pi D^2 K_c C n \qquad (12.55)$$

Because the rainfall droplet size distribution is not uniform, a more general form for Λ is obtained by integrating over the size distribution of droplets, $N(D_p)$:

$$\Lambda = \frac{W_{bc}}{C} = \int_0^\infty \pi D_p^2 K_c N(D_p) dD_p \qquad (12.56)$$

Because raindrop size distributions are non-Gaussian, application of eqn 12.55 using the mean droplet diameter can lead to large errors in Λ. The droplet size distribution can have a large effect on the scavenging coefficient, depending on the relative proportion of small droplets.

The "wet removal" lifetime of soluble gases is a function of the local scavenging rate and the frequency and duration of clouds and precipitation. It is also a function of the nature of the cloud cover and precipitation type. These complicating factors make detailed prediction of wet removal difficult. As a general example, Brasseur et al. (1998) used the method outlined above in a three-dimensional tropospheric model. They reported zonally averaged HNO_3 wet removal lifetimes in the lower troposphere of 0.5–2 days in the Tropics, a few days in mid-latitudes, and several days to weeks in polar regions.

12.6.3   Within-cloud removal of gases

The rate of transfer of a soluble species from interstitial air to cloud droplets can be described in a similar manner to that associated with the transfer of

soluble species to falling drops. Because cloud droplets are small, mass transfer will be much more rapid than that associated with falling raindrops. The scavenging time for HNO_3 is on the order of seconds, much faster than the characteristic times associated with the formation of precipitation and the movement of air through clouds (Seinfeld & Pandis 1998). Thus, within-cloud scavenging can usually be considered an equilibrium process.

The relative effectiveness of rainout and washout depends on the solubility of the gas (in addition to the droplet size distribution and cloud dynamics). If washout is insignificant compared to rainout, then the flux at the cloud base and the flux at the ground will be the same. A comparison of cloud-base and ground-level fluxes of two species with different solubilities is shown in Fig. 12.10. In this example, rainout dominates over washout for species that are only slightly soluble ($H^* = 10^3$). This is because only a small amount of the species enters the liquid phase within the cloud and the gas-phase mixing ratio is not affected by the presence of the cloud. As the droplets fall, they are already saturated with respect to the levels of trace gas below the cloud and therefore little additional dissolution occurs. Washout can be important for highly soluble species, however. Because the gas-phase mixing ratio is substantially altered inside the cloud, droplets are undersaturated with respect to the air below the cloud. Additional dissolution can occur as the droplets fall.

12.6.4   Within-cloud removal of particles

Aerosol scavenging within clouds is the results of both nucleation of cloud droplets from condensation nuclei and scavenging of excess aerosols by the cloud droplets. Most of the aerosol mass is incorporated during the nucleation process. Collection of interstitial aerosol by cloud droplets is a slow process and is not a significant aerosol removal mechanism.

12.6.5   Below-cloud removal of particles

As a raindrop sweeps out a volume of air during its fall, it collects particles in this volume. The air-

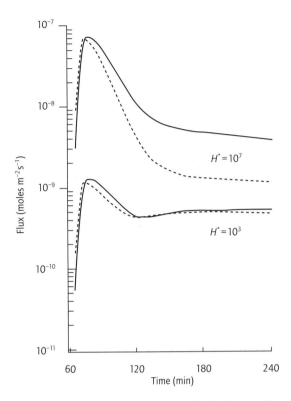

Fig. 12.10 Comparison of cloud-base (dashed line) and ground-level (solid line) flux of soluble species removed by precipitation. (From Xing & Chameides 1990, with kind permission from Kluwer Academic Publishers.)

The collection of particles by falling droplets, like the collection of particles on surface elements, involves inertial tendencies as well as Brownian motion. Large particles will be collected as they are unable to follow the streamlines. Small particles will follow streamlines around the droplet, but may come into contact with the droplet due to Brownian motion. Thus, the scavenging of large and small particles is relatively efficient. Particles in the size range 0.1–1 μm are not collected efficiently. The minimum in the collection efficiency is sometimes referred to as the "Greenfield Gap." Phoretic effects are also important in aerosol scavenging. Electric effects enhance the scavenging, particularly for small particles. Thus, thunderstorms may lead to more efficient removal of small particles than nonelectrified rain events (Pranesha & Kamra 1997).

Collection efficiencies are generally much less than unity. Particles with diameter 0.01 μm are collected with an efficiency of about 0.001–0.01. Particles with diameter 1 μm are collected with efficiencies of about 0.0001–0.001. The collection efficiency increases with diameter for particles larger than about 1 μm and approaches unity for particles larger than 10 μm (Seinfeld & Pandis 1998).

Slinn (1982) used dimensional analysis and experimental data to formulate an expression for the collection efficiency:

flow around the droplet is perturbed such that some particles will follow streamlines around the droplet and will not be collected. Whether a particle is collected or not is described in terms of a collection efficiency, E, which is the fraction of particles of diameter d_p that are collected by a droplet of diameter D_p. The scavenging coefficient for a particular particle diameter is given by

$$\Lambda_p(d_p) = \int_0^\infty \frac{\pi}{4} D_p^2 U_t(D_p) E(D_p, d_p) N(D_p) dD_p \quad (12.57)$$

where $N(D_p)$ is the number distribution of the droplets (Seinfeld & Pandis 1998).

$$E = \frac{4}{ReSc}[1 + 0.4Re^{1/2}Sc^{1/3} + 0.16Re^{1/2}Sc^{1/2}]$$

$$+ 4\frac{d_p}{D_p}\left[\frac{\mu_a}{\mu_w} + (1 + Re^{1/2})\frac{d_p}{D_p}\right] + \left[\frac{St - S^\star}{St - S^\star + \frac{2}{3}}\right]^{2/3}$$

$$(12.58)$$

where μ_a is the viscosity of air, μ_w is the viscosity of water, d_p is the diameter of the aerosol, D_p is the diameter of the droplet, the Reynolds (Re) number is $D_p U_t \rho_a / 2\mu_a$, the particle diffusivity is given by eqn 12.33, and S^* is given by

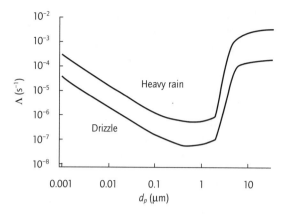

Fig. 12.11 Particle scavenging coefficient as a function of particle size. (Reprinted from *Atmospheric Environment* (Garcia Niento *et al.* 1994). Copyright 1980, with permission from Elsevier Science.)

$$S^{\star} = \frac{1.2 + \dfrac{1}{12}\ln(1 + Re)}{1 + \ln(1 + Re)} \qquad (12.59)$$

Using eqns 12.57 and 12.58 along with a specified droplet size distribution and rainfall rate (ignoring phoretic effects) allows the scavenging coefficient to be determined as a function of aerosol size. Figure 12.11 shows the scavenging coefficient as a function of particle size calculated for two rainfall rates. There are three distinct regions evident in Fig. 12.11: below 2 μm, 2–5 μm, and above 5 μm. Within each region the dominant mechanism is Brownian motion, interception, and impaction, respectively.

Although eqn 12.57 captures the essence of the mechanisms responsible for below-cloud capture, modeling wet removal is complicated by the nature of precipitating storms (convection) and difficulties in specifying droplet size distributions. Wet deposition is often highly parameterized in large-scale chemical transport models due to data limitations and difficulties in modeling subgrid-scale processes (processes that occur on spatial scales much smaller than the model grid). Schemes that parameterize the scavenging coeffi-cient as a constant or a simple latitude-dependent function are commonly used (Lee & Feicher 1995). Despite these simplifications global models are able to simulate the wet removal of ^{210}Pb (a pro-duct of ^{222}Rn decay that becomes irreversibly attached to sub-micrometer aerosols) relatively well (Lee & Feicher 1995; Giannakopoulos *et al.* 1999). Lee and Feicher (1995) found that the spread of results obtained from using four different wet deposition schemes was about a factor of two. Giannakopoulos *et al.* (1999) found that convective processes are an important aspect of scavenging and that the performance of even simple scavenging schemes could be improved by coupling the scheme to vertical transport in the model.

REFERENCES

Bache, D.H. (1986) On the theory of gaseous transport to plant canopies. *Atmospheric Environment* **20**, 1379–88.

Baldocchi, D.D. (1988) A multi-layer model for estimat-ing sulfur dioxide deposition to a deciduous oak forest canopy. *Atmospheric Environment* **22**, 869–84.

Baldocchi, D.D. (1992) On estimating HNO$_3$ deposition to a deciduous forest with a Lagrangian random-walk model. In Schwartz, S.E. & Slinn, W.G. (eds), *Precipita-tion Scavenging and Atmosphere–Surface Exchange*, Hemisphere Publishing, Washington, DC.

Baldocchi, D.D., Hicks, B.B. & Camara, P. (1987) A canopy stomatal resistance model for gaseous deposi-tion to vegetated surfaces. *Atmospheric Environment* **21**, 91–101.

Baldocchi, D.D., Hicks, B.B. & Meyers, T.P. (1988) Mea-suring biosphere–atmosphere exchanges of biologi-cally related gases with micrometeorological methods. *Ecology* **69**, 1331–40.

Beverland, I.J., Milne, R., Boissard, C., O'Neill, D.H., Moncrieff, J.B. & Hewitt, C.N. (1996) Measurement of carbon dioxide and hydrocarbon fluxes from a sitka spruce forest using micrometeorological techniques. *Journal of Geophysical Research*, **101**, 22,807–15.

Brasseur, G.P., Hauglustaine, D.A., Walters, S. *et al.* (1998) MOZART, a global chemical transport model for ozone and related chemical tracers 1. Model description. *Journal of Geophysical Research* **103**, 28,265–89.

Businger, J. & Oncley, S. (1990) Flux measurement with conditional sampling. *Journal of Atmospheric Oceanic Technology* **7**, 349–52.

Businger, J.A., Wyngaard, J.C., Izumi, Y. & Bradley, E.F. (1971) Flux-profile relationships in the atmospheric surface layer. *Journal of the Atmospheric Sciences* **28**, 181–9.

Caffrey, P.F., Ondov, J.M., Zufall, M.J. & Davidson, C.I. (1998) Determination of size-dependent deposition velocities with multiple intrinsic elemental tracers. *Environment Science and Technology* **32**, 1615–22.

Cellier, P. & Brunet, Y. (1992) Flux-gradient relationships above tall plant canopies. *Journal of Agricultural and Forest Meteorology* **58**, 93–117.

Dabberdt, W.F., Lenschow, D.H., Horst, T.W., Zimmerman, P.R., Oncley, S.P. & Delany, A.C. (1993) Atmosphere-surface exchange measurements. *Science* **260**, 1472–81.

Desjardins, R.L. (1977) Energy budget by an eddy correlation method. *Journal of Applied Meteorology* **16**, 248–50.

Finkelstein, P.L., Ellestadm, T.G., Clarke, J.F. *et al.* (2000) Ozone and sulfur dioxide dry deposition to forests: Observations and model evaluation. *Journal of Geophysical Research* **105**, 15,365–77.

Fowler, D. & Duyzer, J.H. (1989) Micrometeorological techniques for the measurement of trace gas exchange. In Andreae, M.O. & Schimel, D.S. (eds), *Exchange of Trace Gases Between Terrestrial Ecosystems and the Atmosphere*. Wiley, New York.

Frossling, N. (1938) The evaporation of falling drops. *Gerlands Beitrage Geophysical* **52**, 170–216.

Gao, W., Wesely, M.L. & Lee, I.Y. (1991) A numerical study on the effects of air chemistry on fluxes of NO, NO$_2$, and O$_3$ near the surface. *Journal of Geophysical Research* **96**, 18,761–9.

Garcia Niento, P.J.G., Garcia, B.A., Diaz, J.M.F. & Braña, M.A.R. (1994) Parametric study of selective removal of atmospheric aerosol by below-cloud scavenging. *Atmospheric Environment* **28**, 2335–42.

Garratt, J.R. & Hicks, B.B. (1973) Momentum, heat, and water vapour transfer to and from natural and artificial surfaces. *Quarterly Journal of the Royal Meteorological Society* **99**, 680–7.

Giannakopoulos, C., Chipperfield, M.P., Law, K.S. & Pyle, J.A. (1999) Validation and intercomparison of wet and dry deposition schemes using ^{210}Pb in a global three-dimensional off-line chemical transport model. *Journal of Geophysical Research* **104**, 23,761–84.

Giorgi, F. & Chameides, W.L. (1985) The rainout parameterization in a photochemical model. *Journal of Geophysical Research* **90**, 7872–80.

Guenther, A., Baugh, W., Davis, K. *et al.* (1996) Isoprene fluxes measured by enclosure, relaxed eddy accumulation, surface layer gradient, mixed layer gradient, and mixed layer mass balance techniques. *Journal of Geophysical Research* **101**, 18,555–67.

Hall, B.D. (1997) Measurements of the dry deposition of peroxides to forests. PhD dissertation. Washington State University.

Hall, B.D., Claiborn, C.S. & Baldocchi, D. (1999) Measurement and modeling of dry deposition of peroxides. *Atmospheric Environment* **33**, 577–89.

Hanna, S.R., Briggs, G.A. & Hosker, R.P. (1982) *Handbook on Atmospheric Diffusion*. Technical Information Center, US Department of Energy, Washington, DC.

Horvath, L., Nagy, Z. & Weidinger, T. (1998) Estimation of dry deposition velocities of nitric oxide, sulfur dioxide, and ozone by the gradient method above short vegetation during the TRACT campaign. *Atmospheric Environment* **32**, 1317–22.

Kaimal, J.C. & Finnigan, J.J. (1994) *Atmospheric Boundary Layer Flows: Their Structure and Measurement*. Oxford University Press, New York.

Lee, H.N. & Feicher, J. (1995) An intercomparison of wet precipitation scavenging schemes and the emission rates of ^{222}Rn for the simulation of global transport and deposition of ^{210}Pb. *Journal of Geophysical Research* **100**, 23,253–70.

Lind, J.A. & Kok, G.L. (1986) Henry's Law determinations for aqueous solutions of hydrogen peroxide, methylhydroperoxide, and peroxyacetic acid. *Journal of Geophysical Research* **91**, 7889–95.

McMillan, R. (1988) An eddy correlation technique with extended applicability to non-simple terrain. *Boundary Layer Meteorology* **43**, 231–45.

Majewski, M., Desjardins, R., Rochette, P., Pattey, E., Selber, J. & Glotfelty, D. (1993) Field comparison of an eddy accumulation and aerodynamic-gradient system for measuring pesticide volatilization fluxes. *Environment Science and Technology* **27**, 121–8.

Meyers, T.P. & Paw U, K.T. (1987) Modeling the plant canopy micrometeorology with higher-order closure principles. *Journal of Agricultural and Forest Meteorology* **41**, 143–63.

Meyers, T.P., Huebert, B.J. & Hicks, B.B. (1989) HNO$_3$ deposition to a deciduous forest. *Boundary Layer Meteorology* **49**, 395–410.

Meyers, T.P., Hall, M.E., Lindberg, S.E. & Kim, K. (1996) Use of the modified Bowen-ration technique to measure fluxes of trace gases. *Atmospheric Environment* **30**, 3321–9.

Meyers, T.P., Finkelstein, P.L., Clarke, J., Ellestad, T.G. & Sims, P.F. (1998) A multilayer model for inferring dry deposition using standard meteorological measurements. *Journal of Geophysical Research* **103**, 22,645–66.

Nicholson, K.W. (1998) The dry deposition of small particles: a review of experimental measurements. *Atmospheric Environment* **22**, 2653–66.

Opiolka, S., Schmidt, F. & Fissan, H. (1994) Combined effects of electrophoresis and thermophoresis on particle deposition onto flat surfaces. *Journal of Aerosol Science* **25**, 665–71.

Padro, J., Den Hartog, G. & Neumann, H.H. (1991) An investigation of the ADOM dry deposition module using summertime O_3 measurements above a deciduous forest. *Atmospheric Environment* **25**, 1689–704.

Padro, J., Neumann, H.H. & Den Hartog, G. (1992) Modeled and observed dry deposition velocity of O_3 above a deciduous forest in the winter. *Atmospheric Environment* **26**, 775–84.

Pleim, J.E., Finkelstein, P.L., Clarke, J.F. & Ellestad, T.G. (1999) A technique for estimating dry deposition velocities based on similarity with latent heat flux. *Atmospheric Environment* **33**, 2257–68.

Pranesha, T.S. & Kamra, A.K. (1997) Scavenging of aerosol particles by large water drops 3. Washout coefficients, half-lives, and rainfall depths. *Journal of Geophysical Research* **102**, 23,947–53.

Raupach, M.R. (1979) Anomalies in flux-gradient relationships over forests. *Boundary Layer Meteorology* **16**, 467–86.

Sehmel, G.A. (1980) Particle and gas deposition, a review. *Atmospheric Environment* **14**, 983–1011.

Seinfeld, J.H. & Pandis, S.H. (1998) *Atmospheric Chemistry and Physics: From Air Pollution to Climate Change*. Wiley, New York.

Sillman, S., He, D., Rippin, M.R. *et al.* (1998) Model correlations for ozone, reactive nitrogen, and peroxides for Nashville in comparison with measurements: implications for O_3–NO_x–hydrocarbon chemistry. *Journal of Geophysical Research* **103**, 22,629–44.

Slinn, W.G.N. (1982) Predictions for particle deposition to vegetative canopies. *Atmospheric Environment* **16**, 1785–94.

Stull, R.B. (1988) *An Introduction to Boundary Layer Meteorology*. Kluwer Academic Publishers, Norwell, MA.

Sutton, M.A., Burkhardt, J.K., Guerin, D., Nimitz, E. & Fowler, D. (1998) Development of resistance models to describe measurements of bi-directional ammonia surface–atmosphere exchange, *Atmospheric Environment* **32**, 473–80.

Thom, A.S. (1972) Momentum, mass, and heat exchange of vegetation. *Quarterly Journal of the Royal Meteorological Society* **98**, 124–34.

Webb, E.K., Pearman, G.I. & Leuning, R. (1980) Correction of flux measurements for density effects due to heat and water vapor transfer. *Quarterly Journal of the Royal Meteorological Society* **106**, 85–100.

Wesely, M.L. (1988) *Improved Parameterization for Surface Resistance to Gaseous Dry Deposition in Regional-scale Numerical Models*. EPA/600/3-88/025. US Environmental Protection Agency Report, Washington, DC.

Wesely, M.L. (1989) Parameterization of surface resistances to gaseous dry deposition in regional-scale numerical models. *Atmospheric Environment* **23**, 1293–304.

Wesely, M.L. & Hicks, B.B. (1977) Some factors that affect the deposition rates of sulfur dioxide and similar gases on vegetation. *Journal of Air Pollution Control Association* **27**, 1110–16.

Williams, R.M. (1982) A model for the dry deposition of particles to natural water surfaces. *Atmospheric Environment* **16**, 1933–8.

Xing, L. & Chameides, W.L. (1990) Model simulations of rainout and washout from a warm stratiform cloud. *Journal of Atmospheric Chemistry* **10**, 1–26.

Zhang, L., Padro, J. & Walmsley, J.L. (1996) A multi-layer model versus single-layer models and observed O_3 dry deposition velocities. *Atmospheric Environment* **30**, 339–45.

Part 2

Problems, Tools, and Applications

13 Global Air Pollution Problems

ATUL K. JAIN AND KATHARINE A.S. HAYHOE

13.1 INTRODUCTION

Prior to the Industrial Revolution, the Earth's climate was entirely governed by natural influences. Records preserved in ice cores and ancient sediments reveal large fluctuations in global climate. The Milankovitch theory holds that long-term natural variations in temperature change were primarily governed by 26,000 to 100,000-year cycles in solar flux due to the Earth's orbital variations. Over shorter time periods, temperature was also modified by other internal factors and relationships within the Earth–atmosphere–ocean system. Comparison of historical records of temperature and greenhouse gases (Fig. 13.1) reveals that: variations in temperature are well correlated with variations in major greenhouse gases from biogenic sources, such as carbon dioxide (CO_2) and methane (CH_4); temperature variations were also affected by periodic high dust episodes related to increases in volcanic activity or widespread arid conditions.

Until just a few decades ago, it was believed that human presence on this planet was insufficient to affect the natural environment (e.g. Keller 1998). The truth is that, commencing in the 1600s and rapidly accelerating after the Industrial Revolution of the 1700s and 1800s, human activities began to intervene in the natural world (Pollack *et al.* 1998). At the beginning of this new millennium, emissions from human activities now have the potential to jeopardize life on this planet. Global temperatures currently exceed any global

average temperatures recorded over the last half-million years. Worldwide monitoring has provided concrete evidence that human activities such as fossil fuel combustion, agriculture development, waste generation, synthetic chemicals production, biomass burning, and changes in land use are significantly altering levels of radiatively and chemically active greenhouse gases (GHGs) and aerosols in the atmosphere. These changes have resulted in two major global pollution problems of increasing difficulty to humanity. The first is the depletion of stratospheric ozone, commonly known as the "ozone hole." The problem is that of greenhouse gas induced temperature rise, known as "global warming." Although the science of these two problems is interlinked—the chemicals that attack the ozone layer also contribute to global warming—these two problems are essentially distinct in terms of primary causes, impacts and possible solutions.

In the 1970s, several pioneering studies (Molina & Rowland 1974; Stolarski & Cicerone 1974) drew attention to the accumulation of industrially produced chlorofluorocarbons (CFCs) and other chlorinated and brominated halocarbons in the stratosphere. At stratospheric altitudes, these compounds destroy the ozone shield that protects life on Earth from harmful solar ultraviolet radiation. The original theory—long since verified—states that the stability of halocarbons enables them to transport chlorine and bromine from the Earth's surface to the stratosphere, where the ozone layer is situated. As solar radiation strikes

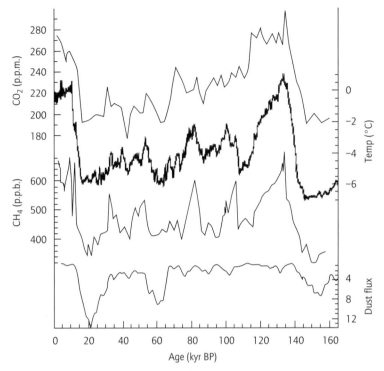

Fig. 13.1 Climate and atmospheric composition records from the Vostok ice core (east Antarctica) covering the past 160,000 years. These include CO_2 and CH_4 greenhouse gas records, which are closely tied to Antarctic temperature variations over the last full glacial/interglacial climate cycle (Lorius *et al.* 1985; Barnola *et al.* 1987; Jouzel *et al.* 1987; Chappellaz *et al.* 1990). Temperature data are plotted as deviations from the present day mean annual temperature. Also included is the record of the flux of dust to the area (shown on an inverted scale for comparison purposes) (Petit *et al.* 1990).

the halocarbons in the stratosphere and photo-chemical reactions occur, the chlorine and bromine are released. Each of these highly reactive molecules is capable of destroying thousands of stratospheric ozone (O_3) molecules before being converted to less reactive forms. Since the beginning of the twentieth century, halocarbons have been manufactured for use in many industrial and commercial products. Their low reactivity (inertness) and long lifetimes of hundreds to thousands of years have made them attractive chemicals for use as foam blowing agents, coolants, propellants, fire retardants, and other industrial applications. The links between human production of these compounds, the build-up in their stratospheric concentrations, and the recent decrease in stratospheric O_3 have been firmly established by numerous observational and modeling studies, as summarized in WMO (1999).

Greenhouse warming is a natural phenomenon that occurs due to naturally occurring GHGs in the Earth's atmosphere. These GHGs absorb much of the Earth's infrared radiation that would otherwise escape to space. The trapped radiation warms the lower atmosphere and the Earth, keeping the Earth's surface over 30°C warmer than it would be otherwise. This phenomenon, known as the "greenhouse effect," enables life to survive on this planet. The major greenhouse gases contributing to this natural effect are those with significant natural sources: water vapor (H_2O), carbon dioxide (CO_2), methane (CH_4), nitrous oxide (N_2O), and tropospheric ozone (O_3).

The problem of global warming is what is known as the **enhanced** greenhouse effect. Since the beginning of industrial times, the greenhouse effect has been enhanced by steadily increasing GHG concentrations in the atmosphere. These concentrations have been increasing due to emissions from human-related activities. Increasing concentrations alter the radiative balance of the Earth and result in a warming of the Earth's surface. The main GHGs affected by human activities are CO_2, CH_4, N_2O, tropospheric O_3, and CFCs (mainly CFC-11 and CFC-12), which are also responsible for stratospheric O_3 loss. Aerosols resulting from human activities, particularly from fossil fuel combustion, can scatter or absorb solar and infrared radiation, cooling and/or heating the Earth's surface. Emissions, concentrations, and radiative effects of these gases and aerosols are the topics addressed in this chapter.

Human-induced changes in climate and in the stratospheric ozone layer impact the entire Earth system, from terrestrial and marine ecosystems to the health and welfare of humanity. International agreements have already begun to control the use of halocarbons, paving the way for the recovery of stratospheric ozone. However, reducing emissions of the other GHGs that contribute to the enhanced greenhouse effect is a far more complex problem than that of reducing halocarbons alone. Measures imposing substantial reductions in CO_2 and other GHG emissions would severely limit global energy use. If GHG emissions are not limited, there are strong scientific grounds to believe they will cause significant global warming in the next few decades. Such a warming would be global, but with regional impacts of varying severity. Concomitant changes are expected to occur in regional patterns of temperature, precipitation, and sea level, with resultant impacts on most societal, economic, and environmental aspects of existence on this planet. With increasing world population, impacts can be expected not only in the areas of agriculture and water resources, affecting food security, but also via the sea-level rise affecting coastal settlements, including a number of the world's major cities. These impacts are expected to impinge on the economy and the welfare of every nation.

This chapter provides the latest scientific understanding of these two problems—stratospheric O_3 loss and global warming—that have been created by the growing influence of human activities on the Earth–atmosphere system. We begin by providing the historical and current greenhouse gas concentrations and presenting their impact on climate and ozone. We then describe their conceivable future trends. We conclude this chapter with the

discussion of the potential impacts of climate and stratospheric ozone changes and current policy considerations.

13.2   HISTORICAL EVIDENCE OF THE IMPACT OF HUMAN ACTIVITIES ON CLIMATE

13.2.1   *Increased greenhouse gas and aerosol concentrations*

Analysis of air bubbles trapped in ice cores reveals that, prior to the Industrial Revolution, atmospheric concentrations of heat-trapping GHGs were relatively constant. Since the mid-1700s, however, the world has become increasingly industrialized. Dramatic increases in the use of carbon-containing fossil fuels such as coal, oil, and natural gas have occurred to meet the demand for energy. Synthetic chemicals have been manufactured to supply important products. At the same time, population has grown exponentially and agricultural activities have developed accordingly. This has resulted in additional energy use and deforestation, particularly in tropical regions. Taken together, increased fossil fuel burning, chemical production, and agriculture have caused emissions of GHGs and aerosol precursors to rise rapidly, resulting in a significant increase in atmospheric levels of these same GHGs and aerosols. The most conclusive evidence for such a connection between human activities and atmospheric change comes from a comparison between historical emissions from energy use and agricultural activities, and the observed atmospheric concentrations of various greenhouse gases. For example, Fig. 13.2 shows fossil fuel related carbon emissions, while Fig. 13.3 shows the observed increasing concentrations for CO_2.

Following is a brief discussion of the observed trends in atmospheric concentrations and the estimated current emissions for the most important GHGs—CO_2, CH_4, N_2O, tropospheric O_3, CFC-11, and other halocarbons—and aerosols. There are also several gases, such as carbon monoxide (CO), nonmethane hydrocarbons (NMHC), and nitric oxides (NO_x), that are not important GHGs in

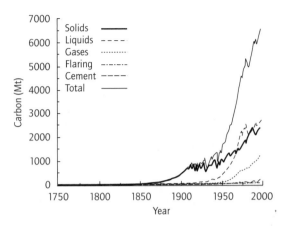

Fig. 13.2 Global CO$_2$ emissions from fossil fuel burning, cement production, and gas flaring for 1751–1997 (Marland *et al.* 2000).

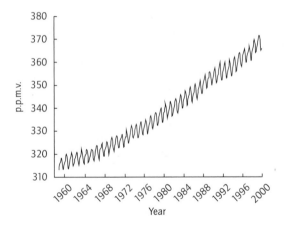

Fig. 13.3 Observed monthly average CO$_2$ concentration (p.p.m.v.) from Mauna Loa, Hawaii (Keeling & Whorf 1999). Seasonal variations are primarily due to the uptake and production of CO$_2$ by the terrestrial biosphere.

themselves, but are important precursors of tropospheric O$_3$ and are discussed in that section.

Water vapor

Water vapor (H$_2$O) is the most important GHG in the atmosphere, responsible for much of the natu-

rally occurring greenhouse effect. Water vapor absorbs infrared (IR) radiation over a large range of wavelengths, leaving only an atmospheric window region between 9 and 12 μm through which IR radiation emitted by the Earth can escape to space.*

Water vapor concentration is highly variable, ranging from over 20,000 p.p.m.v. in the lower tropical atmosphere to only a few p.p.m.v. in the stratosphere (Peixoto & Oort 1992). On a global scale, tropospheric concentrations of water vapor are determined by climate parameters such as evaporation, condensation, and precipitation rates, which are in turn affected by temperature and atmospheric dynamics. As such, tropospheric water vapor is not directly affected by human emissions, but is determined internally within the climate system. However, a human-induced climate warming will affect water vapor in the lower atmosphere. As temperatures rise, more water vapor will evaporate into the atmosphere. In this way, water vapor is projected to be a significant **positive feedback**† to global warming.

Water vapor concentrations in the upper troposphere and lower stratosphere are very low. However, due to the lower temperatures at that altitude, the effectiveness of water vapor as a greenhouse gas is enhanced. The main sources of stratospheric water vapor are transport upwards from the troposphere and the oxidation of methane.

Carbon dioxide

Carbon dioxide (CO$_2$) is the most important GHG produced by human activities. Based on measurements from air bubbles trapped in ice cores, preindustrial CO$_2$ concentration was about 280 p.p.m.v.

---

* The definition of a non-H$_2$O greenhouse gas is that it has at least one strong absorption band in the atmospheric window region where H$_2$O is not absorbing.

† A positive feedback acts to enhance the original forcing. In the case of water vapor, increased temperature due to global warming will enhance evaporation, raising water vapor levels in the lower atmosphere. Since water vapor is a greenhouse gas, more water vapor in the atmosphere means more absorption of infrared radiation, adding to the initial warming effect.

Accurate, real-time measurements of CO_2 began in 1958 (Fig. 13.3), showing that the annually averaged atmospheric CO_2 concentrations have risen from 316 p.p.m.v. in 1959 to 368 p.p.m.v. in 1999 (Keeling & Whorf 1999). CO_2 concentrations exhibit a seasonal cycle due to the uptake and release of atmospheric CO_2 by terrestrial ecosystems.

Atmospheric CO_2 is primarily affected by three main human activities: (i) the burning of fossil fuels, which releases the carbon contained in the fuels into the atmosphere, where it combines with oxygen to form CO_2; (ii) the manufacturing of cement, which releases the carbon contained in limestone into the atmosphere; and (iii) changes in land use, mainly due to deforestation, which releases the carbon contained in biomass into the atmosphere. Annual global emissions from fossil fuel burning and cement production in 1997 are estimated to be 6.6 Pg C yr^{-1}, with 0.2 Pg C yr^{-1} from cement production (Marland *et al.* 2000). Net land-use flux, made up of a balance of CO_2 emissions due to deforestation versus CO_2 uptake due to regrowth on abandoned agricultural land can be estimated based on land-use statistics and simple models of rates of decomposition and regrowth. The annual flux of carbon from land-use change for the period 1990–5 has been estimated to be 1.6 Pg C yr^{-1}, 1.7 Pg C yr^{-1} in the Tropics and a small sink in temperate and boreal areas (Houghton 2000).

In contrast to other GHGs, CO_2 is not removed from the atmosphere by chemical reactions with other atmospheric gases. For this reason, atmospheric CO_2 does not have a specific lifetime. Instead, it is part of a cycle whereby carbon is transferred between Terrestrial, oceanic, and atmospheric reservoirs over time scales ranging from tens to thousands of years.

Methane

On a molecule basis, methane (CH_4) is approximately 66 times more effective as a greenhouse gas than CO_2 (WMO 1999). When its effectiveness as a GHG is combined with the large increase in its atmospheric concentration, methane becomes the second most important GHG contributing to global warming. Based on ice core measurements

and direct measurements begun in 1978, atmospheric CH_4 concentration increased steadily from its preindustrial concentration of about 0.7 p.p.b.v. to 1.75 p.p.m.v. by 1990, a factor of 2.5 increase (Rasmussen & Khalil 1981; Dlugokencky *et al.* 1998; Etheridge *et al.* 1998). The current total source strength for methane is about 600 Tg CH_4 yr^{-1}, with approximately 55% of CH_4 emissions being produced by human-related sources, while 45% is natural in origin (Houghton *et al.* 2001). Human-related sources are mainly biogenic in origin and related to agriculture and waste disposal, including cows and other ruminants, rice paddies, biomass burning, animal and human waste, and landfills. CH_4 is emitted naturally by wetlands, termites, other wild ruminants, ocean, and hydrates.

Over 90% of CH_4 is removed from the atmosphere through reactions with the hydroxyl radical (OH). Reactions with OH oxidize CH_4 to carbon monoxide, which in turn reacts with OH, lowering OH levels and slowing the removal of the original CH_4. For this reason, although the chemical lifetime of CH_4 is 8.9 years (WMO 1999), due to the CH_4–OH–CO feedback cycle, emissions of CH_4 from human activities are removed from the atmosphere over an adjustment time of approximately 12 years. The remaining 10% of CH_4 is removed through uptake by soil or transport into the stratosphere.

From preindustrial times to the beginning of the 1990s, the increase in CH_4 was well correlated with increases in population and agriculture (Khalil & Rasmussen 1996). However, over the past decade, this correlation became less distinct, with an overall slowing of the growth rate and several abrupt fluctuations over short time periods (Fig. 13.4, lower panel). Although the cause of the observed global decline in methane growth is still not well understood, the short-term variations are thought to be linked to chemical- and temperature-related impacts due to the Pinatubo eruption in 1991, while the longer-term decline in the growth rate may signal a decoupling between traditional CH_4 sources and actual emissions (Bekki *et al.* 1994; Hogan & Harriss 1994; Dlugokencky *et al.* 1998). Figure 13.4 (upper panel)

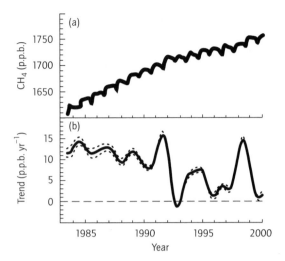

Fig. 13.4 (a) Globally averaged atmospheric CH_4 concentrations (p.p.b.v.) derived from NOAA Climate Monitoring Diagnostic Laboratory air sampling sites (Dlugokencky *et al.* 1998). The continuous line is a deseasonalized trend curve fitted to the data. The dashed line is a model-estimated trend fit to the globally average values. (b) Atmospheric CH_4 instantaneous growth rate (p.p.b.v. yr^{-1}), the derivative with respect to the trend curve shown in (a).

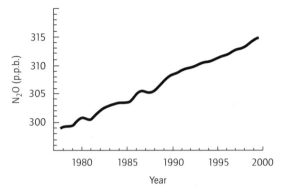

Fig. 13.5 Global annual average N_2O concentrations since 1978 (Elkins *et al.* 1998).

also shows the seasonal variations in CH_4 concentrations with a peak in late summer, due to strong emissions from temperature-dependent northern wetlands and variations in the removal of CH_4 from the atmosphere by hydroxyl (OH) radicals.

Nitrous oxide

Nitrous oxide (N_2O), also known as laughing gas, is the third most important GHG produced by human activities. On a molecule basis, N_2O is about 200 times more efficient than CO_2 in absorbing infrared radiation. N_2O levels continue to grow in the global atmosphere since preindustrial times. Ice core measurements show that prior to the preindustrial time, concentrations of N_2O were relatively stable at about 270 p.p.b.v. (Machida *et al.* 1995). Since then the mean global atmospheric concentrations of N_2O have increased 20% to 314 p.p.m.v. by 1998 (Elkins *et al.* 1998; Prinn *et al.*

1998). Direct measurements of the global mean N_2O concentrations for the period 1984–98 with the seasonal cycle removed show a continuous increase in the N_2O concentration at a rate of 0.8 p.p.b.v. yr^{-1} (Fig. 13.5; Elkins *et al.* 1998).

N_2O is produced from a wide variety of microbial sources in soils and water. Based on recent estimates, total global N_2O emissions are about 16.4 Tg N yr^{-1}, of which 43% of emissions arise from human-related activities, while 57% are of natural origin. Major anthropogenic sources of N_2O include fertilized cultivated soils, biomass burning, industrial sources, and cattle and feed lots. The main natural sources for N_2O are the ocean, tropical soils such as wet forests and dry savannas, and temperate soils such as forests and grasslands.

Atmospheric N_2O is primarily removed in the stratosphere by photolysis and reaction with electronically excited oxygen atoms.

Halocarbons

Halocarbons are unique in that they contribute to both global warming and stratospheric ozone depletion. These gases are important GHGs because they have strong infrared absorption lines in the atmospheric window region. Due to their relatively high concentrations, the most potent halocarbons in the current atmosphere are the chlorofluorocarbons $CFCl_3$ (CFC-11) and CF_2Cl_2 (CFC-12). A molecule of CFC-11 and CFC-12 is about 12,400 and 15,800 times more effective, respectively, at absorbing infrared radiation than an additional molecule of CO_2 (Jain *et al.* 2000).

In addition to their radiative properties, halocarbons are extremely long-lived. There are no major removal mechanisms for these gases in the troposphere, so most halocarbons are transported directly to the stratosphere. In the stratosphere, halocarbons are broken down by high-energy radiation from the Sun, releasing the chlorine and bromine they contain. These highly reactive forms of Cl and Br can then catalyze thousands of O_3-destroying reactions before being converted to less reactive compounds.

Since the beginning of the twentieth century, halocarbons have been manufactured for use in many industrial and commercial products. Their inertness and long lifetimes have made them attractive chemicals for use as foam blowing agents, coolants, propellants, fire retardents, and other industrial applications. With the exception of the naturally occurring and short-lived emissions of CH_3Cl and CH_3Br from volcanic and oceanic activity, essentially all the halocarbons in the atmosphere are man-made. The ozone-depleting halocarbons with the largest potential to influence both climate and the ozone layer are chlorofluorocarbons (CFCs), mainly CFC-11 ($CFCl_3$), CFC-12 (CF_2Cl_2), and CFC-113 ($CF_2ClCFCl_2$). Less effective ozone-depleting substances include hydrochlorofluorocarbons (HCFCs) and the hydrofluorocarbons (HFCs) that contain fluorine instead of chlorine. The chemical link between these compounds and stratospheric O_3 destruction was first suggested by Molina and Rowland (1974) and has since been confirmed by numerous observational and modeling studies (WMO 1999). In addition, it is now clear from measurements in polar Firn air that there are no natural sources of CFCs, HCFCs, or HFCs (Butler *et al.* 1999).

CFC-11 and CFC-12 have the largest atmospheric concentrations, at 0.26 and 0.53 p.p.b.v., respectively. Tropospheric concentrations of both of these gases were increasing at about 4% per year through the 1980s and in the early 1990s. However, their growth rates have now slowed appreciably (Elkins *et al.* 1998), and have even started to decrease in the case of CFC-11, as shown in Fig. 13.6 (WMO 1999). This decrease is due to the banning of the use of CFCs and number of other ozone

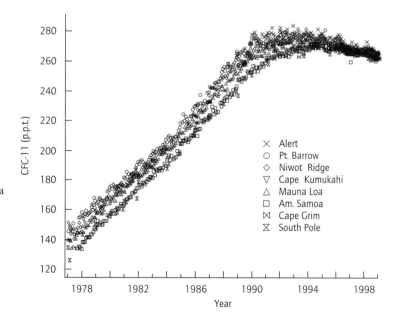

Fig. 13.6 Observed CFC concentrations over time, showing a sharp rise over the past few decades, and the beginning of a recent decline. Based on measurements from NOAA Climate Monitoring and Diagnostics Laboratory (http://www.cmdl.noaa.gov/hats/graphs/graphs.html; updated from Elkins *et al.* 1993).

depleting compounds by 1996 under the Montreal Protocol and its amendments. Atmospheric concentrations of several other halocarbons have until recently been growing at even faster rates than CFC-11 and CFC-12. For example, the concentration of CFC-113 ($C_2F_3Cl_3$) was increasing by about 10% per year in the early 1990s but has also slowed greatly, with a current concentration of about 0.08 p.p.b.v. Recent measurements indicate that abundances of CH_3CCl_3 and HCFC-22 (CHF_2Cl) have been increasing at about 0.5 and 10% per year from their current concentrations of 0.16 and 0.10 p.p.b.v., respectively (WMO 1999, and the references therein).

Bromine is even more effective at destroying ozone than is chlorine. Although total chlorine loading in the atmosphere peaked in 1997, bromine is still increasing slightly (WMO 1999). For these reasons, bromine-containing halons, most notably CF_3Br (Ha-1301) and CF_2ClBr (Ha-1211), have caused concern, especially in light of their early rapid increases in atmospheric concentration that now seem to be leveling off. Primary destruction of these compounds also occurs through photolysis in the stratosphere, resulting in long atmospheric lifetimes. These compounds, however, have small atmospheric concentrations, about 2 p.p.t.v. (Houghton *et al.* 1996; Butler *et al.* 1998; WMO 1999). Hence, their contribution to global warming through the absorption of infrared radiation is considered minimal.

The halocarbons whose production has been banned due to environmental concerns had many important uses to humanity. Therefore, alternatives to these chemicals were required to meet expanding worldwide needs for refrigeration, air conditioning, energy efficient insulation, plastic foams, solvents, and aerosol propellants in, for example, medical products. The proposed replacements, particularly HFCs (hydrofluorocarbons), retain many of the desirable properties of CFCs; however, as a result of the addition of one hydrogen into their molecular structure, they have much shorter lifetimes in the atmosphere, reducing their contribution to global warming. In addition, HFCs do not contain chlorine; therefore these chemicals have no potential for ozone depletion.

Because of their increasing use in recent years, HCFC concentrations continue to rise almost exponentially. Figure 13.7 shows the HFCs with the largest measured atmospheric abundances, along with some major HCFCs, the latter being controlled under the Montreal Protocol and its amendments.

Other perfluorogenated species such as, CF_4, C_2F_6, and sulfur hexafluoride (SF_6), have very long lifetimes (greater than 1000 years) and significant infrared absorption lines in the atmospheric window region. CF_4 and C_2F_6 are by-products released to the atmosphere during the production of aluminum. SF_6 is mainly used as a dielectric fluid in heavy electrical equipment. Most of the current emissions of C_2F_6 and SF_6 arise from human activities. Harnisch and Eisenhauer (1998) have shown

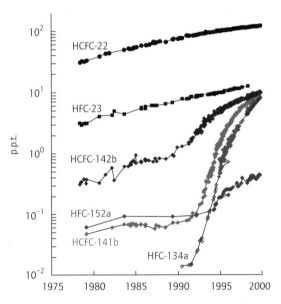

Fig. 13.7 HFC-23, -152a, -134a, and HCFC-22, -142b, and -141b concentrations at Cape Grim, Tasmania, for the period 1978–99. The different symbols show data from different networks: SIO (filled circles), NOAA-CMDL (open diamonds, Montzka *et al.* 1994, 1996a, b, 1999), UEA (filled diamonds and squares, Oram *et al.* 1995, 1996, 1998, 1999), and AGAGE (open circles, Miller *et al.* 1998; Prinn *et al.* 1998; Sturrock *et al.* 1999, 2001).

that CF_4 and SF_6 are naturally present in fluorites, and outgassing from these materials leads to natural background concentrations of 40 p.p.t.v. for CF_4, and 0.01 p.p.t.v. for SF_6. However, at present, human emissions of CF_4 exceed natural emissions by a factor of 1000 or more.

Atmospheric concentrations of CF_4 and SF_6 are increasing as shown in Figs 13.8 and 13.9. Current concentrations of CF_4 probably exceed 70 p.p.t. (Harnisch *et al.* 1999) and global average concentration of SF_6 were 3–4 p.p.t. (Maiss *et al.* 1996; Maiss & Brenninkmeijer 1998). Because of their long lifetime, comparatively small emissions of these species will accumulate and lead to a signifi-

cant impact on climate over the next several hundred years. For this reason, these species are included as part of the "basket" of GHGs to be reduced under the Kyoto Protocol (UN 1997).

Stratospheric ozone

Over 90% of the ozone in the atmosphere is located in the stratosphere (Fig. 13.10, vertical profile of O_3), where it has two important effects. First, stratospheric ozone absorbs the Sun's radiation, heating the stratosphere and causing stratospheric temperature to increase with height, which has a cooling effect on the Earth's surface. Second, stratospheric ozone protects the Earth's surface from harmful UV-B radiation from the Sun, which would otherwise have adverse effects on human, animal, and plant life.

Stratospheric ozone is produced by the photo-dissociation of oxygen by UV radiation, a process unaffected by human emissions. Ozone is then removed in catalytic mechanisms involving free radicals such as NO_x, HO_x, and ClO_x. In the past, stratospheric O_3 production and destruction were in balance, resulting in a constant protective ozone layer. However, human emissions of CFCs and other halocarbons have altered the balance by increasing concentrations of reactive Cl and Br. Both substances are extremely effective catalysts in the O_3 removal cycles that can turn over thousands of times before being converted to less reactive forms.

The increase in stratospheric Cl and Br loading has resulted in significant amounts of O_3 depletion since the "ozone hole" was first observed over Antarctica in the 1970s. This hole has since grown to nearly complete elimination of Antarctic ozone in the lower stratosphere every spring since 1992 (WMO 1995, 1999). Globally, total ozone has decreased by 5%, with largest decreases occurring in both hemispheres at mid- to high latitudes in the lower stratosphere (WMO 1999). Although it has not yet been confirmed observationally, combined stratospheric Cl and Br is expected to have peaked before 2000 based on tropospheric measurements. However, the recent World Meteorological Organization assessment of stratospheric O_3 (WMO

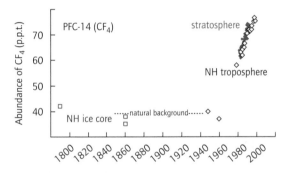

Fig. 13.8 Concentration of CF_4 over the past 200 years as measured in tropospheric air (Harnisch *et al.* 1999), stratospheric air, and ice cores (Harnisch *et al.* 1996).

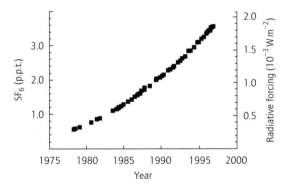

Fig. 13.9 SF_6 concentrations measured at Cape Grim, Tasmania, since 1978 (Maiss & Brenninkmeijer 1998).

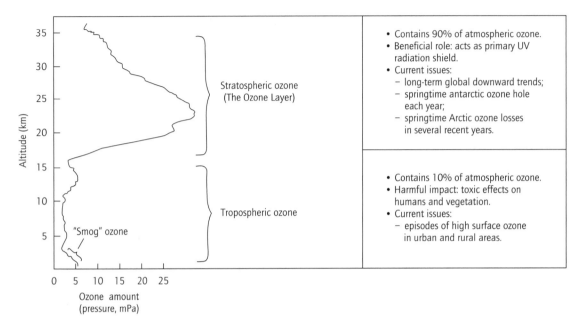

Fig. 13.10   Vertical profile of O_3 in the troposphere and stratosphere (WMO 1999).

1999) stresses that the O_3 layer is currently in its most vulnerable state. Maximum O_3 depletion is not estimated to occur for another one to two decades. Although stratospheric halocarbons are expected to return to pre-1980 levels by 2050, O_3 recovery will be affected by stratospheric background concentrations of aerosols, CH_4, N_2O, water vapor, and other compounds. This leads to uncertainty in projections of future O_3 concentrations for a given Cl/Br level, as the correlation may not be the same as in the past.

Tropospheric ozone

Approximately 10% of the atmosphere's O_3 is located in the troposphere (Fig. 13.10, vertical profile of O_3). In contrast to beneficial effects of stratospheric ozone, lower-level ozone is an important greenhouse gas, contributing to global warming. At high levels such as occur downwind of many urban areas, O_3 also has a toxic effect on human, animal, and plant life. However, as O_3 photolysis is the main source of OH, the atmosphere's cleansing

agent, O_3 is also the primary driver of photochemical processes that remove pollutants from the troposphere.

Ozone is not emitted directly from human activities; however, we are responsible for emissions of a number of gases that are "ozone precursors," i.e. they are involved in chemical reactions that produce ozone. These gases include NO_x, CO, CH_4, and other nonmethane hydrocarbons. Increased emissions of these gases are thought to have significantly raised tropospheric O_3 levels since preindustrial times (Crutzen 1995; Brasseur et al. 1998; Wang & Jacob 1998). Numerous observational and modeling studies have shown that O_3 levels are strongly dependent on background concentrations of these species, as determined by regional emissions, seasonal meteorology, and photochemical activity (Crutzen et al. 1999; Fuglestvedt et al. 1999; Brühl and Crutzen 2000; Lelieveld and Dentener 2000; Mauzerall et al. 2000). NO_x levels are particularly crucial as approximately 50% of tropospheric O_3 is produced by the photolysis of NO_2. The other 50% of tropospheric O_3 results from downward transport from the stratosphere. Due to its short lifetime and inhomogeneous distribution, NO_x is often the limiting species on O_3 production. It is estimated that only 10% of potential tropospheric O_3 production is currently being realized, with a net production of O_3 in high-NO_x areas and a net destruction of O_3 in low-NO_x regions (see Wuebbles *et al.* 2000 for a detailed treatment of tropospheric O_3 chemistry).

Over the past century, ozone production in photochemical smog, such as occurs in the atmospheric boundary layer over many urban areas, has increased steadily. Although smog-induced ozone has major health impacts, it is not a significant contributor to climate. However, changes in tropospheric O_3 above the boundary layer do have the potential to impact climate (Lacis *et al.* 1990). Observation of O_3 trends over the past decade have shown increases over western Europe and Japan (both of which contain many urban areas with high NO_x levels), little change in the Tropics where natural O_3 production is high, and a small decrease over Canada (WMO 1999). Rising levels of CH_4, NO_x, CO, and NMHCs as well as stratospheric O_3 loss will enhance O_3 production, while increasing temperature and water vapor from global warming could increase or decrease O_3 levels. Due to the complexity of the factors influencing tropospheric O_3 abundances, the amount and even the sign of future tropospheric O_3 change is uncertain—although recent modeling studies project a net increase over the next few decades (Brasseur *et al.* 1998; Collins *et al.* 2000; Lelieveld & Dentener 2000; Stevenson *et al.* 2000).

Aerosols

Aerosols differ from GHGs in two primary aspects: (i) they can have either a cooling or a warming effect on climate, depending on aerosol composition; and (ii) they have short lifetimes of only a few days to weeks. Due to this short lifetime, aerosols are inhomogeneously concentrated over areas such as the northern hemisphere continents. This patchy distribution leads to a regionally dependent climate impact, in contrast to the longer-lived, well mixed greenhouse gases.

A number of major types of aerosols are recognized to currently affect climate to a significant degree. The most important type, sulfate aerosols, has a net cooling effect on climate. Sulfate aerosols are not directly produced from human activities. Instead, combustion of fossil fuels containing sulfur results in emissions of sulfur dioxide (SO_2), the precursor for sulfate aerosols. In the atmosphere, SO_2 is quickly oxidized to sulfuric acid, which in turn condenses onto cloud droplet and aerosol particle surfaces to form sulfate aerosols. Coal-fired electricity generation is one of the most important sources of anthropogenic aerosols, generating a major portion of global emissions (Haywood & Boucher 2000, and references therein; Benkovitz *et al.* 1996; Houghton *et al.* 1996). Other human-related emissions arise from combustion of other fossil fuels and biomass burning. Ocean sea salt also contributes approximately 15% of global atmospheric sulfur in the form of DMS. Finally, sulfate and other aerosol types are episodically mass-produced by volcanic eruptions. Under sufficiently violent eruptions, large amounts of aerosols of mixed composition can be injected directly into the stratosphere. In the stratosphere, aerosol lifetime is extended from days to months and even years. As bands of volcanic aerosols circle the Earth, they exert a significant impact on the Earth's climate. Their radiative effects also impact natural emissions and uptake of greenhouse gases including CO_2 and CH_4. The Mt Pinatubo eruption, for example, is thought to be responsible for large fluctuations in the growth rate of CO_2, CH_4, CO, and other gases, enhanced stratospheric O_3 destruction, and a short-term cooling at northern latitudes (Bekki *et al.* 1994; Dlugokencky *et al.* 1998; Wuebbles *et al.* 2000).

Carbonaceous aerosols can have either a warming or a cooling effect on climate, depending on whether they contain black carbon (BC) or organic carbon (OC). BC contained in soot forms the nuclei of infrared-absorbing carbonaceous aerosols that have recently been identified as significant contributors to positive radiative forcing. Major anthropogenic sources of BC are split evenly, with coal and diesel combustion accounting for approximately 50% of emissions, while biomass burning makes up the remainder (Cooke & Wilson 1996; Liousse et al. 1996; Cooke et al. 1999). Combustion also results in emissions of OC, with the ratio between black and organic emissions being primarily dependent on the combustion temperature. Since most large-scale combustion facilities operate at very high temperatures, industry is a major source of BC, while domestic fuel use and biomass burning are responsible for the majority of OC emissions. Organic aerosols exhibit great variety, which makes it difficult to fully characterize their impacts on atmospheric chemistry and climate (Jacobson et al. 2001a, b).

Other human and natural aerosols, including nitrate and mineral dust aerosols, for example, also affect the radiative balance of the atmosphere.

In the past, aerosol emissions were relatively constant, punctuated by occasional volcanic eruptions that have been detected by dust deposits in glacial and polar ice cores (Petit et al. 1990) and correlated with historical records dating back thousands of years. However, the beginning of the Industrial Revolution and large-scale fossil fuel use signaled a significant increase in sulfate, BC, and OC aerosols. Estimates of historical energy use and fuel sulfur content show that sulfur emissions have increased from only a few megatons of sulfur (Mt S) in 1850 to approximately 75 Mt S by 1990 (Dignon & Hameed 1989; Mylona 1996; Lefohn et al. 1999), with recent reductions due to power plant sulfur controls in developed countries (Lefohn et al. 1999). Since emissions of BC and OC are tied to fossil fuel use and biomass burning, concentrations of these aerosols are also estimated to have increased over the past century (Tegen et al. 2000).

13.2.2   Changes in radiative forcing

Clear evidence has been presented in the previous section that the atmospheric concentrations of a number of radiatively active GHGs have increased over the past century as a result of human activity. By infrared absorption, they increase the heat trapping ability of the atmosphere, driving climate change. As discussed above, human-related activities have also been responsible for increases in aerosol particles in the atmosphere, which can have either heating or cooling effects, depending on the aerosol type. A change in the concentrations of GHGs and aerosols will change the balance between the incoming solar radiation and the outgoing infrared radiation from the Earth. This change in the planetary radiation budget is termed the "radiative forcing" of the Earth's climate system. Changes in the Earth's surface temperature are approximately linearly proportional to the radiative forcing inducing those changes, although there is some nonlinearity induced by the sensitivity of climate response to height, latitude, and the nature of the forcing (e.g. Hansen et al. 1997).

Greenhouse gases

Recent assessments of the direct radiative forcing due to the changes in greenhouse gas concentrations are generally in good agreement, determining an increase in radiative forcing of about $2.43\,W\,m^{-2}$ from the late 1700s to the present time (Fig. 13.11; Houghton et al. 2001). Of the $2.43\,W\,m^{-2}$ change in radiative forcing from greenhouse gases over the past two centuries, approximately $0.5\,Wm^{-2}$ occurred within the past decade. By far the largest effect on radiative forcing has been the increasing concentration of carbon dioxide, accounting for about 64% ($1.46\,W\,m^{-2}$) of the total forcing (Jain et al. 2000). The radiative forcings due to changes in concentrations of CH_4, N_2O, and CFCs and other halocarbons since the late 1700s are 0.50, 0.15, and $0.34\,W\,m^{-2}$, respectively (Fig. 13.11).

Stratospheric ozone

Over the past few decades, stratospheric ozone depletion in middle to high latitudes has been primarily responsible for the observed $-0.6°C$ decade^{-1} cooling of the lower stratosphere, resulting in a negative forcing on climate (WMO 1999). Recent estimates (WMO 1999; Houghton et al. 2001) place the forcing due to the loss of stratospheric ozone at $-0.2 \pm 0.1\,W\,m^{-2}$ from 1979 to 1997 and

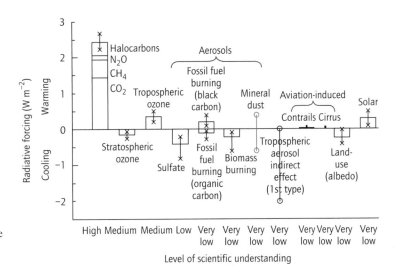

Fig. 13.11 IPCC (Houghton *et al.* 2001) estimated change in globally and annually averaged anthropogenic radiative forcing (W m^{-2}) resulting from number of agents from preindustrial time (1750) to the present (about 2000). The error bars with an "x" delimiter indicate an estimate of the uncertainty range. The error bars with an "o" delimiter denote a forcing for which no central estimate can be given, due to large uncertainties.

-0.35 W m^{-2} from preindustrial times to the present. The later estimate is equivalent to approximately 30% of estimated GHG forcing over the same period; thus, projected stratospheric O$_3$ recovery over the coming century is likely to enhance global warming (WMO 1999).

Tropospheric ozone

Observational evidence for tropospheric ozone increases over the past century is weak. However, as discussed previously, recent modeling studies that account for emissions and concentrations of ozone precursors (e.g. Crutzen 1995; Brasseur *et al.* 1998; Wang & Jacob 1998) estimate a net increase in tropospheric O$_3$ since preindustrial times. Additional calculations indicate that an increase in ozone over recent decades (as suggested by some measurements) should add to the warming effect in radiative forcing from greenhouse gases emitted by human activities (Wang *et al.* 1993; Hauglustaine *et al.* 1994). Estimates of the change in radiative forcing since the 1700s for tropospheric ozone are positive, at 0.35 ± 0.15 W m^{-2} (Houghton *et al.* 2001).

Aerosols

Tropospheric aerosols that are thought to have a

substantial anthropogenic component include sulfate, black carbon, organic carbon, mineral dust, and nitrate aerosols. Due to their short lifetime, the geographic distribution, plus the diurnal and seasonal patterns of the radiative forcing from aerosols, are quite different from that of greenhouse gases. Human-related aerosols appear to have played a significant role in explaining the discrepancy between observed temperature changes and that expected from greenhouse gases (Santer *et al.* 1995; Houghton *et al.* 1996). Anthropogenic aerosols influence the radiative budget of the Earth–atmosphere system in two different ways. The first is the direct effect, whereby aerosols scatter and absorb solar and thermal infrared radiation, thereby altering the radiative balance of the Earth–atmosphere system. The second is the indirect effect, whereby aerosols modify the microphysical and hence the radiative properties and lifetimes of clouds.

Direct aerosol radiative forcings

Sulfate aerosols. Radiative forcing from sulfate aerosols is centered primarily over the industrialized regions in the northern hemisphere midlatitudes. Current estimates for the globally averaged direct effect due to scattering of solar radiation on radiative forcing from changes in sulfate aerosols during the industrial period range from –0.26 to –0.82 W m^{-2} (Kiehl & Briegleb 1993; Boucher & Anderson 1995; Haywood & Ramaswamy 1998; Haywood & Boucher 2000; Kiehl et al. 2000; Houghton et al. 2001). Houghton et al. (2001) estimates the direct effect of sulfate aerosols as –0.4 W m^{-2}, with an uncertainty range of a factor of two. The estimated spatial distribution of the forcings shows strongest radiative forcings over industrial regions of the northern hemisphere. The seasonal cycle is strongest in the northern hemisphere summer when the insolation is the highest (Haywood & Ramaswamy 1998).

Fossil fuel black carbon. In the past few years, there have been a number of global estimates of radiative forcing due to BC aerosol from fossil fuel burning. Based on the calculations discussed in the recent IPCC assessment (Houghton et al. 2001), the estimate for the global mean radiative forcing for fossil fuel BC aerosols was +0.2 W m^{-2}, with a range +0.1 W m^{-2} to +0.4 W m^{-2}. However, recent studies by Jacobson (2001a, b) find a global (fossil fuel + biomass burning) direct forcing ranging from 0.31 W m^{-2} for an external mixture through 0.55 W m^{-2} for a multiple-distribution coated core, up to 0.62 W m^{-2} for internally mixed, coated-core BC.

Fossil fuel organic carbon. Penner et al. (1998) and Grant et al. (1999) modeled the radiative forcing due to an internal and external mixture of fossil fuel BC and OC and found that, in contrast to BC, the radiative forcing due to OC from fossil fuels for the external mixture case was much higher than the internal mixture case (–0.04 W m^{-2}). Thus –0.04 W m^{-2} represents the lower limit for the strength of the forcing due to fossil fuel OC. On the basis of these calculations, IPCC (Houghton et al. 2001) estimated the radiative forcing due to fossil fuel OC of –0.06 W m^{-2} with an uncertainty of at least a factor of three.

Biomass-burning BC and OC. The annual mean radiative forcing due to biomass-burning aerosol where BC and OC are combined has been esti-

mated to be –0.3 W m^{-2} (Hobbs et al. 1997; Penner et al. 1998; Iacobellis et al. 1999; Grant et al. 1999). On the basis of these studies, IPCC (Houghton et al. 2001) estimated the radiative forcing due to biomass burning aerosols of –0.2 W m^{-2}, of which approximately –0.4 W m^{-2} is due to OC and +0.2 W m^{-2} due to BC, with the uncertainty associated with each of the components estimated as at least a factor of three.

Mineral dust aerosol. Although mineral dust particles are usually large, they can be lifted to high altitudes in the troposphere. At high altitudes, in addition to their cooling effect through reflecting solar radiation, dust may exert a significant warming effect as well through the absorption of infrared radiation. The global mean radiative forcing in the solar part of the spectrum is negative, due to the scattering nature of mineral dust aerosol at these wavelengths, and positive in the thermal infrared where they absorb and reradiate. Sokolik and Toon (1996), Hansen et al. (1998), and Miller and Tegen (1998) suggest that the solar radiative forcing is likely to be of a larger magnitude than the terrestrial infrared radiative forcing, giving a net radiative forcing. However, a net positive forcing cannot be ruled out. For this reason, a tentative range of –0.6 W m^{-2} to +0.4 W m^{-2} has been adopted by the latest IPCC assessment (Houghton et al. 2001).

Nitrate aerosol. Although IPCC (Houghton et al. 1994) highlighted the significance of human-related nitrate aerosols, there have been few studies of the direct radiative effect. Van Dorland et al. (1997) and Adams et al. (2001) estimated a radiative forcing of approximately –0.03 and –0.19 W m^{-2} respectively for ammonium nitrate. However, as acknowledged by the authors, many of the assumptions implicit within the calculations make this a first-order estimate only.

Volcanic aerosol. Sulfate and other aerosols emitted by volcanoes are unique, in that they are often injected directly into the stratosphere, where their radiative cooling effects are strong and long lasting. Estimates of radiative forcing vary widely for each volcanic eruption, depending on the amount of aerosols produced and the height and latitude at which they are expelled. For example, the eruption of Mt Pinatubo in 1991 is estimated to have caused an immediate forcing of $-4\,W\,m^{-2}$ immediately following the eruption and as much as $-1\,W\,m^{-2}$ up to two years following (Hansen *et al.* 1992).

Indirect aerosol radiative forcing

There are, in general, two types of aerosol indirect effects. First, an increase in aerosols results in an increase in droplet concentration and a decrease in droplet size for a given liquid water content. In the second type of indirect effect, a reduction in cloud droplet size affects the precipitation efficiency, which tend to increase the liquid water content, the cloud lifetime, and the cloud thickness. Because of the inherent complexity of the aerosol indirect effect, GCM estimates for the aerosol indirect effect are very uncertain. IPCC (Houghton *et al.* 2001) and Haywood and Boucher (2000) reviewed and discussed the various estimates for the globally averaged aerosol indirect forcing available in the literature. Based on IPCC (Houghton *et al.* 2001), the estimated radiative forcing from these effects may vary between zero and $-2\,Wm^{-2}$. Not much emphasis should be given to the exact bound of this interval due to a very low level of scientific understanding of this forcing.

Solar variability

Changes in the solar energy output reaching the Earth constitute the primary external forcing on the climate system. The Sun's energy output is known to vary by small amounts over the 11-year cycle associated with sunspots. There are also indications that the solar output may vary by larger amounts over longer time periods (Hoyt & Schatten 1993). Slow variations in the Earth's orbit over time scales of multiple decades to thousands of years have varied the solar radiation reaching the Earth, and are thought to have affected climate in the past, including the formation of the ice ages. The IPCC (Houghton *et al.* 2001) estimated forcings due to solar variability at about $0.3\,W\,m^{-2}$, with a lower limit of $0.1\,W\,m^{-2}$ (Foukal & Lean 1990) and an upper limit of $0.5\,W\,m^{-2}$ (Nesme-Ribes *et al.* 1993).

Combined effects

Based on the above discussion, Fig. 13.11 shows the global, annual-mean averaged radiative forcing estimates from 1750 to present for the greenhouse gases, aerosols, and forcings discussed above. The height of the rectangular bar denotes a mid-range or best guess estimate of the forcing, while the vertical line about the bar is an estimate of the uncertainty range. This figure clearly shows a large uncertainty in the exact amount of forcing. This is particularly for aerosols because of the difficulty in quantifying the indirect effects of aerosols, and the overall changes in aerosol and ozone concentrations. Because of the hemispheric and other inhomogeneous variations in concentrations of aerosols, the overall change in radiative forcing could be much greater or much smaller at specific locations about the globe, with the largest increase in radiative forcing expected in the southern hemisphere (where aerosol content is smallest).

13.2.3 Past changes in temperature

Observed land and ocean surface temperature

One of the main evidences of global climate change is an increase in historical surface temperatures over land and ocean. Figure 13.12 shows the global average surface temperature increase (i.e. the average of the land surface air temperature and sea surface temperature) of $0.6 \pm 0.2°C$ since 1860 (Jones *et al.* 2001). As indicated in this figure, the increase in temperature is by no mean a uniform one. Instead, the increase is seen to have occurred in two distinct periods: from 1910 to 1945, and from 1976 to the present, with a rate of temperature increase for both periods of $0.15°C$ decade^{-1}. The rate of increase in sea surface temperature over the period 1950–93 is about half that of the mean land-surface air temperature. This most recent period of warming has been almost global, but the largest increases in temperature have occurred over the mid- and high latitudes of the continents

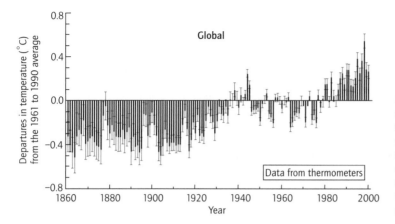

Fig. 13.12 Combined global and annual land surface air and sea surface temperature anomalies relative to the period 1861–1999 as calculated using an optimum average of UKMO ship and buoy and CRU land surface air temperature data taken from Jones *et al.* (2001).

in the northern hemisphere, and the warming has been greater over land compared to oceans.

On a shorter time scale, the 1980s and 1990s have brought some unusually warm years for the globe as a whole. The five warmest years since the beginning of the record in 1861, in terms of global annual average, all occurred in the 1990s. In addition, it is very likely that 1998 was the warmest year in the instrumental record since middle of the nineteenth century (Fig. 13.12). The remarkably consistent month-by-month global warming of 1998 is discussed in Karl *et al.* (2000).

There are of course uncertainties in the temperature records. For example, during this time period recording stations have moved, and techniques for measuring temperature have varied. Also, marine observing stations are scarce. Despite these uncertainties, confidence that the observed variations are real is increased by virtue of the trend and the shape of the changes being similar when different selections of the total observations are made. For instance, the separate records from the land and sea surface and from the northern and southern hemispheres are in close accord (Jones *et al.* 2001; Houghton *et al.* 2001).

Observed temperature above the surface layer

Balloon-based observational temperature records began in 1958. The initial global network of balloon stations consisted of about 540 stations, which grew to about 800 stations by the 1970s. In addition, during the past 25 years or so observations have been available from satellites orbiting around the Earth. Their great advantage is that they automatically provide data with global coverage, a factor often lacking in other data sets. The length of the record from satellites, however, is a comparatively short period in climate terms. Satellite-based time series are compared below with balloon-based and surface-based records.

Figure 13.13 shows that the balloon-based temperature changes since 1958 for the lowest 8 km of the atmosphere and at the surface are in good agreement, with a warming of about 0.1°C decade^{-1}. However, since the beginning of the satellite record in 1979, the temperature data from both satellite and weather balloons show a warming in the global middle-to-lower troposphere at the rate of approximately 0.05 ± 0.05°C decade^{-1} (Houghton *et al.* 2001). As suggested by IPCC (Houghton *et al.* 2001), about half of the observed difference in warming since 1979 is likely to be due to the combination of the differences in spatial coverage of the surface and tropospheric observations, and the physical effects of the sequence of volcanic eruptions and a substantial El Niño that occurred within this period (Houghton *et al.* 2001). The remaining difference is very likely real and not an observing bias. It arises primarily due to differences in the rate of temperature change over

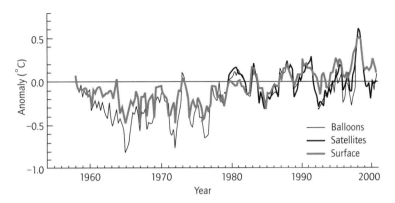

Fig. 13.13 Time series of seasonal temperature anomalies of the troposphere based on balloons (HadRT2.0, 1958–99) and satellites (MSU 2LT, 1979–99), in addition to the surface (CRU + MOHSST, 1958–99). The difference between the balloon and satellite time series is presented in the inset, which indicates that it is very likely that the surface has warmed relative to the troposphere (Houghton *et al.* 2001, and references therein).

the tropical and subtropical regions, which were faster in the lowest 8 km of the atmosphere before about 1979, but which have been slower since then. There are no significant differences in warming rates over mid-latitude continental regions in the northern hemisphere. In the upper troposphere, no significant global temperature trends have been detected since the early 1960s. In the stratosphere, as shown in Fig. 13.13, both satellites and balloons show substantial cooling, punctuated by sharp warming episodes of one to two years long that are due to volcanic eruptions.

13.3 FUTURE OUTLOOK OF CLIMATE AND OZONE CHANGES

Assessment of the future climate and stratospheric ozone responses to the GHG emissions requires mathematical models based on a set of fundamental physical and chemical principles governing the Earth–atmosphere system. A hierarchy of models is available for such studies. These models begin with one-dimensional models where altitude is the only dimension considered, or two-dimensional models that consider both altitude and latitude, both of which are based on box model concepts. Large, three-dimensional coupled atmosphere–ocean general circulation models

(GCMs) of the Earth's climate system have been the reference standard for global change research. GCMs, although extremely important to our understanding of climate and stratospheric ozone and how they might change into the future, are too computationally intensive for extended studies. However, the assessment of the impact of human activities on the future climate requires the investigation of a large number of alternative scenarios. Unfortunately, current and foreseeable computational resources are insufficient to use the best, most complex, representation of the climate system in such studies for more than a few limited analyses. For this reason, a new type of approach known as "integrated assessment" has been developed, with integrated assessment models (IAMs) being increasingly used for climate and ozone analysis due to their simplicity, relatively transparency, and computational efficiency. In the following we briefly discuss the purpose and components of an IAM.

13.3.1 *Integrated assessment modeling (IAM)*

IAM provides an integrated view of human interaction with the physical world. Rather than attempting to use a range of multidimensional and complicated expert models, IAMs build on the knowledge achieved by each individual scientific discipline. The individual components of IAM are often simplified based on the more complex models to allow all the main components of the Earth system to be dealt with in one framework. There is no doubt that a fully integrated model (IAM) of the important processes and relationships between these processes can provide an important tool to evaluate policy alternatives associated with global climate change. In addition, to explore all the possible scenarios and the effects of assumptions and approximations in the model more thoroughly, simpler models such as IAMs are often used to examine the time-dependent changes. These models are constructed to give results similar to the more complex GCMs. Simple models represent only the most critical processes, by tying them to the results of complex models and/or observations. As a result of these simplifications, they are relatively easy to understand and inexpensive to run, so that multiple diagnostic tests can be executed. However, the challenge in such modeling is to adequately represent the complexity of current understanding in the processes affecting emissions of greenhouse gases and other radiatively important constituents, the processes determining the resulting climate and ozone changes, and the processes affecting resulting impacts, plus representation of the feedbacks between many of these processes.

As shown in Fig. 13.14, there are four key components to an IAM. The *Human Activities* and *Ecosystems* modules cover population, technology, and economic changes that contribute to energy use; shifts in land use prompted by the evolution of human societies; and the elements leading to emissions of greenhouse gases. The *Atmospheric Composition* and *Climate* sections deal with the modification of climate induced by these emissions. The impact of climate change is dealt with through interaction between the *Human Activities* and *Ecosystems* components.

13.3.2　The integrated science assessment model

We are currently developing an integrated science assessment model (ISAM) for integrated assessment modeling in order to better represent the spa-

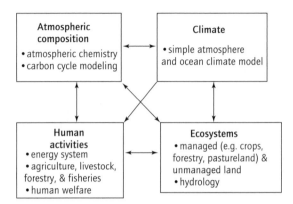

Fig. 13.14　Main components of a full-scale Integrated Assessment Model (IAM).

tial variations and processes relevant to evaluating biogeochemical cycles, determining atmospheric composition, and projecting the resulting global and regional climatic effects. Integrating a process-level understanding of regional impacts into an integrated assessment (IA) model will help to improve the understanding of climate change impacts and extend the range of issues that can be addressed in an IA framework.

The current version of ISAM consists of several submodels, including a terrestrial biosphere–ocean carbon cycle model and two-dimensional (latitude and height) models for chemical transport, radiative forcing, and climate. Connections to the economic and social sciences have been achieved through Pacific Northwest National Laboratory's (PNNL) global change integrated assessment modeling (GCAM), which consists of two parallel but interrelated model development programs, MiniCAM and PGCAM. In this study we use a reduced form version of our ISAM for future emission scenario analysis, as described next.

Here, we use the reduced form ISAM model of Jain *et al.* (1994) and Jain and Bach (1994) to estimate the relationship between the time-dependent rate of GHG and aerosol emissions and quantitative features of climate including GHG concentrations, equivalent effective stratospheric chlorine (EESC), and ozone, global temperature, and sea level.

Carbon cycle

ISAM's global carbon cycle component simulates CO_2 exchange between the atmosphere, carbon reservoirs in the terrestrial biosphere, and the ocean column and mixed layer (Jain *et al.* 1994, 1995, 1996; Kheshgi *et al.* 1996, 1999). The model consists of a homogeneous atmosphere, an ocean mixed layer and land biosphere boxes, and a vertically resolved upwelling-diffusion deep ocean. Ocean and land-biosphere components of this model are described by Jain *et al.* (1995) and Kheshgi *et al.* (1996), while model parameters are discussed by Kheshgi *et al.* (1999). The model takes into account important feedbacks, such as the fertilization and climatic feedback processes. The ocean component of carbon cycle model takes into account the effects of temperature on CO_2 solubility and carbonate chemistry (Jain *et al.* 1995), whereas the biospheric component takes into account the temperature effect through the temperature-dependent exchange coefficient between terrestrial boxes.

Methane cycle

In its reduced form as used in this study, atmospheric CH_4 concentrations are calculated by simulating the main atmospheric chemical processes influencing the global concentrations of CH_4, CO, and OH, using a global CH_4–CO–OH cycle model (Jain *et al.* 1994; Jain & Bach 1994; Kheshgi *et al.* 1999). The removal rates of CH_4 and CO take into account oxidation by OH, soil uptake, and stratospheric transport (Jain & Bach 1994). Reaction with OH radical is responsible for up to 90% of the tropospheric CH_4 sink. In ISAM the concentration of OH is determined by the photochemical balance between the total tropospheric production of OH and the loss rate due to reaction with CH_4, CO, and NMHC (Jain & Bach 1994). The production rate of OH is based on the NO_x emissions and CH_4 concentrations as discussed in Kheshgi *et al.* (1999).

Other GHGs

In its reduced form, ISAM past and future atmospheric concentrations for N_2O and halocarbons are calculated by a mass balance model as described by Bach and Jain (1990).

Climate model

ISAM calculates temperature based on a reduced-form energy-balance climate model of the type used in the 1990 IPCC assessment (Harvey *et al.* 1997). Thermohaline circulation is schematically represented by polar bottom-water formation, with the return flow upwelling through the one-dimensional water column to the surface ocean from where it is returned to the bottom of the ocean column as bottom water through the polar sea. The climate component of ISAM calculates the perturbations in radiative forcings from CO_2 and other GHGs based on updated seasonal and latitudinal GHG radiative forcing analyses (Jain *et al.* 2000).

The response of the climate system to the changes in radiative forcing is principally determined by the climate sensitivity, ΔT_{2x}, defined as the equilibrium surface temperature increase for doubling of atmospheric CO_2 concentration. This parameter is intended to account for all climate feedback processes not explicitly dealt with in the model, particularly those related to clouds and water vapor and related processes. Recent atmosphere–ocean general circulation model (AOGCM) estimates for ΔT_{2x} range from 1.5 to 4.5°C (Houghton *et al.* 2001).

Chlorine and bromine loading and ozone changes

In order to relate emissions of various stratospheric ozone-depleting substances to actual stratospheric ozone depletion, the equivalent effect stratospheric chlorine (EESC) concept (WMO 1999), a simple measure of the chlorine and bromine loading in the atmosphere, is used here. EESC is an index developed to represent potential damage that a given mixture of ozone-depleting substances could create in the stratospheric ozone layer. Because of the impact of transport and other processes on stratospheric O_3, EESC should not be regarded as a perfect measure of the expected future ozone change. Instead, EESC can be thought of as a primary stratospheric forcing mechanism with the attribute that increasing chlorine and bromine loading will decrease ozone (WMO 1999).

The concept of EESC combines the knowledge gained from atmospheric observations with analysis from atmospheric models, in order to evaluate the total amount of chlorine and bromine transferred from the troposphere to the stratosphere, where halocarbons can react with ozone. EESC is directly proportional to the surface emissions of the halocarbons, their reactivity as reflected in their atmospheric lifetimes, and the number of chlorine and bromine atoms released per molecule. Model calculations and laboratory measurements indicate that bromine is much more reactive with ozone than chlorine (WMO 1999). In order to represent bromine loading equivalent to chlorine loading, EESC includes a multiplicative factor, α. Determining the value of α is complicated, as it varies with altitude, latitude, and season in some cases. Here we adopt the WMO (1999) recommended value of 60 for α in representing globally averaged ozone loss.

Based on the satellite and ground-based measurements, WMO (1999) estimated that the depletion of the mean midlatitude (60° S to 60° N) total column ozone between 1979 and 1997 was ~5.5%. In this study, the relationship between future Cl/Br loading and total column ozone change is scaled based on the observed trends in total column ozone and estimated EESC over the period 1979–97. (Ozone depletion first became statistically observable in about 1979.)

Importance of greenhouse feedbacks in estimating future changes in climate

There are a number of important feedback processes occurring in the Earth–ocean–atmosphere system that could significantly modify future CO_2 and other GHG concentrations in the warmer world. Some of these are discussed briefly here. These feedbacks have the potential to be either positive (amplifying the initial change) or negative (dampening them).

As increasing GHG concentrations alter the Earth's climate, changing climate and environmental conditions in turn react back on the carbon cycle and atmospheric CO_2. For example, temperature change affects the growth, disturbance, and respiration of plants and soils in the terrestrial component of the carbon cycle model. Therefore, net emissions of CO_2 from terrestrial ecosystems will be elevated if higher temperature (resulting from CO_2 and other GHG increases) increases respiration at a faster rate than photosynthesis. This would be a positive feedback, resulting in more CO_2 emissions and a further increase in temperature. Climate can also affect the ocean's role in the global carbon cycle. If the ocean became warmer, its net uptake of CO_2 could decrease because of changes in the chemistry of CO_2 in the sea water, biological activities in the surface water, and the rate of exchange of CO_2 between the surface layers and the deep ocean. This would also be a positive feedback, as increasing temperatures would lead to more CO_2 build-up in the atmosphere, again enhancing the initial temperature rise.

In the reduced form of ISAM, the carbon cycle's influence on climate feedback due to the radiative effects of CO_2, other GHGs, and aerosols is accounted for using a "climate sensitivity" parameter as discussed above. Figure 13.15 shows the ISAM-estimated CO_2 concentrations with and without climate feedback for a reference scenario IS92a scenario (Leggett *et al.* 1992). In the climate feedback case, climate sensitivity ranged from 1.5 to 4.5°C for a doubling of CO_2 concentration. This approach yields a lower bound on uncertainties in the carbon cycle and climate. For the reference case, a "best-guess" climate sensitivity of 2.5°C was assumed. The modeled response without the climate feedback leads to a 2100 CO_2 concentration of 682 p.p.m. for the reference case. With climate feedback, the 2100 CO_2 concentration in the reference case was 723 p.p.m., about 6% higher than without climate feedback case. The ranges of 164 p.p.m. or –12%/+11% (about the reference case) in the 2100 CO_2 concentration indicate that there is significant uncertainty about the future CO_2 concentrations due to any one pathway of changes in emissions.

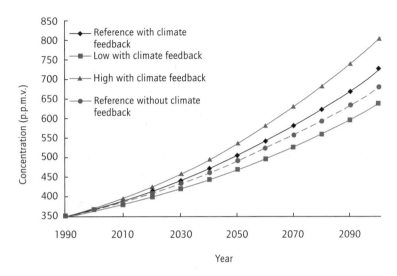

Fig. 13.15 Integrated Science Model (ISAM) estimated CO$_2$ concentrations for the IS92a scenario, with and without climate feedback. The range of values for the climate feedback case was obtained with the range of climate sensitivities from 1.5 to 4.5°C. The reference climate feedback case estimated based on the climate sensitivity of 2.5°C.

As discussed above, observed CH$_4$ concentrations vary seasonally (Fig. 13.4), mainly because CH$_4$ emissions from natural wetlands and rice paddies are particularly sensitive to temperature and soil moisture. Biogenic emissions from these sources are significantly larger at higher temperatures and increased soil moisture; conversely, a decrease in soil moisture would result in smaller emissions. Note that the CH$_4$ component of reduced form ISAM does not take these climate feedbacks into account due to its coarse horizontal resolution.

Chemical feedbacks on atmospheric lifetimes for non-CO$_2$ GHGs

The atmospheric lifetime of a GHG is another important factor when estimating future GHG concentrations and hence the impact of GHG emissions on the Earth's climate. Because of their very long lifetimes (Table 13.1), CO$_2$, CFCs, HCFCs, and N$_2$O are removed slowly from the atmosphere. Hence, their atmospheric concentrations take decades to centuries to adjust fully to a change in emissions. In contrast, some of the halocarbon replacement compounds (HFCs) and CH$_4$ have relatively short atmospheric lifetimes. This enables their atmospheric concentrations to re-

Table 13.1 Lifetime estimates for various greenhouse gases.

Species	Lifetime (years)
CO$_2$	Variable (50 to hundreds)
CH$_4$	9 (8–10)
N$_2$O	120
CFC-11, CFC-12	50 to hundreds
Other CFCs, HCFCs, HFCs	0.01 to hundreds
SF$_6$ and PFCs	Thousands

spond fully to emission changes within a few decades.

For all GHGs except CO$_2$ and H$_2$O, lifetime is calculated based on the balance between emissions and the chemical reactions in the atmosphere. In case of CO$_2$, the lifetime is determined mainly by the slow exchange of carbon between the surface and deep ocean, and the atmosphere and terrestrial biosphere. GHGs containing one or more H atoms (e.g. CH$_4$ and HFC) are removed primarily by reactions with hydroxol radicals (OH), which takes place mainly in the troposphere. The GHGs N$_2$O, PFCs, CF$_6$, CFCs, and halons do not react with OH in the troposphere. These gases are destroyed in the stratosphere or above, mainly by solar ultraviolet radiation (UV) at short wavelengths (<240 nm).

How do these chemical feedbacks affect the future concentrations of a gas? An example is the previous IPCC estimates for future CH_4 concentrations (Houghton *et al.* 1996), which chose to ignore the changing emissions of short-lived gases, such as CO, VOCs (volatile compounds) and NO_x, that affect OH and hence CH_4 removal. As discussed above, the concentration of OH and hence CH_4 depends critically on these compounds (Derwent 1996). To illustrate the potential importance of these short-lived gases, we used our ISAM to calculate the CH_4 increase resulting from a full IS92a scenario (Leggett *et al.* 1992). The scenario includes the projected growth in NO_x, CO, and VOC emissions. As a result of increases in NO_x and tropospheric O_3, the calculated CH_4 concentrations were actually 20% lower than the IPCC projection (Houghton *et al.* 1996) of 3.6 p.p.m. in 2100 (Fig. 13.16), which prescribed constant NO_x, CO, and NMHC emissions at 1990 rates.

These calculations clearly suggest that integrated assessment and other modeling studies should take into account a number of factors, including: the complexity of current understanding of the processes affecting emissions of greenhouse gases and other radiatively important constituents; the processes determining the resulting climate changes; the processes affecting resulting impacts; and the feedbacks between many of these processes.

13.3.3 Future emission projections for major greenhouse gases and aerosols

In order to study the potential implications on climate from further changes in human-related emissions and atmospheric composition, a range of scenarios for future emissions of greenhouse gases and aerosol precursors has been produced by the IPCC Special Report on Emission Scenarios (SRES: Nakicenovic *et al.* 2000). These scenarios are for use in modeling studies to assess potential changes in climate over the next century for the current IPCC international assessment of climate change. Four different narrative story lines, A1, A2, B1, and B2, were developed to describe the relationships between emission-driving forces and their evolution. All the scenarios based on the same story line constitute a scenario family. None of these scenarios should be considered as a prediction of the future, but they do illustrate the effects of various assumptions about economics, demography, and policy on future emissions. In this study we investigate six SRES "marker" scenarios labeled A1B, A1T, A1F, A2, B1, and B2 as examples of the possible effect of greenhouse gases on climate. Of the six scenarios, three (A1B, A1T, A1F) are chosen from the A1 story line and scenario family, and one each from other three story lines and scenario families. Each scenario is based on a narrative, describing alternative future developments in economics, technical, environmental, and social dimensions. These scenarios are no more or less likely than any other scenarios but they have received the closest scrutiny. Details of these storylines and the SRES process can be found elsewhere (Nakicenovic *et al.* 2000). These scenarios are generally thought to represent the possible range for a business-as-usual situation where there have been no significant efforts to reduce emissions to slow down or prevent climate changes. In order to provide direct comparison with the results of

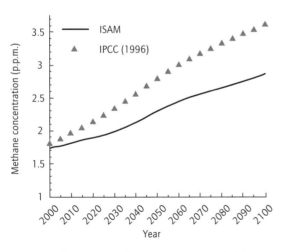

Fig. 13.16 Comparison of Integrated Science Model (ISAM) estimated and the IPCC (Houghton *et al.* 1996) estimated atmospheric CH_4 concentrations for IS92a scenario.

previous IPCC reports (Houghton *et al.* 1996) we have also compared the results of SRES scenarios with the IPCC IS92a scenario.

Figure 13.17a–d shows the anthropogenic emissions for four of the most important gases of concern to climate change: CO_2, CH_4, N_2O, and SO_2. Carbon dioxide emissions span a wide range, from nearly five times the 1990 value by 2100 to emissions that rise and then fall to near their 1990 value. N_2O and CH_4 emission scenarios reflect these variations and have similar trends. However, global sulfur dioxide emissions in 2100 have declined to below their 1990 levels in all SRES scenarios, because rising affluence increases the demand for emission reductions to improve local air quality. Note that sulfur emissions, particular-

ly to mid-century, differ fairly substantially between the scenarios. Also, the SRES scenarios for sulfur emissions are much smaller than IS92a emission estimates, largely as a result of increased recognition worldwide of the importance of reducing sulfate aerosol effects on human health and ecosystems.

Emissions for halocarbons and related greenhouse gases controlled under Montreal Protocol and its Amendments are not covered under the SRES scenarios. For these gases we consider only a single baseline scenario, following the most recent regulations of the Montreal Protocol, developed by the WMO (1999) assessment. The emissions for the HFCs plus SF6 and PFC are based on SRES scenarios.

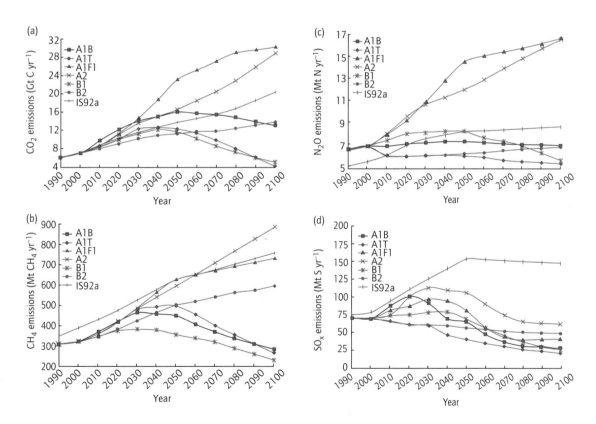

Fig. 13.17 Anthropogenic emissions in the six SRES marker scenarios and IS92a scenario for CO_2, CH_4, N_2O, and SO_2. Note that the SRES emission values are standardized such that emissions from 1990 to 2000 are identical in all scenarios.

Projection of future concentrations

Figure 13.18a–f shows the ISAM-calculated global mean GHG concentrations from 1990 to 2100 for the range of SRES and IS92a scenarios discussed above. Note that for the major greenhouse gases CO_2 and N_2O, the 2100 concentrations in the SRES A2 and A1F1 scenarios lie above IS92a, while the other scenarios lie below the IS92a scenario. In the case of CH_4, the concentrations for scenarios A1F, B1, and B2 lie above the IS92a scenario, while concentrations are lower than IS92a for the other three scenarios. The various scenarios lead to substantial differences in projected greenhouse gas concentration trajectories. The SRES emission scenarios imply the following changes in greenhouse gas concentrations from 1990 to 2100: CO_2 changes range from 20 to 165% (Fig. 13.18a); CH_4 (Fig. 13.18b) from –22 to +110% (Fig. 13.18c); and N_2O from 14 to 49% (Fig. 13.18d). In the case of halocarbons (not shown here), the concentrations first increase before the trend reverses, due to the projected complete phase-out of halocarbon emissions.

Projection of future changes in radiative forcing

Figure 13.19 shows the derived globally averaged radiative forcing as a function of time for these scenarios, relative to the radiative forcing for the preindustrial background atmosphere. For sulfate aerosols, the direct and indirect radiative forcings were calculated on the basis of sulfur emissions contained in SRES and IS92a scenarios as discussed in IPCC (Houghton *et al.* 2001), except for the aerosol direct forcing due to biomass burning. The radiative forcing due to aerosols from biomass burning, OC and BC direct aerosol forcings, the radiative forcings for all GHGs, and direct and indirect aerosol forcing are estimated based on the method described in IPCC (Houghton *et al.* 2001). The direct radiative forcing for biomass burning emissions of OC and BC over the period 1990–2100 was assumed to remain constant at 1990 levels. The contribution from aerosols is probably the most uncertain part of future radiative forcing

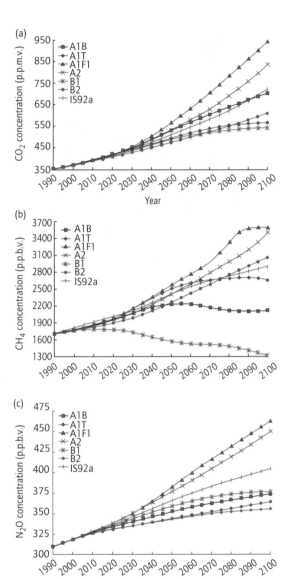

Fig. 13.18 Integrated Science Model (ISAM) estimated atmospheric (a) CO_2, (b) CH_4, and (c) N_2O concentrations from 1990 to 2100 for the six SRES marker and IS92a scenarios.

projections. The calculated radiative forcing from greenhouse gas increases for the preindustrial period to 2100 is about $5\,W\,m^{-2}$ for IS92a. Figure 13.19 shows that for the SRES scenarios the 2100

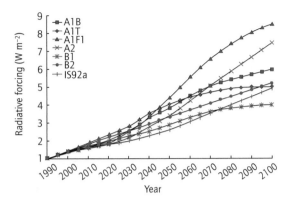

Fig. 13.19 Integrated Science Model (ISAM) estimated radiative forcing from 1990 to 2100 for the six SRES and IS92a scenarios.

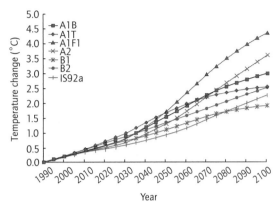

Fig. 13.20 Integrated Science Model (ISAM) estimated global mean temperature change since 1990 for the six SRES and IS92a scenarios. These calculations use a climate sensitivity of 2.8°C for a doubling of CO_2, which appears to best represent the climate sensitivity of current climate models (Houghton *et al.* 2001).

value ranges between 4 and 8.5 W m^{-2}. The range of values represents a sizable increase over the 1.0 W m^{-2} derived for 1990, implying a significant warming tendency. The negative forcing due to tropospheric aerosols offsets some of the greenhouse gas positive forcing in all scenarios.

Future temperature changes

The reduced form ISAM has been used in this analysis to estimate the global mean temperature changes for the IPCC IS92a and various SRES scenarios described earlier. The temperature change results for all scenarios are shown in Fig. 13.20. These calculations have been done for the IPCC (Houghton *et al.* 2001) "best-guess" climate sensitivity estimate of 2.8°C for a doubling of CO_2 (Houghton *et al.* 2001). The IS92a scenario results in a 2.8°C temperature change over the period 1990–2100. For the SRES scenarios the 1990–2100 temperature change ranges between 1.9 and 4.4°C. Note that the temperature changes in the A1T, B1, and B2 scenarios are comparable to IS92a even though their 2100 carbon emissions are lower. This is because of more realistic assumptions for sulfur emissions reductions, which result in dramatically lower sulfur dioxide emissions in the SRES scenarios, and thus a smaller amount of sulfate cooling.

Future development of chlorine and bromine loading and ozone

The future evolution of EESC for the WMO (1999) baseline scenario is shown in Fig. 13.21. Note that the EESC reaches its maximum value of 3.35 p.p.b.v. in year 1995, which is more than doubled compared with the 1970 value of 1.45 p.p.b.v.; after 1995 the EESC recover slowly to levels observed before the extensive human-related emissions in recent decades. Figure 13.21 shows that the largest contribution to EESC comes from CFCs, followed by carbon tetrachloride and methyl chloroform. In 1995, CFCs, CCl_4, CH_3CCl_3, halons, emissions of CH_3Br (anthropogenic), and HCFCs contributed about 40, 11, 11, 9, 4, and 1%, respectively. The natural emissions for CH_3Cl and CH_3Br contributed about 12% each to the total EESC. After 1995, the contribution of all classes of halocarbon, except HCFCs and halons, will decrease. The HCFC contribution will increase from 1% in 1995 to a maximum of about 3% in 2015. The halon contribution is expected to increase from 9% in 1995 to a maximum of about 12% in 2011.

Figure 13.22a and b compares the calculated EESC and total column ozone change for the no protocol scenario and the 1997 Montreal Scenario,

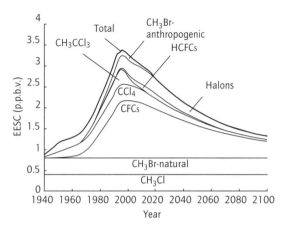

Fig. 13.21 Integrated Science Model (ISAM) estimated past and future equivalent effective stratospheric chlorine (EESC) loading due to various halocarbon emissions following the WMO (1999) baseline scenario, which includes the 1997 Montreal Protocol.

i.e. the WMO (1999) baseline scenario. As noted by WMO (1999), the assumptions that go into these extreme scenarios are subject to considerable uncertainties; hence, the calculated effects should be viewed as mere estimates. Figure 13.22a shows that without any regulation, the EESC by 2050 would be likely to have increased by about four times the 1995 levels, while Fig. 13.22b shows that the ozone depletion at mid-latitudes would have correspondingly decreased by about 45% by 2050, while decreasing further with time. Note that under the current regulations provisions, calling for the complete phase-out of CFCs and halons and other halocarbons, the EESC and the ozone reduction trend is reversed at around 1995, although it is not until about 2040 that the 1979 levels of EESC and ozone are obtained.

13.4 POTENTIAL IMPACTS OF STRATOSPHERIC OZONE AND CLIMATE CHANGES

13.4.1 Impacts of stratospheric ozone depletion

Both ground-based and satellite data clearly confirm that significant depletion of stratospheric O_3

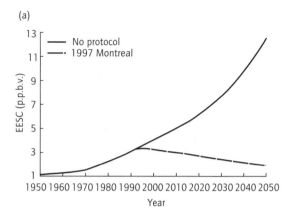

(a)

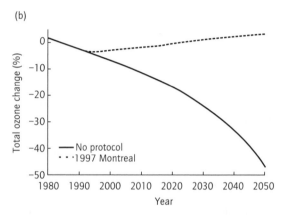

(b)

Fig. 13.22 Integrated Science Model (ISAM) estimated past and future (a) equivalent effective stratospheric chlorine (EESC) loading, and (b) total column ozone change for the no protocol and the 1997 Montreal scenarios.

has occurred over the past two decades. Modeling studies estimate that this depletion is likely to continue past 2050. The impacts due to O_3 loss in the Earth's protective ozone layer have already been observed, and it is estimated that some impacts will remain with us over much of the next century (WMO 1999). But what are these impacts, and how exactly does the ozone layer, over 10 km up in the stratosphere, affect the Earth's surface?

Stratospheric O_3 depletion affects the Earth's surface in two main ways. First, a decrease in stratospheric O_3 allows more UV-B radiation from the Sun to reach the Earth. The relationship between stratospheric O_3 and UV-B radiation incident at the surface is well documented (WMO 1999). Increases in UV radiation due to stratospheric O_3 loss are currently 4–7% at northern and southern hemisphere mid-latitudes, 22% during Arctic spring, and 130% during Antarctic spring (UNEP 1999). Increased UV-B radiation reaching the troposphere and the Earth's surface is responsible for the following impacts.

Human health

Exposure to higher UV-B radiation levels (as opposed to less harmful UV-A radiation) leads to increased risk of eye disease and damage, including snowblindness and cataracts. Ultraviolet radiation can also affect the human immune system, limiting resistance to some infectious diseases and increasing the severity of autoimmune and allergic responses in sunlight. Increases in skin and eye cancer are expected to peak around the middle of the century, with an additional one case per 10,000 due to higher UV levels (UNEP 1999).

Terrestrial ecosystems

Ecosystem response to increased UV-B levels is difficult to quantify, as the response is modified by a number of other changing factors, including temperature, available water and nutrients, and even atmospheric CO_2 levels. UNEP (1999) estimates that while some crops have protection against UV, others will be damaged. The effects of UV radiation may build up from year to year, affecting the timing of plants' life cycles, their ability to protect themselves against disease, and the quality of food produced.

Aquatic ecosystems

The net impact of increased UV-B radiation on marine and freshwater ecosystems is harmful, with polar ecosystems (the area of greatest O_3 loss and UV increase) being most strongly affected. UV radiation has negative impacts on phytoplankton and zooplankton, the foundations of the marine food chain. Large decreases in these tiny organisms could seriously disrupt the larger organisms further up the chain. UV-B is also known to directly affect sea grasses, coral, macroalgae, and other aquatic species that spend at least part of their lives near the surface. A link to climate change is established through phytoplankton that take up CO_2 from the atmosphere. Reductions in phytoplankton therefore imply a slower uptake of CO_2 by the ocean, enhancing global warming.

Air quality

UV-dependent tropospheric photolysis rates are projected to increase with stratospheric O_3 depletion. Photolysis is the primary formation mechanism for OH formation, the atmosphere's "cleansing agent" responsible for removing pollutants from the lower atmosphere. A 1% decrease in stratospheric O_3 is estimated to increase tropospheric OH production by 0.7–1% (WMO 1999). Thus, stratospheric O_3 loss may actually help to clean out many primary pollutants (i.e. substances emitted directly from human activities). However, increased photolysis will also increase O_3 levels in high-NO_x (urban) areas, while decreasing O_3 in low-NO_x (marine and rural) areas. In the lower atmosphere, O_3 is a secondary pollutant. At high concentration levels it is detrimental to human health, often leading to "exercise warnings" in major cities during summer. In the troposphere, O_3 is also a potent GHG. The net sign of tropospheric O_3 change resulting from stratospheric O_3 depletion is uncertain and local changes will be difficult to detect due to the dependence of O_3 production and destruction on many other variables (Tang *et al.* 1998).

The second way in which stratospheric ozone depletion affects the Earth's surface is through its impact on climate. Stratospheric O_3 loss has caused a net negative forcing on climate that has counteracted approximately 30% of the forcing due to GHGs over the past two decades. However, in the future, O_3 recovery will remove this negative forcing, leading to a more rapid increase in radiative forcing than that due to GHG emissions alone.

13.4.2   Impacts due to climate change

In the previous sections, we briefly discussed the climate and ozone changes that we could expect in the future due to current and future human activities. There are many uncertainties in our predictions, particularly with regard to the timing, magnitude, and regional patterns of climate change. Nevertheless, scientific studies have shown that human health, ecological systems, and socioeconomic sectors (e.g. hydrology and water resources, food and fiber production, and coastal systems, all of which are vital to sustainable development) are sensitive to changes in climate as well as to changes in climate variability. Recently, a great deal of work has been undertaken to assess the potential consequences of climate change (Houghton *et al.* 2001; McCarthy *et al.* 2001; Metz *et al.* 2001). This recent study, like many previous studies, has assessed how systems would respond to climate change resulting from an arbitrary doubling of equivalent atmospheric CO_2 concentrations. Here we restrict our discussion to only a brief overview. For those interested, potential climate change impacts are discussed in great detail in Volume III of the most recent IPCC third assessment report (McCarthy *et al.* 2001).

Ecosystems

Ecosystems both affect and are affected by climate. As carbon dioxide levels increase, the productivity and efficiency of water use by vegetation may also increase. Warming temperatures will cause the composition and geographic distribution of many ecosystems to shift as individual species respond to changes in climate. As vegetation shifts, this will in turn affect climate. Vegetation and other land cover determine the amount of radiation absorbed and emitted by the Earth's surface. As the Earth's radiation balance changes, the temperature of the atmosphere will be affected, resulting in further climate change. Other likely climate change impacts from ecosystems include reductions in biological diversity and in the goods and services that ecosystems provide society.

Water resources

Climate change may lead to an intensification of the global hydrological cycle and can have major impacts on regional water resources. Reduced rainfall and increased evaporation in a warmer world could dramatically reduce runoff in some areas, significantly decreasing the availability of water resources for crop irrigation, hydroelectric power production, and industrial/commercial and transport uses. In light of the increase in artificial fertilizers, pesticides, feedlot excrement, and hazardous waste dumps, the provision of good quality drinking water is anticipated to be difficult.

Agriculture

Crop yields and productivity are projected to increase in some areas and decrease in others, especially in the Tropics and subtropics, which contain the majority of the world's population. The decrease may be so severe as to cause increased risk of hunger and famine in some locations that already contain many of the world's poorest people. These regions are particularly vulnerable, as industrialized countries may be able to counteract climate change impacts by technological developments, genetic diversity, and the maintenance of food reserves.

Livestock production may also be affected by changes in grain prices due to pasture productivity. Supplies of forest products such as wood during the next century may also become increasingly inadequate to meet projected consumption due to both climatic and nonclimatic factors. Boreal forests are likely to undergo irregular and large-scale losses of living trees because of the impact of projected climate change. Marine fisheries production is also expected to be affected by climate change. The principal impacts will be felt at the national and local levels.

Sea-level rise

In a warmer climate, sea level will rise due to two factors: (i) the thermal expansion of ocean water as it warms; and (ii) the melting of snow and ice from mountain glaciers and polar ice caps. Over the past century, the global-mean sea level has risen about 25 cm (Houghton *et al.* 2001). Over the next century, ISAM projects a further increase of 30–50 cm in global-mean sea level for the SRES scenarios of greenhouse gas emissions discussed above (Fig. 13.23). These calculations are based on the "best guess" climate sensitivity of 2.8°C. A sea-level rise in the upper part of this range could have very detrimental effects on low-lying coastal areas. In addition to direct flooding and property damage or loss, other impacts may include coastal erosion, increased frequency of storm surge flooding, salt water infiltration and hence pollution of irrigation and drinking water, destruction of estuarine habitats, damage to coral reefs, etc.

Health and human infrastructure

Climate change can impact human health through changes in weather, sea level, and water supplies, and through changes in ecosystems that affect food security or the geography of vector-borne

diseases. The IPCC (McCarthy *et al.* 2001) study dealing with human health issues found that most of the possible impacts of global warming would be adverse.

In terms of direct effects on human health, increased frequency of heat waves would increase rates of cardiorespiratory illness and death. High temperatures would also exacerbate the health effects of primary pollutants generated in fossil fuel combustion processes and increase the formation of secondary pollutants such as tropospheric ozone. Changes in the geographic distribution of disease vectors such as mosquitoes (malaria) and snails (schistosomiasis) and changes in life-cycle dynamics of both vectors and infective parasites would increase the potential for transmission of disease. Nonvector-borne diseases such as cholera might increase in the Tropics and subtropics because of climatic change effects on water supplies, temperature, and microorganism proliferation. Concern over climate change effects on human health is legitimate. However, impacts research on this subject is sparse and the conclusions reached by the IPCC (McCarthy *et al.* 2001) are still highly speculative.

Indirect effects from climatic changes that decrease food production would reduce overall global food security and lead to malnutrition and hunger. Shortages of food and fresh water and the disruptive effects of sea-level rise may lead to psychological stresses and the disruption of economic and social systems.

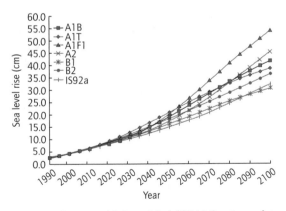

Fig. 13.23 Integrated Science Model (ISAM) estimated sea-level rise (cm) for the six SRES and IS92a scenarios using a 2.8°C climate sensitivity.

13.5 PATHWAYS TO POLICY CONSIDERATIONS

13.5.1 *Policy for the control of stratospheric ozone depletion*

Long-term observations of ozone over Antarctica were the first to reveal the presence of a 50% loss of the springtime total ozone column in the early 1980s (Farman *et al.* 1985). Pioneering chemistry studies by Molina and Rowland (1974) had already suggested a link between emissions of human-produced CFCs and stratospheric O_3 loss. The strong correlation between CFC emission records and the observed mid- and high-latitude declines in total ozone, and the dangerous and immediate nature of stratospheric O_3 depletion impacts on human health and natural ecosystems, prompted a rapid international response.

The Montreal Protocol on Substances that Deplete the Ozone Layer (UN 1987) and its subsequent Amendments – London 1990, Copenhagen 1992, Montreal 1997 – have severely restricted the production and use of ozone-depleting substances. The success of these international agreements has already been observed: the total abundance of ozone-depleting compounds is thought to have peaked before 2000, and a reduction to pre-1980 CFC levels is projected to occur around the middle of this century, leading to a subsequent recovery in total column O_3 levels (Fig. 13.23b).

A number of factors contributed to the success of these agreements to limit ozone-depleting substances. First, most of the production and use of these substances was contained within a relatively narrow range of chemical industries. Thus, regulations could aim at these specific targets (a far easier task than attempting to control energy use, for example, which cuts across all sectors of the economy). Second, it was possible to find viable alternative chemicals with lower ozone depleting and global warming potentials in a short enough time and at a low enough cost to enable the rapid phase-out of CFCs and other harmful substances. Third, rapid action produced a rapid response. Within a decade, tropospheric CFC abundances were declining, and stratospheric O_3 recovery is expected to begin within another decade or two.

The effectiveness of international agreements on ozone depletion has many lessons that can be applied to climate change policy. However, the very factors that contributed to the success of these efforts also serve to highlight the difficulties surrounding climate change policy. In contrast to CFCs and related chemicals, global warming is the result of emissions of a number of greenhouse gases arising from sources tied to all sectors of the human economy—industrial, commercial, transportation, and residential. Specific targets for emission reductions are difficult to resolve. The tabling of any potential legislation is fraught with pitfalls, as targeting one particular source, sector, or even region can lead to questions of equity and fairness. Second, energy use is responsible for over 70% of the observed increase in radiative forcing from preindustrial times to the present (Houghton *et al.* 2001). However, the strong dependence of the world's economy on fossil fuels and the absence of large-scale, economically affordable sources of renewable energy severely limit the amount by which fossil fuel use can be reduced. Given the current lack of viable energy alternatives and the devastating economic and social consequences of severe energy rationing, it is highly unlikely that fossil fuel use can be eliminated or even significantly reduced over the next century. Finally, climate change is a long-range problem. Due to the long lifetime of the gases involved, their slow build-up in the atmosphere, and the response time of the climate system to these forcings, the majority of impacts resulting from climate change have yet to be observed (Metz *et al.* 2001). In the absence of clear, unmistakable, and serious impacts on human society, it is difficult to mobilize the public and governmental support needed to successfully prevent the future climate change that will be caused by current human activities. Unfortunately this short-sighted approach could prove dangerous, as decades of modeling and observational studies have made it clear that in order to prevent serious harm to human life and the environment, GHG emission reductions must occur *before* the most serious impacts become evident. The most dangerous impacts of climate change are not expected to occur until prevention is too late and adaptation is the only option left. For these reasons the world may find climate change to be the global crisis of the new century.

13.5.2 *International actions in response to global warming*

Worldwide concern over climate change and its potential consequences has led the world community to consider international actions to address this issue. These actions fall into two broad categories: an adaptive approach, in which people migrate and/or change their living conditions to adapt to a new environment; and a preventive or "mitigation" approach, in which attempts are made to minimize anthropogenic global climate change by removing its causes. While it is not our intention here to consider or examine the range of possible policy options, it is important to discuss recent international activities that have resulted in a number of recommendations for emission reductions.

In Rio de Janeiro in 1992, the United Nations Framework Convention on Climate Change (UNFCC) agreed to call for the "stabilization of greenhouse gas concentrations in the atmosphere at a level that would prevent dangerous anthropogenic interference with the climate system" (UN 1992). While specific concentration levels and time paths to reach stabilization for greenhouse gases were not stated, analyses of illustrative scenarios for future CO_2 concentrations have given some guidance as to what is required to reach CO_2 stabilization at various levels (Enting *et al.* 1994; Wigley *et al.* 1996). Figure 13.24 shows the calculated allowable emission levels over time that would ultimately stabilize atmospheric CO_2 at levels ranging from 350 to 750 p.p.m.v. These calculations were made with the carbon cycle component of our ISAM model, discussed above. From this figure it is clear that, regardless of stabilization target, global CO_2 emissions initially continue to increase, must reach a maximum some time in this century, and eventually begin a long-term decline that continues through the remainder of the analysis period.

While these reductions are projected to lead to

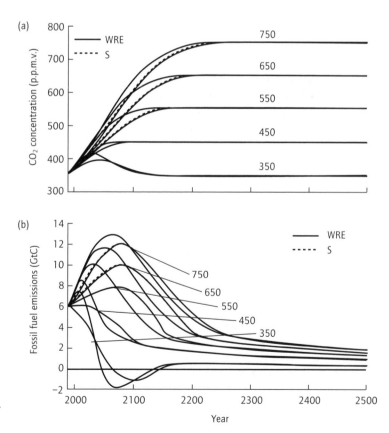

Fig. 13.24 (a) CO_2 concentration stabilization profiles and (b) associated ISAM estimated fossil CO_2 emissions. The "S" and "WRE" pathways are defined in Enting *et al.* (1994) and Wigley *et al.* (1996).

measurable decreases in the rise of CO_2 concentrations, no specific commitments to achieve this goal were made until the December 1997 meeting of the Conference of Parties to the FCCC in Kyoto, Japan (UN 1997). At that meeting, developed nations agreed for the first time to reduce their emissions of greenhouse gases by an average of 5.2% below 1990 levels. Emission targets range from a return to baseline year emissions for most eastern European countries up to an 8% reduction for the European Union. Emission limits for the United States under the Kyoto Protocol consist of a 7% reduction below baseline year emission levels. The baseline year relative to which emission reductions are determined is 1990 for CO_2, CH_4, and N_2O, and the choice of either 1990 or 1995 for HFCs, PFCs, and SF_6. Mitigation actions can include reductions in any of six greenhouse gases: CO_2, CH_4, N_2O, halocarbons (HFCs), perfluorocarbons (PFCs), and sulfur hexaflouride (SF_6).

However, should this protocol enter into force in the USA (a doubtful project as the US government has recently (2001) withdrawn from the Kyoto process, even though the USA is currently responsible for 25% of the world's greenhouse gas emissions), and even if its terms were renewed throughout the remainder of the twenty-first century, it would not achieve the goal of the UNFCC (Wigley *et al.* 1996). This is due to the fact that Kyoto only legislates emission controls for developed or industrialized nations. In the past, a move toward industrialization has been accompanied by an enormous increase in greenhouse gas emissions. Although emissions from the developed countries listed in the Kyoto Protocol currently account for the majority of global greenhouse gas emissions, most developing nations are already moving toward industrialization. If their relationship between greenhouse gas emissions, fossil fuel use, and industrialization follows the paths of other developed nations, emissions from currently developing nations are projected to equal emissions from currently developed nations by 2020 and far surpass them by the end of the century. Thus, emissions from developed nations will make up a smaller and smaller part of the climate change problem as we proceed further into the coming century. For this reason, Kyoto controls on currently developed countries are not enough if we want to prevent dangerous climate change impacts. At the same time, countries in the process of industrialization have the right to be allowed to develop into industrialized nations with higher standards of living and greater wealth. The challenge facing the world community today is how to allow nations the right of development while successfully preventing "dangerous anthropogenic interference with the climate system." The real problem of stabilizing GHG forcing will require much larger reductions that can only be fully supplied by CO_2 emissions (Hoffert *et al.* 1998) from all nations. The future emphasis on CO_2 emission reductions from developed and developing countries highlights the importance of technologies, such as efficiency improvements or carbon capture and sequestration, that provide mechanisms by which fossil fuels can continue to play an important role in future global energy systems without concurrent emissions growth.

13.6   CONCLUSIONS

There is clear evidence that human activities are already having an impact on the Earth's radiation budget, global climate, and stratospheric ozone layer. Potential impacts are projected to be widespread, affecting human health, ecological systems, and socioeconomic sectors. As discussed in this chapter, it is now clear that a certain degree of climatic and ozone changes are now inevitable due to mankind's past and present activities. However, this conclusion has led not to inaction but to the growing consensus that only a concerted action can help to solve the major environmental challenges that still lie ahead.

In its first attempt to confront a global environmental problem, the international community took action to protect the ozone layer through the Montreal Protocol. All nations contributing to the depletion of stratospheric ozone have agreed to control their production of ozone-destroying compounds. As a consequence, abundances of chlorine and other ozone-depleting substances in the atmosphere are decreasing, and stratospheric ozone is expected to begin its recovery within two decades.

In contrast, the problem of global warming is being caused by greenhouse gases emitted from activities and energy resources that lie at the core of human society. Moreover, in tackling the problem of global warming, many of the challenges that lie ahead grow out of existing uncertainties. Measurements and modeling results are plagued with uncertainties, not surprising when one considers how difficult it is to measure and simulate the many interactions of the Earth's complicated climate system. We know from our experience that further research will reduce the uncertainties in some areas. Due to inherent complexity, in other areas the uncertainties may increase and even new ones may be created. Because of these difficulties, it is extremely difficult to develop a global policy to control global warming. However, a precautionary approach to the problem of global warming is warranted on the basis of its potential impact and the scale of the response that is necessary if serious impacts are to be avoided. In order to preserve its future, the world needs concrete action *now* to avert future changes in climate and associated consequences. In that sense, the Kyoto agreement (UN 1997) was an important first step in the right direction. Kyoto initiated a policy of small steps over the next decade based on a balanced portfolio of policy options. This portfolio begins with "no-regret" strategies that make economic and environmental sense whether or not there is a climate change problem. It is unavoidable, however, that as a matter of precaution and prudence, global action will eventually have to go beyond such a "no-regret" policy. Current consensus is that the global community must follow a two-pronged strategy: namely, to conduct research to narrow down uncertainties in our knowledge, while at the same time taking precautionary measures in response to current knowledge.

ACKNOWLEDGMENTS

This work is supported in part by the Integrated Assessment Program, Biological and Environmental Research (BER), US Department of Energy (DOE-DE-FG02-01ER63069).

REFERENCES

Adams, P.J., Seinfeld, J.H., Koch, D., Mickley, L. & Jacob, D. (2001) General circulation model assessment of direct radiative forcing by the sulfate–nitrate–ammonium–water inorganic aerosol system. *Journal of Geophysical Research* **106**, 1097–111.

Bach, W. & Jain, A.K. (1990) CFC greenhouse potential of scenarios possible under the Montreal Protocol. *International Journal of Climatology* **10**, 439–50.

Barnola, J.M., Raynaud, D., Korotkevich, Y.S. & Lorius, C. (1987) Vostok ice core provides 160,000-year record of atmospheric CO_2. *Nature* **329**, 408–14.

Bekki, S., Law, K.S. & Pyle, J.A. (1994) Effect of ozone depletion on atmospheric CH_4 and CO concentrations. *Nature* **371**, 595–7.

Benkovitz, C.M., Scholtz, M.T., Pacyna, J. *et al.* (1996) Global gridded inventories of anthropogenic emissions of sulfur and nitrogen. *Journal of Geophysical Research* **101**, 29,239–53.

Boucher, O. & Anderson, T.L. (1995) GCM assessment of the sensitivity of direct climate forcing by anthropogenic sulfate aerosols to aerosol size and chemistry. *Journal of Geophysical Research* **100**, 26,117–34.

Brasseur, G., Kiehl, J., Muller, J.-F. *et al.* (1998) Past and future changes in global tropospheric ozone: Impact on radiative forcing. *Geophysical Research Letters* **25**, 3807–10.

Brühl, C. & Crutzen, P.J. (2000) NO_x-catalyzed ozone destruction and NO_x activation at midlatitudes to high latitudes as the main cause of the spring to fall ozone decline in the Northern Hemisphere. *Journal of Geophysical Research* **105**, 12,163–8.

Butler, J.H., Montzka, S.A., Clarke, A.D., Lobert, J.M. & Elkins, J.W. (1998) Growth and distribution of halons in the atmosphere. *Journal of Geophysical Research* **103**, 1503–11.

Butler, J.H., Battle, M., Bender, M.L. *et al.* (1999) A record of atmospheric halocarbons during the twentieth century from polar firn air. *Nature* **399**, 749–55.

Chappellaz, J., Barnola, J.M., Raynaud, D., Korotkevich, Y.S., & Lorius, C. (1990) Atmospheric CH_4 record over the last climatic cycle revealed by the Vostok ice core. *Nature* **345**, 127–31.

Collins, W.J., Stevenson, D.S., Johnson, C.E. & Derwent, R.G. (2000) The European regional ozone distribution and its links with the global scale for the years 1992 and 2015. *Atmospheric Environment* **34**, 255–67.

Cooke, W.F. & Wilson, J.J.N. (1996) A global black carbon aerosol model. *Journal of Geophysical Research* **101**, 19,395–409.

Cooke, W.F., Liousse, C., Cachier, H. & Feichter, J. (1999) Construction of a $1° \times 1°$ fossil fuel emission data set for carbonaceous aerosol and implementation and radiative impact in the ECHAM4 model. *Journal of Geophysical Research* **104**, 22,137–62.

Crutzen, P. (1995) Overview of tropospheric chemistry: developments during the past quarter century and a look ahead. *Faraday Discussion* **100**, 1–21.

Crutzen, P.J., Lawrence, M.G. & Poschl, U. (1999) On the background photochemistry of tropospheric ozone. *Tellus* **A51**, 123–46.

Derwent, R.G. (1996) The influence of human activities on the distribution of hydroxyl radicals in the troposphere. *Philosophical Transactions of the Royal Society London, Series A* **354**, 1–30.

Dignon, J. & Hameed, S. (1989) Global emissions of nitrogen and sulfur-oxides from 1860 to 1980. *Journal of the Air Pollution Control Association* **39**, 180–6.

Dlugokencky, E.J., Masarie, K.A., Lang, P.M. & Tans, P.P. (1998) Continuing decline in the growth rate of the atmospheric methane burden. *Nature* **393**, 447–50.

Elkins, J., Thompson, T., Swanson, T. *et al.* (1993) Decrease in the growth rates of atmospheric chlorofluorocarbons-11 and -12. *Nature* **364**, 780–3.

Elkins, J.W., Butler, J.H., Hurst, D.F. *et al.* (1998) Halocarbons and Other Atmospheric Trace Species Group/ Climate Monitoring and Diagnostics Laboratory (HATS/CMDL) (http://www.cmdl.noaa.gov/hats).

Enting, I.G., Wigley, T.M.L. & Heimann, M. (1994) Future emissions and concentrations of carbon dioxide: key ocean/atmosphere/land analyses. CSIRO Division of Atmospheric Research Technical Paper No. 31, CSIRO Australia.

Etheridge, D.M., Steele, L.P., Francey, R.J. & Langenfelds, R.L. (1998) Atmospheric methane between 1000 AD and present: evidence of anthropogenic emissions and climatic variability. *Journal of Geophysical Research* **103**, 15,979–93.

Farman, J.C., Gardiner, B.G. & Shanklin, J.D. (1985) Large losses of total ozone in Antarctica reveal seasonal ClO_x/NO_x interaction. *Nature* **355**, 207–10.

Foukal, P. & Lean, J. (1990) An empirical-model of total solar irradiance variations between 1874 and 1988. *Science* **247**, 556–8.

Fuglestvedt, J.S., Bernsten, T.K., Isaksen, I.S.A., Mao, H., Liang, X.-Z. & Wang. W.-C. (1999) Climatic forcing of nitrogen oxides through changes in tropospheric ozone and methane: global 3D model studies. *Atmospheric Environment* **33**, 961–77.

Grant, K.E., Chuang, C.C., Grossman, A.S. & Penner, J.E. (1999) Modeling the spectral optical properties of ammonium sulfate and biomass aerosols: Parameterization of relative humidity effects and model results. *Atmospheric Environment* **33**, 2603–20.

Hansen, J., Lacis, A., Ruedy, R. & Sato, M. (1992) Potential climate impact of Mt. Pinatubo eruption. *Geophysical Research Letters* **19**, 215–18.

Hansen, J., Sato, M. & Ruedy, R. (1997) Radiative forcing and climate response. *Journal of Geophysical Research* **102**, 6831–64.

Hansen, J., Sato, M., Lacis, A., Ruedy, R., Tegen, I. & Matthews, E. (1998) Climate forcings in the Industrial Era. *Proceedings of the National Academy of Science* **95**, 12,753–8.

Harnisch, J. & Eisenhauer, A. (1998) Natural CF_4 and SF_6 on Earth. *Geophysical Research Letters* **25**, 2401–4.

Harnisch, J., Borchers, R., Fabian, P. & Maiss, M. (1996) Tropospheric trends for CF_4 and CF_3CF_3 since 1982 derived from SF_6 dated stratospheric air. *Geophysical Research Letters* **23**, 1099–102.

Harnisch, J., Borchers, R., Fabian, P. & Maiss, M. (1999) CF_4 and the age of mesospheric and polar vortex air. *Geophysical Research Letters* **26**, 295–8.

Hauglustaine, D., Granier, C., Brasseur, G. & Megie, G. (1994) The importance of atmospheric chemistry in the calculation of radiative forcing on the climate system. *Journal of Geophysical Research* **99**, 1173–86.

Harvey, D., Gregory, J., Hoffert, M. *et al.* (1997) *An Introduction to Simple Climate Models Used in the IPCC Second Assessment Report*. Cambridge University Press, Cambridge.

Haywood, J.M. & Boucher, O. (2000) Estimates of the direct and indirect radiative forcing due to tropospheric aerosols: a review. *Reviews in Geophyshics* **38**, 513–43.

Haywood, J.M. & Ramaswamy, V. (1998) Global sensitivity studies of the direct radiative forcing due to anthropogenic sulfate and black carbon aerosols. *Journal of Geophysical Research* **103**, 6043–58.

Hobbs, P.V., Reid, J.S., Kotchenruther, R.A., Ferek, R.J. & Weiss, R. (1997) Direct radiative forcing by smoke from biomass burning. *Science* **275**, 1776–8.

Hoffert, M.I., Caldeira, K., Jain, A.K. *et al.* (1998) Energy implications of CO_2 stabilization. *Nature* **395**, 881–4.

Hogan, K.B. & Harriss, R.C. (1994) Comment on "A

dramatic decrease in the growth rate of atmospheric methane in the northern hemisphere during 1992" by Dlugokencky *et al. Geophysical Research Letters* **21**, 2445–6.

Houghton, J.T., Meira Filho, L.G., Callander, B.A. *et al.* (eds) (1994) *Climate Change 1994. Radiative Forcing of Climate Change and An Evaluation of the IPCC IS92 Emission Scenarios.* Intergovernmental Panel on Climate Change. Cambridge University Press, Cambridge.

Houghton, J.T., Meira Filho, L.G., Callander, B.A., Harris, N., Kattenberg, A. & Maskell, K. (eds) (1996) *Climate Change 1995. The Science of Climate Change.* Cambridge University Press, Cambridge.

Houghton, J.T., Ding, Y., Griggs, D.J., Noguer, M., van der Linden, P.J. & Xiaosu, D. (eds) (2001) *Climate Change 2001. The Scientific Basis.* Cambridge University Press, Cambridge.

Houghton, R.A. (2000) A new estimate of global sources and sinks of carbon from land-use change. *EOS 81*, supplement s281.

Hoyt, D.V. & Schatten, K.H. (1993) A discussion of plausible solar irradiance variations, 1700–1992. *Journal of Geophysical Research* **98**, 18,895–906.

Iacobellis, S.F., Frouin, R. & Somerville, R.C.J. (1999) Direct climate forcing by biomass-burning aerosols: impact of correlations between controlling variables. *Journal of Geophysical Research* **104**, 12,031–45.

Jacobson, M.Z. (2001a) Global direct radiative forcing due to multicomponent anthropogenic and natural aerosols. *Journal of Geophysical Research* **106**, 1551–68.

Jacobson, M.Z. (2001b) Strong radiative heating due to the mixing state of black carbon in atmospheric aerosols. *Nature* **409**, 695–7.

Jain, A.K. & Bach, W. (1994) The effectiveness of measures to reduce the man-made greenhouse effect: the application of a climate-policy-model. *Theoretical and Applied Climatology* **49**, 103–18.

Jain, A.K., Kheshgi, H.S. & Wuebbles, D.J. (1994) Integrated science model for assessment of climate change model. Paper presented at the Air and Waste Management Association's 87th Annual Meeting, Cincinnati, OH, June 19–24.

Jain, A.K., Kheshgi, H.S., Hoffert, M.I. & Wuebbles, D.J. (1995) Distribution of radiocarbon as a test of global carbon cycle models. *Global Biogeochemical Cycles* **9**, 153–66.

Jain, A.K., Kheshgi, H.S. & Wuebbles, D.J. (1996) A globally aggregated reconstruction of cycles of carbon and its isotopes. *Tellus* **48B**, 583–600.

Jain, A.K., Briegleb, B.P., Minschwaner, K. & Wuebbles, D.J. (2000) Radiative forcings and global warming potentials of thirty-nine greenhouse gases. *Journal of Geophysical Research* **105**, 20,773–90.

Jones, P.D., Folland, C.K., Horton, B., Osborn, T.J., Briffa, K.R. & Parker, D.E. (2001) Accounting for sampling density in grid-box surface temperature time series. *Journal of Geophysical Research* **106**, 3371–80.

Jouzel, J., Lorius, C., Petit, J.R. *et al.* (1987) Vostok ice core: a continuous isotope temperature record over the last climatic cycle (160,000 years). *Nature* **329**, 403–7.

Karl, T.R., Knight, R.W. & Baker, B. (2000) The record breaking global temperatures of 1997 and 1998: evidence for an increase in the rate of global warming? *Geophysical Research Letters* **27**, 719–22.

Keeling, C.D. & Whorf, T.P. (1999) Atmospheric CO_2 records from sites in the SIO air sampling network. In *Trends: A Compendium of Data on Global Change.* Carbon Dioxide Information Analysis Center, Oak Ridge National Laboratory, Oak Ridge, TN.

Keller, C.F. (1998) Global warming: an update. Las Alamos National Laboratory report.

Khalil, M.A.K. & Rasmussen, R.A. (1996) Atmospheric methane over the last century. *World Resources Review* **8**, 481–92.

Kheshgi, H.S., Jain, A.K. & Wuebbles, D.J. (1996) Accounting for the missing sink with the CO_2 fertilization effect. *Climatic Change* **33**, 31–62.

Kheshgi, H.S., Jain, A.K., Kotamarthi, R. & Wuebbles, D.J. (1999) Future atmospheric methane concentrations in the context of the stabilization of greenhouse gas concentrations. *Journal of Geophysical Research* **104**, 19,183–90.

Kiehl, J.T. & Briegleb, B.P. (1993) The relative roles of sulfate aerosols and greenhouse gases in climate forcing. *Science* **260**, 311–14.

Kiehl, J.T., Schneider, T.L., Rasch, P.J., Barth, M.C. & Wong, J. (2000) Radiative forcing due to sulfate aerosols from simulations with the National Center for Atmospheric Research Community Climate Model, Version 3. *Journal of Geophysical Research* **105**, 1441–57.

Lacis, A.A., Wuebbles, D.J. & Logan, J.A. (1990) Radiative forcing of climate by changes in the vertical distribution of ozone. *Journal of Geophysical Research* **95**, 9971–81.

Lefohn, A.S., Husar, J.D. & Husar, R.B. (1999) Estimating historical anthropogenic global sulfur emission patterns for the period 1850–1990. *Atmospheric Environment* **33**, 3435–44.

Leggett, J., Pepper, W.J., Swart, R.J. *et al.* (1992) Emissions scenarios for the IPCC: an update. In *Climate Change 1992: The Supplementary Report to the IPCC Scientific Assessment*. Cambridge University Press, Cambridge.

Lelieveld, J. & Dentener, F.J. (2000) What controls tropospheric ozone? *Journal of Geophysical Research* **105**, 3531–51.

Liousse, C., Penner, J.E., Chuang, C., Walton, J.J., Eddleman, H. & Cachier, H. (1996) A global three-dimensional model study of carbonaceous aerosols. *Journal of Geophysical Research* **101**, 19,411–32.

Lorius, C., Jouzel, J., Ritz, C., Merlivat, L., Barkov, N.E. & Korotkevich, Y.S. (1985) 150,000-year climatic record from Antarctic ice. *Nature* **316**, 591–5.

McCarthy, J.J., Canziani, O.F., Leary, N.A., Dokken, D.J. & White, K.S. (eds) (2001) *Climate Change 2001. Impacts, Adaption and Vulnerability*. Cambridge University Press, Cambridge.

Machida, T., Nakazawa, T., Fujii, Y., Aoki, S. & Watanabe, O. (1995) Increase in the atmospheric nitrous oxide concentration during the last 250 years. *Geophysical Research Letters* **22**, 2921–4.

Maiss, M. & Brenninkmeijer, C.A.M. (1998) Atmospheric SF6: trends, sources, and prospects. *Environmental Science and Technology* **32**(20), 3077–86.

Maiss, M., Steele, L.P., Francey, R.J. *et al.* (1996) Sulfur hexafluoride: a powerful new atmospheric tracer. *Atmospheric Environment* **30**, 1621–9.

Marland, G., Boden, T.A. & Andres, R.J. (2000) Global, Regional, and National CO_2 Emissions. In *Trends: A Compendium of Data on Global Change*. Carbon Dioxide Information Analysis Center, Oak Ridge National Laboratory, US Department of Energy, Oak Ridge, TN.

Mauzerall, D.L., Narita, D., Akimoto, H. *et al.* (2000). Seasonal characteristics of tropospheric ozone production and mixing ratios over East Asia: a global three-dimensional chemical transport model analysis. *Journal of Geophysical Research* **105**, 17,895–910.

Metz, B., Davidson, O., Swart, R. & Pan, J. (eds) (2001) *Climate Change 2001. Mitigation*. Cambridge University Press, Cambridge.

Miller, B., Huang, J., Weiss, R., Prinn, R. & Fraser, P. (1998) Atmospheric trend and lifetime of chlorodifluoromethane (HCFC-22) and the global tropospheric OH concentration. *Journal of Geophysical Research* **103**, 13,237–48.

Miller, R. & Tegen, I. (1998) Climate response to soil dust aerosols. *Journal of Climatology* **11**, 3247–67.

Molina, M.J. & Rowland, F.S. (1974) Stratospheric sink for chlorofluoromethanes: chlorine-atom catalyzed destruction of ozone. *Nature* **249**, 810–14.

Montzka, S.A., Myers, R.C., Butler, J.H. & Elkins, J.W. (1994) Early trends in the global tropospheric abundance of hydrochlorofluorocarbon-141b and -142b. *Geophysical Research Letters* **21**, 2483–6.

Montzka, S.A., Butler, J.H., Myers, R.C. *et al.* (1996a) Decline in the tropospheric abundance of halogen from halocarbons: implications for stratospheric ozone depletion. *Science* **272**, 1318–22.

Montzka, S.A., Myers, R.C., Butler, J.H. *et al.* (1996b) Observations of HFC-134a in the remote troposphere. *Geophysical Research Letters* **23**, 169–72.

Montzka, S.A., Butler, J.H., Elkins, J.W., Thompson, T.M., Clarke, A.D. & Lock, L.T. (1999) Present and future trends in the atmospheric burden of ozone-depleting halogens. *Nature* **398**, 690–4.

Mylona, S. (1996) Sulphur dioxide emissions in Europe 1880–1991 and their effect on sulphur concentrations and depositions. *Tellus* **48B**, 662–89.

Nakicenovic, N., Alcamo, J., Davis, G. *et al.* (2000) *IPCC Special Report on Emissions Scenarios*. Cambridge University Press, Cambridge.

Nesme-Ribes, E., Ferreira, E.N., Sadourny, R., Le Treut, H. & Li, Z.X. (1993) Solar dynamics and its impact on solar irradiance and the terrestrial climate. *Journal of Geophysical Research* **98**, 18,923–35.

Oram, D.E., Reeves, C.E., Penkett, S.A. & Fraser, P.J. (1995) Measurements of HCFC-142b and HCFC-141b in the Cape Grim air archive: 1978–1993. *Geophysical Research Letters* **22**, 2741–4.

Oram, D.E., Reeves, C.E., Sturges, W.T., Penkett, S.A., Fraser, P.J. & Langenfelds, R.L. (1996) Recent tropospheric growth rate and distribution of HFC-134a (CF_3CH_2F), *Geophysical Research Letters* **23**, 1949–52.

Oram, D.E., Sturges, W.T., Penkett, S.A., Lee, J.M., Fraser, P.J. & McCulloch, A. (1998) Atmospheric measurements and emissions of HFC-23 (CHF_3). *Geophysical Research Letters* **25**, 35–8.

Oram, D.E., Sturges, W.T., Penkett, S.A. & Fraser, P.J. (1999) Tropospheric abundance and growth rates of radiatively-active halocarbon trace gases and estimates of global emissions. In *IUGG 99: Abstracts*. International Union of Geodesy and Geophysics.

Peixoto, J.P. & Oort, A.H. (1992) *Physics of Climate*. American Institute of Physics, New York.

Penner, J.E., Chuang, C.C. & Grant, K. (1998) Climate forcing by carbonaceous and sulfate aerosols. *Climate Dynamics* **14**, 839–51.

Petit, J.R., Mounier, L., Jouel, J., Korotkevich, Y.S.,

Kotlyakov, V.I. & Lorius, C. (1990) Paleoclimatological and chronological implications of the Vostok core dust record. *Nature* **343**, 56–8.

Pollack, H., Huang, S. & Shen, P.Y. (1998) Climate change revealed by subsurface temperatures: a global perspective. *Science* **282**, 279–81.

Prinn, R.G., Weiss, R.F., Fraser, P.J., Simmonds, P.G., Alyea, F.N. & Cunnold, D.M. (1998) The ALE/GAGE/AGAGE database, DOE-CDIAC World Data Center. Dataset No. DB-1001.

Rasmussen, R.A. & Khalil, M.A.K. (1981) Atmospheric methane (CH4): trends and seasonal cycles. *Journal of Geophysical Research* **89**, 11,599–605.

Santer, B.D., Taylor, K.E., Wigley, T.M.L., Penner, J.E., Jones, P.D. & Cusbasch, U. (1995) Towards the detection and attribution of an anthropogenic effect on climate. *Climate Dynamics* **12**, 77–100.

Sokolik, I.N. & Toon, O.B. (1996) Direct radiative forcing by anthropogenic airborne mineral aerosols. *Nature* **381**, 681–3.

Stevenson, D.S., Johnson, C.E., Collins, W.J., Derwent, R.G. & Edwards, J.M. (2000) Future tropospheric ozone radiative forcing and methane turnover — the impact of climate change. *Geophysical Research Letters* **27**, 2073–6.

Stolarski, R.S. and Cicerone, R.J. (1974) Stratospheric chlorine: a possible sink for ozone. *Canadian Journal of Chemistry* **52**, 1610–15.

Sturrock, G.A., O'Doherty, S., Simmonds, P.G. and Fraser, P.J. (1999) In situ GC-MS measurements of the CFC replacement chemicals and other halocarbon species: the AGAGE program at Cape Grim, Tasmania. In *Proceedings of the Australian Symposium on Analytical Science*, Melbourne, July.

Sturrock, G.A., Porter, L.W. & Fraser, P.J. (2001) In situ measurement of CFC replacement chemicals and other halocarbons at Cape Grim: the AGAGE GC-MS program. In Tindale, N.W., Derek, N. & Francey, R.J. (eds), *Baseline Atmospheric Program (Australia) 1997–98*. Bureau of Meteorology and CSIRO Atmospheric Research, Melbourne.

Tang, X., Madronich, S., Wallington, T. & Calamari, D. (1998) Changes in tropospheric composition and air quality. *Journal of Photochemistry and Photobiology B* **46**, 83–95.

Tegen, I., Doch, D., Lacis, A.A. & Sato, M. (2000) Trends in tropospheric aerosol loads and corresponding impact on direct radiative forcing between 1950 and 1990: a model study. *Journal of Geophysical Research* **105**, 26,971–89.

UN (1987) *Montreal Protocol on Substances that Deplete the Ozone Layer*. United Nations, New York.

UN (1992) *Framework Convention on Climate Change*. United Nations, New York.

UN (1997) *Kyoto Protocol to the United Nations Framework Convention on Climate Change*. United Nations, New York.

UNEP (1999) *Environmental Effects of Ozone Depletion: 1998 Assessment*. UNEP, Geneva.

Van Dorland, R., Dentener, F.J. & Lelieveld, J. (1997) Radiative forcing due to tropospheric ozone and sulfate aerosols. *Journal of Geophysical Research* **102**, 28,079–100.

Wang, W.C., Zhuang, Y.C. & Bojkov, R.D. (1993) Climate implications of observed changes in ozone vertical distributions at middle and high latitudes of the northern hemisphere. *Geophysical Research Letters* **20**, 1567–570.

Wang, Y. & Jacob, D.J. (1998) Anthropogenic forcing on tropospheric ozone and OH since preindustrial times. *Journal of Geophysical Research* **103**, 31,123–35.

Wigley, T.M.L., Richels, R. & Edmonds, J. (1996) Economic and environmental choices in the stabilization of atmospheric CO_2 concentrations. *Nature* **379**, 240–3.

WMO (1995) *Scientific Assessment of Ozone Depletion: 1994*. Global Ozone Research and Monitoring Project, Report No. 37. WMO, Geneva.

WMO (1999) *Scientific Assessment of Ozone Depletion: 1998*. Global Ozone Research and Monitoring Project, Report No. 44. WMO, Geneva.

Wuebbles, D.J., Hayhoe, K.A.S. & Kotamarthi, R. (2000) Methane in the global environment. In Khalil, M.A.K. (ed.), *Atmospheric Methane: Sources, Sinks, and Role in Global Change*. Springer-Verlag, New York.

14 Regional-Scale Pollution Problems

CRISPIN J. HALSALL

14.1 INTRODUCTION

It is only in the past 40 years that evidence has arisen to show that air pollution poses a problem in areas well removed from obvious sources such as cities and associated industry. Pollutants in ambient urban air will be rapidly dispersed and diluted downwind of the conurbation. This fact, following the catastrophic smog events of the early 1950s in London, led industry—particularly the power industry—to build taller stacks to aid the dispersal of "smoke" away from the local environment. The observation that pollutants were carried by the prevailing winds and remained airborne for long enough to affect areas very far removed from the immediate source was only made in the 1960s when the effects of acid precipitation became apparent on Scandinavian lakes. This was the first time that transboundary air pollution was recognized as a significant hazard, which prompted legislation in the form of the United Nations Economic Commission on Europe (UNECE) to instigate the Convention on Long-range Transboundary Air Pollution (CLRTAP). The convention was the first internationally legally binding instrument to deal with problems of air pollution on a broad regional basis. It was signed in 1979 and entered into force in 1983. The first protocol under this convention focused on sulfur dioxide (SO_2), and in the past 20 years a number of protocols have been ratified for a variety of pollutants. (Further details of specific protocols can be found at the UNECE website http://www.unece.org/env/lrtap/). Table 14.1 provides a summary of these protocols along with the year that they were ratified. It is the implementation of these protocols (at the national level) that has driven regional-scale air pollution management.

The term "regional air pollution" (as opposed to urban or global effects) is a phrase coined largely through work on acidic species, as well as on photochemical oxidants such as ozone (O_3). The term "regional" implies the widespread dispersal of primary and secondary air pollutants, the latter formed from the oxidation or reaction of primary pollutants released from a variety of local or point sources. Perhaps the best examples include the oxidation of SO_2 to sulfate (SO_4^{2-}), or the oxidation of nitrogen dioxide (NO_2) to nitrate (NO_3^-). The formation of nitrate and sulfate particulates is a relatively slow process, which occurs during prolonged atmospheric transport, resulting in deposition of acidic species well away from the point of release. The occurrence of particulate material released either directly to the atmosphere or formed through reaction contributes to reduced visibility or haze, and regional air pollution is often distinguished by an easy reference such as visibility. Perhaps the best example is "Arctic haze," a regional air pollution phenomenon brought about by long-range transport of air pollutants, that was first observed by pilots in the 1950s and is the focus of further attention later in this chapter.

The dimensions to regional-scale air pollution

Table 14.1 UNECE Convention on Long-range Transboundary Air Pollution.

Protocol	Year	Summary
Reduction in sulfur	1985	30% reduction in S emissions by 1993 (base year 1980)
Control of NO_x*	1988	Reduction in NO_x emissions by 9% (base year 1987)
Control of VOC†	1991	30% reduction in VOC by 1999 (base year between 1984 and 1990)
Further reduction in sulfur	1994	Introduction and setting of critical loads
Control on heavy metals	1998	Reduce emissions for Cd, Pb, and Hg to below 1990 levels
Persistent organic pollutants	1998	Eliminate discharges and emissions of 11 pesticides, two industrial compounds and three by-products
Abatement of acidification, eutrophication and ozone	1999	Emission rulings for 2010 for NO_x, VOC and NH_3, i.e. cut NO_x emissions by 41% compared to 1990

* Nitric oxide (NO) and nitrogen dioxide (NO_2).
† Volatile organic compounds.

are often difficult to define. However, it is generally accepted that on a spatial scale the area affected ranges from the size of a country up to the subcontinental level. Often "regional scale" is used in a context to define a certain populated or industrial area where air pollution problems are well characterized, such as the Los Angeles or Athens basins. However, under scientific criteria regional air pollution remains at the synoptic scale (thousands of kilometers). The vertical extent of the pollution is usually confined to the atmospheric boundary layer (~500–2000 m), with incursions into the free troposphere depending on the prevailing meteorology. Regional pollution problems are generally considered over the period of days to months, i.e. on a seasonal time frame. Incidents such as ground-level ozone during the summer or wintertime haze in the Arctic are good examples of this, but this is clearly dependent on the nature and source of pollution. Criteria defining temporal and spatial scales of air pollution are presented in Table 14.2. These definitions are from Borrell *et al.* (1997) as part of EUROTRAC, a European framework organization set up in 1988 with the aim of coordinating research on chemical pollutants in the troposphere. The regional air pollution criteria were developed with photooxidants in mind but are applicable to other types of pollutant.

Table 14.2 Scales of air pollution (adapted from Borrell *et al.* 1997).

	Vertical scale (km)	Horizontal scale (km)	Time scale (days)
Local or urban	2	100	1
Regional or continental	2	2,500	5–100
Global	10	20,000	100 to years

14.2 MONITORING FRAMEWORKS

The main framework of the UN-ECE CLRTAP was the setting up and financing of the EMEP program (Cooperative Programme for Monitoring and Evaluation of the Long-range Transport of Air Pollution in Europe). The EMEP program relies on three main areas to fulfill its role, namely collection of emission data, measurements of air and precipitation quality, and modeling of atmospheric transport and deposition of air pollutants. Through the combination of these three elements, EMEP regularly reports on emissions, concentrations, and/or deposition of air pollutants, as well as the quantity and significance of transboundary fluxes and related exceedances to critical loads and threshold levels. The modeling program has gone some way to help shape emission reduction targets for some of the later protocols. EMEP now operates some 100 sampling stations across 32 European countries. In the USA, air pollution on both local and regional scales falls under the remit of the Environmental Protection Agency, largely through the 1990 Clean Air Act. Air pollution monitoring, assessment of source emissions, and air quality criteria for key pollutants now all fall under EPA control. With regard to regional air pollution the Ambient Monitoring Technology Information Center (AMTIC), operated by the EPA's Monitoring and Quality Assurance Group (MQAG), contains information and files on ambient air quality monitoring programs, as well as details on monitoring methods and information on air quality trends. Both the US EPA and EMEP maintain sophisticated websites containing relevant monitoring data, as well as model results and predictions for regional air quality and deposition estimates.

14.3 THE REGIONAL OZONE PROBLEM

Photochemical oxidants, of which O_3 is the most important, are secondary air pollutants, formed by the reaction of volatile organic compounds (VOCs) and carbon monoxide (CO) with nitrogen oxides (NO_x) in the presence of sunlight. High levels of O_3 in the atmospheric boundary layer (ground-level O_3), along with other photooxidants such as hydrogen peroxide (H_2O_2) and peroxyacetyl nitrate (PAN), are therefore mainly a summertime phenomenon in Europe and North America.

Photochemical smog and associated photooxidants were first studied in the Los Angeles basin of southern California in the late 1940s. The occurrence of ground-level O_3 at concentrations greatly above background levels was established in most industrialized countries in the following decades (e.g. Cox *et al.* 1975; Isaksen *et al.* 1978; White *et al.* 1983; Lui *et al.* 1987; Dollard *et al.* 1995). With the establishment of continent-wide monitoring networks such as the European EMEP

program and the remit granted to the US EPA, levels of ground-level O_3 are now routinely monitored and interpolated, and the extent of the pollution is assessed on a regional basis. Typical European mean rural boundary layer O_3 concentrations vary widely, but generally fall between 20 and 40 p.p.b.v. (parts per billion by volume) on an annual basis. Levels during the summer months, however, are higher, between 50 and 60 p.p.b.v., with episodes often lasting a number of days where concentrations may be well in excess of 80 p.p.b.v. Figure 14.1 shows the average annual maxima in O_3 for the period 1988–94 for EMEP, with monitoring stations plotted as their latitudinal position. This serves to illustrate the elevated concentrations observed in central Europe due to photochemical episodes, where concentrations may reach ~100–150 p.p.b.v., due to stable high-pressure weather systems and an accumulation of precursor compounds. These episodes may typically last from 2 to 10 days. The lower values for the Iberian peninsula reflect the cleaner Atlantic air masses that impact on these sites.

O_3 is not produced or emitted directly by polluting sources, but formed downwind via a complex reaction of VOC and NO_x in the presence of sunlight; a photochemical reaction that is discussed in further detail in the following sections. Consequently, O_3 can be formed over rural areas. In Europe the highest summertime concentrations of O_3 are found in the southeast of the continent, with lower concentrations to the northwest. Typically, the lowest concentrations occur during the winter, with a maximum during early summer.

14.3.1 Human health

An important issue surrounding photochemical oxidants is the human health effects associated with acute exposure during photochemical smog events (PORG 1993). Exposure to high levels of O_3 can result in respiratory disorders through the irritation of the mucous membranes, often accompanied by an inflammatory response in the lining of the lungs. General sensitization can occur particularly for asthmatics or people suffering from other respiratory-related allergies. Detailed studies of

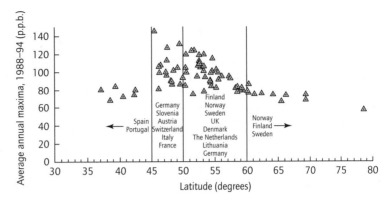

Fig. 14.1 Average of 1988–94 annual maxima at EMEP O$_3$ monitoring sites (plotted according to latitude). (From PORG 1997.)

the dose–response relationship for O$_3$ have shown that both the concentration and duration of exposure are important factors affecting aspects of lung function (Larson *et al.* 1991, cited in PORG 1997). Of importance is the available evidence on general population health effects in view of exposure to O$_3$ over a wide area. It is suggested that those who live in areas of repeated high O$_3$ concentrations are more likely to suffer from asthma and have generally depressed lung function (Schwartz 1989). Several studies have shown clear links between hospital admissions for asthma/respiratory illnesses and levels of O$_3$. However, it is often difficult to distinguish the effects of O$_3$ from those of other air pollutants (e.g. Oehme *et al.* 1996). The World Health Organization (WHO) has summarized a large number of studies from North America and Europe to examine exposure–response relationships. This work calculated a coefficient for hospital admissions for respiratory disorders, based on O$_3$ alone, to be 0.003810 p.p.b.$^{-1}$ O$_3$. This allows actual quantification of hospital admissions due to exposure to O$_3$ at the total population or regional level. Large uncertainties arise, however, once assumptions are used based on whether a threshold effect or nonthreshold effect is followed, and this can result in the generation of a wide range of admission numbers. Nevertheless, the World Meteorological Organization (WMO) has set guidelines for the protection of human health of between 75 and 100 p.p.b. as maximum hourly exposure values. Similarly, individual countries have developed their own criteria based on international guidelines. For instance, in the UK, the Air Quality Guidelines for O$_3$ have been set at 50 p.p.b. for an eight-hour running average. In the USA, the National Ambient Air Quality Standard set by the Environment Protection Agency for O$_3$ is 120 p.p.b. as a one-hour average, although a secondary standard of 80 p.p.b. for an eight-hour running average has still to be implemented.

14.3.2 Effects on vegetation

Large areas in both Europe and North America are exposed to levels of O$_3$ that exceed a threshold value or critical level whereby potential reductions in crop yield may occur. In addition, O$_3$ may also be contributing to forest decline in central Europe (Herman *et al.* 2001). Acute O$_3$ exposure results in visible damage to the foliar parts of plants when rates of O$_3$ uptake (over hours or days) exceed rates of detoxification. Symptoms of acute injury on broad-leafed plants include bleaching, chlorosis, and necrosis spots, while in conifers banding and tip necrosis of the needles are often observed (Krupa *et al.* 1989). Chronic responses to O$_3$ exposure are often more subtle. This results from both high and low rates of O$_3$ uptake throughout the growth season or life cycle of the plant. While visible signs of plant injury may occur through chronic exposure, reductions in growth or primary productivity are the real area of concern.

The benchmark for chronic effects on vegetation is based on the critical levels approach, whereby concentrations of O_3 above a certain level are likely to cause an adverse effect on vegetation, such as reduced crop yield (Grunhage *et al.* 1999). Establishing an appropriate critical level has been the focus of much scientific investigation, with particular attention on crop species. The first critical levels were derived from American studies, such as the National Crop Loss Assessment Program, which established the first long-term (or "growing season") O_3 critical level of 25 p.p.b. as a seasonal seven-hour mean (Heck *et al.* 1988). Concern over the use of a mean value—which may hide the effects of high O_3 levels—has resulted in the assessment of different O_3 indices. For example, Lee *et al.* (1988) obtained data from a wide range of studies investigating different crop types in a variety of locations over a number of years. They demonstrated that those indices that emphasized peak concentrations or accumulated exceedances of a threshold concentration gave a better fit to crop yield data than mean concentrations alone. This gave rise to a variety of long-term experiments to investigate threshold values, perhaps the most extensive being the European open-top chamber programme (EOTCP), which ran over five years (1987–91) in nine countries. Chamber measurements were conducted on a variety of crop types exposed to varying concentrations of O_3. For the crop species studied, yield reduction was highly correlated with cumulative exposure above a threshold of either 30 or 40 p.p.b. There was less fit with the data when the 50 p.p.b. threshold was utilized. As background levels of O_3 are typically ~25–30 p.p.b. in Europe, it was considered that the value of 40 p.p.b. would be the most appropriate criterion. The exposure index is therefore referred to as AOT 40, or accumulated exposure over a threshold of 40 p.p.b. The time period over which the exposure could be assessed was selected as three months, effectively the summer months of May, June, and July (relevant to northern Europe when O_3 concentrations are at their highest), and also the average length of the EOTCP wheat experiments.

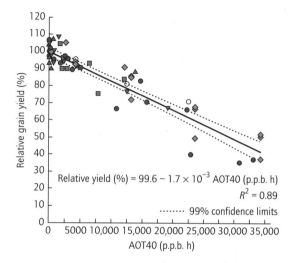

Fig. 14.2 Relative grain yield of spring wheat and O_3 exposure as AOT 40 over three months, based on data from various European and US open-chamber experiments (represented by the different symbols). (From Fuhrer 1996.)

Figure 14.2 shows the relationship between relative grain yield and AOT 40 exposure over three months and includes both European and US data. The critical level for agricultural crops has now been selected as 3000 p.p.b. h (above 40 p.p.b. O_3), accumulated during daylight hours over the three-month period, which from Fig. 14.2 gives rise to an approximate reduction in wheat yield of 10%. Efforts are now directed at mapping AOT 40 exposures across Europe to assess those areas where ambient O_3 exposures exceed critical levels; an approach that will influence control strategies for reducing the emissions of O_3 precursors. AOT 40 exposure values have also been assessed for other vegetation types; for example, the critical ozone level for forest trees has been defined at an AOT 40 of 10 p.p.b. h during daylight hours over a six-month growing season. However, there is evidence to suggest that some species of tree are more sensitive to O_3 exposure and could suffer significant damage even below this critical level (Van der Heyden *et al.* 2001).

Concern over the uncertainties surrounding AOT 40 as an acceptable criterion on which to base a critical level have been highlighted by Fowler *et al.* (1999). In the EOTCP chamber studies, the measured reduction of crop yield under controlled conditions was based on crops grown under optimum conditions (i.e. well watered crops) (Manning & Krupa 1992). The crops were also subject to a continuous flow of air with a stable concentration of O_3. This is often in contrast to the situation in the crop field, where large fluctuations in O_3 may occur over a short time period, affecting both O_3 deposition and plant uptake (Krupa & Kickert 1989). Ongoing studies are focusing on the relationship between measured O_3, deposition, and plant uptake at the micrometeorological scale and the subsequent chronic response of the plant.

14.3.3 Understanding the chemistry behind ozone formation

The formation of ozone in the polluted atmosphere is through a complex series of reactions, but in essence relies on the fact that oxidation of nitric oxide (NO) to nitrogen dioxide (NO_2) can occur without the net consumption of O_3, a process that *does* occur in the unpolluted background atmosphere. During daylight hours NO_2 undergoes photolysis to NO and singlet oxygen (O), the latter reacting with oxygen (O_2) to form O_3. In turn O_3 reacts with NO to form NO_2. The process leads to the "photostationary steady state," in which O_3 formation is matched by O_3 loss. This process is typical of the clean background atmosphere where levels of NO as well competing oxidants to O_3 are low. The following reactions illustrate this process:

$$NO_2 \rightarrow NO + O \qquad (14.1)$$

$$O + O_2(+M) \rightarrow O_3 \qquad (14.2)$$

$$NO + O_3 \rightarrow NO_2 + O_2 \qquad (14.3)$$

In the polluted atmosphere, however, the oxidation of NO to NO_2 can occur via competitive reaction with oxidants other than O_3, namely the peroxy radicals of HO_2 and RO_2. These peroxy radicals are formed by the reaction of the hydroxyl radical (OH) with VOCs. The oxidation of NO can therefore be represented by:

$$HO_2 + NO \rightarrow OH + NO_2 \qquad (14.4)$$

$$RO_2 + NO \rightarrow RO + NO_2 \qquad (14.5)$$

Since the conversion of NO to NO_2 as a result of reactions 14.4 and 14.5 does not consume O_3, the subsequent photolysis of NO_2 (reaction 14.1) results in net O_3 formation (via reaction 14.2), and the photostationary steady state is disrupted.

It must be pointed out that the above description is a much simplified form of a complex series of reactions. Importantly, the VOC or hydrocarbon—usually represented as RH (i.e. alkane)—undergoes a range of oxidation steps resulting in alkoxy radicals (RO), which in turn result in the formation of carbonyls ($R_{-H}O$). These in turn can be further oxidized (ultimately to CO_2), resulting in further O_3 formation with each step. The O_3 in turn can be photolyzed, resulting in more OH, causing a chain propagating process. The governing factor in the oxidation of VOCs are the ambient concentrations of hydroxy (OH), peroxy (HO_2), and alkoxy radicals (RO_x). To illustrate, Fig. 14.3 presents the schematic of photochemical oxidant production in the polluted atmosphere (PORG 1997). It must be pointed out, however, that the cycle of reactions in Fig. 14.3 represents the oxidation of a simple generic alkane to its first generation oxidized product $R_{-H}O$. In fact, the cycle occurs for many different VOCs present in the polluted atmosphere and is repeated for their oxidized products—a process involving many different rates and products and hence yields of O_3. In an attempt to quantify O_3 yield, the photochemical oxidant creation potential (POCP) has been derived for the major hydrocarbon species present in the atmosphere, based on their observed concentration and rates of reaction with OH (the dominating oxidation process) (Derwent *et al.* 1996). Elegant descriptions of photooxidant and O_3 chemistry are

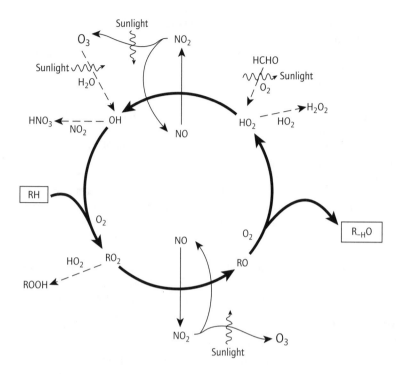

Fig. 14.3 Photochemical oxidant production in a polluted atmosphere (where NO > 20 p.p.t.v.). Note this is a single cycle for a generic saturated hydrocarbon "RH." In an unpolluted atmosphere where levels of NO are <20 p.p.t.v., any RO_2 formed from hydrocarbons may react with HO_2, thus terminating the chain reaction that leads to O_3 formation. (Reprinted from *Environmental Pollution* (Fowler *et al.* 1999). Copyright 1999, with permission from Elsevier Science.)

provided in the excellent reviews of Fowler *et al.* (1999) and Kley *et al.* (1999).

14.3.4 The importance of VOCs and NO_x concentrations in O_3 formation

With regard to regional O_3 pollution, attention must now be turned toward the O_3 precursors of VOCs and NO_x. In the urban atmosphere during incidents of high sunlight (usually associated with stable anticyclonic conditions), the formation of O_3 is highly sensitive to levels of VOCs and NO_x present in the atmosphere. For instance, in the urban atmosphere levels of primary pollutants such as NO, emitted directly from nearby sources such as vehicle exhausts, are high. These high levels of NO will react not only with HO_2 and RO_2, but also with O_3. This has the effect of actually *reducing* O_3 in the polluted urban atmosphere,

notably for periods of the day when NO levels are high, such as the morning and afternoon rush hours. Levels of O_3 may only start to increase once the polluted air mass is advected out of a conurbation and levels of NO start to decrease. The sensitivity of O_3 formation is therefore often expressed in terms of the relative ratio of VOC and NO_x concentrations. Situations may arise whereby the VOC/NO_x ratio is low, resulting in VOC-limited conditions, whereby a reduction in "RH" will result in a decrease in the reaction chain length apparent in Fig. 14.3, resulting in a subsequent decrease in O_3. Conversely, a reduction in NO_x relative to VOCs may result in NO_x-limited conditions, resulting in plentiful HO_2 and RO_2 radicals that may give rise to further O_3 yield. Obviously, further reduction in NO_x in a NO_x-limited area would ultimately result in O_3 decline, but a decline in a highly NO_x-polluted environ-

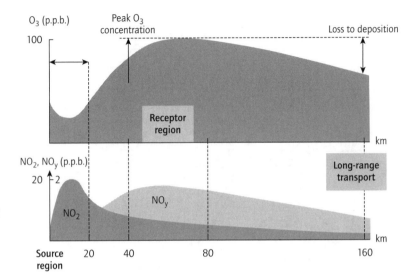

Fig. 14.4 Schematic diagram
showing the urban–rural
boundary layer and the
formation of O_3 and the
oxidation of NO_x to NO_y.
(Reprinted from *Environmental
Pollution* (Fowler *et al.* 1999).
Copyright 1999, with
permission from Elsevier
Science.)

ment may give rise to extra O_3. The VOC/NO_x ratio has been estimated to lie in the range of 1–10 for the polluted boundary layer over Europe, and in some rural areas due to the emission of biogenic VOCs may increase to 10–20, a situation resulting in NO_x-limited conditions and favoring O_3 formation (Kley *et al.* 1999). Figure 14.4 displays a modeled two-dimensional cross-sectional profile of O_3 and NO_x concentrations in a transect starting upwind of an urban area and extending some 150 km downwind. The formation of O_3 is at its highest in the downwind rural area. This graph effectively illustrates the potential for O_3 production over a wide area, particularly given the multitude of urban centers across Europe and North America. Similarly, the decline in NO_x with increasing distance from the urban source is a result of further oxidation, leading to the formation of acidic species such as nitric acid (HNO_3). Collectively these species are termed NO_y, and further examples are given in Fig. 14.5.

In order to predict O_3 levels, photochemical oxidant models have been developed in both North America and Europe over the past 20 years and vary according to level of complexity. Essentially, these describe how the primary emitted pollutants are dispersed and chemically transformed as they are transported away from major source areas. The response in O_3 concentrations to changes in precursor emissions (i.e. NO_x and VOCs) are often the goal of these models, and may be used to assess the impact of regulatory polices. These models operate over varying degrees of spatial resolution (gridded scales) and rely on the accuracy of emission inventories in order to predict ground-level air concentrations, deposition fluxes and AOT 40 values. As with other air pollutants (see Section 14.4.4), both Lagrangian and Eulerian modeling approaches have been utilized for regional photochemical oxidant modeling. The Lagrangian models are generally less computationally demanding, whereby chemical production and removal of pollutants occur within a parcel of air advected over the surface along a trajectory. In the Eulerian approach the chemistry and transport are determined over a fixed grid for a region of interest; the chemistry and various processes are solved simultaneously, akin to numerical weather models.

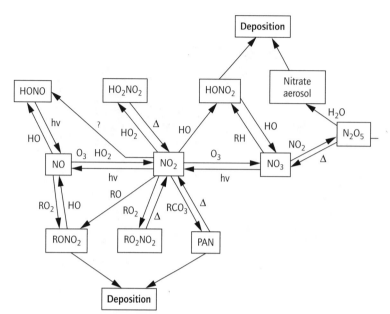

Fig. 14.5 A schematic illustrating the complex chemistry of nitrogen oxides in the atmosphere. (From Sturges 1991, fig. 6.2; with kind permission from Kluwer Academic Publishers.)

As an example, EMEP operates the Lagrangian ozone model, which follows parcels of air along 98-hour long trajectories, with emissions of NO_x, VOC, SO_2, and CO taken from the underlying grid according to national emission inventories. Trajectories are calculated every six hours to 740 arrival points covering the whole of Europe, for time periods of up to six months, using a polar-stereographic grid with a size of $150 \times 150\,km$ at $60°\,N$. This grid size is satisfactory for quantifying the transboundary exchange of O_3. However, it actually hides large "within-square" variability. This has led to the development of the EMEP Eulerian photochemistry model, which has been designed as an extension of the EMEP Eulerian acid deposition model. The grid scale in this model is on a much higher resolution of $50 \times 50\,km$. In general, both modeling approaches are able to reproduce the broad features of O_3 episodes, such as the maximum concentrations and regional distribution. However, comparison between daily maximum observed O_3 concentrations for selected monitoring stations and the model output from

the Eulerian model generally show a better correlation (Simpson *et al.* 1998). Plate 14.1 (facing p. 180) shows a comparison between the Lagrangian and Eulerian models for AOT 40 values generated over the period April–September 1996. While the spatial resolution is clearly different, the models also differ in other respects, such as initial boundary conditions and the amount of chemistry actually detailed.

14.4 DEPOSITION OF NITROGEN AND SULFUR ACROSS EUROPE: ACIDIFICATION AND EUTROPHICATION

The term "acid rain" gained wide use during the 1970s when the "acidification" of freshwater lakes in Scandinavia, upland waters in the UK and Adirondack lakes in the northeastern USA were linked to deposition of acidic species released from urban/industrial areas. The major component to this acid precipitation was sulfate (SO_4^{2-}), released initially as SO_2 from combustion sources using fuels containing relatively high sulfur content (i.e. coal and oil). The contribution from NO_x to overall acidity has in the past been less than that of sulfur for Europe and North America. However, with the decline of sulfur emissions, NO_x has become the dominant contributor in relative terms to acidification.

14.4.1 Atmospheric behavior of sulfur and nitrogen

In the atmosphere, SO_2, NO_x, and NH_3 undergo (further) oxidation, ultimately to form sulfuric and nitric acids. SO_2 may be oxidized directly by the OH radical to form sulfur trioxide (via an adduct that reacts with O_2), which in turn rapidly reacts with water to form sulfuric acid. This is outlined in reactions 14.6–14.8.

$$SO_2 + OH \rightarrow HSO_3 \qquad (14.6)$$

$$HSO_3 + O_2 \rightarrow SO_3 + HO_2 \qquad (14.7)$$

$$SO_3 + H_2O \rightarrow H_2SO_4 \qquad (14.8)$$

Importantly, SO_2 is also soluble in water, existing in the aqueous phase as the bisulfite ion (HSO_3^-). Aqueous-phase oxidation occurs through several routes, including reaction with dissolved O_3 or, more importantly, via hydrogen peroxide (H_2O_2), which is readily soluble in water. Oxidation of SO_2 in the aqueous phase is an important process within cloud droplets and it is estimated that between 48 and 84% of all SO_2 conversion to H_2SO_4 in the troposphere occurs within clouds (Langner & Rodhe 1991). The rate of oxidation of SO_2 by these various processes shows a wide variation. For example, the gas-phase oxidation via OH is on the order of 10 days, time for significant transport away from the initial source areas. However, the lifetime of SO_2 is actually on the order of a day or so, once other oxidation processes are taken into account, as well as removal via wet and dry deposition. Nevertheless, the formation of oxidation products such as sulfate is sufficiently slow to allow for wide atmospheric dispersal, with subse-

quent deposition of sulfate species occurring over a wide regional area.

Nitrogen oxide chemistry is complex and has been highlighted in Section 14.3. Figure 14.5 presents a variety of reactions that NO_x may undergo in the atmosphere. Many of these reactions are reversible, unlike those of sulfur, with the reactions dependent on temperature, sunlight, and oxidant concentrations, changes to which may favor the reverse of a reaction and formation of the parent NO_x again. Figure 14.5 is best interpreted by focusing on the three rows. The top row compounds are the inorganic nitrates formed from reactions of the middle row with O_3 and the reactive OH and HO_2 radicals. The bottom row comprises the generic organic nitrates, formed from the reaction of the middle row NO_x with alkyl radicals such as RO and RO_2 (produced via the reaction of hydrocarbons with HO_x). The most common and widely studied compound of the organic nitrates is PAN, although the occurrence and levels of other organic nitrates in both the polluted and unpolluted atmosphere are of growing interest (Fischer *et al.* 2000; Kastler *et al.* 2000). The formation of HNO_3 can occur directly via reaction of NO_2 with OH during daylight, or through the reaction of NO_2 with O_3 at night to form nitrogen trioxide (NO_3), which in turn can react with NO_2 to form nitrogen pentoxide (N_2O_5). This ultimately reacts with water to produce HNO_3. The atmospheric lifetime of NO_x through these reactions and via deposition is approximately 1–2 days, again allowing sufficient time for significant transport on a regional basis.

14.4.2 Contemporary emissions of sulfur and nitrogen

Within Europe the emissions of sulfur compounds and their ensuing deposition is dominated by anthropogenic sources. These emission sources are well characterized, with the signatory countries on the various UNECE Protocols producing a yearly emission inventory for the various pollutant types (http://projects.dnmi.no/~emep/index.html). Emissions of sulfur (as SO_2) across Europe ("European Community") were estimated to be ~12 Mt for 1995, compared to a similar amount for nitrogen (as NO_2). Naturally occurring sulfur compounds, notably dimethyl sulfide (DMS) emitted from oceans, contribute significantly to the global emission inventory, but apart from coastal areas are not considered to be significant for Europe. Emissions of nitrogen are dominated by both NO_x, emitted via traffic and stationary sources, and ammonia (NH_3), originating from livestock manure and its handling. Combined emissions of NO_2 and NH_3 greatly outweigh the emissions of SO_2. In addition, significant amounts of NO may be released from agricultural land, although detailed emissions for the whole of Europe are not yet available, and may be a significant source of NO_x in agricultural areas (Skiba et al. 1992). Estimated emissions of NH_3 are shown for the whole of Europe in Plate 14.2 (facing p. 180). These are based on the EMEP 50×50 km grid scale and are estimated for 1999. Release of reduced nitrogen such as NH_3 contributes to eutrophication as well as acid deposition. Emission estimates are therefore required in order to implement the critical load concept (outlined in the next section), based around the chemical buffering capacity of a catchment soil.

14.4.3 Acidification and eutrophication

It is now well known that deposition of protons, in association with sulfur and nitrogen species, has damaged ecosystems across Europe and the northeastern USA. The first effects of acid precipitation were observed on aquatic systems, resulting in the decline of natural fish stocks. The effect of lowered pH in catchment waters is the mobilization of metal ions from soil/rock strata, with particular concern over aluminum, which may be toxic to certain fish species at lowered pH. In addition, critical nutrients such as phosphate (PO_4^{2-}) (which is found at trace levels naturally) may be lost as they precipitate out with aluminum ions entering a lake system. The overall effect of acidification of natural waters is a reduction in species richness. Not only are certain fish species vulnerable but also amphibians and invertebrates. Loss of the latter may even affect bird populations as they

migrate to new areas from which to find an insect food source.

Damage to vegetation via acidic species is also well established and has been reviewed by Wellburn (1994). Direct effects on the foliar parts of vegetation have been observed under both simulated and real acid rain conditions, where impact of rainwater with lowered pH results in visible damage to the leaves of sensitive crop types, such as radishes, beets, and soya beans. The degree of injury is dependent on many factors, notably the concentration of acidity in the rain and the duration of contact, but it is also dependent on the temperature, humidity, wind turbulence, and leaf morphology. The effect of acidity can be exacerbated through the evaporation of water from retained droplets on the leaf surface, resulting in concentrating the acidity within the water drop and promoting damage to the waxy cuticle of the leaf (Unsworth 1984). On conifer needles cracking of the thin waxy plugs covering stomata may occur, allowing ingress of pollutants and pathogens into the leaf and water loss, and increasing the effects of frost damage.

It is unlikely, however, that these effects alone can be the cause of the large-scale declines observed in forested regions in both Europe (Krause et al. 1986) and North America (Johnson et al. 1986). Acid deposition can also disrupt the soil nutrient status due to changes in soil chemistry. Many of the essential elements or nutrients, particularly the cations of potassium (K^+), calcium (Ca^{2+}), and magnesium (Mg^{2+}), may be leached out of the soil via ion exchange, through increasing inputs of hydrogen ions (H^+). In effect, continual input of acid deposition reduces the buffering capacity of a soil, although this is highly dependent on the soil type and the water percolation rate. In a naturally acidic forest soil with low cation content, a large loss of nutrient ions may occur even though the pH change is minimal. Importantly, as the buffering capacity of the soil in the form of available Ca^{2+} and Mg^{2+} becomes exhausted, this may result in the release of Al^{3+} ions in exchange for H^+ ions (Matzner & Prenzel 1992). This mobilization of Al^{3+} ions has been suggested as toxic to the uptake mechanisms of fine root hairs and to the fungi in mycorrhizal association around the roots. This process of catchment degradation by increased acidic input through sulfur and nitrogen may contribute to forest decline, as well as a decline in other habitat types.

The shift away from sulfur in air pollution emissions to nitrogen-based pollutants, such as NO_x and NH_3, has led to increased deposition (both wet and dry) of nitrogen species onto sensitive ecosystems. High emissions of NH_3 are evident in areas with intensive animal husbandry, such as parts of Belgium and the Netherlands (see Plate 14.2, facing p. 180). Inorganic nitrogen is an essential nutrient for plant growth. However, the increased loading of nitrogen through atmospheric deposition can be detrimental to sensitive habitats. As a consequence, detrimental effects can occur on the major species comprising that habitat. A good example of this is presented by Wellburn (1994) for a typical forest system and is outlined here. Initially, increased nitrogen loading to a forest may result in a growth spurt, usually reflected in an increase in foliar vegetation, but which subsequently leads to an imbalance between proteins and carbohydrates. Valuable minerals such as K^+, Mg^{2+}, and PO_4^{2-} may be utilized in order to address this imbalance, whereby excess nitrogen can be lost as a range of toxic by-products, including amines, amides, and ammonium derivatives. These in turn can promote parasitic attack on the tree surfaces. Translocation of carbohydrates from the roots to other parts of the tree will have the effect of disrupting the symbiosis that often exists between the roots and specific species of fungi, resulting in reduced microbial activity in the soil. The diminishing effects on the roots and enhanced foliar vegetation (leaves) may render the trees more susceptible to wind-blow damage or excess water loss, effectively increasing the overall stress on the tree.

The inputs of excess nitrogen through deposition associated with regional air pollution may be leading to eutrophication within various ecosystems, such as grasslands (Wilson *et al.* 1995) and coastal waters (Jaworski *et al.* 1997), and is a growing area of concern. A comprehensive review by Smith *et al.* (1999) presents the impacts of eutrophication on freshwater, marine, and terrestrial ecosystems.

14.4.4   Critical loads and atmospheric deposition

The critical load concept is based on the limits of tolerance that a habitat can have to a certain pollutant type—above this limit substantial ecological harm could occur. Critical loads are essentially estimates of the quantities of pollutants that a habitat or ecosystem can absorb without ecological harm. The concept has been developed with acidification in mind, and limits of tolerance for sulfur and nitrogen acidity have been estimated for most natural and seminatural areas within Europe, based on soil quality (acid neutralizing capacity), vegetation type, etc. The 1994 UNECE Protocol on Further Reduction in Sulfur Emissions (see Table 14.1) incorporated the critical load concept as the basis for setting a new reduction target (Jenkins *et al.* 1998). In order for critical loads to be used, target loads need to be set for different areas in order to try to halt the acidification process. Therefore, target loads can be either higher or lower than the critical load values and may be determined by political agreement, taking into account socioeconomic factors as well as scientific findings. A review of the calculation procedure for target loads in the Netherlands has been presented by Van der Salm and de Vries (2001). They investigate soil quality criteria used to assess critical loads to Dutch forests, and investigate the effect of introducing a second criterion on the existing critical load calculations.

Application of critical loads (for either acidification or eutrophication) requires accurate assessments of pollutant deposition in order to estimate the exceedance of these loads for a certain area. Models such as the Regional Acidification Information and Simulation (RAINS) model can predict exceedances and allow the effect of emission reduction strategies to be assessed for a target area (Amann & Klaassen 1995). However, local variations in deposition may have a large effect on the accuracy of such models. This has been investigated by examining spatial variability in deposition across an EMEP grid square (150×150 km), through both statistical evaluation and comparing model results with actual deposition measurements. From their findings, Hirst et al. (2000) estimate the exceedance of acidification critical loads in Europe to be approximately double those estimated when local variation in deposition is ignored.

Understanding the deposition of sulfur and nitrogen species has therefore been at the forefront of scientific efforts to investigate the effects of regional air pollution. Importantly, it is necessary to define the removal rates of pollutants from the atmosphere in order to predict deposition (or more appropriately transboundary exchange) for both nitrogen and sulfur species. For Europe the modeling of the relationships between emissions (sources) and deposition (receptors) has been carried out under the EMEP framework, effectively to assess the impact of policy aimed at curbing pollutant emissions. Both Lagrangian and Eulerian models have been developed for the quantitative description of the distribution and deposition of sulfur and nitrogen. A recent example is EURAD (European Acid Deposition Model), an Eulerian model intended for the simulation of acidification, nutrification as well as photooxidants over Europe (Borrell et al. 1997). The model encompasses the whole troposphere and the lower stratosphere and covers the whole of Europe in a grid of 80×80 km or less. The model utilizes a sophisticated meteorological model taking meteorological data from the European Center for Medium Range Weather Forecasting. In addition, EURAD can provide the boundary conditions on which to run a nested model, the latter operating at much finer spatial resolutions. A nested model is one that runs inside another, and a good example of this is the European Modeling of Atmospheric Constituents (EUMAC) Zooming model (EZM). This details the meteorological and photochemical phenomena on an urban scale with grids as small as 5×5 km and can operate as part of EURAD (Moussiopoulos 1994).

This type of approach to modeling, i.e. setting the boundary conditions using a larger-scale model such as EURAD and a finer-scale model to depict local-scale processes, has also proved useful for the modeling of deposition to sensitive receptors (i.e. certain habitats), where assessments of critical loads is crucial. One such approach has been utilized in the Netherlands to assess the dry deposition of SO_2, using a local-scale deposition model incorporated in an EMEP long-range transport model (Erisman & Baldocchi 1994). Figure 14.6 illustrates this scheme for calculating fine-scale deposition fluxes using a combination of nested models.

14.5   ARCTIC HAZE

So far this chapter has focused on the industrialized regions of Europe and North America. Obviously these regions contain major sources of air pollution and the effects of these pollutants on a regional scale have been observed and reasonably well characterized. By no means are other parts of the globe free from regional air pollution problems: one only has to think of the forest fires of Indonesia in 1997, or the growing number of reports of regional-scale air pollution to come out of the People's Republic of China (e.g. Zhang et al. 1998; Tao & Feng 2000). However, an area that generally receives less attention is the Arctic. This region experiences the haze phenomenon; that is, pollution that occurs in the lower atmosphere of the Arctic, which builds up during the winter months and is usually observed during March/April after polar sunrise. This has been the subject of scientific study over the past 20 years or so and dispels the myth that the Arctic is a pristine area beyond the reach of industrialized pollution. It is now well established that intrusion of pollutants to this region occurs on a seasonal basis, with polluted air being carried into the Arctic via long-range atmospheric transport (LRT) from southerly source regions. Aside from the aesthetic impact of Arctic haze, the deposition of pollutants and their subsequent effects on the fragile polar ecosystem are an area of real concern. This has resulted in international

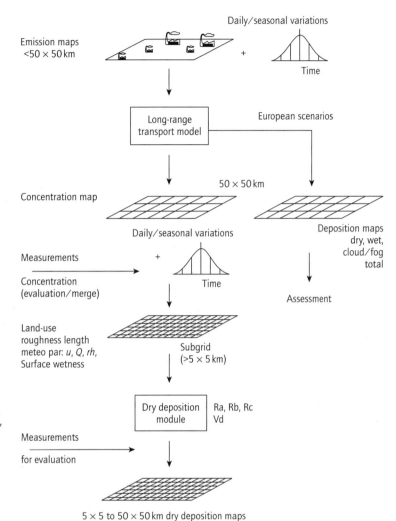

Fig. 14.6 Flow diagram depicting the incorporation of a high resolution model within a coarse spatial model, resulting in the calculation of dry deposition of SO₂ to simple surfaces. Note that R_a, R_b, R_c relate to "resistances" based on empirical data for deposition modeling to different surfaces, i.e. vegetation. (From Erisman & Baldocchi 1994; © Munksgaard International Publishers Ltd, Copenhagen.)

programs to investigate and understand pollutant impact on the ecosystem, the foodchain, and ultimately the indigenous Arctic peoples (e.g. AMAP 1997).

Unlike the Antarctic, the Arctic is entirely surrounded by industrialized regions with high population densities and associated air pollution problems. Furthermore, the northern polar region also contains its own industrialized areas, notably in Fenno-Scandinavia as well as Russia, and sup-

ports indigenous peoples who rely on its various habitats for their survival and culture. Point pollution events, such as hydrogen bomb tests in the Russian Arctic (1960s), oil spills, and mining activity, are ongoing and fairly frequent events. However, much of the pollution to impact the Arctic is believed to be atmospherically derived through LRT (AMAP 1997; CACAR 1997). The term "Arctic haze" was coined in the 1950s by US pilots, who noted a brown haze at different layers in the lower troposphere during polar sunrise (Mitchell 1956; Raatz 1984). This problem was already known to native people, who called it "poo-jok" or dark haze. Subsequent scientific investigation revealed that the haze was derived through transport of airborne pollutants from source regions further south, particularly Eurasia (Rahn & McCaffrey 1980). The haze consists of combustion products such as carbonaceous particulate matter, heavy metals, and basic air pollutants such as SO_x and NO_x. Its sources and composition have been the subject of a review by Barrie (1986).

14.5.1   Pollutant transport into the Arctic and spatial distribution

The prevailing near-surface meteorology during the Arctic winter is largely to account for the haze phenomenon. During winter in the northern hemisphere, near surface Arctic meteorology is dominated by four semipermanent atmospheric pressure systems: two centers of low pressure, situated over the Atlantic and Pacific oceans (Icelandic and Aleutian lows), and two high-pressure systems, situated over the North American and Eurasian landmasses (Raatz 1991). During winter the high-pressure systems develop and broaden, resulting in intense airflow between the mid-latitudes and the Arctic. In the summer these pressure systems diminish, with a corresponding reduction of airflow into the Arctic. To illustrate this, Fig. 14.7 displays a stylized, typical winter surface-pressure map for the northern hemisphere, highlighting the four dominant pressure cells, along with the accompanying airflow movements. Note that the air mass circulation of the Asiatic "high" and the Icelandic "low" complement each other, resulting in enhanced air mass movement into the Arctic and subsequent transfer of airborne pollutants from mid-latitudinal sources. Due to these well developed pressure systems, the wintertime circulation in the Arctic and the mid-latitudes is intense, whereas during the summer, due to weakening pressure gradients, the near-surface circulation is considerably slower. The occurrence of pollutants in the Arctic air mass is brought about by "episodic" air mass movement favored by the prevailing meteorology outlined above. These air masses carry pollutant burdens that are representative of the sources over which the air parcel passed and result in the rapid and efficient transfer of both gaseous and particle pollutant loads to the Arctic.

An important feature of the lower Arctic troposphere is the existence of intense temperature inversions. These persist over the whole Arctic region during the winter and over snow- and ice-covered surfaces during the summer. The increase in temperature with height effectively serves as a barrier to vertical turbulent mixing for pollutants that enter the Arctic. In winter, surface inversions form during anticyclonic conditions as a result of energy loss from the snow-covered surfaces in the absence of incoming solar radiation. Typically, surface inversions may extend up to 2000 m accompanied by a temperature increase of ~10°C from the ground to the inversion top. These inversions may affect the distribution of pollutants in a particular air mass. For example, in a well mixed air mass, pollutants in the upper layers may become decoupled from the pollutants in the lower inversion layer, which are then subject to different removal rates such as dry deposition. This can result in the occurrence of pollutant "bands" in the lower troposphere. Pollutant bands may also be enhanced by the occurrence of lifted inversions. These occur when a turbulent mixing layer separates the inversion base from the ground. The occurrence of lifted inversions, however, is more frequent during the summer than the winter, when the Arctic region is impacted by warm moist air.

Probably the most important factor influencing Arctic haze is the slow removal rates of pollutants once transported into the Arctic. The lack of light over the winter months, coupled to the lack of precipitation and strong temperature inversions, reduces the dispersal of pollutants, resulting in long atmospheric lifetimes of air pollutants in this region. The spatial distribution of Arctic haze is widespread across the whole Arctic region, with levels of SO_4^{2-} (the major aerosol component) generally showing higher levels on the Eurasian side of the Arctic and concentrations declining toward

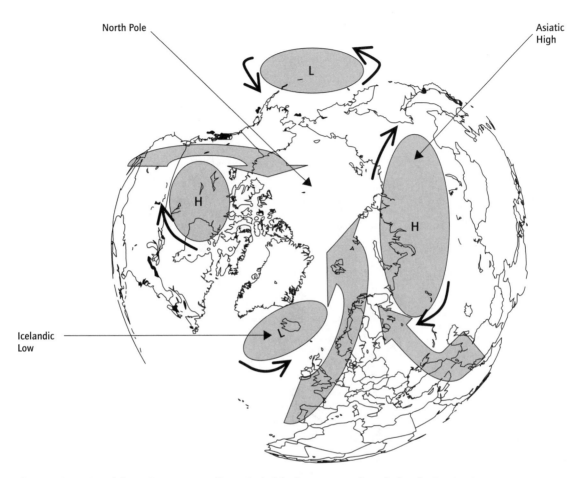

Fig. 14.7 Location of the major pressure cells, typical of the lower atmosphere during the Arctic winter.

the high Arctic and northern Canada. A minimum or "saddle" is observed in the region of the Greenland–North American Arctic, which falls between the influence of the Eurasian and eastern North American sources. Increased precipitation in this area due to the storminess of the eastern Canadian and Greenland Arctic also has the effect of decreasing the pollutant loading. Figure 14.8 shows the spatial distribution of the arithmetic mean SO_4^{2-} air concentrations ($\mu g\,m^{-3}$) for the period January–April 1980. Superimposed are the three main sources of air for the North American airshed.

14.5.2 Physical and chemical properties of Arctic haze

Arctic haze is actually a mixture of gases and aerosol, although most of the research has focused on the aerosol fraction due to its role in reducing visibility. The following information is summarized from Sturges (1991) and describes the typical "haze" composition. Aerosol particles are most numerous in the diameter range (D_p) of 0.005–0.2 μm, with the total number concentration ranging from ~10 to 4000 cm^{-3} and a geometric average of 200–300 cm^{-3}. The shape of the number size dis-

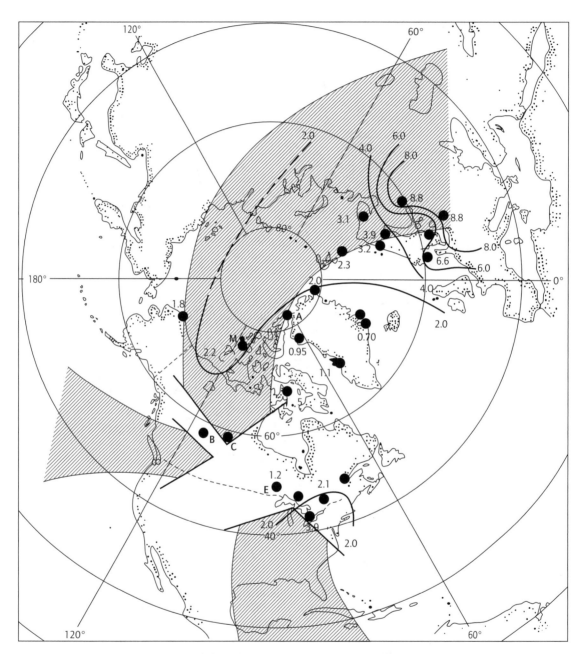

Fig. 14.8 A stylized diagram representing the three major sources of air for the North American airshed (Arctic, Pacific, Caribbean), including the spatial distribution of SO_4^{2-} air concentrations ($\mu g\,m^{-3}$) for the period January–April. (From Sturges 1991, fig. 6.5; with kind permission from Kluwer Academic Publishers.)

tribution is highly variable below 0.1 μm, largely due to the presence (or absence) of nucleation mode particles formed from the condensation of reactive gases such as SO_2. Most of the aerosol occurs in the **accumulation mode size range** (D_p 0.1–1 μm) responsible for light scattering and the related haze. Above 0.2 μm the number concentration decreases rapidly with increasing diameter, although the mass (or volume) of aerosol during the winter is actually concentrated in the larger particles (D_p 0.1–0.2 μm), with a significant contribution from the **coarse particle mode** (D_p 1–10 μm). The number concentration for this size range, however, is approximately four orders of magnitude less than the accumulation mode. The presence of the **giant particle mode** (D_p > 10 μm) in haze aerosol has also been reported. Both the coarse and giant particle modes are believed to originate from aeolian crustal material such as soil and clay minerals, as well as sea salt to a lesser extent.

The major component of haze aerosol is SO_4^{2-} and to a lesser degree ammonium bisulfate (NH_4HSO_4). Most of the sulfur (~75%) enters the Arctic as SO_2, while the remainder (~25%) is already present as the oxidized SO_4^{2-}. This is important with respect to the nature of the pollution, as oxidation of SO_2 to SO_4^{2-} requires oxidants in the form of OH for gas-phase oxidation or H_2O_2 or O_3 for dissolved aqueous-phase oxidation. These oxidants, however, are largely absent during the Arctic winter due to the lack of sunlight driving photochemical processes, with a result that the ratio of SO_2/SO_4^{2-} remains high throughout the winter. Following polar sunrise and the increase in incident sunlight, concentrations of SO_2 decrease more rapidly than those of SO_4^{-2}. Gas-phase oxidation of SO_2 to SO_4^{2-} during the spring months has been inferred through numerous measurements, where high SO_2 levels have been associated with nucleation mode particles, derived through the oxidation of SO_2 to H_2SO_4. The average composition of the soluble fraction of aerosol (both fine and coarse particle sizes) associated with the haze is displayed in Fig. 14.9. Non-sea-salt sulfate and its associated ammonium ion dominate the mass of fine particles, while sea salt (Na^+, Cl^-) dominates

the coarser-size particles (particularly at Arctic coastal sites). The sulfates contribute significantly to the accumulation mode particles that are involved in the scattering of sunlight and hence the reduction in visibility resulting in haziness.

Ratios of SO_4^{2-} with trace metals such as vanadium (V) have allowed comparisons to other continental aerosols to distinguish source types and regions. Barrie and Hoff (1984) used the ratios of SO_4^{2-}/V and SO_4^{2-}/SO_2 in both Arctic and mid-latitude source regions as an input to a simple

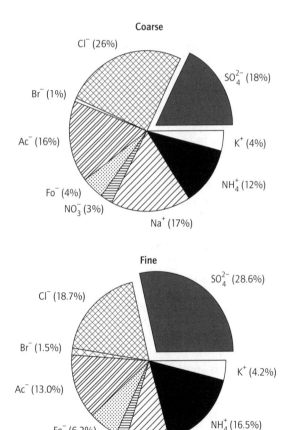

Fig. 14.9 Average composition of coarse (2–10 μm) and fine (<2 μm) aerosols at Barrow, Alaska, in March and April 1986. (From Sturges 1991, fig. 6.7; with kind permission from Kluwer Academic Publishers.)

chemical transport model and found that the mean oxidation rate of SO_2 between sources and the Canadian Arctic was 0.04–0.1% h^{-1} during the darkest part of winter (December to February) and increased to 0.2% h^{-1} in April after polar sunrise.

Reactive nitrate (NO_3^-) makes a far smaller contribution to haze aerosol than sulfate, and on a molar basis can be an order of magnitude lower than concentrations of sulfur oxides. A seasonal trend is apparent that is similar to that of sulfate, with high levels during the winter, although a similar spring maximum does not occur. Importantly, much of the inorganic NO_x is actually present as NO_y, particularly at higher altitudes, with a large percentage of oxidized nitrogen occurring as organic forms such as alkyl nitrates, peroxy-nitrate and PAN.

Carbonaceous material is also a major component of Arctic haze. This has been reported as soot, elemental carbon, or graphitic carbon, although these labels imply some structural form to the carbon. Therefore, it is more appropriate to term this carbonaceous material simply as "black carbon." Like many of the other haze constituents, levels are highest during the winter months and closely match those of SO_4^{2-}, with a peak in concentrations occurring between January and March (Hopper *et al.* 1994). Good correlations have been observed between black carbon and CO_2 at several locations in the Arctic, indicating the influence of combustion sources on haze pollution. Combustion-derived trace organic compounds have also been observed, such as the polycyclic aromatic hydrocarbons (PAHs). These semivolatile persistent organic pollutants have been observed in both gaseous and particle phases during the Arctic winter (Halsall *et al.* 1997). Examination of the reactive/stable isomers for selected PAHs (i.e. B[a]P/B[e]P) reveal a ratio that is close to unity throughout much of the winter season, a ratio that is not dissimilar to urban observations. This reveals that depletion of the reactive isomer (through either OH attack or some particle surface-mediated process) is greatly reduced during the dark winter period. Removal from the Arctic atmosphere is therefore likely to occur largely through deposition to Arctic surfaces rather than by degradation in the atmosphere, resulting ultimately in the exposure of Arctic flora and fauna to these toxicologically relevant contaminants.

14.5.3   Deposition

Deposition of airborne pollutants in the Arctic has been assessed largely through snow pack and ice-core sampling. Direct atmospheric sampling of precipitation is difficult, particularly during the winter, due to the low levels of precipitation (especially for the high Arctic) and the problems associated with blowing snow. There have been numerous snow pack chemistry surveys conducted in different regions of the Arctic, which have been reviewed by Barrie (1986). These earlier surveys revealed that surface winter snow in many parts of the Arctic has a relatively low pH (4.5–5.6), with the predominant anions being SO_4^{2-} and NO_3^-. The snow pack is therefore slightly acidic, with most of the acidity accounted for by SO_4^{2-} and NO_3^-. In an earlier study by Rahn and McCaffery (1979), the concentrations of various trace elements were determined in the winter snowpack of northern Alaska. It was found that the snow pack was enriched in zinc (Zn), cadmium (Cd), and mercury (Hg) relative to the crustal material and that anthropogenic sources were most likely to account for this.

Ice cores have also demonstrated the importance of the atmospheric deposition of anthropogenic pollution. Cores taken on the Agassiz ice cap on Ellesmere Island in the Canadian Arctic show that the conductivity and acidity of the ice are well correlated, and undergo a strong seasonal variation that mirrors that of Arctic air pollution (e.g. Barrie 1985; Peters *et al.* 1995). Deposition trends of PAHs taken from the Greenland ice cap have shown a dramatic increase in concentration over the past 100 years that correlates well with world petroleum production. PAH concentrations relative to plant fatty acids (C_{20}–C_{32}) demonstrate that contributions of anthropogenic PAHs have significantly increased since the 1930s (Kawamura *et al.* 1994).

14.6 CURRENT TRENDS AND UNCERTAINTIES IN REGIONAL AIR POLLUTION

Scientific advances with respect to air pollution, particularly in predictive modeling, have clearly shaped recent UNECE Protocols as well as national strategies for improving air quality (the two being closely linked). Importantly, monitoring networks are now well established in both Europe and North America, with 10-year or longer data sets existing for many air pollutants. Indeed, the list of pollutants has also increased, with VOCs now included to an extent in the EMEP program, as well as a selection of persistent organic pollutants. For Europe, long-term monitoring data have been subject to trend analysis. For O_3 there appears to be both decreasing and increasing trends depending on location. For example, in the Netherlands a downward trend of 0.4–1.6% per year was evident for nine sites between 1981 and 1994, which is in contrast to the other continental EMEP sites (PORG 1997). The monitoring station at Preilla, Lithuania, actually shows a clear upward trend of 2.6% per year between 1982 and 1993. This apparent difference in trends for different parts of Europe may be down to several factors. Since the 1980s controls have been implemented to reduce the O_3 precursors of VOC and NO_x (see Table 14.1). These controls vary from country to country, but for some countries it is estimated that there have been significant reductions in these precursors that will influence peak O_3 levels (Berge *et al.* 1994). The controls on anthropogenic releases of VOCs are likely to have been more effective than those on NO_x, affecting the VOC/NO_x ratio in certain parts of Europe and hence the rate of O_3 formation. Furthermore, from study of trends in O_3 and their precursors, it is also evident that large-scale prevailing meteorology plays a major role in influencing the year-on-year trend. A poor summer typified by cloudy conditions and relatively low temperatures will result in lower average O_3 concentrations than a warmer summer (PORG 1997). Generally, however, peak ozone concentrations in Europe are declining, while long-term average concentrations are increasing.

Importantly, in order to assess the future trends for O_3, emission inventories for precursors need to be improved, particularly on the spatial scales on which photooxidant models can operate. Speciated VOC inventories are clearly lacking over most of Europe and the quantitative role played by biogenic VOCs such as isoprene and monoterpenes on O_3 yield is an ongoing area of research. Detailed emissions inventories for the other regional pollutants outlined in this chapter (such as NH_3) are also required to assess ongoing acidification and nutrification. At the other end of the scale, deposition estimates at the fine spatial scale need to be improved, particularly with respect to the critical loads concept. An essential prerequisite for this approach is that the pollutant load (deposition) can be quantified with accuracy and over a sufficient resolution in both time and space. This is particularly important with respect to model development for dealing with complex terrain such as hills and forests, which have a significant influence on the deposition of airborne pollutants.

Finally, it is only controls implemented to abate regional air pollution in temperate latitudes that will result in the decline of air pollution in remote locations such as the Arctic. The focus for this region has been to quantify the contribution of various air pollutants from source areas such as Russia. In a recent study, potential source contribution functions (PSCF) have been developed by combining aerosol data with air parcel back trajectories to identify source areas and preferred transport pathways. For the winter, high PSCF values for black carbon as well as for the aerosol light scattering coefficient ($\lambda = 450\,nm$) were related to industrial sectors in Eurasia (Polissar *et al.* 2001).

REFERENCES

Amann, M. & Klaassen, G. (1995) Cost-effective strategies for reducing nitrogen deposition in Europe. *Journal of Environmental Management* **43**, 289–311.

AMAP (1997) *Arctic Pollution Issues: A State of the Arctic Environment Report*. Arctic Monitoring and Assessment Programme, Oslo.

Barrie, L.A. (1985) Atmospheric particles: their physical/chemical characteristics and deposition processes relevant to the chemical composition to glaciers. *Annals of Glaciology* 7, 100–8.

Barrie, L.A. (1986) Arctic air pollution: an overview of current knowledge. *Atmospheric Environment* **20**, 643–63.

Barrie, L.A. & Hoff, R.M. (1984) The oxidation rate and resident time of sulphur dioxide in the arctic atmosphere. *Atmospheric Environment* **18**, 2711–22.

Berge, E., Styve, H. & Simpson, D. (1994) *Status of the Emission Data at MSC-W*. EMEP MSCW Report 2/95. The Norwegian Meteorological Institute, Oslo.

Borrell, P., Builtjes, P.J.H., Grennfelt, P. & Hov, O. (eds) (1997) *Photo-oxidants, Acidification and Tools: Policy Applications of EUROTRAC Results*. The Report of the EUROTRAC Application Project. Springer-Verlag, Berlin.

CACAR (1997) *Canadian Arctic Contaminants Assessment Report. Northern Contaminants Program*. Department of Indian and Northern Affairs, Ottawa.

Cox, R.A., Eggleton, A.E.J., Derwent, R.G., Lovelock, J.E. & Pack, D.H. (1975) Long-range transport of photochemical ozone in north-western Europe. *Nature* **255**, 118–21.

Derwent, R.G., Jenkin, M.E. & Saunders, S.M. (1996) Photochemical ozone creation potentials for a large number of reactive hydrocarbons under European conditions. *Atmospheric Environment* **30**, 189–200.

Dollard, G., Fowler, D., Smith, R.I., Hjellbrekke, A.-G., Uhse, K. & Wallasch, M. (1995) Ozone measurements in Europe. *Water, Soil and Air Pollution* **85**, 1949–54.

Erisman, J.W. & Baldocchi, M. (1994) Modeling dry deposition of SO_2. *Tellus* **46B**, 159–71.

Fischer, R.G., Kastler, J. & Ballschmiter, K. (2000) Levels and patterns of alkyl nitrates and halocarbons in the air over the Atlantic Ocean. *Journal of Geophysical Research* 105, 14,473–94.

Fowler, D., Cape, J.N., Coyle, M. *et al.* (1999) Modeling photochemical oxidant formation, transport, deposition and exposure of terrestrial ecosystems. *Environmental Pollution* **100**, 43–55.

Fuhrer, J. (1996) The critical levels for effects of ozone on crops, and the transfer to mapping. In Karenkampi, L. & Skarby, L. (eds.) *Critical Levels for Ozone in Europe: Testing and Finalising the Concept*. University of Kuopio, Kuopio.

Grunhage, L., Jager, H.J., Haenel, H.D., Lopmeier, F.J. & Hanewald, K. (1999) The European crtical levels for ozone: improving their usage. *Environmental Pollution* **105**, 163–73.

Halsall, C.J., Barrie, L.A., Fellin, P. *et al.* (1997) Spatial and temporal variation of polycyclic aromatic hydrocarbons in the Arctic atmosphere. *Environmental Science and Technology* **31**, 3593–9.

Herman, F., Smidt, S., Huber, S., Englisch, M. & Knoflacher, M. (2001) Evaluation of pollution-related stress factors for forest ecosystems in central Europe. *Environmental Science and Pollution Research* **8**, 231–42.

Heck, W.W., Taylor, O.C. & Tingey, D.T. (1988) *Assessment of Crop Loss from Air Pollutants*. Elsevier, New York.

Hirst, D., Kåresen, K., Høst, G. & Posch, M. (2000) Estimating the exceedance of critical loads in Europe by considering local variability in deposition. *Atmospheric Environment* **34**, 3789–800.

Hopper, J.F., Worthy, D.E.J., Barrie, L.A. & Trivett, N.B.A. (1994) Atmospheric observations of aerosol black carbon, carbon dioxide and methane in the high Arctic. *Atmospheric Environment* **28**, 3047–54.

Isaksen, I.S.A., Hov, O. & Hesstvedt, E. (1978) Ozone generation over rural areas. *Environmental Science and Technology* **12**, 1279–84.

Jaworski, N.A., Howarth, R.W. & Hetling, L.J. (1997) Atmospheric deposition of nitrogen oxides onto the landscape contributes to coastal eutrophication in the north-east United States. *Environmental Science and Technology* **31**, 1995–2004.

Jenkins, A., Helliwell, R.C., Swingewood, P.J., Sefton, C., Renshaw, M. & Ferrier, R.C. (1998) Will reduced sulphur emissions under the second sulphur protocol lead to recovery of acid sensitive sites in the UK? *Environmental Pollution* **99**, 309–18.

Johnson, A.H., Friedland, A.J. & Dushoff, J.G. (1986) Recent and historic red spruce mortality: evidence of climatic influence. *Water, Air and Soil Pollution* **30**, 319–30.

Kastler, J., Jarman, W. & Ballschmiter, K. (2000) Multifunctional organic nitrates as constituents in European and US urban photo-smog. *Fresenius Journal of Analytical Chemistry* **368**, 244–9.

Kawamura, K., Suzuki, I., Fujii, Y. & Watanabe, O. (1994) Ice core record of polycyclic aromatic hydrocarbons over the last 400 years. *Naturwissenschaften* **81**, 501–5.

Kley, D., Kleinmann, M., Sanderman, H. & Krupa, S. (1999) Photochemical oxidants: the state of the science. *Environmental Pollution* **100**, 19–42.

Krause, G.H.M., Arndt, U., Brandt, C.J., Bucher, J., Kenk, G. & Matzner, E. (1986) Forest decline in Europe:

development and possible causes. *Water, Air and Soil Pollution* **31**, 647–88.

Krupa, S.V. & Kickert, R.N. (1989) Ambient ozone (O_3) and adverse crop response. *Environmental Reviews* **5**, 55–77.

Krupa, S.V., Tonneijck, A.E.G. & Manning, W.J. (1989) Ozone. In Flagler, R.B. (ed.), *Recognition of Air Pollution Injury to Vegetation: A Pictorial Atlas.* Air and Waste Management Association, Pittsburgh.

Langner, J. & Rodhe, H. (1991) A global three-dimensional model of the tropospheric sulfur cycle. *Journal of Atmospheric Chemistry* **13**, 255–63.

Larsen, R.I., McDonnell, W.F., Horstmann, D.H. & Folinsbee, L.J. (1991) An air quality data analysis system for interrelating effects, standards and needed source reductions, part II. A log-normal model relating human lung function decrease to O_3 exposure. *Journal of Air and Waste Management Association* **41**, 455–9.

Lee, E.H., Tingey, D.T. & Hogsett, W.T. (1988) Evaluation of ozone exposure indices in exposure–response modelling. *Environmental Pollution* **54**, 43–62.

Liu, S.C., Trainer, M., Fehsenfeld, F.C. *et al.* (1987) Ozone production in the rural troposphere and the implications for regional and global ozone distribution. *Journal of Geophysical Research* **92**, 4191–207.

Manning, W.J. & Krupa, S.V. (1992) Experimental methodology for studying the effects of crops and trees. In Lefohn, A.S. (ed.), *Surface Level Ozone Exposures and Their Effects on Vegetation.* Lewis Publishers, Chelsea, MI.

Matzner, E. & Prenzel, J. (1992) Acid deposition in the German Solling area: effects on soil solution chemistry and Al mobilization. *Water, Air and Soil Pollution* **61**, 221–34.

Mitchell, M. (1956) Visual range in the polar regions with particular reference to the Alaskan Arctic. *Journal of Atmospheric Physics* (special supplement), 195–211.

Moussiopoulos, N. (1994) *The EUMAC Zooming Model: Model Structure and Applications.* EUROTRAC ISS, Garmisch-Partenkirchen.

Oehme, F.W., Coppock, R.W., Mostrom, M.S. & Khan, A.A. (1996) A review of the toxicology of air pollutants: toxicology of chemical mixtures. *Veterinary and Human Toxicology* **38**, 371–7.

Peters A.J., Gregor, D.J., Teixeira, C.F., Jones, N.P. & Spencer, C. (1995) The recent depositional trend of polycyclic aromatic hydrocarbons and elemental carbon to the Agassiz Ice Cap, Ellesmere Isand, Canada. *Science of the Total Environment* **160/1**, 267–79.

Polissar, A.V., Hopke, P.K. & Harris, J.M. (2001) Source regions for atmospheric aerosol measured at Barrow, Alaska. *Environmental Science and Technology* **35**, 4214–26.

PORG (1993) *Ozone in the United Kingdom.* United Kingdom Photochemical Oxidants Review Group. Department of the Environment, London.

PORG (1997) *Ozone in the United Kingdom.* United Kingdom Photochemical Oxidants Review Group. Department of the Environment Transport and Regions, London.

Raatz, W.E. (1984) Observations of "arctic haze" during the "Ptarmigan" weather reconnaissance flights, 1948–1961. *Tellus* **36B**, 126–36.

Raatz, W.E. (1991). The climatology and meteorology of Arctic air pollution. In Sturges, W.T. (ed.), *Pollution of the Arctic Atmosphere.* Elsevier Science, Harlow.

Rahn, K.A. & McCaffrey, R.J. (1979) Compositional differences between arctic aerosol and snow *Nature* **280**, 479–80.

Rahn, K.A. & McCaffrey, R.J. (1980) On the origin and transport of the winter Arctic aerosol. *Annals of the New York Academy of Sciences* **338**, 486–503.

Schwartz, J. (1989) Lung function and chronic exposure to air pollution: a cross sectional analysis of NHANES II. *Environmental Research* **50**, 309–21.

Skiba, U., Hargreaves, K.J., Smith, K.A. & Fowler, D. (1992) Fluxes of nitric and nitrous oxides from agricultural soils in a cool temperate climate. *Atmospheric Environment* **26A**, 2477–85.

Simpson, D., Altenstedt, J. & Hjellbrekke, A.G. (1998) The lagrangian oxidant model: status and multi-annual evaluation. In *Transboundary Photo-oxidant Air Pollution in Europe: Calculations of Tropospheric Ozone and Comparison with Observations* (EMEP MSC-W Status Report No. 5–29). The Norwegian Meteorological Institute, Oslo.

Smith, V.H., Tilman, G.D. & Nekola, J.C. (1999) Eutrophication: impacts of excess nutrient inputs on freshwater, marine and terrestrial ecosystems. *Environmental Pollution* **100**, 179–96.

Sturges, W.T. (ed.) (1991) *Pollution of the Arctic Atmosphere.* Elsevier Science, London.

Tao, F.I. & Feng, Z.W. (2000) Critical loads of SO_2 dry deposition and their exceedence in South China. *Water, Air and Soil Pollution* **124**, 499–538.

Unsworth, M.H. (1984) Evaporation from forests in cloud enhances the effects of acid deposition. *Nature* **312**, 262–4.

Van der Heyden, D., Skelly, J., Innes, J. *et al.* (2001) Ozone exposure thresholds and foliar injury on forest plants in Switzerland. *Environmental Pollution* **111**, 321–31.

Van der Salm, C. & de Vries, W. (2001) A review of the calculation procedure for critical acid loads for terrestrial ecosystems. *Science of the Total Environment* **271**, 11–25.

Vestreng, V. (2001) Emission data reported to UN-ECE/EMEP: evaluation of the spatial distribution of emissions. EMEP/MSC-W. Note 1/01. ISSN 0332-9879.

Wellburn, A. (1994) *Air Pollution and Climate Change: The Biological Impact*, 2nd edn. Longman Scientific & Technical, New York.

White, W.H., Patterson, D.E. & Wilson, W.E. Jr (1983) Urban exports to the nonurban troposphere: Results from Project MISTT. *Journal of Geophysical Research* **88**, 10,745–52.

Wilson, E.J., Wells, T.C.E. & Sparks, T.H. (1995) Are calcareous grasslands in the UK under threat from nitrogen deposition. An experimental determination of a critical load? *Journal of Ecology* **83**, 823–32.

Zhang, Y., Shao, K., Tang, X. & Li, J. (1998) The study of urban photochemical smog pollution in China. *Acta Scientiarum Naturalium Universitatis Pekinensis* **34**, 392–400.

15 Urban-Scale Air Pollution

JES FENGER

15.1 INTRODUCTION

Urban regions are by their nature concentrations of humans, materials, and activities. They therefore exhibit both the highest levels of pollution and the largest targets of direct impacts. Air pollution is, however, enacted on all geographic and temporal scales, ranging from strictly "here and now" problems related to various aspects of human health and well-being, over regional phenomena like acidification and forest die-back with a time horizon of decades, to global phenomena, which in the course of centuries have impacts for humans and nature over the entire globe.

Urban-scale air pollution is a vast subject, with different socioeconomic aspects in different parts of the world—sometimes even within a specific region. And it cannot be treated in isolation, but must be seen in connection with air pollution on other scales—both when large-scale phenomena influence the urban air quality and when urban sources, via long-range transport, give rise to large-scale pollution impacts.

15.1.1 Urban air pollution as a chain of events

In a systematic treatment pollution can be conceived as a chain of decisions and processes. An example for urban-scale air pollution is shown in Fig. 15.1.

The chain starts with a series of (often unplanned) decisions about activities related to space heating, traffic, industry, etc. These activities lead to emission of pollutants, which are transported, dispersed, and sometimes transformed in the atmosphere. Depending upon the conditions the result is levels (concentrations) of pollutants that have generally unwanted, impacts. The impacts are evaluated and result in attempts at mitigation, which may be applied at every link of the chain in the form of reduction of emissions, facilitation of dispersion, or counteraction of impacts.

On the urban scale there is often, but not always, a reasonable relation between local emissions and resulting local pollution levels. In contrast to regional and global pollution, efforts to mitigate urban-scale air pollution may therefore generally have direct and observable effects.

15.1.2 Air pollution through the ages

Initially air pollution was an indoor phenomenon, caused by open fires without controlled venting. As an example it has been shown (Skov *et al.* 2000) that the indoor exposure to nitrogen oxides and organic compounds in a reconstructed Iron Age house must have been comparable to, or even higher than, that in a modern city. Similar observations have been made in present-day dwellings in the developing world (Smith 1988, 1993). Ambient urban air pollution, however, is as old as cities, and literature as well as historical records testify that the problems were extensive. Attempts at regulation have been documented for at least 2000 years (e.g. Mamane 1987). Some problems may even have been underestimated, since generally people

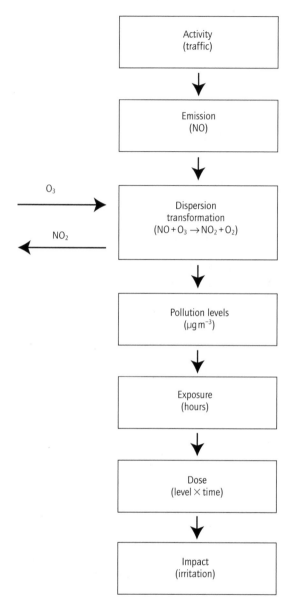

Fig. 15.1 Pollution is a chain of decisions and processes. This also applies to urban air pollution, and is here shown for traffic emission of NO in a North European City. In practice many such chains interact.

were less critical about their living conditions, and they had no means of evaluating long-term impacts of, for example, carcinogens. Further, many of the records concern aesthetic impacts in the form of smell and soiling, which are not deleterious to health in themselves.

Finally, it should be recognized that up to the Second World War many people had an ambivalent attitude toward pollution, which to some extent was perceived as a symbol of wealth and growth. Thus advertisements showed pictures of fuming chimney stacks (Fig. 15.2) and automobiles with visible exhaust — images hardly anyone would cultivate today!

Semiquantitative evaluations of early urban air pollution have been attempted in various ways, e.g. via records of material damage and impacts on human health and vegetation. Also, simplified dispersion modeling is possible when the consumption of fuels and raw materials within a confined area is reasonably well known (Brimblecombe 1977, 1987).

Some direct measurements of air pollutants were carried out by scientists and amateur enthusiasts in the nineteenth century, but systematic and official investigations with continuous time series are of fairly recent date.

15.1.3 *Global growth and urbanization*

Human societies as such were first formed with the advent of agriculture about 10,000 years ago. The global population was then hardly more than five million. Until the start of industrialization and improved hygiene in the nineteenth century the population slowly grew to one billion, but then the rate increased dramatically. At the beginning of the twentieth century the number had doubled to two billion. Just after the Second World War the world population was about 2.5 billion, and in the past 50 years it has more than doubled and has now passed six billion. Some projections suggest that the number will level off at about 10 billion in the second half of this century, with the sharpest rise in the (now) developing regions, especially Africa.

Since 1950 global urbanization, defined as the

Fig. 15.2 Advertisement for a British exhibition in Copenhagen 1932. Obviously before flue gas cleaning became fashionable.

fraction of people living in settlements above 2000 inhabitants, has risen from below 30 to 45%. In the more developed countries it is now around 75% and in the less developed 37% (Population Reference Bureau 1999). In 1950 there were only eight cities with inhabitant numbers above five million, in 1990 there were about 35 (WHO/UNEP 1992), and now there are probably more than 40, about half of them situated in East Asia. About 20 urban regions each have populations above 10 million people. By 2015 the number may have risen to 26, and for the foreseeable future the rate of megacity growth in the developing world will far outpace that in the developed countries (Lynn 1999).

Today cities cover 2% of the Earth's surface, and account for roughly 78% of the carbon emissions from human activities, 76% of the industrial wood use, and 60% of the water tapped for use by people (O'Meara 1999). These global trends will continue in the next decades (WHO/UNEP 1992). Particu-

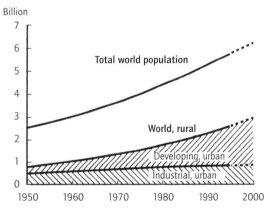

Fig. 15.3 The increase since 1950 of the total world population and the urban population in developed and developing countries. (Based on material compiled in Brown *et al.* 1998.)

larly in the developing countries, there is significant migration of people from the countryside to the towns—both because of the mechanization of farming and because of opportunities in new industries and public services. This may lead to a further growth in urbanization (UNEP 1997). In China alone more than 100 million people have been reported to move around in search for work, and in Beijing the authorities appear to actively prevent their permanent settlement. In Asia, Latin America, and Africa this urbanization has been accompanied by the proliferation of slums and squatter settlements (UN 1997a).

Regions with high birth rates and extensive immigration are faced with environmental problems due to unplanned urban growth and emerging megacities (Lynn 1999). In some cases the situation is aggravated by polluting industries, which have been transferred from industrialized countries with stricter environmental legislation and higher wages. On the other hand there has been a tendency in the industrialized world that relatively fewer people live in the inner areas of the cities (OECD 1995b). The consequences are expanding road systems and increasing commuting traffic, which may counteract improvements in automobile technology.

15.1.4 Global energy consumption, production, and emissions

World production of automobiles has increased by a factor of five since 1950 (Fig. 15.4) and the total fleet is now above 500 million. In addition to this there has been a substantial production of motorbikes, nearly half in China, with an estimated fleet of 80 million (OECD 1995a).

In the same period industrial production has increased by a factor of 10 and global energy consumption by nearly a factor of five (UN 1997b and earlier reports). Since the major part of the energy has been produced by fossil fuels, and to a minor extent by biofuels, initially without flue gas cleaning, the global emissions of air pollutants have increased correspondingly. Since 1950 the global emission of sulfur oxides has more than doubled, and the emission of nitrogen oxides increased by a

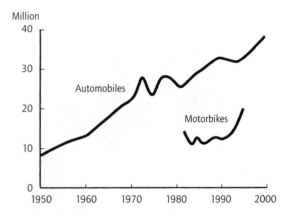

Fig. 15.4   The world production of automobiles (upper curve) and motorbikes (lower curve). (Based on material compiled in Brown *et al.* 1998.)

factor of four. A substantial part of these emissions occur in urban areas.

15.2   POLLUTANTS AND SOURCES

To date about 3000 different anthropogenic air pollutants have been identified, most of them organic compounds (including organometals). Combustion sources, especially motor vehicles, emit several hundred different compounds, but only for about 200 of them have the impacts been investigated. Ambient concentrations are known for a much smaller number and time series for even fewer.

The pollutants can be divided into two groups (Wiederkehr & Yoon 1998): the traditional **major air pollutants** (MAP, comprising sulfur dioxide, particles, nitrogen dioxide, carbon monoxide, lead, and ozone); and the **hazardous air pollutants** (HAP, comprising chemical, physical, and biological agents of different types). The HAPs are generally present in the atmosphere in much smaller concentrations than the MAPs, and they often appear more localized (typically in urban areas or near industries), but they are—due to their high specific activities—nevertheless toxic or hazardous. Both in basic investigations and in abatement strategies HAPs are difficult to manage, not only because of their low concentrations but also because they are in many cases not identified.

Compounds that increase the greenhouse effect (carbon dioxide, methane, nitrous oxide) or deplete the ozone layer (e.g. CFCs) are not urban pollutants *per se*, but control of their emissions is related to emission of other pollutants, and their impacts have a bearing on urban quality of life.

15.2.1 Characteristics of pollutants

Sulfur dioxide (SO$_2$) is the classical air pollutant associated with sulfur in fossil fuels. Emission can be successfully reduced using fuels with low sulfur content, e.g. natural gas or oil instead of coal. In larger plants in industrialized countries desulfurization of the flue gas is an established technique.

Nitrogen oxides (NO$_x$) are formed by oxidation of atmospheric nitrogen or nitrogen compounds in the fuel during combustion. The main part, especially from automobiles, is emitted in the form of the nontoxic nitric oxide (NO), which is subsequently oxidized in the atmosphere to the secondary and toxic pollutant NO$_2$. Emissions can be reduced by optimization of the combustion process (low NO$_x$ burners in power plants and lean burn motors in motor vehicles) or by means of catalytic converters in the exhaust.

Carbon monoxide (CO) is the result of incomplete combustion. Emissions can be reduced by increasing the air/fuel ratio, but with the risk of increasing the formation of nitrogen oxides. In automobiles the most effective reductions are carried out with catalytic converters.

Particulate matter (PM) is not a well defined entity. Originally it was determined as soot or "black smoke," for which there is a European Union (EU) air quality limit value (Edwards 1998). Later the concept of total suspended particulate matter (TSP) was introduced, but since 1990 size fractionating has been attempted by measurements of PM$_{10}$ (particles with aerodynamic diameter less than 10 μm). The major part of PM$_{10}$ may have a natural origin (e.g. sea spray or desert and soil dust), and it is therefore important to also measure PM$_{2.5}$ or even, when the appropriate technology has been developed, PM$_1$.

Usually particles are grouped in three modes: ultrafine, fine, and coarse. The ultrafine particles are formed by chemical reactions or are condensed from hot vapor, e.g. from diesel exhaust, and they coagulate into fine particles (Whitby & Sverdrup 1980). Defined as having an aerodynamic diameter less than 0.1 and 2.5 μm respectively (UNEP/WHO 1994), the ultrafine and fine particles are thus predominantly of anthropogenic origin. Coarse particles, on the other hand, are often of natural origin (dust, seaspray, pollen, or even insects). In determinations of TSP the coarse particles dominate with their high mass, almost irrespective of their relative number.

In urban atmospheres the actual size spectra show quantitative differences with, for example, more pronounced mass peaks for fine particles in suburban sites and for coarse particles near sea coasts (Fig. 15.5).

The application of different measuring techniques complicates the evaluation of changes in pollution levels. To some extent it is possible to establish relationships between the concentrations of fine and coarse particles relevant to epidemiological studies (Wilson & Suh 1997), but only under well defined conditions. Measurements from Erfurt in the former East Germany

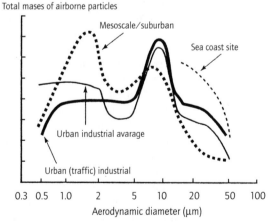

Total mases of airborne particles

Fig. 15.5 Schematic size distribution of particulate matter in various atmospheres. (From UNEP/WHO 1994). The corresponding deposition in the human respiratory system is shown in Fig. 15.13.

show that the level of PM$_{2.5}$ has been reduced substantially since the reunification of Germany and the subsequent introduction of updated technology. Nevertheless, the amount of even smaller particles has increased and so has the total number of particles, indicating a change in major sources (Tuch *et al.* 1997).

A further complication is that the chemical composition of particles is not well known, and that health impacts may be due to other pollutants adsorbed on them—typically heavy metals or less volatile organic compounds.

Emissions of particles of anthropogenic origin can be reduced by use of cleaner fuels, better combustion techniques, and a series of filtration or impaction technologies.

Volatile organic compounds (VOCs) as air pollutants are the result of incomplete combustion of fuels, are formed during combustion, or are due to evaporation. Some industrial processes and the use of solvents result in the emission of VOCs. In urban air the most important compounds are benzene and the series of polyaromatic hydrocarbons (PAH), but 1,3-butadiene, ethene, propene, and a series of aldehydes have also received attention (Larsen & Larsen 1998).

Biogenic VOCs, emitted from vegetation, do not pose a health risk in themselves, and the sources in cities are modest, but they must be taken into account in relation to regional photochemical air pollution, which in turn may influence urban air quality.

Lead as an additive to gasoline has by and large been phased out in the major part of the industrialized world, but it is still used in many developing countries and economies in transition, where emissions from industrial activities are also significant. Other heavy metals of interest as air pollutants include cadmium, nickel, and mercury, all with industrial sources.

Ozone is a secondary pollutant formed in photochemical reactions between VOCs and nitrogen oxides. Apart from small sources of indoor pollution like photocopiers and printers, it has no direct sources, and it can therefore only be controlled via the primary pollutants. Other constituents of photochemical air pollution, e.g. aldehydes and

PAN (peroxy-acetyl-nitrate), are not regulated separately.

15.2.2 *The main source groups*

Since air pollution is mostly related to combustion or evaporation, the different sources emit by and large the same compounds but in varying proportions, under different conditions, and with different time patterns (Stanners & Bourdeau 1995). There is also geographic variation, thus, for example, in eastern Europe SO$_2$ from space heating plays a relatively more important role compared to western and southern Europe, and in southern Europe the contribution of SO$_2$ from traffic is relatively high due to the use of diesel oil with a high sulfur content.

Space heating from small individual units was originally the main source of SO$_2$, CO, soot, and organic compounds in cities, but now—with the introduction of cleaner fuels, better combustion technology, and district heating—has diminishing emissions in industrialized countries.

Power generation with fossil fuels, in some cases combined with district heating, is, despite cleaner fuels and flue gas cleaning, still an important source of SO$_2$ (more than half) and NO$_x$ (about one-fifth) in Europe. In general, however, the impacts on urban air quality are relatively modest, since many plants are located in rural areas and equipped with high stacks.

Incineration of waste in larger plants has some similarities to power production, and is in many cases used for district heating. Due to the very mixed fuels, the emission of heavy metals and toxic organic compounds (e.g. dioxin) must be considered.

Industry is indirectly responsible for all the emissions related to energy use. Depending upon the production, emission of organic compounds and heavy metals may be significant. If the emissions arise from diffuse sources they may be difficult to control and have relatively large local impacts.

Use of solvents in households, crafts (painting), and minor industries gives rise to the evaporation of organic compounds. A more responsible use, including a switch to water-based coatings, can by and large solve the problem.

Traffic, and especially individual motorized traffic in the form of automobiles and motorbikes, is the dominant source of air pollution in most urban areas (Schwela & Zali 1999), not only in terms of local emissions, but also in terms of resulting pollution levels, since the emissions take place at low height and often in street canyons. In Europe all mobile sources account for nearly 70% of the emissions of CO, more than 60% of the NO_x, and more than 30% of the VOCs. Polyaromatic hydrocarbons (PAH) are mainly emitted from traffic (Nielsen 1996; Nielsen *et al.* 1999) and some have only recently been noticed (Enya *et al.* 1997).

The almost complete removal of lead as an additive to gasoline (Nriagu 1990) has not been completely without risk of side-effects. Changes in the composition of the gasoline to increase the octane number may increase the emission of aromatic hydrocarbons, including benzene. Benzene concentrations have increased in many urban atmospheres with the introduction of catalytic converters (Richter & Williams 1998). An alternative additive, MTBE (methyl-tert-butyl ether), not only increases the octane number, but also improves the combustion and thus reduces the emissions of carbon monoxide and hydrocarbons. It is, however, an air pollutant, causing both immediate eye and respiratory irritation and long-term risk of cancer. More important may be the contamination of soil and groundwater, especially around gasoline filling stations (transmedia pollution). In Denmark MTBE is only used for 98 octane gasoline.

Particles in the urban atmosphere, and especially small particles from diesel exhaust, are receiving increasing attention (Marico *et al.* 1999; Maynard & Howard 1999; Schwela & Zali 1999).

15.2.3 *Emission inventories*

Emissions are measured or calculated individually only for large point sources, such as power plants. In most cases emission inventories are carried out on the basis of emission factors. These factors may express the emitted amount of a given compound for a given activity or use of fuel. A typical emission factor might thus be: "$0.013\,kg\,CO\,GJ^{-1}$, when natural gas is used in district heating plants" (for details see Chapter 17).

National inventories of emissions as they are carried out, for example, in the European CORINAIR database (CORINAIR 1996) are used in international negotiations and may for lack of better information suggest general trends in air pollution levels. For studies of **transboundary air pollution** a spatial resolution of 50 km is normally sufficient (EMEP/MSC-W 1998, 1999).

For urban areas, however, more detailed investigations are necessary, with time and space resolutions relevant to the applied scale. Proxy data for larger areas can be generated on the basis of information on, say, traffic patterns (Friedrich & Schwarz 1998), but detailed investigations of individual streets (Berkowicz 1998) must be based on actual traffic counts. Of special importance is the time-dependent traffic density, with more or less pronounced rush hours (Fig. 15.6).

15.3 FROM EMISSION TO POLLUTION LEVELS

Emission of air pollutants from urban sources determine to a large extent the urban air quality. However, advection, dispersion, and to some extent simultaneous transformation are also important. Therefore, the impacts of different sources and their relative importance cannot be evaluated on the basis of emission inventories alone.

In the design of cost-effective abatement strategies (e.g. Krupnick & Portney 1991) it must be realized that the relations between emissions and resulting concentrations (so-called "immissions") are by no means simple. Measurements are still the foundation of understanding, but application of mathematical modeling, and also of physical modeling in wind tunnels, is of increasing importance in urban air pollution management. As a consequence, numerous techniques have been developed for different spatial scales, ranging from entire regions down to individual streets. Some models only describe dispersion or have simple

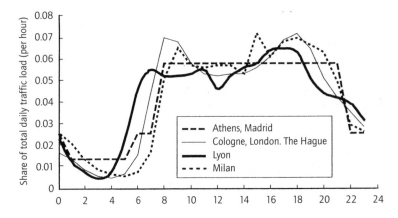

Fig. 15.6 Hourly profiles of traffic density for a series of European cities in 1990. (From European Commission 1996b.)

reaction schemes; more sophisticated models comprise a large number of interacting reactions. Such models can be further developed to form full decision support systems (e.g. Dennis *et al.* 1996).

15.3.1 *The urban climate*

An urban area differs from the surrounding rural region in important ways, which influence its climate, the possibility of dispersion of air pollutants, and their impacts. A city is generally darker than its surroundings and thus has a higher energy absorption (lower albedo). This effect is amplified by the rougher surface created by buildings, which trap solar radiation. Finally, various energy-consuming activities add to the heating, which may result in temperatures a few degrees above the surrounding rural areas.

At low wind speeds this so-called "urban heat island" can give rise to a circulation where the pollution is trapped. This situation can be aggravated by the topography, e.g. if the city is situated in a valley, which further restricts horizontal mixing. At higher wind speeds the urban area acts as the source of a plume, with elevated pollution levels downstream (Fig. 15.7).

The relative humidity in the city center is generally lower—partly because of the elevated temperature, partly because of enhanced runoff of precipitation in sewers and drains. On the other hand, the general instability generated by the heat island and the turbulence due to the rough surface may increase rainfall (Goldreich 1995).

15.3.2 *Urban-scale dispersion and transformation*

Dispersion in the urban area

The importance of dispersion was recognized with the invention of chimneys, and meteorological conditions played a crucial role in a series of pollution disasters (Brimblecombe 1987). Dispersion mechanisms have received special interest with the increasing traffic in built-up areas, where street canyons exhibit special flow patterns (Fig. 15.8). Consistent with this, both measurements and dispersion calculations (e.g. Berkowicz 1998) have shown that the wind direction is important for long-term pollution levels; therefore, in areas with a dominant wind direction – in northern Europe from the west – the orientation of the individual street may result in significantly different levels at the two sides. In streets parallel to the wind direction traffic emissions may give concentrations which built up downstream.

Wind speed is also of importance. Figure 15.9 demonstrates the significance of wind speed for the resulting air pollution in a street canyon, where a fairly strong wind ($8\,m\,s^{-1}$) is seen to nearly halve the concentration of NO_2 at rush hours.

The overall significance of the climatological conditions is clearly demonstrated in a com-

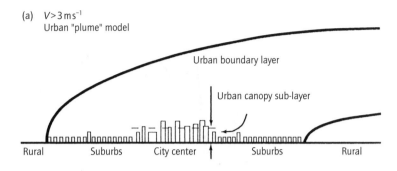

(a) $V > 3\,\mathrm{m\,s^{-1}}$
Urban "plume" model

Urban boundary layer

Urban canopy sub-layer

Rural   Suburbs   City center   Suburbs   Rural

(b) $V > 3\,\mathrm{m\,s^{-1}}$
Urban "dome" model

Urban dome

Rural   Suburbs   City center   Suburbs   Rural

Fig. 15.7 Urban heat island in the case of (a) moderate to strong wind and (b) weak wind. (From Mestayer & Anquetin 1995; with kind permission from Kluwer Academic Publishers.)

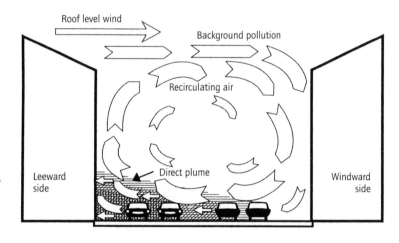

Roof level wind

Background pollution

Recirculating air

Leeward side

Direct plume

Windward side

Fig. 15.8 Vortex in a street canyon with a roof level wind at a right angle to the direction of the street. Even at moderate wind speeds (3–4 m s^{-1}) short-term pollution concentrations may be 5–6 times higher at the leeward site. For wind directions parallel to the street, the "vortex effect" disappears, but may be replaced with a "tunnel effect," where concentrations build up in the wind direction.

parison between air quality in Copenhagen and Milan. Since the frequency of low wind speeds is considerable higher in Milan than in Copenhagen, Milan has much higher pollution levels for comparable emissions (Vignati *et al.* 1996). In recent studies by Cocheo *et al.* (2000) the urban levels of benzene measured in six European cities were found to decrease drastically with increased wind speed.

Chemical reactions

During dispersion the pollutants interact chemically (e.g. Finlayson-Pitts & Pitts 1997) and for the urban atmosphere reactions between nitrogen oxides, organic compounds, and ozone are the most important. Photochemical smog with formation of ozone and other oxidants was first recognized in Los Angeles in the mid-1940s as an urban phenom-

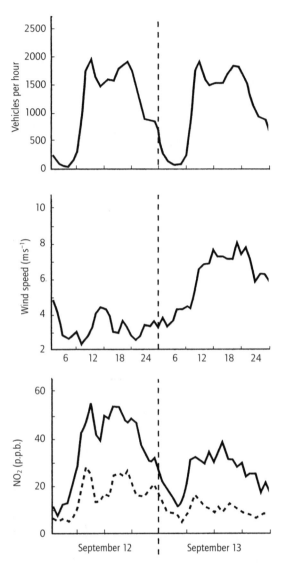

Fig. 15.9 Relations between traffic intensity, wind velocity, and NO$_2$ concentrations on two consecutive weekdays in 1997, measured at Jagtvej, a busy street surrounded with buildings in Copenhagen. The concentrations are measured both at the roadside (upper curve) and at a nearby rooftop (lower curve). (Based on material from Kemp *et al.* 1998.)

enon related to automobile exhaust in a subtropical, topographically confined region.

Photochemical smog is now observed in many parts of the world, but with distinctly different patterns. In the south of Europe, cities like Athens and Rome may experience a "summer smog" of the Los Angeles type, but in many cases it is a large-scale phenomenon (Guicherit & van Dop 1977). In cities in the northern part of Europe, one of the predominant reactions is a reduction of ozone by nitric oxide in automobile exhaust (Fig. 15.1) to form oxygen and nitrogen dioxide:

$$NO + O_3 \rightarrow NO_2 + O_2$$

As a result ozone levels are generally lower at ground level in the streets than at roof level or in the surrounding countryside. Urban ozone levels are higher during weekends with low traffic and may be practically nil during some pollution episodes (Fig. 15.10). Note also in Fig. 15.10 that the concentration of NO follows the traffic intensity at rush hours and weekends much more closely than NO$_2$, the concentration of which is largely determined by the available O$_3$, supplied from outside the city. A disappointing consequence of this reaction pattern is that a given reduction in the emission of nitrogen oxides (NO$_x$), e.g. by the introduction of catalytic converters on automobiles, will generally result in less than the corresponding reduction in the concentrations of NO$_2$ (Palmgren *et al.* 1996).

These mechanisms often lead to the formation of elevated ozone concentrations downstream from the city (city or urban plume). In a similar phenomenon on a larger scale, elevated ozone concentration in central Europe due to extended high-pressure events can, via long-range transport, be detected in northern Europe. Here concentrations normally build up over several days—often in parallel with rising temperatures, high stability, and solar radiation.

Since ozone is a secondary pollutant it can only be regulated via the primary pollutants. Long-range chemical/transport models can demonstrate the effects of changes in the emissions and ozone concentrations, and they clearly indicate that a

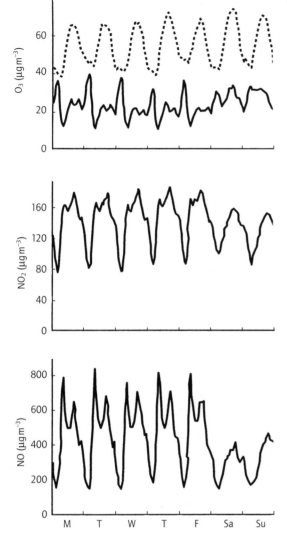

Fig. 15.10 Average weekly variations for ozone and nitrogen oxides at H.C. Andersens Boulevard in Copenhagen. For comparison a typical variation at a rural site is shown. (Based on HLU 1994.)

concerted international effort is necessary. In computer experiments it has thus been shown that in the hypothetical situation where all Danish emissions were reduced to zero, the average ozone

levels in Denmark would go *up* slightly (Zlatev *et al.* 1996).

Formation of particles

As stated in Section 15.2.1 particles in urban air can have many forms. Primary particles can be fairly large soot particles that cause soiling (Section 15.4.2), but are of minor importance for human health (Section 15.4.1). Less volatile organic compounds from, in particular, diesel exhaust may condense and form smaller particles or may add to small soot particles. The actual formation of secondary particles is predominantly a result of oxidation of sulfur dioxide to sulfate and nitrogen oxides to nitrate. The most important oxidizing agent is hydroxyl, OH-radicals being formed in photochemical reactions. Sulfuric acid vapor nucleates on its own or with water molecules to form a fine aerosol in the nanometer range. Nitric acid may, for example, react with ammonia and form ammonia nitrate particles. Although these secondary particles may be formed in the urban atmosphere they are spatially more uniformly distributed, with smaller urban to rural gradients than primary particles in traffic emissions.

15.3.3 Relations between air pollution on different scales

In more open spaces (parks, squares, residential areas) the pollution levels take the form of an urban background, with increasing impacts from more distant sources. Recent applications of mesoscale computer models have also demonstrated that the regional component is important, especially in areas with a complex landscape such as coastal regions; thus studies in the Mediterranean region and southern Europe have indicated that in certain periods the urban areas may be significantly affected by sources located hundreds of kilometers away (Kallos 1998).

Figure 15.11 shows as an example and symbolically the average impacts of different sources on the urban levels of nitrogen oxides.

Growing global consumption of fossil fuels leads to energy-related emissions of carbon dioxide

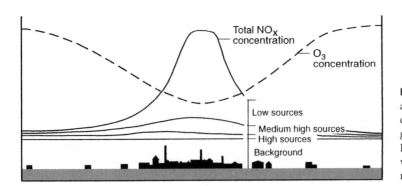

Fig. 15.11 A cut through an urban area with contributions to the concentration of nitrogen oxides at ground level. In cities in northern Europe the concentration of ozone varies inversely with that of nitrogen oxides.

(e.g. Ellis & Tréanton 1998) and may eventually, via the enhanced greenhouse effect, result in climate changes with impacts on all human activities and natural ecosystems. One of the results of the UN conference on environment and development in Rio de Janeiro in 1992 was an action plan for the attainment of a sustainable global development in the twenty-first century—the so-called Agenda 21. As a consequence many cities and administrative units in the industrialized world have embarked on local programs, and more than 290 European cities have signed the Aalborg Charter of European Cities and Towns towards Sustainability. Noteworthy in this connection is The International Council for Local Environmental Initiatives (ICLEI 2000), whose purpose is to achieve and monitor improvements in global environmental conditions through cumulative local actions. Although political attention emphasizes climate protection and thus reduction of emission of the ultimate product of combustion, CO_2, attempts to save energy may also improve urban air quality (e.g. Pichl 1998).

On the other hand, climate changes may alter the dispersion of urban air pollutants via a change (possibly an increase) in the occurrence of anticyclonic conditions, during which the dispersion is limited. This possibility is discussed by the Intergovernmental Panel of Climate Change (IPCC), but no definite conclusions have been reached so far. Clearly, the links between climate change, urban air quality, and human exposure to air pollutants are extremely complex and cannot as yet be quantified.

15.3.4 Outdoor and indoor air quality

Indoor air quality is largely determined by indoor sources—not only in occupational environments, but also in offices, libraries, museums (Brimblecombe 1990), and private homes with emissions from open fires, evaporation from synthetic building materials and solvents, use of detergents, smoking, etc. However, in urban areas ambient air pollution often has a significant influence, which must be taken into account in the evaluation of various impacts.

Generally the contributions from outdoor sources are reduced, and especially in buildings with recirculation and fresh air intake through filters. In a study in the National Museum of Denmark (Schmidt et al. 1999) it was found that elements of mainly antropogenic origin (sulfur, vanadium, and lead, and elements of mainly soil origin silicon, cadmium, and titanium) have the same time pattern in an unventilated room as in the outdoors, but with indoor concentrations reduced to about one-fifth. In a ventilated room the levels were reduced to about 1%. Nitrogen dioxide was reduced to one-half in the unventilated room and to one-third in the ventilated.

A crucial issue is the impact of outdoor particles on indoor particle concentrations. Theoretical calculations (Wallace 1996) indicate that without indoor sources and for a typical air exchange rate of $0.75\,h^{-1}$ the indoor/outdoor ratio should be 0.65 for fine particles and 0.43 for coarse.

15.3.5 Air quality indicators

The complex nature of air pollution, especially with respect to health impacts in cities, has prompted attempts to define so-called indicators (Wiederkehr & Yoon 1998), which condense and simplify the available monitoring data to make them suitable for public reporting and decision-makers. The OECD (1998) has applied major pollutants measured in a specified way as indicators for the total mix of pollutants.

In another type of synthesis the OECD (1998) has aggregated monitoring data from various regions (western Europe, USA, Japan) to demonstrate overall trends in pollution levels. The results of such an exercise should be treated with some caution, and in particular comparisons between different cities are dangerous (see Section 15.6.1). Yearly averages, however, appear to represent the general developments in specific cities reasonably well (see Fig. 15.27).

The weighed means of concentrations of several pollutants relative to guideline values have also been used (Kassomenos *et al.* 1999). Again comparisons between different cities are dangerous—if not impossible—due to differences in the mix of pollutants, in the monitoring networks, and in the climate-related impacts.

15.4 URBAN-SCALE IMPACTS

Urban air pollution has a series of impacts on human health and well-being, materials, vegetation (including urban agriculture), and visibility. These impacts depend in the first instance on the relevant pollution levels, but also on other factors, such as climate, lifestyle, and the possibility of interaction between different components. For short recent reviews with references see Fenger *et al.* (1998, Chapters 18–21).

15.4.1 Human health and well-being

The impacts on human morbidity and mortality of air pollution have been unambiguously documented during acute episodes in the past. Today

impacts on human health and well-being are more subtle—at least in developed countries—but they are still the main concern with urban air pollution. They also by and large determine abatement strategies (Bascom *et al.* 1996; Holgate *et al.* 1999; WHO 2000).

In some cases the abatements have had impressive effects. Thus the phasing-out of lead in gasoline in the industrialized countries (Nriagu 1990) has resulted not only in a drastic reduction in ambient lead concentrations, but also in significant decreases in blood lead levels (typically by a factor of three) in the relevant populations, converging to a "natural level" of $3 \mu g \, dl^{-1}$ (Thomas *et al.* 1999).

It should be emphasized, however, that the health of urban populations is determined by many other factors related to urban living, which blur the picture (Phillips 1993). The most important confounder is active and passive smoking, which has large regional and cultural variability, but in general appears to have health impacts well above those of general air pollution.

The human respiratory system

Air pollutants can have impacts on humans through various routes, e.g. when heavy metals are transferred to food and water. Some compounds (ozone, nitrogen dioxide, aldehydes) act directly as eye irritants, but the dominant route is via the respiratory system (Fig. 15.12).

For gaseous pollutants the rate of uptake depends upon solubility. SO_2 is soluble and normally more than 95% is deposited in the upper airways, whereas NO_2 and O_3 penetrate deeper into the lungs. The penetration of insoluble particles (soot, flyash, mineral dust) is highly dependent upon their size (Fig. 15.13).

Even for a well defined exposure the uptake depends on respiration rate, which increases with physical activity. The eventual effects depend upon the individual sensitivity, with the elderly, children, and asthmatics being especially sensitive to acute respiratory impacts.

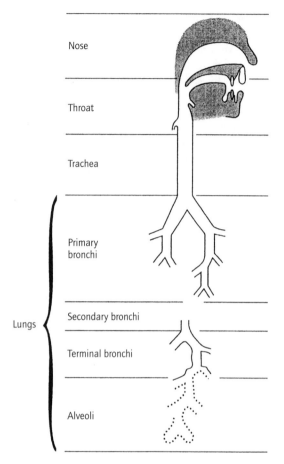

Fig. 15.12 Simplified structure of the human respiratory system.

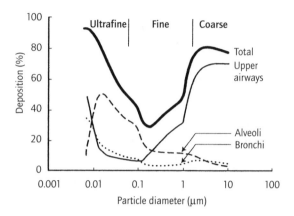

Fig. 15.13 Deposition of particles in the human respiratory system. Note that coarse particles are predominantly caught in the upper airways, whereas fine and especially ultrafine particles penetrate deep into the lungs.

Exposure

Until recently most studies of air pollution were centered on determination of ambient levels, which have formed the basis of legislation and abatement strategies. Equally important, however, is the extent to which people are actually exposed to the measured or calculated pollution levels. So far the pollution exposure of the population has in many cases been assessed on the basis of crude assumptions, e.g. that the levels observed at a single or a few monitoring stations are representative of the exposure of the population in a larger urban area. Some results of such an evaluation (WHO 1995) are summarized in Section 15.6.3.

Direct personal monitoring demonstrates that the levels *an individual* is exposed to vary drastically during the day, and that the ambient, outdoor air quality does not adequately describe the actual exposure in sufficient detail (an example is shown in Fig. 15.14). A further complication is that people appear to stir up "personal clouds" of particle-laden dust from their surroundings and therefore may experience exposure to fine particles about 60% greater than classical monitoring would suggest (Renner 2000).

A realistic evaluation of *population exposure* therefore requires statistical information, where the time and activity pattern of the entire population must be taken into account. Thus Danish studies (Andersen 1988) indicate that on average the population spent only one hour per day outdoors, one hour commuting and the remaining 22 hours indoors. More recent studies (Jensen 1998) apply a Geographic Information System (GIS) to combine air pollution data calculated with the OSPM model (Berkowicz 1998) with population data available from administrative databases.

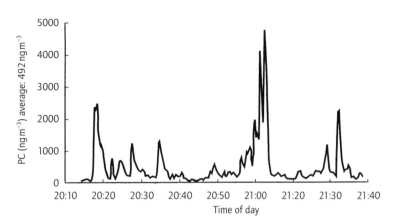

Fig. 15.14 Particle concentration measured on a trip from Aeropuerte Internacional to a hotel in the south of Mexico City. The measurements were based on photoelectric charging and emphasize polycyclic aromatic hydrocarbons (PAH) from combustion of organic material. (Reprinted from *Atmospheric Environment* (Zhiqiang *et al.* 2000). Copyright 2000, with permission from Elsevier Science.)

In view of the large fraction of time spent indoors, especially in cities with a cold climate, the relations between outdoor pollution and related indoor levels (see Section 15.3.4) must be taken into account. That is not always simple; as described in Section 15.3.2, the general pollution level for given emissions decreases with increasing wind speed. For benzene, however, this is not reflected convincingly in personal exposure (Cocheo *et al.* 2000), and in some cases population exposures were found to exceed the average ambient urban concentration. It is speculated that the reason may be that people are generally outdoors when the ambient levels are high and indoors when they are low, and that the indoor environment may store pollution with an outdoor origin (a sort of "flywheel effect").

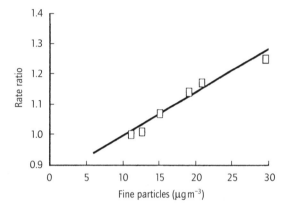

Fig. 15.15 Estimated adjusted mortality-rate ratios and fine particle pollution levels (PM$_{2.5}$) in six US cities. The rate for the least polluted city is defined as 1. (From Dockery *et al.* 1993.)

Particles

Soot and TSP have always been considered a health risk, but in the past decade fine particles have received special attention—partly because of increasing traffic, partly because of the development of adequate monitoring techniques (Pope *et al.* 1995; Maynard & Howard 1999; Section 15.2.1).

In an often cited investigation of the association between air pollution and mortality in six US cities Dockery *et al.* (1993) reported a strong correlation between the concentrations of small particles (PM$_{2.5}$) and mortality rate ratios (Fig. 15.15). For other pollution parameters (TSP, SO$_2$, sulfate,

O$_3$, and acidity) some increase with pollution levels was observed.

In a larger study, Pope *et al.* (1995) linked ambient air pollution data from 151 US metropolitan areas to the survival or death of 0.55 million persons. After correction for a series of confounders a significant association between fine particulate exposure and survival emerged. Adjusted relative risk ratios had a value of 1.17 between the highest and lowest polluted area.

In view of the surprisingly large effect the results of these investigations should not be uncritically transferred to other countries with different age distribution, lifestyle, climate, and pollution mix. Nevertheless, taking together the two studies, the WHO estimates a relative risk of 1.1 per long-term exposure to $10\mu g$ $PM_{2.5}$ m^{-3}. With a series of further assumptions, Brunekreef (1997) calculates a corresponding loss in life expectancy of more than one year for Dutch men.

Numerous other epidemiological studies of short-term and long-term effects of air pollution have shown that fine particles at the present levels are responsible for significant pulmonary impacts, especially for people already suffering from respiratory and cardiopulmonary diseases. Schwartz et al. (1996) have reported that a $10\mu g\,m^{-3}$ increase in $PM_{2.5}$ two-day mean was associated with a 1.5% increase in total daily mortality. Somewhat larger increases were found for deaths caused by chronic obstructive pulmonary disease (3.3%) and ischemic heart disease (3.3%).

Notwithstanding this extensive documentation, some important problems remain to be satisfactorily elucidated. First, the statistical procedures are debated. The question of how to relate daily mortality with longer-term mortality effects (McMichael et al. 1998) and especially how to filter out a possible "harvesting effect" (Zeger et al. 1999) arises. Second, the assessment of human exposure is still being developed (Mage et al. 1999). Finally, it must be admitted that epidemiologic studies provide little (or no) information on the underlying impact mechanisms, which are still poorly understood (Brunekreef 1999).

Some recent European studies

In the APHEA (Air Pollution and Health, a European Approach) the effects of several air pollutants in a total of 15 European cities in 10 countries were investigated. In western European cities it was found that an increase of $50\mu g\,m^{-2}$ in sulfur dioxide or black smoke was associated with a 3% increase in daily mortality. For PM_{10} it was 2%. The rises in cardiovascular and respiratory mortality were 4 and 5% respectively (Katsouynnia et al. 1997).

The daily hospital admissions for respiratory

diseases (Anderson et al. 1997; Spix et al. 1998) were found to increase, most significantly, for elevated levels of ozone, although the underlying statistics have later been criticized (Hasford & Fruhmann 1998).

In the Swedish SHAPE (Stockholm Study on Health Effects of Air Pollution and Their Economic Consequences) (Bellander et al. 1999) it has been calculated how many hospital admissions can be avoided in the county of Stockholm if all levels of PM_{10} and nitrogen dioxide are reduced to the levels in the fringe areas (reduction of 5 and $16\mu g\,m^{-3}$ respectively). On this basis it is estimated that 700 of the yearly admissions of about 65,000 for respiratory, cardiac, and circulatory diseases could be avoided.

It should be noted that Stockholm is a fairly clean city (similar to the least polluted in the US studies) and a complete removal of all particle emissions would only increase the average lifespan by about 2 months. Nevertheless this is about double the effect of a hypothetical prevention of all deadly traffic accidents.

In the so-called EXPOLIS (Air Pollution Exposure Distributions of Adult Urban Populations in Europe) study (Jantunen et al. 1998), personal exposure was determined by monitors and diary records in six European cities (Helsinki, Bilthoven, Prague, Basel, Grenoble, and Milan).

Economic evaluation of health impacts

Cost–benefit analyses including valuation of human health are controversial, with moral and ethical complications. Nevertheless, this has been attempted in many cases related to air pollution — often with the explicit aim of demonstrating that control is worthwhile considering the reduced loss of labor force, cost of health care, etc. In a recent study the European Commission (1998) performed an economic evaluation of the air quality targets put forward in 1996 (see Section 15.5.1). The overall conclusion was that in the case of PM_{10} the benefits exceed the costs by a factor of 100–200, that for SO_2 and NO_2 the benefit–cost ratios are in range of 1–10, but that in the case of lead the costs appear to exceed the benefits.

15.4.2 *Material damage*

The most conspicuous impact of urban air pollution is the degradation of materials, which leads to direct economic loss, failure of equipment, and deterioration of irreplaceable cultural artifacts. The local urban climate (Section 15.3.1) is important, since temperature and relative humidity control the moisture layer on surfaces in the absence of precipitation. This effect partially counteracts the higher pollution levels in the center of a city (Kucera & Fitz 1995). This can also explain the noticeable impacts in fairly clean Nordic cities compared to cities in the south.

Soiling

Particulate pollution has different effects: the soiling itself represents an aesthetic loss. Cleaning can be expensive (Newby *et al.* 1991) and may further impose mechanical wear. Soiling, especially with hygroscopic compounds (e.g. sulfates), facilitates the formation of moisture. Finally, some compounds act as catalysts for various reactions — notably oxidation of SO_2 and NO_x.

Corrosion of metals

In dry air and at normal temperatures oxygen reacts with most metals and in some cases forms protective layers of oxides, which inhibit further corrosion. Aluminum is covered with Al_2O_3 and iron with a mixture of oxides. Copper is covered with an attractive verdigris, which in a pure atmosphere mainly consists of basic copper carbonates. In a polluted atmosphere a less protecting layer of sulfates is formed.

In humid air the corrosion is generally a more rapid electrochemical process in a moist layer on the metal surface, where, for example, iron is converted to rust (FeOOH), which is peeled or washed off. Pollutants will dissolve in the moist layer (Fig. 15.16) and sulfur dioxide, for example, will be oxidized to sulfate:

$$SO_2 + O_2 + 2e^- \rightarrow SO_4^{2-}$$

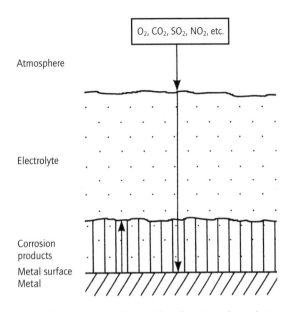

Fig. 15.16 Corrosion of a metal surface is an electrolytic process in a thin humid layer. The attack need not be uniform. Often it appears in the form of "pit corrosion," with more complex reactions in a confined area with a typical diameter of a few millimeters.

Simultaneously, the metal (e.g. iron) will be dissolved:

$$Fe \rightarrow Fe^{2+} + 2e^-$$

Ferrous sulfate can be further oxidized to ferric sulfate and converted to rust. This liberates sulfate, which can act once more. At the same time the rust maintains the humid corrosion layer.

The lower levels of SO_2 in the industrialized countries have increased the relative significance of other atmospheric pollutants. Most important is NO_2. Its corrosive reactions are not fully understood and it may not have impacts *per se*. It appears, however, to increase the effect of sulfur dioxide — possibly by facilitating its oxidation to sulfate (Arroyave & Morcillo 1995).

Stone materials

Stone used as building material and for monuments has a wide range of composition, texture,

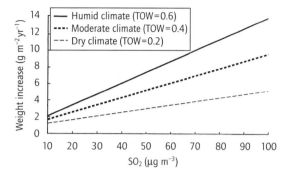

Fig. 15.17 Weight increase of sandstone calculated from a dose–response relation showing the effect of different climates on the deterioration rate. TOW is the so-called time of wetness, here defined the fraction of time, when the temperature is above 0°C and the relative humidity above 80%. (From Tidblad & Kucera 1998.)

and structure. Most important is the porosity, ranging from 0.5% for the dense granites and marbles to 25% for some limestones and sandstones. The classical pollutant causing degradation is SO_2, which is dissolved in the moisture layer to form sulfite and eventually is oxidized to sulfate. This attacks calcium carbonate $(CaCO_3)$ and forms gypsum $(CaSO_4 \cdot 2H_2O)$. The process initially results in a weight increase (Fig. 15.17), but in time the gypsum is washed off by rain.

In porous materials, notably soft sandstones used for ornaments, hygroscopic salts can be formed inside the stone. During changes of temperature and humidity various transformations result in mechanical stress, which enhances degradation. This also means that degradation can continue after the pollution has been reduced or stopped.

Organic materials

Degradation of organic materials is mainly related to ozone, which attacks double bonds in unsaturated polymers. This may lead either to chain-scissoring (with loss of tensile strength) or to cross-linking (with loss of elasticity). Thus accelerated wear of tires in California was an early sign of photochemical air pollution.

The exposure to ozone results in fading and cracking of outdoor paints and other coatings, but the impacts may be difficult to distinguish from direct action from sunlight. However, indoor effects have also been observed. Accelerated laboratory exposure to a mixture of photochemical oxidants leads to extensive fading of artists' colorants, and humidity enhances the effect (Grosjean *et al.* 1993). It appears that in some museums without proper air conditioning, especially in the Tropics, serious fading may result within a few months.

Economic evaluation of material damage

Material damage is better documented than health impacts, and experiments are less controversial. Nevertheless, economic evaluations are uncertain due to a lack of accurate materials inventories. Various estimates of maintenance costs (Mayerhofer *et al.* 1995; Kucera *et al.* 1996; Tidblad & Kucera 1998) suggest total European damage in the order of several billion euro per year. Thus the annual savings from reduced damage to buildings in Europe of implementing the second sulfur protocol (Section 15.5.1) are estimated at about 10 billion euro, with about two-thirds in urban areas.

Notwithstanding the uncertainties in such evaluations, it appears that the savings from reduced material damage may balance a considerable part of pollution abatement costs. On the other hand, the impacts can be significantly reduced or even prevented by a modification or replacement of materials, e.g. when metals are alloyed or replaced by plastic.

A special problem is posed by damage to irreplaceable cultural monuments and art, the reason being partly that the value is often related to a thin layer of decoration and not to mechanical strength. Attempts to evaluate the benefit derived in the form of pleasure from such objects (contingent valuation) suggest that the economic loss of damage is in the same order of magnitude as that of damage to trivial materials.

15.4.3 Urban ecosystems

The natural environment in urban areas is of importance for various reasons (Ashmore 1998, and references therein). First, it has utilitarian value. Many city dwellers grow food for their consumption, and it is estimated that a total of 800 million people are engaged in urban agriculture—mainly in Asia, Africa, and Latin America. Commercial production of high-value crops is normally found in the outskirts, but in densely populated regions like the Netherlands more intensive agriculture activity is incorporated into urban planning.

Second, vegetation has aesthetic value and trees may improve the urban climate by offering shade and cooling by transpiration, and by influencing deposition. It appears that urban trees generally reduce ozone concentrations in cities, but tend to increase average ozone concentrations in the surrounding area (Novak *et al.* 2000). It has even been argued that urban vegetation may act as a sink for many pollutants. On the other hand, trees tend to reduce the local dispersion of pollutants.

Third, sensitive elements may provide useful bio-indicators of trends in air quality. The diversity of lichen has in many countries been used to indicate levels of SO_2. Lichens have also been used for mapping deposition of metals, which they accumulate effectively.

Early observations testify to significant damage to urban vegetation. Although urban air quality has improved considerably in recent years there is evidence of continuing impacts (Ashmore *et al.* 1998).

15.4.4 Visibility

The most visible air pollutant is black smoke, which acts as a precursor to fogs (an important ingredient in early English detective novels). Although the London "Pea souper" and other spectacular events (Fig. 15.18) are things of the past, reduction of visibility may still be a nuisance.

It is pleasant when the air is clear and it is advantageous in many ways, including road safety. A reduction in visibility is therefore not only unsavory, it may also be dangerous. What determines whether we can see an object of a reasonable size is not so much its radiance as its contrast—normally defined as the relative difference between the lu-

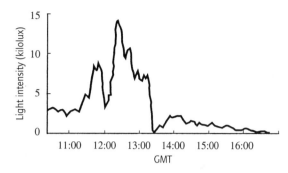

Fig. 15.18 The light intensity in London on the "Dark Day," January 16, 1955. Here the smoke concentration at ground level was not particular high, but a higher layer practically blocked all sunlight. (From Brimblecombe 1987.)

minosity of the object and that of the background. When light passes through the atmosphere it is absorbed and scattered by molecules and particles, and consequently the contrast is gradually reduced. For contrasts below 0.02–0.05 the object becomes for most people indistinguishable from the background, and the corresponding distance is termed the visibility.

In practice visibility is determined by light scattering on particles, and there are different processes, which depend upon the ratio between the wavelength and the particle size. Thus not only the contrast, but also the color mix, of sight is changed. For relatively large particles like fog the scattering is largely independent of the wavelength and results in a diffuse white light, which may reduce visibility to nearly zero.

When the relative humidity becomes so large that particles adsorp water their size will increase drastically and so will the scattering—especially for water soluble compounds like sulfates and nitrates.

The number of particles in urban air is typically highest in the morning and afternoon, not only because of rush hours, but also because of a low-lying mixing layer. At the same time the humidity may be high and thus the visibility low (Fig. 15.19).

In clean dry atmospheres, as in some natural parks, visibility may be as far as 200 km. In

European cities (without direct fog) visibility may be down to 2 km. It is thus rarely a practical, only an aesthetic, problem. In the case of more serious episodes pollution may build up over several days (Section 15.3.2).

15.5   MEANS OF MITIGATION

As indicated in Fig. 15.1 the impacts of urban air pollution can be mitigated at all points of the "pollution chain." Some possibilities have already been mentioned. The classic solution, with dispersion from high stacks (Fig. 15.20), tends to only transfer the problem. With the growth and eventual merging of urban areas this option is only acceptable in combination with advanced flue gas cleaning.

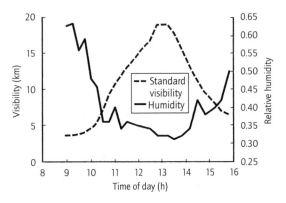

Fig. 15.19   Variation of visibility as a function of humidity in Vienna on a day with stagnant air. Note that the humidity and the visibility vary inversely. (From Horvath 1998.)

The problem with increasing traffic and its emissions can in principle be attacked in two ways: on the technical level by reductions of the individual emissions; and on the economic and planning level by a reduction of activities, either by reducing the need for transport as such (promoting walking and bicycling) or by promoting public transport. Some solutions based on traffic planning, however, are of debatable value. They improve the air quality in city centers, but transfer problems to the outskirts and may even increase total activity.

15.5.1   Legislation and technical solutions

Past experiences have with depressing clarity shown that existing technical possibilities and recommended management practices will not be implemented unless legally or economically enforced. Air quality expressed as pollutant concentrations is controlled by limit values. With the possible exception of ozone limit values, they are only relevant for pollution levels in urban or industrial areas. The scientific foundation is experiments on humans or animals and epidemiological investigations. The results are evaluated by the WHO and expressed in the form of guidelines (WHO 2000) that are subsequently used as the basis for legally binding limit values.

Most countries have established such limit values for the major air pollutants and use, in addition, guideline values for a series of other compounds. Most important are the limits in the EU and the USA, which in many cases have served as models for other regions. US emission standards, especially for motor vehicles (Faiz *et al.* 1996), have been used in this way.

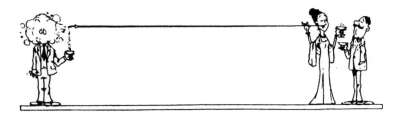

Fig. 15.20   The policy of high stacks only transferred the problems. (An early cartoon from *Euroforum*.)

Air quality in the EU

In the EU the setting of limit values is a multistep process (Edwards 1998) involving a system of EU directives, the first being adapted in 1980. Since 1996 a framework directive has provided a basic structure, and daughter directives lay down limit values and proscribe dates for attainment, methods for measurements, etc., which are mandatory throughout the EU. These directives are subsequently ratified in the individual member states in the form of national legislation (Edwards 1998). This system has so far covered sulfur dioxide, particulate matter, nitrogen dioxide, benzene, carbon monoxide, and lead. Threshold values for ozone for the giving of information and warnings to the public are also regulated by EU directives. Other pollutants for which legislation is considered important include polyaromatic hydrocarbons, cadmium, arsenic, nickel, and mercury.

Vehicle emissions in the EU

The most direct means of improving air quality is of course through regulation of emissions. The EU legislation on vehicle emissions and fuel quality standards has evolved greatly since the first directive in 1970. The early legislation had the dual purpose of reducing pollution and avoiding barriers to trade due to different standards in different member states. It is now aiming at meeting air quality targets. The "Auto Oil I Programme," undertaken by the European Commission in conjunction with industry (European Commission 1996a), has set up targets for a series of traffic-related pollutants and assessed different technologies and fuel quality standards. The target for nitrogen dioxide was in full compliance with the new WHO guideline of $200 \mu g\,m^{-3}$ as a maximum one-hour average.

By means of models with simplified chemical reactions, the impacts, compared to 1990, on the air quality in seven representative European cities (Athens, Cologne, London, Lyon, Madrid, Milan, and The Hague) have been evaluated (European Commission 1996b). Already agreed measures (e.g. the introduction of three-way catalytic converters) were expected to reduce pollution from ve-

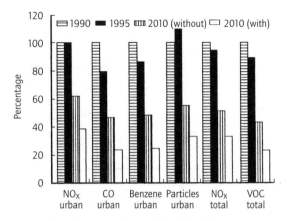

Fig. 15.21 Expected impacts of the European Auto Oil Programme on emission of air pollutants in the European Union and specifically in urban areas. The diagram shows calculated emissions in 1990 and 1995 and projected emissions in 2010 without and with the programme in force. (Based on EC Commission 1996a.)

hicles by 40–50% in 2010 compared to 1990. The Auto Oil Programme will increase the reduction to 70% — even for the expected higher traffic volume (Fig. 15.21).

It appeared that the objectives for carbon monoxide ($10\,mg\,m^{-3}$, one hour maximum), benzene ($10 \mu g\,m^{-3}$ annual mean), and particulate matter ($50 \mu g\,m^{-3}$ 24-hour average) would be met by 2010, and the NO_2 objective should be met in most of the Union in 2010. In some cities, like Athens, however, further action would be needed. A more stringent target value for benzene of $2.5 \mu g\,m^{-3}$, preferred by several member states, would be exceeded in all investigated cities except The Hague.

In an ongoing second program nontechnical measures for local use in areas with high pollution levels, such as road pricing, traffic management, and scrapping schemes, will be evaluated. Further, the resulting reductions in pollution levels will be calculated in more detail. An example from a Danish site is shown in Fig. 15.22.

Stationary sources in the EU

EU legislation on industrial air pollution has al-

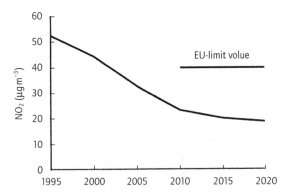

Fig. 15.22 The yearly average of NO₂ at Jagtvej in Copenhagen, a street with high buildings and a present traffic of about 22,000 per day, which is not expected to rise. For 1995 the value is based on measurements. Later values are calculated with a street pollution model coupled to a simple model for urban background pollution (Berkowicz 2000). The EU limit value of $40\,\mu g\,m^{-3}$ will easily be met.

ways taken air quality limit values into account and required the operators to use "best available technology, not entailing excessive costs." A Council Directive from 1996 on integrated pollution prevention and control (IPPC) has the purpose of reducing pollution to the environment as a whole, avoiding transfer from one medium to another (Edwards 1998).

The United States Clean Air Act

In the USA the first federal air pollution legislation was enacted in 1955, and in 1970 the administration was transferred to the new US Environmental Protection Agency (EPA). The Clean Air Act was first passed by Congress in 1967 as the Air Quality Act, which was amended in 1990. Under the act, the EPA set national air quality standards (NAAQS) for six pollutants: sulfur dioxide, nitrogen dioxide, carbon monoxide, particulate matter, and lead.

The standards are reviewed regularly, and new proposals were put forward in 1997. For ozone the new EPA standard allows no more than 0.08 p.p.m. ($160\,\mu g\,m^{-3}$) as an eight-hour average. For

particles a separate standard is set for $PM_{2.5}$ (Brown 1997).

Long-range transport and urban air pollution

The United Nations Economic Commission for Europe (UNECE), comprising all European countries and Canada and the USA, has been an important forum for east–west discussions of air pollution. It was also in the ECE that the so-called Geneva Convention on long-range, transboundary air pollution was established and signed in 1979. After it had been ratified by a sufficient number of member states it came into force in 1983. A series of related protocols set targets for reductions of national emissions of sulfur dioxide, nitrogen dioxide, and volatile organic compounds.

These protocols are all aimed at protecting natural systems. A new multipollutant–multieffect "Gothenburg Protocol" comprehensively addresses acidification, eutrofication, and photochemical air pollution. However, since the largest part of the relevant emissions occur in urban areas, the necessary reductions have direct impacts on urban air quality. Thus the decreases in levels of sulfur dioxide in the 1980s are related to the use of natural gas, low-sulfur fuel oil, and desulfurization (or alternative sources of energy), necessary to comply with international agreements. Attempts to reduce ecological impacts of large-scale photochemical air pollution will directly influence urban ozone levels—especially in the north of Europe.

In other parts of the world, notably in east Asia, unregulated local air pollution emissions have increasing regional and transboundary impacts.

15.5.2 Economic incentives

Attempts to reduce urban driving by various types of economic incentives have had some success, but they are often opposed by trade. It must also be considered that the placing of restrictions *in* cities may promote the growth of big shopping centers, hotels, and office buildings *outside* the cities, where they can offer free parking space and other facilities, often resulting in an increase in total traffic volume (OECD 1995b). In such cases improvement in urban air quality is paid for with more pollution on a larger geographic scale.

Taxes

Taxes can be imposed on the purchase of a vehicle or as an annual tax related to vehicle weight, energy consumption, or emissions. Generally, the impact on traffic pollution is a reduction, but it must be taken into account that purchase taxes discourage the replacement of an outdated automobile fleet. Fuel taxes will in the long run lead to more effective driving and increase the demand for smaller and more efficient automobiles. The total effect on energy consumption and emissions is much higher (and more logical) than that of any other form of taxation.

The basic problem with taxes is that the real political purpose often appears to be a means to provide additional revenue rather than protection of the environment. In addition, attempts to reduce transport contrast with the general belief that mobility in society is a prerequisite for economic growth.

Parking and driving restrictions and fees

More rational types of economic incentive are aimed at reducing traffic in sensitive areas. Driving in city centers can thus be reduced by restricting the number of parking places or by increasing the fees for their use.

Pricing of the infrastructure in the form of **toll roads** is used in several areas. Typically, a fee is paid for the use of highways, but pay stations can also form a ring around a town to tax all traffic in the urban area.

A more advanced, in principle more fair, but not yet fully developed, system is **road pricing**. Here the position of each automobile is determined by satellite or by coils in the road network and the information on movements is stored electronically *in* the automobile. This allows detailed taxing according to type of vehicle, different zones, and time of the day, etc., and thus encourages economical driving. The general objection to the system is the

vision of a "big brother society" where all movements of all people are registered. This, however, appears to be only a technical problem.

15.5.3 City planning

The result of a choice between public and individual transportation depends upon the proximity of access to the public system. In an industrialized country a distance of more than 1 km often means the use of an automobile (Fig. 15.23).

The impacts of urban air pollution can therefore be mitigated by constructive city planning. The complete separation of industry and habitation, originally envisaged as an environmental improvement and a reasonable solution in a society with heavily polluting industries, is now outdated and often only leads to increased commuting traffic and congestion. The ideal now is integrated land use, which minimizes transport and thus total urban emissions. Open spaces and parks can be used to improve environmental quality.

In existing cities the possibilities for restructuring are limited, but the construction of ring roads, which lead part of the traffic around the city center, is one of the options (OECD 1995b).

In the industrialized world few cities and urban

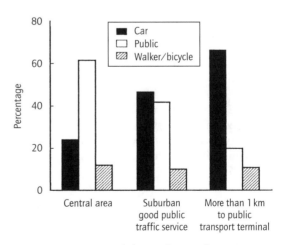

Fig. 15.23 Distance and choice of mean of transportation in the Copenhagen area. (From Nielsen 1997.)

areas can be constructed from scratch, but when possible new concepts of integrating urban planning, building design, and supply of renewable energy should be applied. The climate of the city is also important, and the influence of buildings and street canyons on solar radiation, shade, and wind pattern should be taken into account (Bitan 1992).

In this planning, which to a large extent is planning of traffic, it must be realized that air pollution is not the only environmental impact, and probably not even the most important. It is estimated (European Commission 1995) that on average the external costs of air pollution (not including the greenhouse effect) from transport in the EU amount to 0.4% of the gross national product, compared to 0.2% from noise, 1.5% from accidents, and 2.0% from congestion.

Figure 15.24 shows as an example of a Danish suburb west of Copenhagen, where the urban density increases close to a railway station. In this area public facilities such as a shopping center and schools are located within walking distance. The unintended side-effect of such a solution, where nearly everything is built at one time, is often a homogeneous and uninspiring style. (Apparently very few architects live in such developments.)

Even for a fixed volume of traffic in a given infrastructure the pollution from traffic can be reduced. The emissions per kilometer from a specific automobile depend upon the driving pattern and increase drastically with "stop and start" driving in congested urban areas. Here average speed appears to be a good measure of the number of accelerations and stops and thus low speeds result in high specific emissions. Smooth traffic (secured with speed limits and green waves), which increases the average speed from 20 to 40 km in main flows, may thus reduce the emissions by 30–40% (Fig. 15.25).

15.6   CASE STUDIES

The complex interplay between human activities, technical and legislative development, and natural parameters gives rise to completely different pollution patterns in different cities of the world. However, there are some common features for cities in the industrialized world (mainly exemplified by western Europe), the developing world (mainly east Asia), and economies in transition (eastern Europe), respectively.

15.6.1   Comprehensive records of urban air quality

In most of the industrialized world urban air pollution is now monitored routinely. Since 1974 WHO and UNEP have, within the "Global Environment Monitoring System," collaborated on a project to monitor urban air quality, the so-called GEMS/AIR (UNEP 1991; WHO/UNEP 1992; GEMS/AIR 1996; and a series of related reports). Concentrations of air pollutants in selected countries are also reported yearly by the OECD (1999). A comprehensive presentation of urban air pollution in Europe, based on data from 79 cities in 32 countries (Richter & Williams 1998), has recently been published by the European Environment Agency.

These and similar data give an indication of trends in ambient air quality at the national level and in selected cities. Often, however, the data are based on only a few monitoring stations placed at critical sites, and thus represent micro-environments. It should also be taken into account that the coverage of stations varies from country to country (Larssen & Hagen 1997), and that average values can therefore be differently biased.

Air pollution in the developing countries and in some countries with economies in transition is not documented in detail and longer time series are very rare. In most cases a general trend in air quality can only be estimated on the basis of dubious emission inventories. Data presented in the open literature are seldom up to date and normally concern specific cities, which may not be fully representative. In recent years many governmental and private institutions from the industrialized countries have acted as consultants in developing countries or performed investigations, but not all the efforts have been reported in the open literature.

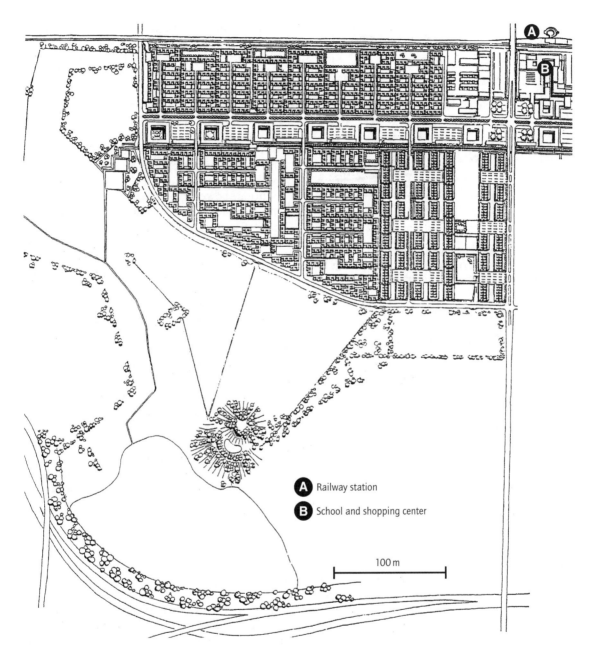

Fig.15.24 Albertslund, west of Copenhagen, is a typical example of an urban development aimed at public transport. (From Gaardmand 1993.)

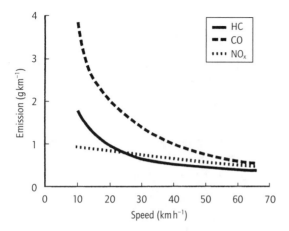

Fig. 15.25 Average speed and emissions for a Danish passenger car around 1990. Although the technology has been continuously improved the tendency is obvious. (From Krawack 1991.)

15.6.2 General pollution development

Seen over longer periods, pollution in major cities tends to increase during the build-up phase, pass through a maximum and then become reduced, as abatement strategies are developed (Mage *et al.* 1996; Goklany 1995). Depending upon the time of initiation of emission control the stabilization and subsequent improvement of the air quality may occur sooner or later in development (WHO/UNEP 1992).

In the industrialized Western world urban air pollution is in some respects in the last stage, with effectively reduced levels of sulfur dioxide and soot. In recent decades, however, increasing traffic has switched the attention to nitrogen oxides, organic compounds, and small particles. In some cities photochemical air pollution is an important urban problem, but in the northern part of Europe it is a large-scale phenomenon, with ozone levels in urban streets normally being lower than in rural areas. Cities in eastern Europe have been (and in many cases still are) heavily polluted. After the recent political upheaval, followed by a temporary recession and subsequent introduction of new technologies, the situation appears to be improving. However, the rising number of private auto-

mobiles is an emerging problem. In most developing countries the rapid urbanization has so far resulted in uncontrolled growth and deteriorating environment. Air pollution levels are still rising on many fronts.

In accordance with this development, the environmental quality in a given country generally depends upon the average income of the inhabitants (Shafik 1994). The availability of safe water and adequate sanitation increases with income, and so does the amount of municipal waste per capita. Air pollution, however, appears initially to increase with income up to a point and then to decrease (Fig. 15.26). Based on more recent data, Grossman and Krueger (1995) estimate that the turning point for different pollutants varies, but in most cases comes before a country reaches a per capita income of US$8000. Although this suggests that global emissions will decrease in the very long term, continued rapid growth over the next several decades must be expected (Selden & Song 1994). Emission of CO_2, the (so far) unavoidable end-product of most energy production still increases in all cases.

Fully in line with this typical development, the World Commission on Environment and Development in its report "Our Common Future" (1987) conceives technological development and rising standards of living as a prerequisite for environmental improvement. Or as Berthold Brecht has put it "Erst kommt das Fressen, dann kommt die Moral" (mistranslated into English: "First development and only later pollution control").

15.6.3 Industrialized (OECD) countries

Europe

Europe is a highly urbanized continent with more than 70% of the population living in cities. Population changes are in most countries modest and partly due to refugees and migration from east to west.

The large resources of coal were the primary source of energy during the Industrial Revolution, but in recent decades oil and gas have been found in the North Sea and in Russia and have been trans-

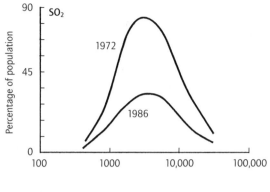

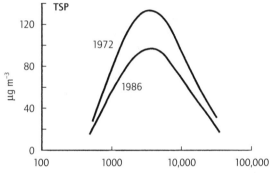

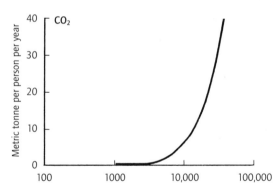

Fig. 15.26 Urban concentrations of SO_2, particulate matter, and CO_2 as a function of per capita income (Shafik 1994). The curves only show general tendencies, there are marked deviations. Thus in the early 1980s Kuwait had both some of the highest incomes and highest pollution levels. (From Smith 1988.)

ported through pipelines to other regions. In western urban areas the consumption of coal in small units, e.g. used for domestic heating, has been replaced with less polluting fuels, resulting in reduced emissions of sulfur dioxide and particles.

The total emission of sulfur dioxide steadily increased from about 5 Mt in 1880 to a maximum of nearly 60 Mt in the 1970s, only interrupted by the Second World War. It peaked in the mid-1970s, but has now been reduced to less than half. For the traffic-related pollutants nitrogen oxides, carbon monoxide, and VOCs, an increase has only recently been reversed (10–15%) by the introduction of three-way catalytic converters (TWC) (EMEP/MSC-W 1998, and related earlier and later reports).

European road traffic currently accounts for the largest part of total CO emissions, more than half of the NO_x-emissions, and a third of the VOC emissions, but only a minor part of sulfur dioxide—in western and northern cities about 5% and in southern cities, where diesel oil has a higher sulfur content, 14%.

It is estimated that in 2010 the CO, VOC, and NO_x pollution from traffic within the EU will be reduced drastically despite the expected traffic increase (Figs 15.21 and 15.22).

The European cities differ in various ways, which influence the relations between emissions and resulting pollution levels. Western Europe is influenced by the predominant westerly wind, bringing moist air from the sea, a climate that also favors long-range transport. In the northern part of Europe the small amount of sunlight favors persistent inversions with poor dispersion conditions. In central and eastern Europe high pressure, with air stagnation and accumulation of local pollution, is frequent. During the summer the climate in the Mediterranean Region likewise favors accumulation of local emissions, whereas during the winter large-scale wind systems are more frequent. The formation of photochemical oxidants depends upon sunlight, which in combination with poor dispersion conditions results in frequent episodes during summer.

The most extensive comprehensive treatment of Europe's environment up to 1992 was given in the so-called Dobris Assessment (Stanners &

Bourdeau 1995) from the European Environment Agency. It has since been updated in a second assessment (EEA 1998). The general air quality in European cities has improved in recent decades — often despite an increase in population density and standard of living — but air pollution is still considered a top priority environmental problem with both urban and large-scale impacts. Its special aspects are treated in more detail in several reports from the agency (e.g. Jol & Kielland 1997).

Smoke. The drastic improvements in urban air quality in the 1960–70s, partly brought about by a change from coal to less polluting fuels for domestic heating, partly by the closing down of polluting industry, has also resulted in a marked reduction in incidence of fog (e.g. Eggleston *et al.* 1992), but the lack of chemical analysis and size fractionation preclude more than a qualitative evaluation of the health impacts.

Sulfur. In many western European cities severe pollution with sulfur dioxide is a thing of the past. In Copenhagen the yearly average of SO_2 concentration was about $80\,\mu g\,m^{-3}$ around 1970 and during the winter about $120\,\mu g\,m^{-3}$. Today it is well below the WHO guideline of $50\,\mu g\,m^{-3}$. In provincial towns it is even lower (Fig. 15.27). The levels in provincial cities are not much higher than those at rural sites, indicating that most of the sulfur dioxide is due to long-range transport.

In accordance with this a more detailed analysis (Kemp & Palmgren 1999) has shown that there are still significant peak values (95th and 98th percentiles), identified as being due to long-range transport from central and eastern Europe. Consistent with this, levels of particulate sulfur (sulfate) show only a slight downward trend up to 1996. In 1997 levels for all types of sulfur pollution were very low, but so far it has not been established whether this was due to reduced emissions or special meteorological conditions.

This development has had several causes. An important aspect is the Geneva Convention on transboundary air pollution, which resulted in reduction of total emissions, but a widespread transition from individual to district space heating produced in large units with high stacks (often as combined heat and power production) has played a role. A further factor is that the oil crisis in the early 1970s resulted in better insulation of buildings. In Denmark the consumption of energy used for space heating was reduced in the period 1972–82 by about 30% despite an increase in area heated of about 20% (ENS 1997).

In 1990, 10 European cities observed exceedances of the long-term WHO-AQG for SO_2 of $50\,\mu g\,m^{-3}$; in 1995 only Katowice and Istanbul did. The short-term guideline of $125\,\mu g\,m^{-3}$ daily average is, however, still exceeded for a few days per year in many countries.

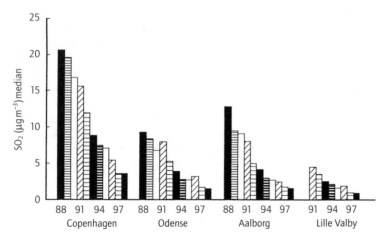

Fig. 15.27 Trends for annual medians based on hourly average concentrations of SO_2 measured in the center of the Danish capital Copenhagen (1.8 million inhabitants), in three major provincial cities, and at a rural station (Lille Valby). (From Kemp & Palmgren 1999.)

Nitrogen oxides. In recent years urban air pollution nitrogen dioxide has shown a downward trend in most European cities, although the short-term WHO guideline (corresponding to $200 \mu g m^{-3}$ as maximum hourly value) is exceeded in cities (EEA 1998).

The situation is complicated by chemical reactions in the urban atmosphere. Since the introduction of catalytic converters on all Danish gasoline automobiles registered after 1990 there has been a marked reduction in the levels of NO in Danish cities. Levels of NO_2, however, have been reduced much less (the median in Copenhagen by about 20%), reflecting the point that emissions of NO are not the limiting factor, but the available O_3 (Kemp *et al.* 1998).

Carbon monoxide. Urban concentrations of carbon monoxide have likewise been reduced since 1990, although exceedances of the eight-hour WHO guideline have been reported from many cities (EEA 1998).

Lead. In countries that have reduced or eliminated lead in gasoline the lead levels have substantially declined (EEA 1998), especially in countries with few or no lead-emitting industries. Since automobile exhaust is emitted at low height and often in street canyons there has been a close correspondence with the lead concentrations in urban air. In Denmark, where lead from gasoline in 1977 accounted for 90% of the national lead emissions, the problem has virtually disappeared. Remaining low lead concentrations of about $20 ng m^{-3}$ are essentially due to long-range transport (Fig. 15.28).

In the early 1980s, 5% of Europe's urban population in cities with reported lead levels were exposed to more than the WHO guideline of $0.5 \mu g$ m^{-3}. At the end of the decade levels above the guideline value were no longer reported from western countries.

Volatile organic compounds. Recently published (EEA 1998) levels of benzene range from a few to more than $50 \mu g m^{-3}$, with the highest values normally found near streets with high traffic volume. Of the reporting cities only Antwerp did not ob-

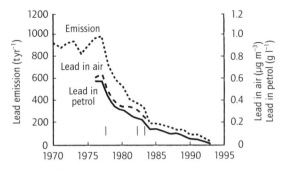

Fig. 15.28 Annual average values for total Danish lead emissions 1969–93, lead pollution in Copenhagen since 1976, and the average lead content in gasoline sold in Denmark. (From Jensen & Fenger 1994.) The dates of tightening of restrictions on lead content are indicated with bars. Lead concentrations for recent years can be found in Kemp *et al.* (1998, 1999). In 1998 they were below $20 ng m^{-3}$.

serve exceedances of the WHO-AQG guideline of $2.5 \mu g m^{-3}$ as a yearly average.

In a study of personal benzene exposure in a series of European cities (see Sections 15.2.2 and 15.4.1) it was shown that the levels decreased from north to south, probably reflecting a decrease in dispersion (Fig. 15.29).

In the 1960s the annual average concentration of BaP was above $100 ng m^{-3}$ in several European cities (WHO 2000). In most developed countries improved combustion technology, a change of fuels, and catalytic converters in motor vehicles have reduced the urban levels to $1–10 ng m^{-3}$. However, urban air pollution by potentially carcinogenic species is still not satisfactorily understood.

Ozone. European ozone levels appear to have increased from about $20 \mu g m^{-3}$ around 1900 to about double now, with the most rapid rise between 1950 and 1970, concurrent with the rise in emissions of primary precursors (Volz & Kley 1988). In the outskirts of Paris the early ozone levels were about 20 $\mu g m^{-3}$, but in the center only $3–4 \mu g m^{-3}$. Since there was no nitric oxide from motor vehicles it

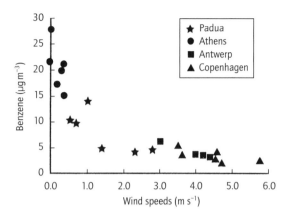

Fig. 15.29 Benzene levels as a function of wind speed in a series of European cities. The same pattern is observed when the levels are plotted against the latitude of the city. (From Cocheo *et al.* 2000.)

is assumed that O_3 was reduced by SO_2 or NH_3 (Anfossi & Sandroni 1997).

Summer smog with high ozone concentrations occurs in many European countries. As an urban phenomenon it is most serious in Athens and Barcelona, with concentrations up to $400 \mu g\,m^{-3}$, but Frankfurt, Krakow, Milan, Prague, and Stuttgart are also affected. Generally the present European ozone concentrations decrease from southeast to northwest.

An EU ozone directive contains a threshold value for giving information to the population of $180 \mu g\,m^{-3}$ and for warning of $360 \mu g\,m^{-3}$. In the Nordic countries it is seldom necessary to give mandatory ozone information and never mandatory warning. Implementation of the VOC protocol is expected to result in a 40–60% reduction in high peak values and 1–4% in annual average O_3 concentrations (EEA 1998).

General trends. According to the OECD (1998) the typical reduction of sulfur dioxide in Western European cities in the period 1988–93 was nearly 40% for traffic sites and about 20% for residential areas. For NO_2 the reductions were 10 and 12% respectively. Levels of CO were in general unchanged.

The urban population exposure has decreased correspondingly. In a WHO (1995) study, a series of data for the period 1976–90 from European urban areas with populations above 50,000 were pooled in two groups: up to and after 1985. Notwithstanding the limited and not fully representative data it appears that in western Europe up to 1985, 58% were living in areas with annual mean concentrations above the WHO air quality guideline of $50 \mu g\,m^{-3}$. After 1985 the fraction has been reduced to 14%. Similar, but smaller, reductions were observed for black smoke and SPM. Only for NO_2 was a slight increase observed (WHO 1995). The number of peak values has also decreased, and the total population experiencing episodes exceeding $250 \mu g\,m^{-3}$ SO_2 decreased during the 1980s from 71 to 33%.

North America

The two countries in North America, the USA and Canada, are among the wealthiest in the world with respect to both natural resources and production. This has previously led to serious urban pollution especially in the USA.

Many cities in the early industrialized USA were, like those in Europe, characterized by heavy smoke, and have subsequently gone through the typical development process. As an example (Davidson 1979), Pittsburgh already had pollution problems in 1804, and they by and large increased until the first effective smoke control in the late 1940s after the city experienced severe pollution due to the Second World War steel production.

As a general measure of the development in air pollution in the period 1970–93, the overall emissions of carbon monoxide, volatile organic compounds, and particulate matter have been reduced by 24, 24, and 78% respectively. A decline in emissions of nitrogen oxides from vehicles has been offset by increased electricity generation, giving an overall increase of 14%, but since the emissions are generally not in urban areas, this has not prevented a reduction of NO_2 levels in cities. Lead as an air pollutant has virtually disappeared, but toxic chemicals are still a problem.

Los Angeles still has the largest ozone problem in the USA. In 1990 the highest one-hour average was $660\,\mu g\,m^{-3}$. Ozone concentrations showed a significant decrease of 30% from 1988 to 1993 in urban residential areas in the USA both as an average and in the most polluted cities. Many counties will find it hard to comply with the new ozone standard of 0.08 p.p.m. (Brown 1997).

According to the OECD (1998) the trends in air quality in city areas in the USA in the period 1988–93 were decreases of the order of 23% for SO_2, 11% for NO_2, and 22% for CO.

In Canada emissions of sulfur dioxide and particulate matter have been reduced significantly since the early 1970s, and lead has virtually disappeared. However, some central Canadian cities still experience unacceptable air quality, with high levels of ozone and particulate matter, especially during the summer.

Japan

Japan is in many respects not typical of east Asia, but more like the western countries. According to the OECD (1998) the trends in air quality in Japanese city areas in the period 1988–93 were decreases of 20% for SO_2 and 26% for NO_2, whereas CO was largely unchanged.

The capital of Japan, Tokyo, is an encouraging example of an industrial megacity where air pollution is controlled. In the 1960s it was heavily polluted due to coal combustion and insufficient emission controls. Concentrations peaked in the late 1960s with annual mean values for SO_2 and TSP up to 200 and $400\,\mu g\,m^{-3}$ respectively (Komeiji *et al.* 1990). By switching the major fuel consumption from coal to oil and installing of dust collectors the annual mean values of SO_2 and TSP were brought well below WHO guidelines in the 1980s (WHO/UNEP 1992), and as a spectacular example the reduction of particulate emissions in the 1970s has resulted in a corresponding increase in visibility (Fig. 15.30).

As in the Western world, stricter control of vehicle emissions has been counteracted by a growth in traffic and the levels of traffic-related pollution have merely been stabilized. However, ozone lev-

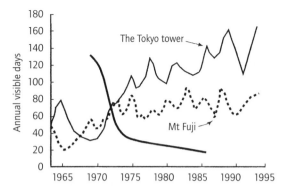

Fig. 15.30 The concentration of particulate pollution in Tokyo (right scale) and the days per year when the Tokyo Tower and Mount Fuji were visible from a suburban site. (From Kurashige & Miyashita 1998.)

els have generally been halved since 1970, when the yearly average was about $80\,\mu g\,m^{-3}$.

15.6.4 *Economies in transition*

In Eastern Europe solid fuels are still used in private houses and industry. Many buildings are badly insulated with large potential savings. Combined heat and power production is limited, partly due to problems with financing and management.

In 1990 the energy intensity of economies in central and eastern Europe (CEE) was about three times higher than that in western Europe; per unit of GDP emissions of NO_x and SO_2 were more than four times higher; and emissions of particles and VOCs were considerably higher (Bollen *et al.* 1996; UNEP 1997). SO_2 emissions were highest in the northern part of the region (Poland), which depends heavily on indigenous coal and lignite (Adamson *et al.* 1996), and lower in the southern CEE, where oil and gas are available.

In general it appears that industrial "hot spots" have shifted from western Europe to the east and southeast, where heavy industry, the use of low-quality fuels, and outdated production technologies have resulted in high emission levels.

Nevertheless, emissions have been reduced in some areas, partly as a result of German reunification in 1990 and the collapse of the Soviet Union in

1991. In 1990–1 distribution problems, ethnic conflicts, and organized crime in combination with outdated technology resulted in a drop in the Soviet Russian gross domestic product of 10–15%. And in the period 1990–6 NO_x emissions dropped by one-third and SO_2 by one-quarter.

A typical Polish example is Katowice (GEMS/AIR 1996), where since the late 1980s many older more polluting industrial plants have been closed down. In the period 1988–92 this — in combination with the introduction of better emission controls — resulted in a reduction of emissions of nitrogen oxides and particulate matter of 22 and 41% respectively. In the same period the air quality was improved by 30 and 44% for SO_2 and NO_2 respectively. Such reductions are also reflected in the investigations of population exposure carried out by the WHO (1995).

Further substantial reductions in emissions are possible, even if only current western European practices are applied (Bollen et al. 1996), but it is a serious problem for CEE to raise funds for economic and technological growth in connection with a transition to a market economy. For some cities, e.g. Krakow, the most effective strategy to improve air quality was found to be a ban on the use of coal — possibly limited to the town center (Adamson et al. 1996).

The southeastern urban areas have a partly outdated automobile fleet and are decades behind in the organization of road traffic. Only recently have efforts been put into the construction of ring highways (e.g. in Budapest and Prague) to reduce unnecessary crossing of the city center. Lead pollution is still found in eastern Europe, notably in Romania and Bulgaria near large uncontrolled metal industries (WHO 1995). The worst environmental problems are probably in the Balkans (Brown 1999), where they are aggravated by armed ethnic conflicts bordering on civil war.

Despite the general improvement in European air quality the WHO short-term air quality guidelines for SO_2 and TSP are often violated during wintertime smog, with the highest exceedances observed in central European cities. A positive development in the entire region can be hoped for with the extension of the EU.

15.6.5   Developing countries

Latin America

Latin America and the Caribbean is the most urbanized region in the developing world, with a rapidly increasing vehicle fleet, which is the dominant source of air pollution (Onursal & Gautam 1997). In Mexico City it accounts for 99% of the CO, 54% of the hydrocarbons, and 70% of NO_x. Leaded gasoline is still permitted in most countries and even in 1995 constituted the entire sale in Venezuela. In other countries it is being phased out; in Mexico City lead concentrations decreased by 80% in the period 1990–4.

Most of the air pollution occurs in major urban centers. In 1994, 43 of them had more than one million inhabitants. Often the situation is aggravated by the cities (e.g. Mexico City and Santiago de Chile) being situated in valleys surrounded by mountains.

Not surprisingly, the most critical air pollutants in Mexico City are ozone and its precursors NO_2, VOCs, and PM. The ambient ozone concentrations have consistently exceeded the Mexican one-hour standard of $220\,\mu g\,m^{-3}$. The highest value ever recorded (in 1992) was $955\,\mu g\,m^{-3}$.

In Santiago the most critical air pollutant is particulates, especially in the colder period (April–September), the principal source being a large number of poorly maintained diesel buses (GEMS/AIR 1996). The concentration of TSP is among the highest in any urban area in the world. In 1995 it reached a one-hour mean of $621\,\mu g\,m^{-3}$, compared with the Chilean standard of $260\,\mu g\,m^{-3}$.

São Paulo in Brazil is the third of the three most polluted cities in the region. There has been some success in attempts to control emissions from the rapidly growing industry, and SO_2 concentrations were reduced substantially during the 1980s (WHO/UNEP 1992), but ambient air quality standards for all traffic relevant pollutants are exceeded.

Africa

For the larger African cities — Cairo, Alexandria,

Nairobi, and Johannesburg—air pollution has been monitored for some years and there is an increasing awareness of the need for air quality management (WHO/UNEP 1992; GEMS/AIR 1996).

Urbanization is, however, increasing rapidly all over Africa and especially in the least developed countries, by up to 5% per year, the driving force being a mixture of population growth, natural disasters, and armed ethnic conflicts. Most African cities have been unable to keep pace with this development and lack adequate industrial and vehicle pollution control.

Much of the urban population growth is in coastal cities, e.g. in the Mediterranean area. So far general air pollution appears to be modest, but urban problems are emerging. In most countries, however, emission inventories are nearly nonexistent, pollution is neither monitored nor controlled, and there are no long-term records of pollution levels and impacts (UNEP 1997).

West Asia

In the past two decades West Asia has been radically transformed, with an urban growth rate above 4% per year. Although the most pressing urban problem seems to be waste management, air pollution is also emerging (UNEP 1997). In many cases protective trade regimes and a lack of environmental regulations have prevented adequate replacement of outdated polluting industries. Fuels with high sulfur content and old inefficient automobiles using leaded gasoline have exacerbated urban air pollution. Recently, however, the situation has been improving. In Oman new industries will be regulated with environmental standards.

East Asia

East Asia contains three of the world's largest countries (China, India, and Indonesia), several minor landlocked states, and a series of island states (including the highly industrialized Japan). The region is 35% urbanized and contains about half of the largest cities in the world. It therefore represents all stages of pollution development. Urbanization is not restricted to the continent

and the major island states, but is also seen as in-migration to the main island on small island states, e.g. in the Maldives. Urban congestion and air pollution is seen as a high-priority problem in many countries, such as China, India, Pakistan, Indonesia, Philippines, and Thailand (UNEP 1997).

Taken as a whole the rapid growth of energy use, combined with extensive use of coal in most of Asia, has resulted in a drastic increase in emissions of sulfur dioxide. Attempts to solve *local* problems by installing taller stacks only transformed them into an extensive *regional* pollution, and the acidification phenomena, known from North America and Europe, are now emerging.

In the important agricultural and industrial region of the Chinese Jiangsu province and Shanghai municipality SO_2 emissions are already high and are projected to double by the year 2010 (Chang *et al.* 1998). Model calculations demonstrate that in large regions the WHO guideline for long-term exposure $(40–60 \mu g\,m^{-3})$ is exceeded, as is in some regions even the one-hour guideline $(350 \mu g\,m^{-3})$. Without drastic measures the short-term guideline will by 2010 be exceeded in a large part of the province for more than 5% of the time. In line with this the new version of the Atmospheric Pollution Control Law passed by the National People's Congress in 1995 calls for reductions of emissions from power plants and other large coal users based on a permit system and emission taxes (Chang *et al.* 1998).

Aerosol analysis in 1987–92 by a privately established network (Hashimoto *et al.* 1994) demonstrated TSP levels for Chinese cities up to averages above $500 \mu g\,m^{-3}$ (Lanzhou), to be compared with a tentative WHO guideline of $120 \mu g\,m^{-3}$ as a one-hour average. Lower values in Seoul (around $70 \mu g\,m^{-3}$) showed a gradual increase.

Some of the highest reported levels of TSP, compared to what is seen in European cities, should, however, be evaluated critically, since not all is of human origin; high concentrations observed in Beijing in the December–April period are partly due to sand storms from the northern desert. Irrespective of the origin these particles are generally larger (Section 15.2.1) and thus have minor health impacts (Section 15.4.1).

A further complication, with problems for abatement strategies, is forest fires, which are generally blamed for causing smog over major cities in the ASEAN region (Hassan *et al.* 1997).

Urban transport is an increasing problem, which has been treated in a series of reports from the World Bank (Walsh 1996; Walsh & Shah 1997). A special aspect is the extensive use of motorbikes (Fig. 15.4), partly because they are the cheapest means of individual motorized transport for the expanding working class, partly because many Asian cities are too crowded to allow a drastic expansion of the automobile fleet. In 1992 motorcycles accounted for 27% of the vehicle fleet in Beijing, and for 65% in Guangzhou. This results in comparatively large emissions in relation to fuel consumption, since most motorcycles have two-stroke engines with poor pollution characteristics.

On a rapid path to industrialization, Taiwan is characterized by limited land and fast economic growth. Consequently, the environmental stress is serious. In the early 1990s the monthly SO_2 averages in Taipei were reported to be about 30 p.p.b. ($80 \mu g\,m^{-3}$), but a gradual decrease is expected concurrent with a reduction in permissible sulfur contents in fuels. The limit for diesel fuel was reduced to 0.05% in 1999. Lead in gasoline is also being phased out (Fang & Chen 1996).

In a series of major cities, where energy production is based on gas or low-sulfur coal (e.g. Bombay, Calcutta, Bangkok), SO_2 is not a serious problem, with average levels about $30 \mu g\,m^{-3}$. In many cases, however, TSP levels are above WHO guidelines.

15.7   CONCLUSIONS

Urban air pollution and its impact on urban air quality are a worldwide problem. This manifests itself differently in different regions, depending upon economic, political, and technological development, upon the climate and topography, and last — but not least — upon the nature and quality of the available energy sources. Nevertheless, a series of general characteristics emerges.

15.7.1   *From space heating to traffic*

Originally urban air pollution was a strictly local problem mainly connected with space heating and primitive industries. In the earlier stages of modern industrialization it was considered unavoidable or even a symbol of growth and prosperity. The situation in the industrialized western world has in most respects proved this viewpoint outdated. Emissions from industry and space heating are by and large controllable, but the urban atmosphere is now in most cities dominated by traffic emissions, with documented impacts on human health. Attention has thereby been shifted from sulfur dioxide and soot to nitrogen oxides, the whole spectrum of organic compounds, and particles of various sizes and composition, which are reported to be carcinogenic and/or cause a significant reduction in life expectancy through respiratory and cardiovascular diseases. These pollutants require much more detailed investigations in the form of chemical analysis and computer modeling.

The nature of material damage has shifted. In many respects the problems are diminishing, with reduced levels of sulfur dioxide, the introduction of more resistant materials, and the filtration of indoor air. On the other hand, the interaction between different pollutants and their impacts on, for example, sensitive electronic equipment are not fully understood.

In principle, control of emissions of sulfur and nitrogen oxides is relatively straightforward when they are related to power production in large plants, which can be compelled to use clean fuels and equipped with proper cleaning technology. Traffic emissions are more difficult to control, since they, in the nature of things, arise from small units. Meeting the increasingly stringent air pollution targets is therefore not an easy task. According to the conclusion of the "Auto Oil Programme," even with the maximum technical package introduced in the EU, not all cities will be able to comply.

The situation in developing countries is mixed. In some major cities in Asia sulfur emissions have been brought under control, e.g. via a transition from coal to natural gas, but particularly in China a rapid growth in energy production based on coal has resulted in increasing sulfur pollution on both urban and regional scales. In the developing countries, however, and in some economies in transition (including eastern Europe) traffic is becoming *the* problem. This is a challenge to city planning in these countries, where the long repressed wishes for private automobiles are difficult to reconcile with environmental protection.

15.7.2 Regional impacts on urban air pollution

The interactions between the cities and their surroundings are becoming increasingly important. With expanding and often merging urban areas and diminishing emissions in the cities proper, pollution levels can to a large extent be determined by long-range transport. The same applies to lead pollution in countries where lead has been removed from gasoline. Another example is photochemical air pollution, which in many cases is a large-scale phenomenon, where emissions and atmospheric chemical reactions in one country may influence urban air quality in another.

15.7.3 Cities as sources of pollution

A more far-reaching problem is the city as a source of pollution. In the past, local problems were attacked by dispersing pollutants from high stacks, but this only resulted in a transfer to a larger geographic scale in the form of acidification and other transboundary phenomena. Now long-lived greenhouse gases, and especially carbon dioxide, threaten the global climate—irrespective of their origin. This problem can only be solved by a general reduction in net emissions.

In the industrialized countries developments in technology and legislation to protect air and water quality have in many cases resulted in improved energy efficiency and emission reductions, although some means of improving urban air quality, such as catalytic converters, which consume energy and emit nitrous oxide (another greenhouse gas), are contrary to this goal. On a global basis, the growing population and its demand for a higher material standard of living have so far counteracted any reductions in total emissions. Therefore, responsibility for the future is both national and global, with many actors, comprising national environmental agencies, international organizations, the World Bank, etc.

Long-term abatement, however, has been intensified in recent years. An example is the ICLEI initiative, with the purpose of achieving and monitoring improvements in global environmental conditions through cumulative local actions.

15.7.4 A comprehensive approach

The realization that traffic is rapidly becoming *the* urban air quality problem in both industrialized and developing countries calls for comprehensive solutions, where traffic-related air pollution is seen in connection with other impacts of traffic, such as noise, accidents, congestion, and general mental stress. As a consequence, technological improvements in the form of less polluting vehicles are not sufficient. Support for infrastructure, where the need for transport is minimized, and where use of public means of transport dominates use of private automobiles and motorbikes, should be encouraged. Unfortunately, most attempts at control will be perceived as a restriction of individual freedom, and they are frequently met with outspoken opposition. Obviously, a change in attitude is called for.

ACKNOWLEDGMENTS

The author thanks colleagues who have provided material and offered advice during the preparation of this chapter—and especially Ole Hertel and Finn Palmgren. Further, he thanks the librarians at the National Environmental Institute, Birgit Larsen and Kit Andersen, and the lithographic artist Britta Munter.

REFERENCES

The literature relevant to urban-scale air pollution is

vast, and the references below are only given as typical examples. Some of the presented data have been used by different authors and are compiled from various sources. Not all references are therefore primary, but they are key to a full documentation. Note that many institutions publish series of reports. New Issues may have appeared after finalization of this chapter.

Adamson, S., Bates, R., Laslett, R. & Pototschnig, A. (1996) *Energy Use, Air Pollution, and Environmental Policy in Krakow. Can Economic Incentives Really Help?* World Bank Technical Paper no. 308. The World Bank, Washington DC.

Andersen, D. (1988) *The Everyday Life of the Danish Population 1987* (in Danish). Institute of Social Science, Copenhagen.

Anderson, H.R., Spix, C., Medina, S. *et al.* (1997) Air pollution and daily admissions for chronic pulmonary disease in 6 European cities: results from the APHEA project. *European Respiration Journal* **10**, 1064–71.

Anfossi, D. & Sandroni, S. (1997) Ozone levels in Paris one century ago. *Atmospheric Environment* **31**, 3481–2.

Arroyave, C. & Morcillo, M. (1995) The effect of nitrogen oxides in atmospheric corrosion of metals. *Corrosion Science* **37**, 293–305.

Ashmore, M. (1998) Impacts on urban vegetation and ecosystems. In Fenger, J. *et al.* (eds), *Urban Air Pollution, European Aspects.* Kluwer Academic Publishers, Dordrecht.

Bascom, R., Bromberg, P.A., Daniel, A. *et al.* (1996) Health effects of outdoor air pollution. *American Journal of Respiratory Critical Care Medicine* **153**, 3–50, 477–98.

Bellander, T., Svartengren, M., Berglind, N., Staxler, L. & Järup, L. (1999) *SHAPE. The Stockholm Study on Health Effects of Air Pollution and Their Economic Consequences.* Vägverket, Borlänge.

Berkowicz, R. (1998) Street scale models. In Fenger, J. *et al.* (eds), *Urban Air Pollution, European Aspects.* Kluwer Academic Publishers, Dordrecht.

Berkowicz, R. (2000) A simple model for urban background pollution. *Environmental Monitoring and Assessment.*

Bitan, A. (1992) The high climatic quality city of the future. *Atmospheric Environment* **26B**, 313–29.

Bollen, J.C., Hettelingh, J.-P. & Maas, R.J.M. (1996) *Scenarios for Economy and Environment in Central and Eastern Europe.* RIVM report no.481505002. Netherlands Institute of Public Health and the Environment.

Brimblecombe, P. (1977) London air pollution. *Atmospheric Environment* **11**, 1157–62.

Brimblecombe, P. (1987) *The Big Smoke. A History of Air Pollution in London since Medieval Times.* Methuen, London.

Brimblecombe, P. (1990) The composition of museum atmospheres. *Atmospheric Environment* **24B**, 1–8.

Brown, K.S. (1997) A decent proposal? EPA's new clean air standards. *Environmental Health Perspectives* **105**, 378–83.

Brown, L.R., Renner, M. & Flavin, C. (1998) *Vital Signs 1998. The Environmental Trends that Are Shaping Our Future.* W.W. Norton & Company, Inc. New York.

Brown, V.J. (1999) The worst of both worlds: poverty and politics in the Balkans. *Environmental Health Perspectives* **107**, A606–13.

Brunekreef, B. (1997) Air pollution and life expectancy: is there a relation? *Occupational and Environmental Medicine* **54**, 781–4.

Brunekreef, B. (1999) All but quiet on the particulate front. *American Journal of Respiratory Critical Care Medicine* **159**, 354–6.

Chang, Y.S., Arndt, R.L., Calori, G., Carmichael, G.R., Streets, D.G. & Sue, H.P. (1998) Air quality impacts as a result of changes in energy use in China's Jiangsu province. *Atmospheric Environment* **32**, 1383–95.

Cocheo, V., Sacco, P., Boaretto, C. *et al.* (2000) Urban benzene and population exposure. *Nature* **40**, 141.

CORINAIR (1996) *Atmospheric Emission Inventory Guidebook.* European Environment Agency, Copenhagen.

Davidson, C.I. (1979) Air pollution in Pittsburgh: A historical perspective. *Air Pollution Control Association* **29**, 1035–41.

Dennis, R.L., Byon, W.D., Novak, J.H., Galuppi, K.J. & Coats, C.J. (1996) The next generation of integrated air quality modeling: EPA's models-3. *Atmospheric Environment* **30**, 1925–38.

Dockery, D.W., Pope, C.A. III, Xu, X. *et al.* (1993) An association beteween air pollution and mortality in six US cities. *New England Journal of Medicine* **239**, 1753–9.

Edwards, L. (1998) Limit values. In Fenger, J. *et al.* (eds), *Urban Air Pollution, European Aspects.* Kluwer Academic Publishers, Dordrecht.

EEA (1998) *Europe's Environment. The Second Assessment.* European Environment Agency, Copenhagen.

Eggleston, S., Hackman, M.P., Heyes, C.A., Irwin, J.G., Timmis, R.J. & Williams, M.L. (1992) Trends in urban air pollution in the United Kingdom during recent decades. *Atmospheric Environment* **26B**, 227–39.

Ellis, J. & Tréanton, K. (1998) Recent trends in energy-related CO_2 emissions. *Energy Policy* **26**, 159–66.

EMEP/MSC-W (1998) *Transboundary Acidifying Air Pollution in Europe*. Research Report no. 66. Norwegian Meteorological Institute, Oslo.

EMEP/MSC-W (1999) *Transboundary Acidifying Air Pollution in Europe*. Research Report no. 83. Norwegian Meteorological Institute, Oslo.

ENS (1997) *Energy Statistics* (in Danish). Danish Energy Agency, Copenhagen.

Enya, T., Suzuki, H., Watanabe, T., Hirayama, T. & Hisamatsu, Y. (1997) 3-Nitrobenzanthone, a powerful bacterial mutagen and suspected human carcinogen found in diesel exhaust and airborne particles. *Environmental Science and Technology* **31**, 2772–6.

European Commission (1995) *Towards Fair and Efficient Pricing in Transport*. COM(95)691. European Communities, Luxembourg.

European Commission (1996a) *The European Auto Oil Programme*. A report by the Directorate Generals for Industry, Energy, and Environment. European Communities, Luxembourg.

European Commission (1996b) *Air Quality Report of the European Auto Oil Programme*. Report of Sub Group 2. European Communities, Luxembourg.

European Commission (1998) *Economic Evaluation of Air Quality Targets for Sulfur Dioxide, Nitrogen Dioxide, Fine and Suspended Particulate Matter and Lead*. European Communities, Luxembourg.

Faiz, A., Weaver, C.S. & Walsh, M.P. (1996) *Air Pollution from Motor Vehicles. Standards and Technologies for Controlling Emissions*. World Bank, Washington, DC.

Fang, S.-H. & Chen, H.-W. (1996) Air quality and pollution control in Taiwan. *Atmospheric Environment* **30**, 735–41.

Fenger, J., Hertel, O. & Palmgren, F. (eds) (1998) *Urban Air Pollution. European Aspects*. Kluwer Academic Publishers, Dordrecht.

Finlayson-Pitts, B. & Pitts, J.N. Jr (1997) Tropospheric air pollution, ozone, airborne toxics, polycyclic aromatic hydrocarbons, and particles. *Science* **276**, 1045–52.

Friedrich, R. & Schwarz, U.-B. (1998) Emission inventories. In Fenger, J. *et al.* (eds), *Urban Air Pollution, European Aspects*. Kluwer Academic Publishers, Dordrecht.

Gaardmand, A. (1993) *Land Use Planning in Denmark 1938–1992* (in Danish). Arkitektens Forlag, Copenhagen.

GEMS/AIR (1996) *Air Quality Management and Assessment in 20 Major Cities*. United Nations Environment Programme and World Health Organization, New York.

Goklany, I.M. (1995) Richer is cleaner. In Bailey, R. (ed.), *The True State of the Planet*. The Free Press, New York.

Goldreich, Y. (1995) Urban climate studies in Israel—a review. *Atmospheric Environment* **29**, 467–78.

Grosjean, D., Grosjean, E. & Williams, E.L. II (1993) Fading of artists' colorants by a mixture of photochemical oxidants. *Atmospheric Environment* **27A**, 765–72.

Grossman, G.M. & Krueger, A.B. (1995) Economic growth and the environment. *The Quarterly Journal of Economics* **110**, 353–377.

Guicherit, R. & van Dop, H. (1977) Photochemical production of ozone in Western Europe (1971–1975) and its relation to meteorology. *Atmospheric Environment* **11**, 145–55.

Hasford, B. & Fruhmann, G. (1998) Air pollution and daily admissions for chronic obstructive pulmonary disease in six European cities: results from the APHEA project. *European Respiration Journal* **11**, 992–3.

Hashimoto, Y., Sekine, Y., Kim, H.K., Chen, Z.L. & Yang, Z.M. (1994) Atmospheric fingerprints of East Asia, 1986–1991. An urgent record of aerosol analysis by the Jack Network. *Atmospheric Environment* **28**, 1437–45.

Hassan, H.A., Taha, D., Dahalan, M.P. & Mahmud, A. (eds) (1997) *Transboundary Pollution and the Sustainability of Tropical Forests*. Asean Institute of Forest Management, Kuala Lumpur.

HLU (1994) *Air Quality in the Copenhagen Area* (in Danish). Environmental Protection Agency, Copenhagen.

Holgate, S.T., Samet, J.M., Koren, H.S. & Maynard, R.L. (eds) (1999) *Air Pollution and Health*. Academic Press, London.

Horvath, H. (1998) Reduction of visibility. In Fenger, J. *et al.* (eds), *Urban Air Pollution, European Aspects*. Kluwer Academic Publishers, Dordrecht.

ICLEI (2000) *Recent Publications*. The International Council For Local Environmental Initiatives (http://www.iclei.org/iclei/icleipub.htm).

Jantunen, M.J., Hänniken, O., Katsouyanni, K. *et al.* (1998) Air pollution exposure in European cities. The "EXPOLIS" study. *Journal of Exposure Analysis and Environmental Epidemiology* **8**, 495–518.

Jensen, F.P. & Fenger, J. (1994) The air quality in Danish urban areas. *Environmental Health Perspectives* **102**(suppl. 4), 55–60.

Jensen, S.S. (1998) Mapping human exposure to traffic air pollution using GIS. *Journal of Hazardous Materials* **61**, 385–92.

Jol, A. & Kielland, G. (eds) (1997) *Air Pollution in Europe 1997*. EEA Environmental Monograph no. 4. European Environment Agency, Copenhagen.

Kallos, G. (1998) Regional/mesoscale models. In Fenger, J. *et al.* (eds), *Urban Air Pollution, European Aspects*. Kluwer Academic Publishers, Dordrecht.

Kassomenos, P., Skouloudis, A.N., Lykoudis, S. & Flocas, H.A. (1999) "Air-quality indicators" for uniform indexing of atmospheric pollution over large metropolitan areas. *Atmospheric Environment* **33**, 1861–79.

Katsouynni, K., Touloumi, G., Spix, C. *et al.* (1997) Short term effects of ambient sulfur dioxide and particulate matter on mortality in 12 European cities: results from time series data from the APHEA project. *British Medical Journal* **314**, 1658–63.

Kemp, K. & Palmgren, F. (1999) *The Danish air quality monitoring programme*. Annual Report for 1998. NERI Technical Report no. 296. National Environmental Research Institute, Roskilde.

Kemp, K., Palmgren, F. & Mancher, O.H. (1998) *The Danish air quality monitoring programme*. Annual Report for 1997. NERI Technical Report no. 245. National Environmental Research Institute, Roskilde.

Komeiji, T., Aoki, K. & Koyama, I. (1990) Trends of air quality and atmospheric deposition in Tokyo. *Atmospheric Environment* **24A**, 2099–103.

Krawack, S. (ed.) (1991) Air pollution from individual and public transport (in Danish). *Miljøprojekt* 165. Danish EPA, Copenhagen.

Krupnick, A.J. & Portney, P.R. (1991) Controlling urban air pollution: a benefit–cost assessment. *Science* **252**, 522–8.

Kucera, V. & Fitz, S. (1995) Direct and indirect air pollution effects on materials including cultural monuments. *Water, Air and Soil Pollution* **85**, 153–65.

Kucera, V., Pearce, D. & Brodin, Y.-W. (eds) (1996) *Economic Evaluation of Air Pollution Damage to Materials*. Swedish EPA, Stockholm.

Kurashige, Y. & Miyashita, A. (1998) How many days can Mt.Fuji and the Tokyo Tower be seen from the Tokyo suburban area. *Journal of the Air and Waste Management Association* **48**, 763–5.

Larsen, J.C. & Larsen, P.B. (1998) Chemical carcinogens. In Hester, R.E. & Harrison, R.M. (eds), *Air Pollution and Health. Issues in Environmental Science and Technology 10*. The Royal Society of Chemistry, Cambridge.

Larssen, S. & Hagen, L.O. (1997) *Air Pollution Monitoring in Europe—Problems and Trends*. Topic report 26. European Environment Agency, Copenhagen.

Lynn, W.R. (1999) Megacities. Sweet dreams or environmental nightmares. *Environmental Science and Technology News* 238A–40A.

Mage, D., Ozolins, G., Peterson, P. *et al.* (1996) Urban air pollution in megacities of the world. *Atmospheric Environment* **30**, 681–6.

Mage, D., Wilson, W., Hasselblad, V. & Grant, L. (1999) Assessment of human exposure to ambient particulate matter. *Journal of the Air and Waste Management Association* **49**, 1280–91.

Mamane, Y. (1987) Air pollution control in Israel during the first and second century. *Atmospheric Environment* **21**, 1861–3.

Marico, M.M., Podsiadlik, D.H. & Chase, R.E. (1999) Examination of size-resolved and transient nature of motor vehicle particle emissions. *Environmental Science and Technology* **33**, 1618–26.

Mayerhofer, P., Weltschev, M., Trukenmüller & Friedrich, R. (1995) A methodology for the economic assessment of material damage caused by SO_2 and NO_x emissions in Europe. *Water, Air and Soil Pollution* **85**, 2687–92.

Maynard, R.L. & Howard, C.V. (eds) (1999) *Particulate Matter. Properties and Effects upon Health*. Bios Scientific Publishers, Oxford.

McMichael, A.J., Anderson, H.R., Brunekreef, B. & Cohen, A.J. (1998) Inappropriate use of daily mortality analyzes to estimate longer-term mortality effects of air pollution. *International Epidemiological Association* **27**, 450–3.

Mestayer, P.G. & Anquetin, S. (1995) Climatology of cities. In Gyro, A. & Rys, F.-S. (eds) *Diffusion and Transport of Pollutants in Atmospheric Mesoscale Flow Fields*. Kluwer Academic Publishers, Dordrecht.

Newby, P.T., Mansfield, T.A. & Hamilton, R.S. (1991). Sources and economic implications of building soiling in urban areas. *Science of the Total Environment* **100**, 347–65.

Nielsen, H.P. (1997) Localization, transport means and urban structure (in Danish). *Byplan* 6/97.

Nielsen, T. (1996) Traffic contribution of polycyclic aromatic hydrocarbons in the center of a large city. *Atmospheric Environment* **20**, 3481–90.

Nielsen, T., Feilberg, A. & Binderup, M.L. (1999) The variation of street air levels of PAH and other mutagenic PAC in relation to regulations of traffic emissions and the impact of atmospheric processes. *Environmental Science and Pollution Research* **6**, 133–7.

Novak, D.J., Civerolo, K.L., Rao, S.T., Sistla, G., Luley, C.J. & Crane, D.E. (2000) A modeling study of the impact of urban trees on ozone. *Atmospheric Environment* **34**, 1601–13.

Nriagu, J.O. (1990) The rise and fall of leaded gasoline. *Science of the Total Environment* **92**, 13–28.

OECD (1995a) *Motor Vehicle Pollution. Reduction Strategies beyond 2010.* Organization for Economic Co-operation and Development, Paris.

OECD (1995b). *Urban Travel and Sustainable Development.* Organization for Economic Co-operation and Development, Paris.

OECD (1998) *Advanced Air Quality Indicators and Reporting.* Organization for Economic Co-operation and Development, Paris.

OECD (1999) *OECD Environmental Data. Compendium 1999.* Organization for Economic Co-operation and Development, Paris.

O'Meara, M. (1999) *Reinventing Cities for People and the Planet.* Worldwatch Paper 147.

Onursal, B. & Gautam, S.P. (1997) *Vehicular Air Pollution. Experiences from Seven Latin American Urban Centers.* World Bank Technical Paper no. 373. World Bank, Washington, DC.

Palmgren, F., Berkowicz, R., Hertel, O. & Vignati, E. (1996) Effects of reduction of NO_x on NO_2 levels in urban streets. *Science of the Total Environment* **189/190**, 409–15.

Phillips, D.R. (1993) Urbanization and human health. *Parasitology* **106**, S93–107.

Pichl, P. (ed.) (1998) *Exchange of Experiences between European Cities in the Field of Clean Air Planning, CO_2 Reduction and Energy Concepts.* Umwelt Bundes Amt, Berlin.

Pope, C.A. III, Dockery, D.W. & Schwartz, J. (1995) Review of epidemiological evidence of health effects of particulate air pollution. *Inhalation Toxicology* **7**, 1–18.

Population Reference Bureau (1999) *World Population Data Sheet.* Population Reference Bureau, Washington, DC.

Renner, R. (2000) Pollution monitoring should get personal, scientists say. *Environmental Science and Technology News* 64–65A.

Richter, D.U.R. & Williams, W.P. (1998) *Assessment and Management of Urban Air Quality in Europe.* EEA Monograph no. 5. European Environment Agency, Copenhagen.

Schmidt, A.L., Bronée, P., Kemp, K. & Fenger, J. (1999) Airborne dust on museum exhibits. Paper presented to

Air Quality in Europe: Challenges for the 2000s, Venice, May 19–21.

Schwartz, J., Dockery, D.W. & Neas, L.M. (1996) Is daily mortality associated specifically with fine particles? *Journal of the Air and Waste Management Association* **46**, 927–39.

Schwela, D. & Zali, O. (eds) (1999) *Urban Traffic Pollution.* E & FN Spon, London.

Selden, T.M., & Song, D. (1994) Environmental quality and development: is there a Kuznets curve for air pollution emissions? *Journal of Environmental Economics and Management* **27**, 147–62.

Shafik, S. (1994) Economic development and environmental quality: an econometric analysis. *Oxford Economic Papers* **46**, 757–73.

Skov, H., Christensen, C.S., Fenger, J., Larsen, D. & Sørensen, L. (2000) Exposure to indoor air pollution in a reconstructed house from the Danish Iron Age. *Atmospheric Environment* **34**, 3801–4.

Smith, K.R. (1988) Air pollution. Assessing total exposure in developing countries. *Environment* **30**, 16–20, 28–30, 33–5.

Smith, K.R. (1993) Fuel combustion, air pollution exposure, and health: the situation in the developing countries. *Annual Review of Energy and the Environment* **18**, 529–66.

Spix, C., Anderson, H.R., Schwartz J. *et al.* (1998) Short-term effects of air pollution on hospital admission of respiratory diseases in Europe: a quantitative summary of APHEA study results. *Archives of Environmental Health* **53**, 54–64.

Stanners, D. & Bourdeau, P. (eds) (1995) *Europe's Environment: The Dobris Assessment.* European Environment Agency. Office for Publications of the European Communities, Luxemburg.

Thomas, V.M., Socolow, R.H., Fanelli, J.T. & Spiro, T.G. (1999) Effects of reducing lead in gasoline: an international experience. *Environmental Science and Technology* **33**, 3942–8.

Tidblad, J. & Kucera, V. (1998) Materials damage. In Fenger, J. *et al.* (eds), *Urban Air Pollution, European Aspects.* Kluwer Academic Publishers, Dordrecht.

Tuch, Th., Heyder, J., Heinrich, J. & Wichmann, H.E. (1997) *Changes of the Particle Size Distribution in an Eastern German City.* Presented at the Sixth International Inhalation Symposium, Hannover Medical School.

UN (1997a) *Report of the United Nations Conference on Human Settlements (Habitat II).* United Nations, New York.

UN (1997b) *Statistical Yearbook 1995.* United Nations, New York.

UNEP (1991) *Urban Air Pollution.* UNEP/GEMS Environment Library No 4. United Nations Environment Programme, Nairobi.

UNEP (1997) *Global Environment Outlook.* United Nations Environment Programme, Nairobi.

UNEP/WHO (1994) *GEMS/AIR Methodology Reviews Volume 3: Measurement of Suspended Particulate Matter in Ambient Air.* United Nations Environment Programme, Nairobi.

Vignati, E., Berkowicz, R. & Hertel, O. (1996) Comparison of air quality in streets of Copenhagen and Milan, in view of the climatological conditions. *Science of the Total Environment* **189/190**, 467–73.

Volz, A. & Kley, D. (1988) Evaluation of the Montsouris series of ozone measurements made in the nineteenth century. *Nature* **332**, 240–2.

Wallace, L. (1996) Indoor particles: a review. *Journal of the Air and Waste Management Association* **46**, 98–126.

Walsh, M.P. (1996) Motor vehicle pollution control in China: an urban challenge. In Stares, S. & Zhi, L. (eds), *China's Urban Development Strategy.* World Bank Discussion Paper no. 353. World Bank, Washington, DC.

Walsh, M. & Shah, J.J. (1997) *Clean Fuels for Asia. Technical Options for Moving toward Unleaded Gasoline and Low-sulfur Diesel.* World Bank Technical Paper no. 377. World Bank, Washington, DC.

Whitby, K.T. & Sverdrup, G.M. (1980) California aerosols: their physical and chemical characteristics. *Advances in Environmental Science and Technology* **9**, 477–517.

WHO/UNEP (1992) *Urban Air Pollution in Megacities of the World.* Blackwell, Oxford.

WHO (1995) *Concern for Europe's Tomorrow: Health and Environment in the WHO European Region.* Wissenschaftliche Verlag, Stuttgart.

WHO (1987/2000) *Air Quality Guidelines for Europe.* World Health Organization, Copenhagen.

Wiederkehr, P. & Yoon, S.-J. (1998) Air quality indicators. In Fenger, J. *et al.* (eds), *Urban Air Pollution, European Aspects.* Kluwer Academic Publishers, Dordrecht.

Wilson, W.E. & Suh, H.H. (1997) Fine particles and coarse particles: concentration relationships relevant to epidemiologic studies. *Journal of the Air and Waste Management Association* **47**, 1238–49.

World Commission on Environment and Development (1987) *Our Common Future.* Oxford University Press, Oxford.

Zeger, S.L., Dominici, F. & Samet, J. (1999) Harvesting-resistant estimates of air pollution effects on mortality. *Epidemiology* **10**, 171–5.

Zhiqiang, Q., Siegmann, K., Keller, A., Matter, U., Sherrer, L. & Siegmann, H.C. (2000) Nanoparticle air pollution in major cities and its origin. *Atmospheric Environment* **34**, 443–51.

Zlatev, Z., Fenger, J. & Mortensen, L. (1996) Relationships between emission sources and excess ozone concentrations. *Computers Mathematics Application* **32**, 101–23.

16 Atmospheric Monitoring Techniques*

ROD ROBINSON

16.1 INTRODUCTION

A diverse range of measurement techniques has been developed to meet the many requirements that exist for atmospheric monitoring. It is not possible in a single chapter to discuss all of these in full technical detail. Therefore, the aim of this chapter is to give an overview of the techniques that are routinely available. It covers techniques for monitoring atmospheric gases, particularly those associated with trace gases and air pollution and the measurement of particulate matter. Techniques for determining meteorological parameters, such as wind, temperature, and pressure, are also addressed.

The monitoring techniques are divided into those that make measurements of the lower troposphere, and those that make measurements of the middle to upper atmosphere (upper troposphere and stratosphere). Included in the latter group are techniques that measure the total column content or concentration profile through the whole atmosphere.

The techniques described range from relatively inexpensive and simple technologies for routine monitoring of ambient air to specialized satellite-borne remote sensing equipment. The intention is to make the reader aware of a wide range of capabilities, and to give some indication of the applications in which these may be appropriate.

16.2 REQUIREMENTS

Measurements are made to fulfill certain requirements, and these drive the development of monitoring techniques. Requirements for monitoring in the lower troposphere are dominated by the regulatory need for routine measurements of air quality and therefore focus on pollutants in ambient air. The requirements for measurements of the middle to upper atmosphere are primarily driven by the needs of atmospheric research, including the provision of data to support climate modeling. The most common measurements in the upper atmosphere are of ozone, water vapor, and those trace-gas species that are active in atmospheric chemistry. Requirements are developed by organizations such as the World Meteorological Organization (WMO) within frameworks such as its Global Atmosphere Watch (GAW).

Factors to be considered when selecting a monitoring method include: the species to be measured, the required measurement uncertainty, time scale, spatial coverage, and monitoring location. Examples of the requirements for measurements include: atmospheric research to aid in the understanding of the transport of pollutants, investigation of chemical reactions, validation of models, and informing policy decisions. Legislative requirements include assessment of the levels of pollutants to meet regulations designed to protect the environment and public health. Short-term local variations may be of interest, such as the effect of traffic density on local air quality, or long-term data, such as the yearly average of criteria pollutants, may be required. There are nearly as many techniques for monitoring as there are reasons to monitor. They range from cutting-edge research instruments to well established standardized methods. Inevitably trade-offs are required; for example, instruments capable of making high time resolution or spatially resolved measurements are expensive and are often not suitable for long-term continuous operation.

16.3   STANDARDIZED METHODS

It is important when choosing a monitoring technique to be able to clearly specify the measurement requirements. As an example, one of the primary reasons for air quality monitoring is to satisfy legislation; this legislation clearly defines the measurement requirements. Such monitoring is targeted at locations close to ground level where there is most likely to be a direct impact on human health. The European Commission has set objectives for air quality in the Air Quality Framework Directive and its Daughter Directives (Commission of the European Communities 1996, 1999, 2000). The species covered under the framework directive are carbon monoxide, sulfur dioxide, nitrogen dioxide, ozone, benzene, lead, mercury cadmium, arsenic, and particulates. The Framework Directive and its Daughter Directives specify the limit values, monitoring strategies, and data quality objectives. The monitoring requirements, and the associated data quality objectives, are categorized into bands, depending on concentrations measured. For example, if concentrations are close to the limit values, continuous monitoring to strict data quality objectives is required; if lower levels are found, then more indicative monitoring may be carried out. At the lowest level of requirements data may be estimated purely from modeling. The measurements are used to calculate average values over defined time periods for comparison with air quality objectives and in such cases the time scale of the measurements is clearly defined.

Comprehensive measurement requirements allow a well informed decision to be made on the appropriateness of proposed monitoring techniques. Monitoring systems can be assessed to ensure compliance with these requirements. A number of approaches may be adopted to assess whether a method meets the data quality requirements. If the method is standardized as a national or international standard—for example, by the European Committee for Standardization, CEN— then appropriate validation may have been carried out as part of the development of the standard. Where commercial instrumentation is being considered, product certification can aid the selection of an appropriate measuring system. Schemes such as the UK Environment Agency's Monitoring Certification Scheme, MCERTS (EA 2000), provide certification of instruments to performance standards. These standards include assessment of characteristics such as linearity, drift, and susceptibility of the instrument to interference from other species. Based on such performance characteristics the measurement uncertainty (ISO 1995, 2001) can be calculated and an assessment of the appropriateness of the monitoring system for a particular application can be made. It is important to include the measurement procedure in the assessment of a method. The use of a certified instrument does not in itself guarantee data quality, as the protocol for the operation of the instrument must also be considered. This includes aspects such as calibration procedures, quality assurance, and quality control procedures and data handling.

16.4   SAMPLING TECHNIQUES

One of the fundamental components of atmospheric monitoring is the technique used to sample the atmosphere (Boubel et al. 1994). The sampling methods used can be broadly divided into three categories: the collection of samples for subsequent analysis, direct sampling with online analysis, and remote sampling.

The collection of samples for subsequent analysis may be passive or active. Passive techniques are those such as diffusive samplers, where the sampler is placed in the atmosphere and the quantity to be measured is collected directly from the ambient air. Passive sampling generally takes place over a relatively long time scale. Passive samplers provide low-cost measurement solutions; for example, a large number of samplers can be deployed to screen a wide area to detect air quality hot spots. Active samplers draw a volume of air into a collecting medium. This may be a canister or a sorbent cartridge. Active sampling techniques have much shorter sampling periods than passive sampling, typically of a few hours duration.

Direct sampling systems are automated measuring systems (AMS) combining an active sampling stage with online analysis. They are generally self-contained units that output a digital or analog signal in concentration units. Typically such systems have sampling periods of a few minutes to an hour. Most automatic monitoring systems produce results that are nearly instantaneous measurements; in general they sample rapidly enough to be considered as continuous monitors. The time series of measurements they produce allow short-term variations in the measured quantity to be observed. By contrast, most systems that are used to collect samples for subsequent analysis do so over an extended period (this is usually necessary in order to collect a sufficient quantity for analysis). They are suitable for measurements with a relatively long averaging period and provide truly continuous sampling during this period. The advantage of such samplers over the instrumental systems is in general their simplicity, small size, and low cost.

The sampling techniques described above are all point sampling methods, sometimes known as *in situ* samplers. They measure the atmosphere at a single localized point. Such systems are often fixed, though they may be mounted on a mobile platform; for example, on balloon payloads.

In contrast, remote samplers measure the concentration of atmospheric compounds at a location, or locations, separate from the sensing equipment itself. Remote sampling techniques used in atmospheric monitoring generally employ optical or microwave radiation to probe the atmosphere, though acoustic techniques are also used. In the case of a remote sampling system the sample is defined by the volume of the atmosphere probed by the sensing beams. Systems that measure the path-integral concentration of a gas along a sensing beam, typically referred to as long path or open path sensors, are included in this category.

16.5 EXPRESSION OF RESULTS

The concentration of gaseous compounds in the atmosphere is usually expressed in one of two ways, as a mass fraction ($mg\,m^{-3}$) or as a volume or molar fraction (parts per million, p.p.m., or parts per billion, p.p.b.). Mass fraction units must be expressed under stated conditions of temperature and pressure, as the volume of a gas depends on these parameters. Current European Air Quality Directives require the measurements of concentration to be given in units of micrograms per cubic meter referenced to a temperature of 293 K and a pressure of 101.3 kPa. It is usual to use the ideal gas law to derive the conversion factor between p.p.m. units (which are invariant under different temperature and pressure) and mass fractions. The general form of the conversion is given in eqn 16.1.

$$\text{Conc}(mg\,m^{-3}) = \frac{\text{Conc(p.p.m.)} \times \text{Molecular weight(g)}}{\text{Molar volume}} \quad (16.1)$$

This assumption of the ideal behavior of the gases is reasonable for the concentrations likely to be observed in ambient air. The nonideal behavior (due to compressibility) of gases such as sulfur dioxide will only be apparent at percentage concentration levels. For sulfur dioxide the compressibility will lead to a change in the molar volume of the sample of roughly 1.5% at a concentration of approximately 50% molar fraction.

16.6 MONITORING AIR QUALITY

This section describes techniques for monitoring

the air that surrounds us, specifically in the planetary boundary layer and lower troposphere. The initial part describes measurement techniques for determining air pollutant concentrations, while the monitoring of meteorological parameters is covered in a later section.

A significant amount of monitoring is carried out to determine and control the emission of pollutants into the atmosphere; such monitoring is not explicitly covered here as the measurements are made at source rather than in the atmosphere. Many of the measurement techniques discussed are appropriate as the basis for the measurement of the emissions of atmospheric pollutants; however, the specific application of the techniques will be different due to the particular measurement problems associated with emissions monitoring.

16.6.1 Wet chemical techniques

Some of the earliest routine measurement techniques were wet chemical methods. These techniques involve collecting a gaseous species by drawing sampled air through a liquid in which the gas to be measured dissolves. The solution is subsequently analyzed, usually after transfer to an analytic laboratory, though some on-line automated systems have been developed. The early techniques for analysis were often based on titration.

One of the earliest national air quality networks was the black smoke and sulfur dioxide monitoring network in the UK. This started daily monitoring in the 1960s. This network utilized a wet chemistry technique for monitoring sulfur dioxide. In this technique the sample gas is drawn through a dilute acidified solution of hydrogen peroxide in Drechsel bubblers. Sulfur dioxide in the sample gas reacts to form sulfuric acid in solution. The concentration of sulfuric acid is then subsequently determined by titration with an alkali. The absorbing solution is acidified to pH 4.5 to encourage the absorption of the strongly acidic sulfur dioxide in preference to compounds such as carbon dioxide. However, the technique is not completely selective to sulfur dioxide, and any other acidic compounds in the atmosphere will act as interfer-

ents. Monitoring stations using this technique continue to be operated in the UK to provide measurements of sulfates in support of acid deposition monitoring. Alternative wet chemical methods that can be used to analyze sulfur dioxide concentrations include ion chromatography, the tetrachloromercurate (TCM) method, and the Thorin method. The standard developed by the International Standardization Organization (ISO) to define the TCM method, ISO 6767:1990, has recently been withdrawn, as has ISO 4221:1980, which defined the Thorin spectrophotometric method.

Additional wet chemical methods will not be described in detail. In general, the wet chemistry techniques have been superseded by automated instrumental methods, particularly for the criteria pollutants identified in the European Air Quality Directives. Automated instruments have the advantage that they require little operator intervention. However, wet chemical techniques can be tailored to very specific measurements, can be very sensitive, and are relatively cheap, though they do require experienced analytic personnel.

16.6.2 Diffusive and permeation samplers

Some of the simplest samplers to deploy are those based on passive sampling techniques. Passive samplers fall into two categories: diffusive samplers and permeation samplers. Passive samplers are placed in the air to be monitored and collect the target gas over a relatively long period by diffusion into the sampling material. They are then sealed and returned to a laboratory for analysis to determine the concentration, or mass, of target species collected. They can be employed for personal exposure monitoring for occupational health, as they can easily be worn as a badge. A disadvantage of these techniques is that because they do not actively draw the air sample onto the sampling medium there is no direct means to measure the sampling volume. How the sample volume is determined depends on the details of the diffusion technique and the design of the sampler.

Diffusive samplers rely on establishing a concentration gradient between the air mass being

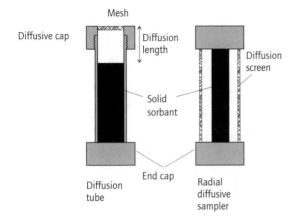

Fig.16.1 Examples of diffusive sampler designs.

sampled and an absorbing medium upon which the gas will be captured. In their most common form the samplers consist of a small diameter tube containing the absorber, as illustrated in Fig. 16.1. The distance between a well defined inlet and the surface of the absorber defines the diffusive length and this establishes the gradient along which diffusion of the gas takes place. A diffusion cap, consisting of a thin mesh, is used to define the inlet. This helps to stop external air movements from disturbing the air mass inside the diffusion tube and also stops the ingress of large bodies. The diffusion of the gas over time is given in eqn 16.2, which is derived from Fick's diffusion law.

$$m = \frac{D(c_1 - c_2)At}{l} \qquad (16.2)$$

where: m is mass of gas transferred along diffusion length; D is diffusion coefficient; c_1 is concentration of gas at entrance to tube; c_2 is concentration of gas at sorbent surface; A is cross-sectional area of tube; t is time; l is diffusion length.

The diffusion rate of a gas depends on the cross-section of the tube, the diffusion length, the diffusion coefficient of the gas, and the concentration difference between the end of the tube and the surface of the absorbent. Ideally the concentration of the gas will be zero at the surface of the absorbent, due to absorption of the gas into the body of the ab-

sorbent. If the absorbent surface becomes saturated such that the number of sites at which molecules can attach are significantly reduced then the concentration of the gas at the surface of the sorbant will be nonzero. This will cause the diffusion rate to reduce over time. For a given ambient concentration the diffusive uptake rate is given by eqn 16.2.

The performance of a diffusive sampler depends critically upon the selection of an appropriate absorbent. If an absorbent with a high efficiency is used the concentration of the target gas at the absorbent surface will remain near zero and the uptake rate of the diffusive sampler will be close to its theoretical value. Diffusion uptake rates for typical sorbents used for sampling hydrocarbons are tabulated in ISO/DIS 16017-2 (1999 Draft).

A further issue with diffusive samplers that affects the uptake rates is known as back diffusion. This occurs if the concentration at the absorbent surface is higher than the ambient concentration, perhaps due to fluctuations in the ambient concentration. In this situation diffusion from the surface of the absorbent to the external air may occur. This issue is investigated in Bartley *et al.* (1987). A diffusive sampler will correctly integrate fluctuating ambient concentrations if the sampling time is significantly longer than the time taken for a single molecule to diffuse onto the absorbent. This is typically of the order of 10 seconds.

An additional influence on diffusive samplers is the wind velocity, which can change the effective diffusion length. At low wind speeds it is possible for a stable layer of air, or boundary layer, to form outside the sampler. The sample gas is depleted in this stable layer and this increases the effective diffusion length of the sampler. At high wind speeds turbulent air may disrupt the air within the tube, effectively reducing the diffusion length. The effect of low wind speeds may be reduced by careful tube design; a long thin tube will have an external boundary layer that will be insignificant compared to the internal tube length, whereas a short, wider diameter, tube will be more sensitive to this effect. For high wind speeds a cap containing a fine mesh will act as a screen to stop turbulent air affecting the air inside the tube.

Precipitation may also affect the sampler by blocking the diffusive opening and it is therefore usual to mount the samplers in some form of shelter to protect them from rain and snow, and to stop direct sunlight from overheating the absorbent. The open ends of the samplers should not be shielded from wind, as this will affect their sampling characteristics. For nitrogen dioxide monitoring the tubes are generally clear plastic. Care must be taken to ensure that the material is opaque to ultraviolet radiation to avoid the photochemical conversion of nitric oxide to nitrogen dioxide occurring within the sampler. For monitoring hydrocarbons the tubes are usually stainless steel or brass, but for labile compounds such as dioxins or strongly polar compounds nonreactive tubes are required.

Permeation samplers use a thin permeable membrane, across which the gas molecules selectively permeate onto a sorbent. Samplers have been developed which use a porous diffusive layer surrounding an inner absorbent tube, also shown in Fig. 16.1. These samplers often have a greater surface area than tube samplers and therefore have faster uptake rates and hence are often used for short-term workplace and personnel exposure monitoring. These samplers may become saturated if used for long-term monitoring.

A number of standard methods have been defined for the design and use of diffusive samplers including ISO (1999). A key component of monitoring methods using diffusive samplers is the use of field blanks. Field blanks are treated identically to the measurement tubes but are not exposed to the atmosphere. The blank samplers are analyzed in exactly the same way as the sample tubes. The results of the blanks give a measurement of the achievable detection limits for the method and act as a quality control check.

Species that can be monitored using passive samplers include nitric oxide, nitrogen dioxide, sulfur dioxide, ozone, benzene toluene ethylbenzene xylenes (BTEX), 1,3-butadiene, and other speciated hydrocarbons (with carbon numbers from 2 to 12). Typical absorbents for hydrocarbons include Tenax-A and Tenax-D, XAD, Chromosorb-160, and active carbon. Sulfur dioxide and nitrogen dioxide tubes use a mesh coated in triethanolamine (TEA) as the sorbent medium, though sodium carbonate or potassium hydroxide can also be used for sulfur dioxide monitoring.

16.6.3 Analysis of collected samples

Exposed samplers are analyzed using a number of techniques; these include capillary column gas chromatography, gas chromatography coupled with mass spectrometry, ion exchange chromatography, and spectrophotometry. The decision on which technique to use will depend on the gas being measured and absorbent used. There are two common techniques for extracting the sample from the absorber: thermal desorption and solvent desorption. In thermal desorption the tube is heated to drive the sample off while clean carrier gas is passed through the tube; in solvent desorption a solvent is passed through the tube, removing the sample due to a closer affinity than the sorbent.

Gas chromatography is generally used for the analysis of tubes used for hydrocarbon measurements. Gas chromatography combined with mass spectroscopy (GC/MS) may also be used, particularly if it is necessary to identify unknown species. Ion-exchange chromatography is usually used for the analysis of nitric oxide, nitrogen dioxide, and sulfur dioxide samples.

Calibration of the analysis stage can be carried out using tubes loaded with a known amount of the gas being measured. These may be prepared by spiking the tubes with a known mass of liquid or by exposing the tubes to a calibration gas. These calibration tubes can be used to assess the desorption efficiency of the tubes. The uptake rates of the tubes can be assessed by exposing tubes in calibrated atmospheres (Martin et al. 2001).

16.6.4 Active sampling

The long averaging times of passive samplers can be a disadvantage for some applications. Active sampling techniques alleviate this by using a pump to draw the sampled air through a trap containing the sorbent medium. This technique has significantly lower averaging times than passive samplers, typically of a few hours. The technique also has the potential to achieve much lower detection sensitivities due to the high sampling volumes. However, there is an increased risk of overloading the sorbent and experiencing break-through. Pumped sorbent tubes are one of the primary methods for short-term monitoring of speciated hydrocarbons. Typical sorbents for this purpose are Tenax, XAD, Chromosorb, and acti-vated carbon. The analysis of pumped tubes is car-ried out in the same way as for diffusive samplers.

16.6.5   Canister sampling

One of the simplest techniques for monitoring the concentration of one or more species in the atmos-phere is to sample the air into a canister and trans-port it to a laboratory for analysis. The canister must be unreactive with respect to the sampled gas; this is not always assured, and for reactive gases this can rule out the technique. The canister may be a rigid sample vessel, or it may be a polyvinyl fluoride (PVF) Tedlar™ bag.* The rigid vessel may be pre-evacuated and opened to draw the sample in, or a pump may be used to fill the ves-sel. In the case of a Tedlar™ bag a pump is usually required, although it is possible to use a Tedlar™ bag inside an evacuated container. In this case the vacuum in the container will expand the Tedlar™ bag, drawing ambient air into it. The pump, if used, will be in the sampling path and so it must be care-fully chosen to ensure it does not affect the sample. Inert pumps are available: one type uses metal bel-lows, other designs use PTFE bellows, and either will be suitable for most ambient measurements.

Subsequent analysis is usually performed using gas chromatography or gas chromatography cou-pled with mass spectrometry. A pre-concentrator stage is often required because, unless the sample has been pumped, it will be at ambient pressure. Cryogenic pre-concentration gas chromatography with an FID detector can achieve very low detec-tion limits, better than 100 parts per trillion for hydrocarbons (H. D'Souza, NPL, private commu-nication 2002).

*Tedlar™ is a trade mark of E.I. du Pont de Nemours and Company.

16.6.6   Automated measuring systems

The systems most commonly used for monitoring gaseous compounds for the assessment of ambient air quality are automated measuring systems (AMSs). These systems provide continuous meas-urements of the concentration of one or more gases. They are usually self-contained units with integral sampling pumps, though in some cases an external (down stream) pump must be supplied. A generic AMS is shown in Fig. 16.2. It is a require-ment of the sampling systems that they do not alter the composition of the sample gas; for this reason the pump is usually situated after the meas-urement or collection stage. Most AMSs draw air through their sample chambers at relatively low flow rates. The sample line or manifold required to bring ambient air to the instrument is often a significant length. At typical AMS sample flow rates the residence time in such a sample line is unacceptable. To overcome this, ambient-air monitoring applications typically use a single large-diameter glass or PFA manifold. A high-volume pump is used to draw ambient air through at a relatively high flow rate, which results in a low residence time for the sampled gas. A number of different measuring instruments may be con-nected to this manifold. These instruments can be connected using much shorter individual sample lines, keeping the overall sample residence time to a minimum.

One of the primary reactions that may occur in a sample line is the oxidation by ozone; for exam-ple, converting nitric oxide to nitrogen dioxide. Careful design of the sample line and regular main-tenance to ensure the line is clean will reduce the effect of these reactions.

The following sections discuss some of the most common techniques used in AMS systems for monitoring specific pollutants in ambient air.

16.6.7   Optical spectroscopy

A number of measurement techniques are based on optical spectroscopy. Some of these are specific to the measurement of a single gas, and are tuned to the characteristic wavelength absorption of the

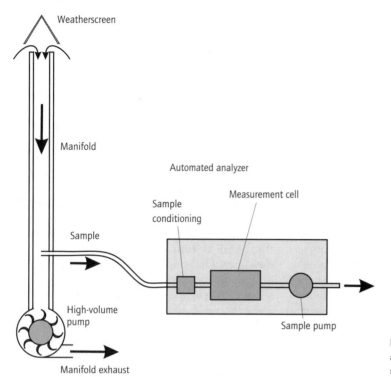

Fig.16.2 Typical configuration for an automated air quality monitoring station.

target gas. Other techniques are more generic and are capable of measuring multiple species. This section describes general spectroscopic techniques.

A gas molecule will absorb electromagnetic radiation at a number of wavelengths that are characteristic of the molecule. A detailed description of the absorption mechanisms can be found in any textbook on molecular physics. Many gases absorb light in the ultraviolet region of the spectrum due to electron absorption. Many molecules will also absorb light at specific wavelengths in the infrared region of the spectrum due to rotational and vibrational absorption mechanisms. Most hydrocarbons have absorption spectra in the infrared spectral region. If an optical beam is passed through an air sample the spectrum of this light will exhibit a number of absorption features, characteristic of the gases in the sample. The degree of absorption at these specific wavelengths is directly proportional to the amount of each absorbing gas

in the sample. Beer–Lambert's Law (eqn 16.3) describes the relationship between the absorption of the transmitted light and the concentration of the species. Analysis of the spectrum of the transmitted light beam therefore allows the identification and quantification of components of the sample gas.

$$I = I_o \exp(\alpha CL) \qquad (16.3)$$

where: I is intensity of transmitted radiation; I_o is intensity of radiation source; α is absorption coefficient of gas; C is concentration of gas over pathlength, L.

Measured absorption features are not, in general, discrete sharp features but are bell-shaped curves exhibiting a characteristic width. Ambient pressure and temperature conditions strongly affect the absorption line shape. Two key broadening mechanisms are Doppler broadening due to the Brownian motion of the molecules, which is related to the temperature of the sample, and collision broadening, the magnitude of which is related to the pressure of the sample (Elachi 1987). Collision broadening, which produces a Lorentz line shape, dominates in the lower atmosphere. Doppler broadening dominates in the upper atmosphere. It is also more apparent in the infrared spectral region because Doppler broadening is proportional to the wavelength. In addition, the instrument used to measure the spectrum will have a particular wavelength resolution. This will smooth out features in the spectrum, resulting in measured line shapes being broadened. The effect of the instrument broadening on a theoretical discrete absorption feature is known as the instrument line function. If it is known then it can be deconvolved from measured line shapes to recover the "true" line shape. This can be important when attempting to resolve two absorption features that overlap in the measured spectrum.

One of the main difficulties in spectroscopy, particularly in the infrared region of the spectrum, is caused by the absorption features of two common, abundant, atmospheric species, water vapor and carbon dioxide. Their absorption features effectively block many regions of the spectrum, and severely limit the choice of absorption lines which can be measured in the free atmosphere. Weaker absorption features of water and carbon dioxide may not block the transmitted light but they can interfere with target gas absorption lines. A key component of any spectroscopic measurement is the choice of wavelength at which to measure a given species. Knowledge of likely interfering gases is required in order to select spectral regions in which the target gas absorption is free from overlapping spectra. A number of commercial packages and libraries of absorption line parameters are available that allow the user to specify the species of interest, and the expected concentrations (Rothman *et al.* 1998). The software can be used to calculate the absorption of the target gas and the likely interference by atmospheric gases. Such tools are invaluable in defining spectroscopic measurements. It is also important for the user to be aware of the wavelengths used by commercial instruments to identify specific species. This can

be particularly important if the instrument is to be used in a situation where previously unconsidered interfering species may be present in high concentrations; for example, in industrial environments. These issues not withstanding, spectroscopy is one of the key techniques for monitoring species in the atmosphere. It has a further advantage: it can be extended from point measurement systems, where the absorption takes place in a measurement cell, to remote sensing or long path systems.

In order to improve measurement sensitivity many spectroscopic techniques use multiple-pass gas cells. These increase the effective path-length of the optical beam through the sample gas, increasing the absorption. Typical designs of multipass cells are White cells, Heriott cells, and astigmatic Heriott cells. White cells are relatively simple to align and can be easily adjusted to give variable path lengths. Astigmatic Heriott cells are designed to achieve high reflection efficiency and are therefore capable of very high path lengths. However, they are more complicated to configure. Path lengths of tens of meters are possible with Heriott cells having physical dimensions of less than 1 m.

16.6.8 *Applications of the spectroscopic technique*

The following sections outline a number of specific techniques that use spectroscopy as the basis for the measurement of gaseous species.

Some techniques use a light source that emits radiation over a broad range, or band, of wavelengths covering the region of interest. After the light passes through the sample the intensity at specific wavelengths is measured. In the case of a scanning spectrometer the wavelength at which the intensity is measured is scanned over the whole wavelength region of interest. In principle such a technique can scan a number of absorption features and therefore measure multiple species. However, in practice such wavelength scanning is slow. Alternatively, spot measurements can be made at a number of specific wavelengths. This is often achieved by filtering the light to only allow specific wavelengths to fall onto a detector, which enables the intensity at specific wavelengths to be measured. The most usual approach is to use a filter wheel that contains a set of optical filters, each letting specific wavelengths pass. The main difficulty with this approach is the availability of optical filters that exhibit sharp narrow band-pass at the desired wavelengths.

Techniques may directly measure the absorption of light at specific wavelengths. Alternatively they may measure the first or higher order differentials of the absorption feature, and hence retrieve the line shape. This may be achieved by varying the filter band-pass wavelengths and measuring the rate of change of the transmitted intensity as the filter pass-band scans across an absorption feature. One such technique employs filters that have a geometric relationship between their pass-band and the angle of the incident light. Oscillating the orientation of the filter causes the wavelength of light passed to change in a sinusoidal manner. Measuring the differential of the intensity with respect to the wavelength allows broad-band filters to be used to detect absorption features which are narrow in comparison with the bandwidth of the filter.

In differential optical spectroscopy (DOAS), the absorption of the optical beam at two (or more) carefully selected wavelengths is measured. In a two-wavelength system, one measurement of the intensity of the transmitted light is made at the on-resonant wavelength. This is chosen to be at a wavelength that is absorbed by the target species. Another measurement is made at the off-resonant wavelength, which is in a nearby spectral region that is not absorbed by the target species. By analyzing the difference in the measured intensity between the two wavelengths the absorption due to the target species can be determined. The principle of DOAS is shown in Fig. 16.3. The off-resonant wavelength is used to correct for the effects of changes in atmospheric conditions and changes in the intensity of the transmitted beam. In many operational systems a number of different wavelengths are used in order to measure more than one species. The wavelengths are generally selected using filters in a filter wheel.

A key technique, currently used in a number of

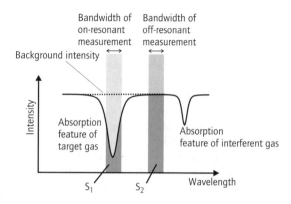

Fig.16.3   The principle of the differential absorption spectroscopy technique.

commercial instruments, is Fourier transform spectroscopy (FTS). This technique is capable of measuring a number of different atmospheric species simultaneously. The technique is based on the principles of the Michelson interferometer. This is shown in schematic form in Fig. 16.4. An optical beam is incident on a beam splitter, which divides the beam into two paths. One beam is reflected off a fixed mirror, the other is reflected off a movable mirror. The two beams are reflected back to the beam splitter and then onto a detector. If the movable mirror is moved a short distance this alters the phase of the two beams, causing them to interfere when they recombine. If the mirror is oscillated, an interferogram will be observed at the detector. The spectrum of the incident radiation can be retrieved from this by taking the inverse Fourier transform of the interferogram.

The measurement of the spectrum is nearly instantaneous. Compared to a wavelength scanning spectrometer, operating over the same resolution and measurement period, the FTS exhibits an improved signal to noise known as the Fellgett advantage (Schnopper & Thompson 1974). This is only strictly true when the FTS noise is independent of the signal strength. This is the case for most infrared applications, where detector noise dominates. Such FTS systems operating in the IR region are commonly known as FTIR systems.

The distance the movable mirror travels deter-

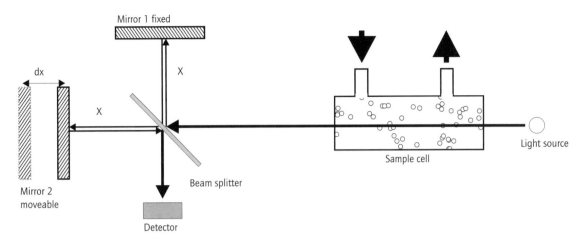

Fig.16.4 Simplified schematic of a Fourier transform spectrometer.

mines the spectral resolution of the measured spectra. The FTS is capable of achieving very high spectral resolution, and commercial instruments are available with resolutions of better than $0.1\,cm^{-1}$. Such instruments are capable of resolving the absorption features of many atmospheric pollutants and trace gases. Software packages exist which will match retrieved spectra against reference spectra to identify species present in the atmosphere. A further advantage of FTS is that the spectrum produced is constant in spectral resolution across the entire spectrum; this is often not the case in grating spectrographs.

Concentration calibration can be achieved from the use of gas cells containing known concentrations of standard gas, or by using reference spectra, obtained under calibration conditions. External reference spectra should be convolved with the instrument function of the FTS in order to be comparable with spectra produced by the FTS. Many systems use an internal helium–neon laser to calibrate the movement of the mirror by observing the interferogram of the narrow-band laser light. Such lasers can also be used to give information on the instrument line function.

In some systems, narrow-band transmitted light is used to probe specific absorption features. These systems generally use laser sources and one of the most common examples of these techniques

is tuneable diode laser spectroscopy (TDLS). Diode lasers are available which exhibit very narrow bandwidth and single-mode operation. The key attribute of diode lasers is that the wavelength of the laser radiation can be changed by varying the current across the diode, and more coarsely by varying the temperature of the diode. Over short-wavelength regions the diode can be tuned continuously and very accurate wavelength control can be achieved. Diodes can be selected such that the laser wavelength can be tuned to span a specific absorption feature of the target gas. In addition, the bandwidth of the laser can be selected, such that it is narrow enough to resolve the spectral feature. The laser radiation is passed through a multi-pass cell through which the sample gas is passed. The multi-pass cell is used to provide a long path length to increase the absorption due to the target gas and therefore to improve the sensitivity and detection limit of the technique. The TDL is scanned across the absorption feature of the target gas. The intensity of the radiation after passing through the sample cell is detected and recorded. The profile of the absorption feature can then be determined, and the concentration of the species calculated by reference to calibration results. Problems can arise with the selection of the diode laser as they tend to vary subtly in their performance. It can be difficult to obtain a laser that has the required wavelength scanning range while maintaining single-mode operation. Problems can also occur with mode jumping, in which the laser suddenly and randomly switches mode, causing the wavelength of the emitted light to jump. A number of research instruments have been developed that utilize the TDLS technique and some commercial instrumentation is also now available.

16.6.9 Ultraviolet absorption spectroscopy

UV absorption is commonly used to measure ozone. Ozone absorbs UV radiation at a number of wavelengths; a typical wavelength used in analyzers is 254 nm. At this wavelength there are no common atmospheric components that have interfering absorption features. The most common configuration uses a sample cell and a reference cell containing ozone-free air (generally this is generated from the ambient air by removing the ozone with a scrubber). The intensity of UV radiation passing through the sample cell and the reference cell is alternately measured using a photomultiplier. A schematic of a UV absorption spectrometer is given in Fig. 16.5. The concentration of ozone in the sample is determined from the difference in intensities between these measurements, by application of Beer–Lambert's Law. Calibration of ozone analyzers is complicated by the difficulty in producing a stable ozone gas mixture in a cylinder. It is common to use an ozone photometer as a transfer standard to calibrate field instruments against traceable laboratory ozone generators.

16.6.10 Nondispersive infrared

The nondispersive infrared (NDIR) technique is one of the primary techniques used to measure carbon monoxide. It is based on the fact that carbon monoxide absorbs IR radiation at specific wavelengths around 4.5 μm. Figure 16.6 shows the detection principle for an NDIR analyzer. The detection element consists of two cells containing carbon monoxide separated by a movable membrane that forms one plate of a capacitor. Two infrared beams fall onto the detector, on each side of the membrane. One beam passes through a reference measurement chamber, the other passes through the sample chamber. The intensity of the two beams is different due to the absorption of one by carbon monoxide in the sample chamber. When the light falls on the detector, carbon monoxide in the detector cells absorbs IR radiation. This causes the carbon monoxide gas to expand, which in turn causes the membrane to move. A chopper wheel alternately blocks each beam, causing the cells in the detector to be alternately exposed to light. This causes the membrane to oscillate. The amplitude of the oscillation is proportional to the difference in intensity of the two beams, which is in turn proportional to the amount of carbon monoxide in the sample gas. This technique is sensitive to water

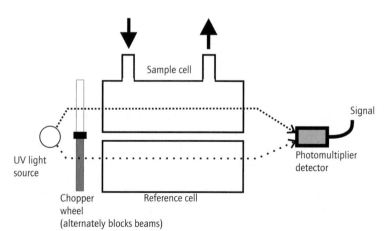

Fig.16.5 Schematic of an ultraviolet spectrometer.

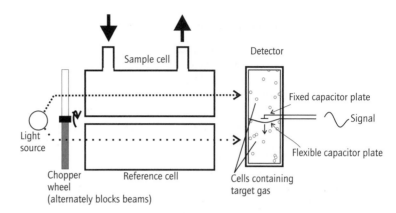

Fig.16.6 Schematic of a nondispersive infrared system.

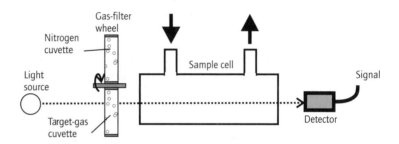

Fig.16.7 Schematic of a gas filter correlation system.

vapor and carbon dioxide, which act as interferents. It is therefore necessary to dry the sampled air before it is analyzed. A cell containing carbon dioxide can be placed in the optical beams to remove the effect of carbon dioxide interference.

16.6.11 *Gas-filter correlation spectroscopy*

Gas-filter correlation (GFC) is a highly specific technique that can be employed to measure species in wavelength regions in which there is the potential for interference from other species. The technique, shown schematically in Fig. 16.7, uses a filter wheel mounting two gas cells or cuvettes. One of these contains nitrogen and the other contains a mixture of nitrogen and the target gas. The filter wheel is placed in the path of an IR beam, which passes through the sample chamber and onto a detector. When the cell containing the target gas is in the path of the IR beam the gas in the cell completely absorbs, and in effect blocks, spe-

cific characteristic wavelengths. Because these wavelengths have been completely absorbed there is no further absorption of the beam by the same gas in the sample chamber. When the nitrogen cell is in the path of the beam all wavelengths pass into the sample chamber, where some will be absorbed by any target gas present. Two signals are produced from the detector, one for each filter position. Both signals will be affected in the same way by the absorption due to species other than the target gas. The only difference between the two signals will be the absorption due to the target gas in the sample chamber. The difference between the intensity of these two signals, normalized by the IR source intensity, is proportional to the concentration of the target gas in the sample chamber. Gas filter correlation can be used for monitoring a number of species in the atmosphere, including nitrous oxide, carbon monoxide, and carbon dioxide. Detection limits of 50 p.p.b. for carbon dioxide measurements can be achieved with this technique.

16.6.12 Ultraviolet fluorescence

Fluorescence techniques are spectroscopic techniques that measure emission spectral features rather than absorption features. The most common use of this technique is for the measurement of sulfur dioxide; such a system is shown in Fig. 16.8. Sulfur dioxide molecules in the sample gas are excited to an unstable energy state by ultraviolet radiation; typically light at a wavelength of 212 nm is used. These energy states decay almost immediately, resulting in the emission of fluorescent radiation. This radiation is detected by a photomultiplier. The output signal from this is normalized for the intensity of the incident UV radiation, and is directly proportional to the concentration of sulfur dioxide in the sample gas. The UVF technique is susceptible to a "quenching" effect due to aromatic hydrocarbons, which absorb the fluorescent radiation and cause the technique to underread the sulfur dioxide concentration. These hydrocarbons are removed from the sample gas stream by a scrubber, known as the hydrocarbon kicker. It is necessary to check the operation of this kicker as part of a quality control procedure. The UVF technique can also be used to measure other species, including hydrogen sulfide. Typically the technique has a detection limit of better then 1 p.p.b. for sulfur dioxide measurements.

16.6.13 Chemiluminescence

Chemiluminescence is the primary technique used to measure nitric oxide (NO), nitrogen dioxide (NO$_2$), and total oxides of nitrogen (NO$_x$). NO$_x$ is, by convention, the sum of nitric oxide and nitrogen dioxide (and does not include nitrous oxide) (ISO 1985). The chemiluminescence technique specifically measures the concentration of nitric oxide based on the detection of light emitted from the reaction between nitric oxide and ozone. Ozone is generated in the system at a high concentration so that the reaction is nitric oxide limited. Therefore the intensity of light detected is proportional to the concentration of nitric oxide in the sample. The concentration of NO$_x$ is measured by reducing all nitrogen dioxide in the sample to nitric oxide using a heated molybdenum catalyst. The nitric oxide concentration in this converted sample is therefore equivalent to the NO$_x$ concentration. The nitrogen dioxide concentration is then determined by subtraction of the nitric oxide value from the NO$_x$ value. Two configurations of analyzer are commonly used. Single-channel systems sequentially switch the gas stream either through the NO$_x$ converter or not, and then into a single measuring cell. Such a system is shown in Fig. 16.9. Two channel systems have separate gas streams, one passing through the NO$_x$ converter, the other not, and two detection chambers. The correct measurement of nitrogen dioxide depends on the operation of the NO$_x$ converter, and the efficiency of the conversion is a key performance characteristic of the systems. Single-channel systems discontinuously measure nitric oxide and NO$_x$ and can potentially give incorrect results if the nitro-

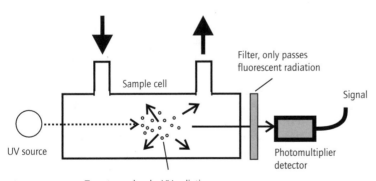

Fig.16.8 Schematic of an ultraviolet fluorescence analyser.

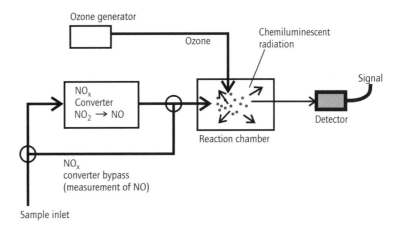

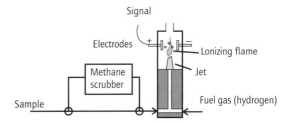

Fig.16.9 A chemiluminescence analyzer for measuring nitric oxide and nitrogen dioxide.

gen dioxide concentration is rapidly changing. This can occur at highly polluted locations, and care should be taken if single-channel systems are to be operated in these locations. In some single-channel instruments the length of the NO_x converter by-pass pipe is chosen to ensure that samples take the same time to travel through the converter or the by-pass. Therefore, consecutive measurements of nitric oxide and NO_x are made on the same air sample, which reduces the influence of rapidly varying concentrations. Chemiluminescence techniques can also be used to measure ammonia and formaldehyde using similar chemiluminescence reactions.

16.6.14 Flame ionization detectors (FID)

The concentration of total organic carbon (TOC), or the total hydrocarbons, in ambient air is usually measured using a flame ionization detector (FID). Figure 16.10 shows a schematic representation of an FID analyzer. The FID works by combining the sample gas stream with hydrogen fuel gas and then burning this in a controlled flame. The hydrocarbon molecules form charged carbon atoms when they burn, and these produce a current across an applied potential. This current is proportional to the number of carbon atoms in the sample gas. Many FID instruments contain a catalytic converter which can be used to remove methane from the sample gas. Such instruments are able to meas-

Fig.16.10 Schematic of a flame ionization detector.

ure the concentration of total hydrocarbons, nonmethane hydrocarbons, and, by subtraction, methane. FID detectors are also used as the detection element for GCs used to monitor hydrocarbons. Care should be taken to ensure that the calibration gas used is in an air matrix as the use of a nitrogen balance gas will affect the response of the FID.

16.6.15 Automated on-line gas chromatographs

A common method for measuring speciated hydrocarbons in ambient air is to use passive or active sampling followed by laboratory analysis using GC or GC/MS as described earlier. However a number of automated analyzers are available for measuring and speciating benzene, toluene, ethylbenzenes, and xylenes (BTEX). A limited number of systems are also available that give automated psuedo real-time measurements of a wider range of speciated hydrocarbons. In general, such systems consist of a traditional GC instrument coupled to a sampling and pre-concentration stage. Typical sampling periods are up to half an hour. The collection stage is then heated and the trapped hydrocarbons are driven off and transferred to the GC using a carrier gas flow. Helium is typically used as the carrier gas. A pre-concentration stage is then often used before the sample enters the GC column, which increases the sensitivity of the technique. The sample is passed into the GC column and measured using a detector, typically an FID. The concentrations of individual hydrocarbons can then be determined and quantified from the resulting chromatogram. There are a number of variations employed by different systems to improve the measurements. These include cooling of the trap and pre-concentrator cryogenically or by using carbon dioxide cooling or Peltier coolers. Cooling improves the trapping efficiency of many sorbants, particularly for lower molecular weight hydrocarbons, and is necessary to measure hydrocarbons with three carbon atoms or less. Some systems employ multiple sampling traps to allow sampling to continue on one trap while the previous sample is being transferred. GCs with more than one column are also used. These allow a broad range of hydrocarbons to be identified by selecting one column to measure lighter hydrocarbons and the other to measure heavy hydrocarbons. The choice of column is critical in automated systems, particularly if a large number of hydrocarbons are to be monitored. The temperature profile used in the GC must be carefully selected to separate the compounds of interest over the range of concentrations expected. In general, automated systems are able to routinely identify BTEX species. However, for measurements of large numbers of hydrocarbons where the chromatography is more complex, off-line post-analysis is often required to achieve acceptable data quality. Calibration is required not only to enable the calculation of concentration from the chromatogram peak areas, but also to determine elution times to allow peak identification. Multiple component gas standards are used to calibrate the systems. Issues can arise with

misidentification of peaks by automated systems, particularly if elution times shift due to column degradation or interference effects.

A network of 13 automated monitoring stations, measuring 25 hydrocarbons on an hourly basis, was operated in the UK from 1992 to 2000. A comprehensive quality control and quality assurance process enabled the network to achieve data quality and data capture targets in line with European requirements.

16.6.16   Particulate monitors

The measurement of suspended particulate matter in the atmosphere is a problem distinct from other, gas sensing, measurements. The measurements required fall into three broad areas: measurement of the mass, chemical analysis, and measurement of the size distribution. A number of techniques have been developed to perform these measurements.

The collection of particulate matter that is in suspension in ambient air requires specialized sampling systems. Particulate matter behaves very differently in a flowing gas stream to gas molecules and ideally samples should be collected isokinetically. Under isokinetic conditions all particles in the ambient sample are transferred into the sampling system. In reality this is rarely achieved, or required, as most particulate sampling is intentionally size-selective and uses specific size-selective sampling heads.

The measurement of the mass of particulate matter may be of total suspended particulates (TSP) or of a particular size fraction. The size fractions are related to particular health implications of fine particulates, which can remain entrained in air through the trachea and travel into the deep lung. The most commonly monitored size fractions include particulate matter with aerodynamic diameters of less than $10\,\mu m$ (PM_{10}), $2.5\,\mu m$ ($PM_{2.5}$), and $1\,\mu m$ (PM_1). These size distributions are realized in practice by size-selective sampling heads, designed to allow only the required particulate matter to pass through. These sampling heads typically use impaction or cyclonic techniques to differentiate between particulate with different aeronautical diameter. The CEN standard EN12341:1999 (Centre for European Normalization 1999) defines the design of a reference PM_{10} sampling head.

A number of techniques are available for measuring the mass of collected particulate matter. These include automated real time systems such as the tapered element oscillating microbalance (TEOM), beta-gauge, and a range of optical system and samplers designed to collect material for off-line gravimetric determination.

Gravimetric samplers are active sampling systems that collect particulate matter onto filters. High-volume samplers draw sample air at roughly $100 \, m^3 h^{-1}$; low-volume samplers typically sample volumes of $1 \, m^3 h^{-1}$. The mass of particulate matter is subsequently determined by weighing the filters. Care must be taken in transporting the filters and in handling and preparation of the filters for weighing. High-volume samplers are in general less convenient to site and transport than low-volume samplers. However, their higher collection rate gives a larger mass of particulate material to weigh and hence potentially improves the uncertainty of the method and allows shorter sampling times. Gravimetric samplers are the basis for most methods used to determine the chemical composition of the particulate matter. Quartz fiber filters are generally used for PM_{10} sampling. These should not be used for sampling material for subsequent analysis for metal content. Glass fiber filters are also applicable for particulate sampling. They are more robust than quartz filters, which can improve weighing uncertainties. There have been some reports that glass filters may exhibit artifacts due to adsorption of reactive species. The flow control systems must be calibrated in the samplers. However, the primary source of uncertainty is in the filter handling and weighing procedures. Other issues can arise due to the loss of volatile compounds from filters. For this reason filters should be weighed as soon as possible after sampling and should not be stored in direct sunlight.

The TEOM system (developed by Rupprecht and Patashnick Co., Inc.) is an on-line automated system that produces real-time measurements of particulate mass. The technique is based on a hollow tapered crystal element, which is free to vibrate at one end. A filter unit is fitted on this end of the crystal element and sample gas is drawn through this filter and through the hollow crystal element. The crystal vibrates at a resonant frequency that changes as the mass of particulate on the filter increases. The resonant frequency is directly proportional to the mass of particulate on the filter. The TEOM measures the integral particulate mass on the filter with time. It is necessary to periodically replace the filter element. The system calibration is checked using pre-weighed filter papers.

Beta-gauge systems collect particulate matter on a moving filter tape. The filter passes between a beta-ray source and an ionization chamber. The ionization chamber is used to measure the beta radiation passing through the filter tape. Particulate matter on the filter absorbs the beta radiation and the intensity measured by the ionization chamber is proportional to the mass of particulate matter deposited on the tape. The system is calibrated using filters with known beta-ray attenuation. Filter tape systems can also be used to collect particulate matter for subsequent determination of the particle size distribution by automated image analysis.

Optical particulate monitoring systems are generally based on the detection of light scattered by particulate matter passing through an optical beam. One of the main applications of these systems is to make real-time measurements of the size distribution of particles in the sample gas. If the mass of collected particulate is required then this must be determined by off-line weighing of particulates collected on an additional in-line filter. The scattering effect of particulates is dependant on surface qualities as well as size and this can cause inaccuracies in the measurements. Some configurations of this technique effectively measure a single particle at a time, allowing particle size counting and imaging. Commercial single-particle mass spectrometers are also available; these are able to provide chemical speciation on a particle-by-particle basis.

The amount of particulate matter in the atmosphere is expressed in mass per volume units (usually $\mu g\,m^{-3}$), which is easily derived from mass measurements if the sample volumes are known or measured.

16.6.17   Sampling multiphase species

In order to sample multiphase species such as poly aromatic hydrocarbons (PAHs), dioxins, and metals, it is necessary to sample both the particulate and gas phases. This is usually achieved by using cartridge samplers in a gravimetric particulate-sampling system. The cartridge samplers include a filter to trap particulate matter and an absorbing medium backing the filter to trap the gaseous phase. This also ensures that any material outgassed from the trapped particulate matter is re-trapped. For PAHs an XAD sorbent is used, and for dioxins a polyurethane foam sorbent is used. Subsequent analysis can be carried out using high-resolution GC/MS.

Most metals monitoring is carried out using dual-phase samplers followed by laboratory analysis. For metals analysis the filters may be analyzed using a number of laboratory techniques, including atomic absorption spectroscopy (AAS) (ISO 1993), X-ray fluorescence spectroscopy (XRF), inductively coupled plasma mass spectroscopy (ICPMS) and high-performance liquid chromatography (HPLC). The choice of analysis technique will be influenced, *inter alia*, by the amount of sample collected, the data quality required, and the levels of metals expected.

Some instrumental systems have been developed to monitor vapor-phase mercury online; these include cold vapor atomic fluorescence spectrometry (CVAFS). Optical techniques can also be used. These include DOAS, which can measure the spectroscopic absorption features of vapor-phase mercury in the ultraviolet spectral region.

16.6.18   Remote sensing in the troposphere

Remote sensing is defined as the detection of a quantity at a location remote from the measuring instrument. In the context of the remote measurement of atmospheric pollutants in the lower tropo-

sphere, this utilizes the interaction of a beam of optical or microwave radiation with species in atmosphere. In many cases, as mentioned in previous sections, optical techniques used as the basis for point monitoring instruments can be adapted for remote sensing configurations. Remote sensing falls into two broad categories, passive and active. Passive remote sensing involves the detection and analysis of radiation incident on the detection system from a remote, uncontrolled source. In general the Sun and other extraterrestrial bodies are used as the source. It follows from this that such techniques are generally used to monitor the concentrations of atmospheric species through the whole atmosphere rather than within the troposphere. Some passive infrared systems have been used for the monitoring of emission sources and elevated plumes within the troposphere. They work by comparing the emissions observed in light passing through the plume with those seen in "clear" air. There are also monitors for the emission spectra of hot gases—for example, in flares and jet engine exhausts—but such techniques are beyond the scope of this book.

Active remote sensors generate and transmit the optical radiation used for detection. Two configurations are commonly used in tropospheric monitoring: open path systems and lidar.

Open path techniques

Open path techniques measure the total concentration of one or more species in the atmosphere within a defined, extended, optical path. Two of the basic configurations for an open path monitoring system are given in Fig. 16.11.

Single-ended (mono-static) systems operate by transmitting an optical beam into the atmosphere to a passive retro-reflector which returns the beam to the detector. In bi-static systems the transmitter and the detector are separated at the two ends of the optical beam. In some situations a mono-static system can be an advantage. They only require a passive retro-reflector at one end and so power and access need only be available at the source/detector location. The retro-reflector can therefore be mounted in hazardous environments or at elevat-

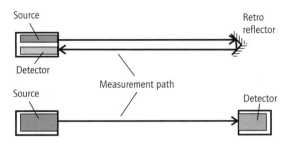

Fig.16.11 Open-path monitor configurations.

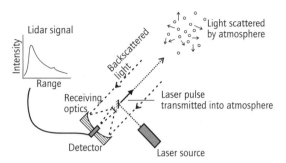

Fig.16.12 Simplified diagram of an aerosol lidar.

ed locations; for example, on a mast. A further advantage is that a single mono-static monitor can be pointed sequentially at different retro-reflectors. This allows a range of monitoring configurations to be employed, including vertical "curtains" of beams and ring fencing of emission sources (Woods *et al.* 1992).

Three of the most common techniques used in open path configurations are differential optical absorption spectroscopy (DOAS), tuneable diode laser spectroscopy (TDLS), and Fourier transform spectroscopy (FTS). The principles of these techniques have been described above. The key benefit of open path systems is that they require no extractive sampling systems and there are no issues of sample contamination. They also provide a spatial average of concentrations in the atmosphere, which in many applications can be more representative. They can also measure relatively large areas and are able to provide a continuous time series of measurements. Calibration of the systems is usually achieved by placing a gas cell containing a known concentration and path length in the measurement beam. In all open path systems it is necessary to determine the background intensity of the beam from wavelengths that are not significantly absorbed, in order to normalize for attenuation of the beam in the atmosphere by other mechanisms such as scattering from fog. A potential problem in open path FTIR systems is the effect of background solar radiation scattered into the measurement path. This results in a spurious spectrum due to the stray radiation. Various techniques have been employed to reduce this effect; one is discussed in Richardson (1997).

Typical DOAS systems monitor sulfur dioxide, nitric oxide, ozone, and BTEX compounds in the ultraviolet and carbon dioxide, methane, and other hydrocarbons in the infrared, with detection limits typically of 1 p.p.b. over a 1 km path length (Casavant & Kamme 1992). The FTS open path system can, in principle, continuously measure the concentration of all species that have significant absorption features over path-lengths of up to 1 km. Care should be taken in interpreting detection limits for open path systems as these systems measure the integrated path concentration. It should be noted that a detection limit of 1 p.p.b. over 1 km implies a uniform distribution of 1 p.p.b. along the entire path length, and is equivalent, for example, to 1 p.p.m. distributed over 1 m. Typical detection sensitivities for an FTS system are 100 p.p.b. km for BTEX and 20 p.p.b. km for ethane.

Lidar

Lidar (light detection and ranging) can be thought of as the optical equivalent of radar. A short high-power pulse of laser radiation is transmitted into the atmosphere along a given line of sight (Measures 1984). A fraction of the light is scattered by particulate matter or aerosols and molecules in the atmosphere. A small component is returned, or backscattered, to the lidar system. The backscattered light is collected by a telescope and measured using high-speed detectors. The most basic configuration for a lidar system is given in Fig. 16.12. It is possible to determine the distance from the lidar system at which light incident on the detector at any given time was backscattered. It is half the distance given by the elapsed time multiplied by the speed of light. The backscattering media are continuous, and therefore light is scattered from all ranges. However, they are not homogeneous and hence different regions will scatter more or less light depending on the variation in the scattering media. Scattering of the light in directions that are off the axis of the transmitted line of sight, and absorption by gases and aerosols, cause the light to be attenuated as it travels through the atmosphere. Therefore at a maximum range the incident light reaching the lidar will fall below the limit of detection. The intensity of the light incident at the detector at an elapsed time t is described by the lidar equation, eqn 16.4.

$$P_x = E_x \frac{D_x}{r^2} B_x(r) \exp\left\{-2\int_0^l [A_x(r') + \alpha_x C(r')] dr'\right\}$$
$$(16.4)$$

where: P_x is the returned intensity at wavelength x; E_x is the transmitted energy at wavelength x; D_x is a range-independent efficiency; r is range from the lidar; B_x is backscatter coefficient of the atmosphere at wavelength x; $C(r)$ is the concentration of an absorbing gas with an absorption coefficient of α_x; A_x is absorption of wavelength x by the all other species in the atmosphere. This equation includes a term describing the absorption due to a target gas, showing how in principle lidar can observe gaseous absorption as well as backscatter. It should be noted that this equation is a simplified form, which neglects the effects of multiple scattering. It is assumed that light scattered off the axis of the beam is not subsequently scattered back to the detector. Based on this equation, the detected signal can be analyzed as a function of time to provide spatial information on the distribution of the scattering and absorbing media in the atmosphere. The limiting spatial resolution possible with a lidar, along the line of sight, is half the pulse length multiplied by the speed of light. In practice the detector speed, acquisition digitization rate, and signal averaging that are necessary to achieve usable

signal to noise will reduce the achievable resolution below the theoretical limit.

In the basic aerosol lidar configuration the detected light is at the same wavelength as the transmitted pulse. Such scattering, which does not change the wavelength of the light, is called elastic scattering. The primary elastic scattering mechanisms are Mie scattering by aerosols with diameters of the same order of magnitude as the wavelength of the incident light and Rayleigh scattering by molecules (Hinkley 1976). In Rayleigh scattering the intensity of light scattered back towards the lidar system has a dependence on wavelength of λ^{-4} and therefore drops off sharply with increasing wavelength. The backscattered component in Mie scattering is of an order of magnitude less than the forward scattering component. However, in the troposphere the number density of particles is sufficiently high to produce a Mie backscattering component which is of the same order as the Rayleigh scattering, at visible wavelengths. In the IR spectral region ($750\,\text{nm} < \lambda$) Mie scattering dominates, though this also decreases with wavelength. In addition in the IR there are strong regions of absorption by atmospheric carbon dioxide and water vapor, leading to a few "windows" in which measurements can be made. These are roughly at wavelengths of 1.5–1.8, 2–2.5, 3.4–4.4, 4.5–5.0 and 8–13 μm, as can be seen from Fig. 16.13, which shows the atmospheric absorption spectra for a typical atmospheric composition.

The basic scattering lidar described above can be used to determine the spatial distribution of aerosols and particulate matter. Such systems can be used to determine the transport and dispersion characteristics of elevated plumes and to monitor and map the distribution of aerosols in the boundary layer. These measurements can be used to measure turbulence and convective processes in the boundary layer. The boundary layer can be observed in the lidar return signal as a region of turbulence followed by a sharp cut-off in the signal from the cleaner air from the free troposphere. A vertical pointing lidar system can be used to automatically measure the height of the boundary layer. Typical aerosol mapping systems have oper-

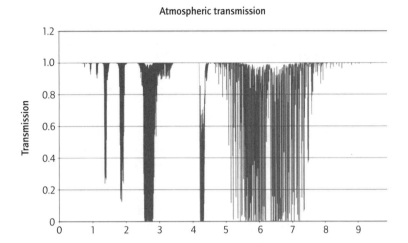

Fig.16.13 Typical atmospheric absorption spectra.

ating ranges of up to 3 km in the troposphere, though some can operate up to ranges 15 km, into the stratosphere.

Raman lidar

Raman lidar operates on a different principle to the basic aerosol lidar described above. It detects light backscattered at a different wavelength from that transmitted. This backscattered light is generated by inelastic Raman scattering from molecules in the atmosphere. Raman scattering occurs at vibrational-rotational frequencies of molecules. For a given incident wavelength a number of Raman-shifted wavelengths may be produced. These may be at wavelengths that are either longer (Stokes) or shorter (anti-Stokes) than the incident light. The wavelength shift is also dependent on the temperature of the scattering molecule. By selecting the wavelength detected, the lidar system can be tuned to measure the backscatter from a particular molecular species. Most Raman lidar systems operate in the UV spectral region.

The major drawback of Raman lidar is the low Raman scattering cross-section, which results in very low return signals. In order to increase signal to noise Raman lidars are often operated at night to reduce the interference from ambient light. Some daylight systems have been developed which use baffles or filters to block out as much ambient light

as possible. Recent work has involved the use of a Fabry–Perot etalon to separate multiple nitrogen Raman wavelengths from the sky background, to enable daylight nitrogen measurements (Arshinov *et al.* 2000). Because it is homogeneously distributed in the atmosphere, measurements of the nitrogen concentration allow determination of the temperature profile of the atmosphere. One of the most common species measured with Raman lidars is water vapour, which is present in the atmosphere at high concentrations and therefore has a relatively strong Raman signal.

A further difficulty with Raman lidars is that the Raman wavelength may suffer attenuation, particularly from absorption by ozone in the UV spectral region. It is often necessary to measure several Raman-shifted wavelengths in order to normalize the signal to remove the effects of attenuation. A common configuration that enables the determination of the water vapour partial pressure is to simultaneously measure Raman wavelengths associated with nitrogen and water vapour. The nitrogen signal is used to normalize the signal to take account of signal attenuation and atmospheric density. This technique has also been used to measure the partial pressure of other species. A further refinement of the technique is to simultaneously measure the elastic scattering due to Mie and Rayleigh scattering. Combining this measurement with the nitrogen signal allows the aerosol backscattering ratio to be determined (Whiteman *et al.* 1992). Systems have been developed that can perform these measurements during the day and night (Goldsmith 1998).

Differential absorption lidar (DIAL)

Differential absorption lidar (DIAL) is a development of the lidar technique that allows the measurement of spatially resolved pollutant concentrations in the atmosphere. It is based on the principle of lidar described above and is operated in a configuration similar to DOAS. The key feature of the measurement technique is that it can be tuned to measure specific gaseous pollutants based on their spectroscopic absorption characteristics. Two laser pulses are transmitted sequentially, at different wavelengths. These are carefully tuned to detect a given species: one lidar wavelength is absorbed by the target species; the other wavelength is selected to be just off the absorption feature. The wavelength separation between these is kept as small as possible. This reduces the influence of differential scattering between the two lidar signals. If these wavelengths are chosen carefully the only differences between the signals will be due to absorption by the target species. The difference in the intensity of these two signals is proportional to the concentration of the target species, due to Beer–Lambert's Law. Light scattered from differ-

ent ranges returns to the detector at different times and so by analyzing the differential absorption at different times a measurement of the spatial distribution of the concentration of the species along the measurement line of sight can be made.

DIAL systems operating in the UV and visible spectral regions can be used to monitor species such as ozone, sulfur dioxide, carbon dioxide, BTEX, and water vapor. Infrared DIAL systems have also been developed (Woods *et al.* 1992). These are capable of measuring a wide range of species, particularly hydrocarbons, water vapor, oxygen, hydrogen chloride, and nitrogen dioxide, with ranges of up to 3 km and range resolutions of better than 10 m. The potential for interference from other absorption lines means that absorption line selection in the infrared region is critical (Ambrico *et al.* 2000). The atmospheric backscatter is weaker in the infrared region and therefore specialized high-power laser sources and detection optics are required. In addition, if the DIAL is to be used to monitor a range of species the lasers must be continuously tuneable (Gardiner *et al.* 1996). These issues have meant that IR DIAL remains a specialized technique. However, commercial systems are operated in the UK; an example of such a system is shown in Fig. 16.14 and described in Robinson *et al.* (1995). Typical detection sensitivities for this system are given in Table 16.1.

By sweeping the optical beam in a vertical or

Fig. 16.14 An infrared DIAL system in operation monitoring hydrocarbon concentrations at an industrial plant. Reproduced with permission of National Physical Laboratory, UK.

Table 16.1   Detection sensitivities for the NPL IR/UV DIAL facility achievable under typical atmospheric conditions.

Ultraviolet DIAL System		Infrared	
Species	Typical detection limit (p.p.b.)	Species	Typical detection limit (p.p.b.)
NO	5	Methane	50
NO_2	10	Ethyne	40
SO_2	10	Ethene	10
O_2	5	Ethane	20
Hg	0.5	Other alkanes	40
Benzene	10	HCl	20
Toluene	10	N_2O	100
Xylene	20		

Quoted detection limits are stated at a range of 200 m from the DIAL.

horizontal plane, a map of the pollutant concentration can be built up. This can be combined with wind measurements to directly measure the emission flux passing through the plane. No dispersion modeling is required as the total concentration of gases is measured. Pollutant mapping can be carried out in urban environments to investigate dispersion and street canyon effects. The use of a dual DIAL system simultaneously operating in both the ultraviolet and infrared spectral regions has enabled the measurement of chemical reactions in airborne plumes.

A number of key papers covering the development of DIAL and other forms of lidar have been compiled in Grant *et al.* (1997).

16.6.19   *Satellite remote sensing of the lower troposphere*

In general satellite instrumentation does not have the required sensitivity and vertical spatial resolution to make measurements of the troposphere. However, some instruments have been developed, such as MOPITT (Measurements of Pollutants in the Troposphere), which was flown on the ERS-2 satellite. This system measured vertical profiles of carbon dioxide and methane in the troposphere. Techniques such as limb sounding do have the potential to allow retrieval of vertical profiles into the troposphere.

16.7   MONITORING METEOROLOGICAL PARAMETERS

Ground-level meteorological measurements (below 10 m) are usually made *in situ* using meteorological equipment mounted on a mast. The WMO has developed a standard specification for meteorological equipment, and a global network of sites produce routine measurements using such equipment. The basic equipment for wind measurements is a cup anemometer and a wind vane. These provide wind speed and direction measurements respectively. Alternatively, a rotating vane mounting a propeller anemometer on the front can measure both direction and speed. Careful calibration of these physical measuring systems is necessary and care should be taken in the transportation of portable equipment, as mechanical damage can easily affect the calibration of the equipment. When selecting an anemometer consideration should be given to the threshold wind speed at which the cups or propeller start to rotate. For many applications the system should be capable of detecting wind speeds of less than $1\,m\,s^{-1}$. In order to measure turbulence and the vertical wind component a measurement of the wind vector is required. Three-axis wind speed measurement using three propellers has been used. However, the vertical component of such measurements is not particularly accurate. The most common, and sensitive, instruments for monitoring three-dimensional wind vectors are sonic anemometers. These systems employ sonic wind speed sensors mounted in three axes to give real-time measurements of the wind vector. The sensors work by measuring the difference in the speed of two sound pulses sent in opposite directions between two points. They report the wind speed in three directions and usually also report the measured speed of sound. They can be used for measurements of turbulence and can be used in combination with gas sensors to make eddy-correlation flux measurements.

Information on the optical scattering properties of the atmosphere can be obtained using an integrating nephelometer, which measures the optical depth of the atmosphere. Recent studies have compared nephelometer results with predictions on aerosol scattering derived from climate models.

It is often necessary to obtain information on the vertical profile of meteorological data. For many air quality and dispersion modeling purposes information is required up to the top of the boundary layer. The most common way to determine profiles of temperature, pressure, and humidity is from balloon sondes. Sondes can make measurements up to altitudes of approximately 15 km. At this altitude the balloon expands to the point of bursting and the sonde parachutes to the ground. Wind measurements have been made in the past by tracking the motion of the balloon using radar-theodolites. These sondes are known as rawinsondes. Current systems employ differential GPS receivers to track the balloon and determine wind speed and direction. There is a worldwide network of weather sonde launching sites and balloons are usually launched regularly, every 12 hours. For some applications, such as short-term local studies, a tethered balloon can be used with local telemetry of the data. These can routinely be flown up to 300 m and can be raised and lowered to gain profile information. These packages can mount standard *in situ* wind monitoring equipment. In addition, most atmospheric research aircraft include meteorological monitoring equipment as a part of their instrument packages.

A number of remote sensing instruments have been developed for monitoring meteorological parameters. These include Doppler lidar systems, correlation lidar, sodar and Doppler sodar, radar, and radio acoustic sounding systems (RASS). Many lidar systems have been developed for wind profiling and Delaval *et al.* (2000) describe an intercomparison between four wind lidar systems.

Aerosol lidars, described above, can be used to monitor turbulence within the boundary layer by observing variations in the aerosol concentration. Doppler lidar operates by measuring the Doppler shift on the scattered light detected by the lidar. The Doppler shift is used to determine the velocity component of the scattering medium (taken to be representative of the air mass velocity) parallel to the lidar beam. Owing to the spatial resolving capabilities of the lidar measurement, these systems are able to measure the velocity profile along the lidar beam. By scanning the direction of the lidar a velocity map can be built up, and measurements of the off-vertical wind velocity can be made.

Correlation lidars are able to measure three-dimensional wind velocity vectors. These systems work by scanning the lidar beam in a cone. Correlation analysis of the lidar signal at each location in the scan is used to track the movement of air masses by the inhomogeneity in the scattering in the atmosphere. The technique is based on the fact that if a particular pattern of scattering material is observed on one line of sight, and then is seen at a later time in another line of sight, the velocity of the air mass containing the scattering material can be determined from the time and spatial difference between the two lines of sight. In actual applications the data reduction is not as simple. However, by observing the correlation between each line of sight, at different time delays, the wind vector can be calculated with a reasonable uncertainty. The technique is not effective if the air being monitored is homogeneous in terms of its scattering properties. The technique determines the velocity components normal to the lidar beam with better accuracy than those parallel to it. However, as described above, Doppler lidar can provide this information.

Sodar systems operate in a similar way to lidar

and radar systems but employ sound waves (sonic detection and ranging). Sodars emit an audible "ping" from vertical pointing loudspeakers; the returned sound is collected using a parabolic reflector and measured with a sensitive microphone. The sound waves are reflected by variations in the refractive index of the atmosphere that have a similar scale to the wavelength of the sound. These are typically temperature variations. Sodars can therefore be used to determine the boundary layer height due to the turbulence caused by the temperature inversion. In addition they can be used to measure turbulence and buoyancy waves in the nighttime boundary layer. Typical sodars have an operational range of 1 km. Careful positioning and shielding of the sodar is necessary to reduce interference from ambient noise. In addition, their audible sound makes them inappropriate for long-term use in urban areas. Doppler sodars measure the frequency shift in the audio spectrum due to the motion of the reflecting air mass. These can typically measure the vertical wind speed profile at ranges of up to 1 km.

Basic "weather" radars are used to monitor and map precipitation by measuring the backscatter of microwave radiation by water droplets. These systems may be operated in vertical (range height indication, RHI), or horizontal (plan position indication, PPI) scanning modes or may be mounted on aircraft or satellite platforms.

Microwave radar can also be used to probe the vertical distribution of moisture fluctuations in the atmosphere. Such systems can measure and map turbulence up to about 10 km and can be used to observe convective flows and other atmospheric dynamics. Radar can be operated in a Doppler mode to measure the vertical wind speed. By measuring off the vertical axis the three-dimensional wind vector can be measured. The Doppler spectrum of the radar return provides information on the turbulence and mixing processes in the atmosphere.

RASS use a radar combined with a sodar system to provide information on the vertical temperature profile. The radar is able to measure the speed of the sound wave due to the enhanced backscatter from the pressure wave produced by the sodar sound pulse. The velocity of the sound wave provides information on the temperature of the air. It is necessary to correct for the effect of any vertical wind component; this can be simultaneously measured by the radar in wind sounding mode. RASS have similar performance to radar systems for monitoring wind velocities and can measure the temperature profile out to the range of the sodar (about 1 km).

Some satellite instruments, such as the wind scatterometer on the ERS satellite are used to monitor surface wind above the oceans. The satellites measure the change in radar reflection from the surface of the ocean caused by wind-induced ripples. Higher wind speeds increase the ripples. These in turn increase the radar return and this enables the wind velocity to be determined.

16.8 MONITORING OF THE MIDDLE TO UPPER ATMOSPHERE

This section discusses techniques for monitoring the middle to upper atmosphere. For the purposes of this discussion this is taken to be monitoring of the upper troposphere and stratosphere, as well as measurements of the total integrated concentration, or column, and vertical profiles of species through the whole atmosphere. Monitoring of the upper troposphere and mesosphere is not covered in detail, as such monitoring is mainly concerned with the interaction of the Earth's atmosphere with solar radiation and particles.

16.8.1 *Requirements*

A number of properties of the atmosphere are routinely measured, such as the radiative solar flux reaching the Earth's surface and the total column of species such as ozone, sulfur dioxide, and nitric oxide. Several networks of monitoring stations, both national and global, produce regular measurements. These include ozone monitoring stations and general meteorological stations. However, many of the drivers for measurements of the atmosphere above the atmospheric boundary layer come from specific atmospheric research programs. These include studies of stratospheric ozone and its chemistry, measurements in support of climate change research, the transport and mixing of trace gas species, water vapor transport, and cloud formation. Key requirements are for measurements of the vertical profile of parameters in the atmosphere, the measurement of trace compounds with sub-p.p.b. concentrations and large-scale mapping of the atmosphere.

The Network for the Detection of Stratospheric Changes is a global network of monitoring stations that measure ozone and trace gas concentrations in the stratosphere. The aim of the network is to provide long-term data sets to enable long-term changes in the stratosphere to be detected, particularly with regard to the ozone layer, to support climate change models, and to provide calibration of satellite observations.

16.8.2   In situ *measurements*

Many measurements of atmospheric properties in the upper atmosphere are made by *in situ* instrumentation. Such instruments generally use techniques similar to those for ground measurements described earlier in this chapter. However, the need to transport the sensors into the upper atmosphere places different design criteria on them. A number of platforms are used to carry *in situ* instruments; these include balloons, rockets, and aircraft. Balloons and rockets require telemetry unless the payload is guaranteed to be recovered intact. Some measurements are made by sampling the atmosphere in canisters that are returned to Earth for analysis in the laboratory. The advantage of *in situ* measurements is that they are analytically very specific and do not require complicated retrieval processes to generate vertical information. An important concern of all *in situ* measurements, as discussed above, is ensuring that the sample is not affected by the sampling system and measuring system before it is analyzed. It is also important, and not necessarily a trivial task, to ensure the systems remain in calibration while aloft.

Regular balloon launches are made with sondes that transmit vertical information on a number of parameters. Balloons of this type are regularly launched to retrieve meteorological data. A meteorological sonde typically measures values of air temperature, air pressure, and relative humidity. Radiosondes transmit their measured data to a ground receiving station. Ozonesondes are an example of an *in situ* measurement technique used to measure vertical profiles of ozone concentration. Ozonesondes are composed of an ozone sensor, a battery, and a small gas pump and can be carried on weather balloons. The balloon radiosonde is used to transmit values of the detector current, detector temperature, and pump speed to a ground receiving station. The sampled air is pumped through a sensor that produces an electrical current. A typical ozone sensor consists of two cells containing potassium iodide solutions. Each chamber has a platinum electrode at its base and the electrodes are connected by current sensor. Air is pumped into one chamber and the ozone reacts with the potassium iodide to produce iodine. This produces a current between the two cells. The current is proportional to the rate of the reaction, which is proportional to the ozone concentration. The current is measured in the sensor and transmitted to the ground. The ozone concentration is directly proportional to the current.

A disadvantage of balloon-borne instrumentation is that the payload is limited in weight and size. Long-term flights are also limited by the power supply that can be carried. The primary disadvantage is, however, the lack of control over the flight path of the balloon. For vertical soundings this may not be a major issue, but for long-term research flights the outcome of the mission may depend on the meteorological conditions at the time of launch.

A number of research aircraft have been developed to carry instrumentation for monitoring the atmosphere. Some of these are remotely operated unpiloted systems, such as the Altus unmanned aerospace vehicle UAV. Others are specifically designed research aircraft or converted commercial aircraft such as the UK Facility for Airborne Atmospheric Measurement (FAAM) BAes146-300 G-LUXE, the DC-8 flown by NASA, and the Lockheed WP-3D Orions flown by NOAA. The UK G-LUXE aircraft (this is the registration name of the plane) has a ceiling of approximately 10 km and a flight duration of 6 hours. This aircraft will replace the Hercules C-130, operated by the UK Meteorological Office, which is now been withdrawn from service. The NASA DC-8 based at Dryden Flight Research Center has a ceiling of nearly 13 km and can carry 13 tonnes of sensing equipment. It is used for atmospheric monitoring campaigns, for sensor development, and to provide *in situ* validation measurements for satellite remote sensors (information obtained from NASA FS-2001-06-050-DFRC fact sheet). NASA also fly ER-2 high-altitude research planes. These are civilian research versions of the US U2-S military reconnaissance planes. The planes are capable of flying atmospheric monitoring payloads up to altitudes of above 21 km. At this altitude they are capable of making *in situ* measurements of the stratosphere.

A number of other meteorological planes are operated worldwide; for example, the European Fleet for Airborne Research (EURFAR) lists 12 operational planes available for research flights in Europe alone.

16.8.3 *Remote sensing techniques*

Many remote sensing techniques used to monitor the whole atmosphere are passive systems. These use solar radiation to make measurements of gases in the upper atmosphere. The most basic instrument of this type is the radiometer. Radiometry is the measurement of the total incident radiative flux at a receiver. In general this is a measurement of the total flux at the Earth's surface, though radiometers are also employed on Earth-observing satellites. Strictly speaking a radiometer is not a remote sensor, as the information obtained is not a measure of a quantity at a remote location. However, it does measure the total effect of radiative transfer processes through the remote atmosphere. A number of specific measuring systems are available; these include pyronometers, which measure the total incident radiation from all angles, and pyrheliometers, which measure the direct normal irradiance (the intensity of light coming directly from the Sun). Ultraviolet radiometers may measure total UV light or may be made specific to UV-A (roughly 386 nm) or UV-B (306 nm) by the use of filters. Radiometers may also measure far infrared (3–50 μm) radiation, which is the wavelength region of emitted thermal radiation from the Earth.

Dobson and Brewer spectrophotometers

Dobson and Brewer spectrophotometers are passive sensors that are used to measure the total amount of ozone, or column, in a vertical path through the atmosphere. They measure the solar radiation at the Earth's surface at four wavelengths. Two of these are in a wavelength region in which ozone absorbs; usually this is in the region of 290–320 nm. Since a number of atmospheric components, such as gases and aerosols, also absorb light in these wavelengths, two wavelengths in a region of the spectrum where ozone does not absorb are simultaneously measured. It is assumed that the effect of absorption by clouds and other atmospheric contaminants is equal across the wavelengths (Komhyr & Evans 1980). By taking the ratio of the two ozone-absorbed wavelengths from the two background intensities, the effects of the clouds and aerosols are cancelled out, leaving the ozone absorption signal.

The method was developed in the 1930s and it is still in use today (Dobson 1957a, b; Dobson & Normand 1962). The light entering the spectrophotometer is split into two beams that fall alternately on a photomultiplier tube. The more intense beam is reduced in intensity by passing it through a piece of glass of varying thickness. The glass wedge is adjusted until both beams give the same signal on the photomultiplier. The position of the glass wedge is noted and the amount of ozone is then derived from calibration data. The Dobson spectrophotometer can be used to determine ozone by measuring light from direct sunlight, diffuse light from clear or cloudy skies, and even reflected sunlight from the Moon. The direct Sun measurements are preferred because the uncertainties of ozone measurements get larger as the amount of light entering the instrument decreases.

The Brewer spectrophotometer uses the same basic technique as the Dobson instrument. However, the Brewer spectrophotometer is automated. Most Brewer instruments are programmed to take measurements at regular observation times. The instrument measures UV light at five wavelengths (306, 310, 313, 317, and 320 nm). The total column ozone amount is calculated by using a more complicated form of the equation used for the Dobson instrument. This equation includes terms for sulfur dioxide, which is an interferent at the wavelengths used. The uncertainty for a total ozone measurement made by a calibrated Brewer instrument is estimated to be ±2.0%. Ozone column content is usually expressed in Dobson units. One Dobson unit is equivalent to a total amount of ozone in the atmosphere that would have a thickness of 1 mm at 273 K and 101.3 kPa (van Roozendael et al. 1998).

A number of other spectrometers, including FTS systems, are used to measure species in the atmosphere by observing absorption lines in the solar spectrum (Hervig et al. 1995; Meier 1997). In general such measurements are made at high altitudes to avoid water vapor and other gaseous absorption present at lower altitudes. Measurements are made of the total absorption, and hence total column content of gases. By analyzing the line shape of individual absorption features it is also possible to retrieve the vertical distribution of species. This is possible because different processes cause different broadening characteristics in absorption lines. At low altitudes pressure broadening dominates, giving rise to the characteristic broad tails of the Lorentz line-shape. At high altitudes the Doppler broadened line-shape dominates. The relative contribution of the different line broadening processes can be determined using numerical analysis if the temperature and pressure profiles in the atmosphere are known or can be assumed. Fitting algorithms are used to determine the vertical distribution of the measured gas that would produce the observed line shape. Such retrieval algorithms are validated and constrained by simultaneous measurements using balloon sondes and other in situ techniques.

Passive microwave systems are used to observe emission spectra from a number of atmospheric species. Many atmospheric species have rotational spectral emission lines in the microwave region. Two of the principle species observed using passive microwave techniques include water vapor and oxygen. Water vapor has emission lines at 22 and 183 GHz, and oxygen has a broad band at 60 GHz and a single line at 118 GHz. Due to its homogeneity, oxygen measurements can be used to determine the temperature profile. Microwave measurements rely on radiative transfer models in order to retrieve vertical concentration profiles. One of the advantages of microwave systems is that they are unaffected by cloud cover. A number of microwave sounders are operated at NDSC measurements sites to measure vertical profiles of water vapor, carbon monoxide, and ozone (De la Noe et al. 1998; Sinnhuber et al. 1998).

Lidar

Active techniques such as lidar are also used to probe the upper atmosphere. Because the systems are used to probe high into the stratosphere very high power lasers are required. Detectors operating in photon-counting mode are needed to observe the faint return signals. Therefore the systems are usually purpose built to specifically measure particular parameters; for example, ozone concentration or partial pressure. Such systems are optimized to measure ozone and are rarely capable of tuning to different wavelengths in order to measure more than one species. Common wavelengths used for the measurement of stratospheric ozone are 308 and 351 nm. The 308 nm beam is absorbed by ozone absorption, while the 351 nm radiation is not. Wavelength selection is particularly important in a stratospheric DIAL system. If a lidar is used to measure stratospheric ozone the light passes twice through any ozone at lower altitudes. It is therefore necessary to select wavelengths such as 308 nm, which are not strongly absorbed by ozone, as otherwise the return signal would be completely absorbed by ozone at lower altitudes. In the troposphere, different, more strongly absorbed wavelengths must be used. Ozone lidars designed for stratospheric monitoring are capable of making measurements from roughly 15 to 50 km in the atmosphere with 1 km resolution. Other lidar systems are used to probe atmospheric density, water vapor concentration, and cloud profiling.

In order to overcome the effect of the optically dense troposphere, lidar systems have been mounted on aircraft. For example, the OLEX system is mounted on the European CITATION II and FA-20 aircraft and is capable of measuring vertical ozone profiles at ranges of up to 26 km with a resolution of 1 km. It operates at wavelengths of 308, 345, and 532 nm. Another example of an airborne DIAL system is the Lidar pour l'Etude des interactions Aerosols Nuages Dynamique Rayonnement et du cycle de l'Eau (LEANDRE II), which has been flown on the French ARAT aircraft. The system is able to measure the water vapor mixing ratio in the troposphere (Bruneau *et al.* 2001) at ranges of up to 5 km with a resolution of 300 m.

Lidar systems have also been launched as a part of rocket borne payloads, to allow *in situ* measurements to be made into the mesosphere. The Transmitter and Receiver of Optical Light (TROLL) system was developed and flown as a part of the Rocketborne Optical Neutral Gas Analyzer with Laser Diodes (RONALD) payload. The rocket payload includes a spectrometer able to make *in situ* measurements of water vapor and carbon dioxide, and electrostatic probes to measure ion densities in the ionosphere. The TROLL system is able to make measurements of the atmospheric density at locations in the atmosphere that are not affected by the bow-shock of the rocket (Eriksen *et al.* 1999).

Satellite remote sensors

Satellite-mounted instrumentation has the potential to provide measurements of the atmosphere with very large spatial, even global, coverage. A large number of Earth observation satellite missions have been flown. Many carry instrumentation to monitor surface conditions, such as land use, sea temperature, and surface altitude. For example, the Along Track Scanning Radiometers, ATSR-1, ATSR-2, and the future AATSR instruments, provide sea surface temperatures. A significant number of instruments have also been flown

to monitor atmospheric parameters. Satellite instruments provide data for estimating atmospheric water vapor, ozone, and trace gas concentrations, aerosol content, clouds dynamics, and precipitation. In addition, satellite data are used to study radiative and dynamic processes within the atmosphere.

Satellite instrumentation used for atmospheric monitoring can be passive or active. Passive instrumentation uses solar, or other external, radiation sources or upwelling infrared radiation, or directly observes emissions from target species. Active systems usually employ either radar or lidar systems. There are four different passive remote sensing techniques based on the viewing geometry of the instrument, as shown in Fig. 16.15. They are nadir viewing, occultation, limb emission, and limb scattering. Most active systems operate in a nadir-viewing configuration.

Passive nadir-viewing techniques observe directly below the satellite and measure solar radiation scattered from the atmosphere. Instruments operating in the ultraviolet are used to measure the total column amounts and profiles of ozone and some trace gases. The total ozone mapping spectrometer (TOMS) instrument measures total column ozone. The TOMS instrument was flown on the Nimbus-7 satellite and gathered data from 1978 to 1988. Similar instruments have continued to provide continuous data on ozone concentrations. Measurements from TOMS have been compared with ozone measurements from Dobson ozone spectrometers and other ground-based techniques in the NDSC network (Nichol *et al.* 1996; Lambert *et al.* 2000).

Other instruments, such as the global ozone monitoring experiment (GOME), are potentially able to measure the vertical profile of ozone, nitrogen dioxide, CFCs, and some trace atmospheric species. This instrument is a nadir scanning spectrometer operating in the visible and ultraviolet wavelengths. The sensor has spectral resolution of 0.4 nm and a field of view of up to 960×80 km. It uses DOAS techniques to recover gas concentrations. Currently the system has been used to determine total column amount of ozone and nitrous oxide. GOME was flown on the 1995 ERS-2 satel-

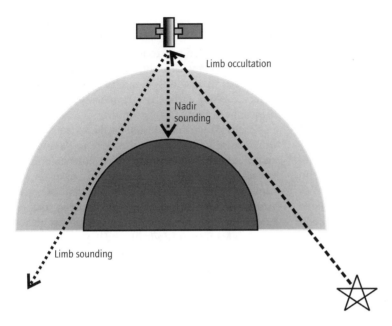

Fig.16.15 Configuration of satellite sensing systems.

lite (Burrows *et al.* 1999) and a second-generation instrument will fly on the future polar-orbiting METOP satellites. In general, passive nadir techniques provide good horizontal resolution, but less resolved vertical information.

The occultation technique uses solar, lunar, or stellar radiation measured through the limb of the atmosphere. As the source is occulted by the Earth, the optical path through the limb passes through the atmosphere at different altitudes. By measuring absorption features this technique can measure vertical profiles of a number of trace gases. Instruments measure the spectra of the source above the atmosphere and compare this with the spectra through the atmosphere. The occultation technique can achieve improved vertical resolution, at the expense of horizontal resolution and coverage, as measurements can only be made during occultations of the light source. Stellar observing instruments can carry out more measurements than solar/lunar instruments, but the reduced intensity of stellar light reduces detection sensitivity. The halogen occultation experiment (HALOE) instrument on board the UARS satellite used solar occultation to measure ozone, hydrogen chloride,

hydrogen fluoride, methane, water vapor, oxides of nitrogen, and aerosols in the infrared. It also measured temperature profiles (Singh & Gross 1996). The Stratospheric Aerosol and Gas Experiments, SAGE I/II, used the limb occultation technique to provide vertical profiles of a number of species together with aerosol extinction. The global ozone mapping by occultation of stars instrument (GOMOS) is a stellar occultation instrument on the European ENVISAT, launched in early 2002. This instrument will measure the spectra of the star in five bands from the ultraviolet to the near infrared. Approximately 100 stars are bright enough to be candidates for measurements.

The limb emission or limb viewing technique measures the radiation emitted thermally in the atmosphere along the line of sight of the instrument. The instrument observes along a tangent through the limb of the atmosphere. These instruments measure thermal infrared radiation emitted from gases in the atmosphere, and can identify trace gas species from the emission spectra. The relatively weak emission spectra are more easily observed in a limb-sounding configuration because there is little interference from the cold background of space and the path length through the upper atmosphere is longer than for nadir viewing systems. The retrieval algorithms for limb emission spectroscopy are discussed in Elachi (1987). The spectroscopy of the atmosphere using far infrared emission (SAFIRE) instrument uses this technique. The michelson interferometer for passive atmospheric sounding (MIPAS) is a limb-sounding FTS system that will measure high-resolution emission spectra. It is on board the European ENVISAT and will monitor a wide range of traces gases in the near infrared, including ozone, water vapor, methane, oxides of nitrogen, and CFCs. The microwave limb sounding (MLS) instrument flown on the UARS satellite observed emission lines from ozone, water vapor, sulfur dioxide, nitric acid, and hyopochlorite (ClO). Froidevaux *et al.* (1996) reports results of validation of the MLS ozone measurements.

The limb scattering technique measures scattered radiation from near horizontal paths through the limb of the atmosphere. This provides vertically resolved measurements, using techniques similar to those used for passive nadir sounding instruments. This allows for continuous coverage through the daylight portion of the upper troposphere and stratosphere. One of the first high-resolution spectrometers to be used for limb sounding was the atmospheric trace molecule spectrometer (ATMOS), which was an FTIR system. The scanning imaging absorption spectrometer for atmospheric cartography (SCIAMACHY) instrument, which is on the ENVISAT payload, will employ this technique.

Most active satellite instruments are either microwave radars or lidars. A crucial step in space borne lidar was made with the 1994 lidar in space technology experiment (LITE) mission flown aboard the shuttle (McCormick 1996). This 10-day experiment made measurements of stratospheric and tropospheric aerosol concentrations using three wavelegth channels at 355, 532, and 1064 nm. It also enabled retrieval of the planetary boundary layer height, and stratospheric temperature and density profiles. LITE results for stratospheric aerosol concentrations have been compared with SAGE-II results (Kent *et al.* 1996), and various

comparisons were made between LITE and airborne and ground-based lidar systems.

Another satellite-borne lidar proving experiment was the ALISSA system, which was operated from the MIR platform from 1996 to 1999 (Chanin *et al.* 2000). This instrument was a Mie lidar that was mainly used to determine cloud top height and to detect optically thin high-altitude cirrus clouds, which were not detectable by previous meteorological satellites.

Following the success of the LITE and ALISSA experiments a number of satellite-borne lidars have been proposed and are under development. One is due to be flown as the main payload in the Atmospheric Dynamics Mission (ADM) Aeolus satellite. This system, the atmospheric laser Doppler instrument (ALADIN) is a Doppler lidar capable of mapping the wind field (Morancais *et al.* 2000). The system will operate at 355 nm. It is an off-nadir pointing instrument that has two detection systems, one designed to measure Doppler shifted Mie aerosol backscatter and the other to measure Doppler shifted Rayleigh molecular backscatter. While both these backscattering mechanisms are elastic the relative motions of molecules and aerosols produce different line broadening effects, and hence detection systems can be tuned to be most sensitive to each type of scattering. Mie scattering is only very slightly broadened due to the motion of the aerosols because of wind variability, whereas Brownian motion of molecules causes considerable broadening of the Rayleigh scattered light. The use of two complementary detection systems allows optimum detection of the backscatter from all altitudes, as molecular scattering is dominant at high altitudes but is attenuated in the troposphere, where Mie scattering dominates. The two detection systems operate on very different principles: the very narrow bandwidth Mie lines will be measured using a heterodyne system, which has a very high resolution; the Rayleigh channel will use a Fabry–Perot etalon to allow it to directly measure the returned lines. The system is designed to achieve a vertical resolution of 1 km with a horizontal resolution of 50 km and an along-track sampling spacing of 200 km.

The geoscience laser altimeter system (GLAS) is a lidar altimeter due to be flown on the Ice, Cloud and Land Elevation Satellite (ICESat), which is scheduled for launch in late 2002. While GLAS is primarily designed to measure ice-sheet topography and associated temporal changes, it is also designed to deliver cloud heights and atmospheric properties such as temperature and density profiles (McCormick *et al.* 2000).

16.9   CONCLUSIONS

This chapter has reviewed a small selection of the vast number of measurement techniques that have been developed to meet the requirements of the atmospheric science community. It has not been possible, in a single chapter, to cover many of the experimental techniques that have been developed to meet specific applications. Without a doubt the future will see the development of smaller, faster sensors, particularly solid state sensors for air quality measurements. More and more laboratory techniques will migrate into reliable field instruments, and satellite-borne lidars and other active sensors will revolutionize atmospheric profiling. Future requirements for air quality monitoring include lower detection sensitivities to meet the requirements of the new air quality targets and systems to monitor a wide range of air toxics and micro-pollutants.

REFERENCES

Ambrico, P.F., Amodeo, A., Girolamo, P.D. & Spinelli, N. (2000) Sensitivity analysis of differential absorption lidar measurements in the mid-infrared region. *Applied Optics* **39**, 6847–65.

Arshinov, Y., Bobrovnikov, S., Serikov, I. *et al.* (2000) Spectrally absolute instrumental approach to isolate pure rotational Raman lidar returns from nitrogen molecules of the atmosphere. In Dabas, A., Loth, C. & Pelon, J. (eds), *Advances in Laser Remote Sensing: Selected Papers Presented at the Twentieth International Laser Radar Conference (ILRC), Vichy, France, 10–14 July 2000.* Edition de l'Ecole polytechnique, Paris.

Bartley, D.L., Deye, G.J. & Woebkenberg, M.L. (1987) Diffusive monitor test: Performance under transient conditions. *Applied Industrial Hygiene* **2**, 119–22.

Boubel, R.W., Fox, D.L., Turner, D.B. & Stern A.C. (1994) *Fundamentals of Air Pollution*, 3rd edn. Academic Press, New York.

Bruneau, D., Quaglia, P., Flamant, C. & Pelon, J. (2001) Airborne lidar LEANDRE II for water-vapor profiling in the troposphere. II. First results. *Applied Optics-LP* **40**, 3068–82.

Burrows, J.P., Weber, M., Buchwitz, M. *et al.* (1999) The global ozone monitoring experiment (GOME): mission concept and first scientific results. *Journal of Atmospheric Science* **56**, 151–75.

Casavant, A.R. & Kamme, C.J. (1992) An inter-comparison of sulfur dioxide, nitrogen dioxide, and ozone with DOAS optical. In *Optical Remote Sensing: Applications to Environmental and Industrial Safety Problems.* Air and Waste Management Association, Pittsburgh.

Centre for European Normalization (1999) *EN 12341:1998 Air Quality. Determination of the PM 10 Fraction of Suspended Particulate Matter. Reference Method and Field Test Procedure to Demonstrate Reference Equivalence of Measurement Methods.* CEN, Brussels.

Chanin, M.-L., Hauchecorne, A., Malique, C., Desbois, M., Tulinov, G. & Melnikov, V., (2000) The ALISSA lidar onboard the MIR platform. In Dabas, A., Loth, C. & Pelon, J. (eds), *Advances in Laser Remote Sensing: Selected Papers Presented at the Twentieth International Laser Radar Conference (ILRC), Vichy, France, 10–14 July 2000.* Edition de l'Ecole polytechnique, Paris.

Commission of the European Communities (1996) Council Directive 1996/62/EC of 27 September 1996 on ambient air quality management and assessment. *Official Journal of the European Communities* **L296**, 55.

Commission of the European Communities (1999) Council Directive 1999/30/EC of 22 April 1999 relating to limit values for sulfur dioxide, nitrogen dioxide, and oxides of nitrogen, particulate matter and lead in ambient air. *Official Journal of the European Communities* **L163**, 41–58.

Commission of the European Communities (2000) Council Directive 2000/69/EC of 16 November 2000 relating to limit values for sulfur dioxide, nitrogen dioxide, and oxides of nitrogen, particulate matter and lead in ambient air. *Official Journal of the European Communities* **L313**, 12–21.

De la Noe, J., Lezeaux, O., Guillemin, G., Lauque, R., Baron, P. & Ricaud, P. (1998) A ground-based microwave radiometer dedicated to stratospheric ozone monitoring. *Journal of Geophysical Research* **103**, 22 147–61.

Delaval, A., Flamant, P.H., Aupierre, M. *et al.* (2000) Intercomparison of wind profiling instruments during the VALID field campaign. In Dabas, A., Loth, C. & Pelon, J. (eds), *Advances in Laser Remote Sensing: Selected Papers Presented at the Twentieth International Laser Radar Conference (ILRC), Vichy, France, 10–14 July 2000*. Edition de l'Ecole polytechnique, Paris.

Dobson, G.M.B. (1957a) Observers' handbook for the ozone spectrophotometer. *Annals of the International Geophysical Year* **5**, 46–89.

Dobson, G.M.B. (1957b) Adjustment and calibration of the ozone spectrophotometer. *Annals of the International Geophysical Year* **5**, 90–113.

Dobson, G.M.B. & Normand, C.W.B. (1962) Determination of the constants etc. used in the calculation of the amount of ozone from spectrophotometer measurements and of the accuracy of the results. *Annals of the International Geophysical Year* **16**, 161–91.

Elachi, C. (1987) *Introduction to the Physics and Techniques of Remote Sensing*. Wiley Interscience, New York.

Environment Agency (2000) *MCERTS Performance Standards for Continuous Emission Monitoring Systems*. Environment Agency, National Compliance Assessment Center, Lancaster.

Eriksen, T., Hoppe, U.P., Thrane, E.V. & Blix, T.A. (1999) Rocketborne Rayleigh lidar for in-situ measurements of neutral atmospheric density. *Applied Optics* **38**, 2605–13.

Froidevaux, L., Read, W.G., Lungu, T. *et al.* (1996) Validation of UARS Microwave Limb Sounder ozone measurements. *Journal of Geophysical Research* **101**, 10017–60.

Gardiner, T.D., Milton, M.J., Molero, F. & Woods, P.T. (1996) Infrared DIAL measurements with an injection-seeded OPO. In Ansman, A., Neuber, R., Rairoux, P. & Wandinger, U. (eds), *Advances in Atmospheric Remote Sensing with Lidar, Selected Papers of the Eighteenth Inernational Laser Rader Conference (ILRC) Berlin*. Springer-Verlag, Berlin.

Goldsmith, J.E.M. (1998) Turn-key Raman lidar for profiling atmospheric water vapor, clouds and aerosols. *Applied Optics* **37**, 4979–90.

Grant, W.B., Browell, E.V., Menzies, R.T., Sassen, K. & She, C. (eds) (1997) *Selected Papers On Laser Applications in Remote Sensing, International Society for Optical Engineering*. SPIE, Bellingham.

Hervig, M.E., Russell III, J.M., Gordley, L.L. *et al.* (1995) Ground-based FTIR spectroscopic absorption measurements of stratospheric trace gases with the sun and moon as light sources. *Journal of Molecular Structure* **347**, 407–16.

Hinkley, E. (ed.) (1976) *Laser Monitoring of the Atmosphere*. Springer-Verlag, Berlin.

International Standards Organization (1985) *ISO 7996 Ambient Air. Determination of the Mass Concentrations of Nitrogen Oxides. Chemiluminescence Method*. ISO, Geneva.

International Standards Organization (1993) *ISO 9855 Ambient Air. Determination of the Particulate Lead Content of Aerosols Collected in Filters. Atomic Absorption Spectroscopy Method*. ISO, Geneva.

International Standards Organization (1995) *Guide to the Expression of Uncertainty in Measurement*. ISO, Geneva.

International Standards Organization (1999) *ISO/DIS 16017-2. Indoor, Ambient and Workplace Air. Sampling and Analysis of Volatile Organic Compounds by Sorbent Tube/Thermal Desproption/Capillary Gas Chromatography. Part 2: Diffusive Sampling*. ISO, Geneva.

International Standards Organization (2001) *ISO/FDIS 14956 (Standard in Draft). Air Quality. Evaluation of the Suitability of a Measurement Procedure by Comparison with a Required Measurement Uncertainty*. ISO, Geneva.

Kent, G.S, Osborn, M.T., Trepte, C.R. & Skeens, K.M. (1996) LITE measurements of aerosols in the stratosphere and upper troposphere. In Ansman, A., Neuber, R., Rairoux, P. & Wandinger, U. (eds), *Advances in Atmospheric Remote Sensing with Lidar, Selected Papers of the Eighteenth Inernational Laser Rader Conference (ILRC) Berlin*. Springer-Verlag, Berlin.

Komhyr, W.D. & Evans, R.D. (1980) Dobson spectrophotometer total ozone measurement error caused by interfering absorbing species such as SO_2, NO_2 and photochemically produced O_3 in polluted air. *Geophysical Research Letters* **7**, 157–60.

Lambert, J.-C., Van Roozendael, M., Simon, P.C. *et al.* (2000) Combined characterisation of GOME and TOMS total ozone measurements from space using ground-based observations from the NDSC. *Advances in Space Research* **26**, 1931–40.

McCormick, M.P. (1996) The flight of the lidar in space technology experiment LITE. In Ansman, A., Neuber, R., Rairoux, P. & Wandinger, U. (eds), *Advances in Atmospheric Remote Sensing with Lidar, Selected Papers of the Eighteenth Inernational Laser Rader Conference (ILRC) Berlin. Springer-Verlag, Berlin.

McCormick, M.P. (2000) A bright future for spaceborne lidars. In Dabas, A., Loth, C. & Pelon, J. (eds), Advances in Laser Remote Sensing: Selected Papers Presented at the Twentieth International Laser Radar Conference (ILRC), Vichy, France, 10–14 July 2000. Edition de l'Ecole polytechnique, Paris.

Martin, N.A., Dedman, S.A., Goody, B.A. & Henderson, M.H., (2001) The dosing of pumped adsorption tubes and diffusion tubes with benzene in a controlled atmosphere test facility. Poster presentation at international conference on Measuring Air Pollutants by Diffusive Sampling, Montpellier, September 26–28.

Measures, R.M. (1984) Laser Remote Sensing: Fundamentals and Applications. Wiley, New York.

Meier, A. (1997) Determination of atmospheric trace gas amounts and corresponding natural isotopic ratios by means of ground-based FTIR spectroscopy in the high Arctic. Reports on Polar Research 236, 311.

Morancais, D., Fabre, F., Berlioz, P., Maurer, R. & Culoma, A. (2000) Spaceborne wind lidar concept for the atmospheric dynamics mission (ALADIN). In Dabas, A., Loth, C. & Pelon, J. (eds), Advances in Laser Remote Sensing: Selected Papers Presented at the Twentieth International Laser Radar Conference (ILRC), Vichy, France, 10–14 July 2000. Edition de l'Ecole polytechnique, Paris.

Nichol, S.E., Keys, J.G., Wood, S.W., Johnston, P.V. & Bodeker, G.E. (1996) Intercomparison of total ozone data from a Dobson spectrophotometer, TOMS, visible wavelength spectrometer, and ozonesondes. Geophysical Research Letters 23, 1087–90.

Richardson, R.L. (1997) Reduction of stray light in monostatic open-path FT-IR spectrometers with a plane correction mirror. Applied Spectroscopy 51, 1254–6.

Robinson, R.A., Woods, P.T. & Milton, M.J. (1995) DIAL measurements for air pollution and fugitive loss monitoring. Proceedings SPIE Air Pollution and Visibility Measurements, 2506.

Rothman, L.S., Rinsland, C.P., Goldman, A. et al. (1998) The HITRAN molecular spectroscopic data base and HAWKS: 1996 edition. Journal of Quantum Spectroscopic Radiation Transfer 60, 665–710.

Schnopper, H.W. & Thompson, R.I. (1974) Fourier spectrometers. In Carleton, N. (ed.), Methods of Experimental Physics. Volume 12, Part A: Astrophysics, Optical and Infrared. Academic Press, London.

Singh U.N. & Gross, M.R. (1996) Validation of temperature measurements from the Halogen Occultation Experiment. Journal of Geophysical Research 101, 10277–85.

Sinnhuber, B.-M., Langer, J., Klein, U., Raffalski, U., Künzi, K. & Schrems, O. (1998) Ground based millimeter-wave observations of Arctic ozone depletion during winter and spring of 1996/97. Geophysical Research Letters 25, 3227–30.

Van Roozendael, M., Peeters, P., Roscoe, H.K. et al. (1998) Validation of ground-based visible measurements of total ozone by comparison with Dobson and Brewer spectrophotometers. Journal of Atmospheric Chemistry 29, 55–83.

Whiteman, D.N., Melfi, S.H. & Ferrare, R.A. (1992) Raman lidar system for the measurement of water vapor and aerosols in the Earth's atmosphere. Applied Optics-OT 31, 3068–82.

Woods, P.T., Partridge, R.H., Milton, M.J., Jolliffe, B.J. & Swann, N.R. (1992) Remote sensing techniques for gas detection, air pollution and fugitive loss monitoring, In Optical Remote Sensing; Applications to Environmental and Industrial Safety Problems. Air and Waste Management Association, Pittsburgh.

17 Emission Inventories

DAVID HUTCHINSON

17.1 INTRODUCTION

An atmospheric emissions inventory is a schedule of the sources of a pollutant or pollutants within a particular geographic area and over a particular period of time. The inventory usually includes information on the amount of the pollutant released from road traffic and other modes of transport, major industrial sources, and average figures for the emissions from smaller sources throughout the area. However, inventories may also be prepared for particular purposes that are more restricted in their scope. Emission inventories are an essential tool in the management of local air quality. While measurements of pollutant concentrations in the atmosphere, carried out by many national governments, regional organizations, and cities around the world, show the extent of air pollution, emission inventories identify the sources and help in the development of air quality improvement strategies.

The first emission inventories were prepared in the 1960s. In the United States, the establishment of the National Ambient Air Quality Standards (NAAQS) created the need for emission inventories, because once there were standards, there was a need to quantify the extent of pollution, and to identify the sources, the type, and the amount of pollution they produced (Cabreza 1999). Two of the earliest inventories prepared in the UK were for Sheffield (Garnett 1967) and Reading (Marsh & Foster 1967). The first London inventory was prepared by the Scientific Branch of the former Greater London Council by David Ball and Sam Radcliffe (1979). This was an inventory of sulfur dioxide (SO_2) emissions.

17.1.1 The United States

In the United States regulations requiring the annual reporting of emissions have been in place since the late 1970s. Emission inventories are used for a wide variety of purposes, but are most often prepared in response to regulation. Emission inventory data are used to evaluate the current state of air quality in relation to air quality standards and air pollution problems, to assess the effectiveness of air pollution policy, and to initiate any changes that may be needed. Individual states may have their own specific inventory requirements, while at the federal level, requirements for emission estimates stem mainly from the Clean Air Act (Eastern Research Group 1999).

The Clean Air Act directs the US Environmental Protection Agency (US EPA) to identify and set National Ambient Air Quality Standards for the most common air pollutants. US EPA uses these "criteria pollutants" as indicators of air quality. These pollutants are:
- ozone (O_3);
- carbon monoxide (CO);
- nitrogen oxides (NO_x);
- sulfur dioxide (SO_2);
- particulate matter with aerodynamic diameter less than or equal to $10\,\mu m$ (PM_{10});

• particulate matter with aerodynamic diameter less than or equal to 2.5 μm (PM$_{2.5}$);
• lead (Pb).
In addition to these pollutants, US EPA also regulates emissions of volatile organic compounds (VOCs) under criteria pollutant programs. VOCs are ozone precursors—they react with nitrogen oxides in the atmosphere to form ozone. VOCs are emitted from motor vehicle fuel distribution, chemical manufacturing, and a wide variety of industrial, commercial, and consumer solvent uses.

The Clean Air Amendment Act 1990 requires the development of "comprehensive, accurate, and current" inventories from all sources. The US EPA's Office of Air Quality Planning and Standards develops and maintains emission estimating tools to support the preparation of emission inventories by federal, state, and local agencies, consultants and industry (Sasnett & Misenheimer 1996). Inventory data quality is important because these data allows us to balance community health risks against pollution control decisions and to determine the type and extent of controls that are needed to prevent health effects caused by poor air quality. There is a close relationship between control strategies and public health. Underestimates in an inventory result in inadequate levels of control to protect public health, whereas overestimates can result in unnecessary and costly industrial controls (Cabreza 1999).

Emission inventories provide the technical foundation for state, local, and federal programs designed to improve or maintain ambient air quality (Eastern Research Group 1999). Specific examples of end uses for emission inventories include:
• meeting Clean Air Act requirements for specific inventories as part of state implementation plans (SIPs);
• tracking progress toward meeting the National Ambient Air Quality Standards and emission reductions;
• determining compliance with emission regulations and setting the baseline for policy planning;
• identifying sources and general emission levels, patterns, and trends in order to develop control strategies and new regulations;

• serving as the basis for the modeling of predicted pollutant concentrations in ambient air;
• providing input for human health risk assessment studies;
• conducting environmental impact assessments for proposed new sources;
• serving as the basis for construction and operating permits;
• serving as a tool to support future emissions trading programs;
• siting ambient air quality monitoring equipment.
A further use, not mentioned by Eastern Research Group but one that is becoming increasingly important, is as a basis for emissions trading. One example is trading within the nine northeastern states participating in the Ozone Transport Commission (see http://www.sso.org/otc/).

17.1.2 Europe

There are a substantial number of international agreements that require the preparation of emission inventories. The Convention on Long Range Transboundary Air Pollutants was adopted in Geneva in 1979 within the framework of the United Nations Economic Commission for Europe (UNECE). States are required to report emissions data to the Executive Body of the Convention in order to meet their obligations to comply with the Convention Protocols and to assess which countries need to adopt the most stringent abatement strategies. The Protocols are:
• the 1984 Protocol on Long-term Financing of the Cooperative Programme for Monitoring and Evaluation of the Long-range Transmission of Air Pollutants in Europe (EMEP);
• the 1985 Protocol on the Reduction of Sulphur Emissions or their Transboundary Fluxes by at least 30%;
• the 1988 Protocol concerning the Control of Nitrogen Oxides or Their Transboundary Fluxes;
• the 1991 Protocol concerning the Control of Emissions of Volatile Organic Compounds or Their Transboundary Fluxes;
• the 1994 Protocol on Further Reduction of Sulphur Emissions;

• the 1998 Protocol on Heavy Metals;
• the 1998 Protocol on Persistent Organic Pollutants (POPs);
• the 1999 Protocol to Abate Acidification, Eutrophication and Ground-level ozone.

The parties to the Convention, including the UK, are required to submit details of their annual national emissions of sulfur dioxide (SO_2), oxides of nitrogen (NO_x), methane (CH_4), nonmethane VOCs (NMVOC), ammonia (NH_3), carbon monoxide (CO), heavy metals, and persistent organic pollutants. Emissions data reported by national governments under the Convention can be found on the UNECE website (http://www.unece.org/env/lrtap/).

The EMEP, established under the Convention, in turn set up a Task Force on Atmospheric Emissions Inventories in 1991 with a variety of objectives, including the preparation of an *Emission Inventory Guidebook* (EMEP/CORINAIR 2002). The aim of this was to increase reporting of emissions by the signatories to the Convention. It also established a 50×50 km grid, known as the EMEP grid, for reporting and analysis purposes, and 11 main source categories. These are:

1 Public power, cogeneration, and district heating plants.
2 Commercial, institutional, and residential combustion plants.
3 Industrial combustion.
4 Production processes.
5 Extraction and distribution of fossil fuels.
6 Solvent use.
7 Road transport.
8 Other mobile sources and machinery.
9 Waste treatment and disposal.
10 Agriculture.
11 Nature.

In 1985 the European Community set up a program for the collection of information on the state of the environment and natural resources in the Community. This programme was given the name CORINE (Coordination d'Information Environmentale), and included a project to gather and organize information on emissions to the atmosphere that were relevant to acid deposition. This is known as CORINAIR. The program included the

preparation of an inventory as well as the preparation of a default emission factor handbook—to be used in the absence of more specific data—computer software, and a standard set of descriptions of both sources and activities. This is known as the Selected Nomenclature for Air Pollution (SNAP). The first CORINAIR inventory was completed in 1990 (Jol 1998).

The United Nations Framework Convention on Climate Change (UNFCCC) also requires that national inventories of anthropogenic emissions of greenhouse gases are to be prepared. The Intergovernmental Panel on Climate Change (IPCC), with the support of OECD and the International Energy Agency, prepared *Greenhouse Gas Inventory Reporting Instructions*, which were first published in 1994 and then revised in 1996 (Houghton *et al.* 1997). There is also advice on *Good Practice Guidance and Uncertainty Management in National Greenhouse Gas Inventories* (Penman *et al.* 2000). From 2000 onwards, parties to the convention must follow the revised UNFCCC reporting guidelines (UNFCCC 2000). There is an on-line searchable database of greenhouse gas inventory data for CO_2, CH_4, N_2O, CO, NO_x, NMVOCs, SO_2 (http://www.unfccc.de/resource/index.html).

The European Environment Agency was set up in 1993 in order to provide the member states of the European Union "with objective, reliable and comparable information at the European level enabling them to take the requisite measures to protect the environment, to assess the results of such measures and to ensure that the public is properly informed about the state of the environment." As part of its first work program the Agency designated five European Topic Centres to address the problems of inland waters, the marine and coastal environment, nature conservation, air quality, and air emissions. The work program for the European Topic Centre on Air Emissions (ETC/AEM) requires the development of emission inventory guidelines at various levels, the compilation of a European emissions inventory, and a review of the CORINAIR methodology. The Agency also took over the editing and publication of the EMEP *Atmospheric Emission Inventory Guidebook* referred to above. The European Environment

Agency now publishes an *Annual European Community emission inventory* (Gugele & Ritter 2002).

In 2001 the European Union adopted the National Emission Ceilings Directive (NECD) setting national emission ceilings for each EU member state as the primary means of implementing its acidification strategy and to make progress on the problem of ground-level ozone. The NECD proposal sets ceilings for national emissions of SO_2, NO_2, NH_4, and VOCs to be attained by 2010.

17.1.3   *The United Kingdom*

As part of the UK government's continuing program of air pollution studies, the Department for Environment, Food and Rural Affairs (DEFRA) has developed a national inventory of air pollution sources and the type and quantity of the pollutants they emit. This is called the National Atmospheric Emissions Inventory (NAEI). The coverage of this inventory has been expanded as emissions of pollutants have grown or as evidence of their adverse effects has accumulated. Initially only emissions of black smoke and SO_2 were estimated but now many more pollutants are covered. They are NMVOCs, SO_2, NO_x, CO, CO_2 and other greenhouse gases (N_2O and CH_4), NH_3, HCl, benzene, 1,3-butadiene, PM_{10}, $PM_{2.5}$, black smoke, and metals (Pb, Cd, Hg, Cu, Zn, Ni, Cr, As, Se, V, Mn). There are also persistent organic pollutants on the UNECE list (currently dioxins, polycyclic aromatic hydrocarbons, polychlorinated biphenyls, lindane, hexachlorobenzene, and pentachlorophenol), as well as hydrofluorocarbons, perflurocarbons, and sulfur hexafluoride. CFCs are currently covered by a separate inventory. The NAEI is maintained for the DEFRA by the National Environmental Technology Center (Goodwin *et al.* 2002). The purpose of the NAEI is:

• to provide an input to discussions with various international bodies, including those mentioned above;

• to provide the data necessary to report on compliance with the UNECE protocols and other international agreements;

• to provide an input to UK policy-making with respect to pollution abatement and control;

• to assist in judging the effectiveness of existing policies;

• to assist in the interpretation of air quality measurements;

• to provide an input to atmospheric dispersion models;

• for general public information.

Although the first urban atmospheric emissions inventories in the UK had been prepared in the 1960s (see above), it was not until 1994 that the then Department of the Environment (now DEFRA) initiated a comprehensive program of urban inventory preparation (Hutchinson 1998). The first urban inventory to be commissioned was for the West Midlands area, covering the city of Birmingham and the adjacent urban local authorities of Coventry, Dudley, Sandwell, Walsall, and Wolverhampton (Hutchinson & Clewley 1996). This was followed by inventories covering:

1 Greater Manchester, an area traditionally associated with the textile industry in the northwest of England but now including power stations and petrochemical industries.

2 Merseyside, including the northwest coastal port city of Liverpool, power stations, and industry.

3 Glasgow, a large center of industry and population in the Scottish lowlands.

4 Portsmouth and Southampton, representing a mixed urban/industrial corridor and major port complex on England's south coast.

5 Bristol, a major city and industrial center in the west of England.

6 Port Talbot, Neath, and Swansea, a corridor of traditional heavy industry in south Wales.

7 Middlesborough and Teesside, another area of traditional heavy and chemical industries associated with a port in the northeast of England.

8 West Yorkshire, comprising the coalescing cities of Leeds, Bradford, and Huddersfield in the north of England.

In addition to these, an emissions inventory was prepared for the Greater London area with funding from the European Commission and the London local authorities (Buckingham *et al.* 1998). The primary purpose of all these inventories is to assist the local authorities in managing air quality and particularly in meeting their obligations under the Environment Act 1995 to review air quality within their areas.

The pollutants and pollutant groups included in the urban inventories are SO_2, NO_x, CO, NMVOCs, CO_2; benzene; 1,3-butadiene; particulate matter less than 10 μm aerodynamic diameter (PM_{10}). Lead is not included in the inventories because airborne lead levels have shown a downward trend since 1981, when the amount of lead in gasoline was reduced from 0.45 to 0.40 g l^{-1} (with a further reduction to 0.15 g l^{-1} in 1985). Catalytic converters have been fitted to all new vehicles since 1993, and leaded gasoline was withdrawn from sale on January 1, 2000.

The NAEI can be described as a "top-down" inventory (see later discussion of "top-down" and "bottom-up" inventories) in which national data are allocated to smaller areas on the basis of the resident population and other appropriate indicators of regional activity. The London and other urban emissions inventories are bottom-up inventories in which local data are used to compile an inventory of local emissions. These can then be aggregated into larger areas or to a whole city. The NAEI and the urban inventories were originally seen as complementary but, as the NAEI has been developed and refined, it has increasingly adopted a bottom-up approach to data assembly wherever this is possible.

17.1.4 *Japan*

The Ministry of the Environment (formerly the Environment Agency of Japan) has monitored emissions of SO_x, NO_x, and particulate matter since 1974. The Ministry's Air Quality Management Division carries out an annual survey in collaboration with local government and consultants. Survey forms are sent to some 70,000 factories and other buildings, which have between them some 174,000 boilers, furnaces, and other facilities controlled under the Air Pollution Control Law that are treated as point sources. Forty different types of fuel and material are covered and 77

categories of industry. Some individual local authorities also carry out their own surveys (Tonooka 1999).

In addition to point sources, emissions from anthropogenic sources, and from small combustion sources such as heating systems in domestic and other buildings, are estimated as area sources by applying emission factors to energy use statistics. Similarly, emissions from mobile sources, including road vehicles, aircraft, railways, and ships, are estimated using energy statistics, transport statistics, and emission factors. The data sources used in emissions estimation are shown in Table 17.1.

As in other countries, the Japanese national atmospheric emissions inventory is used by the Ministry of the Environment to monitor long-term trends in emissions and as a basis for developing policy. It is used, for example, to establish and follow up area-wide total emissions control programs. Where required, in order to study regional variations or for dispersion modeling, the results can be mapped to a 1 kilometer resolution based on a longitude and latitude grid.

17.2 EMISSION INVENTORY PROCEDURES

There are two main approaches that can be followed in estimating emissions, which are often referred to as the "top-down" and "bottom-up" approaches. The top-down approach means that emission estimates are derived from national or regional data on, for example, amount of fuel used, the quantity of chemical produced, or the distance traveled by road vehicles. National or regional data are scaled to the area covered by the inventory using some measure of activity thought to be directly or indirectly related to the emissions in the area of study. Population figures, employment statistics, and sales data are often used. For example, the population in an inventory area can be used as the activity data to estimate VOC emissions from dry cleaning facilities if the total national or regional population and the total quantity of dry cleaning agents used are known. In the alternative bottom-up approach, estimates are made of emis-

Table 17.1 Information sources used in compiling the Comprehensive Survey of Air Pollutant Emissions in Japan (data from Tonooka 1999).

Large category	Middle category	Detailed category	Analytical factors	Fundamental data source
Stationary combustion sources* (point sources)	Registered combustion source†	Manufacturing industry; power plant; waste incinerator; gas supply facility; district heating system; agriculture; buildings‡	Industry type; facility (furnace) type; fuel or material type; regional resolution, i.e. prefecture, municipality, and grid zone; facility capacity size rank; facility install data rank; emission control equipment type	Comprehensive Survey of Air Emissions in Japan by the Japan Environment Agency
Stationary combustion sources (area sources)	Buildings	Offices, hotels, hospitals, schools, etc.§; small combustion equipment¶	Building type; thermal use type; fuel type; regional resolution**; Yearbook; Boiler Association statistics	Comprehensive Survey of Air Emissions in Japan; Survey on Gas Utility Industry; LPG research material
	Residential houses	Small combustion equipment¶; residential house type; regional resolution**	Thermal use purpose type; fuel type; Survey on Gas Utility Industry; LPG research material	Yearbook; family income and expenditure; population census
Mobile sources	Vehicles	Light vehicles; passenger cars; buses; light trucks; light duty trucks; heavy duty trucks; special use vehicles; motorcycles; construction machines, etc.	Vehicle type; fuel type; owner type; road classification; regional resolution††	Motor Vehicle Transport Statistics; Road Traffic Census; Current Survey on the Supply and Demand of Petroleum Products; Emission Factor Survey of Vehicles
	Railways	Diesel powered; other; regional resolution‡‡	Company; rail line	Railway Statistics; Compilation of Emission Factors AP-42 (US EPA)
	Vessels	In harbor area, coastal, or off-shore; vessel type; transportation type or fishery; cruising mode type; regional resolution‡‡	Harbor classification; harbor; Harbor statistics; Current Survey on the Supply and Demand of Petroleum Products	Statistical Survey of Coastwise Vessel Transport; Compilation of Emission Factors AP-42 (US EPA)
	Aircraft	Landing; cruising; aircraft type; LTO cycle; company; regional resolution‡‡	Airport; domestic or international; Statistical Survey of Air Transport	US EPA (1992) Procedures for Inventory Preparation Vol 450/4-81-026 (revised)

* Including noncombustion industrial sources.
† Registered to Japan Environment Agency as Air Pollutant Emission Source under the Air Pollution Control Law.
‡ Only includes boilers large enough to require registration.
§ Buildings other than factories, storage facilities, and residential homes.
¶ Small emission sources including boilers small enough to be exempt from registration.
** Regional resolution of area sources are prefecture, municipal government, and 1×1 km grid system.
†† Regional resolution of motor vehicles on national and prefectural roads is prefecture, municipal government, road link, and 1×1 km grid system. Regional resolution of motor vehicles on city, town, and village roads is prefecture, municipal government and 1×1 km grid system.
‡‡ Regional resolution of railways, vessels and aircraft is the same as for area sources; prefecture, municipal government, and 1×1 km grid system.

Table 17.2 Characteristics of "top-down" and "bottom-up" inventories (extrapolated from Eastern Research Group 1999).

Characteristics of a top-down approach

Typically used to inventory area sources.

Requires minimum resources by grouping like emission sources together and making use of readily available activity and emission data.

Used when: (i) local data are not available; (ii) the cost to gather local information is prohibitive; or (iii) the end use of the data does not justify the cost of collecting detailed site-specific data.

Emission factors or national- or regional-level emissions estimates are used to estimate emissions in a state or county based on a surrogate parameter such as population or employment in a specific sector.

One potential problem with this approach is that an emission estimate will lose some accuracy due to the uncertainty associated with the estimate and the representativeness of the estimate once extrapolated to the local level.

Characteristics of a bottom-up approach

Typically used to inventory point sources; however, it can be used to inventory area sources when resources are available to collect local activity data through a survey effort.

Requires more resources to collect site-specific information on emission sources, activity levels, and emission factors.

Results in more accurate estimates than a top-down approach because data are collected directly from individual sources and not derived from a national or regional estimate.

sions from individual sources and these are then summed for the inventory area. The characteristics of the two approaches are summarized in Table 17.2.

The sources of air pollution emissions included in an inventory are usually classified as:

• point sources, including high-intensity emissions from industrial plants;

• area sources, including emissions from agricultural and other land, and low-intensity emissions from sources such as building heating systems;

• line sources, including roads and railways.

The number of emission sources that have been continuously monitored is generally small in relation to the total number of sources in a city, region, or country. Hence, information on the actual emissions occurring from an individual source is only available in a limited number of cases in any area. Some of the possible ways of estimating emissions are summarized in Table 17.3. The majority of emissions must be estimated from other information such as fuel consumption, vehicle–kilometers traveled (VKT), or some other measure of activity relating to the emissions. Emission factors, derived from the results of the monitoring which has been undertaken, are then applied to the activity data in order to estimate the likely emissions:

$$\text{Activity rate} \times \text{Emission factor} = \text{Emission}$$

The procedure for applying emission factors to information on activities in order to estimate total emissions and emissions by area is illustrated in Fig. 17.1. The processes of assembling an inventory are shown in Fig. 17.2.

For many of the pollutants of concern, the major source of emissions is the combustion of fossil fuels. Consequently, the collection and analysis of fuel consumption statistics plays a significant part in the preparation of emission inventories. However, it is important to consider the differences between consumption and fuel deliveries when making use of the available data. Most of the readily available statistics relate to fuel deliveries, which, in many cases, relate closely to consumption. However, in the case of fuels that may be stockpiled, such as coal, there may be significant differences between delivery and consumption. In the case of transport fuels, there may be significant geographic differences between the point of delivery and where the fuel is used. A striking example is London's Heathrow Airport, to which some 4 billion liters of aviation fuel are delivered annually. Only a small fraction of this fuel is used within the London area, while the remainder is used on journeys to the far corners of the Earth.

Where possible, emission factors derived from measurements undertaken locally should be used because these reflect particular local conditions or industrial practice. In those countries with a long-standing concern over air quality, enough sources have been studied to derive robust emission factors to cover road traffic, industrial processes,

Table 17.3   Methods of estimating emissions (extrapolated from Eastern Research Group 1999).

Continuous emission monitors	Continuous emission monitors measure (with very short averaging time) and record actual emissions during the time of monitor operation. Continuous emission monitoring data can also be used to estimate emissions for different operating and longer averaging times
Source testing	Emission rates are derived from short-term emission measurements taken at a stack or vent. Emission data can then be extrapolated to estimate long-term emissions from the same or similar sources
Material balance	Emissions are determined based on the amount of material that enters a process, the amount that leaves the process, and the amount shipped as part of the product itself
Emission factors	An emission factor is a ratio that relates the emission of a pollutant to an activity level at a plant that can be easily measured, such as an amount of material processed or an amount of fuel used. Given an emission factor and a known activity level, a simple multiplication yields an estimate of the emissions. Emission factors are developed from separate facilities within an industry category, so they represent typical values for an industry, but do not necessarily represent a specific source. Published emission factors are available in numerous sources
Fuel analysis	Emissions are determined based on the application of conservation laws. The presence of certain elements in fuels may be used to predict their presence in emission streams. For example, SO_2 emissions from oil combustion can be calculated based on the concentration of sulfur in the oil. This approach assumes complete conversion of sulfur to SO_2. Therefore, for every kilogram of sulfur (molecular weight $= 32$ g) burned, 2 kg of SO (molecular 2 weight $= 64$ g) are emitted
Emission estimation models	Emission estimation models are empirically developed process equations used to estimate emissions from certain sources. An example emission estimation model is the US EPA TANKS software for estimating volatile organic compound emissions from fixed- and floating-roof storage tanks
Surveys and questionnaires	Surveys and questionnaires are commonly used to obtain facility-specific data on emissions and their sources
Engineering judgment	An engineering judgment is made when the specific emission estimation techniques such as stack testing, material balance, or emission factor are not possible. This estimation is usually made by an engineer familiar with the specific process, and is based on whatever knowledge is available

and other sources. In many other countries, however, the number of emission sources that have been continuously monitored is limited and there are not enough data to derive robust emission factors. In these cases there is no alternative but to make use of factors developed elsewhere. Three of the most often used sources of emission factors are:

1 *EMEP/CORINAIR Emission Inventory Guidebook*, (2002) published by the European Environment Agency.

2 *Compilation of Air Pollution Emission Factors*, generally known as AP-42, prepared by the US Environmental Protection Agency (1995), and many related documents and software available via the CHIEF website (http://www.epa.gov.ttn/chief/)

and the Office of Transportation and Air Quality (formerly the Office of Mobile Sources) website (http://www.epa.gov/otaq/).

3 Revised 1996 IPCC Guidelines for National Greenhouse Gas Inventories (Houghton *et al.* 1997) and the Good Practice Guidance and Uncertainty Management in National Greenhouse Gas Inventories (Penman *et al.* 2000). These relate to emissions of greenhouse gases, including CO_2, CH_4, N_2O, CO, NO_x, NMVOCs, and SO_2, rather than the pollutant particularly associated with industry and traffic.

For the UK, a great deal of useful information on emission sources, emission factors, and procedures is given in the annual reports of the National Atmospheric Emissions Inventory, mentioned

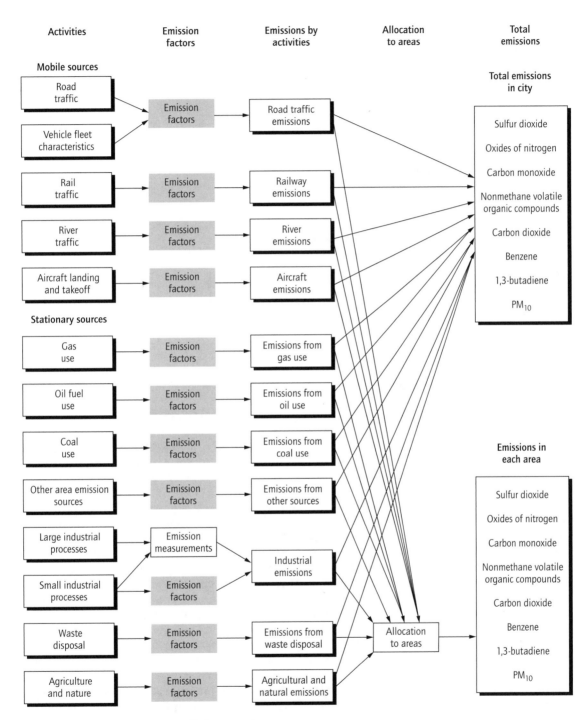

Fig. 17.1   Emission inventory compilation procedure.

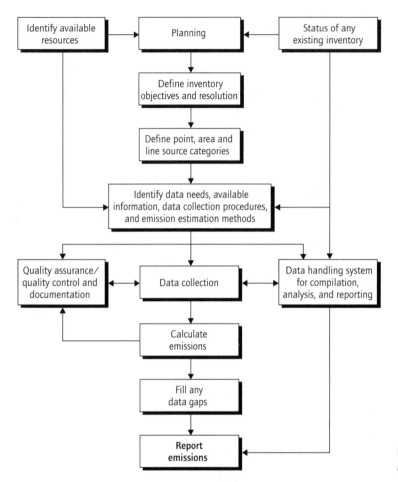

Fig. 17.2 Activities for preparing an inventory.

above (http://www.naei.org.uk/index/php). Further guidance has been provided by the former Department of the Environment, Transport and the Regions (2000a, b) for local authorities that need to prepare inventories as part of their work under the Environment Act 1995. Also in support of local authorities, the former London Research Centre (now absorbed into the Greatre London Authority) and RSK Environment created the *UK Emission Factors Database* (http://www.naei.org.uk/emissions/index.php).

When factors developed elsewhere are used, there may be some doubt as to how appropriate they are. An example is dust from agricultural activity such as the harvesting of cereals. There are no locally derived emission factors in the UK. While there are data on the exposure of agricultural workers to dust as a result of harvesting and other agricultural operations, it is not possible to convert this into emission factors per square kilometer of land. The only available emission factors for harvesting of cereals are those in AP-42 but it is questionable how applicable these are in the UK's generally wetter climate.

It is impossible to discuss all the different procedures that may be applicable to the preparation of emission inventories for countries, regions, cities, and districts in a single chapter. The following sections, which consider emissions from road traffic, other mobile sources, area sources, and point sources, therefore focus on the preparation of inventories for cities.

17.3 EMISSIONS FROM ROAD TRAFFIC

Road traffic is already, or is rapidly becoming, the primary source of many of the pollutants in many parts of the world. Table 17.4 shows the contribution of road traffic to total emissions in the UK and in London. In urban areas, the proportions will generally be higher because they are the main centers of the population and business that give rise to journeys.

The emission of the different pollutants from road vehicles varies according to a wide range of factors. Any one vehicle is quite likely to be some way from the "average" for its type in terms of its emissions at any given time, and any usable emissions factors must be based on the average characteristics of sensible subsections of the vehicle fleet. The following are examples of generic characteristics that can be used to divide the fleet into categories for the derivation of emissions factors:
- the vehicle type (e.g. motorcycle, automobile, or heavy goods vehicles) and further subdivisions based on, for example, engine size;
- the fuel used (gasoline/diesel);
- the vehicle technology (e.g. compliance with emissions standards, including the provision of a catalytic converter).

Table 17.4 Emissions from road traffic as a percentage of all emissions in the UK and London (data from Greater London Authority 2002).

Pollutant	UK 1998	London 1999
Benzene	71	74
1,3-butadiene	85	93
CO	69	94
NO_x	44	58
PM_{10}	20	68
SO_2	1	38

For each of these groups, it should be possible to derive indices describing typical emissions for a variety of driving scenarios (or "drive cycles") under idealized or average conditions.

There are also a number of specific factors affecting the operation of individual vehicles on a day-to-day basis. Examples include:
- traffic conditions (emissions will vary according to the instantaneous speed of the vehicle, road gradients, and the degree of stop/start driving);
- the vehicle condition, such as its age, engine tuning, and operating efficiency;
- individual driver behavior (e.g. gentle as opposed to aggressive or "heavy-footed" driving);
- ambient temperature and climatic conditions.

With the exception of the first of these, the inherently wide degree of unpredictability coupled with the impracticality of characterizing individual journeys means that it is generally not meaningful to attempt to produce emissions factors disaggregated on these bases, and emissions indices produced on the basis of typical driving cycles are assumed to subsume (or "average out") these variations.

However, as vehicle speed is such an important determinant of emissions and information about traffic speeds is often available, it is desirable to refine the generalized driving cycle-based emissions factors to allow disaggregation according to speed. Thus, the typical set of road vehicle emissions factors will detail quantities emitted at selected points over the speed range for a series of vehicle–fuel–technology combinations. Figure 17.3 illustrates the different emission rates for different types of vehicle in the UK fleet in 1997 and, as expected, in 2005.

Many different models have been developed around the world for estimating emissions from vehicles. The first version of the US EPA highway vehicle emission factor model, MOBILE, was published as "look-up tables" in 1978 (US EPA 1999). It is now a computer program that provides average in-use fleet emission factors for three criteria pollutants: VOCs as a precursor to ground level ozone, CO, and NO_x. Data are provided for each of eight categories of vehicle, for any calendar year between 1970 and 2020, and under various condi-

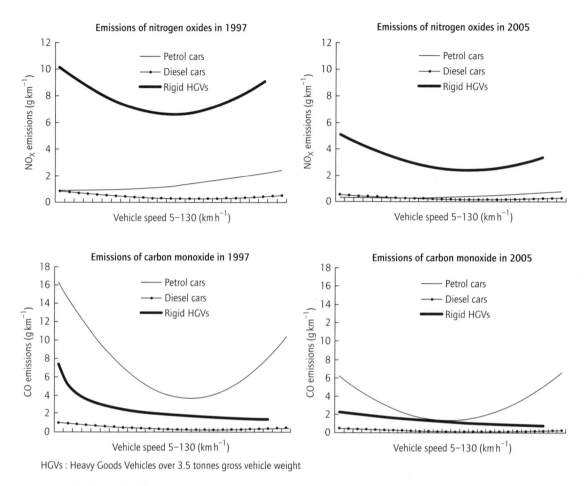

HGVs : Heavy Goods Vehicles over 3.5 tonnes gross vehicle weight

Fig. 17.3 Vehicle speed and emissions.

tions affecting in-use emissions (e.g. ambient temperatures, average traffic speeds, gasoline volatility) as specified by the model year. The model is used by US EPA in evaluating control strategies for vehicles, by states (except California) and other local and regional planning agencies in the development of emission inventories and control strategies for State Implementation Plans under the Clean Air Act, and the development of environmental impact statements. The latest version, MOBILE 6, can be found on the US EPA Office of Transportation and Air Quality (formerly the Office of Mobile Sources) website (http://

www.epa.gov/otaq/ m6.htm), including a User's Guide to MOBILE 6.

The use of the MOBILE model is not confined to the USA. For example, it has been used to estimate emissions from road traffic in Russia and China. There have been no comprehensive vehicle emissions testing programs necessary to build up data on the emission characteristics of the indigenous vehicle fleet in these countries. However, many of the vehicles in use are quite similar to those in use in the USA in the 1970s. In the absence of anything better, the MOBILE model has been used to estimate emissions from them.

COPERT III is a program that has been developed in Europe in order to calculate emissions from road traffic. The emissions calculated include CO, NO_x, VOCs, PM_{10}, N_2O, NH_3, and SO_2, as well as fuel consumption. Emissions from internal combustion engines used in "off-road" applications are also covered. Its initial development dates back to 1987, when the CORINAIR Working Group on Emission Factors for Calculating Emissions from Road Traffic began developing a methodology for estimating emissions from road traffic. COPERT III is an updated version of COPERT 90 in terms of both methodology used and the software supplied. It is supported by a comprehensive manual that is designed to help the users of COPERT III to produce, in a short time, a complete annual emission data set from road transport (and off-road machinery) for a specific country (Kouridis *et al.* 2000; Ntziachristos & Samaras 2000).

There are many similarities between the methods of estimating vehicle emissions adopted in MOBILE and COPERT. The most important difference is that the starting point in MOBILE is the annual vehicle miles traveled (VMT) whereas the starting point in COPERT is the vehicle fuel used. In many countries the only information known for certain is the total quantity of gasoline, diesel fuel, and liquefied petroleum gas (LPG) sold. There is no information on the vehicle miles or vehicle–kilometers traveled (VKT) or, if there is any, it is unreliable (Zachariadis & Samaras 1999).

Vehicle emissions factors are derived from actual test data over a large sample of vehicles, such as in one project undertaken jointly by the Transport Research Laboratory in the UK, the Institut National de Recherche sur les Transports et leur Sécurité (INRETS) in France and TÜV Rheinland in Germany as part of the European Commission's DRIVE research and development program (Jost *et al.* 1992). In this study, a survey of the operating characteristics of automobiles in urban areas was undertaken in six European cities in order to develop 14 typical driving cycles. These driving cycles were then reproduced on chassis dynamometers in each of the participating laboratories and fleets of vehicles tested in order to measure fuel consumption and emissions. The vehicles were selected from three age bands, which reflected the introduction of progressively tighter emission control requirements for gasoline-engined automobiles, and three different engine sizes, as well as diesel-engined automobiles. The vehicles were tested in an "as found" state (i.e. without any prior maintenance or repair) so as to reflect the variation in vehicle condition found "on the road."

The results of this and similar programs provided the basic emissions factors for automobiles and other vehicles used in COPERT and other programs. The form of these factors is illustrated by Fig. 17.3, which shows the relationship between vehicle speed and emissions of NO_x and CO for different types of vehicle. It should be noted that the speed values given relate to complete driving cycles undertaken at the average speed, rather than to instantaneous speeds, since at any given point on the speed axis the automobile could be either accelerating hard or coasting, resulting in very different emissions. This means that these factors can be applied appropriately to average traffic speeds along different roads within an area, although they are not ideal for more microscale studies (for example, of individual road junctions) where automobiles are predominantly either accelerating or idling.

To estimate emissions from road traffic within a city, as opposed to a whole country, we would ideally require data describing the distance (i.e. the vehicle–kilometers) driven by different vehicle types on all roads within the area. Such detailed information is hardly ever available. A method has therefore to be evolved that approaches different elements of the total road traffic emission using the best available data in each case. For this purpose emissions from road traffic can be considered as comprising three categories:

• emissions from traffic on the major road network;

• emissions from traffic on the minor road network;

• emissions associated with trip ends (i.e. trip starts and finishes).

Each of these categories is treated separately, as described in the following sections.

17.3.1   Traffic on major roads

There are two primary sources of road traffic data currently in use for preparing urban emission inventories: traffic surveys and transportation models. Information from traffic surveys is attractive because it relates to *real* traffic on *real* roads, whereas transportation models are a computerized reflection of the actual conditions. However, traffic surveys have the disadvantage that they only provide information relating to the specific survey points, rather than area-wide information. On the other hand, transportation models are available for many large cities and are comprehensive.

In many urban areas the traffic authorities maintain and operate a transport or traffic planning model. Such a model, once it has been "calibrated" so that it accurately reproduces known flows in the recent past, is then used to test the traffic impacts of a variety of future transport infrastructure and policy scenarios. This calibrated or "base" run of the model provides (with appropriate manipulation) a network-based representation of present-day traffic flows on the more significant roads across the area, and is an ideal starting point for the calculation of emissions from traffic on major roads. For example, the London model contains a geographic representation of the major road network consisting of approximately 13,000 individual links (discrete sections of road) within and including the M25 orbital motorway. The network is illustrated in Fig. 17.4. Similar data are available for Tokyo and are referred to later.

Traffic models have seldom been designed with emissions inventories in mind, and before emissions can be calculated, it will probably be necessary to embellish the basic model data in several ways. The principal modifications are likely to be:
1 The very basic vehicle-type breakdowns available from the model may need to be refined into categories suitable for the calculation of emissions. This is done with reference to traffic counts averaged across parts of the study area and vehicle fleet statistics in order to establish, for example, the proportion of the different types of heavy goods vehicles in the flow and the split between gasoline- and diesel-fueled vehicles.

2 The model network may need to be modified to fit wholly within the study area, and to remove extraneous information that may be necessary for traffic modeling but is not necessary in an emissions inventory.
3 To obtain an annual estimate, the hourly flows given by the model will need to be grossed up (by appropriate factors) to annual flow totals. In addition, factors can be produced to describe the variation in traffic flow on an hourly, daily, or seasonal basis. These can be useful in in-depth air quality modeling (for example, studies of the build-up of pollution in individual street canyons).

Following these modifications, emissions factors can be applied directly to calculate the emissions from each vehicle type on each link represented in the model, and from there to produce an annual total for each link and for the entire network. These results can be displayed in a variety of ways, such as on a link-by-link basis, or allocated to the individual areas, such as 1×1 km grid squares, in which they fall.

17.3.2   Traffic on minor roads

Although traffic models may give relatively complete coverage of the more important roads in a city, a significant number of vehicle–kilometers might be expected to be driven on the remaining, so-called "minor," roads. These roads vary widely in character, from commercial access roads to locally significant routes in the suburbs and, on the periphery of the area, roads that are essentially rural in character. However, most of these roads share the characteristic that they are only generally used to access the nearest major road. It is often very difficult to establish a precise figure for this traffic. Minor roads tend to be overlooked by conventional traffic survey programs, which usually concentrate on the busier parts of the network. The result is that there is usually little or no information available to give an indication of link-specific flows, or even the total vehicle–kilometers driven on these roads within an area.

London is fortunate to have an alternative source of data, the Rotating Traffic Census under-

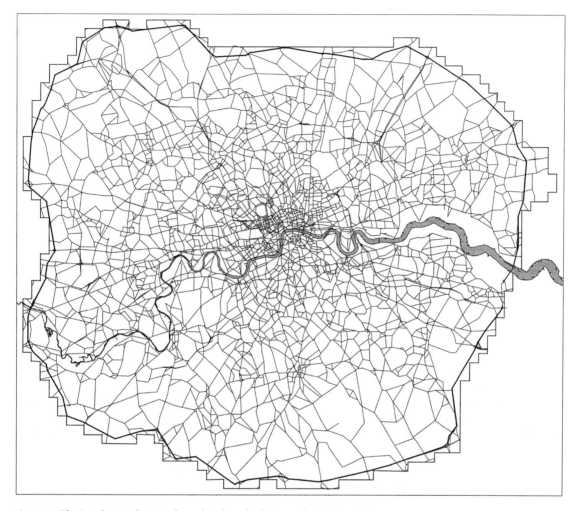

Fig. 17.4 The London road network used in the calculations of vehicle emissions.

taken by the Department for Transport. This provides an estimate of total VKT on both major and minor roads within the London area. It is possible to subtract the total kilometerage accounted for by the traffic model from this total to give a "residue," which can be considered to be the traffic on the minor road network. These additional vehicle–kilometers then need to be apportioned across the study area, the most suitable basis being the relative density of the minor road network in different parts of the city. In the absence of any alternative information it may be necessary to rely on "engineering judgment" (see Table 17.3) in order to estimate minor road traffic.

As it is not generally possible to apportion flows accurately to individual minor roads, the emissions estimates for these roads should be treated as an "area," rather than "line" sources. Similarly, as it is not possible to establish an average traffic speed for each link, a typical urban driving speed must be assumed. Emissions for traffic on minor roads can then be added to those for major roads, to arrive at a total for traffic movements on all roads.

17.3.3 Emissions associated with trip ends

The traffic sources so far considered assume normal vehicle running conditions. There are two circumstances where this assumption needs to be modified. These are, first, emissions associated with the "cold start" phenomenon, which occurs at the start of a trip, and, second, the similar "hot soak" emission, which occurs at the end of a trip once the vehicle engine has been switched off.

When vehicles are first started they take some time for the engine to reach normal operating temperature. During this "cold start" period, emissions are substantially higher than when the engine is "hot." This is true of gasoline-engined vehicles with or without catalytic converters. Catalytic converters do not become effective until both the engine and catalyst have reached their normal operating temperature, and vehicles making short journeys in cold weather conditions may not reach full operating temperature throughout their journey.

To estimate the contribution from cold starts, it is necessary to obtain disaggregate information describing the distribution of trip ends across the study area. As with traffic on major roads, this information is often available from a traffic model, generally in the form of "origin/destination (OD) matrices" for a system of "zones" (small polygons) across the modeled area. These trip ends can be apportioned to individual areas of the city, and the total cold start penalty for each area calculated directly.

This method has the following limitations:

1 All cold start emissions are assumed to occur within the area where the trip starts, whereas it is known that vehicles can take several kilometers to reach normal operating temperature, particularly during cold weather.

2 The complex interrelationships between catalyst performance, ambient temperature, and time since the vehicle was last operated (i.e. the engine temperature when started) are subsumed within the emission factors, which assume average conditions.

Transport surveys generally disregard the shorter stages that may comprise longer trips. These stages result, for example, from stopping to buy a newspaper and refueling the vehicle during a journey to work. Travel surveys treat this a single journey, but this is unsatisfactory from the point of view of emissions because each stage involves either a cold or a hot engine start. The number of trips derived from transport surveys therefore underrepresents the number of engine starts. Neither do they take account of engine starts as a result of the vehicle stalling or very short vehicle movements such as reversing out of a garage. A comparison between the results of instrumented vehicle studies and travel surveys in the USA gives an average of 1.68 starts per trip (Yotter & Pale 1996). A comparison of data from the DRIVE program referred to above and the UK National Travel Survey gives a very similar figure of 1.66 starts per trip (J. Hickman, personal communication 1996). This, or a similar locally derived figure, should be used to convert vehicle trip data to vehicle starts.

The hot soak phenomenon is a type of evaporative emission, arising from the evaporation of hydrocarbons from the engine and fuel system. Such losses occur continuously, both when the vehicle is in use and also when it is parked. Evaporative losses occurring while the vehicle is in use are accounted for in the general road traffic emissions factors for NMVOCs and do not need to be treated separately here. However, evaporation is particularly pronounced when a hot engine is switched off, when the heat of the engine promotes evaporation from the carburettor through the breathers and the air filter. These "hot soak" emissions are closely associated with trip ends, and can be apportioned to the inventory grid using a similar method to cold starts.

The resulting estimates of cold start and hot soak emissions are then combined with the corresponding estimates of emissions for major and minor roads to arrive at a total emission for all road traffic.

17.3.4 Other emissions associated with road traffic

The emissions so far considered arise directly from

the burning of fossil fuels in vehicle engines. There are several other important sources of emissions that are not amenable to estimation on a line-source basis and are therefore treated as area sources. These are:

1 Road traffic gives rise to dust from tire and brake wear, as well as the resuspension of existing dust on the road surface.

2 Significant hydrocarbon emissions are associated with the vehicle refuelling process, both as part of the general fuel production and delivery chain, and also as actual forecourt-based refuelling of individual vehicles. These sources may be included as an area source or as individual point sources, depending upon the available information.

3 Most traffic and transport models do not explicitly include the quite significant numbers of "off-road" vehicles. These range from large pieces of equipment such as tractors, mobile plant, and other works vehicles that do not usually venture onto public roads, to small equipment such as lawn mowers and chain saws. All of these have in common the use of gasoline or diesel as fuel, but can only be estimated as area sources based on their fuel use.

17.4 EMISSIONS FROM RAIL TRANSPORT

As with road transport, an estimate of the emissions is required for each link in the rail network and for each area, such as 1×1 km grid squares, in which they fall. The use, by passenger services, of each link by different type of locomotives and different lengths of train is quantified, during both a peak hour and a nonpeak hour Monday–Friday. These figures are then scaled up to give an annual usage of each link, and the links apportioned to kilometer grid squares or other areas. Emission factors are used to calculate the resultant emissions. For freight traffic, it is necessary to assess the tonne–kilometers of freight moved as the train weight affects the emissions. This is then apportioned to those lines where freight was known to travel and the emissions are calculated.

17.5 EMISSIONS AT AIRPORTS

Airports constitute a collection of diverse emission sources. The principal source is aircraft, but other significant sources include:

1 Ground support equipment, including aircraft tractors, aircraft servicing vehicles and baggage handling vehicles (these vehicles and equipment are treated as an "off road" emissions source).

2 Road traffic delivering or collecting passengers, baggage, and freight, or otherwise serving the airport.

3 Stationary sources, including the airport heating system.

The first two of these are treated as mobile emission sources and the third as a stationary source.

With airports, it is especially important to recognize that the concern is with emissions actually occurring within the vicinity of the airport and not with the total energy use and emissions taking place as a result of the existence of the airport. The largest proportion of an aircraft's emissions generally occur away from the airport during the course of its flight to another airport, within the same country or further afield. The emissions occurring within 1000 m of the ground as a result of the aircraft landing, taxiing, and taking off are calculated in accordance with UNECE/EMEP and CORINAIR conventions. This is referred to as the landing and takeoff, or LTO, cycle and is illustrated in Fig. 17.5.

Information on the types and numbers of aircraft using each airport comes from the airport itself. The emission factors used to calculate the actual emissions have generally been developed by the aircraft engine manufacturers on the basis of data collected during engine certification tests. The International Civil Aviation Organization (1995) compiled and published emission factors based on certification tests for aircraft engines manufactured worldwide. Since its publication, the database has since been further developed in electronic form by UK Defence and Evaluation Research Agency, now QinetiQ, and further information is available on the website (http://www.quinetiq.com/aviation_emissions_

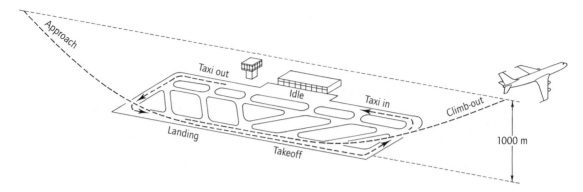

Fig. 17.5   The aircraft landing and takeoff cycle.

Table 17.5   Emissions from a range of common aircraft types (kg) (data from EMEP/CORINAIR 2002).

Aircraft type	Emissions per LTO cycle				
	NO_x	NMVOC	CO	SO_2	CO_2
BAe 146	4.2	0.8	9.7	0.6	1,794
Boeing 727	12.6	6.6	26.4	1.4	4,450
Boeing 737 400	8.3	0.3	11.8	0.8	2,600
Boeing 747 100-300	55.9	35.9	78.2	3.4	10,754
Boeing 747 400	56.6	0.5	19.5	3.4	10,717
Boeing 757	19.7	0.7	12.5	1.3	3,947
Boeing 767 300	26.0	0.2	6.1	1.6	5,094
Boeing 777	53.6	20.5	61.4	2.6	8,073

databank/index.asp). Further information can also be found on the US EPA Office of Transportation and Air Quality website (http://www.epa.gov/otaq/aviation.htm). Emissions from some common aircraft types are shown in Table 17.5 (EMEP/CORINAIR 2002).

Aircraft emit differing amounts and differing proportions of the various pollutants at different stages of the LTO cycle. Aircraft are designed to have adequate power for takeoff when they are fully loaded with passengers, freight, and fuel, even under extreme conditions such as very hot days. During takeoff aircraft engines produce high emissions of NO_x but relatively low emissions of VOCs and CO. Emissions of NO_x are reduced during climb-out. During the approach to landing, NO_x emissions are much lower, while VOC and CO emissions are at the mid-point in the emission range. While taxiing or when engines are idling,

they emit the lowest level of NO_x but the highest levels of VOCs and CO. These variations are illustrated in Fig. 17.6 (Webb & Draper 1996).

When large aircraft are on the ground with their engines shut down, they need power and preconditioned air in order to maintain the aircraft in operation. Large commercial aircraft are fitted with auxiliary power units (APUs) to provide independent power. These units are essentially small jet engines that generate electricity and compressed air. They burn jet fuel and generate exhaust emissions like larger engines. In use, APUs run at full throttle and therefore produce relatively low emissions of VOCs and CO but high NO_x emissions.

17.6   EMISSIONS FROM SHIPPING

Shipping is a relatively small source of emissions

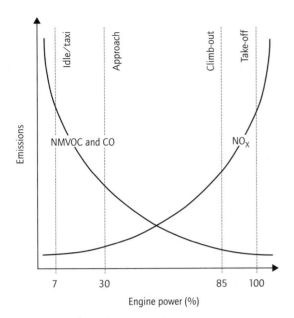

Fig. 17.6 Variations in emissions at different stages in the landing and takeoff cycle.

in many cities, or totally absent, but it may be considered either for overall source comparison or because it constitutes a significant local source of emissions. Where there is shipping, information on the movement of vessels is used in conjunction with emissions factors from the *Marine Exhaust Emissions Research Programme* undertaken by Lloyd's Register (1995). The US EPA is initiating a review of its guidance on developing emission inventories for ocean-going and harbor vessels operating at port areas. As part of this work, it has published the report *Analysis of Commercial Marine Vessels Emissions and Fuel Consumption Data* (Energy and Environmental Analysis 2000). Some further information relating to smaller vessels can be found on the US EPA Office of Transportation and Air Quality website (http://www.epa.gov/otaq/marine.htm).

17.7 AREA EMISSION SOURCES

This section and the following one are concerned with stationary sources of emissions to the atmosphere, including those from building and other heating systems, industrial processes, waste disposal, and natural sources. Some are treated as "area" sources and others as "point" sources. Area sources are emission sources that are distributed over a wide area, such as domestic central heating boilers, whereas point sources are emission sources located at a particular point, such as those from a waste incineration plant, a power station, or a large industrial plant.

The majority of air pollution emissions from stationary sources are the result of the burning of fossil fuels, natural gas, oil fuels, and coal. The use of these fuels has changed considerably over the past 50 years and these changes have had a significant effect on the types of pollution released into the atmosphere.

Throughout the 1950s and 1960s, the majority of energy needs were met by coal and oil, used both directly and to produce electricity. Coal, and later oil, was also used to produce "town gas," but from 1970 onwards "town gas" was steadily replaced in much of western Europe by natural gas from the North Sea for domestic, commercial, and industrial heating. Gas is now the principal fuel for space heating in all sectors, meeting 25% of overall energy demand in Europe (Eurostat 2001). Power stations are one of the largest remaining users of coal, accounting for 77% of coal used in the UK in 1996.

17.7.1 Emissions from fuel combustion

Emissions arising from heating systems in homes and in commercial and industrial buildings are treated as area sources unless they are substantial and occur at a specific location, when they are treated as point sources. Information on fuel use has to be compiled by one of three possible methods:

1 A top-down apportionment of national or regional fuel-use data on the basis of the distribution of dwellings, employment, or other business statistics.
2 A bottom-up estimation based on the number of dwellings and the number of different types of

commercial and industrial businesses in the area. This depends on information being available on the average fuel use by different types of dwelling and commercial and industrial business.

3 Information on the actual quantities of fuel delivered to the area by public utilities and commercial suppliers. This is the most reliable type of information but public utilities and commercial suppliers may regard such information as commercially confidential and be unwilling to provide it.

Throughout the inventory care must be taken to avoid double count emissions. For example, if area fuel supply data are used, point source emissions from fuel combustion must be subtracted from emissions calculated on an area basis.

17.7.2   Area sources of NMVOC emissions

There are a significant number of activities and substances in daily use that give rise to NMVOC emissions but for which it would be very difficult to gather enough information about their use locally to prepare emissions estimates. These activities include industries that are not covered by process regulations (industries controlled by process regulations are generally treated as point sources) and those used by the domestic market. Examples of the former are food production, use of paints and adhesives, and dry cleaning. Examples of the latter are cosmetics, paints, and automobile-care products. However, information may be available from trade sources on the overall use of these substances nationally. This information has been used, for example, to prepare UK national emissions estimates and forecasts, and is discussed in Passant and Lymberidi (1998).

The use of these substances is broadly related to the distribution of population. Because of the difficulty in compiling information on their use locally, these emissions are generally apportioned to the residential and employment population depending on the nature of the emission source. Many sources have to be split between the employment and residential population, such as consumer aerosols, dry cleaning, retail decorative

paint use, and fish frying. Others, such as printing, packaging, photocopying, and bread baking can be allocated on the basis of employment population.

17.7.3   Evaporative emissions from vehicle refueling

In the UK, some 5% of NMVOC emissions originate from vehicle refueling (Passant & Lymberidi 1998). These emissions are in addition to emissions from vehicle exhaust. As diesel fuel is much less volatile than gasoline, the problem of evaporative emissions is generally related to gasoline-fueled vehicles. Refueling is split into two sections: the delivery of fuel to filling station storage tanks (referred to as stage 1); and during vehicle refueling (stage 2). European Union Directive 94/63/EC requires the introduction of stage 1 vapor recovery controls. The implementation of this Directive in the UK has the effect of bringing filling stations under local authority air pollution control, so that, in the future, they can be treated as point sources. In countries where there are no such controls, it may be necessary to treat evaporative emissions from vehicle refueling as an area source.

17.7.4   Emissions from landfill

Landfill sites are widely accepted as a significant source of methane. However, there is a problem in producing emissions factors for landfill sites, as emission rates vary between sites according to composition of the waste, site depth, time since disposal, moisture, temperature, and permeability of the site cap. The data available about individual landfill sites are also often limited. While the locations of sites may be known, records of the start and completion dates, the nature of the material filled, and other data are frequently incomplete. In the past operators did not always keep records and, even though landfill regulations have been tightened in most countries, it is often impossible to reconstruct past events. This makes it difficult to produce accurate estimates of emissions on an area basis.

17.7.5 Construction dust

Construction and maintenance work is a large source of particulate emissions in any urban area. These sources include new building construction, demolition and redevelopment, major building refurbishment, and routine maintenance. Even quite modest operations such as preparing paintwork prior to redecoration produce considerable quantities of fine dust. Dust is also produced as a result of highway maintenance and other civil engineering works. The Quality of Urban Air Review Group (QUARG) estimates the total UK emissions of PM_{10} from these sources in a range from 500 to 18,000 tonnes per annum (Expert Panel on Air Quality Standards 1995). Many of the larger construction works, and larger sources of PM_{10}, by their very nature move site as each project is completed. An emissions inventory may well be unable to track these sources as quickly as they move. Any estimate of emissions from this source may have to be apportioned according to the residential and employment population, or treated as an aggregate total for the city as a whole.

17.7.6 Road surface dust

Emissions of particulate matter from vehicle exhausts, as well as those resulting from brake and tire wear, were considered above. In addition, dust on road surfaces becomes resuspended as a result of air turbulence caused by passing traffic. Emissions vary with the "silt loading" of the road surface, as well as the average weight and volume of traffic. The silt loading refers to the mass of material less than 75 μm in physical diameter on the road surface. The most recent emission factors are those given in AP-42 (US EPA 1995), but when these are applied in northern Europe they result in improbably high emissions. Earlier studies undertaken by Ball and Caswell (1983) in London suggest a very much lower emission rate of $0.1 \, g \, km^{-1}$ based on measurements of the silt loading on a sample of roads in central and suburban London. Ball and Caswell suggested that the emissions of resuspended road dust were comparable to the exhaust emissions at that time.

More research is needed to verify these emission factors.

17.7.7 Agriculture and natural emission sources

Natural source, or biogenic, emissions can make a significant contribution to total VOC and NO_x emissions. Biogenic emission estimates for the USA have been reported at 30,860,000 tons of VOCs and 346,000 tons of NO_x per annum. This is in comparison to estimates of 21,090,000 tons of anthropogenic VOCs and 23,550,000 tons of anthropogenic NO_x estimated for 1990 (Radian Corporation 1996). Estimating emissions of VOCs and NO_x from natural sources is an essential part of preparing an inventory of ozone precursors. A report prepared by the Radian Corporation (1996) presents a standard approach to developing biogenic emission estimates for ozone inventories, and VOC and NO_x emission estimates from the natural sources of lightning and oil and gas seeps.

Analysis of land cover data compiled from satellite images may be important for estimating emissions from agricultural activity. In the UK, the Center for Ecology and Hydrology's (CEH) Land Cover Map (see http://www.ceh.ac.uk/data/lcm/index.htm) provides information on 25 types of land cover, including agricultural land, woodland, and water areas, as well as urban and suburban land, on a 25 meter grid (Fuller *et al.* 1994). This is derived from Landsat Thematic Mapper data. The proportion of each land cover type is used, in conjunction with emission factors, to calculate the emissions from nonurban land uses.

Land cover data are supplemented by data from the Census of Agriculture, which is undertaken every June in the UK. This provides information on crops cultivated and livestock. For example, within the M25 London orbital motorway in 1994 there were 27,289 cattle and calves, 58,674 sheep and lambs, 23,057 pigs, and 608,092 poultry (Ministry of Agriculture, Fisheries and Food 1995). These animals produce 8.8 tonnes of NMVOCs per year, plus 3.4 tonnes of methane. There are also 19,900 hectares of cereals grown in the area. Although it would be possible to use the emission factors given in AP-42 (US EPA 1995) to calculate the particulate emissions from harvesting these cereals, as commented above, it is questionable how applicable these are in the UK's generally wetter climate.

17.8 POINT SOURCE EMISSIONS

Many of the emissions to the atmosphere resulting from industrial processes and the combustion of fossil fuels are not uniformly spread across urban areas but concentrated at particular points. These are the emission sources that are most likely to have continuous emission monitoring equipment fitted, which continuously measures (with very short averaging time) and records the actual emissions, and therefore rely least on the application of emission factors in order to estimate the quantities of pollutants released. They are also the processes that require specific authorization from national or regional regulatory agencies in order to operate.

In the UK the Environmental Protection Act 1990 introduced a single system of control over "prescribed processes" in industry and commerce that is likely to result in the release of pollutants to the environment (air, water, and land). The Environment Agency is responsible for the control of about 2200 processes that are most likely to cause serious pollution. These are referred to as "Part A processes." Local authorities control more than 12,000 less-threatening processes, referred to as "Part B processes." A common system of authorization, enforcement, and public access to information applies to the Environment Agency and local authorities. This system is being progressively replaced by a system of integrated pollution prevention and control, introduced by the Pollution Prevention and Control Act 1999, in order to comply with the requirements of the EU Directive on Integrated Pollution Prevention and Control (96/61/EC).

Both the Environmental Protection Act 1990 and the Pollution Prevention and Control Act 1999 require operators of specified industrial installations to obtain specific authorization, to monitor the release of pollutants, and to submit information in order to demonstrate compliance with the standards set by the Environment Agency. Failure to provide that information constitutes a breach of the authorization, and the Environment Agency can take enforcement action. Information on the actual releases of air pollutants reported to the Agency is available at http://216.31.193.171/asp/introduction.asp. Operators of plants under local authority control are required to obtain specific authorization and to meet specified emission limits, but they are not at present required to report the actual amounts of pollutants released.

While this system of control provides a great deal of information on industrial processes and their location, in the case of Part B processes it does not generally provide the information that is required to compile emissions inventories. In most cases authorizations specify release concentrations, and any monitoring relates to these, rather than total annual emissions. Unlike with Part A processes, there is no requirement for operators of processes to submit details of the quantities of substances released during the preceding year. In London there are comprehensive data for only 59 Part A processes. Emissions from the 750 Part B processes must, in the majority of cases, be estimated using emission factors. This inevitably affects the accuracy of the final emissions inventory.

17.9 A COMPARISON OF THE LONDON AND TOKYO ATMOSPHERIC EMISSIONS INVENTORIES

London and Tokyo are two of the world's larger capital cities, and have many common features. Both cities are capitals of island nations lying off the coast of large continents. Both are the principal cities in developed countries that are members of the OECD. Both cities have well developed public transport systems and the majority of people use them to travel to work (Focas 1998). However, there are also significant differences. London lies at 51° 30′ N (equivalent to just north of Calgary in Canada), while Tokyo lies some 1100 miles closer to the Equator at 35° 40′ N (which is equivalent to half-way between Los Angeles and San Francisco in the USA, the Gulf of Taranto in Italy, or just south of Madrid in Spain). London has a relatively mild climate, benefiting greatly from the warm waters of the Gulf Stream. Summers are significantly hotter and more humid in Tokyo than in London, and this is reflected in the much greater use of domestic air conditioning.

London is a much older city than Tokyo, having developed over almost 2000 years from its Roman foundations. The city grew concentrically, absorbing existing towns, until the 1950s, when further outward expansion was curtailed by the designation of the Metropolitan Green Belt (Clout & Wood 1986). Tokyo, on the other hand, was the insignificant small town of Edo until the shogun Tokugawa Ieyasu moved his government there in 1590 (Nouet 1990). By that time London's population was probably around 220,000. Even in the 1880s it took only two hours to walk across the entire built-up area of Tokyo. In both London and Tokyo, the most rapid physical expansion went hand in hand with the development of the suburban railway system, although Tokyo's expansion was westwards because Tokyo Bay lay to the east. Tokyo's growth continued for longer, stopping only when it reached the mountains, which define the western edge of the Kanto plain.

The different histories of development in the two cities are reflected in the their physical form today. In London the highest densities of development occur at the center of the city, with offices and other commercial buildings, and densities fall steadily toward the peripheral areas of suburban housing. In Tokyo the highest densities of development occur in the east, and fall steadily toward the mountains in the west. Within the mountainous region there is only sparse development in the valley bottoms. These profiles of development are reflected in energy use and emissions, and can be seen in Fig. 17.7.

As explained above, the London inventory was prepared as part of a program of support for local authorities in meeting their obligations under the Environment Act 1995. The pollutants included in the inventory are those covered by the UK national air quality strategy: NO_x, SO_2, CO, NMVOCs, benzene, 1,3-butadiene, and PM_{10} (Department of the Environment, Transport and the Regions 2000b). The inventory also includes CO_2 because of its importance as a greenhouse gas. The area covered by the inventory is the area encircled by the M25 London orbital motorway, which includes the whole of the administrative area of Greater London. The aggregate area is 2466 km^2, with a population of 7.8 million. The base year for the inventory is 1996 but it is being updated. All sources of emissions are covered, including industry, commercial and domestic buildings, road and rail transport, airports, shipping, and off-road mobile sources. A report on the inventory was published in February 1998 (Buckingham *et al.* 1998) and includes maps showing the intensity of emissions on a 1 × 1 km grid.

The Tokyo inventory was compiled by the Institute of Behavioral Sciences (IBS) under contract to Tokyo Metropolitan Government's Bureau of Environmental Protection as part of a program of work on NO_x and PM_{10} abatement. There is no published report in English. Additional data are available for SO_2, CO, total hydrocarbons (THC), and CO_2. The area covered by the inventory includes the whole of the Tokyo metropolitan area. The aggregate area is 2,012 km^2, with a population of 11.77 million. The base year for the inventory is 1995. All sources of emissions are covered, including industry, commercial and domestic buildings, road transport, airports, shipping, and construction equipment. There is no published report but data were provided to the former London Research Centre in order to enable a comparison to be made of the intensity of energy use and emissions in the two cities. This involved the conversion by IBS of the data for Tokyo from a longitude and latitude grid to a 1 × 1 km grid.

The methodologies adopted in compiling the Tokyo and London inventories are very similar. Where there are directly measured data on the emissions from individual industrial and commercial facilities, these data have been used in prefer-

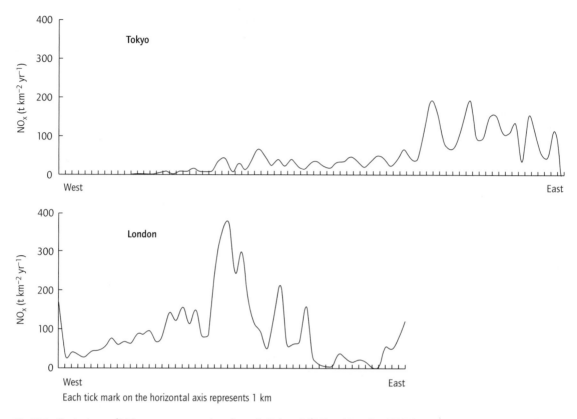

Fig. 17.7 Emissions of NO_x on a cross-section through Tokyo (1995) and London (1996).

ence to emission factors. In the case of Tokyo there are 13,251 registered facilities, which have measured data for NO_x (8658 facilities), SO_2 (5048) and PM_{10} (8427). Although NO_x emissions are only measured at 65% of the registered facilities, these are estimated to account for 90% of total emissions from stationary sources. As noted above, in London there are comprehensive data for only 59 "Part A" processes. Of the 750 facilities operating the less hazardous processes and smaller Part B processes, some 45% fall into the broad category of "coating processes," including the respraying of vehicles, printing, and the manufacture of printing inks and coating materials. At present, the regulating authorities specify the pollutants to be measured on a plant-by-plant basis, and there is no general requirement for all regulated facilities to monitor a standard set of pollutants. For these rea-

sons the London atmospheric emissions inventory relies much more heavily on the use of emission factors than the Tokyo inventory.

Where there are no measured data, emission factors are used in conjunction with data on the amounts of the various fuels used. In many cases the fuel-use data relate to larger areas than the 1×1 km grid. Where this is the case in Tokyo, other data on employment and dwellings have been used to apportion data to grid squares. In London, gas-use data relate to small enough areas to apportion to grid squares directly. So far, only a few *ad hoc* comparisons have been made between emission factors used in the two cities. For example, measured emissions of NO_x from waste incineration plants in Tokyo lie in the range of $0.6–1.06 \, kg \, t^{-1}$ of waste. The default emission factors used in London are $1.0 \, kg \, t^{-1}$ for industrial waste and $1.8 \, kg \, t^{-1}$ for domestic waste. These assume no NO_x abatement measures, whereas many of the Tokyo plants are fitted with a variety of NO_x abatement equipment. The NO_x emission factors for natural gas, which meets almost half of all London's energy needs, are very close—1.7 kg 1000 m^{-3} in Tokyo compared to 1.8 kg 1000 m^{-3} in London. The factors for LPG, of which much more is used in Tokyo than in London, are also close: 2.99–3.98 kg 1000 m^{-3} in Tokyo, compared to 2.27–4.40 kg 1000 m^{-3} in London.

The methodology for estimating vehicle emissions is similar in Tokyo and London but not identical. In London a traffic model, based on origin and destination surveys and on observed traffic flows, has been used to provide estimates of traffic volumes and speeds on each link in a network of major roads throughout the metropolitan area. In Tokyo the estimates are based on the observed traffic flows on the main roads. Vehicles are divided into nine types in the Tokyo and 10 types in the London model. Full account is taken in both cases of the variety of engine sizes, fuels (gasoline and diesel in the case of London, and gasoline, diesel, and LPG in Tokyo), and the age of vehicles in the vehicle fleet. Vehicle age is important in the London inventory because a significant number of vehicles are still operating without catalytic converters. The major roads are then treated as line sources.

The vehicle emission factors used in London are those provided by the National Environmental Technology Centre in agreement with the Transport Research Laboratory. In Tokyo the vehicle emission factors for NO_x, CO, and hydrocarbons are provided by the Japan Environment Agency and for PM_{10}, SO_2, CO_2, and fuel consumption by Tokyo Metropolitan Government. No detailed comparison has yet been made between the London and Tokyo emission factors. However, a simple division of the total NO_x emissions by the total vehicle–kilometers driven in each city gives an average emission rate of 0.8 g km^{-1} in Tokyo and 2.42 g km^{-1} in London, excluding the contribution of cold starts in both cases. Catalytic converters only became mandatory on new gasoline-engined vehicles in the UK in 1993 so in 1996 (the base year for the London inventory) the majority of vehicles were still without catalysts. The significantly higher NO_x emissions per kilometer in London are therefore to be expected.

In both London and Tokyo, there is a significant residue of traffic on minor roads. The aggregate length of minor roads in each grid square is used to apportion this residue. The resultant emissions occurring in each grid square are then treated as area sources.

Table 17.6 compares emissions of key pollutants in London and Tokyo categorized by the prin-

Table 17.6 Emissions in Tokyo and London (t, CO_2 in kt) (data from London Research Centre and Institute of Behavioral Sciences, Tokyo).

	Tokyo					London				
	Power plants & industry	Domestic & commercial	Road traffic	Waste	Total	Power plants & industry	Domestic & commercial	Road traffic	Waste	Total
Sulfur dioxide	2,638	1,131.0	5,963	2,227	11,959	11,296	2,559	5,570	402	19,827
Nitrogen oxides	4,353	10,679	45,423	4,063	64,518	6,587	16,255	114,422	1,859	139,123
Carbon monoxide	3,629	22,228	211,822	23,128	260,807	4,413	4,724	563,129	10	572,276
Carbon dioxide	6,162	13,141	15,893	4,262	39,458	3,670	18,711	10,249	7	32,637
NMVOC	340	3,037	41,867	1,077	46,321	475	1,900	106,367	3	108,745
Benzene	n/a	n/a	n/a	n/a	n/a	34	164	3,369	0	3,566
1,3-butadiene	n/a	n/a	n/a	n/a	n/a	0	0	844	0	844
Fine particles	590	766	4,565	384	6,305	525	238	7,649	491	8,903

n/a, not available.

cipal sources. Some categories of uses have been excluded, as they were not included in the 1×1 km gridded data for Tokyo. These categories include aircraft, shipping, railways in the case of London, and off-road mobile uses.

The use of a 1×1 km grid as the basis for analysis allows a comparison to be made of the relative intensity of emissions use in Tokyo and London. Figure 17.7 shows the intensity of NO_x emissions per square kilometer on a west–east axis through the center of Tokyo and the center of London. In the case of Tokyo the line in Fig. 17.7 passes through Shinjuku and the Otemachi area, two very important business districts containing many company head offices. The line in London passes through the West End, which is London's main commercial and shopping district, as well as the

City, which is the principal financial district and the direct equivalent of Otemachi.

These two diagrams illustrate the differences between the two cities in their physical development and the way that this is reflected in their use of energy and the resultant emissions. In Tokyo buildings and activities are concentrated in the east of the metropolitan area, often referred to as the 23 wards, and decline to a very low level to the west in the Tama area. Within the 23 wards, individual districts such as the Shinjuku and the Otemachi areas stand out as peaks and can clearly be seen in Fig. 17.7. In London buildings and activities are concentrated in the center, declining to both east and west, but less steeply to the west than to the east. Much of the industry that used to exist in east London has now closed. Within the

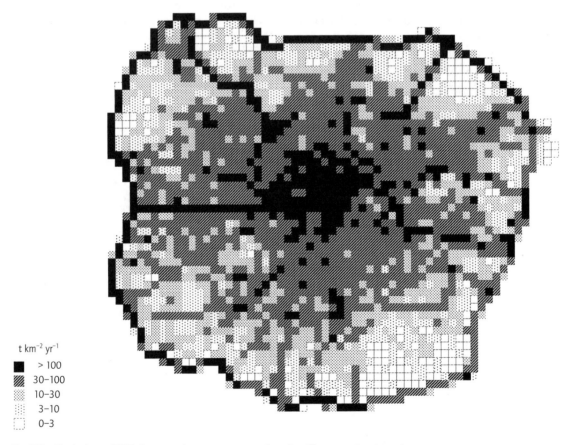

t km^{-2} yr^{-1}
■ > 100
▨ 30–100
▨ 10–30
▤ 3–10
□ 0–3

Fig. 17.8   Emissions of NO_x from stationary sources and road traffic in London (1996).

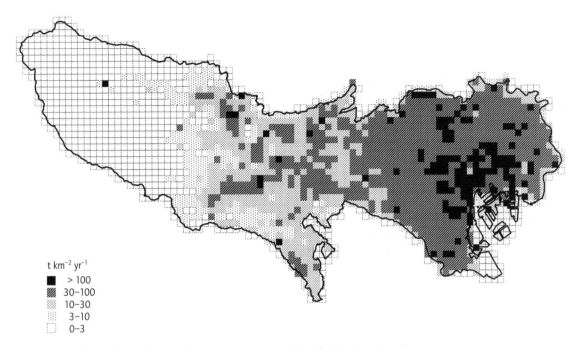

t km^{-2} yr^{-1}
■ > 100
▨ 30–100
▦ 10–30
▨ 3–10
▫ 0–3

Fig. 17.9 Emissions of NO$_x$ from stationary sources and road traffic in Tokyo (1995).

center, the West End and the City stand out as twin peaks. The rise at the extreme western and eastern edges of the graph is due to the heavy traffic using the M25 orbital motorway.

Figures 17.8 and 17.9 show the variations in the intensity of emissions in London and Tokyo in the form of maps. Again the same patterns are evident, with the Tokyo map darker (higher emissions) on the right than on the left. In the case of London the map is darker at the center than at the periphery, apart from the line of the M25 orbital motorway which appears as a dark ring around the outer edge of the map.

It is clear from the simple comparisons presented here that the levels of NO$_x$ emissions in Tokyo are generally lower than those for the corresponding areas in London. This seems likely to be the result of four factors:

1 A higher proportion of energy demand being met by electricity in Tokyo than in London. Emissions associated with electricity used in Tokyo but generated elsewhere are not included.

2 The later introduction of catalytic converters on gasoline-engined vehicles in London, with the result that a significant proportion of vehicles were still operating without catalysts in the inventory base year.

3 More widespread use of NO$_x$ abatement equipment in Tokyo.

4 Greater use of actual NO$_x$ emission measurements, rather than emission factors, in compiling the Tokyo emissions inventory.

The comparison of atmospheric emissions inventories prepared in different countries is inherently difficult because of differences in procedures, data quality and availability, and background assumptions. Nevertheless, such comparisons can draw attention to the factors influencing the levels of emissions and, potentially, provide guides to more effective abatement strategies.

17.10 CONCLUSIONS

The measurements of pollutant concentrations in the atmosphere carried out by many national governments, regional organizations, and cities around the world show the extent of air pollution but they do not identify the sources. The preparation of emission inventories complements the measurements by identifying the sources of the pollutants, and is a prerequisite to the taking of effective action to improve air quality.

This chapter describes procedures that are still being developed, even though the first inventories were prepared in the 1960s. As the understanding of air pollution and its health effects evolves, there are new requirements. The incorporation of a $PM_{2.5}$ standard into the US National Ambient Air Quality Standards required the collection of new sets of data and the development of new emission factors. This process will continue until, perhaps at some time in the far distant future, we can ensure clean air for everyone.

A single chapter does not allow a full discussion of each level of inventory preparation: supranational, national, regional, city, and district. Reference has been made to the procedures in the USA, Europe, and Japan. The discussion of the process of inventory preparation has focused on the city level because that is, perhaps, the most complex. The methodology described seeks to achieve a reasonable balance between accuracy and cost-effectiveness in delivering what is required. Resources of time and money are always limited, and we can always think of ways that things might have been done better. Despite these limitations, or perhaps because of them, the methodology will continue to be improved and refined.

REFERENCES

Ball, D.J. & Caswell, R. (1983) Smoke from diesel-engined road vehicles: an investigation into the basis for British and European emission standards. *Atmospheric Environment* 7, 169–81.

Ball, D.J. & Radcliffe, S.W. (1979) *An Inventory of Sulfur Dioxide Emissions to London's Air*. Research Report 23. Greater London Council, London.

Buckingham, C., Clewley, L., Hutchinson, D., Sadler, L. & Shah, S. (1998) *London Atmospheric Emissions Inventory*. London Research Center, London.

Cabreza, J. (1999) The emission inventory: evaluation, implementation, and future challenges. In *Emissions Inventory: Living in a Global Environment*, Proceedings of a specialty conference, December 8–10, 1998, New Orleans. Air and Waste Management Association, Sewickley, PA.

Clout, H. & Wood, P. (eds) (1986) *London: Problems of Change*. Longman, London.

Department of the Environment (1996) *Digest of Environmental Statistics, No. 18*. HMSO, London.

Department of the Environment, Transport and the Regions (2000a) *Review and Assessment: Estimating Emissions*. LAQM TG2(00). DETR, London.

Department of the Environment, Transport and the Regions (2000b) *The National Air Quality Strategy for England, Scotland, Wales and Northern Ireland: Working Together for Clean Air*. The Stationary Office, London.

Eastern Research Group (1999) *Handbook for Criteria Pollutant Inventory Development: A Beginner's Guide for Point and Area Sources*. EPA-454/R-99–037. United States Environmental Protection Agency, Research Triangle Park, NC (http://www.epa.gov/ttnchie1/brochures/beginner.pdf).

EMEP/CORINAIR (2002) *EMEP/CORINAIR Emission Inventory Guidebook*, 3rd edn. Technical report no. 30. European Environment Agency, Copenhagen (http://reports.eea.eu.int/EMEPCORINAIR3/en/tab_content_RLR).

Energy and Environmental Analysis, Inc. (2000) *Analysis of Commercial Marine Vessels Emissions and Fuel Consumption Data*. Report EPA420-R-00–002. US Environmental Protection Agency, Ann Arbor, MI (http://www.epa.gov/otaq/models/nonrdmdl/c-marine/r00002.pdf).

Eurostat (2001) *Eurostat Yearbook 2001*. Office for Official Publications of the European Communities, Brussels.

Expert Panel on Air Quality Standards (1995) *Particles*. HMSO, London.

Focas, C. (ed.) (1998) *The Four World Cities Transport Study*. The Stationery Office, London.

Fuller, R.M., Groom, G.B. & Jones, A.R. (1994) The land cover map of Great Britain: an automatic classification of landsat thematic mapper data. *Photogrammetric Engineering and Remote Sensing* 60, 553–62.

Garnett, A. (1967) Some climatological problems in urban geography with reference to air pollution. *Transactions of the Institute of British Geographers* 42, 21–43.

Goodwin, J.W.L., Salway, A.G., Dore, C.J., Murrells, T.P. et al. (2002) *UK Emissions of Air Pollutants*

1970–2000. AEA Technology, Culham, Oxfordshire (http://www.naei.org.uk/reports.php).

Greater London Authority (2002) *Cleaning London's Air: The Mayor's Air Quality Strategy.* Greater London Authority, London.

Gugele, B. & Ritter, M. (2002) *Annual European Community CLRTAP emission inventory 1990–2000.* Technical report no. 91. European Environment Agency, Copenhagen. (http://reports.eea.eu.int/technical_report_2002_91/en/tab_content_RLR).

Houghton, J.T., Meira Filho, L.G., Lim, B. *et al.* (eds) (1997) *Revised 1996 IPCC Guidelines for National Greenhouse Gas Inventories*, three volumes. UK Meteorological Office, Bracknell (http://www.ipcc-nggip.iges.or.jp/public/gl/invs1.htm).

Hutchinson, D. (1998) The UK urban emissions inventory program. In *Emissions Inventory: Planning for the Future*, Proceedings of a Specialty Conference, October 28–30, 1997, Research Triangle Park, NC. Air and Waste Management Association, Pittsburgh.

Hutchinson, D. & Clewley, L. (1996) *West Midlands Atmospheric Emissions Inventory.* London Research Center, London.

International Civil Aviation Organization (1995) *ICAO engine exhaust emissions data bank.* Doc 9646-AN/943. International Civil Aviation Organization, Montreal.

Jol, A. (1998) Methods, results and current state-of-play of the European CORINAIR programme. In *Emissions Inventory: Planning for the Future*, Proceedings of a Specialty Conference, October 28–30, 1997, Research Triangle Park, NC. Air and Waste Management Association, Pittsburgh.

Jost, P., Hassel, D., Weber, F.J. & Sonnborn, K.-S. (1992) *Emission and Fuel Consumption Modeling Based on Continuous Measurement.* TÜV Rheinland, Cologne.

Kouridis, C., Ntziachristos, L. & Samaras, Z. (2000) *COPERT III Computer Program to Calculate Emissions from Road Transport: Users Manual.* Technical Report no. 50. European Environment Agency, Copenhagen (http://reports.eea.eu.int/technical_report_No_50/en).

Lloyd's Register (1995) *Marine Exhaust Emissions Research Programme.* Lloyd's Register of Shipping, London.

Marsh, K.J. & Foster, M.D. (1967) An experimental study of the dispersion of the emissions from chimneys in Reading. I: The study of long-term average concentrations of sulfur dioxide. *Atmospheric Environment*, **1**, 527–50.

Ministry of Agriculture, Fisheries and Food (1995) *The Digest of Agricultural Census Statistics: United Kingdom 1994.* HMSO, London.

Nouet, N. (1990) *The Shogun's City: A History of Tokyo.* Paul Norbury, Sandgate.

Ntziachristos, L. & Samaras, Z. (2000) *COPERT III Computer Program to Calculate Emissions from Road Transport: Methodology and Emission Factors.* Technical Report no. 49. European Environment Agency, Copenhagen (http://reports.eea.eu.int/Technical_report_No_49/en).

Passant, N.R. & Lymberidi, E. (1998) *Emissions of Non-methane Volatile Organic Compounds from Processes and Solvent Use.* Report, AEAT-2837 Issue 1. AEA Technology, Culham.

Penman, J., Kruger, D., Galbally, I. *et al.* (eds) (2000) *Good Practice Guidance and Uncertainty Management in National Greenhouse Gas Inventories.* Institute for Global Environmental Strategies, Hayama, Japan (http://www.ipcc-nggip.iges.or.jp/public/gp/gpgaum.htm).

Radian Corporation (1996) *Biogenic Sources Preferred Methods.* EIIP Document Series, Volume V. US Environmental Protection Agency, Research Triangle Park, NC (http://www.epa.gov/ttn/chief/eiip/techreport/volume05/v01.pdf).

Sasnett, S. & Misenheimer, D. (1996) Regulatory requirements for emissions reporting to EPA. *The Emissions Inventory: Programs and Progress*, Proceedings of the conference at Research Triangle Park NC, October 11–13, 1995. Air and Waste Management Association, Pittsburgh.

Tonooka, Y. (1999) *Country Report on Emissions Estimation in Japan.* Report to the Expert Group Meeting on Emissions Monitoring and Estimation, January 27–29. Mimeo.

UNFCCC (2000) *Review of the implementation of commitments and other provisions of the convention: UNFCCC guidelines on reporting and review.* FCCC/CP/1999/7. UNFCCC Secretariat, Bonn (http://www.unfccc.int/resource/docs/cop5/07.pdf).

US Environmental Protection Agency (1995) *Compilation of Air Pollution Emission Factors.* AP-42 5th edn. United States Environmental Protection Agency, Office of Air Quality Planning and Standards, Research Triangle Park, NC (http://www.epa.gov.ttn/chief/).

US Environmental Protection Agency (1999) *Description of the MOBILE Highway Vehicle Emission Factor Model* (http://www.epa.gov/oms/models/mdlsmry.txt).

Webb, S. & Draper, J. (1996) Airport emission inventories—the California FIP. *The Emissions Inventory: Programs & Progress*, Proceedings of the conference at Research Triangle Park, NC, October 11–13, 1995. Air and Waste Management Association, Pittsburgh.

Yotter, E.E. & Pale, A.A. (1996) Activity used in the development of the California starts methodology. *The Emissions Inventory: Programs & Progress*, Proceed-

ings of the conference at Research Triangle Park, NC, October 11–13, 1995. Air and Waste Management Association, Pittsburgh.

Zachariadis, T. & Samaras, Z. (1999) An integrated modeling system for estimation of motor vehicle emissions. *Journal of the Air and Waste Management Association* **49**, 1010–26.

18 **Pollutant Dispersion Modeling**

YASMIN VAWDA

18.1 INTRODUCTION

It is often necessary for operators of industrial processes, developers, or regulatory authorities to quantify the air pollution impact of an existing or proposed scheme (e.g. a power station or a road) for the purposes of licensing or planning. A predictive tool such as an atmospheric dispersion model can be useful in this respect, as it provides a means of calculating air pollution concentrations near ground level in the vicinity of the emitting source, given information about the emissions and the prevailing meteorology. It is the concentration of the pollutant near ground level that is important in determining whether there are any potential ill effects, e.g. on human health; air quality standards and guidelines are set in terms of concentration values, not source rates.

The aim of this chapter is to provide details on what types of model are available, how they should be used, and their limitations. The information required to run dispersion models, the type of results that are produced, and how these can be interpreted are described. The case studies illustrate how the results of various dispersion models can be used in an air quality assessment and highlight some of the problems that are encountered in the modeling approach. A detailed technical description of the theory of atmospheric dispersion is provided in Chapter 10.

Dispersion models take a number of forms, from simple nomograms and spreadsheets to sophisticated computer programs. It is important

that the model is appropriate for the complexity of the study (DoE 1995a). Factors such as model accuracy and validation need to be taken into account (see Case study 5). It is also important to assess the availability of both meteorological and emissions data; some models do not require detailed meteorological data because they make broad assumptions about the prevailing local climate, while others may require quite detailed information on meteorology and emissions.

18.2 EMISSION SOURCES RECOGNIZED BY ATMOSPHERIC DISPERSION MODELS

The three main sources of air pollution modeled are:

1 Emissions from vehicle exhausts on roads, which generally make the greatest contribution to air pollution in urban areas.
2 Controlled industrial, commercial, and domestic emissions from chimneys.
3 Fugitive emissions (e.g. leakages from industrial plant, or particulate matter from mineral extraction schemes; see Chapter 9).

These sources may be divided into three main categories that are recognized by dispersion models:

1 Point sources. These are individual chimneys. The simpler models can treat only one stack at a time (EA 1998), though more sophisticated programs can include a very large number of stacks simultaneously.

2 Line sources. Emissions from road traffic are usually treated as lines, where each straight line represents a segment of road. The simplest models may treat only one straight section of road (Buckland & Middleton 1997), whereas the more advanced programs can handle a very large number of roads, including bridges and street canyons (Benson 1992).

3 Area sources. A cluster of point or line sources (e.g. a large number of vehicles in a parking lot) may be treated as an area source. Similarly, a large city may be split into a number of grid squares (each with both traffic and industrial emissions), and the emissions from each square treated as an array of area sources.

Volume sources and puff releases can also be modeled using more specialized programs, but such applications are less common in the quantification of air pollution impacts, and are not discussed further in this chapter. Roads and industrial chimneys are by far the most commonly modeled sources of air pollution.

The type of model that is used is dependent upon the pollutant released, and the appropriate period over which concentrations will be considered. Most pollutants that are commonly investigated are released as buoyant gases (e.g. sulfur dioxide emissions from a power station chimney), and most dispersion models have been developed to treat the atmospheric dispersion of these hot plumes. Some of the more sophisticated models also incorporate formulae for taking into account deposition processes that may occur in the atmosphere (such as washout of pollutants due to rainfall, or the gravitational settling of particles—see Case study 4). Some models also attempt to account for chemical reactions that may occur during the transport of pollutants in the atmosphere, although most models are suited for unreactive (inert) species only (see Case studies 2 and 6). Moreover, pollutants such as nitrogen dioxide and PM_{10} have both primary and secondary sources, and there are few dispersion models that can reliably calculate secondary pollutant concentra-

tions on a local scale for practical applications. However, a number of empirical relationships are available to describe the conversion of nitric oxide to nitrogen dioxide (Fisher *et al.* 1999), which may be applied to the results of modeling the concentration of total nitrogen oxides.

18.3 ASSESSMENT CRITERIA FOR THE RESULTS OF DISPERSION MODELS

Air quality standards and guidelines for different pollutants are expressed in various averaging periods and percentiles, depending on whether the pollutant has acute effects (which can arise after short periods of exposure) or chronic effects (resulting from exposure over a long period) on human health or vegetation. For example, the air quality objectives that have been set within the UK Air Quality Strategy (DETR 2000a) are based on averaging periods ranging from 15 minutes to one year. Some of these UK air quality criteria are shown in Table 18.1; the results of dispersion models need to be compared against standards of this type. Therefore, it is important that dispersion models calculate air pollutant concentrations in the appropriate averaging periods and statistics. Moreover, special applications of dispersion models (e.g. for assessing odor impacts—see Case study 1) may require estimates of concentrations over very short time scales, i.e. a few seconds.

18.4 METEOROLOGICAL DATA REQUIREMENTS OF ATMOSPHERIC DISPERSION MODELS

A detailed discussion of meteorology is presented in Chapter 10 and a simplified treatment of the parameters that are required by many dispersion models is given below. At the simplest level, highly convective conditions and high wind speeds often give rise to highest ground-level concentrations of pollutants emitted from chimneys. In contrast, calm, low wind speed conditions are

Table 18.1   Examples of UK Air Quality Strategy objectives for the protection of human health.

Pollutant	Concentration	Measured as	Date to be achieved by
Benzene	16.25 μg m⁻³ (5 p.p.b.)	Running annual mean	End of 2003
	5 μg m⁻³ (1.5 p.p.b.)	Running annual mean	End of 2010
1,3-Butadiene	2.25 μg m⁻³ (1 p.p.b.)	Running annual mean	End of 2003
Carbon monoxide	10 mg m⁻³ (8.6 p.p.b.)	Running annual mean	End of 2003
Lead	0.5 μg m⁻³	Annual mean	End of 2004
	0.25 μg m⁻³	Annual mean	End of 2008
Nitrogen dioxide	200 μg m⁻³ (105 p.p.b.) not to be exceeded more than 18 times a year	One-hour mean	End of 2005
	40 μg m⁻³ (21 p.p.b.)	Annual mean	End of 2005
Sulfur dioxide	350 μg m⁻³ (132 p.p.b.) not to be exceeded more than 24 times a year	One-hour mean	End of 2004
	125 μg m⁻³ (47 p.p.b.) not to be exceeded more than three times a year	24-hour mean	End of 2004
	266 μg m⁻³ (100 p.p.b.) not to be exceeded more than 35 times a year	15-minute mean	End of 2005

more likely to give rise to the highest concentrations from road traffic emissions.

The main factors that affect the dilution process are the wind speed, the height at which the substance is emitted, and atmospheric turbulence. Only if the wind direction is roughly aligned between the point of release and the point at which an individual is standing will that individual be subject to any exposure. An individual standing in the open air within an urban area will be subject to the influence of many sources within the area, and hence will be exposed to some extent to traffic fumes, regardless of wind direction.

Many polluting sources are released from tall chimneys; this is to ensure that sufficient mixing and dilution occurs before the pollutants reach the ground, in order to render concentrations harmless. In many cases the emissions are hotter than the surrounding air when released from the chimney, which results in the plume being buoyant. This, together with the velocity of the plume, means that extra dilution occurs before the substance reaches

the ground. The height to which the released material rises depends on atmospheric conditions and dispersion models are designed to take this "plume rise" into account (see Case study 3).

In recent years dispersion models have focused on improving the description of dispersion and turbulence in the atmosphere at heights well above the ground (Weil 1992). This is important for calculating atmospheric concentrations of emissions from tall chimneys, such as those at power stations (HMIP 1996). These new models rely on a better understanding of the lower layers of the atmosphere (Chapter 10). However, many models simplify the issue by categorizing the turbulence of the atmosphere into six or seven stability classes, the "Pasquill" categories A to G. Pasquill category A represents very convective conditions where surface heating generates additional turbulence over that caused by the wind, category D represents neutral conditions and categories F and G represent a very stable, low-turbulence situation.

Case study 1 **An assessment of the odor impact of a proposed sewage treatment plant**

Key features of study

The purpose of the study was to determine whether emissions of hydrogen sulfide from the proposed 15 m chimney at a new sewage treatment plant were likely to give rise to any odor beyond the site boundary, where complaints from local residents could result. The plant was in a rural area, with no other significant sources of hydrogen sulfide, either natural or anthropogenic. Therefore, the background concentration of H_2S was taken to be zero. The assessment criterion was taken to be a "recognizable" odor. The published odor detection threshold for hydrogen sulfide is 0.5 p.p.b. (Woodfield & Hall 1994), which would result in a very faint odor. A "recognizable" odor would be 2–5 times the detection threshold, i.e. 1.0–2.5 p.p.b.

An understanding of the relationship between the mean and fluctuating plume concentrations is important for assessing odors. Over a short time scale (i.e. a few seconds), the unsteady character of the dispersion process (due to constantly varying wind speed and direction) leads to large excursions of peak concentrations from the hourly mean. Most atmospheric dispersion models predict one-hour averages as the shortest averaging period. However, the human nose can readily perceives these peaks in fluctuating concentrations of odorous substances.

Measurements of hourly meteorological parameters representative of the site (from a nearby airport) were available for a three-year period (1997, 1998, and 1999). It is advisable to consider more than a single year of meteorology as the frequency distribution of wind speed, direction, and stability can vary from year to year. A Gaussian dispersion model was used to calculate one-hour mean hydrogen sulfide concentrations. As the perception of an odor can occur over a matter of seconds, the one-hour mean predictions were extrapolated to short-term peaks, by multiplying by a factor of 10 (Hall & Kukadia 1993).

Methodology

The model input parameters are shown in Table 18.2. Receptors for the modeling were located at the site boundary in eight directions from the chimney round the compass at 45° intervals, and at selected properties beyond the site boundary, representative of residential and amenity areas in the neighboring community. The model took into account building downwash effects, as several buildings greater than 6 m in height were proposed adjacent to the chimney. The emission rate of hydrogen sulfide was taken to be the manufacturer's guaranteed limit, to ensure a degree of conservatism in the assessment.

Table 18.2 Emission parameters for proposed sewage treatment plant.

Stack height (m)	15
Stack diameter (m)	0.8
Volumetric flow rate at 20°C ($m^3 s^{-1}$)	7.643
Normalized flow rate ($N m^3 s^{-1}$)	7.121
Exit temperature (°C)	20
H_2S emission concentration – manufacturer's guaranteed limit at 0°C (p.p.b.)	20
Exit velocity ($m s^{-1}$)	15
H_2S emission rate ($g s^{-1}$)	0.000216

Results and assessment

The highest short-term concentration was predicted to occur at a dwelling to the north of the site, with a value of 0.45 p.p.b. The highest short-term concentration at the northern site boundary was predicted to be 0.33 p.p.b. These concentrations were derived by multiplying the model predicted one-hour mean concentrations by 10. These short-term concentrations are well below the environmental criterion of 1 p.p.b. for a "recognizable" odor for H_2S.

Conclusions

A chimney height of 15 m would be adequate to ensure that emissions of H_2S from the sewage treatment plant stack would not give rise to a "recognizable" odor beyond the site boundary. The model results suggest that even a faint odor (i.e. at the detection threshold for H_2S) is unlikely.

This conclusion assumes that the meteorology during 1997, 1998, and 1999 would be representative of future conditions at the site. It is noteworthy that fugitive emission of hydrogen sulfide (e.g. uncontrolled leaks) were not included in the assessment, as their emission characteristics could not be reliably quantified. Moreover, it was assumed that H_2S would behave as an inert species during its dispersion and dilution, and there would be no synergistic effects with other air pollutants that would alter its odorous properties.

Case study 2 **Impact of high nitrogen oxides emissions during start-up conditions at a power station**

Key features of study

Start-up of turbines at power stations can result in temporary emissions of nitrogen oxides (NO_x) far greater than the emission rate during routine operation. Start-up conditions could prevail several times a year, i.e. after maintenance or emergency shut-downs. Start-up emissions need to be modeled to assess the short-term impact of nitrogen dioxide (NO_2) on the community. The national air quality standard for one-hour mean NO_2 concentrations was 287 $\mu g\,m^{-3}$.

Methodology

The turbine manufacturer provided the emissions data shown in Table 18.3. The NO_x emission rate (expressed as an hourly average) was calculated from an emission of 133 $kg\,h^{-1}$ over the first hour (the start-up period).

Table 18.3 Emission parameters for power station during start-up.

Stack height (m)	60
Stack diameter (m)	7
Exit temperature (°C)	88
Exit velocity ($m\,s^{-1}$)	14.2
NO_x emission rate ($g\,s^{-1}$)	36.9

The dispersion model was used to predict one-hour mean NO_x concentrations at a line of receptors (at intervals of 50 m) directly downwind of the stack. A matrix of 20 stability and wind speed combinations (Table 18.4) were used as meteorological inputs to the model. These combinations collectively represent conditions that prevail for 99% of the time at the site (as shown by inspecting a long-term frequency analysis of stability and wind speed recordings for the site).

The dispersion model could not take into account the chemical transformation of the primary-emitted NO to NO_2 in the atmosphere. Therefore, the model is suitable only for predicting

Table 18.4 Predicted one-hour mean ground level NO_x concentrations during start-up.

Pasquill stability category	Wind speed ($m\,s^{-1}$)	Maximum NO_x concentration ($\mu g\,m^{-3}$)	Maximum NO_2 concentration ($\mu g\,m^{-3}$)	Distance downwind at which maximum concentration occurs (m)
A	1	94	47	<50
A	2.5	198	99	<50
B	1	4	2	3500
B	2.5	84	42	200
B	4.5	116	108	150
C	1	<1	<1	3500
C	2.5	12	6	1500
C	4.5	38	19	550
C	6.5	54	27	500
C	9.5	54	27	450
C	15	44	22	400
D	1	<1	<1	3200
D	2.5	2	1	1650
D	4.5	18	9	1150
D	6.5	34	17	850
D	9.5	46	23	750
E	2.5	4	2	4500
E	4.5	10	5	1400
F	1	<1	<1	150
F	2.5	<1	<1	850

(*Continued on p. 508*)

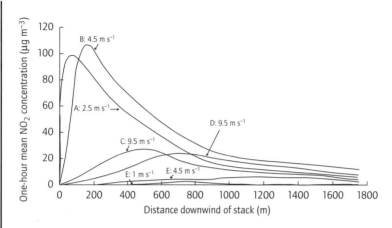

Fig. 18.1 Predicted one-hour mean contribution of stack to ground-level NO_2 concentration ($\mu g\,m^{-3}$) during start-up of turbines. Worst case wind speed in each stability category.

concentrations of total NO_x, and a subsequent allowance must be made to account for how much of this NO_x is in the form of NO_2 at ground level (Janssen *et al.* 1988; Bange *et al.* 1991). Studies have shown that the principal NO_x emitted by power station stacks is nitric oxide (NO). NO is oxidized to NO_2, primarily by reaction with ozone (O_3). The formation of NO_2 proceeds relatively slowly because the available O_3 is rapidly depleted. Additional O_3 enters the plume as it mixes with the surrounding air to continue the reaction, but this is generally a slow process (Janssen *et al.* 1988). Within 5 km of a source, less than 20% of the NO_x is in the form of NO_2, under stable atmospheric conditions. During unstable atmospheric conditions with more mixing, up to 50% of the NO may be converted to NO_2. Many modeling studies assume that 50% of the NO_x in a power station plume is in the form of NO_2 at ground level, and this factor was applied to the modeled NO_x concentrations in this study.

Results and assessment
The maximum predicted NO_x and NO_2 concentrations for each

stability and wind speed combination are shown in Table 18.4, together with the distance downwind of the stack where this peak occurs. The NO_2 results are displayed as concentration profiles in Fig. 18.1. The maximum predicted NO_2 concentration was $108\,\mu g\,m^{-3}$, occurring 150 m downwind of the stack during convective atmospheric conditions (i.e. warm, sunny days) with a wind speed of about $5\,m\,s^{-1}$. The annual mean NO_2 concentration in the area was estimated at 25–$30\,\mu g\,m^{-3}$, based on national mapping of NO_2 concentrations (measured using networks of diffusion tubes). Even when this annual mean was added to the maximum predicted short-term concentration of $108\,\mu g\,m^{-3}$, the air quality standard of $287\,\mu g\,m^{-3}$ for a one-hour mean NO_2 concentration is not approached.

Conclusions
Start-up emissions of NO_x from the power station were unlikely to cause breaches of the relevant air quality standard for NO_2.

For the prediction of long-term pollutant concentrations, such as the annual average, it is often faster for a model to use a statistical summary of meteorological data, which could be in the form of a "Pasquill stability frequency analysis"—this describes the frequency distribution of wind speed, direction, and Pasquill stability category for a number of years. Models in which short-term time-averaged concentrations over periods of about one hour are calculated may not require me-

teorological data (as specific combinations of the key parameters can be user-defined), although reference to the frequency of specific combinations of wind speed, direction, and stability for the area can be very useful for the interpretation of results. Models may also be run to predict a sequence of hourly concentrations using a long sequence of hourly meteorological data. Generally, dispersion models assume that meteorological conditions remain steady over one-hour periods. Statistical and sequential hourly meteorological data sets will give different results when calculating peak one-hour mean concentrations, but should give similar results when used to calculate annual means. Sequential hourly data sets can be a very powerful tool in investigating pollution episodes but only if the emission characteristics corresponding to the period of the episode are known.

18.5 TYPES OF ATMOSPHERIC DISPERSION MODEL

At the simplest level, predictions of air quality may be made using a nomogram, workbook, or spreadsheet, i.e. a "screening" model. Other models are computer-based, using a PC program. The most advanced dispersion models need very detailed meteorological inputs, and may require a workstation or very powerful PC on which to run.

There is a whole class of models based on the Gaussian formulation, in which the vertical and horizontal profiles of the plume concentration follow the shape of a Gaussian function (a symmetric bell-shaped distribution), which has convenient mathematical properties. A Gaussian model also assumes that one of the seven stability categories, together with wind speed, can be used to represent any atmospheric condition when it comes to calculating dispersion. Within each stability category the plume spread has a different dependence on downwind distance (Turner 1994).

Until recently, the Gaussian distribution was considered to be the best mathematical approximation of plume behavior for short averaging periods, and Gaussian models were generally used in most applications. More recently it has been recognized that there are different turbulence and diffusion characteristics within the atmosphere at different heights (see Chapter 10). The so-called "new generation" dispersion models adopt a more sophisticated approach to define vertical profiles of the turbulence parameters, and modify the Gaussian distribution under the more convective (unstable) conditions. These "new generation" models are not restricted to describing plume

shape according to a finite number of dispersion categories. Few Gaussian models can interpolate between discrete dispersion categories.

18.5.1 Screening models

In order to simplify the running of a computer model, some models already contain preset meteorological conditions and certain emissions data, and such models will usually calculate worst-case concentrations. The screening models consist of empirical results from field observations or wind tunnel experiments, or calculations of advanced models under clearly specified conditions. Screening models are a very useful way of quickly gaining an initial impression of the air pollution impact of the scheme under investigation.

18.5.2 Computerized dispersion models

Most dispersion models are computer-based programs suitable for a desktop PC. Their main differences to screening models are that:

1 They require more information on the source, e.g. diurnal traffic flow data for a road rather than simply a peak hour or annual average flow.
2 They can take greater account of hour-by-hour variations in meteorology, e.g. by use of sequential, hourly meteorological data from a representative meteorological observing site.
3 The results may be more accurate than those of screening models, but only if detailed and accurate meteorological and emissions data are available.
4 More than one type of source can be treated, e.g. point, area, and volume sources could be modeled simultaneously.
5 They are usually multisource, i.e. a large number of emission sources can be modeled simultaneously.
6 Their output choices can be more versatile, i.e. a wide range of averaging periods may be modeled.
7 Special effects such as atmospheric photochemistry, complex terrain, and buildings effects can be taken into account by some models.
8 Some of the more advanced models incorporate the most up-to-date treatment of the meteorology

of the atmosphere, if the necessary meteorological parameters are available.

Brief descriptions of some widely used models are given in the Appendix to this chapter for the purpose of illustrating relative capabilities; the list is not exhaustive, and does not indicate any type of approval or endorsement.

18.6 INPUT DATA REQUIREMENTS

All dispersion models need data on the pollutant emission rate. These need not necessarily be exact, as in many cases useful decisions can be made choosing highest likely emissions to represent a worst case (e.g. by assuming that emissions from a chimney are at the limits defined by the regulator, or that traffic on a road is at its peak-hour level— see Case studies 1 and 6 respectively). Most models also require details on how the pollutant is being released, and the environment into which the release occurs. It is also necessary to define the locations at which the impact of the emissions will be predicted (the "receptor" locations).

Where the model assumes that the emissions are not chemically transformed in the atmosphere, then the predicted concentration is directly proportional to the emission rate, i.e. if the emission rate is doubled, the predicted ground-level concentration also doubles. The collation of accurate emissions data is therefore extremely important (Vawda 1998).

18.6.1 Emissions data for chimneys

The emission rate is normally given in grams per second $(g s^{-1})$ or equivalent, convenient units, and may be derived from measurements of pollutant concentrations in the stack gases, plant manufacturers' specifications, or process characteristics such as fuel consumption, composition of fuel or feedstock—emission factors enable one to calculate emissions from such surrogate data. When calculating the emission rate, consideration must also be given as to whether the operation is continuous or "batch" (i.e. the process runs for a specified period of time), or whether there are specific scenarios that

might give rise to higher emissions, such as during start-up (see Case study 2). If a model is being used to predict annual average concentrations, an average or typical emission rate will be adequate. However, if short-term impacts are of interest (e.g. hourly or 15-minute average concentrations), a number of "worst-case" scenarios should also be included.

Some of the release conditions that need to be defined for dispersion models are used to calculate the plume rise, from the buoyancy and momentum of the exhaust gases. These parameters are:
- the height of the stack;
- the diameter of the stack at the exit point;
- the temperature of the emissions at the exit point;
- the velocity of the emissions at the exit point.

18.6.2 Emissions data for roads

The emission rate for a specific section of road is usually required by a dispersion model in grams per meter per second $(g m^{-1} s^{-1})$. The user may have to precalculate the line source emission rate based on a knowledge of individual vehicle type emission factors (DETR 1998). Emissions from individual road vehicles are dependent upon a number of factors, including the type of vehicle and the speed at which the vehicle is traveling. The emissions data required for road traffic emissions modeling include:
- the length of road section;
- the volume of traffic, e.g. vehicles per hour, vehicles per day, or the maximum hourly traffic flow;
- speed of traffic;
- vehicle mix (fraction of gasoline and diesel automobiles, light goods vehicles and heavy goods vehicles, buses, etc.);
- whether the road is in a street canyon, elevated, or in a cutting.

A notable feature of most road traffic models is that they take some account of the extra atmospheric mixing and turbulence caused by the movement of vehicles.

18.6.3 Local environmental conditions

Both nearby buildings and complex topography can have a significant effect upon the dispersion

Case study 3 **Carbon monoxide emissions from two municipal waste incinerators**

Key features of study

A company operates two incinerators (Sites A and B) on opposite sides of a small town in Neverland. Each site has a single chimney. The pollutant of concern from the stacks is carbon monoxide. The Industrial Source Complex (ISC) model was used to predict the cumulative concentrations of carbon monoxide from the two stacks. The predicted concentrations, after making an allowance for background levels, were compared against the relevant air quality criterion for carbon monoxide, to assess the combined air quality impact of the processes.

Methodology

Neverland has no national air quality standard or guideline for carbon monoxide. Therefore, the air quality assessment makes use of guidelines set by the World Health Organization (1987). These have been set for the protection of human health, and represent levels below which there should be no risk of harm. The one-hour mean guideline has been set at $30 \, \text{mg m}^{-3}$ (25 p.p.m.), and the eight-hour mean guideline has been set at $11.4 \, \text{mg m}^{-3}$ (10 p.p.m.).

No ambient monitoring data for carbon monoxide were available specifically for the town in which the two factories were sited. Therefore, use was made of national statistics, which suggested that the annual mean background concentration (i.e. away from the influence of busy roads) was about $0.6 \, \text{mg m}^{-3}$. Even close to busy roads, carbon monoxide concentrations were unlikely to exceed $5 \, \text{mg m}^{-3}$ as an eight-hour mean. The highest one-hour mean carbon monoxide concentration near busy roads during adverse weather conditions was unlikely to exceed $10 \, \text{mg m}^{-3}$.

The Industrial Source Complex (ISC) model was run using one year of sequential hourly meteorological data from the nearest observing station. The model calculated the maximum one-hour mean carbon monoxide concentration as well as the maximum running eight-hour mean. The model allows the effects of aerodynamic downwash on the dispersion of the plume to be taken into account; therefore, building heights, dimensions, and locations were defined in relation to the location of the stacks. The emissions parameters required by the model are shown in Table 18.5 – these were provided by the operator.

The data in Table 18.5 suggest that dispersion and dilution of the emissions from Site A will be less effective than from Site B. This is because the emissions at Site A occur from a shorter stack, and with a lower exit velocity, which would result in less plume rise. Moreover, the mass emission rate from the Site A stack is higher than at Site B, such that higher ground-level concentrations of CO are expected close to Site A.

Results and assessment

The model predictions for the combined carbon monoxide contribution of the two stacks are shown in Table 18.6. The maximum off-site one-hour mean concentration is $6.5 \, \text{mg m}^{-3}$, occurring at the eastern boundary of Site A. The maximum off-site eight-hour mean carbon monoxide concentration was also predicted to occur at this location, with a value of $1.7 \, \text{mg m}^{-3}$. Only the predictions of off-site concentration were reported, as the air quality assessment was concerned with the impact on the community, and not workplace exposure.

The total carbon monoxide concentrations (i.e. predicted contribution from Sites A and B added to the worst-case estimates of

Table 18.6 Maximum predicted carbon monoxide concentration contributed by stacks.

	One-hour mean (mg m^{-3})	Running eight-hour mean (mg m^{-3})
Maximum concentration (occurring at Site A boundary, 300 m east from stack at Site A)	6.5	1.7
Maximum concentration at boundary of Site B (occurring 50 m east of stack at Site B)	1.5	0.5
Maximum concentration at a residential property beyond the site boundary	2.3	1.0

Table 18.5 Emission parameters for the dispersion modeling.

	Stack on Site A	Stack on Site B
Carbon monoxide emission rate (g s^{-1})	185	115
Stack exit temperature of gases (K)	325	325
Stack exit velocity of gases (m s^{-1})	20	28
Stack height (m)	40	60
Stack diameter (m)	0.8	0.3

(Continued on p. 512)

Table 18.7 Total carbon monoxide concentrations compared with air quality guidelines.

	Maximum predicted contribution of the stacks	Worst-case estimate of background level	Total concentration	Air quality guideline
One-hour mean ($mg\,m^{-3}$)	6.5	10	16.5	30
Eight-hour mean ($mg\,m^{-3}$)	1.7	5	6.7	11.4

the background levels) are compared in Table 18.7 against the WHO guidelines. It is clear that even when the impact of the two factories is superimposed on the highest concentrations likely to arise near busy roads, the air quality guidelines are not approached.

Conclusions

The impact of the two incinerators on carbon monoxide concentrations in the local community is not significant, in terms of either incrementing background levels or approaching the relevant health-based air quality criteria.

characteristics of a plume. Many dispersion models contain algorithms to take account of these effects, although it must be accepted that the results should be treated with a greater degree of caution. Buildings may cause a plume to come to ground much closer to the stack than otherwise expected, causing significantly higher pollutant concentrations. As a general guide, building downwash problems may occur if the stack height is less than two and a half times the height of nearby buildings (see Case study 1). To take account of local building effects, models require information on the dimensions and location of the structures in relation to the stack. A number of different building wake algorithms are available.

Plumes can impact directly on hillsides under certain meteorological conditions (resulting in much higher ground-level concentrations that would occur over flat terrain), or valleys may trap emissions during low-level inversions. The more sophisticated models can take account of these terrain effects (though this is an area of active research), and require the input of contour heights in the area surrounding the stack. Terrain effects are unlikely to be significant where the hills have a slope of less than about 10%.

The model user is frequently required to define the "surface roughness," a term that describes the

degree of ground turbulence caused by the passage of winds across surface structures. Surface roughness is much greater in urban areas (due to the presence of tall buildings) than in rural areas. Calculations of dispersion that take account of the greater aerodynamic roughness of the surface structures in urban areas tend to predict higher concentrations closer to the stack than calculations under equivalent conditions that assume typical rural roughness.

The model user needs to define the receptor locations. It is good practice to identify the nearby, sensitive locations (e.g. dwellings, amenity areas) to the source, though many models allow the user to specify a "grid" of receptor locations, such that the results can be presented as isopleths superimposed on a base map of the area (see Case study 4). When setting up a receptor grid it is important to ensure that there are sufficient receptor points to be able to reliably predict the magnitude and location of the maximum concentration; if the grid of receptor points is too widely spaced, the maximum concentration may be missed.

18.6.4   Availability of input data

The results of a dispersion model are only as good as the data that are put in. For example, it would not be

Case study 4 **Nuisance assessment for deposited dust from a proposed opencast coal site**

Key features of study

An opencast coal site can be a significant source of dust, causing soiling of surfaces at properties within 500 m of the boundary. The nuisance impact of a proposed open-cast coal mine on the local community was carried out by modeling the dispersion and deposition of particulates, using the Fugitive Dust Model (FDM). This model can predict dust deposition rates as well as the more common parameter of interest in air quality studies, the ambient concentration of fine particles. Estimates of the existing, background dust deposition rate were in the range 10–50 mg m^{-2} day^{-1} (averaged over a month), with higher peaks close to agricultural tilling. Emission factors for particulates from various dust-generating activities at the site (Table 18.8) were derived from an authoritative database (US EPA 1991).

FDM makes the basic assumption that deposition from the dust plume is proportional to the airborne dust concentration close to the ground. Deposition is treated through two parameters: the gravitational settling velocity and the deposition velocity. The gravitational settling velocity accounts for the removal of particulate matter from the atmosphere due to gravity. Since only the larger particles have sufficient mass to overcome turbulent eddies, this mechanism is significant only for the larger size ranges. The deposition velocity accounts for the removal of particles by all other means, including impaction and adsorption.

The FDM model can treat point, area, and line sources. For the purpose of this study, aggregate dropping, loading onto dumper trucks, and localized scraping of surface material were treated as point sources. Emissions of dust from haul routes and perimeter bunds were treated as line sources, and wind blow from stockpiles was treated as area sources.

Methodology

The village of Little Blakenham lies to the northeast of the proposed site for the opencast coal mine, downwind of the prevailing winds in the area (as shown by a 10-year wind-rose representative of the locality: Fig. 18.2). These long-term meteorological data were input to the FDM model, such that the results of deposited dust concentrations represent a 10-year mean. The emissions data were based on the year of most intense mining activity at the site, to represent a worst case. Examples of the emission factor equations are shown in Table 18.8 (US EPA 1991). Information on site activities (e.g. number of vehicle movements on haul routes, quantities of material handled) necessary for calculating particulate emission rates from the emission factor equations was provided by the developer, and is summarized in Table 18.9. It is acknowledged that the most significant sources of dust at surface mineral sites are mechanical handling operations and haulage of material on unsurfaced site roads (DoE 1995b).

Dust deposition rates were modeled for a number of key properties in the village close to the proposed opencast site, and also over a uniform grid of receptors extending 3 km in all directions from the site with spacings of 10 m.

Results and assessment

A dust deposition rate of 200–350 mg m^{-2} day^{-1} (averaged over a month) is frequently cited in the UK as the level above which nuisance complaints may arise (IIRS 1991). Washington state has adopted a nuisance standard of 187 mg m^{-2} day^{-1}, that in use in West Australia is 133 mg m^{-2} day^{-1}, and the standard in Germany is 350–650 mg m^{-2} day^{-1}. However, it should be noted that these various standards apply to sampling equipment with different efficiencies.

The maximum predicted long-term dust deposition rate at a sensitive location was predicted as 16 mg m^{-2} day^{-1}, occurring at the closest property to the northeast of the opencast site, at a distance of 100 m from the site boundary (Fig. 18.3). Even if this is

Table 18.8 Examples of US EPA emission factor equations for particulates <30 μm.

Vehicle movements on unpaved haul routes	E (kg vehicle^{-1} km^{-1}) = k 1.7 (s/12) (S/48) $(W/2.7)^{0.7}$ $(w/4)^{0.5}$ $(365-p)/365$	k, particle size multiplier, e.g. 0.8 for <30 μm; s, silt content; S, average vehicle speed; W, average vehicle weight; w, number of tyres; p, number of days in year with >0.25 mm of rainfall
Bulldozing/scraping	E (kg h^{-1}) = $(2.6s^{1.2})/M^{1.3}$	s, silt content; M, moisture content
Material handling/transfer	E (kg t^{-1}) = $[k\,0.0016\,(U/2.2)^{1.3}]/(M/2)^{1.4}$	k, particle size multiplier (e.g. 0.8 for <30 μm); M, moisture content; U, wind speed, read directly from meteorological file

(*Continued on p. 514*)

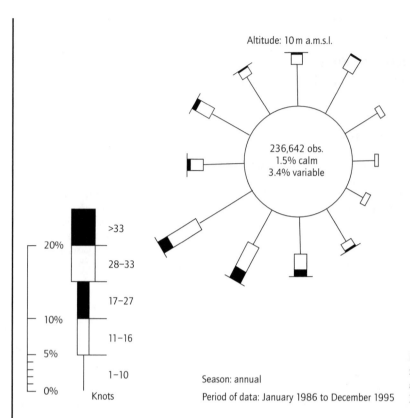

Altitude: 10 m a.m.s.l.

236,642 obs.
1.5% calm
3.4% variable

>33
28–33
17–27
11–16
1–10

20%
10%
5%
0%

Knots

Season: annual

Period of data: January 1986 to December 1995

Fig. 18.2 Ten-year wind rose for meteorological observing station representative of Little Blakenham.

Table 18.9 Operational details and site characteristics for proposed opencast coal site in year of peak activity.

Working hours	7 a.m. to 7 p.m., Monday to Friday 7 a.m. to 1 p.m., Saturday	Working hours	7 a.m. to 7 p.m., Monday to Friday 7 a.m. to 1 p.m., Saturday
Duration of topsoil, subsoil, and overburden removal	4 weeks	Moisture content of topsoil, subsoil, and overburden	8%
Maximum height of overburden stockpile	4 m	Silt content of all materials handled	4%
		Moisture content of coal stockpiles	6%
Time to build overburden stockpile to maximum height	20–25 weeks	Height of coal stockpiles	4 m
		Average weight of vehicles on haul routes	30 t
Area of overburden stockpile	2500 m^2		
Total area of perimeter bunds	2000 m^2	Number of wheels for site vehicles	6
Time to construct perimeter bunds	4 weeks	Average speed of vehicles	20 km h^{-1}
Maximum depth for excavation	30 m	Area of coal stockpiles	1000 m^2
Time to reach maximum depth	30 weeks	Number of haul routes	4
Vehicle movements on haul routes	10 vehicles h^{-1} as a daily average on each haul route	Total length of haul routes	500 m

(Continued)

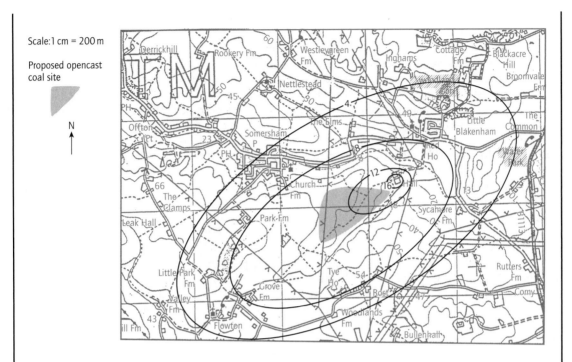

Fig. 18.3 Isopleths of predicted long-term dust deposition rates (mg m^{-2} day^{-1}) contributed by proposed opencast coal mine. (With kind permission of the Ordnance Survey © Crown Copyright. NC/00/971.)

added to a background dust deposition rate of 50 mg m^{-2} day^{-1}, the various nuisance criteria are not exceeded.

Conclusions

The dispersion modeling has shown that operations at the proposed opencast coal site are unlikely to cause soiling nuisance at nearby properties. However, the study did not model the contribution of the opencast site to ambient fine particulate concentrations, which would be necessary for a comparison against health-based air quality standards for airborne PM$_{10}$ concentrations. The study did not attempt to take into account the reduction in emission rates that could be achieved by means of mitigation measures, e.g. good site practices such as watering of haul routes and covering of stockpiles.

meaningful to attempt to predict the highest one-hour mean NO$_2$ concentration close to a road if only daily average traffic data were available; the peak hour vehicle flows and speeds (or at least some estimate of these quantities) are required. Similarly, the contribution of a power station to an observed SO$_2$ episode is unlikely to be indicated by modeling based on the annual average SO$_2$ emission rate; information is required on the variation of SO$_2$ release rates on an hourly basis, coincident with the period of the pollution episode. Advanced models can take account of diurnal, daily, and monthly variations in the emission parameters, based on traffic flow changes and/or industrial output cycles, fuel use scenarios, etc. Such a model may be used if the input data are available at this level of detail.

18.7 OUTPUT DATA
AND INTERPRETATION

The output from dispersion models usually consists of concentration values, which can be presented in many different ways. Concentrations are usually expressed as milligrams or micrograms per cubic meter of air ($mg\,m^{-3}$ and $\mu g\,m^{-3}$ respectively). Concentrations of polluting gases may also be expressed as ratio of the volume of the substance to the volume of air—parts per million (p.p.m.) or parts per billion (p.p.b.).

The concentration of a pollutant will vary almost instantaneously because of the turbulence in the atmosphere. For practical purposes, concentrations calculated by dispersion models are expressed as averages over specified time periods (Beychok 1994). Concentrations are usually calculated over sequences of one-hour periods, each with an associated meteorological condition and, if known, different hourly emission rates. Sequences of calculated hourly average concentrations are not always of direct use in air pollution applications and can be processed in various ways. The annual mean concentration is the simplest long-term output result from dispersion models.

18.8 BACKGROUND AIR QUALITY

Dispersion models can only predict ground-level concentrations arising from the sources that have been input to the model. In all situations, there will be an additional pollutant component arising from those sources that have not been included. These may be local sources (such as nearby roads) or distant sources (in other parts of the country or even neighboring countries). The background component can be significant in determining whether air quality standards are exceeded or not, and must be given careful consideration.

For situations where the annual average concentration is predicted, it is relatively straightforward to include the background component. For example, modeling predictions for a road in a town may indicate a maximum annual average benzene contribution of 1 p.p.b. from the road at the nearest recep-

tor. Data from a nearby automatic monitoring station, but away from the influence of the road being modeled, may indicate that the annual average is about 0.5 p.p.b. (i.e. the urban background concentration). The total concentration can be calculated by simply adding $1 + 0.5 = 1.5$ p.p.b. The annual average concentration modeled from a chimney stack could be treated in exactly the same way.

However, it is more difficult to assess the effects of the background component where peak one-hour concentrations are being considered. In this case, it is necessary to treat low-level (i.e. roads) and high-level (i.e. chimneys) sources in a different way.

For example, the maximum one-hour concentration of NO_x predicted by a dispersion model to arise from the road may be 40 p.p.b. under worst-case atmospheric conditions. These conditions are likely to be the same as those giving rise to an elevated background, e.g. calm, stable conditions during the wintertime. Data from a nearby automatic monitoring station, away from the direct influence of the road, may indicate that the maximum wintertime one-hour concentration is about 75 p.p.b. The total concentration can be estimated by adding $75 + 40 = 115$ p.p.b.

In a different example, the maximum one-hour mean NO_x contribution from a chimney stack may be predicted by a dispersion model to be 50 p.p.b. In this case, the weather conditions that cause the highest concentration from the stack (high wind speed, convective conditions) are unlikely to be the same as those causing an elevated background—during calm, stable conditions the impact of the stack emissions is likely to be very low. The total concentration is therefore estimated by adding the peak one-hour stack contribution to the annual average background, where the latter is again estimated from a suitable urban background monitoring station.

18.9 CHOICE OF
DISPERSION MODEL

The Royal Meteorological Society has published guidelines that seek to promote best practice in the

Case study 5   **Validation of two road traffic dispersion models against ambient monitoring data**

Key features of study

A city council needed to know how reliable the results of two road traffic dispersion models (AEOLIUSF and CAR INTERNATIONAL) would be for predicting carbon monoxide concentrations for the city center. The local authority would then be in a position to choose the better tool in fulfilling its urban planning and regulatory duties with respect to air pollution (DETR 1997). For this reason, a validation study was undertaken of the two models, by comparing model predictions against continuous monitoring data for carbon monoxide, collected over a period of three months at a kerbside site on Straight Street.

Methodology

The available traffic data for Straight Street are shown in Table 18.10. In the absence of further details, it was assumed that these flows were representative of every day in the week over the three-month study period. A constant speed of $15 \, km \, h^{-1}$ was defined for AEOLIUSF and a canyon height of 20 m was assumed for Straight Street.

It is noteworthy that both AEOLIUSF and CAR INTERNATIONAL can treat only a single road, whereas the monitoring data will reflect the influence of emissions from many roads in the city center. Therefore, care must be taken to estimate a realistic urban background concentration for the assessment. The three-month mean background carbon monoxide concentration was estimated at 0.4 p.p.m. from data collated from another continuous monitoring station in the city center, which was distant from all major roads.

Traffic data collated by local authorities often do not include a detailed breakdown of vehicle types, to correspond with the more detailed information on vehicle emission factors that are available (DETR 1998). Therefore, assumptions need to be made on the vehicle mix based on national statistics. The assumptions made for this study were as follows:
- half the HGVs were medium goods vehicles (3.5–7.5 t) and half were 7.5–17 t;
- of the LGVs, half had gasoline engines and half had diesel engines;
- of the automobiles, 80% had gasoline engines with catalytic converters, and 20% had diesel engines.

The carbon monoxide emission factors are shown in Table 18.11. These were the most up-to-date figures for urban situations available from national government (DETR 1998). The Dutch emission factors in CAR INTERNATIONAL were overwritten.

The only meteorological observing station with hour-by-hour readings of the necessary meteorological parameters (as required by AEOLIUSF) was 10 km from the city center (at the local airport), with very different physiography. However, no other suitable meteorological data were available. The format of the meteorological data file required by AEOLIUSF is shown in Table 18.12. CAR INTERNATIONAL requires only an average wind speed over the period of interest; the three-month mean wind speed was found to be $4.3 \, m \, s^{-1}$.

The monitoring station recorded carbon monoxide concentrations as 15-minute averages. These data were processed to derive the three-month mean concentration, the maximum one-hour and eight-hour means, and the 98th percentile of 24-hour, eight-hour, and one-hour means, for comparison against the outputs of AEOLIUSF and CAR INTERNATIONAL.

Table 18.10   Available traffic data (vehicles h^{-1}) for Straight Street.

Hour	Cars	Buses	Light goods vehicles (LGV)	Heavy goods vehicles (HGV)	Motorcycles	Total
08.00–09.00	1185	176	115	33	5	1514
09.00–10.00	906	186	146	42	3	1283
10.00–11.00	667	163	121	68	11	1030
11.00–12.00	670	165	129	61	8	1033
12.00–13.00	664	154	108	50	8	984
13.00–14.00	679	151	107	44	4	985
14.00–15.00	762	164	120	45	5	1096
15.00–16.00	740	166	126	60	6	1098
16.00–17.00	923	186	118	47	7	1281
17.00–18.00	854	179	73	29	4	1139

No traffic data available for period 18.00–08.00 hours.

(Continued on p. 518)

Results and assessment

CAR INTERNATIONAL underpredicted the three-month mean carbon monoxide concentration by 25% (Table 18.13), but gave

Table 18.11 Carbon monoxide emission factors (urban driving cycle) used in CAR INTERNATIONAL and AEOLIUSF models.

Vehicle type	Emission factor ($g\,km^{-1}\,vehicle^{-1}$)
Motorcycles	22
Diesel cars	0.08
Gasoline cars (catalysts)	4.13
Gasoline light goods vehicle (LGV)	13.57
Diesel light goods vehicle (LGV)	0.26
Bus/coach	17.65
Medium goods vehicles (3.5–7.5 t)	4.39
Heavy goods vehicle (HGV) (7.5–17 t)	5.25

good agreement against monitoring data for the 98th percentile of 24-hour means and the 98th percentile of eight-hour means. However, the model underpredicted the 98th percentile of one-hour means.

AEOLIUSF underpredicted the three-month mean concentration (Table 18.13), but this could be due to the assumption of zero traffic over the nighttime period, as traffic data were available only for the daytime. AEOLIUSF also underpredicted the maximum running eight-hour mean concentration and the maximum one-hour mean concentration.

Conclusions

The validation study showed that both models had a tendency to underestimate carbon monoxide concentrations for the city center. The factors that could be responsible for these underpredictions are:

1 The carbon monoxide emission factors published by the government could have been too low, and not suitable for con-

Table 18.12 Example of meteorological data (one day) required by AEOLIUSF.

Year	Month	Day	Hour	Direction	Temperature	Pressure	U_{10}
1997	4	1	0	230	9.5	1013	8.7
1997	4	1	1	230	9.2	1013	9.3
1997	4	1	2	220	9.6	1013	7.7
1997	4	1	3	220	9.1	1013	7.2
1997	4	1	4	230	9.6	1013	7.2
1997	4	1	5	240	8.9	1013	6.2
1997	4	1	6	240	9.4	1013	8.2
1997	4	1	7	260	9.3	1013	7.7
1997	4	1	8	260	9.4	1013	6.7
1997	4	1	9	260	9.8	1013	6.7
1997	4	1	10	270	10.8	1013	8.2
1997	4	1	11	260	10.9	1013	8.2
1997	4	1	12	270	10.5	1013	7.7
1997	4	1	13	240	10.6	1013	7.7
1997	4	1	14	250	10.8	1013	8.2
1997	4	1	15	260	11.0	1013	8.2
1997	4	1	16	270	10.2	1013	7.2
1997	4	1	17	260	10.1	1013	7.2
1997	4	1	18	260	8.9	1013	6.7
1997	4	1	19	250	8.4	1013	5.1
1997	4	1	20	250	8.3	1013	5.1
1997	4	1	21	250	8.8	1013	5.7
1997	4	1	22	250	8.7	1013	7.2
1997	4	1	23	270	8.7	1013	6.2

(Continued)

Table 18.13 Comparison of carbon monoxide modeling results against monitoring data.

Statistic (p.p.m.)	Measured concentration (monitoring data)	Modeling results*	
		AEOLIUSF	CAR INTERNATIONAL
Maximum running eight-hour means	3.3	2.8	–
Maximum one-hour mean	5	4.2	–
Three-month means	1.1	0.7	0.8
98th percentile of 24-hour means	1.8†	–	2.0
98th percentile of eight-hour means	2.4‡	–	2.3
98th percentile of one-hour means	4.9	–	2.7

* Including an allowance for background concentrations.
† 98th percentile of daily means.
‡ 98th percentile of running eight-hour means.

gested, city-center driving conditions. These published emission factors were not speed related.

2 The estimate of the background carbon monoxide concentration may have been too low.

3 The assumptions made on the details of vehicle mix on Straight Street could have been in error. Straight Street may have had a vehicle type distribution very different from that of the national trunk roads.

It would be advisable for the local authority to repeat the validation exercise after making suitable adjustments and refinements to these factors in turn (i.e. a sensitivity analysis), before either model could be used routinely for modeling air quality in the city center for regulatory purposes.

use of mathematical models of atmospheric dispersion (DoE 1995b). Should the results of a screening model suggest that an air quality standard could be approached, it would be prudent to use a computerized Gaussian model. Some areas may be prone to unusual weather patterns (e.g. coastal locations) or there may be complex topographical features (e.g. deep valleys): in these cases, the impact of even small industrial plant may need to be investigated using models that can incorporate these effects. If the polluting potential of a proposed development is high (e.g. NO_x from a large power station in a densely populated urban area), if there are other significant sources nearby (e.g. majors road and an incinerator), and if local background concentrations are known (from

monitoring) to be already close to the air quality standard, then the use of a multisource model, which can treat both chimney and road traffic emissions, would be advisable.

Screening models may be most appropriate when the meteorological data lack the reliability and accuracy that would warrant the use of more sophisticated dispersion models; screening models usually require little or no meteorological input, and therefore cannot take account of local conditions.

When providing meteorological data to the more advanced computerized models, careful consideration needs to be given to the choice of meteorological observing station from which sequential or statistical data sets are purchased;

geographic proximity alone may not be the only criterion. Coastal influences, and the situation with respect to hills and valleys, can have profound effects on the local wind flow and turbulence.

The choice of model depends on the relevant averaging time required of the results. If the objective is to compare the modeling results against some available monitoring data (e.g. to validate the model predictions for a particular situation, and therefore lend it some credibility as a forecasting tool—see Case study 5), then the model results should be for the same averaging periods as the measurements. The receptors for the modeling study must be coincident with the monitoring equipment; accurate geographic referencing of sources and receptors is required. If the purpose of the dispersion modeling is to assess compliance with national air quality standards, the dispersion model should, ideally, generate results for the same averaging periods as the standards. However, only a few of the advanced models can generate percentile statistics. Empirical corrections to predictions for other sampling periods can be applied provided that there are sufficient data from monitoring sites in similar locations (DETR 2000b). A few models have a concentration fluctuations module, which can be used to model averaging periods of less than an hour; for other models, appropriate empirical corrections may be applied to the predicted one-hour mean results to estimate shorter-term concentrations (see Case study 1).

18.10   ACCURACY OF DISPERSION MODELING PREDICTIONS

The accuracy of model predictions depends mostly on the quality of the input data. It is not possible to calculate accurately the concentrations for a specified one-hour time interval because one cannot know with any certainty (unless there are very detailed measurements available) which meteorological and emission conditions applied precisely to that particular interval. Dispersion models have been designed to calculate the average concentration over a large number of time intervals subject to

similar meteorological conditions. There are real fluctuations in concentrations about the average caused by variations in meteorological and emission conditions that are not treated within a model. This issue is not significant when one is using a model to calculate long-term average concentrations over, say, one year. If the prediction of an annual mean value lies within ±50% of the measurement, the results of a dispersion model would usually be deemed to be acceptable. However, dispersion models are less accurate in their predictions of maximum short-term averaged concentrations and high percentile values (see Case study 5). Moreover, most dispersion models have been compiled to overestimate concentrations, in order to ensure a conservative approach for regulatory purposes.

Different models designed to calculate the same quantity will produce different answers. The reason why there is not a single right answer is that model developers have to make choices about the formulae they use and choose from the literature one that is practical, has been tested against measurements (i.e. validated), and is consistent with the other parts of the model; not all model developers will make the same choice. Some will place greater emphasis on one aspect rather than another and this is why models differ.

APPENDIX: LIST OF MODELS

This list of commonly used atmospheric dispersion models is provided as a guide. The list is not exhaustive, and many other models with comparable features are reported in the scientific literature.

Traffic emissions models

The **Design Manual for Roads and Bridges** (DMRB) is a screening model developed by the UK Transport Research Laboratory for single roads, based on a series of graphs and tables. It predicts pollutant concentrations in the averaging periods and statistics directly comparable against UK air quality standards and objectives. The model contains UK vehicle exhaust emission factors, which are speed- and year-related.

Case study 6 **Prediction of short-term NO₂ concentrations close to a road**

Key features of study

A residential property was situated 10 m from the kerb of a busy, straight road. The one-hour mean NO_2 concentrations during the most common weather conditions at the site, during rush hour traffic conditions, were predicted using the CALINE4 dispersion model. CALINE4 incorporates a photochemistry module, which calculates downwind NO_2 concentrations depending upon the distance, wind speed, ambient temperature, and background concentrations of NO, NO_2, and O_3.

Methodology

The input data required by the model are summarized in Table 18.14. The NO_x emission rate is determined by a number of factors, including the vehicle mix along the road during the peak hour (categorized by means of the number of heavy duty vehicles and light duty vehicles) and the vehicle speed during the rush hour. The weighted mean NO_x emission rate is derived from the LDV and HDV emission rates at vehicle speeds of 40 km h⁻¹. The most commonly occurring stability and wind speed combination at the site (based on nearby synoptic observations) was Pasquill category D, with a wind speed of 7 m s⁻¹. A worst-case wind direction directly from the road to the receptor was assumed for the modeling. The background concentrations of NO, NO_2, and O_3 were derived from empirical monitoring data to describe conditions relevant to neutral stability and moderate or high wind speed conditions.

Results and assessment

The predicted NO_2 concentration at the property was 110 p.p.b. There is no need to take separate account of the background level of NO_2 as this was automatically considered by the model. The

Table 18.14   Input data for CALINE4 dispersion model.

Peak hour 2-way flow of light duty vehicles (LDV)	2000 vehicles h⁻¹
Peak hour 2-way flow of heavy duty vehicles (HGV)	150 vehicles h⁻¹
Total peak hour 2-way vehicle flow	2150 vehicles h⁻¹
Peak hour vehicle speed	40 km h⁻¹
Weighted mean NO_x emission rate	4.98 g km⁻¹ vehicle⁻¹
Atmospheric stability	Neutral (Pasquill category D)
Wind speed	7 m s⁻¹
Ambient temperature	15°C
Background concentrations	NO = 10 p.p.b. NO_2 = 25 p.p.b. O_3 = 35 p.p.b.
Wind direction	Perpendicular to road, directly toward receptor

predicted NO_2 concentration exceeds the WHO air quality guideline of 105 p.p.b.

Conclusions

The NO_2 concentration at the property arising from peak-hour NO_x emissions along the road would give cause for concern. If the WHO guideline is exceeded during neutral, moderate wind speed conditions, the margin of exceedance would be even greater during stable, low wind speed weather conditions (e.g. cold foggy mornings and evenings), which could coincide with peak-hour traffic flows. A detailed assessment of the frequency of the worst-case wind direction (i.e. directly from the road to the receptor) would be helpful in assessing the likely frequency of breaches of the air quality guideline.

CAR INTERNATIONAL is a screening model developed by TNO Environmental Sciences and the RIVM National Institute of Public Health and the Environment for Dutch regulatory purposes. The model is widely used in the Netherlands. The model calculates a long-term average concentration and the 98th percentile of one-hour, 24-hour, and eight-hour means.

AEOLIUSF is a screening model that has been developed by the UK Meteorological Office for sin-

gle, street canyon situations. It calculates hour-by-hour concentrations based on hourly meteorology read from a preprocessed data file.

CAL3QHC is a US EPA model designed to handle near-saturated and/or overcapacity traffic conditions and complex intersections. It predicts one-hour mean concentrations.

The **California Line Source Model (CALINE)** was developed by the California Department of Transportation and the US Federal Highways

Agency. It can model junctions, street canyons, parking lots, bridges, and underpasses; it includes a photochemistry model to predict downwind concentrations of NO_2 from NO emitted by vehicle exhausts. CALINE predicts one-hour mean concentrations, and hourly meteorological conditions are user-defined.

The **Point, Area and Line (PAL) Source Model** uses the same Gaussian line source algorithms as CALINE, but has special area source algorithms for treating edge effects more accurately. This facility is useful for modeling parking lots, small areas of cities, and airports.

Industrial source models

D1 Stack Height Calculations is a screening procedure based on nomograms, which estimates a chimney height that would ensure that ground-level concentrations of a pollutant did not exceed a specified standard or guideline for that pollutant for more than about five minutes, under weather conditions that are likely to occur 98% of the time.

Guidance for Estimating the Air Quality Impact of Stationary Sources is a screening method published by the UK Environment Agency that provides precalculated dispersion results for stack emissions expressed as nomograms. These are based on a large number of computations using the ADMS model. The predicted pollutant concentrations are directly comparable against UK air quality objectives.

SCREEN is a US EPA model that uses worst-case meteorological data. It can model a single point, area, or volume source, and can take account of building wake effects. It also has a limited ability to treat terrain above stack height.

R91 is a Gaussian model developed by a Working Group led by the UK National Radiological Protection Board. It has been used in the past by UK regulatory bodies as a reference model. R91 originally consisted of nomograms, but PC versions are commercially available. DISTAR can treat only flat terrain and cannot model building wake effects, although reports by NRPB show how the basic R91 model can include these effects. The

model calculates one-hour, monthly, and annual averages, and it uses statistical meteorological data.

The **Rough Terrain Diffusion Model** (RTDM) is a US EPA Gaussian model capable of predicting short-term concentrations arising from point sources in complex terrain. It calculates one-hour averages only. RTDM requires on-site hourly measurements of turbulence intensity, vertical temperature difference, horizontal wind shear, and wind profile exponents.

The **Industrial Source Complex** (ISC) is a US EPA multisource Gaussian model, capable of predicting both long-term and short-term concentrations arising from point, area, and volume sources. Gravitational settling of particles can be accounted for using a dry deposition algorithm; wet deposition and depletion due to rainfall can also be treated. Effects of buildings can be considered. The model has urban and rural dispersion coefficients.

The **Atmospheric Dispersion Modeling System** (ADMS) is a "new generation" multisource dispersion model from the UK. Specific features include the ability to treat both dry and wet deposition, building wake effects, complex terrain, and coastal influences. ADMS allows the use of point, area, volume, and line sources, and can predict both long-term and short-term (down to one-second mean) concentrations. Urban and rural dispersion coefficients are included, and percentile calculations are possible.

AERMOD is a US EPA update of the ISC model, intended to be the "new generation" model for US regulatory purposes. It contains improved algorithms for convective and stable boundary layers, for computing vertical profiles of wind, turbulence and temperature, and for the treatment of all types of terrain.

The **INDIC AirViro** system differs from other PC-based models in that it requires a UNIX workstation and needs complex physiographic and meteorological configuration by the software supplier. Unlike most Gaussian models which rely upon meteorological information collected from a single site, the INDIC AirViro model describes a pattern of small-scale winds based upon the surface characteristics. The INDIC AirViro model interfaces with a sophisticated emissions database capable of accepting point sources (i.e. stacks), area sources, and line sources (i.e. traffic), and detailed diurnal, seasonal, and production variations of emissions (both traffic and industrial). The INDIC system can be applied using a number of Gaussian model options, and a street canyon model option.

The **Fugitive Dust Model** (FDM) is a US EPA Gaussian model for predicting the dispersion and deposition of particulate matter, e.g. dust from surface mineral workings. Point, line, and area sources can be modeled, using long-term statistical frequency analyses of meteorological conditions. The particle size distribution of the emissions must be user-defined. Emission factors are usually derived from a US EPA database.

REFERENCES

Bange, P., Janssen, L.H.J.M., Nieuwstadt, F.T.M., Visser, H. & Erbrink, H.H. (1991) Improvement of the modeling of daytime nitrogen oxide oxidation in plumes by using instantaneous plume dispersion parameters. *Atmospheric Environment* **25A**, 2321–8.

Benson, P.E. (1992) A review of the development and application of the CALINE3 and 4 models. *Atmospheric Environment* **26B**, 379–90.

Beychok, M.R. (1994) *Fundamentals of Stack Gas Dispersion*. Irvine, Homewood, IL.

Buckland, A.T. & Middleton, D.R. (1997) Nomograms for calculating pollution within street canyons. *Atmospheric Environment* **33**, 1017–36.

DoE (1995a) *The Environmental Effects of Dust from Surface Mineral Workings*. Department of the Environment Minerals Division, London.

DoE (1995b) *Royal Meteorological Society Policy Statement. Atmospheric Dispersion Modeling: Guidelines on the Justification and Use of Models, and the Communication and Reporting of Results*. Department of the Environment, London.

DETR (1997) *Review and Assessment of Air Quality for the Purposes of Part IV The Environment Act*. Department of the Environment, Transport and the Regions, London.

DETR (1998) *The National Air Quality Archive* (http://www.environment.detr.gov.uk/airq/aqinfo.htm).

DETR (2000a) *The Air Quality Strategy for England, Scotland, Wales and Northern Ireland*. Department of the Environment, Transport and the Regions, London.

DETR (2000b) *Review and Assessment: Pollutant Specific Guidance, LAQM.TG4 (2000)*. Department of the Environment, Transport and the Regions, London.

DMRB (1999) *Design Manual for Roads and Bridges*. Department of Transport, London.

EA (1998) *Guidance for Estimating the Air Quality Impact of Stationary Sources*. Report no. GN24 National Center for Risk Analysis and Options Appraisal. Environment Agency, London.

Fisher, B.E.A., Moorcroft, S. & Vawda, Y. (1999) Air quality dispersion modeling: issues which arise during the review and assessment process. *Clean Air* **29**(2), 38–41.

Hall, D.J. & Kukadia, V. (1993) Approaches to the calculation of discharge stack heights for odour control. *Clean Air* **24**(2), 74–92.

HMIP (1996) *Validation of the UK-ADMS Dispersion Model and Assessment of Its Performance Relative to R91 and ISC Using Archived LIDAR Data*. HMIP Commissioned Research, DoE Report No. DoE/HMIP/RR/95/022. DoE/HMIP, London.

IIRS (1991) *Impact on Atmospheric Environment of Proposed Underground Zinc/Lead Mine*. Report R6/6599. Irish Institute for Industrial Research and Standards, Dublin.

Janssen, L.H.J.M., van Wakeren, J.H.A., van Duuren, H. & Elshout, A.J. (1988) A classification of NO oxidation rates in power station plumes based on atmospheric conditions. *Atmospheric Environment* **22**, 43–53.

Middleton, D.R. (1995) *Operation of a New Box Model to Forecast Urban Air Quality: BOXURB*. Meteorological Office, Bracknell.

Turner, D.B. (1994) *Workbook of Atmospheric Dispersion Estimates: An Introduction to Dispersion Modeling*. CRC Press, Boca Raton, FL.

US EPA (1991) *Compilation of Air Pollutant Emission Factors. Volume I, Stationary Point and Area Sources*, 4th edn, supplement B. US Environmental Protection Agency, Washington, DC.

Vawda, Y. (1998) Urban emissions inventories: their uses and limitations in dispersion modeling. *Clean Air* **28**(5), 149–53.

Weil, J.C. (1992) Updating the ISC model through

AERMIC. In *Proceedings of the 85th Annual Meeting of Air and Waste Management Association*. Air and Waste Management Association, Pittsburgh, PA.

WHO (1987) *Air Quality Guidelines for Europe*. World Health Organization Regional Office for Europe, Copenhagen.

WHO (1994) *Update and Revision of the Air Quality Guidelines for Europe*. World Health Organization Regional Office for Europe, Copenhagen.

Woodfield, M. & Hall, D. (1994) *Odour Measurement and Control: An Update*. AEA Technology and DoE, London.

19 Climate Modeling

WILLIAM LAHOZ

19.1 INTRODUCTION

It is well known from the past record that climate change has taken place on a wide range of time scales, the best known example being the alternation of glacial and interglacial periods at intervals of about 100,000 years over the past half million years. This climate variation has a large impact on human activities and the human economy. Occurrence of extreme events in precipitation leading to droughts or floods has always been a cause of concern, and more so in recent years with the increasing pressure on finite food resources. That climate change can be anthropogenic is an important current concern. An example is provided by the increases in greenhouse gas concentrations since preindustrial times (i.e. since about 1750), which have tended to warm the Earth's surface and to produce other changes in climate (IPCC 1996).

The World Meteorological Organization (WMO) has established a number of initiatives to address climate change issues. Two of these are the World Climate Research Program (WCRP) and the Intergovernmental Panel on Climatic Change (IPCC). The WCRP was established to determine: (i) to what extent climate can be predicted; and (ii) the extent of human influence on climate. The remit of the WCRP concerns long-range weather predictions and variability from interannual to multidecadal time scales. The IPCC was established to: (i) assess scientific information on climate change; (ii) assess environmental and

socioeconomic impacts of climate change; and (iii) formulate response strategies.

When modeling climate change it is not enough to just consider the atmosphere. Instead, all components of the Earth's climate system must be considered. These components are: (i) the atmosphere (which comprises the gaseous envelope of the Earth); (ii) the oceans; (iii) the cryosphere (which comprises the continental ice, mountain glaciers, surface snow cover and sea ice); and (iv) the land surface mass (which includes the biomass within or above it). The interactions between different components of the Earth's climate system occur in many ways and over a number of time scales, and climate models must be able to capture them.

A large number of feedback processes may be identified within the interactions which occur within the Earth's climate system. Some act to amplify variations in the system (positive feedback), others act to dampen them (negative feedback). Important examples of simple feedbacks include: (i) ice–albedo feedback; (ii) water vapor–radiation feedback, and (iii) cloud–radiation feedback. These feedbacks also must be considered when modeling climate change.

The questions associated with modeling climate change are two: what can we predict; and how far into the future can we predict? These questions presuppose, first, that there exist modeling tools that are useful for making climate predictions; second, that there are observations, albeit not error-free, that enable the evaluation of predictive skill, and, third, that there are limitations to our predictive capability. The fact that these limitations exist is a key issue of climate modeling, relating to the nature and range of climatic phenomena we can predict using climate models, and the space and time scales over which these phenomena can be predicted.

Variability is a very important aspect of the behavior of the Earth's climate system and has important implications for the detection of climate change, which involves distinguishing between "natural" and "anthropogenic" variability. This natural variability occurs at a variety of space and time scales, and can be purely internal (due to complex interactions between individual components of the Earth's climate system, such as the atmosphere and ocean) or externally driven by changes in solar activity or the volcanic aerosol loading of the atmosphere.

Any "signal" of anthropogenic effects on climate must be distinguished from the background "noise" of climate fluctuations that are entirely natural in origin and occur on a variety of space and time scales. It is difficult to separate a signal from the noise of natural variability in the observations. This is because there are large uncertainties in the evolution and magnitude of both anthropogenic and natural forcings of the Earth's climate system, and in the characteristics of natural variability, which translate to uncertainties in the relative magnitudes of signal and noise. Shine and Forster (1999) give a review of the anthropogenic forcings of climate change.

The models used for climate predictions also exhibit variability. If the models provide a faithful representation of the Earth's climate system, it is reasonable to assume that, if the model is run under present climatic conditions, this variability will be similar to that exhibited by the current climate record. Similarly, one expects that the variability of the model, when run under past climatic conditions, would be similar to that exhibited by the appropriate proxies of the past climate record.

Although climate models may not reproduce all components of the observed variability, there are a number of factors that provide confidence in their ability to simulate important aspects of anthropogenic change in response to anticipated changes

in atmospheric composition (IPCC 1996). These include:

1 The most successful climate models simulate well the important large-scale features of the components of the Earth's climate system.

2 Many climate changes are consistently reproduced by different models in response to greenhouse gases and aerosols and can be explained in terms of physical processes that are known to be operating in the real world.

3 The models reproduce with reasonable fidelity other less obvious climatic variations due to changes in forcing.

4 The model results exhibit natural variability on a wide range of space and time scales that is broadly comparable to that observed.

Climate models are of varying complexity. Some of the most complex models are atmospheric and oceanic general circulation models (AGCMs and OGCMs, respectively). In many instances, AGCMs and OGCMs, developed as separate models, are combined to give coupled atmosphere–ocean general circulation models (CGCMs); these tend to be very complex and require large computer resources to run. All these GCMs aim to provide a picture of the Earth's climate system that is as complete and consistent as possible. Completeness comes from the inclusion of all the significant components of the Earth's climate system. Consistency comes from the correct representation of the interrelationships between these components. The limitations in achieving these goals chiefly arise from: (i) an incomplete understanding of the behavior of the components of the Earth's climate system; and (ii) constraints in available computing power.

GCMs include the physical laws that describe the atmospheric and oceanic dynamics and physics; some AGCMs also incorporate comprehensive photochemistry schemes, and a smaller number are beginning to include biological processes. These models also include representations of land-surface processes, sea-ice related processes, and many other complex processes involved in the Earth's climate system. Many physical processes, such as those related to clouds, take place on much smaller spatial scales than those of the model, and therefore cannot be properly resolved and modeled explicitly. To overcome this limitation, the average effects of these phenomena are included in the model by making use of physically based relationships with the larger-scale variables. This technique is known as parametrization.

There are a number of GCM climate models and details of many of them can be found in Phillips (1994) and references therein. Although these models are continually being updated, e.g. in their vertical extent or in their representation of the dynamics and physics, the versions described in Phillips tend to provide the basis for the latest versions of these models.

Due to the complexity of GCMs it is often difficult to establish cause and effect of phenomena; furthermore, it is often expensive to run the computer experiments required for the comprehensive analysis of these phenomena. To remedy these shortcomings, simpler models (often simplified GCMs) are used to study climate change. Despite the advantages of ease of understanding and of relatively inexpensive computer costs, these simpler models have a significant shortcoming compared to GCMs, i.e. they are less physically realistic and less complete due to the omission of components of the Earth's climate system.

Examples of simpler physically based models include, in increasing order of complexity: energy balance models (which determine the effective radiative temperature of the Earth's climate system by assuming a balance between the incoming solar radiation and the outgoing longwave radiation), radiative–convective models (which determine the vertical distribution of globally averaged temperatures for the atmosphere and the underlying surface), zonally averaged models (which are two-dimensional models capable of simulating vertical and meridional variations in surface and atmospheric properties averaged around latitude circles), and mechanistic models (e.g. GCMs, which exclude the troposphere). Details of many of these simpler models can be found in Cotton and Pielke (1995) and references therein.

There are a number of models used for prediction at the seasonal time scale (which lies at the interface between the time scales for weather and climate prediction) for the Tropics that are based on empirical associations between elements of the general circulation of the atmosphere and ocean derived from historical data. Such associations are used to define predictors of regional circulation or precipitation, a particular example being the Indian summer monsoon. A comprehensive review is given by Hastenrath (1991).

Empirical models based on relationships between antecedent sea surface temperature (SST) anomalies and upper-air circulation anomalies are also used for prediction at the seasonal time scale over North America (e.g. Livezey 1990); similarly, empirical models based on teleconnections between El Niño–Southern Oscillation (ENSO) events and the European region are used for prediction at the seasonal time scale over Europe (e.g. Fraedrich & Muller 1992).

In this chapter we focus on GCMs and their components. A detailed description is provided in Section 19.2.

Our knowledge of the Earth's climate system ultimately comes from observations. Although observations have uncertainties and biases, they are the "truth" against which theories and models must be confronted and evaluated. Examples of observing platforms include ground-based instruments, balloons, aircraft, and remote sounding satellites. Each platform has advantages and disadvantages. For example, ground-based instruments have high spatial resolution but poor global coverage, whereas satellite observations have low spatial resolution but good global coverage. For observing instruments aboard satellite platforms, nadir sounders have good horizontal resolution but poor vertical resolution, whereas limb sounders have poor horizontal resolution but good vertical resolution. General details of observing platforms and their characteristics can be found in Daley (1991) and references therein.

Predictions of the variability of the Earth's climate system require an understanding of the variability in the observations representing the "truth." This requires observations that are of high quality (i.e. have small errors and biases), consistent (i.e. there is uniformity in the observing system characteristics), and long term (i.e. the results have statistical significance). Global coverage at spatial and temporal resolutions that allow the representation of significant climatic phenomena are further requirements. The heterogeneity of observations, the changes in observing systems, the lack of a significant global observing system before the International Geophysical Year (IGY) of 1957 and the lack of significant satellite observations before the 1970s make the above requirements difficult to meet.

Concerns about these difficulties led Bengtsson and Shukla (1988) and Trenberth and Olson (1988) to propose reanalysis of past atmospheric observations using a fixed, up-to-date data assimilation system that would attempt to combine, in an optimal manner, information from observations and a sophisticated AGCM. Since then a number of reanalyses have been developed and used extensively to study issues such as climate change and stratospheric ozone (O_3) depletion. Examples of existing reanalyses include a 15-year reanalysis for the period 1979–93 by the European Centre for Medium-range Weather Forecasts (ECMWF) (Gibson *et al.* 1997) and a 40-year reanalysis from 1957 by the National Centers for Environmental Prediction (NCEP) in collaboration with the National Center for Atmospheric Research (NCAR) (Kalnay *et al.* 1996).

A detailed description of how the predictive skill of climate models is evaluated and of the role observations play in this evaluation is provided in Section 19.3.

Both the climate models used for prediction and the observations used to evaluate the predictive skill of these models have limitations. A few of these have already been identified: shortcomings in our understanding of the behavior of the components of the Earth's climate system, constraints in available computing power, and uncertainties in observations. Other limitations include the nonlinear nature of the Earth's climate system and uncertainties in the initial conditions and external forcings required to run climate models. All these limitations are important, but a fundamental one is nonlinearity, which can cause the Earth's climate system to exhibit chaotic behavior and be

very sensitive to initial conditions (Lorenz 1963); this limitation profoundly influences the way we predict climate change.

Useful tropospheric weather forecasts can be made using AGCMs for periods of up to 10 days. These forecasts simulate the evolution of weather systems and describe the associated weather. In the lower stratosphere, on the other hand, AGCMs tend to have higher forecast skill than in the mid-troposphere. This can be ascribed to the flow regime in the lower stratosphere being dominated by lower wavenumbers than in the mid-troposphere. This difference between the mid-troposphere and the lower stratosphere illustrates the impact of space scales on the predictive skill of a model.

Similarly, the different time scales in weather forecasts and climate simulations have an impact on the predictive skill of the model. In climate simulations, unlike weather forecasts, the interest is in the longer time scales. which, depending on the situation of interest, can range from the seasonal time scale to the multidecadal time scale. The longer time scales mean that the internal component of the natural variability in the model (which often exhibits the signature of chaotic behavior) tends to have a larger influence on the variability of the model in relation to the influence exerted by the initial conditions and the boundary conditions. This means that at these longer time scales there is a marked tendency for the model to exhibit a nonunique response to the prescribed SST forcing. (Note that there is still an element of predictability in the atmosphere at these time scales due to the slow-varying nature of the ocean and, consequently, the forcing SSTs; this allows the possibility of significant predictive skill at these longer time scales.) Consequently, the deterministic approach often used in weather forecasts is in general not very useful for climate prediction and a statistical approach is required. In this approach, it is the statistics of the Earth's climate system that are of interest rather than the day-to-day evolution of weather. These statistics include measures of variability as well as mean conditions, and are taken over many weather systems and for several months, years, or more.

When a model is used for climate prediction it is first run for many simulated decades without any changes in external forcing in the Earth's climate system. The quality of the simulation can then be assessed by comparing statistics of the mean climate, the annual cycle, and the variability on different time scales with observations of the present climate. This is the control run. The model is then run with changes in external forcing, e.g. with changing greenhouse concentrations. This is the perturbation run. The differences between the perturbation and control runs provide an estimate of the consequent climate change due to changes in that forcing factor. This strategy is intended to simulate changes or perturbations to the Earth's climate system.

An increasingly important aspect of this statistical approach is the idea of ensembles or multiple realizations of the Earth's climate system (e.g. Palmer & Anderson 1994). In this approach, several members of the ensemble are constructed so as to capture a significant fraction of the model's variability. The advent of increased computing power makes this approach feasible and it is rapidly becoming the standard in climate prediction. (Note that because of increased computing power, this ensemble approach is also becoming the standard in numerical weather prediction, or NWP.) Typically, the ensemble members are obtained by starting the model runs from initial conditions one day apart. This provides a rough estimate of the model's variability due to small perturbations in the initial conditions, and also can provide a way of estimating the random variability in the model. Climatic conditions that are highly predictable will tend to be associated with low spread in the characteristics of the ensemble members, whereas conditions that are highly unpredictable will tend to be associated with high spread in the characteristics of the ensemble members. This is related to the concept that repeatability in climatic conditions is associated with high predictive skill. The extent to which the estimates of climate prediction are reliable will depend on ensemble size, and a number of studies have investigated this issue, especially in the context of seasonal prediction (e.g. Buizza & Palmer 1998).

It is important to realize that although there are limitations to our predictive capability, the intelligent use of climate models and observations can still be of value in quantifying the climate response to anthropogenic climate forcing. This quantification must be made within the framework of a statistical paradigm rather than that of a deterministic paradigm. In this statistical paradigm, multiple realizations of the Earth's climate system are obtained and it is their statistical properties, in particular the mean or "climatology" of the ensemble, the standard deviation or "variability" of the ensemble, and the large-scale spatial averages and the long-term temporal averages, that are important.

More details of how climate models are used for climate prediction, including how the modeling experiments are set up and how we analyze the results, is provided in Section 19.4. Details of the latest results concerning the current state of climate change predictions are provided in Section 19.5. Future avenues for the field of climate prediction are outlined in Section 19.6.

19.2 THE MODELING TOOLS

A fundamental assumption of the modeling approach to studying the Earth's climate system is that models provide a realistic representation of this system and can be used as a surrogate laboratory for testing hypotheses. In this section we focus on GCMs and describe their constituent parts; how models are evaluated to see if they provide a realistic representation of the Earth's climate system is described in Section 19.3; how models are used for testing hypotheses is described in Section 19.4.

GCMs attempt to include all components of the Earth's climate system. These components are: (i) the atmosphere; (ii) the oceans; (iii) the cryosphere; and (iv) the land surface mass. In this section we discuss these components and, in the case of the atmosphere, discuss a number of subcomponents, including dynamics, radiation, parametrizations, and chemistry. Figure 19.1 is a schematic representation of how these components and subcomponents interact.

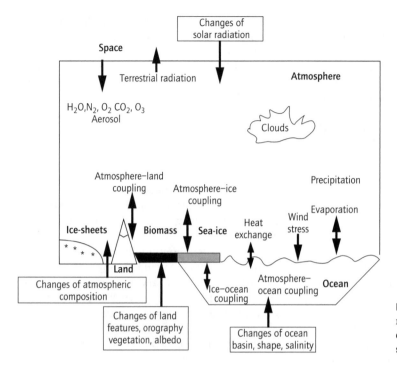

Fig. 19.1 Schematic of the relationship between the components of the Earth's climate system.

19.2.1 Atmosphere

Dynamics

Climate modeling requires a discrete representation of the physical laws that describe the atmospheric and oceanic dynamics and physics. This is because, in general, the equations embodying these physical laws do not have analytic or closed solutions, but can be integrated by discrete numerical methods on computer platforms of varying complexity.

Typically, the continuous equations discretized in AGCMs are the primitive equations, which are an approximation to the Navier–Stokes equations (Andrews *et al.* 1987). The discretization of these equations requires a choice of representation in the horizontal and vertical space dimensions. The choice of horizontal coordinates is not trivial because of the sphericity of the Earth. Grids covering the entire spherical surface with more or less uniformly spaced grid points have been proposed and used. Another approach is to combine two sets of regular polyhedra to cover the Earth's surface.

Nowadays, however, it is more common to use the longitude–latitude grid based on the spherical coordinates.

The primitive equations are often solved in a coordinate system in which geometric height is replaced by a new vertical coordinate that is a function of the hydrostatic pressure. A difficulty with pure pressure coordinates is that constant-pressure surfaces can intersect the lower boundary of the domain of the AGCM (the Earth's surface). A solution is to introduce sigma coordinates, where the pressure is normalized by the surface pressure. An improvement to sigma coordinates is the use of hybrid coordinates, where the vertical coordinate changes from a pure sigma coordinate to a pure pressure coordinate over the vertical range of the domain of the AGCM, with a transition region where a mixture of pure sigma and pure pressure coordinates are used. The specification of the surface pressure depends on the choice of model orography, which, in turn, will depend on the horizontal resolution of the AGCM. Similarly, the land–sea mask in the model will also depend on the horizontal resolution of the AGCM.

There are two conceptually different methods to represent continuous functions on digital computers: as a finite set of grid-point values or as a finite set of series-expansion functions. The grid-point approach is used in conjunction with finite-difference methods. Although the easiest way to implement grid-point methods involves using a grid with uniform resolution, it is often advantageous to use nested grids when the features of interest span a relatively large range of space scale. This technique, called adaptive mesh refinement, is beginning to be implemented in AGCMs, but is not commonly used in climate modeling. Series-expansion methods that are potentially useful in geophysical fluid dynamics include the spectral method, the pseudospectral method, and the finite-element method. The spectral method plays an important role in AGCMs, in which the horizontal structure of the numerical solution is often represented as a truncated series of spherical harmonics. The discrete representation of the continuous equations can either be fixed with respect to the atmospheric flow (this approach is termed Eulerian) or follow the atmospheric flow (in fluid dynamics this approach is termed Lagrangian; as the method employed in climate modeling requires periodic interpolation to a grid, it is termed semi-Lagrangian). Details of these methods can be found in, for example, Durran (1999).

The computation of space and time derivatives in a numerical scheme presents different sets of practical problems. After the nth step of the numerical integration, the numerical solution will be known at every point on the spatial grid, and several grid-point values may be easily included in any finite-difference approximation to the space derivatives. Thus, it is easy to construct high-order centered approximations to space derivatives. In contrast, storage considerations dictate that the numerical solution be retained at as few time levels as possible, and the only time levels available are those from previous iterations. Thus, higher-order finite-difference approximations to the time derivative tend to be one-sided. Despite these differences between space and time

differences in numerical schemes, one must not assume that they are completely independent. Indeed, techniques such as the Lax–Wendroff method (see Durran 1999) cannot be properly analyzed without an understanding of the relationship between space truncation error and time truncation error.

Time difference formulae used in the numerical solution of partial differential equations (PDEs) are related to the methods used to integrate ordinary differential equations (ODEs). In comparison with typical ODE solvers, the methods used to integrate PDEs are of very low-order. There are two basic reasons for this.

1 The approximation of the time derivative is not the only source of finite difference error in the solution of PDEs; other errors arise via the approximation of the space derivatives. Often the largest errors in the solution are introduced via the numerical evolution of the space derivatives, making it pointless to expend effort in developing higher-order time difference schemes.

2 The practical limitations on computational resources often leave no other choice.

Although information from several earlier time levels can be incorporated into the time difference scheme, the large storage requirements of AGCMs preclude the use of data from more than two earlier time levels. Thus, climate modelers have tended to use two- or three-time-level schemes. Durran (1999) discusses the general family of three-time-level schemes, including the leapfrog and Adams–Bashforth schemes. A feature of these two schemes is the presence of a so-called unphysical "computational mode"; Durran discusses ways of handling it. Currently, AGCMs tend to use two-time-level schemes because: (i) provided that they converge to the physical solution they do not have a computational mode; and (ii) they incur smaller computational costs (i.e. they are more efficient). Examples of two-time-level schemes that do not have a computational mode include the forward explicit and the backward implicit schemes. The former is unstable, whereas the latter is stable. A difference scheme is stable if its solutions remain bounded for all sufficiently small time steps (Haltiner and Williams (1980) provide a more rigorous definition for this term, as well as for other terms concerning the properties of numerical schemes).

The equations associated with NWP and climate modeling permit both relatively slow meteorological modes and relatively fast inertial gravity waves. The latter are important for the "geostrophic adjustment" process whereby real and spurious imbalances between the wind and mass fields created by, for example, observation errors, mountains, and point heat sources are eliminated. The inertial gravity waves impose a stringent computational stability condition on the timestep Δt, and require a much smaller Δt than that needed to maintain a high degree of numerical accuracy for the meteorological modes with respect to the time truncation error. This can be dealt with in AGCMs by the use of a semi-implicit time-stepping scheme. In such a scheme the fast gravity waves are treated implicitly and the remainder explicitly.

Numerical schemes can have exponential amplification of the solution of the difference equation (this is known as computational instability), and this must be avoided. A necessary condition for computational stability is the Courant–Friedrichs–Levy (CFL) criterion:

$$|(c\Delta t)/(\Delta x)| \leq 1$$

where c is a characteristic speed of the system (e.g. zonal wind), and Δt and Δx are the time and space discretization steps, respectively. The CFL criterion can be interpreted as requiring that the numerical dependence of a finite-difference scheme include the domain of dependence of the associated PDE.

Numerical schemes can generate unphysical small-scale, high-frequency waves (often termed "noise"), which contaminate the desired physical solution. This noise can be removed in a number of ways. A common one is to introduce an explicit artificial diffusion term that is scale-selective, i.e. can discriminate between different space scales. Alternatively, a time-difference scheme can implicitly achieve selective damping as a function of frequency. A combination of explicit and implicit damping can provide the best means of filtering small-scale, high-frequency waves.

When computing horizontal derivatives by means of finite differences at some grid point, many of the data come from adjacent points, which suggests that it would be more efficient if the variables are staggered in the horizontal. Five different arrangements (termed Arakawa A–E) of the dependent variables (mass and velocity) have been considered (Arakawa & Lamb 1977). The Arakawa B and C grids (Fig. 19.2) have the most desirable properties for AGCMs; for example, they are best at adjusting the initial conditions to a balanced state.

The domain over which the primitive equations are solved is finite. In the case of climate modeling this requires a bottom and a top boundary. The bottom boundary of an AGCM is straightforward; it is the Earth's surface: ocean, ice, or land (see Sections 19.2.2 to 19.2.4). The top boundary of an AGCM, however, has to be placed at an arbitrary location. In the past, most AGCMs had a top boundary at 10 hPa (about 30 km), reflecting the fact that the main focus was the study of the troposphere. Recently, however, it has been recognized that the stratosphere, and in particular O_3 (which mainly resides in the stratosphere), is important, for both NWP and climate modeling. The main reasons are: (i) the desire to perform ultraviolet (UV) forecasting using O_3 information; and (ii) the desire to include in a comprehensive way the impact of stratospheric O_3 changes in climate modeling. As a result, the top boundary of a number of AGCMs is now typically located between 0.1 hPa (about 65 km) and 0.01 hPa (about 80 km), with some having an even higher top boundary. Notwithstanding this, the majority of climate change experiments are still performed with AGCMs with an upper boundary at 10 hPa, mainly due to computer cost considerations.

There are a number of technical issues concerning the treatment of boundary conditions located within the body of a fluid, as is the case for the top boundary in a AGCM. In particular, the conditions imposed at the edge of the domain should mimic the presence of the surrounding fluid. The boundary conditions should, therefore, allow outward-

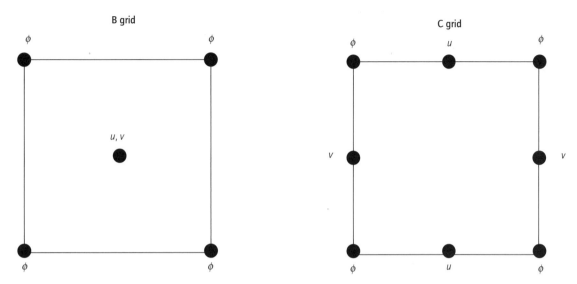

Fig. 19.2 Arakawa B horizontal grid (left) and Arakawa C horizontal grid (right): u and v denote, respectively, the zonal and meridional components of the wind field; ϕ denotes the temperature and geopotential height fields.

traveling disturbances to pass through the boundary without generating spurious reflections that propagate back toward the interior. Details of these technical issues can be found in Durran (1999).

When computing vertical derivatives it is advantageous also to stagger the variables, as this leads to considerable savings in storage space and computation time. Two different arrangements are commonly used in AGCMs. These are the Lorenz grid and the Charney–Phillips grid (Fig. 19.3). Durran (1999) discusses the energy-conservation properties of vertical differencing schemes.

Semi-Lagrangian schemes are of considerable interest because they can be more efficient than competing Eulerian schemes, as they can use a longer time-step. Another advantage of the semi-Lagrangian approach is its ease of use with nonuniform grids. Furthermore, semi-Lagrangian schemes avoid the main source of nonlinear instability in most geophysical wave-propagation problems because the nonlinear advection terms appearing in the Eulerian form of the momentum equations are removed when these equations

are formulated in a Lagrangian representation. Following on from the pioneering work of Robert (1981, 1982), semi-Lagrangian semi-implicit methods are one of the most used architectures in NWP models and climate models. An extensive review of the application of semi-Lagrangian methods to atmospheric problems is provided by Staniforth and Côte (1991).

The characteristic that distinguishes the spectral method from other series-expansion methods is that the expansion functions form an orthogonal set. We focus on the use of spherical harmonic functions as they are often used in climate models. The two-dimensional distribution of a scalar variable on the surface of the sphere can be approximated by a truncated series of spherical harmonics. Spherical harmonics can also be used to represent three-dimensional fields defined within a volume bounded by two concentric spheres if, for example, grid points are used to approximate the spatial structure along the radial coordinate and divide the computational domain into a series of nested spheres. Let λ be the longitude, θ the latitude, and define $\mu = \sin \theta$. If ψ is a smooth function of λ and μ, it can be represented by a convergent ex-

Fig. 19.3 Lorenz vertical grid (left) and Charney–Phillips vertical grid (right): v and θ represent, respectively, the horizontal components of the wind field and the potential temperature (proportional to temperature); $\dot{\sigma}$ denotes the rate of change of the vertical coordinate. Continuous lines denote the model full-levels; dashed lines represent the model half-levels.

pansion of an infinite sum of appropriately weighted spherical harmonic functions, where each spherical harmonic function $Y_{m,n}(\lambda, \mu) = P_{m,n}(\mu)\exp(im\lambda)$ is the product of a Fourier mode in λ and an associated Legendre function in μ. n is summed from $|m|$ to ∞; m is summed from $-\infty$ to ∞. Durran (1999) discusses the properties of spherical harmonics and Legendre functions.

In all practical applications the above infinite series must be truncated to create a numerical approximation where the infinite limits of the sum over m become $m = -M$ and M, and the upper (infinite) limit of the sum over n becomes $N(m)$. The so-called "triangular truncation," in which $N(m) = M$, is unique among the possible truncations because it is the only one that provides uniform spatial resolution over the entire surface of the sphere.

Despite its elegant properties, the triangular truncation may not be optimal in cases where the typical scale of the approximated field has a systematic variation over the surface of the sphere. This is the case, for example, for geopotential height anomalies over the Earth's atmosphere, which are much weaker in the Tropics than at middle latitudes. A variety of alternative truncations have, therefore, been used in low-resolution $(M < 30)$ AGCMs. The most common of these is the so-called "rhomboidal truncation," in which $N(m) = |m| + M$ in the truncated version of the spherical harmonics. The set of indices (m, n) retained in triangular and rhomboidal truncations are shown in Fig. 19.4.

Other truncations in which $N(m)$ is a more complex function of m have also been proposed in order to improve the efficiency of low-resolution climate models. Currently, there does not seem to be a clear consensus about which truncation is most suitable for use in low-resolution AGCMs. The triangular truncation is, however, the universal choice in high-resolution GCMs.

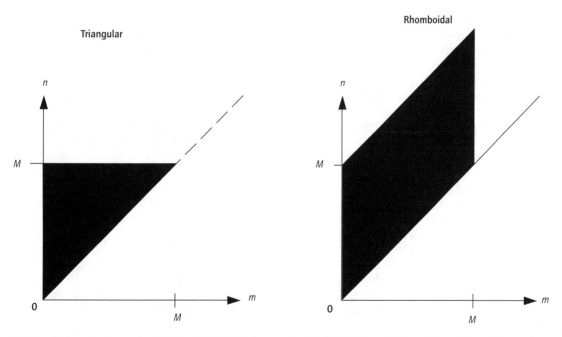

Fig. 19.4   Triangular truncation (left) and rhomboidal truncation (right). Shading indicates the position of the $m \geq 0$ half-plane in which the m and n indices are retained in an Mth-order truncation.

An advantage of using spherical harmonics is that derivatives such as the Laplacian are easily computed using the known properties of the expansion functions. Another advantage is that explicit finite-difference methods applied in a grid-point model can require very small time steps to maintain stability if the grid points are distributed over the sphere on a uniform latitude–longitude grid. This time step restriction arises because of the convergence of the meridians near the pole and the CFL criterion (the "pole problem"). This difficulty can be avoided by the use of spherical harmonics (Durran 1999); in grid-point models it is often treated by Fourier filtering. Another approach is to use a rotated pole.

A disadvantage of spherical harmonics (as well as general spectral methods) with respect to grid-point methods is the occurrence of unphysical Gibbs oscillations. Another disadvantage is the lack of highly efficient algorithms to transform from the spectral formulation to grid-point space (typically a latitude–longitude grid). The transform from the set of spectral coefficients to points on a latitude–longitude grid can be computed using discrete Fourier transforms (Durran 1999). The inverse transform is accomplished using the properties of spherical harmonics and evaluating a Fourier transform using the technique of fast Fourier transforms (FFTs). The integrals arising in this procedure can be computed using Gaussian quadrature techniques (Durran 1999). Provided that one avoids aliasing error (see below), it is possible to evaluate these integrals without introducing errors beyond those associated with the original truncation of the spherical harmonics at some finite wavenumber M. A practical complication when using spherical harmonics in an AGCM is that the calculations of the model physics (e.g. radiation — see later) and of nonlinear quantities are performed in grid-point space (this tends to be done on a "Gaussian grid"), with the result that in these cases the model's algorithm must switch frequently from a spectral formulation to a grid-point formulation and back.

Aliasing error arises because the product of two or more truncated spherical harmonic expansions contains high-order Fourier modes in λ and high-order functions in μ that are not present in the original truncation. As discussed by Durran (1999), to avoid aliasing error at truncation to wavenumber M, the computation of the coefficients in the spherical harmonic expansion requires the evaluation of polynomials of degree $3M$ for a triangular truncation and of degree $5M$ for a rhomboidal truncation. The exact evaluation by Gaussian quadrature of the integral of this polynomial to calculate the spherical harmonic coefficients, requires a minimum of $(3M + 1)/2$ meridional grid points for a triangular truncation and a minimum of $(5M + 1)/2$ meridional grid points for a rhomboidal truncation.

Two additional considerations that arise in using spherical harmonics for practical applications are: (i) the evaluation of derivatives with respect to the meridional coordinate; and (ii) the representation of the vector velocity field. In (i), binary products are computed in grid-point space and the result is transformed in spectral space, where the properties of the expansions are used to evaluate the derivatives within the truncation error of the spectral approximation. In (ii), the velocity components u and v (zonal and meridional components, respectively) are not conveniently approximated by spherical harmonics because artificial discontinuities in u and v are present at the poles unless the wind speed at the pole is zero. This difficulty is commonly avoided by replacing the prognostic equations for u and v by prognostic equations for the vorticity and divergence and by rewriting all remaining expressions involving u and v in terms of the transformed velocities $U = u \cos\theta$ and $V = v \cos\theta$ (where θ is latitude). Both U and V are zero at the poles and free of discontinuities (see Durran 1999 for details).

Ideally, the numerical techniques used in AGCMs should have the following attributes: (i) consistency, i.e. the scheme approaches the corresponding continuous differential equation as the space and time increments in the scheme approach zero; (ii) accuracy (generally, the aim is to have second-order accuracy, i.e. the space truncation error is $(\Delta x)^2$); (iii) stability; (iv) convergence to the physical solution; (v) efficiency; (vi) conservation of physical quantities such as mass, energy, and momentum; and (vii) monotonicity, i.e. the absence of unphysical overshoots and undershoots in, for example, tracer fields or humidity. Some of these requirements are linked. For example, there is a theorem by Lax which states that a linear, stable, consistent scheme must converge. These requirements are discussed in more detail in Haltiner and Williams (1980).

The above attributes are an ideal and the numerical schemes implemented often have conflicting requirements. Examples include the following.

1 Semi-Lagrangian schemes are efficient and stable for large Δt, but are not fully conservative.

2 The semi-implicit treatment of gravity waves is stable for large Δt, but is not accurate. This is not generally a problem if the gravity waves are physically unimportant, but can cause problems such as spurious orographic resonances.

3 Spectral methods give a high order of spatial accuracy and convergence but are associated with unphysical Gibbs oscillations.

4 Total variation diminishing (TVD) schemes used for advection schemes are monotonic but reduce to first-order accuracy near extremes. Within the TVD approach, so-called flux limiters are used to preserve monotonicity in the scheme and avoid spurious ripples. Examples of flux-limiters that are used in AGCMs include the "superbee" and Van Leer limiters.

5 A linear monotonicity preserving method is at most first-order accurate (this is Godunov's theorem). Therefore, monotonic schemes of higher order have to be nonlinear even for linear problems (e.g. for the advection of tracers).

Details of the current parameters in the dynamical formulation of two state-of-the-art AGCMs—the UK Meteorological Office (UKMO) Unified Model (UM) and the ECMWF model— can be found in, respectively, Cullen (1993) and Simmons *et al.* (1999). The UM is a grid-point model, whereas the ECMWF model is a spectral model. The UM tends to be used for climate change studies; the ECMWF model tends to be used for seasonal forecasting.

Radiation

In the Earth's atmosphere, the character of the circulation of the air masses depends strongly on the magnitude and distribution of the net diabatic heating rate. In the troposphere, the net diabatic heating rate is dominated by the imbalance between two large terms, i.e. transfer of heat from the surface and thermal emission of radiation to space. Latent heat is a major component of the flux from the surface to the atmosphere, and clouds play a major role in the emission of radiation to space.

In the stratosphere and mesosphere the situation is simpler. Net heating depends on the imbalance between local absorption of solar UV radiation and infrared (IR) radiative loss. In this region, O_3 is the dominant absorber (and contributor to the heating of the atmosphere) and carbon dioxide (CO_2) is the dominant emitter (and contributor to the cooling of the atmosphere). IR emission by O_3 and water vapor and solar absorption by water vapor, molecular oxygen, CO_2, and nitrogen dioxide play secondary roles. The distribution of the radiative sources and sinks due to these gases exerts a strong control on the large-scale seasonally varying mean temperature and zonal wind fields of the middle atmosphere. IR emission also provides an important mechanism for damping dynamically forced temperature variations.

In the climate modeling literature, solar radiation is commonly referred as shortwave (SW) radiation and thermal radiation (typically IR radiation) as longwave (LW) radiation. In contrast to LW radiation, SW radiation produces only heating because the Earth's atmosphere does not emit at those wavelengths. Clouds tend to have a net heating effect in the LW (as they emit at these wavelengths) and a net cooling effect in the SW (as they reflect or scatter radiation at these wavelengths). The net radiative effect of different cloud types depends on their properties and altitude and, in general, is not fully understood (Harries 1996).

The radiation code in AGCMs solves the equations of radiative transfer and converts the model input parameters into LW and SW heating rates, which, in turn, modify the temperature distribution. The algorithms use diffuse (i.e. with contributions from all directions) irradiances. The irradiance (or flux) is the integral of the component of intensity (or radiance) normal to a surface over the half-space of solid angle in the positive direction (upward irradiance) or the opposite half-space of solid angle in the negative direction (downward irradiance). The net flux in the positive direction is the difference between the upward flux and the downward flux. The heating rate is then computed by taking the divergence of the net flux (rate of change of the net flux with respect to pressure or with respect to optical depth) integrated through an atmospheric column with appropriate boundary flux conditions.

This integration takes account of the optical depth of the atmospheric column and the source terms. For LW radiation, in the absence of scattering, and for local thermal equilibrium (LTE), Kirchhoff's Law holds and this source term is given by the Planck function. The conditions for LTE are satisfied when energy transitions are dominated by molecular collisions, which is the case for the most important radiatively active gases at pressures greater than 0.1 hPa (at altitudes below about 65 km). At higher altitudes LTE breaks down because the interval between collisions is no longer short compared to the lifetime of excited states associated with absorption and emission, and Kirchhoff's Law is no longer valid.

The most important property of the SW spectrum is the solar constant, which represents the flux of radiant energy integrated over wavelength reaching the top of the atmosphere at the mean Earth–Sun distance. The solar constant is weakly variable, having a value of about $1370 \, W \, m^{-2}$ according to recent satellite measurements. When calculating SW heating rates, solar insolation must be taken into account via the solar zenith angle (SZA; this is $0°$ when the Sun is directly overhead and $90°$ when it is at the horizon).

The algorithms that calculate LW heating rates divide the LW part of the electromagnetic spectrum (typically $0–3000 \, cm^{-1}$) into bands. Spectroscopists use the unit of cm^{-1} to denote wavelength; thus $1 \, cm^{-1}$ (1 wavenumber) is equivalent to a wavelength of 1 cm and a frequency of 30 GHz. These bands are selected to include the contribu-

tions from the important gases such as CO_2 (e.g. the $15 \, \mu m$ band; $667 \, cm^{-1}$ band), O_3 (e.g. the 9.6 and $14 \, \mu m$ bands) and water vapor (e.g. the $6.3 \, \mu m$ band). Most of the LW energy emitted by the Earth's surface is absorbed in the troposphere. Only in the atmospheric window at wavelengths of $8–12 \, \mu m$ is absorption weak enough for much of the LW radiation to pass freely through the atmosphere. The $9.6 \, \mu m \, O_3$ band is the only strong absorber at those wavelengths; most of that absorption takes place in the stratosphere, where O_3 concentrations are large.

Similarly, the algorithms that calculate SW heating rates divide the SW part of the electromagnetic spectrum (typically $50,000–1000 \, cm^{-1}$, or $0.2–10 \, \mu m$) into bands. Typically, gaseous absorption by water vapor, CO_2, O_3, nitrous oxide, methane, and oxygen is included. A description of the SW spectrum for O_3 and oxygen can be found in Salby (1996).

In principle, the absorption characteristics of the band determine the corresponding transmission function (defined in Salby 1996). The transmission function varies from 0 to 1 (a transmissivity of 1 indicates no absorption; a transmissivity of 0 indicates full absorption). In practice, however, even individual absorption bands are so complex as to make direct calculations impractical. Instead, the transmission function is evaluated with the aid of band models that capture the salient features of the absorption spectrum in terms of spectroscopic properties like mean line strength, line spacing, and line width.

The radiation algorithms (LW or SW) used by climate models can be very different (see Phillips (1994) for a description of the radiation algorithms used in the climate models that participated in the Atmospheric Modeling Intercomparison Project, AMIP). There are three methods commonly used in climate models to parametrize gaseous absorption in a spectral band: (i) the look-up table method (e.g. Chou et al. 1995); (ii) analytic band models (ABMs; e.g. Brieglieb 1992), and (iii) the k-distribution approach (e.g. Lacis & Oinas 1991).

In the UKMO climate model, the k-distribution approach is used for the LW and SW. The perceived advantages of the k-distribution approach are: (i) scattering can be treated more accurately; and (ii) the total computation amount is linearly proportional to the number of layers, rather than the quadratic dependence of the look-up tables and the ABMs.

Three distinct methods for generating the coefficients for the k-distribution can be identified. First, in the exponential-sum fitting of transmissions (ESFT) technique, the observed or predicted transmissions of each gas in each band and at various absorber masses are used to fit a sum of exponentials, with each exponent equal to a unique coefficient k times the absorber mass (Lacis & Hansen 1974). Second, probability distributions of k values from ABMs have been used, from which the coefficients are expressible in terms of ABM parameters (Lacis & Oinas 1991). This method, however, requires too much computation to be suitable for climate model studies. Third, k-distributions have been created from spectroscopic databases (e.g. Chou *et al*. 1995), and with suitable averaging of k values the coefficients can be found. All three methods assume perfect correlation in wavenumber space of the k values at different pressures, so the integral over wavenumber to find the total broad-band transmission can be replaced by an integral over the cumulative probability distribution of k values. This procedure has the advantage that the cumulative probability distribution is monotonic, whereas the dependence of the absorption coefficient with wavenumber is not; consequently, this integral is relatively straightforward to compute. The line-transmission data are derived typically from established molecular databases such as HITRAN92 (Rothman *et al*. 1992).

The calculation of realistic cloud properties is an important task for a climate model radiation algorithm. Considerable effort has been expended over recent years in developing numerically fast yet accurate algorithms.

In the SW both water clouds and ice clouds must be considered. Approximations termed "thin averaging" and "thick averaging" are often used; these denote approximations that are exact in the limiting case of, respectively, optically thin clouds and optically thick clouds. Edwards and Slingo

(1996) find that for radiation algorithms that employ only a few SW bands, the error in calculating cloud radiative properties is minimized if thick averaging is used for water clouds and thin averaging for ice clouds.

High clouds exert the largest influence of any cloud type on the outgoing LW radiation (OLR). A particular problem with high clouds concerns cirrus clouds, which, due to their low optical depth, inhomogeneity, and composition of nonspherical ice particles, are difficult to observe and model. In the LW it is usually assumed that the effects of absorption are dominant compared with those of scattering, which may therefore be ignored. Calculations of the influence of clouds on LW fluxes are, therefore, often performed with emissivity schemes, which can produce reasonable agreement with measurements of broad-band fluxes (Francis *et al*. 1994). However, the calculations reported by, for example, Ritter and Geleyn (1992) show that scattering may sometimes have a significant effect on broad-band fluxes and heating rates.

Atmospheric aerosols play an important role in the Earth's radiation budget (Hobbs & McCormick 1988). They scatter and absorb solar radiation, causing an increase in the planetary albedo and a reduction in the amount of radiation that reaches the surface. The impact on thermal radiation is generally smaller, although it can be significant (Lacis & Mishchenko 1995). Anthropogenic aerosols are also believed to reduce the magnitude of global warming resulting from the increasing concentrations of greenhouse gases (Houghton *et al*. 1995).

Aerosols have often been ignored in AGCMs, except recently in models used for climate change studies. For example, only seven of the 30 models contributing to AMIP included any representation of aerosols (Phillips 1994). Independent evidence that aerosols need to be included in AGCMs has come from comparisons with data from the Earth Radiation Budget Experiment (ERBE; Harrison *et al*. 1990) and the Scanner for Radiation Budget (ScaRaB; Kandel *et al*. 1998). For example, the clear-sky planetary albedo in the UKMO UM was lower than that from ERBE and ScaRaB and the distribution of the deficit suggested that the lack of aerosols made a substantial contribution. As discussed by Cusack *et al.* (1998), the inclusion of aerosols in the UKMO UM in a formulation where spatially averaged quantities are used (rather than the full temporal and spatial variation of the aerosol optical properties) has the potential to alter the surface and the top of the atmosphere (TOA) radiation budgets in AGCMs, and to improve the agreement of these quantities with measurements significantly.

As interest in climate modeling increased, and awareness of how poorly understood and observed radiation was increased correspondingly, the Intercomparison of Radiation Codes in Climate Models (ICRCCM) project evolved and ultimately culminated in a comprehensive report (see *Journal of Geophysical Research 96*, 8921–9157, 1991). Radiation codes for individual climate models are typically tested by off-line comparison with high-resolution line-by-line models such as GENLN2 (Edwards 1988).

Parametrizations

In this subsection we discuss the following processes, which are parametrized in AGCMs: clouds, the planetary boundary layer (PBL), moist convection, nonconvective precipitation, and gravity wave drag (GWD). The interaction of radiation and clouds was briefly discussed above; in this subsection we discuss cloud schemes in climate models. Land surface processes are discussed in Section 19.2.4. As it involves the need to represent the effect of unresolved small-scale processes on the larger, explicitly resolved scales, the development of parametrization schemes requires an understanding of atmospheric processes at scales smaller than about 200 km in the horizontal and about 1 km in the vertical. Parametrizations require the following properties: (i) they must be physically correct; (ii) they must be dimensionally correct; (iii) they must be invariant under coordinate transformations; and (iv) they must be consistent with the budgets and constraints (e.g. they must not produce negative humidities). Note that as computing power increases and, associated with this development, the resolution of climate mod-

els increases, some parametrizations may become redundant.

Clouds. The recognition that small changes in cloud forcing may have a big impact on climate change, and the desire to have improved precipitation forecasts in NWP models, have led to the introduction of more complex stratiform cloud parametrizations in the current generation of climate models (e.g. Tiedtke 1993). In these parametrizations cloud cover and condensed water contents are predicted, allowing cloud radiative properties to be calculated interactively rather than being prescribed. These predictive schemes are termed "prognostic"; schemes where the cloud radiative properties are prescribed are termed "diagnostic." However, even with these advances the representation of clouds in climate models still remains an area of great uncertainty.

One area of uncertainty in current AGCM cloud parametrizations is the estimation of precipitation from clouds in the temperature range in which ice and water are observed to exist within the same cloud layer (these clouds are termed "mixed-phase clouds"). Studies suggest that sensitivity of modeled climate to SST changes is highly dependent upon the specification of the temperature at which cloud condensate is treated as ice. Given the uncertainty in estimates of climate change from GCMs due to the specification of ice precipitation, it seems vital that more observationally based parametrizations of mixed-phase clouds are developed. An example of such an approach is discussed by Gregory and Morris (1996).

Planetary boundary layer. The PBL extends about 2 km above the Earth's surface. This is the lower part of the troposphere, where the direct influence from the surface is felt through turbulent exchange with the surface. Boundary-layer parametrizations may be divided into two classes according to the vertical resolution of the model close to the ground. If the vertical resolution is such that, at most, one level lies within the boundary layer, so-called "bulk" parametrizations must be used to represent the boundary layer. Alternatively, the boundary-layer structure may be explicitly (albeit crudely) resolved by locating several levels within the lowest 2 km of the model atmosphere. Details of boundary layer schemes can be found in Garratt (1992) and references therein.

In the simplest bulk approach, the surface fluxes are calculated from a basic surface drag law using either the wind at the lowest model level or a wind extrapolated from more than one level. The drag coefficients and turning of the wind through the boundary layer may depend on the underlying surface and the stability of the lowest model layer. Turbulent fluxes either vanish in the free atmosphere or are treated by simple eddy diffusivities.

In explicit boundary-layer models the lowest level is chosen to be within a few tens of meters from the ground, and surface fluxes are determined using either a logarithmic wind profile or, more generally, Monin–Obukhov similarity theory. Above the lowest level, turbulent vertical fluxes are represented as the product of eddy diffusivities and the vertical gradients of the explicitly resolved fields. This diffusivity is often computed from the local wind shear and a mixing length that decreases linearly with height from the lowest model level to vanish just above the top of the PBL. The diffusivity also depends on the local stability, measured by the Richardson number; some schemes use a mixing length that does not vanish in the free atmosphere. So-called "higher-order closure" schemes in which diffusivities are related to an additional predicted variable, the local turbulent energy, have also been used.

Moist convection. Convective parametrization schemes aim to represent the so-called "apparent convective source" and "apparent moisture sink" due to convection (e.g. Gregory 1995), in terms of large-scale atmospheric variables predicted by a model. In doing so they must first predict the vertical distribution of latent heating and transport properties due to convection, through the use of simple cloud models. Second, they must predict the overall magnitude of the energy release from the convection. There are a number of further requirements: (i) the resultant thermodynamic

structure must be realistic; (ii) the schemes must be able to represent the mean distribution of precipitation accurately; and (iii) they must have a realistic representation of atmospheric variability.

Currently, there are four types of convective parametrization schemes: (i) moist convective adjustment schemes (Smagorinsky 1963); (ii) Kuo-type schemes (Kuo 1974); (iii) Betts–Miller adjustment schemes (Betts 1986); and (iv) mass flux convection schemes (e.g. Gregory & Rowntree 1990). We focus on mass flux convection schemes as, with increasing computing power, they are becoming more popular and are used in many NWP and climate models. Furthermore, they provide a physical understanding of how local convection affects the large-scale atmosphere.

In the mass flux scheme it is assumed that within some area (taken to be that associated with a grid point in a numerical model), a fraction is covered by cloud. After mathematical manipulation the convective heating can be expressed in terms of cloud variables and large-scale variables. The latter are available from the grid-point fields of a numerical model, while the former are usually obtained from a one-dimensional steady state entraining plume model of the cloud. A similar procedure is implemented for the convective drying. Details of the manipulations can be found in Gregory (1995). Application of the mass flux theory in a numerical model is a two-stage process. First, a cloud model must be used to estimate the vertical distribution of the cloud quantities. Second, the magnitude of the mass flux at the cloud base must be determined, usually by some reference to the large-scale structure and forcing; this is the "closure" problem.

Two approaches have been suggested for using cloud models to estimate the quantities required to implement mass flux schemes.

1 A spectral cloud ensemble (Arakawa & Schubert 1974), in which several cloud models are used to represent the different cloud types within a grid box of an AGCM.

2 The bulk cloud model (Yanai *et al.* 1973), where a single cloud model is used.

Approach 2 is often used in NWP and climate models as it is cheaper and easier to implement than approach 1.

In recent years several mass flux schemes have been updated to incorporate the effect of convective-scale downdraughts (as they are an important component of many deep convective systems). They are represented by inverted entraining plumes, adiabatic warming through the descent of air being compensated by cooling due to the evaporation of precipitation.

The production of precipitation within clouds is governed by complex microphysical processes. These are poorly represented in most convection schemes. Although the latent heat of freezing is included, little distinction is made between the formation of rain and ice precipitation. The mass flux scheme of Gregory and Rowntree (1990) provides an illustration of some of the schemes currently used.

Nonconvective precipitation. The treatment of nonconvective precipitation (also termed "large-scale precipitation") is typically one of the simpler elements of an overall parametrization scheme. It is computed after other dynamical and physical processes that change the temperature and water vapor content and, in general, involves the condensation (with associated latent heat release) of sufficient water vapor to keep the relative humidity (RH) below a fixed threshold value. This value, which varies between climate models, generally is in the range 80–100%. Values lower than 100% are regarded as representing the small-scale nature of much precipitation, which may occur at points within a grid square, even if in the mean the grid square is not saturated. Too low a threshold in the boundary layer may, however, lead to unrealistic amounts of precipitation from the lowest level above a wet surface.

In the simplest schemes, all condensed moisture falls instantly to the ground, but the evaporation of precipitation may also be taken into account (as can also happen in the convective case). A range of further refinements may also be adopted, including an explicit representation of liquid water.

Gravity wave drag. Gravity waves (GWs) have relatively small space scales that are not resolved at the typical grid scale of the AGCMs used for climate studies. These GWs transfer mean horizontal momentum from the ground to levels in the atmosphere or from one atmospheric layer to another. Flow over topography can generate stationary GWs that break nonlinearly in the troposphere and lower stratosphere. Such waves transfer momentum from the Earth's surface to the breaking region, and this process is thought to act as a significant drag on the eastward mean winds in the mid-latitude troposphere. It is known that AGCMs run without any attempt to impose such a drag tend to produce unrealistically strong mid-latitude eastward surface winds. Other processes (e.g. convection, jet stream instabilities) can produce GWs with nonzero horizontal phase speeds, which act to transfer mean momentum between the troposphere and the stratosphere/mesosphere region. AGCMs are known to produce unrealistic simulations of the extratropical stratospheric/mesospheric circulation unless some account is taken of the effects of these GWs. In particular, the AGCMs have large cold biases (the "cold pole" problem) and unrealistically strong polar night jets. If credible climate simulations (and predictions of climate response to anthropogenic change) are to be obtained, a physically justifiable parametrization of the momentum transport due to unresolvable GWs needs to be formulated. This issue is now recognized as one of the most important challenges in dynamical meteorology.

There are three main GW drag (GWD) parametrizations now being tested: (i) the "Lindzen" scheme (Lindzen 1981); (ii) the "Hines" scheme (e.g. Hines 1993), and (iii) the "Fritts and Lu" scheme (Fritts & Lu 1993). Each is essentially one-dimensional and explicitly considers statistically steady conditions. Each makes assumptions concerning the spectrum of upward-propagating GWs at the lowest level considered (typically near the tropopause) and then uses a simplified treatment of the dynamics to compute the propagation and dissipation of the GWs through the middle atmosphere (and hence the mean-flow modifications induced by the waves). A number of AGCMs use the crude "Rayleigh friction" (RF) parametrization in lieu of elements of the GWD parametrization;

typically, in the RF parametrization the winds in the upper stratosphere and mesosphere are damped toward zero using a damping coefficient that depends on model level. Details of some of the latest developments in GWD schemes and of GW observations can be found in Hamilton (1996).

Chemistry

As climate models become more realistic, the need to take account of atmospheric chemistry has become increasingly apparent (Houghton *et al.* 1992, and references therein). Of the main radiative species, O_3 is possibly the most complex, as it is controlled by both dynamical and photochemical processes with a wide range of time scales in the atmosphere. There are also many chemical feedbacks on climate. For example, chlorofluorocarbons (CFCs) both have a greenhouse warming effect (Houghton *et al.* 1992) and deplete stratospheric O_3 (Rowland & Molina 1975). Thus increasing CFCs may have a smaller impact on surface temperature than originally thought since the direct greenhouse warming may be partially offset by the cooling associated with the lower O_3 levels just above the tropopause (Ramaswamy *et al.* 1992). The lower stratosphere region is also one of the most difficult to model accurately because of the presence of heterogeneous chemical reactions on polar stratospheric clouds (PSCs) (Crutzen & Arnold 1986).

Because of their computational expense, few climate models have been run with full chemistry. At the time of the 1995 WMO ozone assessment, three-dimensional models in the literature included mechanistic models with full chemistry, low-resolution AGCMs with full chemistry, and more complete AGCMs with simplified chemistry. Computing power has only relatively recently reached the stage where AGCMs can be used with fairly complete stratospheric chemistry schemes. The implementation of tropospheric chemistry schemes in AGCMs is not as advanced as that of stratospheric schemes, mainly due to the complexity of tropospheric chemistry and the need to include effects such as deposition (see Chapter 6). Coupled chemistry–climate models have not yet reached the level of maturity and confidence of climate models.

The 1999 ozone assessment (WMO 1999) provides a table of three-dimensional chemistry–climate models. Multiyear integrations with AGCMs with comprehensive chemistry have been or currently are being run with the Hamburg model (Dameris *et al.* 1998), the Météo-France Centre National de Recherches Météorologiques (CNRM) model (Lefèvre *et al.* 1994), and the UKMO chemistry–climate model (Austin *et al.* 1997).

The Goddard Institute for Space Studies (GISS) climate model (e.g. Rind *et al.* 1998) and the EMÉRAUDE climate model used formerly at Météo-France (e.g. Mahfouf *et al.* 1993) have been run with a simple representation of stratospheric chemistry. However, because the chemical and dynamical transport time scales for O_3 are roughly comparable in the stratosphere, using such simplified chemistry has limitations.

Because many climate models generate an unrealistically cold lower stratosphere, there is the prospect that a fully coupled chemistry–climate model could increase the amount of PSCs and produce excessive O_3 depletion by heterogeneous processes. This situation could, for example, produce a current Arctic O_3 hole where one does not presently occur. A common remedy is to correct the bias in polar temperatures by comparing the model temperature climatology with observations.

Typical stratospheric photochemistry schemes in coupled climate–chemistry models include a number of photochemically active tracers; some of these are treated as photochemical families, e.g. O_y (total odd oxygen). In general the species are assumed to be in the gas phase, except for those which are in the solid phase in PSCs (such as water vapor and nitric acid). A further scheme can be included to represent the heterogeneous chemistry taking place on PSCs. In addition to the transported species, a number of species are derived typically from the reservoirs (e.g. HOCl) by photostationary state and conservation assumptions; examples include O_3 and ClO. Photolytic reactions are included and their rates are commonly determined from the altitude, O_3 column, and SZA using a precomputed look-up table. The photochemical data would be taken from an established database (e.g. DeMore *et al.* 1994).

19.2.2 Ocean

There are a number of important differences between the atmosphere and the ocean which have a bearing on the way oceans are modeled. Water is very much denser than air. Thus, the air–sea interface is very stable because of the strength of the gravitational restoring force when it is displaced from its equilibrium position. Because of the stability of the interface, the two media do not mix in any significant way, and transfer of properties between them must take place via a well defined interface. The existence of this interface affects the radiation balance because it reflects radiation.

There is also a discontinuity in optical properties at the ocean surface, which affects the radiation balance. Unlike the atmosphere, the ocean absorbs solar radiation very rapidly; typically, 80% is absorbed within the top 10 m. LW radiation is absorbed even more rapidly by the ocean, with the result that the emission and absorption of LW radiation takes place from a very thin layer, less than 1 mm thick.

Because of the difference in the densities of water and air, the ocean mass is very much greater (about 270 times) than that of the atmosphere. This also implies a large difference in heat capacity. As the specific heat capacity of water is four times that of air, a 2.5 m depth of water has the same heat capacity per unit area as the whole depth of the atmosphere. The large heat capacity of the ocean is important for seasonal changes. Although in the long term each hemisphere loses by radiation as much heat as it receives, this is not true of an individual season. The excess heat gained in summer is not transported to the winter hemisphere, but is stored in the surface layers (to about 100 m depth) of the ocean and returned to the atmosphere in winter. Because of this ability to store heat, the SSTs change by much smaller amounts than the land surface, which cannot store much

heat. Thermal storage in the ocean is also important at longer time scales, and therefore is of significance for climate variability.

By the time of IPCC 1990 (Houghton *et al.* 1990), the state-of-the-art AGCMs used to reproduce the large-scale features of the current climate were coupled to simple representations of the uppermost layer (typically 50–100 m deep) of the ocean (termed "mixed-layer" or "slab" ocean models). These AGCMs were used principally to determine the "equilibrium" climate response to increased concentrations of atmospheric greenhouse gases (in particular CO_2), this being the response obtained when the model has come into statistical equilibrium with the prescribed and enhanced levels of, for example, CO_2. A similar model formulation was used to search for human influences on the climate (e.g. Santer *et al.* 1996). (Note that, due to the relatively slow change in SSTs, many AGCMs are used in a formulation in which the ocean forcing is accomplished via prescribed SSTs, either from climatology or from observations.)

It is increasingly being recognized that the type of climate model that holds out the prospect of dealing satisfactorily with the complexity of the Earth's climate system, its current state, and future evolution, both globally and regionally, is a CGCM. CGCMs are based on the mathematical equations that describe the motions and physical processes in the atmosphere and oceans, and how the various components of the Earth's climate system interact with each other.

The results from the most recent CGCM transient-response, climate-change experiments, which included the effects of both greenhouse gases and aerosols, made a major impact on the issues of detection and attribution of climate change. Pattern-based studies, in which modeled climate responses were compared with observed (geographic, seasonal, and vertical) patterns of atmospheric temperature change began producing more convincing evidence for the attribution of anthropogenic influence on climate (Houghton *et al.* 1995).

CGCMs are now being run with factors, both natural and anthropogenic, that will have influenced the climate since 1860: natural variability (which is inherent in the model), changes in solar output and volcanoes, and anthropogenically induced changes in the concentrations of greenhouse gases (including O_3) and sulfate aerosols. A particular difficulty is that there are many possible combinations of these forcings that could yield the same simulation of the observed global mean temperature change. Some combinations are more plausible than others, but relatively few data exist to constrain the range of possible solutions. The detection of attribution of climate change are discussed in more detail in Section 19.4.

CGCMs have a tendency to drift away from a realistic climatology, and in many this is counteracted by applying prescribed artificial fluxes of heat, fresh water, and, sometimes, momentum at the ocean–atmosphere interface in addition to the normally calculated surface exchanges there. The motivation for this approach is that the modeled climate response to a perturbation may be incorrect if the simulation of the current climate has significant errors. However, this approach does not have a good physical basis. In the face of this, modelers have sought to justify that these flux adjustments do not cause any gross distortion to the climate responses obtained with them.

Over the past few years a number of changes have been introduced to achieve CGCMs that produce stable and realistic long climate simulations without the use of flux adjustments. An example is the UKMO Hadley Centre CGCM (see references in Carson 1999). We describe this model for illustrative purposes.

In this model, the atmospheric component is similar to the UKMO UM described in Cullen (1993). The ocean component is based on the modeling framework established by Bryan and Cox (1967) and Bryan (1969). In this framework, the prediction of currents is carried out using the Navier–Stokes equations with three basic assumptions: (i) the Boussinesq approximation is adopted, in which density differences are neglected except in the buoyancy term; (ii) the hydrostatic assumption is made; and (iii) closure is attained by adopting the turbulent viscosity hypothesis in which stresses exerted by scales of motion too small to be resolved by the grid are represented as enhanced molecular mixing. The temperature and salinity are calculated using conservation equations, again using a turbulent mixing hypothesis for closure. The equations are linked by a simple equation of state. Horizontal mixing of tracers based on the Gent and McWilliams (1990) scheme parametrizes the effects of the finer-scale ocean eddies. The sea-ice model uses a simple thermodynamic scheme and contains parametrizations of ice drift and leads (Cattle & Crossley 1995). The techniques used for computational efficiency include the elimination of high-speed external GWs by the "rigid-lid" approximation.

The atmosphere and ocean components in the CGCM are coupled once a day. The atmospheric model is run with fixed SSTs through the day and the various forcing fluxes are accumulated each atmospheric model time-step. At the end of the day these fluxes are passed to the ocean model, which is then integrated forwards in time. The updated SSTs and sea-ice fields are then passed back to the atmospheric model. (Note that in an AGCM using forcing SSTs, the SSTs would tend to be updated at less frequent time intervals, with five days being standard.)

19.2.3 Cryosphere

Concerning climate and anthropogenic activities, there are five key elements of the terrestrial cryosphere: (i) seasonal snow cover; (ii) sea ice; (iii) the ice sheets; (iv) mountain glaciers; and (v) permafrost.

Seasonal snow cover responds rapidly to atmospheric dynamics on time scales of days and longer. Globally, the seasonal heat storage in snow is small. The primary and large effect is exerted by the high albedo of snow-covered surfaces. In northern winter, at its maximum, snow covers 50% of the land surface and 10% of the ocean surface. Only in recent years have sophisticated modeling techniques been used to look at the influence of snow cover globally.

Sea ice plays a complex role in the Earth's climate system on time scales of seasons and longer. Its seasonal cycle of extent has a similar, though smaller, effect on the surface heat balance to that of snow on land. Sea ice also acts as a barrier to the exchange of moisture and momentum between the atmosphere and ocean. In some regions, sea ice is related to the formation of deep water masses and may play a role in long-term climate change. In AGCMs used to study the current climate, the sea-ice concentration often is specified from climatological data. Within a grid box it is common for the distribution of sea ice to be discrete (there is either no sea ice or the sea-ice cover is complete), regardless of the nature of the cover.

The ice sheets of Antarctica and Greenland are quasi-permanent topographic features (on the shorter climate time scales) and, because of their high albedo, act as elevated cooling surfaces for the atmospheric heat balance. They contain 80% of the Earth's fresh water. Changes in their volume could cause large changes in sea level. AGCMs are the most complex models in a hierarchy of models used to study the climate sensitivity of ice sheets and glaciers; this hierarchy includes "degree-day" models, local energy balance models, regional energy balance models, regional climate models, mesoscale meteorological models, and AGCMs. The main problem in modeling ice sheets with AGCMs is that their typical horizontal resolution (about 100 km) is too coarse to capture elements of the surface regime such as the ablation zone of the Greenland ice sheet. Notwithstanding these shortcomings, the future of AGCMs for use in mass balance research is perceived to be very promising (e.g. Ohmura et al. 1996). In particular, the coarseness of grids has been considerably improved by the introduction of high-resolution grids; this approach is likely to become more common with the increased use of massive parallel computers.

Mountain glaciers are a small part of the cryosphere in volume and surface area. Their relatively rapid movement and high rates of accumulation and ablation (and, therefore, rapid response to climate change) may provide a significant contribution to sea level changes. Because valley glaciers and small ice caps react more swiftly to changing ambient conditions than do the large ice sheets, they can be used as climate indicators; this has

been demonstrated by the dramatic worldwide retreat of valley glaciers during the twentieth century (Oerlemans 1994).

Permafrost is a manifestation of past and present climate, changing significantly on time scales of centuries and longer. It affects surface ecosystems and river discharge into the ocean, which, especially along the estuaries and vast shelf areas of Eurasia, influences the convective regime of the ocean.

The details of how one models the cryosphere become important in paleoclimate studies, where an attempt is made to reconstruct past climates given different solar forcings (associated with the Milankovitch cycles). A key premise behind paleoclimate studies is that the quality of the model representation of components of the Earth's climate system can be evaluated against the paleoclimate record.

The paleoclimate record suggests that over the past three million years or so, the northern part of the northern hemisphere has been subjected to a continuous cycle of glaciations and deglaciations with corresponding sea level changes of approximately 100 m (e.g. Marsiat 1994). Consequently, there is a large interest in modeling the different aspects of the ice age cycle. There are basically three kinds of model that are used in studying the ice age cycle:

1 Models that focus on modeling the ice sheets themselves and include highly simplified forms of the "net-balance function," which depends on snowline parametrization (e.g. Birchfield & Grumbine 1985).

2 Models intended to simulate instantaneous climates in the past or mechanisms supposed to act during some parts of the glacial to interglacial cycle, which frequently use paleoclimatically reconstructed data sets as boundary conditions. This class includes energy balance models (e.g. Hyde et al. 1989) and AGCMs (e.g. Lautenschlager & Herterich 1990).

3 Models that asynchronously couple a climate model to one or several ice sheet models (Marsiat 1994).

In Section 19.3 we discuss the evaluation of paleoclimate models.

19.2.4 Land surface

Land surface schemes are a key component of NWP models and GCMs, calculating the surface to atmosphere fluxes of heat and water, and updating the surface and sub-surface variables which affect these fluxes. In this capacity they are intimately linked to PBL parametrizations and aspects of the cryosphere such as snow cover (see Section 19.2.3). Off-line tests have demonstrated that apparently small differences among these schemes can lead to a wide spread in the results (Henderson-Sellers *et al.* 1996), but it is unclear how this spread is affected by coupling the schemes to a GCM. Furthermore, very little has been published on the impact of land surface process representation on GCM simulations of climate sensitivity to increasing greenhouse gases.

A second generation of land surface schemes were developed in the mid-1980s to address some of the limitations of the earlier "bucket" model. These included a scheme of intermediate complexity developed for use in the then UKMO model and later included in the UKMO UM. A scheme termed MOSES (Meteorological Office Surface Exchange Scheme; Cox *et al.* 1999) has been designed to remove some of the limitations of the earlier UKMO model. We discuss the MOSES scheme for illustrative purposes.

Besides calculating water and energy fluxes, MOSES calculates vegetation to atmosphere fluxes of CO_2, incorporating the direct physiological effect of atmospheric CO_2 concentrations on both photosynthesis and stomatal conductance, which may provide a significant additional climate forcing under enhanced CO_2 (Sellers *et al.* 1996). The calculation of surface CO_2 fluxes is a key step towards the incorporation of a fully dynamic biosphere within AGCMs, and which would enable the full range of climate-vegetation feedback to be included consistently in climate change predictions (Betts *et al.* 1997).

The MOSES scheme has the following components: (i) evaporation fluxes; (ii) surface energy balance; (iii) canopy conductance and primary conductivity, including leaf photosynthesis models; (iv) soil thermodynamics and soil water phase change, including the soil thermal conductivity; (v) soil hydrology; and (vi) surface parameters.

The total moisture flux from the land surface is made up of evaporation from the canopy store, transpiration by vegetation, bare soil evaporation, and sublimation from the snow surface. The surface energy balance at the land–atmosphere interface requires a balance between the net radiation at the surface, and a sum of terms involving the sensible heat flux, the evaporation flux, the ground heat flux, and latent heat effects. The subsurface temperatures are updated using a discretized form of the diffusion equation. However, MOSES includes two significant new features: the soil thermal characteristics are realistic functions of soil moisture content (liquid water and ice), and the soil water phase changes are simulated and the associated latent heat is included in the subsurface calculations. The soil is divided into several vertical layers. The soil hydrology is based on a finite difference approximation to the Richards equation; the prognostic variables of the model are the total moist content within each layer. The surface parameters can be soil dependent or vegetation dependent. Geographically varying fields of the vegetation parameters are currently derived by assigning typical values to the 23 land-cover classes of the Wilson and Henderson-Sellers (1985) land-cover data set. Effective values for the soil parameters are calculated from the grid-box average fractions of sand, silt, and clay using regression techniques. The sand, silt, and clay values are area-weighted averages derived by assigning typical fractions to the three textural classes distinguished within the Wilson and Henderson-Sellers data set.

Further details of the scheme, as well as an assessment of its impact on AGCM climate simulations with fixed SSTs, can be found in Cox *et al.* (1999).

19.3 EVALUATION OF THE MODELING TOOLS

Evaluation of the climate models involved in climate change studies involves three key ideas:

1 Models, despite their complexity, have a number of recognized shortcomings.

2 The representation of the "truth" (in this case, the state of the atmosphere) by observations is not error-free.

3 Predictions should be model-independent and algorithm-independent.

There are a number of ways in which climate models can be evaluated. These include: (i) comparison with theoretical predictions; (ii) comparison with simpler (and better understood) models; (iii) assessment of the physical and dynamical consistency of the model fields; (iv) comparison with other (ideally independent) climate models of similar complexity; (v) reconstruction of current climate and comparison with available data; (vi) reconstruction of past climate and comparison with the paleoclimate record; (vii) observations; and (viii) analyses of the atmosphere commonly derived using a data assimilation algorithm that attempts to combine in an optimal manner data from a GCM and from heterogeneous observations.

Comparison with theoretical predictions is not much used to evaluate climate models, chiefly due to the complexity of climate models, and the fact that approximations to the state of the atmosphere that are amenable to analytic solution are invariably too simple to capture the key ingredients included in a climate model. Comparison with simpler models is occasionally carried out to assess the behavior of climate models. An example would include removing the troposphere from an AGCM and representing it by a forcing field of geopotential height at 100 hPa (typically derived from the AGCM or from atmospheric analyses). The stratosphere in both models would be compared to assess model consistency (by forcing the simpler model with fields from the AGCM) or to assess the model's representation of the troposphere (by forcing the simpler model with fields from the analyses). An assessment of the physical and dynamical consistency of the climate model would typically be one of the first checks carried out on a climate model; commonly, this would involve confirmation from visual inspection of plots of model fields that relations such as thermal wind

balance are not violated. The rest of this section focuses on items (iv) to (viii) above.

A number of model intercomparison projects have been established over the past years. Examples include AMIP, the Palaeoclimate Model Intercomparison Project (PMIP), and the Coupled Model Intercomparison Project (CMIP); details of these projects can be found at the website for the Program for Climate Model Diagnosis and Intercomparison (PCMDI; http://www pcmdi.llnl.gov/PCMDI.html). As AMIP serves as a prototype for the intercomparison of other climate models, such as those for paleoclimate, the ocean, and the coupled ocean–atmosphere system, we focus on AMIP for illustrative purposes. However, we will mention PMIP when discussing the use of the paleoclimate record to evaluate climate models.

AMIP was established in 1990 as an initiative of the Working Group on Numerical Experimentation (WGNE) of the WCRP in order to provide international standards for evaluating the performance of AGCMs used in climate modeling. In particular, it was set up to: (i) document the comparative performance of AGCMs in the simulation of current climate; (ii) facilitate the identification of systematic model errors via coordinated model diagnosis, evaluation, and intercomparison; and (iii) provide a benchmark against which sensitivity experiments and revised model versions could be evaluated in the interests of model improvement.

AMIP has expanded from a few modeling groups to the point where most AGCMs are active participants. Much of the analysis of the AMIP results has been carried out by several diagnostic subprojects that focus on specific aspects of the AGCMs's performance. At present, the second Atmospheric Model Intercomparison Project (AMIP II) is under way. AMIP II includes 35 AGCMs from modeling groups in Europe (the 18 member states that are part of ECMWF), the UK, France, Germany, Russia, the USA, Canada, Australia, Japan, China, and South Korea; some of these models are different versions of the same AGCM. A number of the AGCMs participating in AMIP II have a good representation of the stratosphere. Typically, the AGCMs participating in AMIP had model lids in the neighborhood of 10 hPa; in AMIP II a number of AGCMs have model lids at or above 0.1 hPa. AMIP II has 25 ongoing diagnostic subprojects, including: (i) the evaluation of the synoptic to intraseasonal variability; (ii) the evaluation of simulations of the stratospheric circulation; and (iii) the evaluation of the simulation of snow cover. AMIP II also has three ongoing experimental subprojects, including: (i) the use of multiple model realizations; and (ii) the use of different model resolutions.

Among the more important results from AMIP is the recognition that the performance of AGCMs must be carefully documented under standard conditions if their systematic errors are to be reliably identified and subsequently reduced. This is well illustrated by the reports presented by the various diagnostic subprojects (AMIP 1995).

Examples of these reports include that by Slingo *et al.* (1996) on the simulation of the intraseasonal oscillation in 15 AGCMs, and that by Gaffen *et al.* (1995) on the simulation of tropospheric water vapor in 28 AGCMs. The Slingo *et al.* report concluded that there was a wide range of skill in the AGCMs, and that no model successfully captured the dominance of the intraseasonal oscillation in the observations. The Gaffen *et al.* report concluded that the models tended to underestimate precipitable water over wide geographic locations and appeared to overestimate the poleward flux of moisture. Gaffen *et al.* also concluded that a consensus of the models (as defined by the median or average of the 28 AGCMs considered) gave a better simulation of the observed humidity than the individual AGCMs.

The above examples illustrate a number of features concerning models and model intercomparisons.

1 Typically, no model performs best. By and large, each model will have strengths and weaknesses which will be manifested in a given intercomparison.

2 There are elements of the Earth's climate system that, typically, most models do not simulate as well as would be desired. This is often manifested in model biases, and the presence of such biases across a spectrum of models tends to suggest fundamental deficiencies in the model formulations.

3 An ensemble of different models where one takes the median or average of the ensemble often has higher skill than the individual models. This fact is currently being exploited by the leading European NWP agencies to investigate ways of improving the predictive skill of their seasonal forecasts.

Climate change studies search for signals of anthropogenic influence in elements of the current Earth's climate system such as the temperature structure (e.g. Santer *et al.* 1996; Tett *et al.* 1999). The skill of the climate model simulation is evaluated by comparison with observations; in the case of the Santer *et al.* (1996) study, the observations were radiosonde temperature analyses that spanned the period 1963–87. This evaluation typically involves comparing model and observed changes in, for example, temperature, using methods such as the "centered correlation statistic" (Santer *et al.* 1995), and the "optimal fingerprinting" algorithm (e.g. Allen & Tett 1999). Note that when comparing climate models against observations care must be taken to avoid spurious trends in the observations (Hurrell & Trenberth 1997).

Typically, when the climate model simulation of temperature changes is evaluated against observations, combinations of the main anthropogenic radiative forcings of climate are considered; these include changes in well mixed greenhouse gases (e.g. CO_2 and methane), tropospheric sulfate aerosols, and stratospheric O_3. The main natural radiative forcings of climate, such as changes in solar irradiance and in stratospheric aerosols (due to volcanic activity), are also included to distinguish between the natural and the anthropogenic radiative forcings of climate. Simulations using different combinations of radiative forcings are then evaluated against observations using one of the methods mentioned above. Three key issues in both current climate simulations and future climate change scenarios are: (i) ensuring that the model's climate drift does not dominate over the signal of climate change; (ii) the distinction between natural and anthropogenic climate change; and (iii) the need for predictions to be model-independent. Further details as to how these issues are dealt with in climate modeling are presented in Section 19.4.

The PMIP was set up to coordinate and encourage the systematic study of AGCMs and to assess their ability to simulate large climate changes such as those that occurred in the distant past. PMIP goals include identifying common responses of AGCMs to imposed paleoclimate boundary conditions, understanding the differences in model responses, comparing model results with paleoclimate data, and providing AGCM results for use in helping in the analysis and interpretation of paleoclimate data. Initially, PMIP has focused on the mid-Holocene (6000 years before present, BP) and the last glacial maximum (21,000 years BP), because climatic conditions were very different at those times and because relatively large amounts of paleoclimate data exist for these periods. The major forcing factors are also relatively well known for these periods. For the mid-Holocene the forcing factors include present-day SSTs (if prescribed), a different insolation from the present (due primarily to the precession of the equinoxes), and CO_2 concentrations of 280 p.p.m. (parts per million). For the last glacial maximum the forcing factors include a reconstruction of the SSTs (if prescribed; CLIMAP 1981), a different insolation from the present, CO_2 concentrations of 200 p.p.m., and a reconstruction of the ice sheets (Peltier 1994).

One of the goals of PMIP is to determine which results are model-dependent. The PMIP experiments are limited to studying the equilibrium response of the atmosphere (and surface characteristics like snow cover) to changes in the boundary conditions (e.g. insolation, the distribution of ice sheets, the CO_2 concentration). Control experiments for these PMIP experiments are: (i) present-day conditions with present SSTs, present insolation, and a CO_2 concentration of 345 p.p.m.; and (ii) preindustrial conditions, with SSTs and insolation as for the present, but a CO_2 concentration of 280 p.p.m.

In the mid-Holocene, the northern hemisphere extratropical latitudes received larger incoming solar radiation during the summer season than at present, allowing for a warming of the continents.

The terrestrial data coverage for this period is relatively complete and often accurately dated (e.g. COHMAP 1988). These data permit a relatively detailed comparison of the simulated climate response of AGCMs over the continents to the known changes in the insolation pattern. In particular, the mid-Holocene experiment is understood to be useful in evaluating ground hydrology parametrizations, which strongly influence continental climates.

The Last Glacial Maximum is characterized by large changes in the surface boundary conditions (ice sheet extent and height, SSTs, albedo, sea level) and atmospheric CO_2 concentrations, but only minor changes in the insolation pattern. This period is important for understanding how ice sheets and lowered CO_2 concentrations influence climate. Data for both boundary conditions and model evaluation are relatively abundant for this period. Over the oceans SSTs have been reconstructed (e.g. CLIMAP) and over the continents this period has been extensively studied (e.g. COHMAP). Among the climatic features of interest in this period are the simulated changes in the northern hemisphere tropospheric jet stream location and associated changes in the storm-tracks.

Despite the fact that observations are not error-free and have biases, we regard them as the representation of the true atmospheric state, with appropriate estimates of the error and the bias (note that, for example, biases may be difficult to establish, especially when there is a paucity of independent observations to evaluate the target observations). Consequently, the ideal is to compare climate models against observations; furthermore, these observations must be of high quality, consistent, and sufficiently long term, and must have global coverage at spatial and temporal resolutions that allow the representation of significant phenomena. However, concerning the use of observations to evaluate climate models, there are two key issues that must be taken into account: (i) observations are heterogeneous; and (ii) the data record of observations with, for example, high-quality and global coverage, is not very long, typically extending from the IGY in 1957 to the present day. Issue (i) can be dealt with using the technique of data assimilation (see below). Issue (ii) imposes a limit on the aspects of climate change that can be addressed; for example, information on the multi-decadal natural climate variability crucial to the climate change detection problem is impossible to obtain from the relatively short (about 40 years long) radiosonde temperature record. The optimization and integration of atmospheric and other Earth observation, analysis, and modeling systems, as well the development of new observation capabilities, will address (ii) (see NRC 1998).

Data assimilation (Daley 1991) is a sophisticated data processing method that allows the exploitation of Earth observation (EO) data to the full. It is now the cornerstone of operational weather forecasting, but it is equally applicable to the exploitation of data from research satellites such as NASA's Upper Atmosphere Research Satellite (UARS), launched in September 1991, and the European Space Agency's (ESA) Envisat, launched in 2002.

Data assimilation of atmospheric, oceanic, and land surface measurements is a particular application of techniques used in optimal control theory to control the evolution of a dynamical system (e.g. a AGCM) by the incorporation of observational data. The goal is to produce an optimal estimate of the evolution of the system that is consistent, within the errors, with both the governing equations of the system and the observational data. The advantages of data assimilation for atmospheric applications include: (i) extraction of the maximum amount of information consistent with the errors in the measurements; (ii) a self-consistent picture of the evolving state of the atmosphere; (iii) measurements of one parameter provide information on others via the model equations; and (iv) statistics on the errors of the measurements and errors in the model forecasts.

A number of reanalyses (the 15-year ERA data set; the 40-year NCEP/NCAR data set) have been developed using data assimilation techniques. They are used extensively to study issues such as climate change and stratospheric O_3 depletion, and to evaluate climate models. The analyses themselves are often compared against other analyses, or against independent observations, to evaluate their dependence on the numerical model in the assimilation system. Ideally, analyses should be as independent as possible from the model, especially when these analyses are used to evaluate the model.

19.4 USE OF THE MODELING TOOLS

The use of models to evaluate climate change requires three things: (i) methods of ensuring that the model's climate drift does not dominate over the signal of climate change; (ii) methods of estimating the natural variability of the Earth's climate system; (iii) objective measures for evaluating the model's predictive skill.

CGCMs are currently the best tools available for understanding the current climate and making predictions for future climate change. However, they typically show a climate drift when simulations are run from initial conditions created from climatological estimates of the current state of the atmosphere and ocean. (These initial conditions can be obtained from atmospheric analyses or from a model run, and from established databases of climatological ocean temperatures and salinity.) Climate change experiments using CGCMs need to start close to the model's equilibrium so that the drift does not dominate the climate change signal. The process of achieving this equilibrium state is termed "spin-up."

One method to achieve a spun-up initial state is to start from an estimate of the current climate and run the CGCM from there until the drift is small enough to be acceptable. This method has the drawback of being expensive in terms of computer time (typically, several hundred years of simulation are required). Because the equilibration time scale of the ocean is much longer than that of the atmosphere (hundreds of years compared with several years), it may appear reasonable to treat the two components of the CGCM system separately; however, interactions and feedbacks between the two components ensure that the coupled equilibrium is not equivalent to the two independent equilibria, and a drift still occurs when the two components are recoupled. A faster spin-up method involves using climatic data from the atmosphere component of a CGCM run to drive a long ocean-only run with a long time-step. By allowing much of the ocean spin-up to occur in an accelerated ocean-only phase, significant savings in computer time can be achieved.

A climate change experiment can be performed to address equilibrium or transient issues. In the former, the experimenter is interested in the new equilibrium climatic state attained by the model when changes in natural and/or anthropogenic forcings are imposed; attaining this state may take several years of simulation. In the latter, the experimenter is interested in how the climatic state of the model changes when changes in natural and/or anthropogenic forcings are imposed; changes in the climatic state may begin to become evident after a relatively short period of the simulation.

AGCMs (which are cheaper to run than CGCMs) are often used to study current climate, with the forcing SSTs provided from observations. For studies of future climate change it is impossible to use observed SSTs, and it is difficult to avoid using a CGCM. Nevertheless, SSTs derived from a different (but consistent) CGCM run, or SSTs derived from an AGCM coupled to a mixed ocean layer model, are sometimes used to force an AGCM to study future climate change. The advantage of this approach is the lower cost of the simulation compared to that using a CGCM; this lower cost also means that an ensemble of runs may be performed, and benefit can be made of the key advantage of the ensemble approach, namely that the squared error of the ensemble forecast is smaller than the mean squared error of all the individual forecasts (Brankovic et al. 1990).

When performing an ensemble of climate runs, the spread arises from the chaos inherent in the solutions of the nonlinear equations describing the Earth's climate system (Lorenz 1963). The ensemble average provides the best forecast. However, even if the climate model were perfect, the mean accuracy of the ensemble average forecast is only that represented by the ensemble average and the individual climate realizations, because the real world runs through the "climate experiment" only once. This reflects the fact that ensembles with low spread tend to indicate high predictive skill, whereas ensembles with high spread tend to indicate low predictive skill.

When setting up an ensemble of climate runs one must decide: (i) how many members it should have; and (ii) how the members are to be initialized. Issue (i) is largely determined by computing cost constraints, which means that ensemble size is typically less than 10. Issue (ii) is commonly accomplished by initializing each ensemble member from climate states one day apart, everything else being the same. In this way, the experimenter can capture the uncertainties associated with the initial state, which as demonstrated by Lorenz (1963) can lead to chaotic behavior.

To test hypotheses of climate change forced by natural and/or anthropogenic forcings, a control scenario is needed; this is provided by the natural variability of the climate model. Typically, the natural variability of a climate model is estimated by performing long (e.g. multicentury) control runs with constant forcing parameters (natural and/or anthropogenic). Such simulations provide estimates of the internally generated natural climate variability on a range of space and time scales. Typically, evaluation of the estimates of model natural variability is done by comparing the patterns in model variability on a range of spatial and time scales against those in observations of, for example, surface temperature (Tett et al. 1997).

A common overall approach has emerged to the detection of anthropogenic climate change. A detection statistic is defined and evaluated in an observational data set. This might be: (i) a global mean quantity (Stouffer et al. 1994); (ii) a model versus observation pattern correlation (Mitchell et al. 1995); (iii) the observed trend in pattern correlation (Santer et al. 1996); or (iv) a form of optimized fingerprint (Allen & Tett 1999, and references therein). The same detection statistic is then evaluated, treating sections of a control run of a climate model (in which the forcing is constant) as "pseudo-observations" to provide an estimate of the distribution of that statistic under the null-hypothesis of no anthropogenic change. If the observed value of the chosen statistic lies outside preassigned confidence level (typically 95 or 99%), then detection is claimed with a 0.05 (at the 95% confidence level) or 0.01 (at the 99% confidence level) probability of a false positive result (technically termed a "type-1" error). This approach to quantifying the risk of error requires that the model's simulation of internal climate variability (the natural variability) be realistic.

Typically, one wishes to discriminate between "signal" and "noise" when analyzing time series of the climate record (e.g. temperatures). While the meaning of signal and noise varies with context, there will always be a nonzero probability of incorrectly identifying noise as a deterministic trend or oscillation given limited data. The acceptable probability of such a "type-1" error must be specified, being the "nominal level" of any statistical test. Allen and Smith (1996) discuss the use of sophisticated statistical techniques to analyze records of climate data; in particular, they discuss how to extract physically meaningful oscillations and how to construct hypotheses for testing against "white noise" or "red noise."

Hasselmann (1997) distinguishes between "detection" of anthropogenic climate change and "attribution." The former seeks to rule out, at a certain confidence level, the possibility that an observed change is due to internal variability alone; the latter seeks to demonstrate that the observed change is consistent with the predictions of a climate model subjected to a particular forcing scenario and inconsistent with all physically plausible alternative causal explanations. Formal attribution is a much more demanding aim than detection. In fact, as Hasselmann observes, attribution is a logical impossibility unless physical arguments are used to confine attention *a priori* to a relatively small number of alternative hypotheses. The attribution framework proposed by Hasselmann and implemented by Hegerl *et al.* (1997) also relies heavily on model-simulated climate variability because "consistent" and "inconsistent" are formally defined as "within the bounds of variability as defined by a particular model."

Model simulations of internal climate variability have a number of shortcomings. At the simplest level, the known sources of error of variability in the observational record (e.g. observation errors) are not represented in current climate models. Even if these additional sources are included in the model, it will always be the case that variability on small space and time scales is likely to be underrepresented in the model's finite representation of the climate system. However, one does not require a model simulation of internal variability to be accurate in every respect for the model to be used for uncertainty analysis in climate change detection and attribution. In principle, only those aspects of model behavior that are relevant to the detection and attribution problem need to be realistic. For example, if the chosen detection statistic is the global mean temperature, then all that is required is an estimate of the variability of this quantity on the relevant time scales. The difficulty lies in determining which aspects of model variability are crucial to a particular detection or attribution problem and developing quantitative measures of model adequacy.

Simple checks, such as the comparison of global mean power spectra, can identify gross deficiencies in model variability, but the problem of how to remove the (presumed, but incompletely known) anthropogenic signal from the historical record prior to computing a power spectrum remains. Proxy and incomplete observations of the preindustrial period can help, but separating low-frequency climate variability from slow changes in the relationship between proxy observations and the climate variables they are supposed to represent remains a problem. There is also the intrinsic difficulty that paleoclimate observations are sparse, so paleoclimate reconstructions of any climate index must be contaminated with high spatial wavenumber components of variability that climate models are known to simulate poorly and that, it is hoped, are irrelevant to climate change detection.

The other problem with global mean power spectra is that a deficiency in the model's internal variability may not appear in the global mean while having a significant impact on the chosen detection statistic; this is necessarily true if a "centered" statistic is used, which is defined to be independent of the global mean. Recognizing this, Hegerl *et al.* (1996) use a linear response model to estimate and remove the anthropogenic signal from the historical record and then use the residual as an estimate of natural variability. While this is an improvement on simple power spectra, the approach relies on the adequacy of a simple linear model for both the pattern and amplitude of the anthropogenic signal.

Allen and Tett (1999) propose simple consistency checks for detection of model adequacy based on standard linear regression techniques, which can be applied to both the space–time and frequency domain approaches to optimal detection. The key advantage of the regression-based approach over detection schemes based on pattern correlation is that it provides information on relative amplitudes of response patterns in model and observations; correlations convey no amplitude information. These consistency checks aim to ensure that uncertainty estimates based on model-simulated variability are not demonstrably inaccurate.

As mentioned in Section 19.3, the forcings commonly considered in climate change experiments include anthropogenic forcings—changes in well mixed greenhouse gases (e.g. CO_2), sulfate aerosols, and stratospheric O_3—and natural forcings: changes in stratospheric volcanic aerosols and solar irradiance. Focusing on temperature changes, increases in CO_2 are characterized by a pattern of stratospheric cooling and tropospheric warming, and hemispherically symmetric temperature changes; increases in sulfate aerosol are characterized by a hemispherically asymmetric response, with increased cooling and reduced warming in the northern hemisphere (where the forcing is largest); decreases in stratospheric O_3 are characterized by stratospheric cooling, with maximum cooling at high latitudes in both hemispheres. Increases in stratospheric volcanic aerosols are characterized by stratospheric warming. Most of the energy associated with solar variability is deposited below the tropopause, but it is conceivable that the small forcing in the stratosphere can alter significantly stratospheric temperatures and winds.

A common approach when evaluating climate change is to assume that the observations (y) may be represented as the sum of simulated responses or signals (x_i, modified by an amplitude β_i) and internal climate variability (u, assumed to be normally distributed). The amplitude β_i represents the amount by which the ith signal has to be scaled to give the best fit to the observations. The optimal fingerprinting algorithm (a form of multivariate regression) is then used to estimate the amplitudes β_i and the uncertainty ranges. See Tett *et al.* (1999) for further details.

To test the robustness of the model results a number of approaches are taken. One approach is to use data from different models to explore the sensitivity of the results to model-dependent uncertainties in the definition of an anthropogenic signal. These uncertainties arise from model differences in, for example, the physical parametrizations and the model resolution (see Section 19.2). Santer *et al.* (1996) follow this approach. Another approach is to carry out sensitivity studies in which the analysis procedure is changed. Tett *et al.* (1999) follows this approach. In both approaches, independence of the results to the model used or the algorithm used is an indication of robustness.

19.5 LATEST RESULTS

At the time of the IPCC 1995 report, the most important results related to the issues of detection and attribution of climate change were:
1 The limited available evidence from proxy climate indicators suggests that the twentieth-century global mean temperature was at least as warm as that of any other century since at least 1400 CE. Data prior to 1400 CE are too sparse to allow the reliable estimation of global mean temperature.
2 Assessments of the statistical significance of the observed global mean temperature trend over the twentieth century have used a variety of new estimates of natural internal variability, and externally forced variability. These estimates are derived from instrumental data, paleoclimate data, simple and complex climate models, and statistical models fitted to observations. Most of the studies have detected a significant change and show that the ob-

served warming trend is unlikely to be entirely natural in origin.

3 More convincing recent evidence for the attribution of anthropogenic effects on climate is emerging from pattern-based studies, in which the modeled climate response to combined forcings by greenhouse gases and anthropogenic sulfate aerosols is compared with observed geographic, seasonal, and vertical patterns of temperature change. These studies show that such pattern correspondences increase with time, consistent with the increase in strength of the anthropogenic signal. Furthermore, the probability is very low that these correspondences could occur by chance as a consequence of natural internal variability only. The observed vertical patterns of temperature change are also inconsistent with those expected for solar and volcanic forcing.

4 Our ability to quantify the anthropogenic influence on global climate is currently limited because the expected signal is still emerging from the "noise" of natural variability, and because there are uncertainties in key factors. These include: the magnitude and patterns of long-term natural variability; the time-evolving pattern of forcing by, and response to, changes in concentrations of greenhouse gases and aerosols; and land surface changes. Notwithstanding these uncertainties, the balance of evidence suggests that there is a discernible anthropogenic influence on global climate.

Projections of future anthropogenic climate change (which focus on changes in global mean temperature and sea level) depend, among other things, on the scenarios used to force the climate model. Such scenarios typically include emissions of both greenhouse gases and aerosol precursors. At the time of IPCC 1995, the projections of global mean temperature and sea-level changes did not come directly from CGCMs, chiefly due to their computational expense. Instead, simple upwelling diffusion-energy balance models were used to interpolate and extrapolate the CGCM results. These models, used for similar tasks in previous IPCC reports, were calibrated to give the same globally averaged temperature response as the CGCMs.

Note that the IPCC 1995 report speaks of "pro-jections" and not predictions to emphasize that the climate simulations performed do not represent attempts to forecast the most likely (or "best estimate") evolution of the future climate. The projections are aimed at estimating responses of the climate system to possible forcing scenarios.

19.5.1 Global mean surface temperature response

Taking account of the range in the estimate of climate sensitivity (the likely equilibrium response of global surface temperature to a doubling of the equivalent CO_2 concentration), 1.5–4.5°C, with a best estimate of 2.5°C, and the full set of emission scenarios considered in IPCC 1995, the models project an increase of global temperature between 0.9 and 3.5°C. In all cases the average rate of warming would likely be greater than any seen in the past 10,000 years, but the actual annual to decadal changes would include considerable natural variability. Due to the thermal inertia of the oceans, the global mean temperature would continue to increase beyond the year 2100 even if concentrations of greenhouse gases were stabilized at that time.

19.5.2 Global mean sea-level response

Taking account of the range in the estimate of climate sensitivity and ice melt parameters, and the full set of emission scenarios, the models predict an increase of global mean sea level of between 13 and 94 cm. During the first half of the twenty-first century, the choice of emission scenario has relatively little effect on the projected sea-level rise due to the large thermal inertia of the ocean–ice–atmosphere climate system, but that choice has increasingly larger effects in the later part of the twenty-first century. Furthermore, because of the thermal inertia of the oceans, sea level would continue to rise for many centuries beyond 2100 even if concentrations of greenhouse gases were stabilized at that time. The projected rise in sea level is mainly due to thermal expansion as the oceans warm, but also due to increased melting of glaciers.

In the projections for sea-level change, the combined contributions of the Greenland and Antarctic ice sheets are projected to be relatively minor over the twenty-first century. However, the possibility of large changes in the volumes of these ice sheets (and, thus, in sea level) cannot be ruled out, although the likelihood is considered to be low.

Future changes in sea level will not occur uniformly around the globe. Recent CGCM experiments suggest that the regional responses could differ considerably, due to differences in heating and circulation changes. Furthermore, geological and geophysical processes cause vertical land movements and thus affect relative sea levels on local and regional scales.

Tides and wave and storm surges could be affected by regional climate changes, but future projections are, according to IPCC 1995, highly uncertain.

19.5.3 Spatial patterns

All model simulations, whether they are forced with increased concentrations of greenhouse gases and aerosols, or with increased greenhouse gas concentrations alone, show the following projected changes in temperature and precipitation: (i) generally greater surface warming of the land than of the oceans in winter; (ii) a minimum warming around Antarctica and in the northern North Atlantic, which is associated with deep oceanic mixing in these areas; (iii) maximum warming in high northern latitudes in late autumn and winter associated with reduced sea ice and snow cover; (iv) little warming over the Arctic in summer; (v) little seasonal variation of the warming in low latitudes or over the southern circumpolar ocean; (vi) a reduction in diurnal temperature range over land in most seasons and most regions; (vii) an enhanced global mean hydrological cycle; (viii) increased precipitation in high latitudes in winter.

Although there is less confidence in simulated changes in soil moisture than in those of temperature, some of the results concerning soil moisture are determined more by changes in precipitation and evaporation than by the response of the land surface scheme of the climate model. All model simulations, whether they are forced with increased concentrations of greenhouse gases and aerosols, or with increased greenhouse gas concentrations alone, produce predominantly increased soil moisture in high northern latitudes in winter. Over the northern continents in summer, changes in soil moisture are sensitive to the inclusion of aerosol effects.

In response to increasing greenhouse gases, most models show a decrease in the strength of the northern North Atlantic oceanic circulation, further reducing the strength of the warming around the North Atlantic. The increase in precipitation at high latitudes decreases the surface salinity, inhibiting the sinking of water at high latitudes, which drives the oceanic circulation.

19.5.4 Changes in variability and extremes

Small changes in the mean climate or climate variability can produce relatively large changes in the frequency of extreme events; a small change in the variability has a stronger effect than a similar change in the mean.

A general warming tends to lead to an increase in extremely high temperature events and a decrease in extremely low temperatures (e.g. frosts). New modeling results reinforce the view that variability associated with the enhanced hydrological cycle translates into prospects for more severe droughts and/or floods in some places and less severe droughts and/or floods in other places. By contrast, conclusions based on modeling results regarding extreme storm events are very uncertain. The representation of tropical storms in climate models is not yet mature enough to allow a proper assessment of the changes in their frequency that could be associated with climate change. Several CGCMs indicate that the ENSO-like variability they simulate continues with increased CO_2. Associated with the increase of tropical SSTs as a result of increased greenhouse concentrations, there could be enhanced precipitation variability associated with ENSO events in the increased CO_2 climate, especially over the tropical continents.

19.6 FUTURE AVENUES

The Earth's climate system sciences have developed an enormous capability over the past century to help society anticipate atmospheric and climatic phenomena and events. Progress continues today as improved observational and remote sensing capabilities provide more accurate resolution of the Earth's climate system processes. Furthermore, enhanced physical understanding, new modeling strategies, and powerful computers combine to provide improved simulations and predictions of the Earth's climate system.

The National Research Council (part of the US National Academy of Sciences) states as imperatives for atmospheric science (NRC 1998): (i) improve observation capabilities; and (ii) develop new observation capabilities. The NRC makes the following atmospheric research recommendations:

1 Resolve interactions at atmospheric boundaries and among different scales of flow (including the interaction of the atmosphere with other components of the Earth's climate system).

2 Extend forecasting to new areas (including climate variability and key chemical constituents).

3 Initiate and extend studies of issues related to: climate, weather, and health; management of water resources in a changing climate; and rapidly increasing emissions to the atmosphere.

The NRC also recommends: the development of a strategy to provide atmospheric information; ensuring access to atmospheric information; and assessing the benefits and costs associated with this information.

Climate research aims to understand the physical, chemical, and biological basis of climate and climate change in order to predict climate variability on seasonal to decadal and longer time scales, to assess anthropogenic influence on the climate, and to determine the role of climate change in affecting human activities and the environment. The NRC identifies the following scientific goals in climate change studies:

1 Understand the mechanisms of natural climate variability on time scales from seasons to centuries, and assess their relative importance.

2 Develop climate change prediction, application, and evaluation capabilities.

3 Predict future changes in the Earth's climate system and relate them to human activities.

The highest priority strategies for pursuing these goals are: (i) create a permanent climate observing system; (ii) extend the observational climate record via the development of integrated historical and proxy data sets; (iii) continue and expand diagnostic efforts and process studies to elucidate key climate variability and change processes; (iv) construct and evaluate models (increasingly to be used in an ensemble approach) that are increasingly comprehensive and take advantage of increasing computing power, and that incorporate all the major components of the Earth's climate system.

REFERENCES

Allen, M.R. & Smith, L.A. (1996) Monte Carlo SSA: detecting irregular oscillations in the presence of colored noise. *Journal of Climate* **9**, 3373–404.

Allen, M.R. & Tett, S.F.B. (1999) Checking for model consistency in optimal fingerprinting. *Climate Dynamics* **15**, 419–34.

AMIP (1995) *Proceedings of the First International AMIP Scientific Conference, Monterey, California.* WMO/TD-No. 732. WMO, Geneva.

Andrews, D.G., Holton, J.R. & Leovy, C.B. (1987) *Middle Atmosphere Dynamics.* Academic Press, London.

Arakawa, A. & Lamb, V.R. (1977) Computational design of the basic dynamical processes of the UCLA general circulation model. *Methods in Computational Physics* **17**, 173–265.

Arakawa, A. & Schubert, W.H. (1974) Interaction of a cumulus cloud ensemble with the large-scale environment. Part I. *Journal of the Atmospheric Sciences* **31**, 674–701.

Austin, J., Butchart, N. & Swinbank, R. (1997) Sensitivity of ozone and temperature to vertical resolution in a GCM with coupled stratospheric chemistry. *Quarterly Journal of the Royal Meteorological Society* **123**, 1405–31.

Bengtsson, L. & Shukla, J. (1988) Integration of space and in situ observations to study global climate change. *Bulletin of the American Meteorological Society* **69**, 1130–43.

Betts, A. (1986) A new convective adjustment scheme. Part I: Observational and theoretical basis. *Quarterly Journal of the Royal Meteorological Society* **112**, 677–91.

Betts, R.A., Cox, P.M., Lee, S.E. & Woodward, F.I. (1997) Contrasting physiological and structural vegetation feedbacks in climate change simulations. *Nature* **387**, 796–9.

Birchfield, G.E. & Grumbine, R.W. (1985) "Slow" physics of large continental ice sheets and underlying bedrock and its relation to the Pleistocene Ice Ages. *Journal of Geophysical Research* **83**, 11294–302.

Brankovic, C., Palmer, T.N., Molteni, F., Tibaldi, S. & Cubasch, U. (1990) Extended-range predictions with ECMWF models: time-lagged ensemble forecasting. *Quarterly Journal of the Royal Meteorological Society* **116**, 867–912.

Brieglieb, B.P. (1992) Longwave band model for thermal radiation in climate studies. *Journal of Geophysical Research* **97**, 11475–85.

Bryan, K. (1969) A numerical method for the study of the circulation of the World Ocean. *Journal of Computational Physics* **4**, 347–76.

Bryan, K. & Cox, M.D. (1967) A numerical investigation of the oceanic general circulation. *Tellus* **19**, 54–80.

Buizza, R. & Palmer, T.N. (1998) Impact of ensemble size on ensemble prediction. *Monthly Weather Review* **126**, 2503–18.

Carson, D.J. (1999) Climate modeling: achievements and prospects. *Quarterly Journal of the Royal Meteorological Society* **125**, 1–27.

Cattle, H. & Crossley, J. (1995) Modeling Arctic climate change. *Philosophical Transactions of the Royal Society of London* **A352**, 210–13.

Chou, M.-D., Ridgway, W.L. & Yan M.M.-H. (1995) Parametrizations for water vapor IR radiative transfer in both the middle and lower atmospheres. *Journal of the Atmospheric Sciences* **53**, 1203–8.

CLIMAP (1981) *Seasonal Reconstructions of the Earth's Surface at the Last Glacial Maximum*. Map Chart Series MC-36. Geological Society of America, Boulder, CO.

COHMAP (1988) Climatic changes of the last 18,000 years: observations and model simulations. *Science* **241**, 1043–52.

Cotton, W.R. & Pielke, R.A. (1995) *Human Impacts on Weather and Climate*. Cambridge University Press, New York.

Cox, P.M., Betts, R.A., Bunton, C.B., Essery, R.L.H., Rowntree, P.R. & Smith, J. (1999) The impact of a new land surface physics on the GCM simulation of climate and climate sensitivity. *Climate Dynamics* **15**, 183–203.

Crutzen, P.J. & Arnold, F. (1986) Nitric acid cloud formation in the cold Antarctic atmosphere: a major cause of the springtime ozone hole. *Nature* **324**, 651–5.

Cullen, M.J.P. (1993) The unified forecast/climate model. *Meteorological Magazine* **122**, 81–94.

Cusack, S., Slingo, A., Edwards, J.M. & Wild, M. (1998) The radiative impact of a simple aerosol climatology on the Hadley Centre atmospheric GCM. *Quarterly Journal of the Royal Meteorological Society* **124**, 2517–26.

Daley, R. (1991) *Atmospheric Data Analysis*. Cambridge University Press, New York.

Dameris, M., Grewe, V., Hein, R., Schnadt, C., Brühl, C. & Steil, B. (1998) Assessment of the future development of the ozone layer. *Geophysical Research Letters* **25**, 3579–82.

DeMore, W.B., Sander, S.P., Golden, D.M. *et al.* (1994) *Chemical Kinetics and Photochemical Data for Use in Stratospheric Modeling*. Evaluation Number 11. JPL 94-26.

Durran, D.R. (1999) *Numerical Methods for Wave Equations in Geophysical Fluid Dynamics*. Springer-Verlag, New York.

Edwards, D.P. (1988) Transmittance and radiance calculations: the Oxford line-by-line model GENLN2. In Lenoble, J. & Geleyn, J.-F. (eds) *International Radiation Symposium* (IRS '88). Deepak Publications, Hampton, VA.

Edwards, J.M. & Slingo, A. (1996) Studies with a flexible new radiation code. I: Choosing a configuration for a large-scale model. *Quarterly Journal of the Royal Meteorological Society* **122**, 689–719.

Fraedrich, K. & Muller, K. (1992) Climate anomalies in Europe associated with ENSO extremes. *International Journal of Climatology* **12**, 25–31.

Francis, P.N., Jones, A., Saunders, R.W., Shine, K.P., Slingo, A. & Sun, Z. (1994) An observational and theoretical study of the radiative properties of cirrus: some results from ICE '89. *Quarterly Journal of the Royal Meteorological Society* **120**, 809–48.

Fritts, D.C. & Lu, W. (1993) Spectral estimates of gravity wave energy and momentum fluxes. Part II. Parametrization of wave forcing and variability. *Journal of the Atmospheric Sciences* **50**, 3695–713.

Gaffen, D.J., Rosen R.D., Salstein, D.A. & Boyle, J.S. (1995) Validation of humidity, moisture fluxes and soil moisture in GCMs: report of AMIP diagnostic subproject 11. Part 2: Humidity and moisture flux fields. In Gates, W.L. (ed.) *Proceedings of the First International*

AMIP Scientific Conference Monterey, California. WMO/TD-No. 732. WMO, Geneva.

Garratt, J.R. (1992) *The Atmospheric Boundary Layer*. Cambridge University Press, Cambridge.

Gent, P.R. & McWilliams, J.C. (1990) Isopycnal mixing in ocean circulation models. *Journal of Physical Oceanography* **20**, 15–155.

Gibson, J.K., Kållberg P., Uppala, S., Hernández, A., Nomura, A. & Serrano, E. (1997) *ECMWF Re-analysis Project Report. 1. ERA Description*. European Centre for Medium-range Weather Forecasts, Reading.

Gregory, D. (1995) The representation of moist convection in atmospheric models. In *Parametrization of Sub-grid Scale Physical Processes*. ECMWF Seminar Proceedings. European Centre for Medium-range Weather Forecasts, Reading.

Gregory, D. & Morris, D. (1996) The sensitivity of climate simulations to the specification of mixed phase clouds. *Climate Dynamics* **12**, 641–51.

Gregory, D. & Rowntree, P.R. (1990) A mass flux convection scheme with representation of cloud ensemble characteristics and stability dependent closure. *Monthly Weather Review* **118**, 1483–506.

Haltiner, G.J. & Williams, T.R. (1980) *Numerical Prediction and Dynamic Meteorology*. Wiley, New York.

Hamilton, K. (ed.) (1996) *Gravity Wave Processes and Their Parametrization in Global Climate Models*. Springer-Verlag, New York.

Harries, J.E. (1996) The greenhouse Earth: a view from space. *Quarterly Journal of the Royal Meteorological Society* **122**, 799–818.

Harrison, E.F., Minnis, P., Barkstrom, B.R., Ramanathan, V., Cess, R.D. & Gibson, G.G. (1990) Seasonal variation of cloud radiative forcing derived from the Earth Radiation Budget Experiment. *Journal of Geophysics Research* **95**, 18687–703.

Hasselmann, K. (1997) Multi-pattern fingerprint method for detection and attribution of climate change. *Climate Dynamics* **13**, 601–11.

Hastenrath, S. (1991) *Climate Dynamics of the Tropics*. Kluwer Academic Publishers, Amsterdam.

Hegerl, G.C., von Storch, H., Hasselmann, K., Santer, B.D., Cubasch, U. & Jones, P.D. (1996) Detecting greenhouse gas-induced climate change with an optimal fingerprint method. *Journal of Climate* **9**, 2281–306.

Hegerl, G.C., Hasselmann, K., Cubasch, U. *et al.* (1997) On multi-fingerprint detection and attribution of greenhouse gas and aerosol forced climate change. *Climate Dynamics* **13**, 613–34.

Henderson-Sellers, A., McGuffie, K. & Pitman, A. (1996) The project for intercomparison of land-surface param-

etrization schemes (PILPS): 1992–1995. *Climate Dynamics* **12**, 849–59.

Hines, C.O. (1993) The saturation of gravity waves in the middle atmosphere. Part IV: Cutoff of the incident wave spectrum. *Journal of the Atmospheric Sciences* **50**, 3045–60.

Hobbs, P.V. & McCormick M.P. (eds) (1988) *Aerosols and Climate*. Deepak, Hampton, VA.

Houghton, J.T., Jenkins, G.J. & Ephraums, J.J. (eds) (1990) *Climate Change. The IPCC Scientific Assessment*. Cambridge University Press, Cambridge.

Houghton, J.T., Callander, B.A. & Varney S.K. (eds) (1992) *Climate Change 1992. The Supplementary Report to the IPCC Scientific Assessment*. Cambridge University Press, New York.

Houghton, J.T., Meira Filho, L.G., Bruce, J. *et al.* (eds) (1995) *Climate Change 1994. Radiative Forcing of Climate Change*. Cambridge University Press, Cambridge.

Hurrell, J.W. & Trenberth, K.E. (1997) Spurious trends in satellite MSU temperatures from merging different satellite records. *Nature* **386**, 164–7.

Hyde, W.T., Crowley, T.J., Kim, K.Y. & North, G.R. (1989) A comparison of GCM and energy balance model simulations of seasonal temperature changes over the past 18,000 years. *Journal of Climate* **2**, 864–87.

IPCC (1996) *Climate Change 1995. The Science of Climate Change. Summary for Policymakers and Technical Summary of the Working Group I Report*. IPCC.

Kalnay, E., Kanamitsu, K., Kistler, R. *et al.* (1996) The NCEP/NCAR 40-Year Reanalysis Project. *Bulletin of the American Meteorological Society* **77**, 437–71.

Kandel, R., Viollier, M., Raberanto, P. *et al.* (1998) The ScaRab earth radiation budget dataset. *Bulletin of the American Meteorological Society* **79**, 765–83.

Kuo, H.L. (1974) Further studies of the parametrization of the influence of cumulus convection on large-scale flow. *Journal of the Atmospheric Sciences* **31**, 1232–40.

Lacis, A.A. & Hansen, J.E. (1974) A parametrization for the absorption of solar radiation in the Earth's atmosphere. *Journal of the Atmospheric Sciences* **31**, 118–33.

Lacis, A.A. & Mishchenko, M.I. (1995) Aerosol forcing of climate. In Charlson, R.J. & Heintzenberg, J. (eds) *Climate Forcing, Sensitivity, and Response*. Wiley, Chichester.

Lacis, A.A. & Oinas, V. (1991) A description of the correlated k distribution method for modeling non-gray gaseous absorption, thermal emission and multiple scattering in vertically inhomogeneous atmospheres. *Journal of Geophysical Research* **96**, 9027–63.

Lautenschlager, M. & Herterich, K. (1990) Atmospheric response to ice age conditions: climatology near the Earth's surface. *Journal of Geophysical Research* **95**, 22547–57.

Lefèvre, F., Brasseur, G.P., Folkins, I., Smith, A.K. & Simon, P. (1994) Chemistry of the 1991–92 stratospheric winter: three-dimensional model simulations. *Journal of Geophysical Research* **99**, 8183–95.

Lindzen, R.S. (1981) Turbulence and stress owing to gravity wave and tidal breakdown. *Journal of Geophysical Research* **86**, 9707–14.

Livezey, R.E. (1990) Variability of skill of long-range forecasts and implications for their use and value. *Bulletin of the American Meteorological Society* **71**, 300–9.

Lorenz, E.N. (1963) Deterministic nonperiodic flow. *Journal of the Atmospheric Sciences* **20**, 130–41.

Mahfouf, J.F., Cariolle, D., Royer, J.-F., Geleyn, J.-F. & Timbal, B. (1993) Response of the Météo-France climate model to changes in CO_2 and sea surface temperature. *Climate Dynamics* **9**, 345–62.

Marsiat, I. (1994) Simulation of the northern hemisphere continental ice sheets over the last glacial-interglacial cycle: Experiments with a latitude–longitude vertically integrated ice sheet model coupled to a zonally averaged climate model. *Palaeoclimate* **1**, 59–98.

Mitchell, J.F.B., Johns, T.C., Gregory, J.M. & Tett, S.F.B. (1995) Climate response to increasing levels of greenhouse gases and sulphate aerosols. *Nature* **376**, 501–4.

NRC (1998) *The Atmospheric Sciences: Entering the Twenty-first Century*. National Research Council. National Academy Press, Washington, DC.

Oerlemans, J. (1994) Quantifying global warming from the retreat of glaciers. *Science* **264**, 243–5.

Ohmura, A., Wild, M. & Bengtsson, L. (1996) Present and future mass balance of the ice sheets simulated with a GCM. *Annals of Glaciology* **23**, 187–93.

Palmer, T.N. & Anderson, D.L.T. (1994) The prospects for seasonal forecasting—a review paper. *Quarterly Journal of the Royal Meteorological Society* **120**, 755–93.

Peltier, W.R. (1994) Ice age paleotopography. *Science* **265**, 195–201.

Phillips, T.J. (1994) *A Summary Documentation of the AMIP Models*. Program for Climate Model Diagnosis and Intercomparison (PCMDI) Report No. 18. PCMDI, Livermore, CA.

Ramaswamy, V., Schwarzkopf, M.D. & Shine, K.P. (1992) Radiative forcing of climate from halocarbon-induced global stratospheric ozone loss. *Nature* **355**, 810–12.

Rind, D., Shindell, D., Lonergan, P. & Balachandran, N.K. (1998) Climate change and the middle atmosphere. Part III: The doubled CO_2 climate revisited. *Journal of Climate* **11**, 876–94.

Ritter, B. & Geleyn, J.-F. (1992) A comprehensive radiation scheme for numerical weather prediction models with potential applications in climate simulations. *Monthly Weather Review* **120**, 303–25.

Robert, A. (1981) A stable numerical integration scheme for the primitive meteorological equations. *Atmosphere and Oceans* **19**, 35–46.

Robert, A. (1982) A semi-Lagrangian and semi-implicit numerical integration scheme for the primitive meteorological equations. *Journal of the Meteorological Society of Japan* **60**, 319–24.

Rothman, L.S., Gamache, R.R., Tipping, R.H. *et al.* (1992) The HITRAN molecular database: editions of 1991 and 1992. *Journal of Quantitative Spectroscopy and Radiative Transfer* **48**, 469–507.

Rowland, F.S. & Molina, M.J. (1975) Chlorofluoromethanes in the environment. *Reviews of Geophysics and Space Physics* **13**, 1–36.

Salby, M.L. (1996) *Fundamentals of Atmospheric Physics*. Academic Press, London.

Santer, B.D., Taylor, K.E., Wigley, T.M.L., Penner, J.E., Jones, P.D. & Cubasch, U. (1995) Towards the detection and attribution of an anthropogenic effect on climate. *Climate Dynamics* **12**, 77–100.

Santer, B.D., Taylor, K.E., Wigley, T.M.L. *et al.* (1996) A search for human influences on the thermal structure of the atmosphere. *Nature* **382**, 39–46.

Sellers, P.J., Bounoua, L., Collatz, G.J. *et al.* (1996) Comparison of radiative and physiological effects of doubled atmospheric CO_2 on climate. *Science* **271**, 1402–6.

Shine, K.P. & Forster, P.M.D.F. (1999) The effect of human activity on radiative forcing of climate change: a review of recent developments. *Global and Planetary Change* **20**, 205–25.

Simmons, A.J., Untch, A., Jakob, C., Kållberg, P. & Undén, P. (1999) Stratospheric water vapor and tropical tropopause temperatures in ECMWF analyses and multi-year simulations. *Quarterly Journal of the Royal Meteorological Society* **125**, 353–86.

Slingo, J.M., Sperber, K.R., Boyle, J.S. *et al.* (1996) Intraseasonal oscillations in 15 atmospheric general circulation models: results from an AMIP diagnostic subproject. *Climate Dynamics* **12**, 325–57.

Smagorinsky, J. (1963) General circulation experiments with the primitive equations. I: Basic experiment. *Monthly Weather Review* **99**, 99–164.

Staniforth, A.N. & Côte, J. (1991) Semi-Lagrangian integration schemes for atmospheric models—a review. *Monthly Weather Review* **119**, 2206–23.

Stouffer, R.J., Manabe, S. & Vinnikov, K.Y. (1994) Model assessment of the role of natural variability in recent global warming. *Nature* **367**, 634–6.

Tett, S.F.B., Johns, T.C. & Mitchell, J.F.B. (1997) Global and regional variability in a coupled AOGCM. *Climate Dynamics* **13**, 303–23.

Tett, S.F.B., Stott, P.A., Allen, M.R., Ingram, W.J. & Mitchell, J.F.B. (1999) Causes of twentieth-century temperature changes near the Earth's surface. *Nature* **399**, 569–72.

Tiedtke, M. (1993) Representation of clouds in large-scale models. *Monthly Weather Review* **121**, 3040–61.

Trenberth, K.E. & Olson, J.G. (1988) An evaluation and intercomparison of global analyses from NMC and ECMWF. *Bulletin of the American Meteorological Society* **69**, 1045–57.

Wilson, M.F. & Henderson-Sellers, F. (1985) A global archive of land cover and soils data for use in general circulation climate models. *Journal of Climate* **5**, 119–43.

WMO (1999) *Scientific Assessment of Ozone Depletion: 1998*. Global Ozone Research and Monitoring Project Report No. 44. World Meteorological Organization, Geneva.

Yanai, M., Esbensen, S. & Chu, J.-H. (1973) Determination of the bulk properties of tropical cloud clusters from large-scale heat and moisture budgets. *Journal of the Atmospheric Sciences* **30**, 611–27.

20 Critical Levels and Critical Loads as a Tool for Air Quality Management

WIM DE VRIES AND MAXIMILIAN POSCH

20.1 INTRODUCTION

20.1.1 The critical load concept and its use in policy-making

In the terminology of air pollution impacts, critical levels and critical loads refer to the concentration levels and deposition loads of air pollutants (SO_2, NO_x, NH_3, and O_3), respectively, below which no adverse direct effects are expected. The concept of critical levels and loads is based on the concept of thresholds. Woodwell (1976) defined an ecological threshold as the "maximum exposure (to toxins) that has no effect." Although there are a number of difficulties with the concept of a (nonzero) threshold for ecosystem damage due to atmospheric pollution (Gorham 1976; EPRI 1991), a Canadian working group on impact assessment proposed in 1983 a "target load" to aquatic ecosystems of 20 kg of wet sulfate per hectare per year (discussed in Gorham et al. 1984). This value is still used in Canadian environmental policy, whereas it was in Europe where the concept of critical loads and levels has been developed further during the past 15 years and influenced international agreements of the past seven years.

It was for work under the 1979 Convention on Long-range Transboundary Air Pollution (LRTAP) that the critical loads/levels concept was adopted and further developed. While the earlier protocols to the Convention (1985 on sulfur, 1988 on nitrogen oxides, 1991 on volatile organic compounds) were standstill agreements or required flat-rate emission reductions, the 1994 Sulphur Protocol was the first to consider ecosystem vulnerability in terms of critical loads for the formulation of reduction requirements. The signing of this protocol was a first climax of the work of the Task Force on Mapping Critical Levels/Loads, which was established in 1989 under the Working Group on Effects (WGE) of the Executive Body of the LRTAP Convention. Under this task force, critical load data from individual countries are collected, collated and mapped by the Coordination Centre for Effects (e.g. Posch et al. 1999) and provided to the relevant UNECE bodies under the LRTAP Convention as well as the European Commission to formulate emission reduction strategies in Europe.

20.1.2 Scientific discussions on the topic

The scientific discussion on critical loads (depositions) started at a workshop organized by the Nordic Council of Ministers (NMR) in 1986 in Sundvollen (Norway) and provided, for the first time, estimates of critical loads of sulfur and nitrogen for forest soils, groundwater, and surface waters (Nilsson 1986). The first workshop on critical loads held under the auspices of the United Nations Economic Commission for Europe (UNECE), which provided the permanent secretariat for the LRTAP Convention, was organized in 1988 by the NMR at Skokloster (Sweden). At this workshop, the still-valid definition of a critical load as "the quantitative estimate of an exposure to one or more pollutants below which significant harmful effects on specified sensitive elements of the environment do not occur according to present knowledge" was brought up (Nilsson & Grennfelt 1988).

As the role of nitrogen in the acidification of soils and surface waters gained increasing attention at the end of the 1980s in both the scientific and policy arenas, a workshop was organized by the NMR and the US EPA on that topic in Copenhagen in 1988 (Malanchuk & Nilsson 1989). The purpose of that workshop was to review the state of science on the role of nitrogen in the acidification of the environment. Finally, the foundation for the mapping of critical loads in the ECE countries was laid at a UNECE workshop held in 1989 in Bad Harzburg (Germany), resulting in a manual for mapping critical levels and loads, which has been updated periodically (UNECE 1996a). Furthermore, in a workshop on critical loads of nitrogen organized by the NMR in Lökeberg (Sweden) in 1992, recommendations for deriving critical loads of nitrogen and their exceedances were elaborated (Grennfelt & Thörnelöf 1992). Remaining open questions were discussed at a UNECE workshop in Grange-over-Sands (United Kingdom) in 1994, organized by the UK Department of Environment (Hornung *et al.* 1995).

Parallel to the development of critical loads, critical levels (concentrations), defined as "concentrations of pollutants in the atmosphere above which direct adverse effects on receptors, such as human beings, plants, ecosystems or materials, may occur according to present knowledge" (UNECE 1996a), were developed. Critical levels for sulfur dioxide, nitrogen oxides, ammonia, and ozone were first defined at UNECE workshops in Bad Harzburg (Germany) in the late 1980s and were further elaborated at a UNECE workshop in Egham (United Kingdom) in 1992 (Ashmore & Wilson 1994). Since critical levels for the concentrations of sulfur and nitrogen species are not exceeded on a large scale in Europe, the further development concentrated on critical levels for ozone. A first UNECE workshop concentrating solely on ozone critical levels was held in 1993 in Bern (Switzerland) at which a long-term critical level expressed as accumulated concentration above a threshold (AOT) was adopted (Fuhrer &

Achermann 1994). This concept was evaluated and finalized at a workshop in Kuopio (Finland) in 1996 (Kärenlampi & Skärby 1996).

Critical loads for heavy metals are a more recent development. After a pilot study carried out in the Netherlands (Van den Hout *et al.* 1999), a draft manual for the calculation and mapping of heavy metals (De Vries and Bakker 1996) was discussed and amended at a UNECE workshop in Bad Harzburg (Germany). To date, these updated manuals (De Vries & Bakker 1998; De Vries *et al.* 1998) are used by several European countries to explore the usefulness of the critical loads concept for heavy metals. The approach has, however, not yet been used to calculate and map critical loads for heavy metals in support of European pollution reduction policy. The 1998 protocol on heavy metal emission abatement under the LRTAP Convention is based on flat rate reductions using best available abatement techniques, ignoring differences in susceptibility of receptors to metal input. The ultimate aim is to use a critical load approach as a next step in abating heavy metal emissions under the Convention, as it may lead to more cost-efficient emission reductions.

20.1.3 *Outline of the chapter*

This short historical outline describes only the development within the UNECE context, and cannot discuss the many developments and applications in various countries. In the following sections the methods for calculating critical levels and loads, as described in the above-mentioned documents, are presented and discussed, together with extensions and new developments we believe are useful and should be used. We first give an overview of critical (concentration) levels of SO_2, NO_x, NH_3, and O_3 related to direct impacts on the vegetation canopy, including a method to link those critical levels to critical loads (Section 20.2). The remaining part is dedicated to critical loads related to soil-mediated indirect effects of nitrogen, acidity, and heavy metals. We first briefly review the methods used to calculate critical loads for terrestrial ecosystems (Section 20.3), followed by an overview of impacts, methods, and calculation examples of critical loads for nitrogen (Section 20.4), acidity (Section 20.5), and heavy metals (Section 20.6). Subsequently, we summarize how critical loads are used in policy assessments by international bodies such as the UNECE (Section 20.7). In Section 20.8 a discussion and evaluation of the various approaches is given.

20.2 CRITICAL LEVELS OF AIR POLLUTANTS

Critical (concentration) levels of SO_2, NO_x, NH_3, and O_3 have been derived for forests from a compilation of literature on dose (i.e. pollutant concentration × duration of exposure) response relationships (e.g. Ashmore & Wilson 1994; UNECE 1990, 1996a). Values are presently defined independent of location, tree species and/or soil type. The only differentiation relates to exposure duration, i.e. "acute" effects due to short-term exposures (one-hour to one-day mean values) and "chronic" effects due to long-term exposures (annual mean values). Critical concentration levels presented in this chapter are based on this literature information (UNECE 1990, 1996a).

20.2.1 Sulfur and nitrogen compounds

A summary of critical levels of S and N compounds related to acute effects by short-term exposures and to chronic effects by long-term exposures is given in Table 20.1. The critical values are related to adverse effects on all vegetation types except SO_2, where the values refer to forest ecosystems and natural vegetation.

Table 20.1 Critical levels ($\mu g\,m^{-3}$) used for short-term and long-term exposures to SO_2, NO_x, and NH_3 (based upon UNECE 1990 and Ashmore & Wilson 1994; for a summary see also UNECE 1996a).

Exposure	SO_2	NO_x	NH_3
Short-term (one day; for NO_x four hours)	70	95	270
Long term (one year)	20	30	8

For NO_x and NH_3, separate critical levels have not been set for different vegetation classes because of a lack of information. The sensitivity is, however, thought to decrease according to (semi) natural vegetation greater than forests greater than crops. Critical levels are related to both growth stimulation in response to the fertilizer effect of nitrogen and adverse physiological effects at toxic levels. The critical level for NO_x is based on the sum of NO and NO_2 concentrations. The knowledge to establish separate critical levels for the two gases is still lacking, even though there is evidence that NO is more phytotoxic than NO_2 (Morgan et al. 1992; UNECE 1996a).

20.2.2 Ozone

Critical levels for ozone have been defined by the sum of the hourly ozone concentrations during daylight hours (defined as the time with global radiation $>50\,W\,m^{-2}$) above a threshold value of 40 parts per billion (p.p.b.) during the growing season. The threshold of 40 p.p.b. is based on a reduction in annual biomass increment. The use of daylight hours only is based on the observation that ozone uptake during nighttime is negligible due to the closure of stomatal pores (Ashmore & Wilson 1994). This accumulated ozone exposure over a threshold of 40 p.p.b., called AOT40, is expressed in p.p.b. hours or p.p.m. hours.

For forests, a critical AOT40 of 10,000 p.p.b. h (10 p.p.m. h) accumulated over a six-month growing season (April–September) has been adopted, independent of tree species. Therefore, the following caveats should be kept in mind (Kärenlampi & Skärby 1996):

1 The value is based on data for beech, since deciduous trees are considered more sensitive to ozone than conifers.

2 A cut-off concentration of 30 p.p.b. may be more appropriate in the Nordic countries. In this case, however, the accumulated dose might include background ozone.

3 The six-month growing season starting in April leads to an underestimation for Mediterranean countries and an overestimation for the Nordic countries.

20.2.3 Relationship between critical levels and critical loads

Critical loads or critical deposition levels can be derived from critical concentration levels by first multiplying the critical level by a dry deposition velocity (v_d), leading to an estimate of the critical dry deposition, and subsequently multiplying the results with a total/dry deposition ratio (r_{td}). Estimates of v_d and r_{td} depend on the geographic region and the surface roughness. Examples of estimated critical loads from given critical levels as a function of both parameters are given in Table 20.2. The critical load ranges cover the lower and upper limits of both the dry deposition velocity (v_d) and the total/dry deposition ratio (r_{td}). The critical loads given in Table 20.2 are generally much higher than those related to soil-mediated impacts of S and N compounds.

20.2.4 Uncertainties

There are several assumptions that were made in deriving critical levels that should be considered to assess their reliability. First, critical levels are based on dose–response functions using statistical methods, such as regression curves, to extrapolate the findings in a restricted range of concentrations. Furthermore, critical levels for one pollutant have been set in the context of a pollutant mixture. Available experimental results show that pollutant combination effects are mainly additive (sum of individual effects) or synergistic (more than additive). Synergism is specifically known to occur between NO_x and other pollutants, such as SO_2 and O_3. The definition of the critical level for NO_x thus assumes that SO_2 and O_3 are present at concentrations close to their critical level (Ashmore & Wilson 1994). Finally, critical levels are influenced by climatic conditions (temperature, relative humidity), soil properties, and growth factors (leaf age, developmental stage). Mäkelä and Schöpp (1990), for example, suggested that the long-term critical level for SO_2 should equal $15\,\mu g\,m^{-3}$ in areas where the effective temperature sum above 5°C equals 1000 degree-days.

20.3 METHODS TO DERIVE CRITICAL LOADS FOR TERRESTRIAL ECOSYSTEMS

With respect to the assessment of critical loads, a first distinction can be made between a direct empirical approach, an indirect soil model based approach, and an integrated model-based approach (Fig. 20.1). A description of those methods, including a further distinction in deterministic and probabilistic methods (see Fig. 20.2), is given below.

20.3.1 Empirical approaches

Deterministic approaches

In the deterministic empirical approach, critical loads are derived from observed relationships

Table 20.2 Critical loads of S and N compounds derived from critical levels for short-term and long-term exposures to SO_2, NO_x, and NH_3 as a function of dry deposition velocity and total/dry deposition ratio.

Compound	Critical level ($\mu g\,m^{-3}$)	Velocity (v_d) ($m\,s^{-1}$)	Total/dry ratio (r_{td})	Critical load ($mol\,ha^{-1}\,yr^{-1}$)
SO_2	20	0.010	1.5–2.5	1480–2460
NO_x	30	0.002	1.5–2.5	740–1230
NH_3	8	0.012	1.5–2.5	2670–4450

Note that the multiplication of a critical level in $\mu g\,m^{-3}$ with a dry deposition velocity in $m\,s^{-1}$ gives a deposition rate in $\mu g\,m^{-2}\,s^{-1}$, which in turn has to be multiplied by the number of seconds in a year (31,536,000) and divided by the molar weight multiplied by 100 to get a critical load in $mol\,ha^{-1}\,yr^{-1}$. Molar weights equal 64 for SO_2, 38.4 for NO_x (assuming a critical level of $18\,\mu g\,m^{-3}$ for NO_2 and $12\,\mu g\,m^{-3}$ for NO) and 17 for NH_3.

between atmospheric deposition and effects on "specified sensitive elements" within an ecosystem (ecosystem status) by correlative or experimental research. Such an approach has been used for nitrogen in view of adverse impacts on biodiversity, since nitrogen has a dominating influence on the species diversity of terrestrial vegetation. Critical N loads have, for example, been estimated by comparing the N deposition on grass-dominated and heather-dominated heathlands (e.g. Liljelund & Torstensson 1988) or by experimental investigation of the biomass development of grasses in heathlands as a function of N input (e.g. Roelofs 1986). Tables with empirical critical

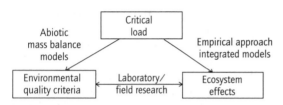

Fig. 20.1 Research methods to derive critical loads (each arrow indicates a relationship that can be assessed by correlative research, process research, and/or model research).

load values and estimates of their reliability have been compiled over the years, both for acidity critical loads and for critical loads of nitrogen as a nutrient (e.g. in UNECE 1996a).

Probabilistic approaches

The drawback of deterministic empirical critical (nitrogen) loads is that they do not give equal protection percentages to the species occurring in different ecosystems. The concept of risk assessment may be used in this perspective as an alternative, because it provides a framework to achieve more standardization in the assessment of protection levels for different environmental problems (e.g. Latour *et al.* 1994). Most progress in assessing and quantifying ecological risks has been made in the field of toxicological stress. There the maximum tolerable concentration (MTC) is chosen as the environmental concentration of a compound (e.g. heavy metal) at which (theoretically) 95% of the species are fully protected. MTCs are calculated by extrapolation of "no observed effects concentrations" (NOEC levels) for single species to an ecosystem, mostly assuming a log-logistic or log-normal distribution of species sensitivities (Slooff

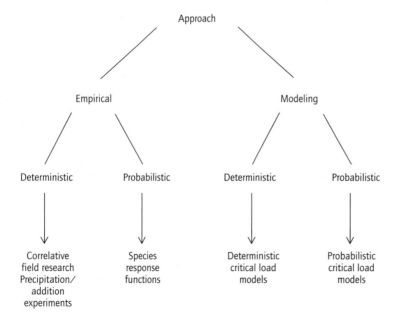

Fig. 20.2 Overview of possible empirical and model-based deterministic and probabilistic approaches to derive critical loads.

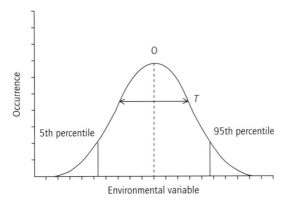

Fig. 20.3 Probability of the occurrence of a species as a function of an environmental variable (species–response function). Between the 5th and 95th percentiles, species are considered protected.

1992). At the species level the risks are assessed based on the species–response function, which describes the occurrence probability of a species as a function of an environmental variable. The species–response function can be characterized by its optimum (O) and standard deviation. Latour *et al.* (1994) used the 5th and 95th percentiles of the species–response curves as NOEC-like measures for the risk at the species level (Fig. 20.3). The 5th percentile corresponds to a reduced occurrence probability due to "limitation," the 95th percentile to one due to "intoxication." Species are considered protected between the 5th and 95th percentiles of a given environmental variable.

An example of a probabilistic empirical model for deriving critical N loads is the MOVE (multiple-stress model for vegetation; Latour & Reiling 1993). MOVE predicts the occurrence probability of about 700 species as a function of three abiotic soil factors: soil acidity, nitrogen availability, and soil moisture. With regression statistics the occurrence probability of a species can be calculated for each combination of the soil factors or for each factor separately (species–response function). Since combined samples of vegetation and environmental variables are rare, the indication values of plant species by Ellenberg (1983) are used to assess the abiotic soil conditions. Combined samples of vegetation with environ-

mental variables are used to calibrate Ellenberg indicator values with quantitative values of the abiotic soil factors. Latour *et al.* (1994) used MOVE to derive critical N loads for fertilized grasslands based on NOECs for 275 species.

20.3.2 *Deterministic soil models*

General approach

In a model-based approach, critical loads are derived with models using critical limits for element concentrations or element ratios in the ecosystem, based on dose–response relationships between these critical limits and the ecosystem status. Using this approach, a critical load of nitrogen, acidity, or heavy metals equals the load causing a concentration in a compartment (soil, groundwater, plant, etc.) that does not exceed a critical limit, thus preventing "significant harmful effects on specified sensitive elements of the environment." Consequently, the selection of critical limits is a step of major importance in deriving a critical load. In defining a critical load one aims at long-term protection of the ecosystem. In this context, protection can be defined as the situation where: (i) no further net accumulation of nitrogen or heavy metals or loss of exchangeable base cations occurs; or (ii) accumulation of nitrogen, acidity, or heavy metals stays below critical limits in defined compartments. The first, so-called precautionary principle, approach implies that the present nitrogen, acidity, or metal status is considered the critical limit above which no further increase is accepted. In the second effect-based approach, the critical limits are based on adverse effects on (parts of) the ecosystem and the nitrogen, acidity, or metal concentrations should stay below those limits. To date, soil chemical criteria (e.g. critical nitrate, aluminum or metal concentrations, or critical base cation to aluminum ratios) have mainly been used to derive critical loads with simple steady-state models. The largest uncertainty in these calculations remains the relation between the critical limits and the "harmful effects," as discussed below.

Apart from assessing reliable critical limits, a major challenge in quantifying critical loads is to

Table 20.3 Examples of critical limits used to derive critical loads.

Element	Soil	Soil solution	Ground water	Surface water	Plant
Nitrogen	$\Delta N = 0$	$N < 0.02$–$0.04\,mol_c\,m^{-3}$	$NO_3 < 0.4$–$0.8\,mol_c\,m^{-3}$	$N < 0.16\,mol\,m^{-3}$	$N < 18\,g\,kg^{-1}$ $N/K < 4\,g\,g^{-1}$
Acidity	$\Delta BS = 0$	$Al/(Ca + Mg + K)$ <0.5–2.0	$pH < 6.0$	$pH < 5.3$–6.0	–
	$\Delta Al_{ox} = 0$	–	$Al < 0.02\,mol_c\,m^{-3}$	$Al < 0.003\,mol_c\,m^{-3}$	–
Metals	$\Delta M = 0$	$Cd < 2.0\,mg\,m^{-3}$	$Cd < 0.1$–$2.0\,mg\,m^{-3}$		–

transform complex empirical findings and conceptual insight into robust ecosystem models, requiring a minimum of data, while still accounting for the most important factors controlling critical loads. In general, critical loads can be calculated either by steady-state models or by dynamic models with different degrees of complexity, as discussed below. The models derived are all deterministic, but a probabilistic approach is possible by combining them with a Monte Carlo simulation analyses, including ranges in input data instead of fixed values.

Critical limits

Critical loads, i.e. maximum deposition levels of nitrogen, acidity, or heavy metals, are derived by setting a limit to the leaching of nitrogen, acidity, or metals, either directly (dissolved concentrations) or indirectly, using soil chemical criteria combined with process descriptions to relate them to dissolved concentrations. An assessment of those limits is thus crucial. In the so-called precautionary principle approach, no further net accumulation of nitrogen or heavy metals or loss of exchangeable base cations or aluminum is allowed. This implies the need for a present value for the concentration of nitrogen, metals, exchangeable base cations, or Al compounds in the soil. Critical limits for the soil solution or for other compartments, such as plants, based on defined effects, are needed in the effects-based approach. In Table 20.3 an overview is given of relevant critical limits for nitrogen, acidity, and heavy metals (limited to cadmium as an example).

Steady-state models

At present critical loads for nitrogen and acidity are mostly calculated with a steady-state single-layer soil model based on a simple mass balance approach. This approach is also the standard for calculating critical loads for heavy metals. These models calculate deposition levels, which avoid the violation of a chosen soil chemical criterion in a steady-state situation. Therefore, processes with a finite time scale, such as cation exchange and sulfate adsorption, are not included. The models include a simple description of the inputs (deposition), outputs (leaching), and permanent sources and sinks of major ions within the rooting zone. For nitrogen, the processes involved are net retention of N by net uptake, denitrification, and net N immobilization. For acidity, processes involved are net retention of N as above, net input of base cations (weathering minus net uptake), and release of Al from silicates and Al hydroxides. Ions involved include sulfate, nitrate, base cations, and aluminum, and their mass balances are combined with a charge balance of those ions in the soil leachate. For heavy metals, it includes mainly adsorption, but other processes, such as weathering and net uptake, can also be included.

In steady-state one-layer soil models the soil is considered a single homogeneous compartment with a depth equal to the root zone. This implies that internal soil processes (such as weathering and uptake) are evenly distributed over the soil profile, and all physicochemical constants are assumed uniform in the whole profile. Furthermore, the simplest possible hydrology is assumed: the annual water flux leaving the root zone equals the annual precipitation minus evapotranspiration. For acidity, this model (called the SMB model) has been and is widely used to produce maps of critical loads of S and N on a European scale (Posch *et al.* 1995, 1999). The methods for computing critical loads for acidity in lakes are similar to those for soils (since soils in the terrestrial catchment influence lake water chemistry), augmented by simple formulations of in-lake processes such as sulfur and nitrogen retention (Posch *et al.* 1997; Henriksen & Posch 2000). For heavy metals, the lake models deviate more strongly from the soil models (De Vries *et al.* 1998). However, these models are not considered in this chapter, which focuses on terrestrial ecosystems.

It is also possible to use multilayer models including element cycling (litterfall, mineralization, and uptake). An example for acidification is the MACAL model (De Vries *et al.* 1994b). Multilayer models have also been constructed for heavy metals (De Vries & Bakker 1998). They are, however, not often used because of their high demand for input data. The steady-state one-layer soil models are primarily developed for applications on a large regional scale, for which data are scarce. In this chapter we thus concentrate on that type of models to discuss the critical load concept.

Dynamic soil models

Dynamic soil models simulate the same processes as steady-state soil models, but include additional processes that play a role on a finite time scale. For nitrogen and heavy metals it includes a dynamic description of nitrogen and metal retention (immobilization or adsorption). Additional processes in dynamic soil acidification models are cation exchange and sulfate adsorption. These models can also be used to calculate critical loads by running the model until a steady state is reached. By trial and error the (constant) deposition level is calculated that fulfills a chosen chemical criterion (e.g. Warfvinge & Sverdrup 1995, where critical acid loads for Sweden are derived with the dynamic model SAFE). Critical loads calculated in this way are equal to those derived by steady-state models, if the process-

es in both models are modeled in the same way. Dynamic soil models can furthermore be used to derive so-called target loads by considering a finite time period (e.g. one forest rotation) in which the system is allowed (or has) to reach a chemical criterion. Unlike the critical load, the present acidification status of the soil system and the magnitude of time-limited processes influence the target load.

Finally, dynamic soil models can also be used to derive the time period before the system reaches a chosen soil chemical criterion for a given deposition scenario. Depending on the present soil status, the model thus calculates the time period before risk increases or before the system starts to recover. Dynamic models are most commonly used in this context (see Cosby *et al.* 1989; de Vries *et al.* 1994c). Some of the more widely used simple dynamic soil models are MAGIC (Cosby *et al.* 1985), SAFE (Warfvinge *et al.* 1993), and SMART (De Vries *et al.* 1989). As with the steady-state models, there are also multilayer dynamic models, but those models are hardly used to calculate critical loads because of their data requirements.

20.3.3 *Integrated soil–vegetation models*

Deterministic models

A major drawback of steady-state soil models is the neglect of biotic interactions. For example, vegetation changes are mainly triggered by a change in N cycling (N mineralization; Berendse *et al.* 1987). Furthermore, the enhancement of diseases by elevated N inputs, such as heather beetle outbreaks, may stimulate vegetation changes (Heil & Bobbink 1993). Consequently, dynamic soil–vegetation models, which include such processes, have a better scientific basis for the assessment of critical N loads. Examples of such models are CALLUNA (Heil & Bobbink 1993) and ERICA (Berendse 1988). The model CALLUNA integrates N processes by atmospheric deposition, accumulation, and sod removal, with heather beetle outbreaks and competition between species, to establish the critical N load in lowland dry heathlands (Heil & Bobbink 1993). The wet heathland model ERICA incorporates the competitive relationships between the species *Erica* and *Molinia*, the litter production from both species, and nitrogen fluxes by accumulation, mineralization, leaching, atmospheric deposition, and sheep grazing.

Neglect of biotic interactions also limits the derivation of critical N loads related to forest damage. Even the derivation of critical loads based on a critical foliar N concentration (see Table 20.3) is impossible with a steady-state soil model, unless purely empirical relationships between N concentrations in the plant and in the soil solution are applied. At present there are several integrated forest-soil models that are potentially useful for a more scientifically based derivation of critical N loads. Examples are the models NAP (Van Oene 1992) and ForSVA (Oja *et al.* 1995).

Probabilistic models

An example of a probabilistic integrated dynamic soil–vegetation model is SMART-MOVE. This model predicts the occurrence probability of plant species in response to scenarios for acidification, eutrophication, and desiccation. The model consists of a soil module (SMART; De Vries *et al.* 1989) and a vegetation module (MOVE; Latour & Reiling 1993; see Fig. 20.4).

The SMART model predicts changes in abiotic soil factors indicating acidification (pH), eutrophication (N availability), and desiccation (moisture content) in response to scenarios for acid deposition and groundwater abstraction. Recently, nutrient cycling (litterfall, mineralization, and uptake) has been incorporated into the model (Kros *et al.* 1995). MOVE predicts the occurrence of species (see Section 20.3.1). Since the species–response functions are based on Ellenberg indicator values,

a calibration of these indication values to quantitative values of the abiotic soil factors is necessary to link the soil module to the vegetation module (see Section 20.3.2). The advantage of SMART-MOVE is that the two approaches to assess critical loads, i.e. (i) the exceedance of critical values for ion concentrations and ion ratios in soil water, and (ii) the analysis of changes in species composition, are combined in a consistent framework.

20.4 CRITICAL LOADS OF NITROGEN

20.4.1 *Impacts of nitrogen and critical limits*

Impacts

The impact of N on an ecosystem depends on its nitrogen status, since N is a nutrient that may be either in short supply or in excess. Since the beginning of the 1980s several authors (e.g. Ulrich *et al.* 1979; Ellenberg 1983, 1985, 1991; Nihlgård 1985; Tamm 1991; Gundersen 1992) have hypothesized that elevated inputs of N lead to vegetation changes, as well as damage to trees. A graphical representation of the hypothesis of the effects induced by elevated atmospheric N deposition on a forest ecosystem going from N limitation to N excess is shown in Fig. 20.5.

In systems with low N status, an elevated input of N will increase forest growth until a certain threshold level. Observations of increased tree growth of European forests (Spiecker *et al.* 1997) may be the effect of increased N inputs. Below the threshold level for reduced forest growth, however, changes in the ecosystem are observed; in particular, the forest biodiversity may gradually change toward more nitrophilic species (Ellenberg

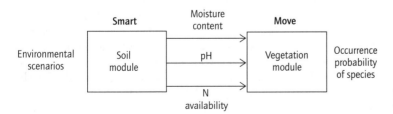

Fig. 20.4 Schematic presentation of the SMART-MOVE model.

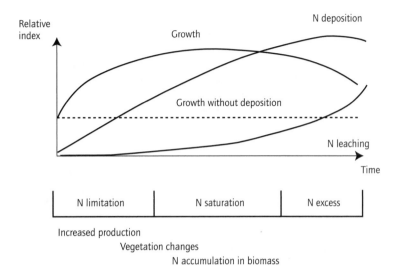

Fig. 20.5 Hypothesis for the responses of temperate forest ecosystems to increased N deposition. The time scale (*x*-axis) for these changes may differ widely between ecosystems and regions. (Adapted from Gundersen 1992.)

1985; Bobbink *et al.* 1995, 1998). A thorough review of the impacts of N inputs on the species diversity of terrestrial ecosystems in general, i.e. ombotrophic bogs and wetlands, heathlands, species-rich grasslands, and forests, including empirical critical N loads related to vegetation changes, has been given in Bobbink *et al.* (1998).

In forested plots with a continuous high N input, essential resources other than N may periodically limit primary production, especially when the canopy reaches its maximum size and N utilization efficiency decreases. The ecosystem then approaches "N saturation" (Aber *et al.* 1989). A forest ecosystem leaching NO_3^- (or NH_4^+) is saturated, but it may still respond to N additions and accumulate a considerable amount of N in the biomass. At the stage of "N saturation" or "N excess," the ecosystem may be destabilized by the interaction of a number of factors (Erisman & De Vries 1999). These are:

1 An increased water stress as a result of increased canopy size, increased shoot/root ratio, and loss of mycorrhizal infection (De Visser 1994).

2 Root damage due to acidification caused by climatically controlled pulses of nitrification (Matzner 1988).

3 Nutrient deficiencies or nutrient imbalances

(Nihlgård 1985; Roelofs *et al.* 1985; Schulze 1989), since the increase in canopy biomass causes an increased demand for base cation nutrients (Ca, Mg, K), whereas the uptake of these cations is reduced by increased dissolved levels of NH_4 and Al (Boxman & Van Dijk 1988), a loss of mycorrhiza or root damage (Schulze 1989).

4 Accumulation of N in foliage (e.g. as amino acids), which may affect frost hardiness (Aronsson 1980) and the intensity and frequency of insect and pathogenic pests (Popp *et al.* 1986; Roelofs *et al.* 1985).

In addition, the nitrate leaching to ground water, such that NO_3^- concentrations in (shallow) ground water exceed the current EU drinking water standard of $50\,\text{mg}\,\text{l}^{-1}$ (e.g. Boumans & Beltman 1991), has to be considered.

Experiments with decreased N deposition at N saturated sites, after building a roof below the canopy to prevent N inputs into the soil, showed an immediate decrease in nitrate leaching (Boxman *et al.* 1995; Bredemeier *et al.* 1995; Wright & Rasmussen 1998). This shows that N saturation is a reversible process in chemical terms. In ecological terms, an improvement is to be expected as well, since several species returned during the period with low N input.

Critical limits

Critical limits for N or NO_3^- in soil solution, ground water, surface water, and plants were given in Table 20.3. A critical N concentration in soil solution related to vegetation changes is difficult to assess. Literature data indicate that vegetation changes may take place in a situation where N leaching hardly increases above natural background values (Van Dam 1990). It is the increase in N availability through enhanced N cycling that triggers vegetation changes (e.g. Berendse et al. 1987). In calculating critical N loads, however, the loss of nitrogen from the ecosystem should be accounted for, even though one should use a natural background value rather than a critical value. Nitrate concentrations in stream water of nearly unpolluted forested areas in Sweden are as low as 0.02 $mol\,m^{-3}$ (Rosén 1990). However, the total N concentration, including NH_4^+ and organic N, is likely to be slightly higher and a natural N concentration range of 0.02–0.04 $mol\,m^{-3}$ seems plausible. With a precipitation excess of 200–400 $mm\,yr^{-1}$, this gives a leaching rate of 40–160 $mol\,ha^{-1}\,yr^{-1}$ (0.5–2 kg $ha^{-1}\,yr^{-1}$), which is a common range of natural N losses from an ecosystem.

Critical limits for nitrate in ground water can be derived from the EU drinking water standard of 50 $mg\,l^{-1}$ (about 0.8 $mol_c\,m^{-3}$). However, countries may wish to use different values based on safety aspects. For example, in the Netherlands a target value of 25 $mg\,l^{-1}$ (0.4 $mol_c\,m^{-3}$) is used for NO_3^-. For surface water a target value of 2.2 $mg\,l^{-1}$ (0.16 $mol_c\,m^{-3}$) is applied to avoid algal growth.

Critical limits have also been derived in relation to nutrient imbalances. Examples are a critical NH_4/K ratio of 5 $mol\,mol^{-1}$ in the soil solution, based on a strong decrease in the uptake of Ca and Mg above this value in a laboratory experiment with two-year-old Corsican pines (Boxman et al. 1988) or a critical N/K ratio in foliage of 4 $g\,g^{-1}$ (11 $mol\,mol^{-1}$) related to K deficiency symptoms (Van den Burg et al. 1988). For most coniferous tree species, an N concentration in the needles of 16–20 $g\,kg^{-1}$ is considered optimal for growth. At these levels the sensitivity to frost and fungal diseases, however, increases too. In a fertilization experi-

ment in Sweden, it was found that frost damage to the needles of Scots pine strongly increased above an N concentration of 18 $g\,kg^{-1}$ (Aronsson 1980). At this N level, the occurrence of fungal diseases such as *Sphaeropsis sapinea* and *Brunchorstia pinea* also appears to increase.

20.4.2 *Empirically derived critical loads related to biodiversity and forest condition*

Until now, critical N loads for terrestrial (and aquatic) ecosystems, related to changes in vegetation and fauna, have generally been derived by a deterministic empirical approach. These empirical loads are based on an extensive but inhomogeneous summary of field studies and large-scale laboratory (greenhouse) experiments (Bobbink et al. 1992, 1995, 1998). These include: (i) precipitation experiments with NH_4 (e.g. on small-scale heathland or soft water ecosystems in a greenhouse); and (ii) correlative field studies between N deposition and species diversity, using present geographic differences or historical data on N deposition and species decline (De Vries & Latour 1995). An overview of empirical data for critical N loads on terrestrial and aquatic ecosystems derived by Bobbink et al. (1998) is given in Table 20.4. The values are all related to vegetation changes with the exception of forests, where they also include impacts of nutrient imbalances.

For forests, empirical critical loads have also been derived in relation to a critical N concentration in the needles of 18 $g\,kg^{-1}$, above which the sensitivity to frost and fungal diseases increases (Section 20.4.1). De Vries et al. (2000a), for example, used a relationship found by Tietema and Beier (1995) between N concentrations in foliage and N leaching rates for a number of intensively monitored plots (NITREX sites). Applying the critical N concentration in this relationship led to a critical N leaching rate of 20 $kg\,ha^{-1}\,yr^{-1}$. Adding estimated data for N uptake, long-term N immobilization, and denitrification (see eqn 20.1 below) thus led to a critical load generally close to 30 $kg\,ha^{-1}\,yr^{-1}$. A more direct approach is to use empirical relationships between atmospheric N deposition and N concentrations in foliage. Figure 20.6 gives an

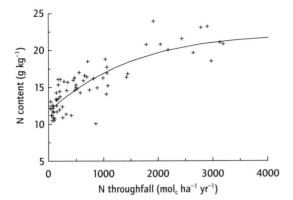

Fig. 20.6 Relationship between nitrogen contents in first year needles of pine and nitrogen deposition at 68 plots in Europe, showing the possibility of setting an empirical critical nitrogen load.

example of such a relationship for Scots pine at 68 intensive monitoring plots in Europe (De Vries *et al.* 2000b). Nearly 70% of the variation in foliar N concentration could be explained by atmospheric deposition, using a nonlinear relationship. Using this relationship, a critical N concentration in the needles of $18 \, g \, kg^{-1}$ was reached near a deposition level of $20 \, mol \, ha^{-1} \, yr^{-1}$ (Fig. 20.6). Both critical loads are in the range given in Table 20.4 for tree health.

20.4.3 Critical loads related to groundwater quality based on a steady-state soil model

Soil model used to calculate critical loads

Critical loads of total N for terrestrial ecosystems can be derived with a simple model of the N balance. A mass balance including all nitrogen fluxes (in $mol_c \, ha^{-1} \, yr^{-1}$) in an ecosystem reads:

$$N_{td} = N_{gu} + N_{im} + N_{de} + N_{vo}$$
$$+ N_{er} + N_{ad} + N_{le} - N_{fi} \qquad (20.1)$$

where the subscript td refers to total deposition, fi to fixation, gu to growth uptake, im to net immobilization, de to denitrification, vo to volatilization, er to erosion, ad to adsorption, and le to leaching.

Adsorption of N can, however, be neglected, as this process only plays a temporary role. Furthermore, even in the short term, adsorption of N can generally be neglected, since: (i) N mainly occurs as NO_3^- except in the topsoil; and (ii) the preference of the adsorption complex for NH_4^+ is mostly low, especially in (acid) sandy soils. Volatilization of N can play a role in grazed woodlands and in areas with frequent forest fires, whereas N removal by erosion may play a role at extremely steep slopes. However, in most cases these N fluxes are negligible. N fixation is also small in most forest and heathland ecosystems except for N-fixing species, such as red alder (De Vries 1993). By assuming N fixation and N loss by volatilization and erosion and N adsorption to be negligible, eqn 20.1 can be written as:

$$N_{td} = N_{gu} + N_{im} + N_{de} + N_{le} \qquad (20.2)$$

Equation 20.2 implies that any N input exceeding the net N uptake by forest growth, a critical long-term immobilization rate, and a natural denitrification rate will lead to an increase in leaching of N, ultimately leading to vegetation changes or even nitrate pollution of groundwater.

When defining the critical load via eqn 20.2 it is implicitly assumed that all terms on the right-hand side do not depend on the deposition of nitrogen. This is unlikely to be the case and thus all quantities should be taken "at critical load." However, to compute "denitrification at critical load" one needs to know the critical load, the very quantity one is trying to compute. The only way to avoid this circular reasoning is to establish a functional relationship between deposition and the sink of N, insert this function into eqn 20.2 and solve for the deposition (to obtain the critical load). This has been done for denitrification. In the simplest case it is linearly related to the net input of N by (De Vries 1993; UNECE 1996a):

$$N_{de} = f_{de}(N_{td} - N_{gu} - N_{im}) \qquad (20.3)$$

where f_{de} ($0 \le f_{de} \le 1$) is the so-called denitrification fraction, which has been formulated as a function of soil type (de Vries *et al.* 1994a). This equation is based on the assumption that the excess N input leaches as NO_3^-. In most (forest) soils, N below the root zone (at 1 m depth) is indeed dominated by NO_3^-, even in the Netherlands, with extremely high NH_4^+ inputs (De Vries 1994), and therefore it is reasonable to assume that NH_4^+ leaching is negligible. This formulation implicitly assumes that immobilization and growth uptake are faster processes than denitrification. From eqns 20.2 and 20.3 a critical N load, $CL_{nut}(N)$, can be derived:

$$CL_{nut}(N) = N_{gu} + N_{im,crit} + N_{le,crit}/(1 - f_{de})$$
$$(20.4)$$

where $N_{im,crit}$ stands for a critical (long-term acceptable) N immobilization and $N_{le,crit}$ for a critical level of N leaching. In wet forest and heathland soils, deep ground water and surface water denitrification is generally not negligible and should be accounted for.

The nitrogen uptake in the critical load equation is the net growth uptake, i.e. the net uptake by vegetation that is needed for long-term average growth. Nitrogen input by litterfall and nitrogen removal by maintenance uptake (needed to resupply nitrogen to leaves) is not considered here, assuming that both fluxes are equal in a steady-state situation. Thus the net uptake is equal to the annual average removal in harvested biomass. This can be estimated from the nitrogen concentration in stems (and branches and leaves, if they are removed, too) in the removed biomass. Dividing that biomass (estimated, for example, from yield tables) by the rotation time gives the annual net uptake flux of nitrogen. Care should be taken not to use biomass data from sites with luxurious growth due to increased nitrogen deposition. As with uptake, N_{im} denotes the long-term net immobilization (accumulation) of nitrogen in the soil, i.e. only the continuous build-up of stable C/N compounds in forest soils. Using data from Swedish forest soil plots, Rosén *et al.* (1992) estimated the annual nitrogen immobilization since the last glaciation at 0.2–0.5 kg N ha^{-1} yr^{-1}. Values between 0.5 and 1.0 kg N ha^{-1} yr^{-1} are currently recommended for the critical loads work under the LRTAP Convention (UNECE 1996a).

Example of a model calculation

Figure 20.7 shows critical nitrogen loads calculated with the steady-state soil model as a function of the net N retention (the sum of net N uptake and long-term net N immobilization), the precipitation excess, and the denitrification fraction. A critical NO_3^- concentration of 0.4 mol$_c$ m^{-3} was used, being a target value in several countries (Table 30.3). The N retention rate varied between 250 (typical for a poorly growing pine forest) and 1000 mol ha^{-1} yr^{-1} (typical for a well growing deciduous forest). The denitrification fraction varied between 0.1 (typical for a well drained sandy soil) and 0.5 (typical for a loamy/clayey soil).

Figure 20.7a illustrates the (large) influence of the precipitation excess on the critical N load in those systems. In the most commonly encountered range in precipitation excesses, varying from 200 to 400 mm yr^{-1}, the critical N load varies between approximately 1500 and 4000 mol ha^{-1} yr^{-1}. Only at a low precipitation excess are the critical N loads in the same range as those derived with respect to species diversity (compare Table 20.4). The ranges in critical N loads are much less influenced by the commonly encountered ranges in N

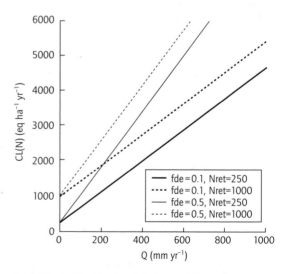

Fig. 20.7 Critical loads of nitrogen with respect to groundwater pollution as a function of precipitation excess, N retention and the denitrification fraction.

Table 20.4 Summary of empirical critical loads of N (in kg N ha^{-1} yr^{-1}) for (semi-)natural freshwater and terrestrial ecosystems (after Bobbink *et al.* 1998).

Ecosystem	Critical load	Indication
Species-rich grasslands		
Calcareous grasslands	15–25†	Increase in tall grass, decline in species diversity
Neutral–acid grasslands	20–30†	Increase in tall grass, decline in species diversity
Montane–subalpine grasslands	10–15‡	Increase in tall graminoids, decline in species diversity
Heathlands		
Lowland dry heathlands	15–20*	Transition heather to grass, functional change
Lowland wet heathlands	17–22†	Transition heather to grass
Species-rich heathlands	10–15†	Decline in sensitive species
Arctic and alpine heathlands	5–15‡	Decline in lichens, mosses and evergreen dwarf shrubs, increase in grasses and herbs
Forests: trees		
Coniferous trees (acidic, managed)	10–15†	Nutrient imbalance (low nitrification)
	20–30†	Nutrient imbalance (moderate to high nitrification rate)
	15–20*	Changes in ground flora
Deciduous trees forests (acidic, managed)	15–20†	Nutrient imbalance; increased shoot–root ratio
Forests: ground vegetation		
Acidic coniferous forests	7–20*	Changes in ground flora and mycorrhizas; increased N leaching
Acidic deciduous forests	10–20‡	Changes in ground flora and mycorrhizas
Acidic unmanaged forests	7–15‡	Changes in ground flora; increased N leaching
Calcareous forests	15–20‡	Changes in ground flora
Wetlands		
Mesotrophic fens	20–35†	Increase in tall graminoids, decline in diversity
Shallow soft-water lakes	5–10*	Decline in isoetid species
Ombotrophic (raised) bogs	5–10†	Decrease in sphagnum and subordinate species, increase in tall graminoids

* Reliable; † quite reliable; ‡ best guess.

retention (Fig. 20.7b), due to the large influence of the leaching rate term, especially at high precipitation excess.

Using a critical limit of 0.02–0.4 mol$_c$ m^{-3} related to vegetation changes leads to much lower critical loads. In this case the critical load is close to the range in N retention, which ranges mostly between 250 and 1000 mol ha^{-1} yr^{-1}, an empirical range that is commonly encountered (see Table 20.4). Conversely, using the EU quality criterion of 0.8 mol$_c$ m^{-3} leads to higher critical loads. In most countries, nitrate pollution of groundwater is therefore not an issue, with the exception of countries with a high nitrogen load related to intensive animal husbandry, such as the Netherlands and parts of Germany and Denmark.

20.4.4   *Critical loads related to biodiversity based on integrated soil–vegetation models*

The CALLUNA model (Heil & Bobbink 1993) has been calibrated with data from field and laboratory experiments in the Netherlands. Atmospheric nitrogen deposition has been varied between 5 and 75 kg N ha^{-1} yr^{-1} in steps of 5–10 kg N during different simulations. From these simulations it became obvious that the critical N load to avoid changes from dwarf shrubs to grasses is 1000–1400 mol$_c$ ha^{-1} yr^{-1}. Berendse (1988) simulated the development of lowland wet heathland after sod removal, using the ERICA model. Using the biomass of *Molinia* with respect to *Erica* as an indicator, his results suggested a critical N load of

Table 20.5 Comparison of empirical data for critical N loads on terrestrial ecosystems derived by Bobbink *et al.* (1998) and the probabilistic integrated soil–vegetation model SMART/MOVE using a protection percentage of 80% (data from Albers *et al.* 2000).

| Ecosystem | Critical load (mol N ha^{-1} yr^{-1}) | |
	Empirical data	SMART/MOVE
Coniferous forests (acidic, managed)	500–1400	1300 (poor sandy soils)
Deciduous forests (acidic, managed)	700–1400	1300 (poor sandy soils)
		1600 (rich sandy soils)
Forests (calcareous)	1100–1400	1200 (loess soils)
		2100 (marine clay soils)
		2500 (calcareous dunes)
Lowland dry heathlands	1100–1400	1100*
Lowland wet heathlands	1200–1600	1400
Acid grassland	1100–1400	900 (sand)
		1050 (peat)
Dune grasslands (noncalcareous)	700–1100	900 (wet grassland)
		900 (nutrient poor dune valley)
		650 (dry grassland)
		1100 (dune heathland)
Peat lands	400–700	400*
Mesotrophic peatlands and swamps	1400–2500	1050 (wet grassland on peat)
		2400 (wetland)

1200–1600 mol$_c$ ha^{-1} yr^{-1} for the transition of lowland wet heathland into a grass-dominated sward (Berendse 1988).

Van Hinsberg and Kros (1999) derived critical loads for nitrogen related to species diversity for the most common terrestrial ecosystems in the Netherlands with the probabilistic integrated SMART-MOVE model. A comparison of their results with empirical data for several ecosystems is given in Table 20.5. In general, the results with SMART-MOVE were within the range of empirical critical N loads. In situations where reliable empirical critical loads were derived, SMART-MOVE results were updated (indicated with an asterisk in Table 20.5).

20.4.5 Uncertainties

Uncertainties in the calculated critical N loads are mainly due to uncertainties in the assumed critical limits, input data, and model assumptions, as described below.

Critical limits and input data

The choice of the critical N leaching rate strongly affects the critical deposition levels of N. For groundwater, those criteria are, however, fixed by legally binding critical limits. The nearly negligible N leaching rate that is often taken to calculate critical N loads related to vegetation changes might not be appropriate in certain situations. Empirical data on vegetation changes in forest (e.g. Bobbink *et al.* 1995, 1998) are slightly higher than the calculated critical N deposition levels, although the results are of a similar order of magnitude. The same is the case with the critical N loads related to effects on trees.

The effect of the uncertainty in the input data can be quantified directly. On a regional scale, the uncertainty in the various N fluxes in the critical load equation (eqn 20.4) is generally within 50%, leading to a resulting uncertainty in critical N loads of the same order of magnitude (De Vries *et al.* 1994a).

Model assumptions

Uncertainties related to the description of N dynamics in the steady-state model result from: (i) neglecting N fixation, which is important for trees such as red alder; (ii) neglecting NH_4^+ fixation, which may play a role in clay soils; (iii) assuming that nitrification is complete, while it is likely to be inhibited at high C/N ratios; (iv) the simple description of net N immobilization; and (v) neglecting the interaction between net N uptake and a change in soil conditions (De Vries 1994). Even though the dynamics of the N transformation processes are strongly simplified, the resulting fluxes for net N uptake, N immobilization, and denitrification seem plausible in view of available data on these processes for forest soils (De Vries *et al.* 1994a).

20.5 CRITICAL LOADS OF ACIDITY

20.5.1 *Impacts of acidity and critical limits*

Impacts

The indirect soil-mediated acidifying impacts of S and N deposition include the loss of base cations from the soil and the release of soluble toxic aluminum. In the 1980s, several authors (e.g. Ulrich *et al.* 1979; Hutchinson *et al.* 1986) considered soil acidification to be responsible for forest decline, since Al^{3+} is toxic to plant roots (Cronan *et al.* 1989; Marschner 1990; Mengel 1991; Cronan & Grigal 1995). This direct relationship has been questioned since then, because there is no clear relationship between forest crown condition and high Al concentrations or Al/base cation ratios (e.g. Hendriks *et al.* 1997). Nevertheless, numerous studies, both in the laboratory and in the field, have shown that high concentrations of Al relative to (divalent) base cations such as Ca and Mg have a negative influence on mycorrhizal frequency, root elongation, and root uptake (see Sverdrup & Warfvinge 1993 for an overview). It may affect fine root growth, thus inhibiting the uptake of water and base cations, causing deficiencies of these nutrients for forest trees (notably Mg).

Roelofs *et al.* (1985) and Schulze (1989) suggested that acidification of soil and excessive N inputs caused nutrient imbalances in the soil and the plants. This coincided with field observations and foliar analyses that deficiencies of Mg and K caused yellowing of needles of Norway spruce (Bosch *et al.* 1983; Zöttl & Mies 1983). Roberts *et al.* (1989) concluded that spruce decline in central Europe mainly results from foliar Mg deficiency due to: (i) an increased Mg demand induced by an increased growth in response to elevated N inputs; and (ii) inhibition of Mg uptake caused by soil acidification (a decrease in exchangeable magnesium, ammonium accumulation, and aluminum mobilization). In general, N contributes to both soil acidification and eutrophication (see later), and the two processes lead to the imbalance in nutrient availability for plant growth (e.g. Heij & Erisman 1997).

Apart from impacts on forest, acid deposition may affect the species diversity of terrestrial vegetation, due to a decrease in pH (Van Hinsberg & Kros 1999), cause pollution of groundwater for drinking water supply due to increased hardness and increased concentrations of Al mobilized from the soil (e.g. Boumans & Van Grinsven 1991), and lead to loss of fish populations caused by a decrease in pH and an increase in labile Al in surface waters (e.g. Hultberg 1988).

Critical limits

Critical limits for Al/(Ca + Mg + K) ratios in soil solution, pH, and Al in either groundwater or surface water were given in Table 20.3. Sverdrup and Warfvinge (1993) derived critical limits for Al/(Ca + Mg + K) ratios for many vegetation types, based on a literature review of numerous laboratory studies, relating those ratios to decreased root biomass and inhibited nutrient uptake. For groundwater, some countries use a critical pH of 6.0 (e.g. Sweden, as reported in Sverdrup *et al.* 1990), whereas other countries use critical Al concentrations based on the EU drinking-water standard of $0.2\,mg\,l^{-1}$ or $0.02\,mol_c\,m^{-3}$ (e.g. the Netherlands, as reported in De Vries 1993). In surface waters the pH should preferably be above 6.0 and certainly above

5.3, whereas the Al concentration should be less than $0.003 \, \text{mol}_c \, \text{m}^{-3}$ to avoid the death of various fish species (Hultberg 1988).

20.5.2 Soil model used to calculate critical loads of acidity

Calculation of critical loads of acidity

Critical loads of acidity, induced by deposition of N and S, can be derived from the steady-state charge balance for the ions in the soil leachate (in $\text{mol}_c \, \text{ha}^{-1} \, \text{yr}^{-1}$) leaving the root zone (modeled as a single homogeneous layer):

$$H_{le} + Al_{le} + BC_{le} + NH_{4,le} + NO_{3,le}$$
$$+ Cl_{le} + HCO_{3,le} + RCOO_{le} \quad (20.5)$$

where $RCOO_{le}$ is the leaching flux of the sum of organic anions. Neglecting OH^- and CO_3^{2-} (a reasonable assumption even for calcareous soils), the alkalinity or ANC (acid neutralizing capacity) can be defined as:

$$ANC_{le} = HCO_{3,le} + RCOO_{le} - H_{le} - Al_{le} \quad (20.6)$$

A steady-state situation with respect to acidification implies a constant pool of exchangeable base cations (BC). Consequently, the following mass balance holds for base cations:

$$BC_{le} = BC_{td} + BC_{we} - Bc_{gu} \quad (20.7)$$

Note that BC_{td} and BC_{we} include all four base cations $(BC = Ca + Mg + K + Na)$, whereas sodium is not taken up by vegetation $(Bc = BC - Na)$. For sulfur the mass balance reads:

$$S_{le} = S_{td} - S_{gu} - S_{im} - S_{re} - S_{ad} \quad (20.8)$$

where the subscript re refers to reduction. An overview of S cycling in forests by Johnson (1984) suggests that the net (growth) uptake, immobilization, and reduction of SO_4^{2-} are generally insignificant. Sulfate adsorption occurs especially in Fe- and Al-oxide rich subsurface horizons (Johnson et al. 1979, 1982; Johnson & Todd 1983). However, when one is deriving a long-term critical load, the effect of adsorption must be neglected, since this

phenomenon is of only temporary importance (several decades). Even in the short term, sulfate adsorption is negligible in most European forest ecosystems, as shown by various budget studies given in Berdén et al. (1987). Furthermore, since sulfur is completely oxidized in the soil profile, S_{le} equals $SO_{4,le}$, and eqn 20.8 simplifies to:

$$SO_{4,le} = S_{td} \quad (20.9)$$

Combining eqns 20.7, 20.8, and 20.9 and the N balance derived in Section 20.4.3 (eqn 20.2; $NO_{3,le} = N_{le}$) yields for the charge balance (eqn 20.5):

$$S^*_{td} + N_{td} = BC^*_{td} + BC_{we} - Bc_{gu}$$
$$+ N_{gu} + N_{im} + N_{de} - ANC_{le} \quad (20.10)$$

where the asterisk denotes Cl-corrected quantities, assuming that chloride comes only from sea spray. It is assumed that there are no sources or sinks of chloride in the soil compartment, and therefore leaching equals deposition. Knowledge of the deposition terms, weathering, and net uptake of base cations, as well as nitrogen uptake, immobilization and denitrification, allows calculation of the ANC leaching, and thus assessment of the acidification status of the soil. Conversely, critical loads of acidity (sum of S and N) can be computed by defining a critical (or acceptable) ANC leaching, $ANC_{le,crit}$, which is set to avoid "harmful effects" on a "sensitive element of the environment" (e.g. damage to fine roots).

$$CL(A) = CL(S+N) = BC^*_{td} + BC_{we} - Bc_{gu}$$
$$+ N_{gu} + N_{im} + N_{de} - ANC_{le,crit} \quad (20.11)$$

Using also the equation for the deposition-dependent denitrification (eqn 20.3), one obtains for the critical loads of sulfur, CL(S), and acidifying nitrogen, CL(N):

$$CL(S) + (1 - f_{de})CL(N)$$
$$= BC^*_{td} + BC_{we} - Bc_{gu}$$
$$+ (1 - f_{de})(N_{im} + N_{gu}) - ANC_{le,crit} \quad (20.12)$$

Note that these critical loads of S and N are not unique; every pair of depositions (N_{td}, S_{td}) that fulfills eqn 20.12 is a critical load. However, when comparing S and N deposition to critical loads one has to bear in mind that the nitrogen sinks cannot compensate for incoming sulfur acidity, i.e. the maximum critical load of sulfur is given by:

$$CL_{max}(S) = BC^*_{td} + BC_{we} - Bc_{gu} - ANC_{le,crit}$$
$$(20.13)$$

This expression has been used to derive the critical deposition of S—used in the negotiations over the 1994 Sulphur Protocol—by multiplying it by the so-called sulfur fraction (e.g. Hettelingh *et al.* 1995). Furthermore, if

$$N_{td} \leq N_{im} + N_{gu} = CL_{min}(N) \quad (20.14)$$

all deposited N is consumed by uptake and immobilization, and sulfur can be considered alone. The maximum amount of allowable N deposition (in the case of zero S deposition) is given by (see eqns 20.12–20.14):

$$CL_{max}(N) = CL_{min}(N) + CL_{max}(S)/(1 - f_{de})$$
$$(20.15)$$

Apart from denitrification, differences in critical load values for forest soils, groundwater, and surface water in the same area are mainly due to differences in the weathering rate and the critical alkalinity leaching flux. The areal weathering rate, expressed in $mol_c\,ha^{-1}\,yr^{-1}$, is determined by the parent material and the considered depth of the soil profile. This depth equals the average thickness of the root zone for forest soils, the unsaturated zone for phreatic groundwater, depth to groundwater extraction for deep groundwater, and the thickness of the catchment for surface water (Sverdrup *et al.* 1990).

Calculation of the critical ANC leaching

Defining a critical chemical value is the most crucial step in calculating critical loads, since this quantity links deposition levels to a "harmful effect." For acidity, the critical leaching of ANC has to be specified. Criteria for [Al] generally refer to the inorganic Al concentration, which is toxic to roots. Assuming that organic anions are completely bound by Al, eqn 20.6 simplifies to:

$$ANC_{le,crit}Q([HCO_3] - [H]_{crit} - [Al]_{crit}) \quad (20.16)$$

where Q is the percolation flux $(m\,yr^{-1})$. The concentration of HCO_3 can be derived by an equilibrium equation describing the dissociation of CO_2 according to:

$$[HCO_3][H] = K_{CO_2}p_{CO_2} \quad (20.17)$$

where K_{CO_2} is the dissociation constant and p_{CO_2} is the partial pressure of CO_2. The concentration of bicarbonate can be neglected in calculating critical loads for acid forest soils, but not in groundwater and surface water, where the critical pH is above 5.

The concentration of Al can be derived by an equilibrium equation describing the dissolution of aluminum (hydr)oxides. In its simplest form, this is modeled by a gibbsite equilibrium:

$$[Al] = K_{gibb}[H]^3 \quad (20.18)$$

This equation is used in the critical load calculations and in many dynamic soil models, including the SMART model. Thus, specifying a critical pH (e.g. $pH_{crit} = 4.2$) the critical ANC leaching can be easily computed. Alternatively, specifying a critical aluminum concentration (e.g. $[Al]_{crit} = 0.2\,mol_c\,m^{-3}$), $ANC_{le,crit}$ can also be computed via eqn 20.18. In several situations, a critical Al concentration has to be derived indirectly from other criteria, including a critical Al/(Ca + Mg + K) ratio, a critical base saturation, or an acceptable Al weathering rate, as discussed below.

20.5.3 Calculation examples based on the use of a steady-state soil model

Impacts on tree roots/forest condition

Calculation of the critical ANC leaching. The most common critical chemical value used in the

European critical loads work is the molar ratio of base cations to aluminum in soil solution. The critical $(Ca + Mg + K)/Al = Bc/Al$ ratio is used to calculate a critical aluminum leaching via:

$$Al_{le,crit} = 1.5 \frac{Bc_{le}}{(Bc/Al)_{crit}}$$

with

$$Bc_{le} = Bc_{td} + Bc_{we} - Bc_{gu} \qquad (20.19)$$

where the factor 1.5 arises from the conversion of the molar Bc/Al ratio to equivalents. Note that weathering generally includes Na, but this is taken care of by correction factors, e.g. $Bc_{we} = 0.85\,BC_{we}$. Multiplying by Q and using eqns 20.16–20.18, the critical ANC leaching is obtained. The same procedure can be used to compute $ANC_{le,crit}$ from a critical Ca/Al or (Ca + Mg)/Al ratio. The standard value used is $(Bc/Al)_{crit} = 1$, but in Sverdrup and Warfvinge (1993) values for many different receptors (plant species) are compiled.

Calculation example. Figure 20.8 shows critical loads for acidity calculated with the steady-state soil model as a function of the net base cation input (base cation deposition and base cation weathering minus net base cation uptake), the precipitation excess, and the critical Al/(Ca + Mg + K) ratio. The net base cation input was assumed to vary between 250 (typical for poor sandy soils with a low weathering rate) and $1000\,mol_c\,ha^{-1}\,yr^{-1}$ (typical for loamy/clayey soils with a high weathering rate) in areas where the net base cation uptake and the base cation deposition are approximately equal. In producing the graphs, a net N removal (the sum of net N uptake, long-term N immobilization, and denitrification) of $500\,mol_c\,ha^{-1}\,yr^{-1}$ was assumed. The acid neutralization rate, which stands for the sum of the net base cation input and the net N removal (i.e. all terms in the critical load equation, eqn 20.11, with the exception of the critical leaching rate of ANC), thus ranges between 750 and $1500\,mol_c\,ha^{-1}\,yr^{-1}$.

Figure 20.8a illustrates the (large) influence of the critical Al/(Ca + Mg + K) ratio in the most commonly encountered range of 0.5–2.0 (see also Table 20.3). At the lowest Al/(Ca + Mg + K) ratio, the critical acid load varies between approxi-

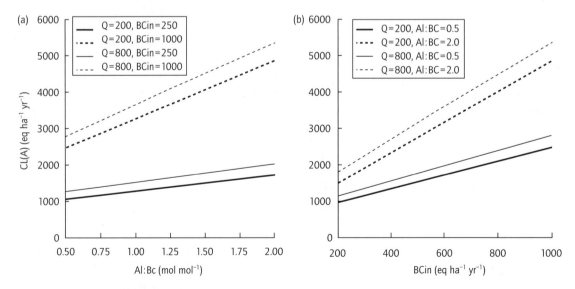

Fig. 20.8 Critical loads of acidity with respect to impacts on ecosystems (roots), groundwater pollution as a function of the quality criteria used, the net base cation input, and the precipitation excess.

mately 1000–3000 mol_c ha^{-1} yr^{-1}. At higher critical ratios, the critical acid load increases strongly, specifically at the higher net base cation input rates. This input rate increases not only the acid neutralization rate, but also the acceptable leaching rate of ANC, since the concentration of base cations in a steady-state situation is calculated as the net base cation input divided by the precipitation excess. This effect is more clearly illustrated in Fig. 20.8b. The impact of the precipitation excess on the leaching rate of ANC (and thus the critical load) is relatively small, since it only influences the leaching of protons. On a Europe-wide scale, it is mainly the noncalcareous sandy soils with a low base cation weathering rate that have critical loads for acidity below the present loads. In many ecosystems, the present load is below the critical load related to elevated Al/(Ca + Mg + K) ratios.

Sustainable base cation pools

Calculation of the critical ANC leaching. Base saturation, i.e. the fraction of base cations on the cation exchange complex, is an indicator of the acidity status of a soil and one may want to keep this pool to avoid nutrient deficiencies in the long term. In the following we show how to link base saturation with ANC, thus allowing the use of a present base saturation for critical load calculations based on the precautionary principle.

Exchange reactions are often described by Gaines–Thomas equations. Lumping together all base cations, the following equations describe the Al–BC–H exchange in the SMART model:

$$\frac{f_{Al}^2}{f_{BC}^3} = K_{AlBC}\frac{[Al]^2}{[BC]^3} \qquad (20.20)$$

$$\frac{f_H^2}{f_{BC}} = K_{HBC}\frac{[H]^2}{[BC]} \qquad (20.21)$$

where f_X is the exchangeable fraction of ion X. Furthermore, charge balance requires that

$$f_{Al} + f_{BC} + f_H = 1 \qquad (20.22)$$

From the equations for Al–BC and H–BC exchange and the gibbsite equilibrium (eqn 20.18) one can derive the following equation:

$$f_{Al} = Kf_H^3 \quad \text{with} \quad K = K_{gibb}\sqrt{K_{AlBC}/K_{HBC}^3} \qquad (20.23)$$

which allows the expression of the base saturation as function of f_H alone:

$$f_{BC} = 1 - f_H - Kf_H^3 \qquad (20.24)$$

Using eqns 20.20 and 20.23 to replace f_{BC} in this equation yields a relationship between the Bc/Al ratio and the exchangeable fraction of H (with [Bc] as additional parameter; for simplicity we assume no Na exchange, i.e. BC = Bc):

$$Kf_H^3 + \left(K(Bc/Al)\right)^{2/3}\left([Bc]/K_{AlBC}\right)^{1/3}f_H^2 + f_H = 1 \qquad (20.25)$$

The last two third-order equations establish a relationship between base saturation and the Bc/Al ratio. This Bc/Al ratio can then be related to an ANC value as described above.

Calculation example. Figure 20.9 shows critical loads for acidity calculated with the steady-state soil model as a function of the net base cation input, the precipitation excess, and the base saturation. The variation in net base cation input was discussed above. Again a net N removal of 500 mol_c ha^{-1} yr^{-1} was assumed. The values used are $K_{AlBc} = 1$ $mol l^{-1}$ and $K = 3.16$ (derived from $log_{10}(K_{gibb}) = 8$, $log_{10}(K_{HBC}) = 5$ and eqn 20.23). More information on the background to these values is given in Posch and De Vries (1999).

Figure 20.9a shows that the calculated critical acid load decreases with an increase in base saturation at a given net base cation input. At a base saturation between 30 and 70%, the critical acid load approximately equals the acid neutralization rate (the sum of the net base cation input and the net N removal), being equal to 750 and 1500 mol_c ha^{-1} yr^{-1} (see above). At a base saturation above 70%, the critical acid load increases strongly,

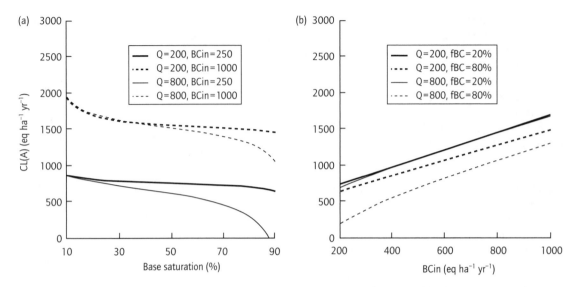

Fig. 20.9 Critical loads of acidity with respect to sustainable soil quality as a function of the base saturation, the net base cation input, and the precipitation excess.

specifically at the higher precipitation excess value. This is because an increase in base saturation is related to an increase in pH, which in turn is related to the generation of acidity by the dissociation of CO_2 (eqn 20.17). Furthermore, the critical load clearly increases below a base saturation of 30%, specifically when the net base cation input is high. This is because below this value an increasing part of the protons or Al is not exchanged with base cations any longer, but leached to groundwater. Below a base saturation of approximately 20–30%, it is therefore more useful to apply the criterion that the pool of readily available aluminum should stay constant (see below). Note, however, that it is likely that the base cation weathering (and thereby the net base cation input) is higher in soils with a high base saturation. Consequently, the upper lines in Fig. 20.9a are more likely for soils with a low base saturation (e.g. below 30%), whereas the lower lines are more likely for soils with a high base saturation (e.g. above 70%).

In Fig. 20.9b, the influence of the net base cation input is further illustrated. As stated above, this input rate increases not only the acid neutralization rate, but also the acceptable leaching rate of

ANC, as it affects the concentration of base cations in a steady-state situation. The effect on the leaching rate of ANC is, however, very limited, illustrated by the nearly one-to-one relationship between critical load of acidity and net base cation input. Compared to the critical loads related to impacts on roots, those related to sustainable base cation pools are generally much lower.

Sustainable aluminum pools

Calculation of the critical ANC leaching. A critical ANC value can also be calculated by aiming at a negligible depletion of Al-hydroxides (Al depletion criterion), being a relevant precautionary approach in acid soils with a low base saturation. This approach is based on the idea that using an ANC limit based on a critical Al concentration or Al/Bc ratio may imply that the accepted rate of Al leaching is greater than the rate of Al mobilization by weathering of primary minerals. The remaining part of Al has to be supplied from readily available Al pools, including Al hydroxides. This causes depletion of these minerals, which might induce an increase in Fe buffering that in turn leads to a decrease in the availability of phosphate (De Vries 1994). Furthermore, the decrease of those pools in, for example, podzolic sandy soils may cause a loss in the structure of those soils.

Negligible depletion of Al hydroxides is achieved when Al leaching equals mobilization of Al from primary minerals. $[Al]_{crit}$ can thus be calculated as

$$[Al]_{crit} = rBC_{we}/Q \qquad (20.26)$$

where r is the equivalent stoichiometric ratio of Al to BC in the congruent weathering of silicates (primary minerals). The other terms in the ANC (eqn 20.16) can again be derived from the equilibrium equations discussed above (eqns 20.17 and 20.18).

Calculation example. Figure 20.10 shows calculated critical loads of acidity with respect to sustainable soil quality for soils with a low base saturation as a function of the precipitation excess, weathering rate, and the stoichiometry of weathering. The net base cation input was assumed to vary between 100 and 500 $mol_c\,ha^{-1}\,yr^{-1}$, a range

typical for poor sandy soils with a low base saturation. Furthermore, as with the previous examples, a net N removal of 500 $mol_c\,ha^{-1}\,yr^{-1}$ was assumed, leading to a net acid neutralization rate of 600–1000 $mol_c\,ha^{-1}\,yr^{-1}$, since the net base cation uptake was assumed to be equal to the base cation deposition.

The critical acid load clearly increases with an increase in the base cation weathering rates and the stoichiometric ratio of Al to BC in the weathering (Fig. 20.10a). This follows directly from eqn 20.26. The weathering rate increases not only the acid neutralization rate, but also the acceptable leaching rate of ANC as described in this equation. This is further illustrated in Fig. 20.10b. Compared to the critical loads related to sustainable base cation pools, those related to sustainable Al pools are generally higher.

Groundwater quality

Critical loads related to an acceptable groundwater quality can directly be derived from a critical limit for the pH or Al concentration and application of eqns 20.16–20.18. Results thus obtained using a

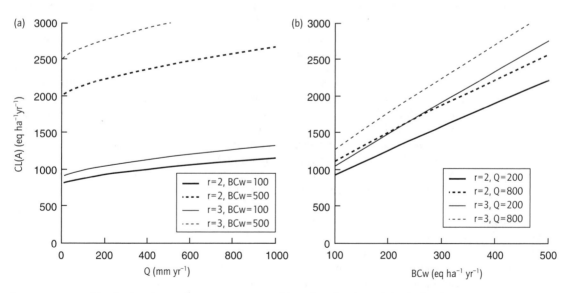

Fig. 20.10 Critical loads of acidity with respect to sustainable soil quality for soils with a low base saturation as a function of the precipitation excess, weathering rate, and stoichiometry of weathering.

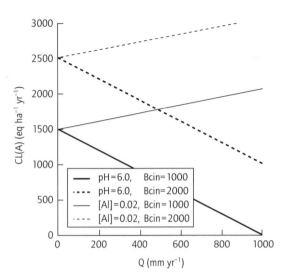

Fig. 20.11 Critical loads of acidity with respect to groundwater pollution as a function of precipitation excess, the net base cation input, and the quality criteria used.

Table 20.6 Comparison of critical loads for acidity N and S calculated for the Solling spruce site with models of different complexity.

Type of model	Complexity	Name	Critical load ($mol_c ha^{-1} yr^{-1}$) N	S
Abiotic mass balance	Simple	SMB	1030	1350
	Complex	PROFILE	820	1120
Integrated abiotic–biotic	Simple	NAP	1580	1560
	Complex	ForSVA	960	1230

critical Al concentration of $0.02\,mol_c\,m^{-3}$ (e.g. applied in the Netherlands) and a pH of 6.0 (e.g. applied in Sweden) are shown in Fig. 20.11. In this example, the net base cation input was assumed to vary between 1000 and $2000\,mol_c\,ha^{-1}\,yr^{-1}$, since groundwater is extracted at greater depth, thus allowing more neutralization by weathering. This range is typical for sandy soils where groundwater is extracted at a depth between 20 and 50 cm (De Vries 1994). Again, a net N removal of $500\,mol_c$ $ha^{-1}\,yr^{-1}$ was assumed, leading to a net acid neutralization rate of $1500–2000\,mol_c\,ha^{-1}\,yr^{-1}$.

The most obvious result is that the critical load for acidity increases with the precipitation rate when the critical Al concentration of $0.02\,mol_c$ m^{-3} is used, whereas the reverse is true when using a critical pH of 6.0. This is because, in the latter case, the dissociation of CO_2 causes an extra generation of acidity, which is larger at increased leaching rates (ANC is positive), whereas in the first case ANC is negative (compare eqns 20.16–20.18). The figure shows that a pH above 6.0 implies that the critical loads for acidity will be low, specifically when the precipitation excess is high. Critical

loads for acidity related to the Al concentration criterion are of the same order of magnitude as those related to the Al depletion criterion (1500–$3000\,mol_c\,ha^{-1}\,yr^{-1}$).

20.5.4 *Critical loads related to biodiversity based on integrated soil–vegetation models*

Apart from nitrogen, Van Hinsberg and Kros (1999) also derived critical loads for acidity related to species diversity for the most common terrestrial ecosystems in the Netherlands with the SMART-MOVE model. Results are described in Albers *et al.* (2000). In general critical acid loads varied between 500 and $2000\,mol_c\,ha^{-1}\,yr^{-1}$.

The NAP (Van Oene 1992) and FORSVA (Oja *et al.* 1995) models have been used to derive critical loads for acidity (the sum of N and S deposition) for the Solling spruce site (Germany). The criteria that were used to derive a critical load were: (i) optimal growth during a relation period (100 years), while avoiding Mg deficiency (NAP); and (ii) a long-term sustainable biomass production, avoiding toxic Al effects (ForSVA). The critical loads for the sum of N and S thus derived were close to those derived by a steady-state soil model (Table 20.6). More information on this comparison is given in De Vries *et al.* (1995). This is an important conclusion, since much work remains to be done, in both modeling efforts and data collection, before truly integrated forest soil models can be used for assessing critical loads on a large regional scale.

20.5.5 *Evaluation and uncertainties*

As with nitrogen, critical acid loads are subject to (large) uncertainties due to uncertainties in the assumed critical Al/(Ca + Mg + K) ratios in relation to effects, model assumptions, and input data.

Critical limits

Although the critical base cation to aluminum ratio is the most widely used critical chemical value in the ongoing critical loads work, it is not undisputed (see Løkke *et al.* 1996; for a critical review see also Cronan & Grigal 1995). Uncertainties in critical values for the Al/(Ca + Mg + K) ratio, related to direct toxic effects of Al, are mainly due to a lack of knowledge about the effects of Al in the field situation. Values are mainly based on laboratory experiments and the applicability to field situations often seems limited. The uncertainty is also partly due to a natural range in the sensitivity of various tree species to Al toxicity (Sverdrup & Warfvinge 1993). Note also that critical annual average values are generally used, whereas the short-term variation can be large, with peak values in the summer. Finally, the Al/(Ca + Mg + K) ratio is probably irrelevant for peat soils, since Al mobilization hardly occurs in these soils.

If one wants to avoid a decrease in base saturation (or pH), the present base saturation has to be used in the critical load calculations. The only uncertainty in this value is the spatial variability in the field situation. The uncertainty in critical loads related to a required constant pool of readily available Al compounds is mainly due to an uncertainty in the weathering rate. In general, the critical loads based on the precautionary principle are lower than the effect-based critical loads and are more reliable.

Model assumptions

The derivation of critical loads with a steady-state soil model, limited to abiotic effects on the soil only, is questionable. The development of multi-stress models, including interactions of desicca-tion, acidification, and eutrophication on forests and effects of drought, pests, and diseases, is necessary to support the results of such simplified models. A first comparison of critical deposition levels for a Norway spruce stand in Solling (Germany), derived with integrated forest-soil models and simple mass balance models, seems promising in this context (De Vries *et al.* 1995).

An important assumption in the SMB model is the homogeneity of the rootzone in both horizontal and vertical directions. Use of a one-layer model implies that the critical Al/(Ca + Mg + K) ratio refers to the situation at the bottom of the rootzone, whereas most roots occur in the topsoil. Values for the Al/(Ca + Mg + K) ratio generally increase with depth due to Al mobilization, BC uptake, and transpiration. Other assumptions in the one-layer model, such as disregarding sulfate interactions, neglecting complexation of Al with inorganic and organic anions, and a simple hydrology, are likely to be less significant (De Vries 1994).

Input data

Assuming that the model structure is correct, the effect of the uncertainty in the input data can directly be quantified. The uncertainty in the calculated net base cation input (deposition and weathering minus uptake), combined with the uncertainty in Al/(Ca + Mg + K) ratio, which affects the associated acidity leaching, has the largest effect on the calculated critical acid deposition level. Considering the above-mentioned uncertainties in input data, it is likely that the overall uncertainty in critical acid deposition levels varies between ±50%.

20.6 CRITICAL LOADS FOR HEAVY METALS

20.6.1 *Impacts of heavy metals and critical limits*

Impacts

With respect to impacts and risks of heavy metals

on terrestrial ecosystems, a major distinction can be made between impacts on humans who use groundwater for drinking water or who consume crops grown on soil (human toxicological risks), and risks to ecosystems themselves (eco-toxicological risks). The eco-toxicological risks associated with elevated heavy metal concentrations in terrestrial ecosystems include:

1 Reduced microbial biomass and/or species diversity of soil microorganisms and macrofungi, affecting microbial processes such as enzyme synthesis, litter decomposition/mineralization, and soil respiration. A review of these effects is given by Bååth (1989).

2 A decrease in abundance, diversity, and biomass of soil fauna, especially invertebrates such as nematodes and earthworms. A review of these effects is given by Bengtsson and Tranvik (1989).

3 Reduced development and growth of roots and shoots (toxicity symptoms), elevated concentrations of starch and total sugar and decreased nutrient concentrations in foliar tissues (physiological symptoms), and decreased enzymatic activity (biochemical symptoms) of vascular plants, including trees. A review of these effects is given by Balsberg-Påhlsson (1989).

4 Heavy metal accumulation followed by possible effects to essential organs of terrestrial fauna, such as birds, mammals, or cattle on agricultural soils. Those effects are considered important with respect to Cd, Cu, and Hg, since these metals can accumulate in the food chain (Jongbloed et al. 1994).

Concern about the atmospheric input of heavy metals (specifically cadmium and lead but also copper and zinc) to terrestrial ecosystems, such as forests, is specifically related to the impact on soil organisms and the occurrence of bio-accumulation in the organic layer (Bringmark & Bringmark 1995; Bringmark et al. 1998; Palmborg et al. 1998). With respect to copper and zinc, the possible occurrence of deficiencies in forest growth is another relevant aspect. Another concern is related to the leaching of metals (specifically cadmium and mercury) to surface water, with an adverse impact on aquatic organisms and bio-accumulation in fish, thus violating food quality criteria.

In the past, several studies have been carried out to assess critical loads of heavy metals for terrestrial and aquatic ecosystems on national (Bakker 1995) and European scales (Van den Hout et al. 1999). With respect to terrestrial ecosystems, the attention was focused on forests, where metal deposition is the only external source. The critical load approach can also be applied to agriculture, where the load refers to the input by both fertilizers/animal manure (sometimes also sewage sludge) and atmospheric deposition. In several countries, there is also concern about the excess input of heavy metals (specifically cadmium, copper, and zinc) in agriculture (e.g. Moolenaar & Lexmond 1998). An excess of heavy metals may yield agricultural products with unacceptable levels of heavy metals and even reduced crop production (Alloway 1990; Fergusson 1990). In this case, the critical load approach can give insight into necessary changes in management practices of agricultural land.

Critical limits

The most commonly available critical limits for heavy metals are critical total concentrations in the soil. The implicit assumption is that (eco-toxicological) effects are due to metal accumulation in the soil. One of the problems with critical metal concentrations is that the official critical limits for mineral soils are not eco-toxicologically based. This is because background concentrations appear to be higher than maximal permissible concentrations in laboratory toxicity tests. This apparent inconsistency is due to differences in metal availability in these toxicity tests and in the field (e.g. Klepper & Van de Meent 1997). Considering this inconsistency, it is just as appropriate to use the present metal concentrations as the critical limit for mineral soil. The critical load then equals the load that does not lead to further accumulation of metals in the soil.

Critical limits for metal concentrations in the soil solution are presently lacking with respect to direct effects on soil organisms. Critical limits based on laboratory studies are mostly related to total metal concentrations (Bååth 1989; Bengtsson

& Tranvik 1989; Witter 1992; Tyler 1992). Until now, critical metal concentrations in soil solution have only been related to effects on plants. The value for Cd of $2.0\,mg\,l^{-1}$ presented in Table 20.3 is based on LOEC data from laboratory studies with culture solutions reported by Balsberg-Påhlsson (1989), applying a safety factor of 10. The range in Cd concentrations of $0.2–2.0\,mg\,l^{-1}$ for surface water is based on critical limits used in various countries (De Vries *et al.* 1998).

20.6.2 *Soil models used to calculate critical loads for heavy metals*

Calculation of critical loads

Critical loads are derived from a steady-state mass balance model:

$$M_{tl} = -M_{lf} + M_{fu} + M_{ru} - M_{we} + M_{le} \quad (20.27)$$

where all terms are fluxes of metal M (in $mg\,m^{-2}\,yr^{-1}$). The subscript tl refers to the total load by deposition and other loads (e.g. fertilizers), lf to litterfall, fu to foliar uptake (or retention), ru to root uptake, we to weathering, and le to leaching. As in previously described steady-state models, a steady-state situation is assumed, implying that the critical load is to be valid for an infinitely long period. A homogeneously mixed soil compartment with only downward transport of water and metals (no seepage flow, surface runoff and bypass flow) is assumed. An additional assumption is that the soil is in an oxidized state. The model can thus not be applied to very poorly drained soils with groundwater levels near the surface. The reason is that anaerobic conditions violate the equilibrium partitioning concept due to precipitation of metal sulfides (e.g. Janssen *et al.* 1996). Note, however, that under such conditions metals are strongly retained (unless a fluctuating groundwater level causes acidification) and critical loads are likely to be very high. The limitations of the various assumptions are further discussed in De Vries and Bakker (1998) and De Vries (1999).

For some metals (e.g. Pb and Cu), the input by litterfall equals the flux due to foliar uptake, implying that deposition reaching the soil by both throughfall and litterfall can be negative at a negligible input of metals. In that case, the element cycle can be neglected. This is not the case for nutrients, such as Cu and Zn. Nevertheless, it is considered appropriate to neglect the cycling of metals and simply include a net uptake of metals due to, for example, forest growth, in view of the uncertainties in metal adsorption and related leaching. Equation 20.27 can thus be simplified to:

$$M_{tl} = M_{gu} - M_{we} + M_{le} \quad (20.28)$$

A critical metal load can be derived from a critical metal leaching rate, being the product of the percolation flux Q and a critical dissolved metal concentration. The latter value can be either used directly or calculated from a metal concentration in the soil, as described below.

Calculation of the critical metal leaching

When the present metal concentration is used as a criterion, the dissolved metal concentration has to be calculated. Equilibrium processes that determine the partition of metals between various phases are adsorption to the soil and complexation. Adsorption can be included in different ways, varying from a simple linear relationship between the (acid digested) total soil concentration and the total concentration in soil solution, to a nonlinear relationship between the reactive metal concentration in the soil and the dissolved free (uncomplexed) metal concentration or activity. An in-depth discussion of various approaches is given in De Vries and Bakker (1998). For simplicity, we here use a simple nonlinear equilibrium partitioning of total concentrations of metals over the solid phase and the soil solution (lumped expression for adsorption and complexation):

$$ctM_r = K_f[M]^n \quad (20.29)$$

where: ctM_r is reactive concentration of metal M in the soil $(mg\,kg^{-1})$; K_f is nonlinear partition or Freundlich adsorption constant $(mg^{1-n}\,l^n\,kg^{-1})$; [M] is concentration of metal M in the soil solution $(mg\,l^{-1})$; n is the Freundlich exponent.

The Freundlich adsorption constant in eqn 20.29 refers to the relationship between reactive metals (added in laboratory experiments) and dissolved metal concentrations. The relationship between the reactive and total metal concentration can be written as:

$$ctM_r = (1 - frM_{im})ctM_s \qquad (20.30)$$

where: ctM_s is the total concentration of metal M in the soil $(mg\,kg^{-1})$; frM_{im} is the immobile fraction of metal M

From eqns 20.29 and 20.30 the dissolved metal concentration [M] can be calculated from the total metal concentration in the soil ctM_s, thus allowing the assessment of a critical metal leaching rate by multiplication with the percolation flux Q.

Taking the minimum critical leaching rate of the criteria related to the present metal concentration in the soil solid phase and the dissolved metal concentration in the soil solution, the critical load will neither cause an accumulation of metals in the soil (precautionary principle) nor affect soil organisms (effect-based approach).

Target loads

The inclusion of an acceptable net accumulation of metal during a given time period is an approach that has been used in Germany to calculate critical loads for mercury. This approach implies that the steady-state principle is abandoned and eqn 20.28 changes into:

$$M_{tl} = M_{gu} - M_{we} + M_{le} + \Delta M_s \qquad (20.31)$$

where ΔM_s is the change in the metal pool in the soil due to adsorption/desorption $(mg\,m^{-2}\,yr^{-1})$.

Note that this change in metal pool is only related to adsorption/desorption. The concentration of metals in soils is such that mineral precipitation is negligible, unless strongly reduced conditions occur, such as in peat lands. One may use this approach when the period to reach steady state is extremely long. The acceptable annual change in the

metal pool, $\Delta M_{s,acc}$, has to be calculated from the mass balance for the metal concentration in the soil compartment:

$$\rho_s z_s (d/dt)ctM_s = M_{tl} + M_{we} - M_{gu} - Q[M] \qquad (20.32)$$

where: ρ_s is the bulk density of the soil $(kg\,m^{-3})$; z_s is the thickness of the soil layer (m); Q is the percolation flux (precipitation minus evapotranspiration) $(m\,yr^{-1})$; and $M_{le} = Q[M]$. In discretized form this differential equation reads:

$$\rho_s z_s (ctM_{s,t+\Delta t} - ctM_{s,t})/\Delta t$$
$$= M_{tl} + M_{we} - M_{gu} - Q[M] \qquad (20.33)$$

where Δt is the time step selected. Using eqns 20.29 and 20.30 to express the concentration on the right-hand side of eqn 20.33 in terms of the metal concentration and selecting an appropriate time step (e.g. $\Delta t = 1$ yr) allows the computation of the metal concentration at any time and for arbitrary initial concentration $ctM_{s,0}$ and net input $M_{tl} + M_{we} - M_{gu}$. Specifying a time period, T, and the acceptable metal concentration to be reached after that time period, $ctM_{s,acc}$, eqn 20.33 can also be used to iteratively calculate the constant net input (target load) that brings the metal concentration from $ctM_{s,0}$ to $ctM_{s,acc}$ within the specified time period T. A rough approximation of the acceptable annual change in the metal pool is obtained from eqn 20.33 by neglecting leaching altogether and setting $t = 0$ and $\Delta t = T$:

$$\Delta M_{s,acc} = \rho_s z_s (ctM_{s,acc} - ctM_{s,0})/T \qquad (20.34)$$

Note that $\Delta M_{s,acc}$ is negative when the initial metal concentration already present in the soil exceeds the critical metal concentration. Unlike a steady-state approach, a simple dynamic approach always requires the inclusion of soil–soil solution interactions such as adsorption/desorption (partitioning). Use of a critical limit for the soil solution requires the derivation of a related critical limit for the soil, which is needed in the accumulation term. Conversely, use of a critical limit for the soil implies that a related critical limit for the soil solution has to be calculated.

20.6.3 Calculation examples related to impacts on soil quality and soil organisms

The calculation examples illustrate the differences in applying a steady-state approach (critical load) and a dynamic approach (target load), using critical limits for both the soil solid phase (precautionary principle) and the soil solution (effect-based approach), focusing on cadmium.

Critical loads

Figure 20.12 shows critical loads of cadmium with respect to a sustainable soil quality (no Cd accumulation) as a function of the initial cadmium content, the precipitation excess, and the adsorption constant. The weathering and uptake rate of Cd were assumed equal, implying that the critical Cd load equalled the critical Cd leaching rate (see eqn 20.28). The leaching rate was calculated

using an immobile Cd fraction of 0.5 and a Freundlich exponent of 0.45 (De Vries & Bakker 1998). The Freundlich adsorption constant was assumed to vary between 3 and 10. This constant is a function of soil properties including the organic matter and clay content and the pH (e.g. Elzinga *et al.* 1996). The values used are typical for an acid sandy soil (pH = 4) and a near neutral sandy soil (pH = 6).

As expected, the critical load increases linearly with the precipitation excess (Fig. 20.12a). The critical load increases strongly with a decrease in the Freundlich adsorption constant (causing less adsorption but more leaching) and an increase in the initial Cd concentration. This may lead to situations where the leaching is higher than acceptable in view of impacts on, for example, soil (or surface water) organisms. The shaded area in the figure is the area where the critical load exceeds the critical limit related to impacts on soil organisms. This load is simply calculated as a critical Cd concentration of $2\,mg\,m^{-3}$ multiplied by the precipitation excess. Critical loads vary mostly

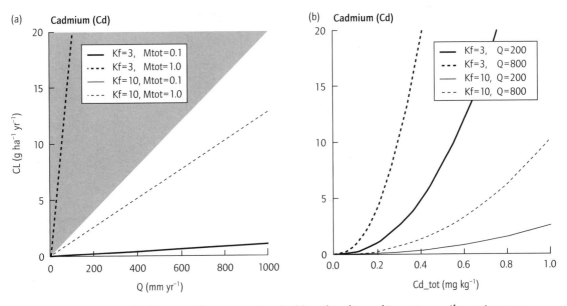

Fig. 20.12 Critical loads of cadmium with respect to sustainable soil quality and impacts on soil organisms as a function of the initial cadmium content, the precipitation excess, and the adsorption constant.

between 1 and 20 g ha^{-1} yr^{-1}, a range commonly encountered for the present loads in Europe (Van den Hout *et al.* 1999).

The influence of the initial Cd concentration in the soil on the calculated critical Cd load is further illustrated in Fig. 20.12b. The figure shows that the differences in Freundlich adsorption constant and precipitation excess become more important at higher initial Cd concentrations. At a Cd concentration of 0.1 mg kg^{-1}, typical for unpolluted sandy soils, the influence is small and the critical load is always below 1–2 g ha^{-1} yr^{-1}.

Target loads

Target loads of cadmium with respect to impacts on soil organisms (using a critical Cd concentration of 2 mg m^{-3}) as a function of the initial cadmium concentration, the precipitation excess, and the adsorption constant are presented in Fig. 20.13 for a time horizon of 100 years. Target loads are calculated by solving eqns 20.32 and 20.33 iteratively. As with the critical loads, the target loads increase with an increase in the precipitation excess, since a

large proportion of the dissolved metal is leached every year, but the influence is not linear any more (Fig. 20.13a). Note that for certain parameter combinations and initial concentrations no positive target load exists below a minimal amount of percolating water. The impact of the initial Cd concentration is now exactly opposite to the critical load. The higher the initial metal concentration in the soil the lower the target load, approaching the critical load related to impacts on soil organisms asymptotically (Fig. 20.13b). The dependence on $ctM_{s,0}$ strongly depends on the combination of the other model parameters.

20.6.4   *Evaluation and uncertainties*

As with nitrogen and acidity, the uncertainty in critical metal loads is mainly determined by the uncertainty in the critical limits for the receptor, in calculation methods, and in input data.

Critical limits

Differences in the literature with respect to criti-

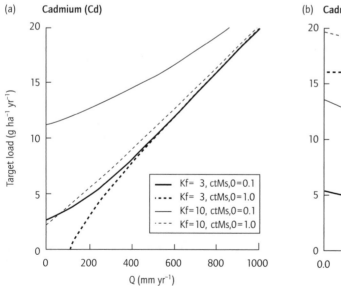

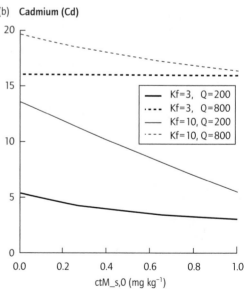

Fig. 20.13   Target loads of cadmium with respect to impacts on soil organisms as a function of the initial cadmium content, the precipitation excess, and the adsorption constant.

cal limits are due to differences in: (i) effects (species) considered; (ii) laboratory (or field) conditions involved, and (iii) extrapolation procedures of single-species toxicity (NOEC) data to a critical value that is assumed to protect an ecosystem sufficiently (Forbes & Forbes 1993; Smith & Cairns 1993). As with nitrogen and acidity, the limits based on the precautionary principle are more reliable than those based on impacts on either soil or surface water organisms or plants.

Calculation methods

The uncertainties related to the calculation methods depend on the type of model used (steady-state or dynamic models). Unlike critical loads, target loads include a definite time target (e.g. 100 years) in relation to what is considered an acceptable input for the future. In general, one can say that the models are not applicable in very wet (anaerobic) circumstances.

Other uncertainties, e.g. due to neglecting the metal cycle, are likely to be less important than the uncertainties due to (unknown) variations in the adsorption constant (see below).

Input data

The influence of input data depend on the critical load model (steady-state or dynamic) used. In general, the uncertainty in the partition (adsorption) constant is the cause of the largest uncertainties in the calculated critical loads for terrestrial ecosystems (De Vries & Bakker 1998). The influence is, however, larger on calculated critical loads with a steady-state model requiring that no further net accumulation of heavy metals occurs than on calculated target loads with a dynamic model, using effect-based critical concentrations in the soil solution, and thus including an acceptable net accumulation in the soil (compare Figs 20.12b and 20.13b). The uncertainty of a steady-state critical load based on a dissolved critical metal concentration is mainly due to the uncertainty in the precipitation excess, apart from the uncertainty in the limit itself.

20.7 THE USE OF CRITICAL LOADS IN POLICY ASSESSMENTS

20.7.1 *Use of critical loads in protocols related to air pollution control*

In the introductory section we sketched the scientific development of the critical load concept and methodology, mostly occurring under the auspices of the LRTAP Convention. This is no coincidence, since articles 2 and 6 of the 1988 Nitrogen Protocol obliged parties to the convention to develop and apply critical loads for determining further emission reductions (UNECE 1996b). As a consequence, critical loads (of sulfur) were used in the integrated assessment and negotiations leading to the 1994 Sulphur Protocol. Furthermore, critical loads of acidity (sulfur and nitrogen), critical loads of nutrient nitrogen, and critical levels of ozone figured prominently in the preparation of the recently signed Protocol to Abate Acidification, Eutrophication and Ground-level Ozone. In this section we summarize the methods used to incorporate critical loads into the integrated assessment framework, which has been used to assist in the formulation of those emission reduction policies.

Critical loads are calculated by most European countries for different receptors (ecosystems), mostly forest soils, but also lakes and seminatural vegetation (heathlands, fens), according to the Mapping Manual (UNECE 1996a) using databases on soils, land cover, and other relevant parameters. Depending on the country, the number of receptors for which critical loads have been computed varies between a few hundred and several hundreds of thousands, and for the whole of Europe the current critical load database held at the Coordination Centre for Effects (CCE) contains various critical load values and auxiliary information for more than 1.3 million receptors (Posch *et al.* 1999). This large number of critical load values requires appropriate methods for presenting them; and since not only the values as such are of interest, but also their geographic location, maps are the obvious choice to display them.

The main purpose of critical loads is to compare

them to deposition estimates, i.e. one is interested in whether critical loads are smaller than the deposition at a given time or whether the deposition exceeds the critical load and thus poses an increased risk for damage to the ecosystem.

20.7.2   Presentation of critical loads in grid cells

Under the LRTAP Convention, deposition estimates are provided by EMEP in the Cooperative Programme for Monitoring and Evaluation of the Long-range Transmission of Air Pollutants in Europe. EMEP's concentration and deposition estimates are provided in $150 \times 150 \, \text{km}^2$ grid cells covering Europe in a polar stereographic projection (EMEP 1998). Consequently, it is only natural to display the critical load information in the same grid system. This is done by: (i) constructing the cumulative distribution function of the critical load values in a grid square, and (ii) computing a desired statistical quantity (e.g. a percentile), which is then mapped.

Let n be the number of critical load values (x_i) in a single grid cell. Sorting these values in ascending order results in a sequence $x_1 \leq x_2 \leq \ldots \leq x_n$. Each value is accompanied by a weight $A_i (i = 1, \ldots, n)$, characterizing the size (importance) of the respective ecosystem. From these normalized weights w_i are computed:

$$w_i = A_i (A_1 + A_2 + \ldots + A_n), i = 1, \ldots, n \quad (20.35)$$

resulting in $w_1 + w_2 + \ldots + w_n = 1$. The cumulative distribution function (cdf) of this set of critical load values is then defined by:

$$F(x) = \begin{cases} 0 & \text{for} \quad x < x_1 \\ W_k & \text{for} \quad x_k \leq x \leq x_{k-1}, \\ & \quad k = 1, \ldots, n-1 \\ 1 & \text{for} \quad x \geq x_n \end{cases} \quad (20.36)$$

where

$$W_k = w_1 + w_2 + \ldots + w_k, k = 1, \ldots, n \quad (20.37)$$

and $F(x)$ is the probability of a critical load being smaller than (or equal to) x, i.e. $1 - F(x)$ is the fraction of ecosystems protected. In Fig. 20.14 an example of a cdf is shown.

All ecosystems in a grid cell are protected if the deposition stays below the smallest critical load value. However, to discard outliers and to account for uncertainties in the critical load calculations, a (low) quantile (percentile) of the cdf is often used to characterize the critical loads in a grid square by a single number. The qth quantile $(0 \leq q \leq 1)$ of a cdf F, denoted by x_q, is the value satisfying

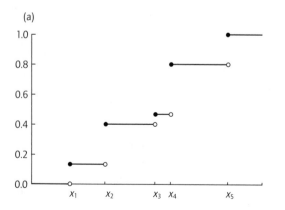

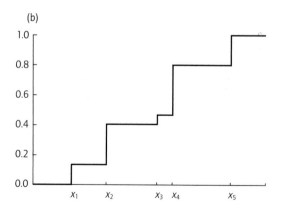

Fig. 20.14   Example of a cumulative distribution function (cdf) for five data points. The filled (open) circles in (a) indicate whether a point is part (not part) of the function. In (b) the same cdf is drawn by connecting all points, the way a cdf is usually displayed.

$$F(x_q) = q \qquad (20.38)$$

which means that x_q, viewed as a function of q, is the inverse of the cdf, i.e. $x_q = F^{-1}(q)$. Percentiles are obtained by scaling quantiles to 100, i.e. the pth percentile corresponds to the $(p/100)$th quantile. The 50th percentile is the median, and the 25th and 75th percentiles are called the lower and upper quartile, respectively. Note that the pth percentile critical load protects $(100 - p)\%$ of the ecosystems.

Computing quantiles, i.e. the inverse of a cdf, which is defined by a finite number of points (critical load values), poses a problem: due to the discrete nature of the cdf a unique inverse simply does not exist. For almost all values of q no value x_q exists so that eqn 20.38 holds; and for the n values $q_i = F(x_i)$ the resulting quantile is not unique: every value between x_i and x_{i+1} could be taken (see Fig. 20.15). Therefore the cdf has to be approximated (interpolated) by a function which allows to solve eqn 20.38 for every q. There is neither a unique approximation nor a single accepted way for calculating percentiles; and in Posch *et al.* (1993) six methods for calculating percentiles are discussed. Note that commonly definitions apply to data with identical weights (i.e. $w_i = 1/n$), but generalization to arbitrary weights is mostly straightfor-

ward. It should be also borne in mind that the differences between different approximation methods vanish when the number of points (critical loads) becomes very large. Here we define the quantile function as (Posch *et al.* 1995):

$$x_q = \begin{cases} x_1 & \text{for} \quad q < W_1 = W_1 \\ x_k & \text{for} \quad W_{k-1} \le q < W_k, \\ & \qquad k = 2, \ldots, n-1 \\ x_n & \text{for} \quad q \ge W_{n-1} \end{cases} \qquad (20.39)$$

An example of a quantile function is shown in Fig. 20.15. The disadvantage of this definition of the quantile function is that it is not continuous, i.e. a small change in q may lead to a significant change in the quantile x_q (jump from x_i to $x_{i\pm1}$). However, none of the disadvantages of a quantile function derived from an interpolated (smoothed) cdf holds: (i) identical values do not lead to ambiguities; (ii) the quantile x_q protects (at least) a fraction q of the ecosystems; and (iii) quantiles of two cdfs are order-preserving, i.e. if $F_1(x) \le F_2(x)$ for all x, then $x_q^{(1)} \le x_q^{(2)}$ for all q (see Posch *et al.* 1995 for examples and further discussion).

It was mostly the 5th percentile of the critical loads in a grid square that was considered "representative" in the integrated assessment leading to the 1994 Sulphur Protocol. In Fig. 20.16 the 5th percentile of the critical loads of sulfur and nitrogen acidity and the critical loads of nutrient nitrogen, calculated according to the methodology described above, are displayed on the EMEP grid covering Europe. It shows that: (i) critical loads are, in general, lower in northern Europe, i.e. ecosystems are more sensitive there; and (ii) critical loads of nutrient nitrogen are smaller than critical loads of acidity almost everywhere in Europe.

20.7.3 *Presentation and use of critical load exceedances in a grid cell*

Once the desired percentile critical load, CL, in a grid square is computed, the excess critical load, *Ex* (also called the critical load **exceedance**), is obtained by subtracting it from the deposition for the same grid square, *Dep* (setting negative exceedances to zero):

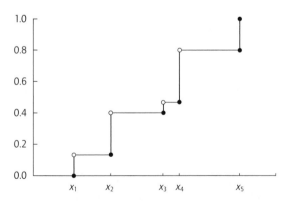

Fig. 20.15 The quantile function (thick vertical lines) of the cdf in Fig. 20.14a (thin horizontal lines). The filled (open) circles indicate whether a point is part (not part) of the function.

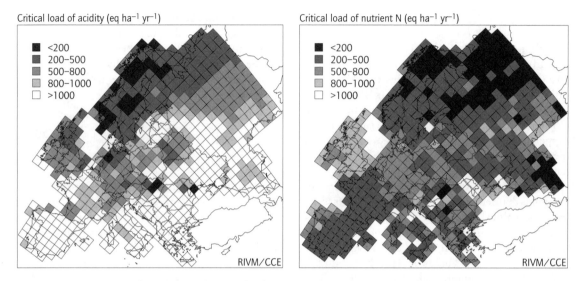

Fig. 20.16   The 5th percentiles of the critical loads of acidity (left) and the critical loads of nutrient nitrogen (right) in each of the $150 \times 150\,\mathrm{km^2}$ EMEP grid cells covering Europe.

$$Ex = \max\{Dep - CL, 0\} \qquad (20.40)$$

Instead of considering only a single percentile critical load in a grid square, all critical load values have to be used to compute the **protection percentage** of the ecosystem area, i.e. the (relative) area within a grid cell in which the respective critical loads are not exceeded. Figure 20.17e and f shows the exceedance of the 5th percentile critical load of nutrient nitrogen (depicted in Fig. 20.16b) and the percent ecosystem area protected, respectively, both for the 1990 total nitrogen deposition.

When comparing present or future deposition scenarios with European critical loads it appeared that nonexceedance could not be reached

everywhere. Thus it was decided by integrated assessment modelers to use uniform percentage reductions of the excess depositions, so-called **gap closures**, for the definition of reduction scenarios. This is illustrated in Fig 20.17a: critical loads and depositions are plotted along the horizontal axis and the (relative) ecosystem area is plotted along the vertical axis. The thick continuous and the thick broken lines are two examples of critical load cdfs (which have the same 5th percentile critical load, indicated by "CL"). "D0" indicates the (present) deposition, which is above critical loads for 85% of the ecosystem area. The difference between D0 and CL is the exceedance in that grid cell. In the protocol negotiations it was decided to

Fig. 20.17   (*facing page*) Left column: cumulative distribution functions (cdf; thick continuous line) of critical loads and different methods of gap closure; (a) deposition gap closure, (b) ecosystem gap closure, and (c) average accumulated exceedance (AAE) gap closure. The thick dashed lines in (a) and (b) depict another cdf, illustrating (i) how different ecosystem protection follows from the same deposition gap closure, or (ii) how different deposition reductions are required to achieve the same protection level. Right column: (d) exceedance of the 5th percentile, (e) percentage of ecosystem area protected, and (f) average accumulated exceedance (AAE) of the critical load of nutrient nitrogen under the 1990 total nitrogen deposition. All values are in $\mathrm{eq\,ha^{-1}\,yr^{-1}}$.

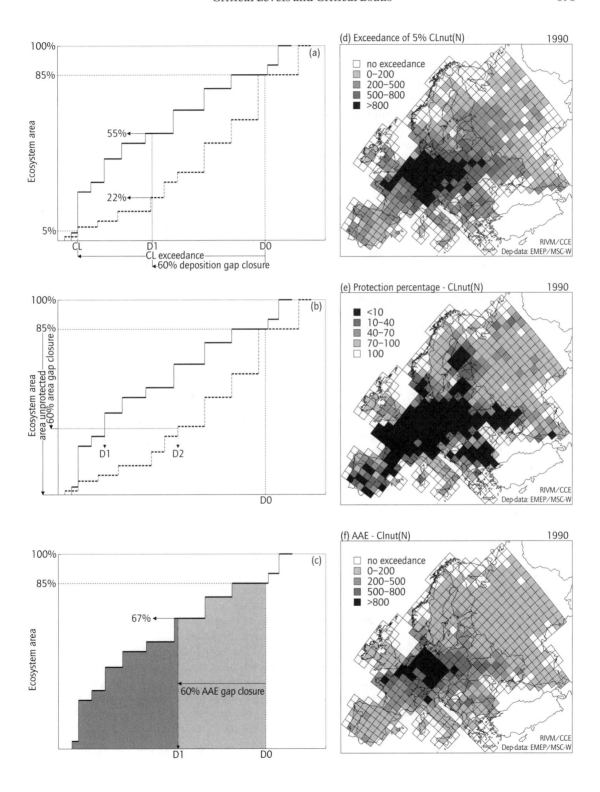

reduce the exceedance everywhere in Europe by a fixed percentage, i.e. to "close the gap" between (present) deposition and (5th percentile) critical load. In Fig 20.17a a **deposition gap closure** of 60% is shown as an example. The figure shows that a fixed deposition gap closure can result in very different improvements in ecosystem protection percentages (55 versus 22%), depending on the shape of the cdf.

To take into account all critical loads within a grid cell (and not only the 5th percentile), one could use an *ecosystem area gap closure* instead of the deposition gap closure. This is illustrated in Fig. 20.17b: for a given deposition D0 to a grid cell the ecosystem area that is unprotected, i.e. with deposition exceeding the critical loads, can be read from the vertical axis. After agreeing to a (percentage) reduction of the unprotected area (e.g. 60%), it is easy to compute for a given cdf the required deposition reduction (D1 and D2 in Fig. 20.17b). The use of the area gap closure becomes problematic, however, if only a few critical load values are given for a grid cell. In such a case the cdf has large discontinuities and (small) changes in deposition may result in either no increase in the protected area at all or large jumps in the area protected.

To remedy the problem with the area gap closure caused by discontinuous cdfs, the **average accumulated exceedance** (AAE) has been introduced. It is defined as

$$AAE = w_1 Ex_1 + w_2 Ex_2 + \ldots + w_n Ex_n \quad (20.41)$$

where Ex_i is the exceedance of the ith ecosystem (eqn 20.40) and w_i its relative weight (eqn 20.35). An example of the AAE, which is the average amount of excess deposition in a grid square (averaged over the total ecosystem area!), is shown in Fig. 20.17f for the critical load of nutrient nitrogen and the 1990 total N deposition.

The AAE is the area under the cdf of the critical loads (the entire grey-shaded area in Fig. 20.17c). Deposition reductions have been negotiated in terms of an **AAE gap closure**, illustrated in Fig. 20.17c: a 60% AAE gap closure is achieved by a deposition D1, which reduces the total grey area by 60%, resulting in the dark gray area; the corres-

ponding protection percentage (61%) can be easily derived. The greatest advantage of the AAE is that it varies smoothly when deposition is varied, even for highly discontinuous cdfs, thus facilitating optimization calculations in integrated assessments.

Another important reason to use the ecosystem area or AAE gap closure is that it can be easily generalized to two (or more) pollutants, which is not the case for a deposition-based exceedance. Such a generalization becomes necessary in the case of acidity critical loads, since both sulfur and nitrogen contribute to acidification. Critical load values are replaced by critical load functions and ecosystem protection percentiles are replaced by protection isolines (for further details see Posch *et al.* 1999).

In the integrated assessment aiding the negotiations for the Protocol to Abate Acidification, Eutrophication and Ground-level Ozone ("multi-pollutant, multi-effects protocol"), the AAE gap closure approach had been chosen to derive emission reduction scenarios for Europe. From the European critical loads data the necessary quantities (protection isolines, AAE-isolines, etc.) were derived by the CCE and passed on to the Task Force on Integrated Assessment Modeling (TFIAM) under the LRTAP Convention, where these data were incorporated into integrated assessment models such as RAINS (Alcamo *et al.* 1990; Schöpp *et al.* 1999). Cost-optimal emission reduction scenarios for environmental targets (gap-closures) of different ambition levels were presented to the negotiators and served as a rational basis for the political decisions on national emission ceilings agreed in the protocol. Critical loads and levels have also been used by the European Commission for designing their acidification and ozone strategies as well as a basis for the negotiations on the latest National Emission Ceilings (NEC) Directive. In conclusion, the concept of critical loads has turned out to be simple enough to allow its application on a pan-European scale and scientifically sound enough to be widely accepted as a means for supporting decision-making with respect to reductions in sulfur and nitrogen emissions.

Table 20.7 Evaluation of various methods for the assessment of critical loads.

Method	Advantage	Disadvantage
Empirical data	Based on empirical relationships between effects and atmospheric (N) deposition Easily applicable for mapping	Not applicable for acidity and metals (no clear empirical relationship) Other effects than acidification and eutrophication may be involved
Steady-state soil models	Simple Easily applicable for mapping	No biotic interactions involved Critical limits are uncertain
Integrated soil–vegetation models	Comprehensive description of the ecosystem Important tool to assess target loads and evaluate scenarios	Model complexity Not easily applicable for mapping

20.8 DISCUSSION AND CONCLUSIONS

In this chapter, we described the methods used to assess critical loads for terrestrial ecosystems, including various quantitative examples. An overview of the advantages and disadvantages of the various methods is given in Table 20.7.

Empirical data and steady-state soil models are easily applicable for mapping. An empirical approach can only derive critical N loads, due to the large influence of N on the species diversity of terrestrial ecosystems. Uncertainties in deterministic empirical data on critical N loads are mainly related to the occurrence of time lags. Harmful effects on ecosystems may already occur before they are visible. The uncertainty in the probabilistic empirical approach described in this chapter is mainly caused by the uncertainty in the relationship between Ellenberg indicator values and N availability.

Uncertainties in critical loads derived by steady-state soil models are determined by the uncertainty in critical limits, model structure, and input data. A systematic overview of these effects has been given. Those models are all relatively simple and neglect interactions with the vegetation. A disadvantage of relatively complex mechanistic models is that input data for their application on a regional scale are generally incomplete and values can only be roughly estimated. Even if the model structure is correct (or at least

adequately representing current knowledge), the uncertainty in the output of complex models may still be large because of the uncertainty of input data. Simpler empirical models have the advantage of a smaller need for input data but the theoretical basis that is needed to establish confidence in the predictions is small, which limits the application of such models for different situations. There is thus a trade-off between model complexity (reliability) and regional applicability. Integrated dynamic soil–vegetation models are relevant to check the results of conventional methods. However, integrated dynamic models are generally complex, limiting their use in mapping.

In general, one can conclude that steady-state soil models do form a relevant tool to calculate critical loads, being an indication of possible adverse effects on terrestrial ecosystems. Indications for such effects become stronger when they are supported by empirical data (specifically the case for nitrogen) or by calculations with integrated soil–vegetation models. Critical loads for heavy metals are likely to be most uncertain due to the uncertainty in critical limits applied.

REFERENCES

Albers, R., Beck, J., Bleeker, A. *et al.* (2000) *Evaluatie van de verzuringsdoelstellingen: de onderbouwing.* RIVM-DLO rapport 725501-001, National Institute of Public Health and the Environment (RIVM), Bilthoven.

Aber, J.D., Nadelhoffer, K.J., Steudler, P. & Melillo, J.M. (1989) Nitrogen saturation in northern forest ecosystems. *Bioscience* **39**, 378–86.

Alcamo, J., Shaw, R. & Hordijk, L. (eds) (1990) *The RAINS Model of Acidification—Science and Strategies in Europe.* Kluwer, Dordrecht.

Alloway, B.J. (1990) *Heavy Metals in Soils.* Blackie and Wiley, Glasgow.

Aronsson, A. (1980) Frost hardiness in Scots pine. II. Hardiness during winter and spring in young trees of different mineral status. *Studia Forest Suecica* **155**, 1–27.

Ashmore, M.R. & Wilson, R.B. (eds) (1994) *Critical Levels of Air Pollutants for Europe.* Background papers prepared for the UN/ECE workshop on critical levels. Egham, March 23–26 1992, Department of the Environment, London.

Bååth, E. (1989) Effects of heavy metals in soil on microbial processes and populations. A literature review. *Water Air and Soil Pollution* **47**, 335–79.

Bakker, D.J. (1995) *The Influence of Atmospheric Deposition on Soil and Surface Water Quality in the Netherlands.* Report R95/013. TNO Institute of Environmental Sciences, Delft.

Balsberg-Påhlsson, A.M. (1989) Toxicity of heavy metals (Zn, Cu, Cd, Pb) to vascular plants. A literature review. *Water Air and Soil Pollution* **47**, 287–319.

Bengtsson, G. & Tranvik, L. (1989) Critical metal concentrations for forest soil invertebrates. A review of the limitations. *Water Air and Soil Pollution* **47**, 381–417.

Berdén, M., Nilsson, S.I., Rosén, K. & Tyler, G. (1987) *Soil Acidification Extent, Causes and Consequences.* Report 3292. Swedish Environmental Protection Board, Solna.

Berendse, F. (1988) *De nutriëntenbalans van droge zandgrondvegetaties in verband met de eutrofiëring via de lucht. Deel 1: Een simulatiemodel als hulpmiddel bij het beheer van vochtige heidevelden.* Centrum voor Agrobiologisch Onderzoek, Wageningen.

Berendse, F., Beltman, B., Bobbink, R., Kwant, R. & Schimtz, M.B (1987) Primary production and nutrient availability in wet heathland ecosystems. *Acta Oecologica* **8**, 265–76.

Bobbink, R., Boxman, D., Fremstad, E., Heil, G., Houdijk, A. & Roelofs, J. (1992) Critical loads for nitrogen eutrophication of terrestrial and wetland ecosystems based upon changes in vegetation and fauna. In Grennfelt, P. & Thörnelöf, E. (eds), *Critical Loads*

for Nitrogen. Nordic Council of Ministers, Copenhagen.

Bobbink, R., Hornung, M. & Roelofs, J.G.M. (1995) The effects of air-borne nitrogen pollution on vegetation-critical loads. In *Updating and Revision of Air Quality Guidelines for Europe.* WHO Europe, Copenhagen.

Bobbink, R., Hornung, M. & Roelofs, J.G.M. (1998) The effects of air-borne pollutants on species diversity in natural and semi-natural European vegetation. *Journal of Ecology* **86**, 717–38.

Bosch, C., Pfannkuch, E., Baum, U. & Rehfuess, K.E. (1983) Über die Erkrankung der Fichte (*Picea abies Karst.*) in den Hochlagen des Bayerischen Waldes. *Forstwissenschaftliches Centralblatt* **102**, 167–81.

Boumans, L.J.M. & Beltman, W. (1991) *Kwaliteit van het bovenste freatische grondwater in de zandgebieden van Nederland onder bos- en heidevelden.* Report 724901001. National Institute of Public Health and Environmental Protection (RIVM), Bilthoven.

Boumans, L.J.M. & Van Grinsven, J.J.M. (1991) *Aluminumconcentraties in het freatische grondwater in de zandgebieden van Nederland onder bos- en heidevelden.* Report 724901002. National Institute of Public Health and Environmental Protection (RIVM), Bilthoven.

Boxman, A.W. & Van Dijk, H.F.G. (1988) *Het effect van landbouw ammonium deposities op bos- en heidevegetaties.* Katholieke Universiteit Nijmegen, Nijmegen.

Boxman, A.W., Van Dijk, H.F.G., Houdijk, A.L.F.M. & Roelofs, J.G.M. (1988) Critical loads for nitrogen with special emphasis on ammonium. In Nilsson, J. & Grennfelt, P. (eds), *Critical Loads for Sulphur and Nitrogen.* Nordic Council of Ministers, Copenhagen.

Boxman, A.W., Van Dam, D., Van Dijk, H.F.G., Hogervorst, R.F. & Koopmans, C.J. (1995) Ecosystem responses to reduced nitrogen and sulfur inputs into two coniferous forest stands in the Netherlands. *Forest Ecology and Management* **71**, 7–29.

Bredemeier, M., Blanck, K., Lamersdorf, N. & Wiedey, G.A. (1995) Response of soil water chemistry to experimental "clean rain" in the NITREX roof experiment at Solling, Germany. *Forest Ecology and Management* **71**, 31–44.

Bringmark, E. & Bringmark, L. (1995) Disappearance of spatial variability and structure in forests floors—a distinct effect of air pollution? *Water Air and Soil Pollution* **85**, 761–6.

Bringmark, L., Bringmark, E. & Samuelsson, B. (1998) Effects on mor layer respiration by small experimental additions of mercury and lead. *The Science of the Total Environment* **213**, 115–19.

Cosby, B.J., Hornberger, G.M., Galloway, J.N. & Wright, R.F. (1985) Modeling the effects of acid deposition: Assessment of a lumped parameter model of soil water and streamwater chemistry. *Water Resources Research* **21**, 51–63.

Cosby, B.J., Hornberger, G.M. & Wright, R.F. (1989) Estimating time delays and extent of regional de-acidification in southern Norway in response to several deposition scenarios. In Kämäri, J., Brakke, D.F., Jenkins, A., Norton, S.A. & Wright, R.F. (eds), *Regional Acidification Models: Geographic Extent and Time Development*. Springer, Berlin.

Cronan, C.S. & Grigal, D.F. (1995) Use of calcium/aluminum ratios as indicators of stress in forest ecosystems. *Journal of Environmental Quality* **24**, 209–26.

Cronan, C.S., April, R., Bartlett, R.J. *et al.* (1989) Aluminum toxicity in forests exposed to acidic deposition. *Water Air and Soil Pollution* **48**, 181–92.

De Visser, P.H.B. (1994) Growth and nutrition of Douglas-fir, Scots pine and pedunculate oak in relation to soil acidification. PhD Thesis, Agricultural University, Wageningen.

De Vries, W. (1993) Average critical loads for nitrogen and sulfur and its use in acidification abatement policy in The Netherlands. *Water Air and Soil Pollution* **68**, 399–434.

De Vries, W. (1994) Soil response to acid deposition at different regional scales. Field and laboratory data, critical loads and model predictions. PhD Thesis, Agricultural University, Wageningen.

De Vries, W. (1999) Approaches and criteria to calculate critical loads of heavy metals for soils and surface waters. In *Proceedings of the Workshop on Effect-based Approaches for Heavy Metals*. Schwerin, Germany.

De Vries, W. & Bakker, D.J. (1996) *Manual for Calculating Critical Loads of Heavy Metals for Soils and Surface Waters. Preliminary Guidelines for Environmental Quality Criteria, Calculation Methods and Input Data*. Report 114. DLO Winand Staring Centre, Wageningen.

De Vries, W. & Bakker, D.J. (1998) *Manual for Calculating Critical Loads of Heavy Metals for Terrestrial Ecosystems*. Report 166. DLO Winand Staring Centre, Wageningen.

De Vries, W. & Latour, J.B. (1995) Methods to derive critical loads for nitrogen for terrestrial ecosystems. In Hornung, M., Sutton, M.A. & Wilson, R.B. (eds), *Mapping and Modeling of Critical Loads for Nitrogen: A Workshop Report*. Institute of Terrestrial Ecology, Huntingdon.

De Vries, W., Bakker, D.J. & Sverdrup, H.U. (1998) *Manual for Calculating Critical Loads of Heavy Metals for Aquatic Ecosystems*. Report 165. DLO Winand Staring Centre, Wageningen.

De Vries, W., Posch, M. & Kämäri, J. (1989) Simulation of the long-term soil response to acid deposition in various buffer ranges. *Water Air and Soil Pollution* **48**, 349–90.

De Vries, W., Reinds, G.J. & Posch, M. (1994a) Assessment of critical loads and their exceedance on European forests using a one-layer steady-state model. *Water Air and Soil Pollution* **72**, 357–94.

De Vries, W., Kros, J. & Voogd, J.C.H. (1994b) Assessment of critical loads and their exceedance on Dutch forests using a multi-layer steady state model. *Water Air and Soil Pollution* **76**, 407–48.

De Vries, W., Reinds, G.J., Posch, M. & Kämäri, J. (1994c) Simulation of soil response to acidic deposition scenarios in Europe. *Water Air and Soil Pollution* **78**, 215–46.

De Vries, W., Posch, M., Oja, T. *et al.* (1995) Modeling critical loads for the Solling spruce site. *Ecological Modeling* **83**, 283–93.

De Vries, W., Reinds, G.J., Klap, J.M., Van Leeuwen, E.P. & Erisman, J.W. (2000a) Effects of environmental stress and crown condition in Europe. III: Estimation of critical deposition and concentration levels and their exceedances. *Water Air and Soil Pollution* **119**, 363–86.

De Vries, W., Reinds, G.J., Kerkvoorde, M. *et al.* (2000b) *Intensive Monitoring of Forest Ecosystems in Europe*. Technical Report 2000a. UNECE EC, Forest Intensive Monitoring Coordinating Institute.

EMEP (1998) *Transboundary Acidifying Air Pollution in Europe. Part 1: Estimated Dispersion of Acidifying and Eutrophying Compounds and Comparison with Observations. Part 2: Numerical Addendum*. EMEP/MSC-W Report 1/98. Meteorological Synthesizing Centre–West, Norwegian Meteorological Institute, Oslo.

EPRI (1991) *The Concept of Target and Critical Loads*. EPRI Report EN-7318. Electric Power Research Institute, Palo Alto, CA.

Ellenberg, H. (1983) Gefährdung wildlebender Pflanzenarten in der Bundes-republik Deutschland—Versuch einer ökologischen Betrachtung. *Forstarchiv* **54**, 127–33.

Ellenberg, H. (1985) Veränderungen der Flora Mitteleuropas unter dem Einfluss von Düngung und Immissionen. *Schweizerische Zeitschift für das Forstwesen* **136**, 19–39.

Ellenberg, H. (1991) Ökologsiche Veränderungen in

Biozönosen durch Stickstoffeintrag. *Berichte aus der ökologischen Forschung* **4**, 75–90.

Elzinga, E.J., van den Berg, B., van Grinsven, J.J.M. & Swartjes, F.A. (1996) *Freundlich isothermen voor cadmium, koper en zink als functie van de bodemeigenschappen, op basis van een literatuuronderzoek*. Report 711501001. National Institute of Public Health and the Environment (RIVM), Bilthoven.

Erisman, J.W. & de Vries, W. (1999) Nitrogen turnover and effects in forests. Contribution to the Welt Forum 2000 Workshop, Slotau, Germany, ECN Report RX 99035.

Forbes, T.L & Forbes, V.E. (1993). A critique of the use of distribution-based extrapolation models in ecotoxicology. *Functional Ecology* **7**, 249–54.

Fergusson, J.E. (1990) *The Heavy Elements. Chemistry, Environmental Impact and Health Effects*. Pergamon Press, Oxford.

Fuhrer, J. & Achermann, B. (eds) (1994) *Critical Levels for Ozone*. FAC-Report 16. Federal Research Station for Agricultural Chemistry and Environmental Hygiene, Liebefeld-Bern.

Gorham, E. (1976) Acid precipitation and its influence upon aquatic ecosystems: an overview. *Water Air and Soil Pollution* **6**, 457–81.

Gorham, E., Martin, F.B. & Litzau, J.T. (1984) Acid rain: ionic correlations in the eastern United States (1980–1981). *Science* **225**, 407–9.

Grennfelt, P. & Thörnelöf, E. (eds) (1992) *Critical Loads for Nitrogen*. Nordic Council of Ministers, Copenhagen.

Gundersen, P. (1992) Mass balance approaches for establishing critical loads for nitrogen in terrestrial ecosystems. In Grennfelt, P. & Thörnelöf, E. (eds), *Critical Loads for Nitrogen*. Nordic Council of Ministers, Copenhagen.

Heij, G.J. & Erisman, J.W. (1997) *Acidification Research in the Netherlands; Report of Third and Last Phase*. Elsevier, Amsterdam.

Heil, G.W. & Bobbink, R. (1993) "CALLUNA": a simulation model for evaluation of impacts of atmospheric nitrogen deposition on dry heathlands. *Ecological Modeling* **68**, 161–82.

Hendriks, C.M.A., Van den Burg, J., Oude Voshaar, J.H. & Van Leeuwen, E.P. (1997) *Relationship between Forest Condition and Stress Factors in the Netherlands in 1995*. Report 148. DLO Winand Staring Centre, Wageningen.

Henriksen, A. & Posch, M. (2000) Steady-state models for calculating critical loads for surface waters—where do we stand today? *Water Air and Soil Pollution*.

Hettelingh J.-P., Posch, M., de Smet, P.A.M. & Downing, R.J. (1995) The use of critical loads in emission reduction agreements in Europe. *Water Air and Soil Pollution* **85**, 2381–8.

Hornung, M., Sutton, M.A. & Wilson, R.B. (eds) (1995) *Mapping and Modeling of Critical Loads for Nitrogen: A Workshop Report*. Institute of Terrestrial Ecology, Huntingdon.

Hultberg, H. (1988) Critical loads for sulfur to lakes and streams. In Nilsson, J. & Grennfelt, P. (eds), *Critical Loads for Sulphur and Nitrogen*. Nordic Council of Ministers, Copenhagen.

Hutchinson, T.C., Bozic, L. & Munoz-Vega, G. (1986) Responses to five species of conifer seedlings to aluminum stress. *Water Air Soil and Pollution* **31**, 283–94.

Janssen, R.P.T., Pretorius, P.J., Peijnenburg, W.J.G.M. & Van den Hoop, M.A.G.T. (1996) *Determination of Field-based Partition Coefficients for Heavy Metals in Dutch Soils and the Relationships of These Coefficients with Soil Characteristics*. Report 719101023. National Institute of Public Health and the Environment (RIVM), Bilthoven.

Johnson, D.W. (1984) Sulfur cycling in forests. *Biogeochemistry* **1**, 29–43.

Johnson, D.W. & Todd, D.E. (1983) Relationships among iron, aluminum, carbon, and sulfate in a variety of forest soils. *Soil Science Society of America Journal* **47**, 792–800.

Johnson, D.W., Cole, D.W. & Gessel, S.P. (1979) Acid precipitation and soil sulfate adsorption properties in a tropical and in a temperate forest soil. *Biotropica* **11**, 38–42.

Johnson, D.W., Henderson, G.S., Huff, D.D. *et al.* (1982) Cycling of organic and inorganic sulfur in a chestnut oak forest. *Oecologia* **54**, 141–8.

Jongbloed, R.H., Pijnenburg, J., Mensink, B.J.W.G., Traas, Th.P. & Luttik, R. (1994) *A Model for Environmental Risk Assesment and Standard Setting Based on Biomagnification. Top Predators in Terrestrial Ecosystems*. Report 719101012. National Institute of Public Health and the Environment (RIVM), Bilthoven.

Kärenlampi, L. & Skärby, L. (eds) (1996) *Critical Levels for Ozone in Europe: Testing and Finalizing the Concepts*. University of Kuopio, Department of Ecology and Environmental Science, Kuopio.

Klepper, O. & Van de Meent, D. (1997) *Mapping the Potentially Affected Fraction (PAF) of Species as an Indicatior of Generic Toxic Stress*. Report 607504001. National Institute of Public Health and the Environment (RIVM), Bilthoven.

Kros, J., Reinds, G.J., De Vries, W., Latour, J.B. & Bollen, M. (1995) *Modeling of Soil Acidity and Nitrogen Availability in Natural Ecosystems in Response to Changes in Acid Deposition and Hydrology.* Report 95. DLO Winand Staring Centre, Wageningen.

Latour, J.B. & Reiling, R. (1993) A multiple stress model for vegetation (MOVE): a tool for scenario studies and standard setting. *The Science of the Total Environment* supplement **93**, 1513–26.

Latour, J.B., Reiling, R. & Slooff, W. (1994) Ecological standards for eutrophication and desiccation: perspectives for a risk assessment. *Water Air and Soil Pollution* **78**, 265–77.

Liljelund, L.E. & Torstensson, P. (1988) Critical load of nitrogen with regards to effects on plant composition. In Nilsson, J. & Grennfelt, P. (eds), *Critical Loads for Sulphur and Nitrogen.* Nordic Council of Ministers, Copenhagen.

Løkke, H., Bak, J., Falkengren-Grerup, U., Finlay, R.D. *et al.* (1996) Critical loads of acidic deposition for forest soils: Is the current approach adequate? *Ambio* **25**, 510–16.

Mäkelä, A. & Schöpp, W. (1990) Regional-scale SO_2 forest-impact calculations. In Alcamo, J., Shaw, R. & Hordijk, L. (eds), *The RAINS model of Acidification — Science and Strategies in Europe.* Kluwer Academic Publishers, Dordrecht.

Malanchuk, J.L. & Nilsson, J. (eds) (1989) *The Role of Nitrogen in the Acidification of Soils and Surface Waters.* Nordic Council of Ministers, Copenhagen.

Marschner, H. (1990) *Mineral Nutrition of Higher Plants.* Academic Press, London.

Matzner, E. (1988) Der Stoffumzatz zweier Waldökosysteme im Solling. *Bericht des Forschungszentrums Waldökosysteme/Waldsterben* **A40**, 1–217.

Mengel, K. (1991) *Ernährung und Stoffwechsel der Pflanze*, 7th edn. Jena.

Moolenaar, S.W. & Lexmond, Th.M. (1998) Heavy metal balances of agro-ecosystems in the Netherlands. *Netherlands Journal of Agricultural Science.*

Morgan, S.M., Lee, J.A. & Ashenden, T.W. (1992) Effects of nitrogen oxides on nitrate assimilation in bryophytes. *New Phytologist* **120**, 89–97.

Nihlgård, B. (1985) The ammonium hypothesis; an additional explanation to the forest dieback in Europe. *Ambio* **14**, 2–8.

Nilsson, J. (ed.) (1986) *Critical Loads for Nitrogen and Sulphur.* Nordic Council of Ministers, Copenhagen.

Nilsson, J. & Grennfelt, P. (eds) (1988) *Critical Loads for Sulphur and Nitrogen.* Nordic Council of Ministers, Copenhagen.

Oja, T., Yin, X. & Arp, P.A. (1995) The forest modeling series ForM-S: applications to the Solling spruce site. *Ecological Modeling* **83**, 207–17.

Palmborg, C., Bringmark, L., Bringmark, E. & Nordgren, A. (1998) Multivariate analysis of microbial activity and soil organic matter at a forest site subjected to low-level heavy metal pollution. *Ambio* **27**, 53–7.

Popp, M.P., Kulman, H.M. & White, E.H. (1986) The effect on nitrogen fertilization of white spruce on the yellow headed spruce sawfly (*Pikonema alaskansis*). *Canadian Journal of Forest Research* **16**, 832–5.

Posch, M. & De Vries, W. (1999) Derivation of critical loads by steady-state and dynamic soil models. In Langan, S.J. (ed.), *Nitrogen Deposition and Its Impact on Natural and Semi-natural Ecosystems.* Kluwer, Dordrecht.

Posch, M., Kämäri, J., Johansson, M. & Forsius, M. (1993) Displaying inter- and intra-regional variability of large-scale survey results. *Environmetrics* **4**, 341–52.

Posch, M., De Smet, P.A.M., Hettelingh, J.-P. & Downing, R.J. (eds) (1995) *Calculation and Mapping of Critical Thresholds in Europe. Status Report 1995.* Coordination Centre for Effects, National Institute of Public Health and the Environment (RIVM), Bilthoven.

Posch, M., Kämäri, J., Forsius, M., Henriksen, A. & Wilander, A. (1997) Exceedance of critical loads for lakes in Finland, Norway and Sweden: reduction requirements for acidifying nitrogen and sulfur deposition. *Environmental Management* **21**, 291–304.

Posch, M., De Smet, P.A.M., Hettelingh, J.-P. & Downing, R.J. (eds) (1999) *Calculation and mapping of critical thresholds in Europe. Status Report 1999.* Coordination Centre for Effects, National Institute of Public Health and the Environment (RIVM), Bilthoven.

Roberts, T.M., Skeffington, R.A. & Blank, L.W. (1989) Causes of type 1 spruce decline. *Forestry* **62**, 179–222.

Roelofs, J.G.M. (1986) The effect of air-borne sulfur and nitrogen deposition on aquatic and terrestrial heatland vegetation. *Experientia* **42**, 372–7.

Roelofs, J.G.M., Kempers, A.J., Houdijk, A.L.F.M. & Jansen, J. (1985) The effect of airborne ammonium sulfate on *Pinus nigra var. maritima* in the Netherlands. *Plant and Soil* **84**, 45–56.

Rosén, K. (1990) *The Critical Load of Nitrogen to Swedish Forest Ecosystems.* Internal Report. Dept of Forest Soils, University of Agriculture Science, Uppsala.

Rosén, K., Gundersen, P., Tegnhammar, L., Johansson, M. & Frogner, T. (1992) Nitrogen enrichment of Nordic forest ecosystems—the concept of critical loads. *Ambio* **21**, 364–8.

Schöpp, W., Amann, M., Cofala, J., Heyes, C. & Klimont, Z. (1999) Integrated assessment of European air pollution emission control strategies. *Environmental Modeling and Software* **14**, 1–9.

Schulze, E.D. (1989) Air pollution and forest decline in a spruce (*Picea abies*) forest. *Science* **244**, 776–83.

Slooff, W. (1992) *RIVM Guidance Document: Ecotoxicological Effect Assessment. Deriving Maximum Tolerable Concentrations from Single-species Toxicity Data.* Report 719102018. National Institute of Public Health and the Environment (RIVM), Bilthoven.

Smith, E.P. & Cairns, J. (1993) Extrapolation methods for setting ecological standards for water quality: statistical and ecological concerns. *Ecotoxicology* **2**, 203–19.

Spiecker, H., Mielikäinen, K., Köhl, M. & Skovsgaard, J.P. (eds) (1997) *Growth Trends in European Forests.* EFI Research Report 5. Springer, Berlin.

Sverdrup, H. & Warfvinge, P. (1993) *The Effect of Soil Acidification on the Growth of Trees, Grass and Herbs as Expressed by the (Ca + Mg + K)/Al Ratio.* Reports in Ecology and Environmental Engineering 2. Dept of Chemical Engineering II, Lund University, Lund.

Sverdrup, H., De Vries, W. & Henriksen, A. (1990) *Mapping Critical Loads. A Guidance Manual to Criteria, Calculation Methods, Data Collection and Mapping.* Nordic Council of Ministers, Copenhagen.

Tamm, C.O. (1991) *Nitrogen in Terrestrial Ecosystems. Questions of Productivity, Vegetational Changes and Ecosystem Stability.* Ecological Studies 81. Springer, Berlin.

Tietema, A. & Beier, C. (1995) A correlative evaluation of nitrogen cycling in the forest ecosystems of the EC projects NITREX and EXMAN. *Forest Ecology and Management* **71**, 143–51.

Tyler, G. (1992) *Critical Concentrations of Heavy Metals in the Mor Horizon of Swedish Forests.* Swedish Environmental Protection Agency, Solna.

UNECE (1990) *ECE Critical Levels Workshop. Final Draft Report.* Umweltbundesamt, Berlin.

UNECE (1996a) *Manual on Methodologies and Criteria for Mapping Critical Levels/Loads and Geographical Areas where They Are Exceeded.* Texte 71/96. Umweltbundesamt, Berlin.

UNECE (1996b) *1979 Convention on Long-range Transboundary Air Pollution and Its Protocols.* United Nations, New York and Geneva.

Ulrich, B., Mayer, R. & Khanna, P.K. (1979) *Die Deposition von Luftverunreinigungen und ihre Auswirkungen in Waldökosystemen im Solling.* Schriften aus der Forstl. Fakultät der Universität Göttingen und der Niedersächsischen Versuchsanstalt, Göttingen.

Van Dam, D. (1990) Atmospheric deposition and nutrient cycling in chalk grassland. PhD Thesis, University of Utrecht, Utrecht.

Van den Burg, J., Evers, P.W., Martakis, G.F.P., Relou, J.P.M. & Van der Werf, D.C. (1988) *De conditie en de minerale-voedingstoestand van opstanden van grove den (Pinus silvestris) en Corsicaanse den (Pinus nigra var. Maritima) in de Peel en op de zuidoostelijke Veluwe najaar 1986.* Instituut voor Bosbouw en Groenbeheer "De Dorschkamp," Wageningen.

Van den Hout, K.D., Bakker, D.J., Berdowski, J. *et al.* (1999) The impact of atmospheric deposition of non-acidifying substances on the quality of European forest soils and the North Sea. *Water Air and Soil Pollution* **109**, 357–96.

Van Hinsberg, A. & Kros, J. (1999) *Eeen normstellingsmethode voor (stikstof)depositie op natuurlijke vegetaties in Nederland. Een uitwerking van de natuurplanner voor natuurdoeltypen.* Report 722108024. National Institute of Public Health and the Environment (RIVM), Bilthoven.

Van Oene, H. (1992) Acid deposition and forest nutrient imbalances: a modeling approach. *Water Air and Soil Pollution* **63**, 33–50.

Warfvinge, P. & Sverdrup, H. (1995) *Critical Loads of Acidity to Swedish Forest Soils: Methods, Data and Results.* Reports in Ecology and Environmental Engineering 5. Dept of Chemical Engineering II, Lund University, Lund.

Warfvinge, P., Falkengren-Grerup, U., Sverdrup, H. & Andersen, B. (1993) Modeling long-term cation supply in acidified forest stands. *Environmental Pollution* **80**, 209–21.

Witter, E. (1992) *Heavy Metal Concentrations in Agricultural Soils Critical to Microorganisms.* Report 4079. Swedish Environmental Protection Agency, Solna.

Woodwell, G.M. (1976) The threshold problem in ecosystems. In Levin, S.A. (ed.), *Ecosystem Analysis and Prediction.* Society for Industrial and Applied Mathematics, Philadelphia, PA.

Wright, R.F. & Rasmussen, L. (eds) (1998) The whole ecosystem experiments of the NITREX and EXMAN projects. *Forest Ecology and Management* **101**, 1–353.

Zöttl, H.W. & Mies, E. (1983) Nährelementversorgung und Schadstoffbelastung von Fichtenökosystemen im Südschwarzwald unter Immissionseinfluß. *Mitteilungen der Deutschen Botanischen Gesellschaft* **38**, 429–34.

21 The Practice of Air Quality Management

BERNARD E.A. FISHER

21.1 INTRODUCTION TO AIR QUALITY MANAGEMENT

The purpose of air quality management is to ensure satisfactory air quality, usually in urban areas. The difference from strategies that rely on measures to control emissions is that air quality management is based on the air quality concentrations that result from all sources in and around an urban area. It is therefore a more appropriate management tool but relies on more detailed assessments of emissions, dispersion, and observations.

The purpose of an air quality strategy is to improve areas of poor air quality and reduce any remaining significant risks to health. At the core of any strategy are air quality standards. The standards should be based on an assessment of the effects of each of the most important air pollutants on public health. Different countries may apply these in different ways but the general principles are the same. This chapter uses the UK as an example but this is only to illustrate principles. In the UK the pollutants within the National Strategy include ozone, benzene, 1,3-butadiene, SO_2, CO, NO_2, particles, lead, and polycyclic aromatic hydrocarbons (PAHs). In setting standards the judgments of experts have been used. Standards have been set at levels of air pollutants at which there would be an extremely small or no risk to human health. There may be differences of detail from country to country.

To implement standards in the UK, Air Quality Objectives have been set. The objectives represent the progress that can be made in a cost-effective manner toward the air quality standards by some future date. This recognizes that improvement in air quality will only come about by planning ahead and by the progressive introduction of improvement measures. Specific objectives for each of the pollutants are listed in Table 21.1. The practice of air quality management requires reviews of air quality in an area to be made and assessments of future air quality to be produced that can be judged against objectives. The review will consider air quality within an area and the assessment will consider whether the objectives will be breached or at risk in 2005. (In the UK ozone is recognized as not generally susceptible to local abatement strategies and is not included in local assessments. Polycyclic aromatic hydrocarbons (PAHs) are an example of a class of trace pollutants that is of concern but for which it is not easy to apply local management techniques because of the difficulty of measurement and applying assessment techniques.) A possible breach of an objective will require the designation of an area, known as an Air Quality Management Area, and the development of Action Plans (DETR 1999a) setting out how to achieve all the air quality objectives listed in Table 21.1.

Air quality management is therefore seen as a fair and logical way to improving air quality. However, the preparation of air quality reviews and assessments relies on technical judgments of matters with greater or lesser degrees of uncertainty, which are not easy to make. The appropriate

Table 21.1 Objectives for the purposes of local air quality management in the UK as at January 2000.

Pollutant	Air quality objective Concentration*	Measured as	Year end to be achieved by
Benzene	$16.25\,\mu g\,m^{-3}$ (5 p.p.b.)	Running annual mean	2003
1,3-Butadiene	$2.25\,\mu g\,m^{-3}$ (1 p.p.b.)	Running annual mean	2003
Carbon monoxide	$11.6\,mg\,m^{-3}$ (10 p.p.m.)	Running eight-hour mean	2003
Lead	$0.5\,\mu g\,m^{-3}$	Annual mean	2004
	$0.25\,\mu g\,m^{-3}$	Annual mean	2008
Nitrogen dioxide	$200\,\mu g\,m^{-3}$ (105 p.p.b.) not to be exceeded more than 18 times a year	One-hour mean	2005
	$40\,\mu g\,m^{-3}$ (21 p.p.b.)	Annual mean	2005
Particles (PM$_{10}$)†	$50\,\mu g\,m^{-3}$ (gravimetric) not to be exceeded more than 35 times a year	24-hour mean	2004
	$40\,\mu g\,m^{-3}$ (gravimetric)	Annual mean	2004
Sulfur dioxide	$350\,\mu g\,m^{-3}$ (132 p.p.b.) not to be exceeded more than 24 times a year	One-hour mean	2004
	$125\,\mu g\,m^{-3}$ (47 p.p.b.) not to be exceeded more than three times a year	24-hour mean	2004
	$266\,\mu g\,m^{-3}$ (100 p.p.b.) not to be exceeded more than 35 times a year	15-minute mean	2005

* The concentration is the air quality standard. Conversions of p.p.b. to $\mu g\,m^{-3}$, and p.p.m. to $mg\,m^{-3}$ at 20°C and 1013 mb.
† Measured using the European gravimetric transfer standard or equivalent. Objectives are subject to revision. Table 21.1 shows the variety of objectives, but does not include those for the protection of vegetation and ecosystem, or for ozone.

authority needs to make judgments according to local circumstances and the information available to it, and this is not a straightforward process. It is readily seen that the objectives appear to be a complex mixture of air pollution units and frequencies of occurrence. This is because they are based on the perceived health effects and not chosen so as to make assessments easy and convenient.

There are various ways of undertaking an air quality review involving various degrees of sophistication. This chapter describes methods by which a start could be made based primarily on desktop studies, using generally available information. The areas of greatest uncertainty are outlined. The intention is that maximum use of available data on air pollution sources and concentrations should be made to complete an air quality review. This is preferable to devoting resources to only part of an assessment, such as an emissions inventory or a monitoring network, without carrying through a complete air quality review. It is always possible to undertake more detailed assessments by doing more monitoring, and to do more modeling using specific information on emissions collected for the purposes of air quality management.

Pollutants such as benzene, 1,3-butadiene, lead, and NO_2 have air quality objectives expressed as long-term averages. The annual mean is the average concentration over a whole year, usually taken to be a calendar year, while the running annual mean represents the average over a 12-month period but starting from any time, and incremented at hourly intervals. The running mean is therefore a slightly stricter standard than the annual mean, although closely related.

The air quality review should consider the air quality at a future date within an urban area. This means that there has to be an appreciation of the source–receptor relationships in an area, based on either models or measurements.

Defining the accuracy of air pollution modeling results is difficult even in very well tried and tested situations, such as the use of dispersion models for single well defined sources. Accepting that perfect accuracy is neither necessary nor achievable, great variations in the accuracy of reviews undertaken by different authorities are not desirable. It is generally recommended that a somewhat pessimistic assessment is performed so that concentrations are accurate to within ±50% most of the time. Confidence can be improved by referring back to measurements as often as possible, using surrogate statistics, such as population or traffic flow, to check on the consistency of the estimated air quality between areas, and taking upper bounds where appropriate to do so.

The air quality review is only the first stage in an assessment. It is likely that monitoring will have to take place in areas identified as air quality management areas. The precise boundaries of these areas are hard to define because of the uncertainty associated with current air quality levels and the uncertainty of future levels. An air quality management area should not be regarded as a region rigidly delineated on a map, although the area has to be marked in a clear way on a map. Instead, it should be thought of as a region chosen to encompass areas where exceedances are likely, such as a local authority area (or areas) or a public highway, for which further consideration of future air pollution levels should be undertaken. By undertaking a review and an assessment, the appropriate authority needs to convince itself that local plans can be implemented in a way that will ensure successful implementation of the National Air Quality Strategy (DoE 1997).

21.1.1 Example of the benefits of air quality management

This chapter contains short sections on categories of air pollution to illustrate the differences in the treatment of pollutants according to their health effects. However, to illustrate how air quality management can assist in the improvement in urban air quality, a preliminary example of the kinds of issues raised is illustrated in work by

Table 21.2 Predicted percentage reduction in annual average urban background concentration (excluding local impact from roads) in London, Bloomsbury, in 1993 if the road transport emissions standards in Berlin were applied in London.

	NO_x	PM_{10}	Benzene	NMVOC	CO
Percentage reduction	12	3	27	21	23

Seika (1996) and Seika *et al.* (1998). The authors compared average concentrations in London and Berlin at a typical central location, not adjacent to a street, where the local impact could be excluded, and at a location within a street. The predicted annual average concentrations in 1993 for NO_x (NO plus NO_2), PM_{10}, benzene, nonmethane volatile organic compounds (NMVOCs), and CO were estimated, assuming that the same road transport emissions applied in London as in Berlin, using a model. Table 21.2 illustrates the results. There would be a significant improvement in the annual average urban air quality in London of about 20–25% for benzene, NMVOCs, and CO if emissions standards in Berlin applied. There would be a slight improvement in NO_x and a very slight reduction in PM_{10}.

These estimates would not be appropriate for an air quality review as required in the UK because: (i) they refer to changes in the annual average estimates, which though related to percentiles, are not expressed in the form defining objectives; (ii) they do not consider concentrations at a future date when improvements in vehicle emissions in London will have taken effect; (iii) they do not consider concentrations at the most exposed locations in London; and (iv) they do not consider all the pollutants considered for an air quality review. In particular, NO_x levels have been estimated, not NO_2. Health effects are associated with NO_2 and not with NO. PM_{10} in this calculation refers to primary particles, that fraction of particulate matter which is emitted directly into the atmosphere as small particles.

Table 21.3 Predicted percentage reduction in annual average urban roadside concentrations in Regent Street, London, and Französische Strasse, Berlin, in 1993 if buses had been converted to compressed natural gas.

	NO_x	PM_{10}	Benzene	NMVOC	CO
Percentage reduction in Regent Street, London	17	24	6	8	6
Percentage reduction in Französische Strasse, Berlin	9	4	1	1	1

Table 21.4 Predicted percentage reduction in annual average urban roadside concentrations in Regent Street, London, in 1993 if taxis had been converted to compressed natural gas.

	NO_x	PM_{10}	Benzene	NMVOC	CO
Percentage reduction	2	12	1	1	1

Seika (1996) considered some other interesting scenarios. He looked at the changes in a major London street (Regent Street) and a major street in Berlin (Französische Strasse) if buses were run on compressed natural gas (see Table 21.3). In this case a considerable improvement in annual average NO_x and PM_{10} concentrations is predicted. If taxis were converted to compressed natural gas, a 10% improvement in annual average PM_{10} concentrations is predicted in Regent Street (see Table 21.4). These are the kinds of interesting scenarios that local air quality management is concerned with.

To undertake assessments an air quality manager needs to have certain data and assessment tools available and there is a danger of information overload. In the UK, data are available from automatic air quality monitoring sites in the National Air Quality Information Archive (see http://www.airquality.co.uk/archive). At early stages in the development of air quality management one is only concerned with reviewing and assessing the air quality in a local authority area. Consideration of future alternative emissions scenarios follows later if an authority needs to consider action plans to bring about improvement.

21.1.3 A staged approach to air quality reviews and assessments

Given the uncertainties associated with air quality management, a staged approach to the review and assessment of air quality is recommended. Figure 21.1 shows the three stages that an authority could follow when undertaking air quality management. Stage 1 is expected to involve simple checklists of polluting sources within an authority area. Stage 2 is expected to involve using simple screening methods to determine whether there are likely to be areas within a local authority boundary where air quality objectives will be breached in 2005. Stage 3 involves more precise and sophisticated assessment methods. At each stage, if concentrations appear to comply with objectives by a wide margin the process can be stopped, as an authority can be confident that objectives will not be exceeded.

Four components of individual concentrations are distinguished. These are: (i) the regional contribution; (ii) the urban background concentration; (iii) the roadside concentration; and (iv) the industrial contribution. The emphasis on these contributions may appear rather obvious, but it should be recognized that (i) and (ii) are largely determined by national policy, (iii) can in part be largely determined by a local administrative body, and (iv) should be determined by the local control agencies and local authorities. The sources affecting the four contributions need to be known if effective plans for improvements are to be drawn up.

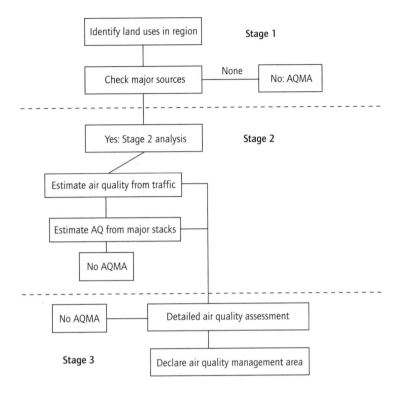

Fig. 21.1 Stages in an air quality management assessment.

21.2 CARBON MONOXIDE

Carbon monoxide is a good pollutant to illustrate the principles of air quality management because its sources and atmospheric behavior are relatively simple to define. Airborne carbon monoxide is produced by incomplete combustion. The major source is road transport, with 91% of emissions in the United Kingdom (88% is associated with gasoline and 3% with diesel). The remainder is emitted by a number of small contributions, such as from industrial combustion and domestic sources (DoE 1995a). Knowledge of the distribution of sources is a necessary starting point to air quality management. However, since emissions per unit area are constrained by the distribution of traffic, the distribution can be summarized as fol-

lows. The average emissions in tonnes per square kilometer are of the order of 200 in urban centers, rising to more than 250 in the center of urban areas in the UK. Outside urban areas the average emission will be much smaller. Spatially disaggregated emissions on a grid basis are available from the urban emissions inventories. Although one expects variations between urban areas there is a consistency with higher emission densities in the larger cities with a higher density of road transport.

Carbon monoxide is degraded slowly in the atmosphere, so that ambient concentrations in urban areas, where the highest concentrations are expected to occur, are controlled by the rate of emission and the subsequent dispersion, or dilution. The air quality standard has been set as 10 p.p.m., measured as a running eight-hour average.

This is a short-term exposure limit and therefore exceedances of the standard, if they occur, will be associated with infrequent episodes of poor dispersion. Maximum concentrations are therefore likely to vary from year to year by ±50%.

Concentrations are expected to be highest at the kerbside in heavy traffic. Concentrations will also be high at locations where dispersion is generally poor, such as in road tunnels or parking lots. There is a regional background concentration of about 0.13 p.p.m., obtained from measurements at a remote site in the British Isles (Mace Head), which is much less than the standard and will have no influence on exceedances. In the UK the urban background CO concentration can be estimated using monitoring data from automatic monitoring stations. Many sites in the network are located near city centers. They have been sited away from major roads, so that measurements are not dominated by the pollution from emissions from vehicles on any one road. Data on the concentrations measured at these sites in recent years suggest maximum eight-hour average concentrations of the order of 4–8 p.p.m., depending on the urban area (summarized in DoE 1995b; more detail in Bower et al. 1997; Broughton et al. 1998). Exceedances of the standard are likely to be infrequent, only possibly occurring next to roads carrying very heavy traffic in road canyon situations.

A typical administrative area will consist of urban areas, for which measurements from sites in other urban areas with a comparable emission density may be used to assess background concentrations, and rural areas, for which the impact of roads can generally be neglected. The assessment should also consider major roads running through urban areas. In this case the maximum eight-hour urban background concentration can be added to the maximum eight-hour CO concentration from the road. Maximum one-hour concentrations at urban background sites tend to be about 20–30% higher than the maximum eight-hour concentration, while annual average CO concentrations are typically about 10% of the maximum eight-hour concentration.

Estimates of the effects of future emissions can be made on the basis of national policy measures to control emissions from vehicles. By incorporating the forecasts of national road transport activity it has been estimated that national emissions of carbon monoxide reduced by 25% in 2000, and are expected to reduce by 40% by 2005, and 50% by 2010, compared with 1995 levels. If these reductions are applied to the monitoring data, estimates of projected maximum eight-hour running mean CO concentrations at UK sites may be made. Concentrations are below the objectives even taking into account the year-to-year variability. Thus CO appears not to be an issue for air quality management in the UK and the same conclusion may be drawn for other countries following similar analysis.

The only caveat to this conclusion would be that concentrations near to major roads should be assessed in more detail. This can be tackled in part by monitoring. However, as concentrations are expected to fall sharply with distance from a road, it is usual and preferable to consider the specific road contribution to overall CO concentrations using dispersion models. These are discussed in more detail in a later section. However, it would seem appropriate to consider major roads defined, say, as major road transport routes in urban areas carrying an average weekday traffic flow in 2005 of more than about 15,000 vehicles per day, and major road transport routes in rural areas carrying an average weekday traffic flow of more than about 50,000 vehicles per day. This definition may have to be altered in the light of experience. One might start by analyzing the roads carrying the largest traffic flows and successively assessing roads carrying lesser traffic flows.

The pattern of distribution of CO concentrations can be explained by reference to simple formulae based on dispersion modeling. The short-term concentration at the center of a circular urban area with a uniform emission density can be estimated using conventional dispersion models (Fisher & Newlands 2000; Fisher & Sokhi 2000). Equation 21.1 has been derived from an area source Gaussian plume model applied to a circular urban area with an assumed uniform emission density. Dispersion is assumed to be subject to slow vertical mixing, a shallow mixing layer, and a very low

Table 21.5 Calculated CO concentrations in some major UK cities.

City	Average emission density of city (t km^{-2} yr^{-1}) CO	Equivalent diameter of city (km)	Maximum calculated eight-hour average concentration (p.p.m.)	Measured max. 1998 eight-hour average concentration (p.p.m.)
Birmingham Center	177	34	3.6	2.6
Birmingham East	177	34	3.6	3.3
Bristol Center	170	8	1.7	3.2
London Hackney	220	56	5.8	6.1
London Bridge Place	220	56	5.8	3.6
London Southwark	220	56	5.8	4.9
London Bloomsbury	220	56	5.8	3.2
Manchester Piccadilly	114	22	1.9	4.4
Manchester Town Hall	114	22	1.9	3.8

wind speed with very variable wind direction fluctuations over an eight-hour period. The urban background CO concentration in the center of the city, C_{urban}, is given by:

$$C_{urban} = 4.28qd^{0.483} \qquad (21.1)$$

where C_{urban} is in p.p.m., q is in kt km^{-2} yr, and d is in kilometers. This shows, as expected, that concentrations are proportional to emission density and dependent, but not strongly dependent, on the diameter of the urban area. This explains the similarity in concentration found at many urban background sites. Applying this formula, CO concentrations in the center of major cities may be estimated (see Table 21.5). Sufficient agreement is found at the most polluted urban background sites for the formula to be used to assess likely exceedances.

Concentrations do not decrease rapidly within an urban area, but once outside of the urban area they decline sharply. Thus, for an initial air quality review the regional contribution to air pollution from sources outside the area of interest can usually be neglected. Thus, estimates of the CO concentrations at a future date at a number of locations in an urban area can be obtained, consisting of the sum of contributions from the urban background concentration and the roadside concentra-

tion. It appears likely that under UK conditions, where road transport is the main source of CO, there is no need to consider designating any Air Quality Management Areas on the basis of CO levels. However, an authority may wish to convince itself of this result. This would take the form of more detailed modeling and monitoring.

21.3 BENZENE

The emission of benzene to the atmosphere arises primarily from the combustion of gasoline and diesel. In the UK 80% is thought to come from mobile sources, which includes gasoline exhausts and diesel engine exhausts. The rest comes from other kinds of sources, such as 7% from gasoline evaporation, 10% from gasoline refining and distribution, and other source categories, none of which when aggregated makes a large contribution to the total (DoE 1995a). A further breakdown according to source type is given by Derwent (1995). Benzene is degraded slowly in the atmosphere (UK PORG 1993), so that concentrations in urban areas, where the highest concentrations are expected to occur, are determined by source strength and dispersion. Highest concentrations are expected to occur near filling stations and adjacent to busy urban streets. Concentrations are lowest in remote rural locations.

Episodes of weather associated with poor dispersion, such as still, cold weather, will be associated with high ambient concentrations. However, the UK Air Quality Standard for benzene is 5 p.p.b. expressed as a running annual average. This is because benzene and 1,3-butadiene, another pollutant considered under air quality management, are the two commonly occurring VOCs that are human carcinogens and thus subject to long-term standards. The long-term average concentration is not strongly affected by the occurrence of meteorological episodes associated with poor dispersion during any 12-month period.

Measurements of benzene, as well as other VOCs, are available from a number of sites in the UK hydrocarbon network, which uses automatic gas chromatography, involving hourly collection and analysis of samples (DoE 1995a; Bertorelli & Derwent 1995). Measurements of benzene using diffusion tubes in London show a clear dependence of concentration on distance from roads. Exceedances of the National Air Quality Standard have been reported at distances out to 100 m from roads. Concentrations of benzene at urban background sites in the UK are around 1 p.p.b. and concentrations of 1,3-butadiene are around 0.2 p.p.b. (note that 1 p.p.b. benzene = 3.24 μg m^{-3} and 1 p.p.b. 1,3-butadiene = 2.21 μg m^{-3}).

In the future, emissions of benzene in the UK are expected to fall by 50%. Significant reductions in emissions from road vehicles are expected through the implementation of vehicle exhaust emission limits, in particular the fitting of catalytic converters to all new gasoline automobiles and the replacement of old automobiles.

In certain areas of the country, industrial emissions of benzene may be significant. However, generally benzene concentrations appear unlikely to exceed the National Air Quality Objective at urban background sites in the center of urban areas in 2005, but diffusion tube measurements in London indicate current exceedances near major roads in the center of urban areas. The measured benzene concentration at a roadside location in Exhibition Road, London, which is subject to moderate traffic flow, was 4.2 p.p.b. between July 1991 and June 1992 (Derwent *et al.* 1995).

In the absence of suitable measurements, an estimate of the benzene concentration may be made in a simple way. The *Digest of Environmental Statistics* (DoE 1995b) suggested that for 1993 the ratio of benzene emissions to total VOC emissions for the whole country was 0.024 by weight. For road transport emissions the *Design Manual for Roads and Bridges* (DETR 1999b) used an average emission ratio of benzene to total hydrocarbons of 0.04. A more detailed benzene emissions inventory of the region may be used if one is available (see example of Buckingham *et al.* 1997a). Assuming that 2.5% of the VOCs is benzene the urban background annual average benzene concentration in the center of the city, C_{urban}, may be estimated from the formula:

$$C_{urban} = 2.6qd^{0.413} \qquad (21.2)$$

where C_{urban} is in p.p.b., q, the VOC emission density of the urban area, is in kt km^{-2} yr^{-1}, and d, the diameter of the urban area, is in kilometers. This again emphasizes that concentrations are proportional to emission density and not strongly dependent on the size of the urban area. In the UK the VOC emission density is typically 60 t km^{-2} in urban areas. Emission densities are much lower in rural areas.

It should be noted that concentrations fall rapidly with distance outside of the urban area and hence the regional contribution to air pollution from sources outside an urban area may be neglected. Equation 21.2 was derived from an area source Gaussian plume model applied to a circular urban area with an assumed uniform emission density, in a similar way to eqn 21.1, but assuming typical meteorological conditions and a uniform wind rose. The annual average concentration was obtained by averaging over a typical frequency of occurrence of dispersion categories for the UK (Fisher & Newlands 2000; Fisher & Sokhi 2000).

The annual average benzene concentration at the edge of major roads may be assessed using a model that depends mainly on the traffic flow along the major roads. Estimates of the benzene concentrations consist of the sum of contributions from the urban background concentration and the roadside concentration. Management of concentrations in the immediate vicinity of a source is determined by the way a process is authorized. For example, emissions of benzene at a gasoline filling station during the replenishment of storage tanks and the refueling of vehicles in the UK are subject to pollution prevention control (Little & Cram 1995).

21.4   1,3-BUTADIENE

When we consider exposure to the general population, the dominant emission of 1,3-butadiene to the atmosphere arises from the combustion of gasoline and diesel (EPAQS 1994c). The emissions in the exhaust gases of motor vehicles are produced by the combustion process itself. The lifetime of 1,3-butadiene in the atmosphere is only a few hours, before it is destroyed by chemical reactions (UK PORG 1993). Rural concentrations are low. A concentration of 0.04 p.p.b. has been reported from West Beckham, Norfolk (Bertorelli & Derwent 1995).

The National Air Quality Standard for 1,3-butadiene is 1 p.p.b., expressed as a running annual average. Measurements in the UK have only been made in ambient air since as recently as 1990. Concentrations of 1,3-butadiene have been reported in 1994 from seven urban monitoring sites (DoE 1995a). These are in London Bloomsbury, London Eltham, Birmingham, Cardiff, Belfast, Middlesbrough, and Edinburgh. The sites are all classified as urban background sites, apart from London Bloomsbury, which is a kerbside site, and London Eltham, which is a suburban site (Bower *et al.* 1997).

By using roadside monitoring data in London, a chemical box model, and a Gaussian line source model, it is possible to derive an average vehicle emission factor for 1,3-butadiene that fits the measurements at a roadside monitoring site in London (Derwent *et al.* 1995). From this analysis the national emission of 1,3-butadiene from gasoline automobiles has been estimated to be 5500 t yr^{-1} in 1992 (Derwent 1995). Estimates for the UK of 1,3-butadiene emissions from sources other than exhaust emissions from motor vehicles are given in the National Air Quality Strategy (DoE 1997).

The ratio of benzene emissions to 1,3-butadiene emissions from road transport is about 10:1 by weight in the United Kingdom. Compliance with the National Air Quality Standard of 5 p.p.b. for benzene, in an area where air quality is largely determined by road transport, therefore corresponds to compliance with a 1,3-butadiene concentration of $5 \times 3.24/(10 \times 2.21) = 0.7$ p.p.b. This is less than the recommended 1,3-butadiene standard. Thus, if it is found that there are no air quality management areas on the basis of a benzene assessment, it is unlikely that there will be a requirement to manage 1,3-butadiene in the area.

In an assessment of 1,3-butadiene one is concerned with long-term average concentrations, which are not strongly affected by the occurrence of meteorological episodes associated with poor dispersion during a 12-month period. Attention needs to be paid to kerbside locations near to roads carrying heavy traffic. Measurements at a roadside location in London (Exhibition Road, which carries a moderate traffic flow) between July 1991 and June 1992 showed an annual average 1,3-butadiene concentration of 0.86 p.p.b. Gasoline-engined motor vehicle 1,3-butadiene emissions declined by about 50% by the year 2000 from their 1992 values, and are expected to decline by 75% by the year 2010.

21.5   LEAD

Except for emissions from local specialized industrial sources nearly all lead in air in the UK used to be from vehicle emissions. The trend in emissions from gasoline-engined road vehicles in the early 1990s is given in the *Digest of Environmental Statistics No. 17* (DoE 1995b) and has reduced since leaded gasoline has been phased out (see www.naei.org.uk/emissions/). Lead compounds, mainly in the form of fine particles of inorganic lead compounds, were emitted from the exhausts of gasoline vehicles using leaded gasoline. A small fraction of the lead would arise from the evaporative losses. Lead in air is now presumed to arise mainly from industrial sources, residual emissions from contaminated vehicle engines, and re-suspension of contaminated soil and dust. Since the lead is attached to fine particles, atmospheric removal processes are generally inefficient and highest concentrations will occur in urban areas or near major roads, where they are controlled by the source strength and the degree of dispersion that has occurred in the atmosphere. Fine particles disperse in the same way as a gas. Highest annual concentrations are expected in confined spaces where mixing is poor regardless of atmospheric conditions, such as in tunnels, at road junctions, and on the central reservations of motorways at times of high traffic flow. Concentrations will be lowest in remote rural locations.

Measurements at a number of locations in the UK are reported in DOE (1995b) and Bertorelli and Derwent (1995). These include a number of sites around industrial works set up to monitor compliance with the EU Directive on lead, a number of "lead in gasoline" sites (two rural, three urban, and three kerbside), and some sites in rural locations. Lead concentrations in air have fallen in recent years in response to the decrease in emissions of lead.

With lead one is concerned with the effect of long-term exposure on human health. The UK national air quality standard is $0.5\,\mu g\,m^{-3}$ as an annual average. Annual averages are not sensitive to the occurrence of particular episodes of poor dispersion, but depend on the general climatology of an area. In the past, before the introduction of unleaded gasoline, concentrations were observed to drop rapidly with distance from major roads and merge with the background within approximately 100 m. Industrial sources that could contribute appreciably to lead levels in their local area should be managed by the control of emissions from industrial sources.

The main focus of a lead assessment used to be on transport emissions close to roads or the locality of lead-emitting industry. In the latter case the DoE (1995a) reports on results from Walsall and Newcastle. Urban background sites in London, Leeds, and Glasgow average about $0.1\,\mu g\,m^{-3}$, well below the national air quality standard, with $0.25\,\mu g\,m^{-3}$ measured at the London Cromwell Road site.

Chamberlain et al. (1979) derived a formula for the annual average lead concentration in urban areas, based on a Gaussian dispersion model and a ground-level source, which may be expressed in the form:

$$C_{urban} = 0.3qd^{0.403} \qquad (21.3)$$

where C_{urban} is the annual average concentration in $\mu g\,m^{-3}$, q is the emission density in $t\,km^{-2}\,yr^{-1}$, and d is the diameter of the urban area in kilometers. This formula is similar to that suggested for estimating benzene concentrations.

Currently and in the future lead concentrations at urban background sites in the center of urban areas are unlikely to exceed standards, given decreasing lead emissions. Only regions containing major point sources of lead emission require a more detailed review of lead levels. Then the stationary source contribution at the point of maximum ground-level concentration of lead may be added to other lead concentrations, to derive the "worst-case" lead concentrations in urban background, urban kerbside, and rural kerbside locations. Since the emissions of lead from gasoline vehicles in the UK have largely disappeared compared with 1995 levels, annual average lead levels should not be significant in urban areas even at the roadside of busy streets.

21.6   NITROGEN DIOXIDE

Oxides of nitrogen are one type of pollutant emitted from road transport. One of the complications of dealing with nitrogen oxides is that there are two forms: nitrogen oxide, NO, and nitrogen dioxide, NO_2. With regard to effects on human health one need only consider NO_2 (EPAQS 1996). Fortunately only a small fraction of NO_x is emitted as NO_2 and the formation of NO_2 from NO is limited by the ozone concentration in the atmosphere. This means that except in unusual circumstances very high concentrations of NO_2 are unlikely to occur.

The urban background concentrations of NO_2 in major cities may be estimated using data from the UK automatic urban network (annual average concentrations typically range from 20 to 30 p.p.b. in urban centers). Annual average concentrations in the middle of cities of various sizes take somewhat similar values. The emission density in UK cities is about $60 t km^{-2}$ in the center of cities, with higher densities in major city centers and much lower concentrations in rural areas (Buckingham & Clewley 1996; Buckingham *et al.* 1997a,b,c). However, NO_2 concentrations from urban background and roadside sources cannot be directly summed. Hence the approach used for CO, lead, and benzene cannot be applied.

In central London urban background concentrations currently exceed the long-term UK national standard of 21 p.p.b., expressed as an annual mean. Outside urban areas concentrations are generally lower than those inside urban areas. The rural concentration is made up of a large number of small contributions from urban and industrial sources within the vicinity and further afield, including emissions from the rest of the country and beyond.

The UK Department of the Environment initiated a large-scale survey of NO_2 using diffusion tube samplers. This national NO_2 diffusion tube network currently consists of approximately 1100 sites. A national overview of the results is available (Stevenson & Bush 1995). The main aim of the diffusion tube survey was to identify areas where additional monitoring might be required to ensure compliance with NO_2 air quality standards. Results from the diffusion tube sampling are useful for obtaining estimates of the long-term (annual) average NO_2 concentration, and can therefore be compared with the long-term UK national air quality objective. This shows that the long-term air quality standard of $40 \mu g m^{-3}$ or 21 p.p.b. (1 p.p.b. $NO_2 = 1.91 \mu g m^{-3}$) is exceeded at many of these sites. However, diffusion tube measurements are not as accurate as results obtained from continuous monitors.

By contrast, the alternative method of assessment is based on the annual average NO_x concentration estimated from NO_x emissions and dispersion models, adjusted for the year 2005. The fraction of NO_x converted to NO_2 needs to be compared with the long-term national air quality objective of 21 p.p.b. This is the key step because of the complications involving the chemistry of NO_2 production. Various models may be used but any model is crucially dependent on the algorithms used to describe the relationship between NO_x and NO_2.

The hourly national air quality objective for NO_2 is 100 p.p.b. or about $200 \mu g m^{-3}$ for the eighteenth highest hourly concentration in a year. An exceedance of the objective depends on the oxidation mechanism. The maximum hourly NO_2 concentrations rarely exceed 100 p.p.b. in urban centers for most years. Chemical reactions can produce NO_2 directly from NO, but it turns out that the reaction is only important in large urban areas when the wind speed is low, typically $1 m s^{-1}$ or less, and concentrations of NO are extremely high. Such episodes of high NO_2 arise during rather exceptional events. In such situations it is not possible to attribute blame to any one source. All sources contributing to the urban pollution levels within the polluted region are in part responsible. These episodes of high NO_2 do occur over London from time to time. The primary driving force is the stagnant air conditions, which do not permit adequate dilution of emissions. The atmosphere in such situations has insufficient capacity to absorb the emitted pollution.

Analysis of the December 1991 episode in London, which was associated with unusually high NO_2 concentrations, showed that measurements could be explained by treating the air over London as a shallow well mixed box in which chemical reactions take place (Derwent *et al.* 1995). Peak NO_2 concentrations are associated not with kerbside monitoring readings but with high concentrations throughout the urban area. This is in contrast to CO and benzene concentrations, which will be high during such episodes, but especially high at roadside locations (EPAQS 1994a,b). National data records (Bower *et al.* 1997; Broughton *et al.* 1998) may be consulted to determine the frequency of NO_2 episodes. Any occurrences should occur in combination with weather conditions associated with low wind speeds.

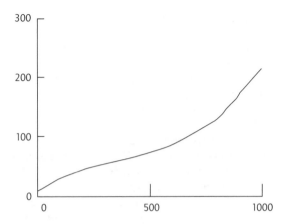

Fig. 21.2 Empirical curve showing relationship between hourly average concentrations of NO_x and NO_2.

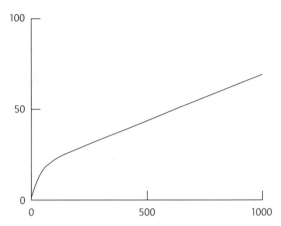

Fig. 21.3 Theoretical curve showing the relationship between hourly average NO_x and NO_2.

An estimate of likely peak hourly NO_2 concentration in 2005 may be made by adding the maximum roadside NO_x and the maximum NO_x from urban sources, and then (i) adjusting the transport emission factors to 2005, and (ii) applying the information on traffic flows in 2005 available from transport planners. An empirical or theoretical curve relating hourly NO_2 to hourly NO_x concentrations must be applied to obtain the maximum hourly NO_2 concentrations in 2005. Two examples are illustrated. Figure 21.2 is an empirical curve based on hourly roadside concentrations in London (Derwent & Middleton 1996), but empirical data when NO_x is greater than 1000 p.p.b. are very sparse. The allowance of up to 18 hours of exceedances suggest that the steeply rising upper part of the curve can be excluded. For very high predicted NO_x concentrations it is preferable to use the theoretical curve in Fig. 21.3, based on the chemical equilibrium between reactions involving nitrogen oxides and ozone.

An estimate of the likely mean annual NO_2 concentration in 2005 may be made in a similar way by adding the mean annual roadside NO_x and the mean annual NO_x from urban sources, and then (i) adjusting the transport emission factors to 2005, and (ii) applying the information on traffic

flows in 2005 available from transport plans. This procedure is somewhat complex. A procedure relating NO_2 to NO_x concentrations must also be applied to obtain the NO_2 concentration in 2005 (Fisher *et al.* 1998).

21.7   SULFUR DIOXIDE

The emission of sulfur dioxide to the atmosphere arises primarily through the combustion of fossil fuels, coal, and oil. There have been significant changes in the pattern of emission in recent years, charted in DoE (1995b). UK SO_2 emissions were at their greatest in 1970, when the annual emission was 6428 kt. In 1993 emissions were estimated to be 3188 kt. This change has come about because of the reduced burning of coal for domestic and industrial purposes in towns. Emissions are now concentrated in a small number of large power stations with tall chimneys located in rural areas, and these will further reduce as a result of the increased use of gas for electricity production. Small and possibly significant emissions arise from road transport, such as diesel vehicles, and from shipping.

National emissions of SO_2 from domestic sources declined from 522 kt in 1970 to 113 kt in

1993 according to DoE (1995b). In 1993 additional emissions from commercial and public services amounted to 88 kt and from road transport to 59 kt. These kinds of emissions, amounting to 260 kt, plus some of the estimated 509 kt from industrial combustion, should be regarded as part of area SO_2 sources in urban centers. Apart from transport sources, the emissions occur some few meters or more above the ground and are discharged through some kind of flue or chimney.

Although a small fraction of the national total, sulfur in vehicle fuel may make a significant contribution to SO_2 concentrations in urban areas. The sulfur content of diesel fuels was restricted to 0.05% or less by weight from 1996 and is reducing further as low-sulfur diesel is promoted widely. The sulfur content of gasoline was typically about 0.04% (EPAQS 1995a) but is also reducing.

Measurements at the Cromwell Road site in London showed evidence that sulfur in gasoline and diesel made some contribution to long-term average kerbside concentrations, but peak concentrations were no higher at this roadside site than at other London urban background sites. If the ratio of CO to SO_2 emissions from road transport is about 100 : 1 an exceedance of the national air quality objective for CO (an eight-hour average concentration set at 10 p.p.m.) corresponds to a SO_2 concentration of $10 \times 1165/(100 \times 2.7) = 43$ p.p.b. SO_2. EPAQS (1995a) recommends that a surrogate one-hour SO_2 concentration of 50 p.p.b. be used when 15-minute averages are not available. Thus compliance with the CO objective more or less ensures that emissions from road transport will not lead to an exceedance of the SO_2 national air quality objective. Air quality reviews for CO and NO_2 should draw attention to areas where road transport is likely to cause exceedances of air quality objectives. The effect of road transport on kerbside SO_2 concentrations is not a priority.

Although SO_2 undergoes chemical reactions in the atmosphere the reaction rate is generally slow. When relating peak concentrations with nearby industrial sources, chemical reactions can be neglected and the SO_2 treated as an inert gas.

Sulfur dioxide and black smoke were the pollutants of greatest concern in the 1950s. An extensive network of monitoring sites was set up in the UK following the Clean Air Act in 1956, forming the so-called National Survey, with up to 1200 monitoring sites mostly located in urban areas. Results from the network were routinely reported in annual summaries. The reductions in concentrations following the decline in concentrations of SO_2 and black smoke has led to a reduction in the network to some 165 sites in the current network of nonautomatic monitoring sites. Sites have been selected to provide representative concentrations in major population centers and long-term trends in concentrations. Because of changes in emissions, past concentrations are no guide to current levels. Some sites have been retained where concentrations have at some time been relatively high and where there may have been a risk of exceeding the EC Directive limit values. During the year April 1993 to March 1994 there were no exceedances of the limit value (Stevenson *et al.* 1995). Data at some rural locations are also reported. (Downing & Campbell 1995; see also airquality.co.uk/archive/).

The appropriate technology at the time when the UK National Survey was set up was to measure SO_2 as a daily average, using a standardized peroxide method. The national air quality objective, designed to protect human health, specifies a SO_2 concentration of 100 p.p.b. $(270 \mu g\,m^{-3})$ when averaged over 15 minutes. The automatic monitoring network makes use of modern instruments that allow 15-minute averages to be reported, although hourly averages only are often reported.

In some areas where concentrations are mainly the result of diffuse, area sources of SO_2 from domestic emissions, and multiple commercial and small industrial sources, a reasonably good relationship exists between daily and 15-minute concentrations. EPAQS (1995a) recommends that the 100 p.p.b. $(270 \mu g\,m^{-3})$ 15-minute standard is equivalent to a maximum daily mean of 28 p.p.b. $(76 \mu g\,m^{-3})$. Thus, in some areas, data from nonautomatic monitoring sites can be used to assess likely exceedances of the SO_2 short-term objective using the daily maximum (Stevenson *et al.* 1995).

Methods for estimating SO_2 concentrations within towns and cities in terms of readily known information and readily estimated parameters were developed when SO_2 was seen as more of a problem. Smith (1973) developed a simple formula that would permit annual average and hourly average concentrations to be estimated. His formula for the annual average concentration, C_{urban}, averaged over the area of a town, may be expressed in the form:

$$C_{urban} = 120qd^{0.341} \qquad (21.4)$$

where q is the emission density of domestic, commercial, and small industrial SO_2 sources in kt $km^{-2}yr^{-1}$ and d is the diameter of the town in kilometers. Lucas (1958) derived a simple formula for the concentration in the center of an urban area, based on Sutton's equation, which may be expressed as

$$C_{urban} = 350qd^{0.378} \qquad (21.5)$$

where q is the emission density of domestic emissions in kt $km^{-2}yr^{-1}$ and d is the diameter of the urban area in kilometers. Both authors emphasized the strong seasonal and daily variation in the emissions, dependent on the use of coal. Lucas applied his results to measurements in London using the winter average of 1954 and 1955, and obtained good agreement for both SO_2 and smoke. For his estimate he used an average winter SO_2 emission density of 200 kt $km^{-2}yr^{-1}$ for domestic sources and obtained a mean winter concentration of 320 $\mu g\,m^{-3}$. Smith (1973) applied his results to measurements of SO_2 in a number of towns in 1970 and also obtained good results for the average SO_2 concentration, using an average emission density of 550 kt $km^{-2}yr^{-1}$ where this included commercial and small industrial sources. At the present time the SO_2 emission density in London is thought to be below about 10 kt $km^{-2}yr^{-1}$ (Buckingham et al. 1997a).

The same kind of approach was used to derive a formula for annual average benzene concentrations in the center of an urban area, referred to Section 21.3. The equivalent formula for SO_2 may be written:

$$C_{urban} = 420qd^{0.413} \qquad (21.6)$$

where C_{urban} is the annual average concentration in $\mu g\,m^{-3}$, q is the emission density in kt $km^{-2}yr^{-1}$, and d is the diameter in kilometers. The formula applied by Chamberlain et al. (1979) to lead concentrations takes a similar form. These formulae all display a somewhat similar dependence of concentration on the diameter of the urban area. However, such relationships refer to long-term average concentrations and therefore cannot be applied to assessments of the UK national air quality objective for SO_2.

In most urban areas the contribution from small diffuse sources within urban areas is not significant. Apart from cities with coal burning the number of 15-minute SO_2 concentrations exceeding 100 p.p.b. is generally less than the objective (35 exceedances). Thus, in most cases an assessment is only concerned with major industrial and commercial sources.

Atmospheric dispersion models are the main tool for assessing the frequency of occurrence of peak short-term SO_2 concentrations. The averaging times appropriate to such models are not always apparent, but models used in the UK generally refer to one-hour averages. Computer models are required, since peak concentrations from an industrial stack depend on a range of factors describing source characteristics, such as stack height, stack diameter, and heat of release, which strongly influence future predicted concentrations. Any dispersion model may be used, provided that it is applied to situations where it has been previously tested against measurements. In some circumstances, such as when the influence of topography is significant, only the most advanced dispersion models are applicable.

21.8   PARTICLES

Emissions of particulate matter are reported in DoE (1995b) in terms of black smoke. This term was introduced to allow for the different soiling capacities of smoke particles from different sources using coal burning as the norm. In contrast, smoke is defined as fine suspended particulate air pollutants (aerodynamic diameter <15 μm) arising from the incomplete combustion of fuel. In recent decades there has been large decrease in the emission of black smoke from domestic sources and an increase from road transport sources. DoE (1988) contains estimates of smoke emissions back to 1960 showing this trend. Estimates were based on the assumption that smoke was a small fraction (dependent on the type of source) of the coal burnt. In the 1950s smoke emissions per household from domestic coal consumption were taken to be approximately equal by weight with SO_2 emissions per household.

Particles in the air arise from other sources as well, loosely categorized as natural, such as sea spray and windblown dust, biological, such as pollen grain and fungal spore fragments, and secondary particles. Secondary particles are formed from gaseous pollutants in the atmosphere. Particles can also grow and combine in the atmosphere. Increasingly attention is being paid to the fraction of airborne particles most likely to be deposited in the lung (Committee on the Medical Effects of Air Pollution 1995a, b). These particles are called PM_{10} (particulate matter less than 10 μm in diameter).

It is not easy to derive a source inventory that includes all types of PM_{10} particles. EPAQS (1995b) contains an emission inventory of PM_{10} particles for the UK that does not include secondary particles, or particles from natural or biological sources. It contains broad similarities with the black smoke inventory but differences in detail, as would be expected. The lack of knowledge regarding total emissions of PM_{10} particles suggests that attempts to relate, and possibly model, total PM_{10} are subject to considerable uncertainty.

Particles that fall within the PM_{10} size fraction will disperse in the atmosphere like a gas. Removal processes do not occur quickly enough to reduce concentrations during the normal time required to transport particles out of an urban area. Ambient concentrations of PM_{10} in urban areas where highest concentrations are expected to occur are controlled by the rate of emission in the urban area and subsequent dispersion.

The appropriate technology at the time when the UK National Survey was set up was to measure black smoke concentrations as a daily average by measuring the blackness caused by particles on a filter when air is drawn through filter paper. Black smoke concentrations measured in this way can be converted into a measure of particulate concentrations by dividing by 0.85. Trends in black smoke concentrations reported in DoE (1995b) reflect the decreasing emissions over the past three decades. Although PM_{10} is measured as part of the UK automatic urban network (Bower *et al.* 1997), information is only available for the past few years. The network contains few sites in rural locations.

Preliminary analyses of PM_{10} and black smoke concentrations (Muir & Laxen 1995; Committee on the Medical Effects of Air Pollution 1995a) in the same urban area suggests an approximate relationship between the two, accepting the limited accuracy of the black smoke measurements at the low concentrations generally found.

Emissions of particles from industrial sources are generally limited by dust control equipment. Discharge rates are set in authorizations for the process and requirements for in-stack monitoring are often imposed. For example, PM_{10} emissions for fossil-fuel power stations for 1993 were approximately 2% of SO_2 emissions (EPAQS 1995b; DoE 1995a).

Major industrial sources individually should not normally lead to exceedances of the particle objectives. However, industrial sources in combination still play an important role in the formation of secondary particles, which are significant on a regional basis, although there is currently no simple method of establishing the relationship between secondary particles and the primary emissions (Airborne Particles Expert Group 1999).

Since major primary sources of PM_{10} exist in urban areas, associated with road transport and in some areas with coal combustion, high concentrations are expected in meteorological episodes when there is poor atmospheric dispersion. In these weather conditions when urban background concentrations are high, the low wind speed and very stable atmospheric conditions will lead to higher concentrations near major roads in the urban area.

Estimates of PM_{10} emissions are likely to be uncertain. National estimates have been subject to large variations (QUARG 1993a, b; EPAQS 1995b), though the contribution from diesel vehicles is thought to be important. Typical urban emission factors are $4\,t\,km^{-2}$ in the center of cities (Buckingham & Clewley 1996; Buckingham et al. 1997a, b, c). Information on PM_{10} emission factors for different categories of vehicles is available from the National Atmospheric Emissions Inventory website at http://www.naei.org.uk/emissions/. Nonexhaust emissions of PM_{10} from brakes and tires make a further (relatively small) contribution, together with an unquantified contribution from dust and PM_{10} resuspended by the movement of vehicles. The latter two source terms are not included in the calculations of PM_{10} from road transport (APEG 1999).

The estimated daily PM_{10} concentrations may be compared with the UK national air quality objective of $50\,\mu g\,m^{-3}$ in 2005, and locations where more than 35 exceedances have occurred should be identified. The procedure to be followed for PM_{10} particles is necessarily associated with considerable uncertainty. An estimate of the annual average urban background concentration of PM_{10} can be estimated from recent measurements, which will include a regional background of secondary particles. It will not be possible to accurately adjust the urban background concentration to 2005 but attempts have been made based on estimates of future regional PM_{10} concentrations.

The expected roadside contribution to PM_{10} in 2005 may be obtained by adjusting the concentration under current conditions to the expected traffic mix and traffic flow in 2005 using dispersion models. The contributions from major roads, urban areas, and the regional PM_{10} concentration can be added together to obtain an estimate of the annual average primary PM_{10} concentration, having first adjusted (i) the road-related components to emission factors in 2005, and (ii) the traffic flows to 2005 from information available from transport plans.

Source–receptor relationships for particle concentration exceedances are very uncertain, but can be estimated by applying empirical factors relating the number of exceedances to annual average concentrations based on current measurements (APEG 1999). Forecasts of future emission factors for road vehicles do not show a large change from current levels, so that use of the same conversion factor may be justified.

21.9   OZONE

Ozone is not a pollutant whose management can be controlled on a local basis; hence it does not form part of a local assessment. However, it is included in the UK National Air Quality Strategy and is regarded as a matter of considerable importance on the regional and international scales.

Ozone occurs naturally in air near ground level. Concern (Advisory Group on the Medical Aspects of Air Pollution Episodes 1995) arises about increased levels of ozone resulting from a complex series of chemical reactions, some of which produce ozone and some of which destroy ozone, involving NO_x and VOCs in the presence of sunlight. All kinds of volatile organic compounds are involved. DOE (1995b) contains total UK national emissions of VOCs according to source type (see also the National Air Quality Information Archive website at http://www.airquality.co.uk/archive/). Each VOC compound has a different propensity for producing ozone (UK PORG 1993). DOE (1995b) also gives national emissions broken down into compound and source type for the 50 most significant VOC species with regard to ozone formation.

The UK Photochemical Oxidants Review Group database (UK PORG 1993) contains a comprehensive list of ozone measurements from remote, rural, and urban monitoring sites in the UK. These data have been used to assess how ozone over the UK as a whole differed from concentrations at a remote rural location according to season and time of day (Derwent & Davies 1994). This analysis showed that ozone is destroyed over the country most of the time. This arises because the ozone production mechanisms are not effective for one reason or another, e.g. no sunlight at night, or because the ozone has been destroyed by deposition at the ground or by chemical mechanisms. Average urban ozone concentrations are lower than those at rural sites, because NO emitted in urban areas destroys ozone in rural air. Ozone is generated in air carrying nitrogen oxides and VOCs away from urban areas, so that ozone concentrations tend to increase in suburban and rural areas.

EPAQS (1994d) recommended an air quality standard for ozone of 50 p.p.b. as an eight-hour running mean concentration. This standard was exceeded at a majority of the UK automatic rural network monitoring sites on some occasions in 1995 (Bower *et al.* 1997). Thus attention focuses on episodes of poor ozone air quality. Most of the sites where no ozone exceedances occurred were at urban center locations.

Complex models of ozone generation have been developed to simulate the movement of air mass trajectories over several days and the accumulation of precursor concentrations. The subsequent chemical reactions leading to high ozone concentrations after three to four days have been successfully represented, leading to predictions that are comparable with measurements (Derwent & Davies 1994). It is only through such models and the consideration of emissions on a regional scale of 1000 km that strategies (Derwent 1995) for reducing peak ozone concentrations during episodes can be evaluated. The impact of emissions in any one local administrative area on the formation of ozone is likely to be small. However, model results indicate that if peak ozone concentrations are to be brought below the national air quality objective, very substantial reductions in hydrocarbon or oxides of nitrogen would be required.

21.10 POLYCYCLIC AROMATIC HYDROCARBONS

Polycyclic aromatic hydrocarbons (PAHs) have not yet been included within the UK National Air Quality Strategy. They, and possibly some heavy metals, will be included in the development of the National Strategy (DETR 1999a). They illustrate the difficulty of including certain categories of air pollutants in a strategy. PAHs are a large group of organic compounds with two or more benzene rings. There are hundreds of individual PAHs that can be present in gaseous form or attached to particulate matter. They are formed as the result of incomplete combustion of organic matter. Major sources include motor vehicles, wood burning, coke production, and coal heating. QUARG (1993a) includes an estimate of PAH emissions by source type. The contribution from domestic coal heating has decreased substantially in urban areas in recent decades and evidence is cited for a decline in atmospheric composition, especially of those PAHs that exist mainly in the vapor phase.

The results from the Toxic Organic Micropollutants Survey (TOMPS) reported in the DoE (1995b) include a common set of ten PAHs in order of increasing molecular weight and decreasing volatility: acenapthene, fluorene, phenanthrene, anthracene, pyrene, benz(a)anthracene, chrysene, benzo(b)fluoroanthene, benzo(a)pyrene, and benzo(ghi)perylene (see Bertorelli & Derwent 1995 for a more detailed summary). It is important that atmospheric concentrations in both the particle and vapor phases are reported. The lighter, less volatile PAHs will tend to be present in the vapor phase, while the heavier, less volatile PAHs will be attached to particles. QUARG (1993a) note that PAHs are mainly carried on respirable particles, and would therefore be collected as PM_{10}.

Several years' worth of data are reported from TOMPS sites (DoE 1995b) in London, Manchester, Cardiff, Middlesbrough, and Lancaster. Total concentrations of PAHs at urban sites show greater variability than expected, and are unexpectedly high at the rural site near Lancaster. QUARG (1993a) remarks that quantitative comparisons between PAH measurements in the UK, including some TOMPS readings, show broadly similar levels, but quantitative comparisons are not very good. Data on individual PAHs are also listed in Bertorelli and Derwent (1995). Further quality assessments and intersite comparisons are needed before it can be established whether spatial variations around the UK are real.

Measurement of PAHs requires long sampling periods (seven days for TOMPS) and lengthy extraction and analysis procedures. Sampling of PAHs is not a routine procedure and not one that local authorities can be expected to undertake. Receptor modeling to identify different categories of sources has been tried but is not an operational procedure.

Analyses of soils and sediments around the world suggest that PAHs are widely distributed. The vapor-phase transport of lighter PAHs is limited by chemical reactions, which vary according to compound, but limit the lifetime to a few hours. Small 1 μm particles carrying heavier PAHs are transported over long distances. Higher concentrations are expected in urban areas where gasoline and diesel transport remain a source, although this is not confirmed by the TOMPS results.

The best known and most measured PAH is benzo(a)pyrene, which is thought to be the most carcinogenic PAH. No safe level can be recommended (WHO 1987), and it is likely that any recommended long-term (annual) concentration level incorporates an acceptable risk. Estimates of risk for PAHs have been based on scaling risk relative to benzo(a)pyrene. A study used a representative sample of 10 PAH compounds to assess the risk of exposure at characteristic locations in the Netherlands: a typical rural site, a typical urban background site, 2 km from a coke-oven plant, near an airport, etc. (Slooff *et al.* 1989). Given the wide uncertainties, the risks at the various outdoor locations were not very different and within a factor of three. The concentration in urban areas of benzo(a)pyrene in the UK is reported to be about $1 \, \mathrm{ng\,m^{-3}}$ (QUARG 1993a), which is similar to that used in the Dutch study and represents a marked decline from past urban concentrations.

Given the lack of knowledge regarding PAH concentrations, the difficulty in relating them to sources, and the variability in measurement, any assessment is likely to be very uncertain. However, an assessment of annual average ground-level benzo(a)pyrene concentrations around large emitting processes could be made using a dispersion model. For major roads running through urban and rural areas the maximum annual average benzo(a)pyrene concentration adjacent to the road could be assessed for a few locations near roads

carrying the highest traffic volumes. From road transport emission factors for benzene and benzo(a)pyrene (Westerholm & Egabäck 1994), annual average concentrations of benzo(a)pyrene may be calculated in the form of annual averages. It should be possible to categorize, albeit qualitatively, where benzo(a)pyrene concentrations are significantly higher than the typical urban level of $1 \, \mathrm{ng\,m^{-3}}$ (EPAQS 1999).

21.11   DISPERSION MODELS FOR LOCAL AIR QUALITY MANAGEMENT

This section summarizes some of the types of air pollution model currently used in the UK (DETR 1998).

21.11.1   R91

This model gets its name from the report number of an NRPB report (Clarke 1979) written by a working group of atmospheric modelers. The nuclear industry was recommended to use the model, so that all parties in regulatory matters were seen to be using the same approach. The R91 report and subsequent ones by the same working group became the standard for the UK nuclear industry. The same uniformity has not been adopted for other industrial sources. Originally the requirement was that all calculations could be done by hand using a calculator, so that the report consisted of graphs and nomograms. Subsequently, the graphs and nomograms have been turned into computer programs. The model is made up of a number of formulae that could be built into source configurations of varying complexity. No generally available code for doing this has been adopted.

The model was designed for use with low-level point sources, but was tested against data from high-level point sources and shown to behave adequately. It is widely used in the UK by the nuclear industry for radiological assessments and has thus acquired status from its widespread use. The model has been used to determine the highest permissible discharge rates from stack arrangements which ensure that the maximum ground-level concentrations comply with specified exposure levels. Both short-term (hourly) and long-term (annual) average concentrations can be calculated. Formulae within R91 are stated explicitly, and these have been used to build the relationships adopted within the GRAM model referred to below.

Special features discussed by the working group included sea breeze and coastal effects, dispersion at low wind speeds, and dispersion around buildings. It was recognized that no single, simple model could cover all situations, and this was the driving force behind the development of the computer model ADMS (see below).

21.11.2 ALMANAC

This model evolved out of the research program of the former Central Electricity Generating Board. In the late 1960s and early 1970s systematic air pollution surveys were carried out around a number of major coal- and oil-fired power stations. This was to establish and confirm that the stack heights being recommended in the new designs would ensure adequate dilution and dispersion. The particular feature of this work was that it was able to investigate a wide range of meteorological conditions and took into account the variability of the atmosphere, which is particularly important for describing dispersion from high-level sources. Much of the time the plume from a tall stack can be above the mixing layer and the pollution is not brought to the ground in the vicinity of the stack.

The results of this research were encapsulated in a dispersion model called ALMANAC, which has been used widely by the power generators in their recent applications to build gas-fired power stations in the UK. The model normally treats only a single stack. Many of the ideas in the model have later been applied in ADMS. Both the highest hourly concentration and the annual average concentration can be calculated using the model. One particularly valuable feature is that the model allows the user to calculate 98th percentiles (the hourly concentration not exceeded on more than 2% of occasions), which is useful when applying European Directives.

21.11.3 ISC

The problem of 98th percentiles does not arise in the USA because regulations do not require concentrations to be expressed in such a way. Dispersion modelers in the USA are quite prepared to use extensive sets of hourly data to generate statistics of pollution concentrations. The ISC (Industrial Source Complex) model has found widespread appeal, especially to users in the UK who are only interested in established models, and require a well accepted and well documented model. There is a long-term version of this model, ISCLT, in which annual averages can be calculated. There is also a short-term version, ISCST, in which hourly average concentrations can be calculated. Many of the national and international guidelines are set in terms of short-term averages, so that a long-term model is not an adequate assessment method on its own.

In recent years there has been considerable research in the USA on dispersion, particularly focusing on the convective boundary layer, where it is felt that existing models such as ISC are inadequate. The US EPA and the American Meteorological Society have been collaborating on the development of a new model, AERMOD, but at the present time this new model has not replaced ISC.

ISC can be used to calculate concentrations from area and volume sources by assuming that the distributed source is equivalent to a virtual point source set some distance further away. It is imagined that the emissions from the virtual point source have mixed in the atmosphere to fill the area or volume of the real source. This method works adequately with small area or volume sources, but can lead to difficulties for wide area sources. The treatment of area sources may therefore not be straightforward with this model.

For multiple point sources there should be no problem in applying ISC, as the model has been widely and extensively used. Care is needed in interpreting results. For example, a high-level source dispersing in stable atmospheric conditions will be predicted to give a maximum ground-level concentration 100 km or more downwind. What this really means is that the plume does not come to the ground until there is a change in weather conditions, e.g. a change from night to day. There is also a danger in interpreting results when sources appear to be aligned in the same wind direction. ISCST will produce high ground-level concentrations, because of the implicit assumption that the plumes from two sources are perfectly aligned. In reality, because of wind direction fluctuations, this will rarely occur.

21.11.4   INDIC system

This is a modeling system that includes data input facilities, output and display options, and a database for processing measurements from a network of air pollution monitors. The output looks very attractive and a lot of effort has been put into the computing aspects of the system. The system comes with a dispersion model that is not dissimilar to the ISC model. However, other dispersion models can be programmed into the INDIC system. The model has to be run on a workstation, rather than a PC.

The database within INDIC has been used to collect and access the monitoring data, which are measured at automatic monitoring sites in UK city centers. The INDIC system is a high-quality solution to air quality management. It is not amenable to short-cut solutions.

21.11.5   ADMS

ADMS (Atmospheric Dispersion Modeling System) has been developed by the Cambridge-based consultancy CERC. The first model was developed under a jointly funded contract in which a number of major companies participated. This contract was for an advanced dispersion model that took into account the most up-to-date understanding of atmospheric dispersion. It was really developed for application to a single tall stack. The model was developed in a modular form to deal with specific features of atmospheric dispersion, such as topography, plume rise, and concentration fluctuations.

The behaviors of the various modules in the models have been tested individually. In addition, validation exercises have been undertaken and published of the behavior of the model. These have focused on the dispersion from large point sources, such as power stations. Particular attention has been paid to dispersion in convective conditions when the plume within the mixing layer can be brought intermittently to the ground close to a tall stack producing high short-term ground-level concentrations. Recently, CERC has developed a version of ADMS that includes multiple sources and deals with urban areas comprising line and area sources. This is now available to users.

21.11.6   Design Manual for Roads and Bridges

Volume 11 of the *Design Manual for Roads and Bridges*, "Environmental Assessment" (DETR 1999b), contains a method that enables carbon monoxide, benzene, particulate, and NO_2 concentrations around a road network to be estimated. The method uses a series of nomograms, such as those in the R91 report. However, the latest version has been generalized to apply to roads in urban areas, though it has been mainly used for screening new road schemes. It is an example of an attempt to produce an accessible tool that nonexperts can use. The current DMRB method includes methods for assessing concentrations in the form required by UK national air quality objectives (see spreadsheet available at http://www.airquality.co.uk/archive/laqm/tools.php).

21.11.7   CALINE4

Among the suite of US EPA dispersion models, there are a number of models for dealing with highways. CALINE4, the latest version of a series of California Air Resources Board line source models, seems to have achieved widest acceptability in the UK. The CALINE4 model is a Gaussian line source model in which the hourly average concentrations in specific meteorological conditions can be calculated. Adjustments are necessary to deal with episodes of low wind speed, stable conditions that persist for several days. However, it contains many features specific to roads, such as queues at traffic junctions and parking lots.

21.11.8 Fluidyn-PANACHE

Fluidyn-PANACHE is potentially a much more powerful model, as it relies on computational fluid dynamics to solve the flow and dispersion around roads and sources. Its graphical interface removes the complexity and hard work normally required to set up and run these kinds of models. Models of this type may be the shape of things to come.

21.11.9 CAR

The Dutch CAR model (Eerens *et al.* 1993) permits calculations of CO, NO_2, benzene, lead, and black smoke up to 30 m from a road. It does not explicitly deal with PM_{10}, but a user-defined air pollutant component can be added. However, the output is in the form of annual averages and 98th percentiles of one-, eight-, and 24-hour average concentrations. Empirical conversion factors from these percentiles to those specified in the UK National Air Quality Strategy is necessary.

21.11.10 GRAM

GRAM (University of Greenwich Review of Air Quality Model) can be used to predict air pollution concentrations near roads, including canyon-like streets in urban areas, where the dispersion of pollutant emissions is restricted and where air pollution problems may arise. The model is based on algorithms (sets of equations) used in the NRPB R91 model for urban dispersion and the USEPA CALINE4 model for roadside dispersion (Fisher & Newlands 2000; Fisher & Sokhi 2000). It illus-

trates the absolute minimum data required to run air quality models. The GRAM model provides concentrations at the appropriate averaging time and frequencies to enable direct comparisons with the UK national air quality objectives to be made.

GRAM allows the user to assess the change that is likely between the present situation and the situation in any year up to 2005 and beyond. GRAM has been designed to, if anything, overestimate pollution concentrations. In this way, the model will enable the user to screen out situations that clearly will meet the UK National Air Quality Strategy in 2005 and retain those that might not, allowing for uncertainty in the estimates. A recent version allows for the uncertainty in input parameters to be assessed using Monte Carlo simulation and for the sensitivity analysis to be performed.

21.11.11 Emission factors for road transport emissions

Apart from the *Design Manual for Roads and Bridges* (DETR 1999b), none of the models discussed above contains information on present and future emissions from road transport, which is an essential part of any air quality calculation. For example, the *Design Manual* contains an average emission factor for CO in $g\,km^{-1}$ for an average 1996 light duty vehicle at a speed of $100\,km\,h^{-1}$, which can be converted to emissions for other speeds, years, and heavy duty vehicles using graphs in the *Design Manual*. Typical emission factors for CO and other pollutants of interest are listed in Table 21.6.

Table 21.6 Emission factors in 1996, 2000, and 2005 for an average light-duty vehicle at an average speed of $100\,km\,h^{-1}$.

Pollutant	Emission rate 1996 ($g\,km^{-1}$)	Emission rate 2000 ($g\,km^{-1}$)	Emission rate 2005 ($g\,km^{-1}$)
Carbon monoxide	4.98	3.34	1.84
Oxides of nitrogen	1.8	1.10	0.47
Total hydrocarbons	0.464	0.274	0.107
Total particulate	0.05	0.035	0.0175
Carbon dioxide	163	163	158

Practical and useful advice on the road vehicle emission factors to be used in models is given on the website http://www.naei.org.uk/emissions/. General information on emission factors is available from the London Research Centre website (http://www.london-research.gov.uk/emission/webhtm.htm). This includes information on cold start emissions.

Considerable differences between predicted and measured concentrations up to a factor of three are to be expected. However, models applied to air quality management should correctly identify those areas at risk of exceedances of national air quality objectives. Models can also be used to investigate how concentrations may vary with changing parameters, and this can still be a useful indicator when making decisions.

21.12   ACCURACY OF AIR POLLUTION MODELS

Defining the accuracy of air pollution modeling results is difficult even in very well tried and tested situations. Accepting that perfect accuracy is neither necessary nor achievable, great variations in the accuracy of air quality assessments are not desirable. A somewhat pessimistic approach to assessment should be taken. This can be achieved by referring back to measurements as often as possible, using surrogate statistics, such as population or traffic flow, to check on the consistency of the estimated air quality between areas, and taking upper bounds where appropriate to do so.

An air quality assessment is only the first stage in air quality management. It is likely that monitoring will have to take place in areas identified as air quality management areas. The precise boundaries of these areas would be hard to define because of the uncertainty associated with current air quality levels and the uncertainty of future levels. An air quality management area should not be regarded as a region precisely delineated on a map, though in practice it may have to be. Instead, it should be thought of as an administrative area, such as a local administrative area (or areas) or a

public highway, for which further considerations of future air pollution levels are required.

There is no straightforward way of specifying the accuracy of dispersion models used in air quality assessments. First, the accuracy depends on the quality of the data used by the model. If the exact levels of emissions are not known, it will not be possible to calculate concentrations accurately. It is not possible to calculate precisely the concentrations for a specified one-hour time interval. This is because one cannot know with any certainty, unless there are very detailed measurements available, which meteorological and emission conditions applied to that particular interval.

Dispersion models have been designed to calculate the average concentration over a large number of time intervals subject to similar meteorological conditions. There are real fluctuations in concentrations about the average caused by variations in meteorological and emission conditions, which are not treated within a model. This problem does not arise when one is using a model to calculate long-term average concentrations over, say, one year. Provided that a dispersion model is used to calculate some form of average concentration, either over time or in more or less identical dispersion conditions, then the experience of model users who have compared average predictions with measurements may be used. If the prediction lies within ±50% of the measurement, a user would not consider that the model has behaved badly. This may be regarded as a rough rule of thumb in assessing model predictions.

Any atmospheric dispersion model will always have a degree of error due to unavoidable inaccuracies in the recorded meteorological data and the simplifications made in the model. In addition, atmospheric mixing has an inherent degree of randomness, due to the turbulent flows that occur. As a result, model validation cannot be expected to show perfect agreement between the calculations and the measurements. If conditions occur many times then it is possible to determine the average plume shape, with a statistical distribution around this average. This is the approach taken in computer models of dispersion. These models can give quite reasonable predictions of the actual dispersion when considering averages over many occasions. Thus the model validations based on infrequently occurring concentrations, such as the highest few hours per year, will rarely match the actual dispersion due to the inherent random error from turbulence, unless based on many years of data, which are rarely available for any site. Because of this, the degree of error in assessments is expected to be greater when calculating the short-term than the long-term air quality objectives.

These features do not undermine the validity of the results from air dispersion modeling, but they do emphasize the need for model validation and caution in drawing conclusions from modeling alone. A convenient measure of the accuracy of the model prediction is to consider predictions at a number of monitoring sites within the study area. The square root of the mean sum of the squared differences between the predicted and measured concentrations at these sites is an estimate of the uncertainty in the prediction. This quantity, the root mean square error, can be used to define a range of values above and below the concentration predicted by a model, in which the actual concentration is expected to lie. From the arguments above, this range is likely to be smaller for air quality objectives based on annual averages and larger for air quality objectives defined by the number of exceedances of short-term average concentrations.

21.13   EXAMPLE OF A METHOD OF ESTIMATING UNCERTAINTY

The accuracy of a model can be estimated by comparing the observed and predicted concentrations at a number of monitoring sites within the study area. For example, given the typical concentrations of urban annual average NO_2 listed in Table 21.7, one can estimate the square of the difference in concentration at each site, and sum these. The square root is the "root mean square difference in concentrations" given by the formula:

$$Rms = \sqrt{\frac{1}{12}\sum_{i=1}^{12}(Measured_i - Predicted_i)^2} \quad (21.7)$$

Table 21.7   Annual average NO_2 concentration (from DETR 1999a).

Site name	Measured concentration (p.p.b.)	Predicted concentration (p.p.b.)
Site 1	39	46.3
Site 2	36.3	37.6
Site 3	31.2	32.8
Site 4	27.2	35.9
Site 5	32.2	32.6
Site 6	31.1	33.1
Site 7	49.9	49.4
Site 8	24.3	30.5
Site 9	36.1	35.4
Site 10	31.6	35.8
Site 11	36.6	41.5
Site 12	25.2	31.9
Root mean square difference in concentration		4.68

The square root of the mean sum of the squared differences between the predicted and measured concentrations at these sites is 4.7 p.p.b. One can therefore assume that the error in the prediction is of this magnitude. More sophisticated treatments are possible depending on circumstances. Concentrations are always positive. Hence there is a tendency for a set of concentration values to be skewed about their mean value and it may be necessary to take logarithms of concentrations before calculating the root mean square concentration differences. Antilogarithms would then need to be taken to obtain the confidence interval in concentration units.

If just one or two monitoring sites are available, it is not possible to make an estimate of the model uncertainty in this way. A check on the model predictions should still be made using any monitoring data available. This would provide an approximate check on the model and, provided that results were within ±50%, would give some confidence. In addition, a sensitivity analysis could be performed using the model, to check on the changes in prediction brought about by changes in the choice of input data. If multiple runs are performed using reasonably realistic input values, a series of output data can be generated and a frequency distribution of predictions constructed. Provided that the measured data, if limited, lie within the range of likely values, bounding say 95% of predictions, then the model could be regarded as acceptable. The frequency distribution can then be used to define an uncertainty associated with the boundaries of the air quality management area. In some cases the uncertainty may be too large to make a decision and further information would need to be collected.

Different dispersion models designed to calculate the same quantities will produce different answers. The reason for this and the reason why there is not a single right answer is that model developers have to make choices about the formulae they should use. Typically the model developer will choose from the literature a formula that is practical (not too many unknown variables and easy to use), has been tested against measurements, and is consistent with the other parts of the model. Not all model developers will make the same choice. Some will place greater emphasis on one aspect than on another and this is why models differ. Ideally, models should be transparent so that all the formulae used by the model are known to anyone who wishes to test or validate the model. It should be possible in principle to reproduce independently the answers from a model from the information supplied with the model. The model validation will not correct imperfections in a model, but by associating a range of values with the model prediction will allow the results to be interpreted appropriately, i.e. large errors in the model validation mean less weight is given to details in the predicted concentration pattern.

21.14 CARBON DIOXIDE

Urban air quality assessments often take account of emissions of carbon dioxide, although it is not strictly relevant to local air quality management. Local administrative bodies are increasingly concerned with sustainable development and encourage energy efficiency. However, improvements in

emissions of local pollutants are not achieved over similar time scales to those for greenhouse gases, and cannot be easily compared. For air quality management one is considering periods of 10 to 20 years. For CO_2 one may be considering improvements over a time scale of 100 years. The usual approach when considering the effect of global emission scenarios is to consider the fraction of CO_2 that is retained in the atmosphere. The approach of Wigley (1991, 1993) has been widely used. This shows the fraction retained as the sum of four exponentially decayed functions. The exponential functions are described by different time scales. The values of parameters are obtained by fitting past trends in concentrations. The overall rate of decay can be compared with a simple box model of CO_2 diffusing into the ocean (Fisher & Macqueen 1980). A rather similar description of the decay is obtained. The key point is that the decay occurs over periods of hundreds of years.

This is illustrated in Fig. 21.4, in which the two models of the fraction of CO_2 retained in the atmosphere are shown. The x-axis shows future

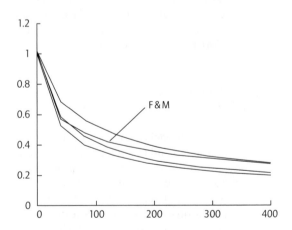

Fig 21.4 Fraction of instantaneous CO_2 emission remaining airborne as a function of time from moment of release from four different models out to 400 years from the point of emission. The curve labeled F&M is based on the diffusion model of Fisher and Macqueen (1980). The other curves are various estimates of Wigley (1991).

years and is drawn from the time of emission to 400 years in the future. Hence, the benefits of improved efficiency and lower CO_2 emissions are long term, whereas those of air quality management are short term.

Increasing interest is being addressed to the effects of some of the pollutants relevant to local air quality management, such as CO and ozone, on climate change, and conversely the effects of climate change on local air quality. However, this takes one beyond the scope of this chapter. The main conclusion from this section is that management of local air quality should not be restricted to the present decade, as specified in current legislation, but directed to possible changes over much longer time scales.

REFERENCES

Advisory Group on the Medical Aspects of Air Pollution Episodes (1995) *Fourth Report: Health Effects of Exposures to Mixtures of Air Pollutants*. HMSO, London.

Airborne Particles Expert Group (1999) *Source Apportionment of Airborne Particulate Matter in the United Kingdom*. Department of the Environment, London.

Bertorelli, V. & Derwent, R. (1995) *Air Quality A to Z*. Department of the Environment, London.

Bower, J.S., Broughton, G.F.J., Willis, P.G. & Clark, H. (1997) *Air Pollution in the UK: 1995*. AEA, London.

Broughton, G.F.J., Bower, J.S., Clark, H. & Willis, P.G. (1998) *Air Pollution in the UK: 1996*. AEA, London.

Buckingham, C. & Clewley, L. (1996) *West Midlands Atmospheric Emissions Inventory*. London Research Centre, London.

Buckingham, C., Clewley, L., Hutchinson, D., Sadler, L. & Shah, S. (1997a) *London Atmospheric Emissions Inventory*. London Research Centre, London.

Buckingham, C., Clewley, L., Hutchinson, D. *et al.* (1997b) *Atmospheric Emissions Inventories for Four Urban Areas: Merseyside, Bristol, Southampton/Portsmouth and Swansea/Port Talbot*. London Research Centre and RSK Environment Ltd, London.

Buckingham, C., Forest, G., Sadler, L., Shah, S. & Wickham, L. (1997c) *Atmospheric Emissions Inventory for Greater Manchester*. London Research Centre and RSK Environment Ltd, London.

Chamberlain, A.C., Heard, M.J., Little, P. & Wiffin, R.D. (1979) The dispersion of lead from motor exhausts.

Philosophical Transactions of the Royal Society London **A290**, 577–80.

Clarke, R.H. (chair) (1979) *A Model for Short and Medium Range Dispersion of Radionuclides Released to the Atmosphere*. NRPB, London.

Committee on the Medical Effect of Air Pollutants (1995a) *Non-biological Particles and Health*. HMSO, London.

Committee on the Medical Effect of Air Pollutants (1995b) *Asthma and Outdoor Air Pollution*. HMSO, London.

Department of the Environment (1988) *Digest of Environmental Protection and Water Statistics*. HMSO, London.

Department of the Environment (1995a) *Air Quality: Meeting the Challenge. The Government's Strategic Policies for Air Quality Management*. HMSO, London.

Department of the Environment (1995b) *Digest of Environmental Statistics No. 17*. HMSO, London.

Department of the Environment (1997) *The United Kingdom Air Quality Strategy*. HMSO, London.

Department of the Environment, Transport and the Regions (1998) *Selection and Use of Dispersion Models: Local Air Quality Management*. HMSO, London.

Department of the Environment, Transport and the Regions (1999a) *The Air Quality Strategy for England, Scotland, Wales and Northern Ireland. A Consultation Document*. HMSO, London.

Department of the Environment, Transport and the Regions (1999b) *Design Manual for Roads and Bridges. Volume 11: Environmental Assessment. Section 3: Environmental Assessment Techniques. Part 1: Air Quality*. HMSO, London.

Derwent, R.G. (1995) Improving air quality in the United Kingdom, *Clean Air*, **25**, 70–94.

Derwent, R.G. & Davies, T.J. (1994) Modeling the impact of NO_x or hydrocarbon control on photochemical ozone in Europe. *Atmospheric Environment* **28**, 2039–52.

Derwent, R.G. & Middleton, D.R. (1996) An empirical function for the ratio of $NO_2:NO_x$. *Clean Air* **26**, 57–60.

Derwent, R.G., Middleton, D.R., Field, R.A., Goldstone, M.E., Lester, J.N. & Perry, R. (1995) Analysis and Interpretation of air quality data from an urban roadside location in central London over the period from July 1991 to July 1992. *Atmospheric Environment* **29**, 923–46.

Downing, C. & Campbell, G. (1995) *Rural Sulphur Dioxide Concentrations in the United Kingdom: 1992*. AEA, London.

Eerens, H.C., Sliggers, C.J. & van den Haut, K.D. (1993) The CAR model: the Dutch method to determine city street air quality. *Atmospheric Environment*, **27B**, 389–99.

Expert Panel on Air Quality Standards (1994a) *Carbon Monoxide*. HMSO, London.

Expert Panel on Air Quality Standards (1994b) *Benzene*. HMSO, London.

Expert Panel on Air Quality Standards (1994c) *1,3-Butadiene*. HMSO, London.

Expert Panel on Air Quality Standards (1994d) *Ozone*. HMSO, London.

Expert Panel on Air Quality Standards (1995a) *Sulphur Dioxide*. HMSO, London.

Expert Panel on Air Quality Standards (1995b) *Particles*. HMSO, London.

Expert Panel on Air Quality Standards (1996) *Nitrogen Dioxide*. HMSO, London.

Expert Panel on Air Quality Standards (1999) *Polycyclic Aromatic Hydrocarbons*. The Stationery Office, London.

Fisher, B.E.A. & Macqueen, J.F (1980) The influence of the oceans on the atmospheric burden of carbon dioxide. *Applied Mathematical Modeling* **4**, 439–48.

Fisher, B.E.A. & Newlands, A.G. (2000) Clarifying the relationship between urban road structure and air quality exceedences using a training model. In Gryning, S.-E. & Batchvarova, E. (eds) *Air Pollution Modeling and its Application XIII*. Kluwer Academic/Plenum, New York.

Fisher, B.E.A. & Sokhi, R.S. (2000) Investigation of road side concentrations in busy streets using the model GRAM: conditions leading to high short-term concentrations. *International Journal of Environmental Technology and Management* **14**, 488–95.

Fisher, B.E.A., Moorcroft, S. & Vawda, Y. (1998) Some issues that arise with air quality reviews and assessments. *Clean Air* **29**, 38–41.

Little, S. & Cram, G. (1995) Atmospheric benzene concentrations near petrol service stations in Middlesbrough. *Clean Air* **25**, 140–7.

Lucas, D.H. (1958) The atmospheric pollution of cities. *Journal of Air Pollution* **1**, 71–86.

Muir, D. & Laxen, D.P.H. (1995) Black smoke as a surrogate for PM_{10} in health studies. *Atmospheric Environment* **29**, 959–62.

Quality of Urban Air Review Group (1993a) *Urban Air Quality in the United Kingdom*. Department of the Environment, London.

Quality of Urban Air Review Group (1993b) *Diesel Vehicle Emissions and Air Quality*. Department of the Environment, London.

Seika, M. (1996) Evaluation of control strategies for improving urban air quality with London and Berlin as examples. PhD Thesis, University of Birmingham.

Seika, M., Harrison, R.M. & Metz, N. (1998) Ambient background model (ABM): development of an urban Gaussian model and its application to London. *Atmospheric Environment* **32**, 1881–91.

Slooff, W., Janus, J.A., Matthijsen, A.F.C.H., Montizaan, G.K. & Ros, J.P.M. (1989) *Integrated Criteria Document PAHs*. National Institute of Public Health and Environmental Protection, Washington, DC.

Smith, F.B. (1973) *A Simple Scheme to Estimate the Average Concentration of a Pollutant within an Urban Area*. Meteorological Office, Bracknell.

Stevenson, K.J. & Bush, T. (1995) *UK Nitrogen Dioxide Survey Results for the First Year 1993*. AEA, London.

Stevenson, K.J., Loader, A., Mooney, D. & Lucas, R. (1995) *UK Smoke and Sulphur Dioxide Networks. Summary Tables for April 1993–March 1994*. AEA, London.

UK Photochemical Oxidants Review Group (1993) *Ozone in the United Kingdom 1993*. Department of the Environment, London.

Westerholm, R. & Egabäck, K.-E. (1994) Exhaust emissions from light-vehicles and heavy-duty: chemical composition, impact of exhaust after treatment, and fuel parameters. *Environmental Health Perspectives* **102** (supplement 4), 13–23.

Wigley, T.M.L. (1991) A simple inverse carbon cycle model. *Global Biogeochemical Cycles* **5**, 373–82.

Wigley, T.M.L. (1993) Balancing the carbon budget. Implications for projections of future carbon dioxide concentration changes. *Tellus* **45B**, 409–25.

World Health Organization (1987) *Air Quality Guidelines for Europe*. WHO Regional Publications European Series No 23, Copenhagen.

The National Air Quality Information Archive is available on the World Wide Web at http://www.airquality.co.uk/archive/

Index

1,3-butadiene 143, 604, 611
  emissions 143, 483

accretion hypothesis 8
accumulation mode 229
acid deposition 384
acid rain 384
acidification 386, 571
activity rate 479
adiabatic cooling 47, 48, 279
advection 285
aerosols 145, 215, 228–54
  absorption 244
  carbonaceous 146
  climate forcing 240–8
  cloud interactions 242
  composition 232
  emissions 233
  monitoring 454
  optical properties 245
  radiative properties 243, 353, 539
  scattering 244
  secondary emissions 234
  size distribution 230
  sources 124–55, 230–8
  sulphate 351
  volcanic emissions 234, 352
air pollution
  economic effects 414
  global 339–75
  legislation 418
  meteorology 255–74
  policy 367–8
  regional scale 376–98
  urban scale 399–438
air quality
  action plans 603
  London 613
  management 562–628

management assessment 607
management in Berlin 605
management in London 605
monitoring station 446
review 606
Air Quality Management Area 603,
  607
aircraft as a measurement platform
  464
airglow 13
Aitken particles 229
albedo 53, 60, 62, 68
alkene oxidation 181
aluminum 577, 580, 582
amino acids 16
ammonia 97
  budget 98
  deposition 324
  sources 98
aqueous phase 211
  photochemistry 219
Arctic haze 65, 388–94
  composition 393
  deposition 394
  particles 393
  properties 391
  source 390
area sources 504
Atmospheric Dispersion Modeling
  System 522, 622
atmospheric dynamics 530
atmospheric energy 35–58
atmospheric stability 49, 51, 279
atmospheric structure 35–58
averaging times 270
Azores high 72

balloon-borne instruments 464
Beer–Lambert Law 43, 446, 460

benzene 143, 427–8, 604, 609
  air quality standard 610
  emissions 483
  urban concentrations 610
bi-directional fluxes 324
biodiversity 572, 584
biogeochemical cycles 90–123
biomarkers 23
biomass burning 236
biosphere 15, 56, 91, 112
black smoke 147
blackbody 159
Boltzmann constant 37, 326
boundary layer 255, 316
  changing conditions 261
  in climate models 540
  complex surfaces 263
  convective 257
  flat terrain 256
  flow 264
  neutral 258
  stable 259
Brewer spectrophotometer 465
Brewer–Dobson circulation 173, 188,
  204
bromoform 113

Cape Grim 167–8, 345, 347
carbohydrate 22
carbon
  cycle 90, 357
  isotope ratio 24
carbon dioxide 18–20, 90, 124–6
  budget 91, 126
  emissions 342, 626
  sources 91
  stabilization 369
carbon monoxide 20, 93, 131, 403,
  604

carbon monoxide (*cont'd*)
  air quality standard 133
  budget 94, 132
  concentrations 133, 608–9
  effects 133
  emissions 483, 511
  at roadside sites 608
  sources 94, 132
  in the UK 607
  in urban areas 609
carbon tetrachloride 109
carbonate 21
carbonyl disulfide 100, 102
carbonyl sulfide 100–2
  chemistry 102
cations in soil 581
cell division 24
Charney–Phillips grid 534
chemiluminescence 452
chlorofluorocarbons 107, 189
  concentrations 108
  emissions 140
  replacement compounds 108, 140, 346
  sources 108
chloroform 110
chloromethane 111
circulation 60
CLAW hypothesis 105
Clean Air Act 137, 615
climate 59–89
  Africa 86
  Antarctic 66
  Arctic 64
  Asia 74
  Australia 82
  –chemistry feedbacks 205, 358–60
  Europe 72
  model evaluation 547–51
  modeling 525–61
  modeling tools 551–4
  North America 74
  polar 62
  south America 83
  sub-tropical 76
  tropical 75
cloud
  condensation nuclei 246–8
  properties 246
  in climate models 540
cold front 302
comets 18
condensation 50, 278
conduction 46

continuity equation 157, 167
convection 46, 296
  in climate models 541
Convention on Long-Range
    Transboundary Air Pollution
    376–7, 420, 474, 562, 591
conveyor belt 303
CORINE program 475
Coriolis force 76, 282, 295, 297–9
Criegee intermediates 180
critical levels 562–602
  ozone 564
  sulfur and nitrogen 564
critical limits 568
  acidity 578
  heavy metals 586
  nitrogen 572
critical loads 387, 562–602
  acidity 577
  assessment 597
  cadmium 589
  exceedances 593
  heavy metals 587
  methods 565
  nitrogen 570, 573–7, 594
  use in policy assessment 591
crop yield reduction 380
cryosphere in climate models 545
cyanobacteria 23–5
cyclones 306

data assimilation 551
denitrification 135
deposition
  aerodynamic resistance 316
  critical loads 387
  dust modeling 515
  modeling 389
  nitrogen 571
  ozone 383
  of particles 324–9, 412
  to plants 318
  resistance model 315, 389
  surface resistance 318
  velocity 315, 323, 325
  Wesley model 319–22
design manual for roads and bridges
    520, 622
dichloromethane 110
dienes 182
differential absorption LIDAR 460
diffusion samplers 442
dimethyl sulfide 100, 103, 234, 246
  chemistry 104–5

oxidation scheme 106
  sources 103
dinitrogen pentoxide 176
dioxin 144
dispersion 255–74, 406
  complex terrain 271–3
  equation 267
  flat terrain 271
  modeling 503–24, 620–3
  modeling, accuracy of 624
  modeling, meteorological data 504
  road traffic modeling 517
dissolution of gases 216
DOAS 448
Dobson spectrophotometer 465
Doppler broadening 45, 446, 469
droplet radius 213
drought 83, 86
dry deposition 314–29
  measurement 327–9

early atmosphere 4
ecosystem effects 566, 575–6
eddy accumulation 328
eddy covariance 327
Ekman layer 289–90
El Niño 78–80
El Niño Southern Oscillation (ENSO)
    80–4
emission
  factors 479, 623
  inventories 405, 473–502
  inventories Europe 474
  inventories Japan 477
  inventories UK 476
emissions
  agriculture 493
  aircraft 490
  airports 489
  area sources 491
  carbon monoxide 511
  from chimneys 510
  cold start 488
  in dispersion models 503
  dust 493
  landfill 492
  London 483, 487
  London and Tokyo compared
    494–9
  National Emissions Ceiling
    Directive 596
  natural 493
  power station 507
  rail 489

road vehicle 483
shipping 490
energy
budget 52–5
consumption 402
transfer 55
equinox 59
ethylene dibromide 113
Eulerian modeling 383, 533
European Environment Agency 475
European Monitoring and Assessment
Programme 379, 384, 388, 474,
592
eutrophication 386
evolution 3, 26

Ferrell cells 60
Fick's Law 443
flame ionization detector 453
forest effects 572, 579
Fourier transform spectroscopy 448

Gaia hypothesis 15, 27
gas chromatography 453
gas constant 35, 276
gas-filter correlation spectroscopy
451
Gaussian plume 268
General circulation models 526–9
global air pollution effects 365–7
global warming international response
368–9
gradient technique 328
gravity waves 196
in climate models 542
greenhouse effect 11, 19, 52, 87, 124,
129, 240–8, 340
greenhouse gases 11, 341
lifetimes 359
future 362
groundwater pollution 580
groundwater quality 582

Hadley circulation 55–6, 60, 85
Halley Bay 190
halocarbons 138, 344
emissions 138
as greenhouse gases 138
lifetimes 139
as ozone depleters 138
halogen
chemistry 184, 223
chlorofluorocarbons 107
cycle 105, 117

organohalogens 107
future concentrations 363
inorganic 114
inorganic chemistry 116
halons 109
heat budget 53
heavy metals 585
Henry's Law 101, 216–19, 320, 330–1
constant 217–18
Herzberg continuum 41
heterogeneous chemistry 211,
238–40
humidity 211
hydrochlorofluorocarbons 108, 346
hydrogen cyanide 127
hydrogen sulfide 100, 102
chemistry 103
sources 103
hydrostatic equation 36
hydroxyl 94, 100, 113, 160, 170, 172

Icelandic low 72
ideal gas 35
IMAGES model 131
indoor air quality 409
industrial source models 522, 621
inner planets 6, 11
integrated assessment modeling
355–6
Intergovernmental Panel on Climate
Change 241, 351, 360, 544, 554
intertropical convergence zone 84–5
inversion 65
irradiance 37
isoprene 94, 131, 328
emissions 95

jetstream 60, 81, 86
Junge layer 201

katabatic wind 69
Kirchhoff's Law 43, 538
Köppen classification 63, 64
Kyoto Protocol 126, 347, 370

La Niña 80
Lagrangian modeling 383, 533
land surface in climate modeling 547
lapse rate 36, 44, 48
dry adiabatic lapse rate 49
saturated adiabatic lapse rate 50
Last Glacial Maximum 550
latent heat 46
of vaporization 47

lead 404, 427, 604, 611–12
LIDAR 457–61, 466
lightning 135
line sources 504
London smog 171
long-range transport 389
Lorenz grid 534
Los Angeles 171

Mace Head 114, 175, 179
mass balance equation 157
material damage 415–17
Mauna Loa 91, 176, 345
mesopause 36
mesosphere 36
meteorology 275–313
monitoring 461–3
methane 17, 92, 127, 162, 343
budget 93
chemistry 161–2
cycle 357
emissions 128–9
sources 93, 127–8
methyl bromide 112
budget 115
methyl chloride 111
budget 115
methyl chloroform 109
methyl iodide 114
budget 115
MIE scattering 458, 469
mineral dust 232, 352
dispersion modeling 513
modeling
climate 525–61
sulfur dioxide 616
molar mass 35
monitoring 439–72
monoterpenes 95–6, 131, 236, 328
monsoon 80, 81, 82
Montreal Protocol 107, 109, 139, 204,
346, 364, 368, 370
mountain atmospheric flow 304
municipal waste incinerator emissions
511

National Air Quality Information
Archive 606, 618
net primary productivity 92
nighttime chemistry 176
nitrate radical 178
nitric acid 177, 204, 239, 331, 408
nitric oxide 98
nitrification 135

nitrite 223
nitrogen cycle 96
nitrogen dioxide 98, 604
  modeling 521
  in urban areas 612–14
nitrous oxide 96, 344
  budget 97
  sources 97
nondispersive infrared analysis 450
North Atlantic
  drift 63
  Oscillation 72
Norwegian cyclone model 300
nuisance assessment 513

occlusion 301
occultation technique 468
oceans in climate models 544
odor assessment 506
organoiodine compounds 114
origin of life 16, 18
outgassing 12
oxides of nitrogen 98, 164, 382, 403
  budget 99, 134
  chemistry 385
  dispersion modeling 507
  emissions 385, 483, 507
  emissions in London 498
  emissions in Tokyo 499
  emissions trends 136
  measurement by
    chemiluminescence 452
  passive sampling 444
  role in atmosphere 134
  sources 134
  UK emissions 135–6
  in urban areas 612–14
oxidizing capacity 159
oxygen 20–9, 140, 408, 618–19
  and evolution 28
ozone
  air quality standard 379, 420
  –alkene chemistry 180
  budget 157, 168–9
  chemistry 380
  column 189
  concentrations 40
  damage 27, 58
  depleting substances 139
  depletion 191, 197–201, 347
  depletion future 204, 364
  deposition 320, 323, 383
  destruction 164
  effects on human health 141, 378

effects on materials 142
effects on vegetation 142, 379
formation 40, 141, 164, 167
isopleth diagram 165
and life 25
measurement by TOMS 467
in the paleoatmosphere 26
profile 26, 40, 157, 174, 347
protection 26
role of hydrocarbons 169
spring maximum 173

paleoclimate 546
paleosol 22
particles
  particulate material 145, 228–54,
    604, 616–18
  sources 146
Pasquill class 264, 505
perchloroethene 110
peroxide 167–8, 221–3, 320, 331
peroxy radical 160, 180
peroxyacetylnitrate (PAN)
  concentration 142
  formation 142, 172
persistent organic pollutants 145
photochemical smog 140, 171
photochemistry 156, 174, 382
  in climate models 543
  rates of photolysis 159, 161
  tropospheric 156–87
  tropospheric budget 170
photostationary state 161
photosynthesis 14, 21, 22, 29, 57, 91
phytoplankton 103, 105, 107, 112,
  131
planets 5
plume 266
  buoyant 267
  dispersion 267
  Gaussian 268
PM_{10} 147, 403, 604, 617
  concentrations 618
  emissions 148, 483, 618
  health effects 147
  sources 148
$PM_{2.5}$ 147, 403
point sources 503
polar high 71
polar stratospheric clouds 201
polar vortex 192
polychlorinated biphenyls 145
polycyclic aromatic hydrocarbons
  144, 603, 619

population 401
potential temperature 48, 278
precipitation 65, 68
pressure 61
  broadening 45
  field, global 70
  profile 35
primitive atmosphere 9

radiation
  absorption 38–40, 538
  balance 59, 75
  blackbody 37, 38–9, 43, 52
  infrared 39–40
  longwave 42
  modeling 537
  net 54
  outgoing 39
  solar 37–8, 158
  terrestrial 38–9
  transfer 42, 45, 56
  ultraviolet 25, 41
radiative forcing 350
  future 362
radical chemistry 160
radiosonde 464
rainwater 212
Raymond scattering 459
remote sensing 248, 308–9, 456,
  463–70, 551
respiratory system 411–12
road pricing 421
Rocky mountains 69
Rossby number 282, 285
Rossby waves 61, 69, 192, 295,
  303–6
rotation of the Earth 59
roughness length 265, 316

Sahel 86
sampling techniques 440
satellites 467, 551
scale height 276
scavenging coefficient 329–31
SCIAMACHY 469
screening models 509
sea breeze circulation 266, 277
sea-level rise 367, 555
sea-salt aerosol 231
sea-surface temperature 73
secondary organic aerosol 234, 246
secondary pollutants 140
semigeostrophic theory 287
sensible heat 47

soil
  dust 146
  models 567
  quality 582, 589
  vegetation models 569
solar flux 15, 35
solar radiation 158
soot 236, 617, 349, 403, 413
South Pole 68
Southern Oscillation 79
Southern Oscillation index 83
space shuttle 250
spectroscopy 447
Stefan–Boltzmann constant 37
Stokes number 326
stratopause 36
stratosphere 36
  chemistry 188–210
  circulation 189
  heterogeneous chemistry 202–4
  oxygen-only chemistry 196
  residual circulation 190
  structure 190
  –troposphere exchange 173, 194
  water vapor 195
street canyon 406
sulfur 99
  budget 100
  chemistry 182, 220, 385
  cycle 99, 183
  emissions 385
sulfur dioxide 99–100, 403, 604, 614
  budget 100
  chemistry 101, 240
  concentrations 137
  deposition 320, 323
  emissions 483
  monitoring 615
  role in atmosphere 136
  sources 137
  UK emissions 137

  urban concentrations 425–6
sulfur hexafluoride 346
sulfuric acid 100–1
Sun 37
sunspot 37
surface reactions 204
synoptic analysis 308
synoptic charts 308
synoptic observations 309
synoptic-scale meteorology 275–313

tapered element oscillating
    microbalance 455
target loads
  cadmium 590
  heavy metals 588
temperature
  changes 353–5, 555
  structure 49
tephigram 281
thermodynamics 48
thermosphere 37
total suspended particulate matter
    403, 413
Toxic Organic Micropollutants Survey
    619
trade winds 77
trichloroethene 110
tropical storms 78
tropopause 36, 194
troposphere 36
  aqueous phase chemistry 211–27

UK Air Quality Objectives 505,
    603–4
UK Air Quality Strategy 505
UKMO climate model 538, 545
uncertainty estimating 625
United Nations Framework
    Convention on Climate Change
    369, 475

units of concentration 441
urban air pollution
  developing countries 430–2
  Europe 424
  Japan 429
  North America 428
urban chemistry 171
urban heat island 406
urban planning 421
urbanization 400

visibility 417
volatile organic compounds 94, 126,
    129, 165, 382, 618
  analysis 453
  anthropogenic emissions 129
  biogenic 130, 169, 404
  lifetimes 97
  sources 97
  UK emissions 130
volcanic gases 12
vorticity 194, 286, 294
  in street canyon 407
Vostok ice core 340

Walker circulation 77, 78, 85
washout ratio 330
water
  content 212
  soluble particles 215
  vapor 342
weather systems 299
weathering 22
wet deposition 329–34
Weybourne atmospheric observatory
    179
wind
  rose 514
  shear 270, 283